STUDENT'S SOLUTIONS MANUAL

TONI GARCIA
Arizona State University

INTRODUCTORY STATISTICS

TENTH EDITION

Neil A. Weiss
School of Mathematical and Statistical Sciences Arizona State University

PEARSON

Boston Columbus Hoboken Indianapolis New York San Francisco
Amsterdam Cape Town Dubai London Madrid Milan Munich Paris Montreal Toronto
Delhi Mexico City São Paulo Sydney Hong Kong Seoul Singapore Taipei Tokyo

Reproduced by Pearson from electronic files supplied by the author.

Publishing as Pearson, 501 Boylston Street, Boston, MA 02116.

ISBN-13: 978-0-321-98928-4
ISBN-10: 0-321-98928-7

3

www.pearsonhighered.com

Contents

CHAPTER 1 SOLUTIONS

Exercises 1.1

1.1 (a) The *population* is the collection of all individuals or items under consideration in a statistical study.

(b) A *sample* is that part of the population from which information is obtained.

1.3 Descriptive methods are used for organizing and summarizing information and include graphs, charts, tables, averages, measures of variation, and percentiles.

1.5 (a) An *observational study* is a study in which researchers simply observe characteristics and take measurements.

(b) A *designed experiment* is a study in which researchers impose treatments and controls and *then* observe characteristics and take measurements.

1.7 This study is inferential. Data from a sample of Americans are used to make an estimate of (or an inference about) average TV viewing time for all Americans.

1.9 This study is descriptive. It is a summary of information on all homes sold in different cities for the month of September 2012.

1.11 This study is descriptive. It is a summary of the annual final closing values of the Dow Jones Industrial Average at the end of December for the years 2004-2013.

1.13 (a) This study is inferential. It would have been impossible to survey all U.S. adults about their opinions on Darwinism. Therefore, the data must have come from a sample. Then inferences were made about the opinions of all U.S. adults.

(b) The population consists of all U.S. adults. The sample consists only of those U.S. adults who took part in the survey.

1.15 (a) The statement is descriptive since it only tells what was said by the respondents of the survey.

(b) Then the statement would be inferential since the data has been used to provide an estimate of what all Americans believe.

1.17 Designed experiment. The researchers did not simply observe the two groups of children, but instead randomly assigned one group to receive the Salk vaccine and the other to get a placebo.

1.19 Observational study. The researchers simply collected data from the men and women in the study with a questionnaire.

1.21 Designed experiment. The researchers did not simply observe the three groups of patients, but instead randomly assigned some patients to receive optimal pharmacologic therapy, some to receive optimal pharmacologic therapy and a pacemaker, and some to receive optimal pharmacologic therapy and a pacemaker-defibrillator combination.

1.23 (a) This statement is inferential since it is a statement about all Americans based on a poll. We can be reasonably sure that this is the case since the time and cost of questioning every single American on this issue would be prohibitive. Furthermore, by the time everyone could be questioned, many would have changed their minds.

(b) To make it clear that this is a descriptive statement, the new statement could be, "Of 1032 American adults surveyed, 73% favored a law that would require every gun sold in the United States to be test-fired first, so law enforcement would have its fingerprint in case it were ever used in a crime." To rephrase it as an inferential statement, use "Based on a sample of 1032 American adults, it is estimated that 73% of American adults favor a law that would require every gun sold in the United States to be test-fired first, so law enforcement would have its fingerprint in case it were ever used in a crime."

1.25 (a) The population consists of all Americans between the ages of 18 and 29.

(b) The sample consists only of those Americans who took part in the survey.

(c) The statement in quotes is inferential since it is a statement about all Americans based on a survey.

(d) "Based on a sample of Americans between the ages of 18 and 29, it is estimated that 59% of Americans oppose medical testing on animals."

Exercises 1.2

1.27 A census is generally time consuming, costly, frequently impractical, and sometimes impossible.

1.29 The sample should be representative so that it reflects as closely as possible the relevant characteristics of the population under consideration.

1.31 (a) Probability sampling consists of using a randomizing device such as tossing a coin or consulting a random number table to decide which members of the population will constitute the sample.

(b) No. It is possible for the randomizing device to randomly produce a sample that is not representative.

(c) Probability sampling eliminates unintentional selection bias, permits the researcher to control the chance of obtaining a non-representative sample, and guarantees that the techniques of inferential statistics can be applied.

1.33 Simple random sampling.

1.35 The acronym used for simple random sampling without replacement is SRS.

1.37 (a) 12, 13, 14, 23, 24, 34

(b) There are 6 samples, each of size two. Each sample has a one in six chance of being selected. Thus, the probability that a sample of two is 2 and 3 is 1/6.

(c) Starting in Line 17 and column 07 (notice there is a column 00), reading single digit numbers down the column and then up the next column, the first digit that is a one through four is a 1. Continue down column 07 and then up column 08. Ignoring duplicates and skipping digits 5 and above and also skipping zero, the second digit found that is a one through four is a 4. Thus the SRS of 1 and 4 is obtained.

1.39 (a) Starting in Line 10 and reading two digits numbers in columns 10 and 11 going down the table, the first two digit number between 01 and 50 is 43. Continuing down the columns and ignoring duplicates and numbers 51-99, the next two numbers are 45 and 01. Then, continuing up columns 12 and 13, the last three numbers selected are 42, 37, and 47. Therefore the SRS of size six consists of observations 43, 45, 01, 42, 37, and 47.

(b) There are many possible answers.

1.41 Dentists form a high-income group whose incomes are not representative of the incomes of Seattle residents in general.

1.43 (a) GLS, GLA, GLT, GSA, GST, GAT, LSA, LST, LAT, SAT.

(b) There are 10 samples, each of size three. Each sample has a one in 10 chance of being selected. Thus, the probability that a sample of three officials is the first sample on the list presented in part (a) is 1/10. The same is true for the second sample and for the tenth sample.

1.45 (a)

E,M,P,L	E,M,L,B	E,P,A,B	M,P,A,B
E,M,P,A	E,M,A,B	E,L,A,B	M,L,A,B
E,M,P,B	E,P,L,A	M,P,L,A	P,L,A,B
E,M,L,A	E,P,L,B	M,P,L,B	

(b) One procedure for taking a random sample of four representatives from the six is to write the initials of the representatives on six separate pieces of paper, place the six slips of paper into a box, and then, while blindfolded, pick four of the slips of paper. Or, number the representatives 1-6, and use a table of random numbers or a random-number generator to select four different numbers between 1 and 6.

(c) 1/15; 1/15

1.47 (a)

F,T	F,G	F,H	F,L	F,B	F,A
T,G	T,H	T,L	T,B	T,A	G,H
G,L	G,B	G,A	H,L	H,B	H,A
L,B	L,A	B,A			

(b) 1/21; 1/21

1.49 (a) I am using Table I to obtain a list of 10 random numbers between 1 and 500 as follows.

I start at the three digit number in line number 14 and column numbers 10-12, which is the number 452.

I now go down the table and record the three-digit numbers appearing directly beneath 452. Since I want numbers between 1 and 500 only, I throw out numbers between 501 and 999, inclusive. I also discard the number 000.

After 452, I skip 667, 964, 593, 534, and record 016.

Now that I've reached the bottom of the table, I move directly rightward to the adjacent column of three-digit numbers and go up.

I record 343, 242, skip 748, 755, record 428, skip 852, 794, 596, record 378, skip 890, record 163, skip 892, 847, 815, 729, 911, 745, record 182, 293, and 422.

I've finished recording the 10 random numbers. In summary, these are:

452	016	343	242	428
378	163	182	293	422

(b) We can use Minitab to generate random numbers. Following the instructions in The Technology Center, our results are 489, 451, 61, 114, 389, 381, 364, 166, 221, and 437. Your result may be different from ours.

1.51 (a) First re-assign the elements 93 though 118 as elements 01 to 26.

Select a random starting point in Table I of Appendix A and read in a pre-selected direction until you have encountered 8 different elements. For example, if we start at the top of the column 10 and read two digit numbers down and then up in the following columns, we encounter the elements 04, 01, 03, 08, 11, 18, 22, and 15. This corresponds to a sample of the elements Cm, Np, Am, Fm, Lr, Ds, Fl, and Bh. Your answer may be different from this one.

(b) We can use Minitab to generate random numbers. Following the instructions in The Technology Center, our results are 8, 2, 9, 20, 24, 19, 21, and 13. Thus our sample of 8 elements is Fm, Pu, Md, Cn, Lv, Rg, Uut, and Db. Your result may be different from ours.

1.53 (a) One of the dangers of nonresponse is that the individuals who do not respond may have a different observed value than the individuals that do respond causing a nonresponse bias in the estimate. Nonresponse bias may make the measured value too small or too large.

(b) The lower the response rate, the more likely there is a nonresponse bias in the estimate. Therefore the estimate will either under or over estimate the generalized results to the entire population.

Exercises 1.3

1.55 Systematic random sampling is easier to execute than simple random sampling and usually provides comparable results. The exception is the presence of some kind of cyclical pattern in the listing of the members of the population.

1.57 Ideally, in stratified sampling, the members of each stratum should be homogeneous relative to the characteristic under consideration.

1.59 (a) Answers will vary, but here is the procedure: (1) Divide the population size, 372, by the sample size, 5, and round down to the nearest whole number if necessary; this gives 74. Use a table of random numbers (or a similar device) to select a number between 1 and 74, call it k. (3) List every 74th number, starting with k, until 5 numbers are obtained; thus, the first number of the required list of 5 numbers is k, the second is k + 74, the third is k + 148, and so forth.

(b) Following part (a) with k = 10, the first number of the sample is 10, the second is 10 + 74 = 84. The remaining three numbers in the sample would be 158, 232, and 306. Thus, the sample of 5 would be 10, 84, 158, 232, and 306.

1.61 (a) Answers will vary, but here is the procedure: (1) The population of size 50 is already divided into five clusters of size 10. (2) Since the required sample size is 20, we will need to take a SRS of 2 clusters. Use a table of random numbers (or a similar device) to select two numbers between 1 and 5. These are the two clusters that are selected. (3) Use all the members of each cluster selected in part (2) as the sample.

(b) Following part (a) with clusters #1 and #3 selected, we would select all the members in cluster 1, which are 1 - 10, and all the members in cluster 3, which are 21 - 30.

1.63 (a) From each strata, we need to obtain a SRS of a size proportional to the size of the stratum. Therefore, since strata #1 is 30% of the population, a SRS equal to 30% of 20, or 6, should be sampled from strata #1. Since strata #2 is 20% of the population, a SRS equal to 20% of 20, or 4, should be sampled from strata #2. Similarly, a SRS of size 8 should be sampled from strata #3 and a SRS of size 2 should be sampled from strata #4. The sample sizes from stratum #1 through #4 are 6, 4, 8, and 2 respectively.

(b) Answers will vary following the procedure in part (a).

1.65 Stratified Sampling. The entire population is naturally divided into subpopulations, one from each lake, and random sampling is done from each lake. The stratified sampling is not with proportional allocation since that would require knowing how many fish were in each lake.

1.67 Systematic Random Sampling. Kennedy selected his sample using the fixed periodic interval of every 50th letter, which is the similar to the method presented in procedure 1.1.

1.69 Cluster Sampling. The clusters of this sampling design are the 46 schools. A random sample of 10 clusters was selected and then all of the parents of the nonimmunized children at the 10 selected schools were sent a questionnaire.

1.71 (a) Answers will vary, but here is the procedure: (1) Divide the population size, 500, by the sample size, 10, and round down to the nearest whole number if necessary; this gives 50. (2) Use a table of random numbers (or a similar device) to select a number between 1 and 50, call it *k*. (3) List every 50th, starting with *k*, until 10 numbers are obtained; thus, the first number on the required list of 10 numbers is *k*, the second is *k*+50, the third is *k*+100, and so forth (e.g., if *k*=6, then the numbers on the list are 6, 56, 106, ...).

(b) Systematic random sampling is easier.

(c) The answer depends on the purpose of the sampling. If the purpose of sampling is not related to the size of the sales outside the U.S., systematic sampling will work. However, since the listing is a ranking by amount of sales, if k is low (say 2), then the sample will contain firms that, on the average, have higher sales outside the U.S. than the population as a whole. If the k is high, (say 49) then the sample will contain firms that, on the average, have lower sales than the population as a whole. In either of those cases, the sample would not be representative of the population in regard to the amount of sales outside the U.S.

1.73 (a) Number the suites from 1 to 48, use a table of random numbers to randomly select three of the 48 suites, and take as the sample the 24 dormitory residents living in the three suites obtained.

(b) Probably not, since friends are more likely to have similar opinions than are strangers.

(c) There are 384 students in total. Freshmen make up 1/3 of them. Sophomores make up 7/24 of them, Juniors 1/4, and Seniors 1/8. Multiplying each of these fractions by 24 yields the proportional allocation, which dictates that the number of freshmen, sophomores, juniors, and seniors selected should be, respectively, 8, 7, 6, and 3. Thus a stratified sample of 24 dormitory residents can be obtained as follows: Number the freshmen dormitory residents from 1 to 128 and use a table of random numbers to randomly select 8 of the 128 freshman dormitory residents; number the sophomore dormitory residents from 1 to 112 and use a table of random numbers to randomly select 7 of the 112 sophomore dormitory residents; and so forth.

1.75 (a) Answers will vary, but here is the procedure: (1) Divide the population size, 435, by the sample size, 15, and round down to the nearest whole number if necessary; this gives 29. Use a table of random numbers (or a similar device) to select a number between 1 and 29, call it *k*. (3) List every 29th number, starting with *k*, until 15 numbers are obtained; thus, the first number of the required list of 15 numbers is *k*, the second is $k + 29$, the third is $k + 58$, and so forth.

(b) Following part (a) with $k = 12$, the first number of the sample is 12, the second is $12 + 29 = 41$. The third number selected is $12 + 58 = 70$. The remaining twelve numbers are similarly selected. Thus, the sample of 15 would be 12, 41, 70, 99, 128, 157, 186, 215, 244, 273, 302, 331, 360, 389, and 418.

1.77 (a) This is a poll taken by calling randomly selected U.S. adults. Thus, the sampling design appears to be simple random sampling, although it is possible that a more complex design was used to ensure that various political, religious, educational, or other types of groups were proportionately represented in the sample.

(b) The sample size for the second question was 78% of 1010 or 788.

(c) The sample size for the third question was 28% of 788 or 221.

1.79 (a) It is also true for systematic random sampling if the population size divided by the sample size results in an integer for m. The chance for each member to be selected is then still equal to the sample size divided by the population size. For example, suppose the population size is N=10 and the sample size is n=2. The chance that each member in simple random sampling to be selected is 2/10 = 1/5. In systematic random sampling for the same example, m=5. The possible samples of size two are 1 and 6, 2 and 7, 3 and 8, 4 and 9, and 5 and 10. Therefore, the chance that a member is selected is equal to the chance of one of those five samples being selected, which is the same as simple random sampling of 1/5.

(b) It is not true for systematic random sampling if the population size divided by the sample size does not result in an integer for m. For example, suppose the population size is N=15 and the sample size is n=2. After dividing the population size by the sample size and rounding down to the nearest whole number, we get m=7. You would select every 7th member after a random starting place k, between 1 and 7, is determined. If k=1, you would select the first and eighth member. If k=7, you would select the seventh and fourteenth member. In this situation, the last member (fifteenth) can never be selected. Therefore, the last member of the sample does not have the same chance of being selected as any other member in the population.

Exercises 1.4

1.81 (a) Experimental units are the individuals or items on which the experiment is performed.

(b) When the experimental units are humans, we call them subjects.

1.83 (a) The response variable is the characteristic of the experimental outcome that is to be measured or observed.

(b) A factor is a variable whose effect on the response variable is of interest in the experiment.

(c) The levels are the possible values of the factor.

(d) For a one-factor experiment, the treatments are the levels of the factor. For multifactor experiments, the treatments are the combinations of levels of the factors.

1.85 In a one-factor experiment, the number of treatments is equal to the number of levels of the factor. Therefore, there are four treatments.

1.87 (a)

A \ B	b_1	b_2	b_3	b_4
a_1	a_1b_1	a_1b_2	a_1b_3	a_1b_4
a_2	a_2b_1	a_2b_2	a_2b_3	a_2b_4
a_3	a_3b_1	a_3b_2	a_3b_3	a_3b_4

(b) There are twelve combinations of the levels of the factors. Therefore, there are twelve treatments.

(c) Yes, you could have multiplied the number of levels in each factor. There are three levels of factor A and four levels of factor B. Therefore, there are (3)(4) = 12 treatments.

1.89 You can multiply the number of levels in each factor. There are m levels in the first factor and n levels in the second factor. Therefore, there are (m)(n) = $m \times n$ treatments.

1.91 (a) There were three treatments.

(b) The first group, the one receiving only the pharmacologic therapy, would be considered the control group.

(c) There were three treatment groups. The first received only pharmacologic therapy, the second received pharmacologic therapy plus a pacemaker, and the third received pharmacologic therapy plus a pacemaker-defibrillator combination.

(d) The first group (control) contained 1/5 of the 1520 patients or 304. The other two groups each contained 2/5 of the 1520 patients or 608.

(e) Each patient could be randomly assigned a number from 1 to 1520. Any patient assigned a number between 1 and 304 would be assigned to the control group; any patient assigned to the next 608 numbers (305 to 912) would be assigned to receive the pharmacologic therapy plus a pacemaker; and any patient assigned a number between 913 and 1520 would receive pharmacologic therapy plus a pacemaker-defibrillator combination. Each random number would be used only once to ensure that the resulting treatment groups were of the intended sizes.

1.93 (a) Experimental units: the drivers

(b) Response variable: the detection distance, in feet

(c) Factors: two factors - sign size and sign material

(d) Levels of each factor: three levels of sign size (small, medium, and large) and three levels of sign material (1, 2, and 3)

(e) Treatments: the nine different combinations of sign size and sign material resulting from testing each of the three sign sizes with each of the three sign materials

1.95 (a) Experimental units: female lions

(b) Response variable: whether or not the female lion approached the male lion dummy

(c) Factors: length and color of the mane on the male lion dummy

(d) Levels of each factor: two different mane lengths and two different mane colors

(e) Treatments: the four combinations of mane length and color

1.97 (a) Experimental units: the children

(b) Response variable: IQ score

(c) Factor: Whether they were given dexamethasone (control or dexamethasone group)

(d) Levels of each factor: two levels of the single factor (control or dexamethasone group)

(e) Treatments: the two levels of the single factor

1.99 (a) This is a randomized block design. The experiment first blocked by gender. All the experimental units are not randomly assigned among all the treatments.

(b) The blocks are the two genders (male and female).

1.101 (a) Simple random sampling corresponds to completely randomized designs since selection is randomly made from the entire population.

(b) Stratified sampling corresponds to randomized block designs since selection is randomly made from within each strata.

Review Problems for Chapter 1

1. Student exercise.

2. Descriptive statistics are used to display and summarize the data to be used in an inferential study. Preliminary descriptive analysis of a sample often reveals features of the data that lead to the choice or reconsideration of the choice of the appropriate inferential analysis procedure.

3. (a) An *observational study* is a study in which researchers simply observe characteristics and take measurements.

(b) A *designed experiment* is a study in which researchers impose treatments and controls and *then* observe characteristics and take measurements.

4. A literature search should be made before planning and conducting a study.

5. (a) A representative sample is one that reflects as closely as possible the relevant characteristics of the population under consideration.

(b) Probability sampling involves the use of a randomizing device such as tossing a coin or die, using a random number table, or using computer software that generates random numbers to determine which members of the population will make up the sample.

(c) A sample is a simple random sample if all possible samples of a given size are equally likely to be the actual sample selected.

6. (a) This method does not involve probability sampling. No randomizing device is being used and people who do not visit the campus cafeteria have no chance of being included in the sample.

(b) The dart throwing is a randomizing device that makes all samples of size 20 equally likely. This is probability sampling.

7. (a) Systematic random sampling is done by first dividing the population size by the sample size and rounding the result down to the next integer, say m. Then we select one random number, say k, between 1 and m inclusive. That number will be the first member of the sample. The remaining members of sample will be those numbered k+m, k+2m, k+3m, ... until a sample of size n has been chosen. Systematic sampling will yield results similar to simple random sampling as long as there is nothing systematic about the way the members of the population were assigned their numbers.

(b) In cluster sampling, clusters of the population (such as blocks, precincts, wards, etc.) are chosen at random from all such possible clusters. Then every member of the population lying within the chosen clusters is sampled. This method of sampling is particularly convenient when members of the population are widely scattered and is most appropriate when the members of each cluster are representative of the entire population. Cluster sampling can save both time and expense in doing the survey, but can yield misleading results if individual clusters are made up of subjects with very similar views on the topic being surveyed.

(c) In stratified random sampling with proportional allocation, the population is first divided into subpopulations, called strata, and simple random sampling is done within each stratum. Proportional allocation means that the size of the sample from each stratum is proportional to the size of the population in that stratum. This type of sampling may improve the accuracy of the survey by ensuring that those in each stratum are more proportionately represented than would be the case with cluster sampling or even simple random sampling. Ideally, the members of each stratum should be homogeneous relative to the characteristic under consideration. If they are not homogeneous within each stratum, simple random sampling would work just as well.

8. The three basic principles of experimental design are control, randomization, and replication. Control refers to methods for controlling factors other than those of primary interest. Randomization means randomly dividing the subjects into groups in order to avoid unintentional selection bias in constituting the groups. Replication means using enough experimental units or subjects so that groups resemble each other closely and so that there is a good chance of detecting differences among the treatments when such differences actually exist.

9. Descriptive study. It is a summary of the scores of major league baseball games on August 14, 2013.

10. (a) Descriptive study. It is a summary of the responses from those that participated in the poll.

(b) Inferential statement. It is an implied estimate of the responses of all adults in the U.S.

11. Inferential study. The results of a sample are used to make inferences about the age distribution of <u>all</u> British backpackers in South Africa.

12. (a) Descriptive study. It is a summary of the percentages of Jewish children sampled in Israel and Britain that have peanut allergies.

(b) Observational study. The researchers simply observed the two groups.

13. This is an observational study. To be a designed experiment, the researchers would have to have the ability to assign some children at random to live in persistent poverty during the first 5 years of life or to not suffer any poverty during that period. Clearly that is not possible.

14. This is a designed experiment since the researcher is imposing a treatment and then observing the results.

15. Because Yale is a very expensive school, incomes of parents of Yale students will not be representative of the incomes of all college students' parents.

16. (a)

H,Z,C	H,Z,A	H,Z,J	H,C,A	H,C,J
H,A,J	Z,C,A	Z,C,J	Z,A,J	C,A,J

(b) Since each of the 10 samples of size three is equally likely, there is a 1/10 chance that the sample chosen is the first sample in the list, 1/10 chance that it is the second sample in the list, and 1/10 chance that it is the tenth sample in the list.

(c) (i) Make five slips of paper with each airline on one slip. Draw three slips at random. (ii) Make 10 slips of paper, each having one of the combinations in part (a). Draw one slip at random. (iii) Number the five airlines from 1 to 5. Use a random number table or random number generator to obtain three distinct random numbers between 1 and 5, inclusive.

(d) Your method and result may differ from ours. We rolled a die (ignoring 6's and duplicates) and got 2, 5, 2, 6, 4. Ignoring duplicates and numbers greater than five, our sample consists of Horizon, Jazz, and Alaska Airlines.

17. (a) Table I can be employed to obtain a sample of 15 random numbers between 1 and 100 as follows. First, I pick a random starting point by closing my eyes and putting my finger down on the table.

My finger falls on three digits located at the intersection of a line with three columns. (Notice that the first column of digits is labeled "00" rather than "01".) This is my starting point.

I now go down the table and record all three-digit numbers appearing directly beneath the first three-digit number that are between 001 and 100 inclusive. I throw out numbers between 101 and 999, inclusive. I also discard the number 0000. When the bottom of the column is reached, I move over to the next sequence of three digits and work my way back up the table. Continue in this manner. When 10 distinct three-digit numbers have been recorded, the sample is complete.

(b) Starting in row 10, columns 7-9, we skip 484, 797, record 082, skip 586, 653, 452, 552, 155, record 008, skip 765, move to the right and record 016, skip 534, 593, 964, 667, 452, 432, 594, 950, 670, record 001, skip 581, 577, 408, 948, 807, 862, 407, record 047, skip 977, move to the right, skip 422 and all of the rest of the numbers in that column, move to the right, skip 732, 192, record 094, skip 615 and all of the rest of the numbers in that column, move to the right, record 097, skip 673, record 074, skip 469, 822, record 052, skip 397, 468, 741, 566, 470, record 076, 098, skip 883, 378, 154, 102, record 003, skip 802, 841, move to the right, skip 243, 198, 411, record 089, skip 701, 305, 638, 654, record 041, skip 753, 790, record 063.

The final list of numbers is 82, 8, 16, 1, 47, 94, 97, 74, 52, 76, 98, 3, 89, 41, 63.

(c) Using Minitab, our results were the numbers 46, 99, 90, 31, 75, 98, 79, 14, 44, 13, 66, 49, 37, 87, 73, 26, 61, 71, 72, 2. Thus our sample consists of the first 15 numbers 46, 99, 90, 31, 75, 98, 79, 14, 44, 13, 66, 49, 37, 87, 73. Your sample may be different.

18. The statement under the vote is a disclaimer as to the validity of the survey. Since the vote reflects only the responses of volunteers who chose to vote, it cannot be regarded as representative of the public in general, some of which do not use the Internet, nor as representative of Internet users since the sample was not chosen at random from either group.

19. The data in this study were clearly not collected via a controlled experiment in which some participants were forced to do crossword puzzles, practice musical instruments, play board games, or read while others were not allowed to do any of those activities. Therefore, any data relative to these activities and dementia arose as a result of observing whether or not the subjects in the study carried out any of those activities and whether or no they had some form of dementia. Since this would be an observational study, no statement of cause and effect can rightfully be made.

20. The researchers did not impose or manipulate any of the conditions of this study. They didn't decide who had cancer, who didn't have cancer, who had hepatitis B, or who had hepatitis C. This study was an observational study and not a controlled experiment. Observational studies can only reveal an association, not causation. Therefore, the statement in quotes is valid. If the researchers wanted to establish causation, they would need a designed experiment.

21. (a) Answers will vary, but here is the procedure: (1) Divide the population size, 100, by the sample size 15, and round down to the nearest whole number; this gives 6. (2) Use a table of random numbers (or a similar device) to select a number between 1 and 6, call it k. (3) List every 6th number, starting with k, until 15 numbers are obtained; thus the first number on the required list of 15 numbers is k, the second is $k+6$, the third is $k+12$, and so forth (e.g., if $k=4$, then the numbers on the list are 4, 10, 16, ...).

(b) Yes, unless for some reason there is some kind of trend or a cyclical pattern in the listing of the athletes.

22. (a) Each category of "Distance from Plant" should be represented in the sample in the same proportion that it is present in the population of City of Durham's water distribution system. 1310/11707 = 0.112. Thus, 11.2% of the sample of 80 water samples should be from "Less than 1.5 miles", 27.0% from "1.5 - less than 3.0 miles", 24.1% from "3.0 - less than 4.5 miles", 13.6% from "4.5 - less than 6.0 miles", 11.5% from "6.0 - less than 7.5 miles", and 12.5% from "7.5 miles or greater". Multiplying each of these fractions by 80 gives us the sample sizes from each category. These sample sizes will not necessarily be integers, so we will need to make some minor adjustments of the results. The first category should have (11.2/100)(80) = 8.96. The second should have (27/100)(80) = 21.6. Similarly, the third, fourth, fifth, and sixth categories should have 19.28, 10.88, 9.2, and 10 for their sample sizes. We round the six sample sizes from the categories to 9, 22, 19, 11, 9, and 10 respectively. We would now randomly select water samples from each region.

23. (a) This is a designed experiment.

(b) The treatment group consists of the 158 patients who took AVONEX. The control group consists of the 143 patients who were given a placebo. The treatments were the AVONEX and the placebo.

24. (a) Experimental units: tomato plants

(b) Response variable: yield of tomatoes

(c) Factor(s): tomato variety and density of plants

(d) Levels of each factor: The four tomato varieties (Harvester, Pusa Early Dwarf, Ife No. 1, and Ibadan Local) would be the levels of variety. The four densities (10,000, 20,000, 30,000, and 40,000 plants/ha) would be the levels of the density.

(e) Treatments: Each treatment would be one of the combinations of a variety planted at a given plant density.

25. (a) Experimental Units: The children

(b) Response variable: Whether or not the child was able to open the bottle

(c) Factors: The container designs

(d) Levels of each factor: Three (types of containers)

(e) Treatments: The container designs

26. This is a completely randomized design. All of the experimental units (batches of doughnuts) were assigned at random to the four treatments (four different fats).

27. (a) This is a completely randomized design since the 24 cars were randomly assigned to the 4 brands of gasoline.

(b) This is a randomized block design. The four different gasoline brands are randomly assigned to the four cars in each of the six car model groups. The blocks are the six groups of four identical cars each.

(c) If the purpose is to learn about the mileage rating of one particular car model with each of the four gasoline brands, then the completely randomized design is appropriate. But if the purpose is to learn about the performance of the gasoline across a variety of cars (and this seems more reasonable), then the randomized block design is more appropriate and will allow the researcher to determine the effect of car model as well as of gasoline type on the mileage obtained.

CHAPTER 2 SOLUTIONS

Exercises 2.1

2.1 (a) Hair color, model of car, and brand of popcorn are qualitative variables.

(b) Number of eggs in a nest, number of cases of flu, and number of employees are discrete, quantitative variables.

(c) Temperature, weight, and time are quantitative continuous variables.

2.3 (a) Qualitative data result from observing and recording values of a qualitative variable, such as, color or shape.

(b) Discrete, quantitative data are values of a discrete quantitative variable. Values usually result from counting something.

(c) Continuous, quantitative data are values of a continuous variable. Values are usually the result of measuring something such as temperature that can take on any value in a given interval.

2.5 Of qualitative and quantitative (discrete and continuous) types of data, only qualitative yields nonnumerical data.

2.7 (a) The first column consists of *quantitative, continuous* data. This column provides the time that the earthquake occurred.

(b) The second column consists of *quantitative, continuous* data. This column provides the magnitude of each earthquake.

(c) The third column consists of *quantitative, continuous* data. This column provides the depth of each earthquake in kilometers.

(d) The fourth column consists of *quantitative, discrete* data. This column provides the number of stations that reported activity on the earthquake.

(e) The fifth column consists of *qualitative* data since the region of the location of each earthquake is nonnumerical.

2.9 (a) The first column consists of *quantitative, discrete* data. This column provides the ranks of the deceased celebrities with the top 10 earnings.

(b) The second column consists of *qualitative* data since names are nonnumerical.

(c) The third column consists of *quantitative, discrete* data, the earnings of the celebrities. Since money involves discrete units, such as dollars and cents, the data is discrete, although, for all practical purposes, this data might be considered quantitative continuous data

2.11 (a) The first column contains types of products. They are *qualitative* data since they are nonnumerical.

(b) The second column contains number of units shipped in the millions. These are whole numbers and are *quantitative, discrete.*

(c) The third column contains money values. Technically, these are *quantitative, discrete* data since there are gaps between possible values at the cent level. For all practical purposes, however, these are *quantitative, continuous* data.

2.13 The first column contains *quantitative, discrete* data in the form of ranks. These are whole numbers. The second and third columns contain *qualitative* data in the form of names. The last column contains the rating of the program which is *quantitative, continuous.*

2.15 The first column is *quantitative, discrete* since it is reporting a rank. The second and third columns are *qualitative* since make/model and type are nonnumerical. The last column is *quantitative, continuous* since it is reporting mileage.

Exercises 2.2

2.17 A frequency distribution of qualitative data is a table that lists the distinct values of data and their frequencies. It is useful to organize the data and make it easier to understand.

2.19 (a) True. Having identical frequency distributions implies that the total number of observations and the numbers of observations in each class are identical. Thus, the relative frequencies will also be identical.

(b) False. Having identical relative frequency distributions means that the ratio of the count in each class to the total is the same for both frequency distributions. However, one distribution may have twice (or some other multiple) the total number of observations as the other. For example, two distributions with counts of 5, 4, 1 and 10, 8, 2 would be different, but would have the same relative frequency distribution.

(c) If the two data sets have the same number of observations, either a frequency distribution or a relative-frequency distribution is suitable. If, however, the two data sets have different numbers of observations, using relative-frequency distributions is more appropriate because the total of each set of relative frequencies is 1, putting both distributions on the same basis for comparison.

2.21 (a)-(b)

The classes are presented in column 1. The frequency distribution of the classes is presented in column 2. Dividing each frequency by the total number of observations, which is 5, results in each class's relative frequency. The relative frequency distribution is presented in column 3.

Class	Frequency	Relative Frequency
A	3	0.60
B	1	0.20
C	1	0.20
	5	1.00

(c) We multiply each of the relative frequencies by 360 degrees to obtain the portion of the pie represented by each class. The result using Minitab is

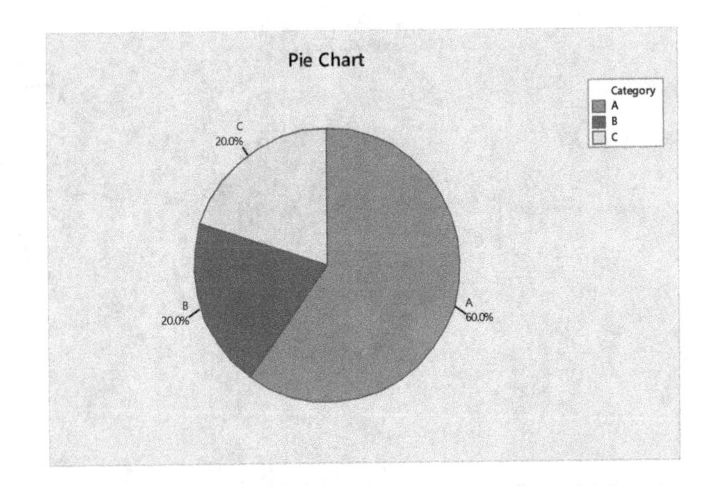

(d) We use the bar chart to show the relative frequency with which each class occurs. The result is

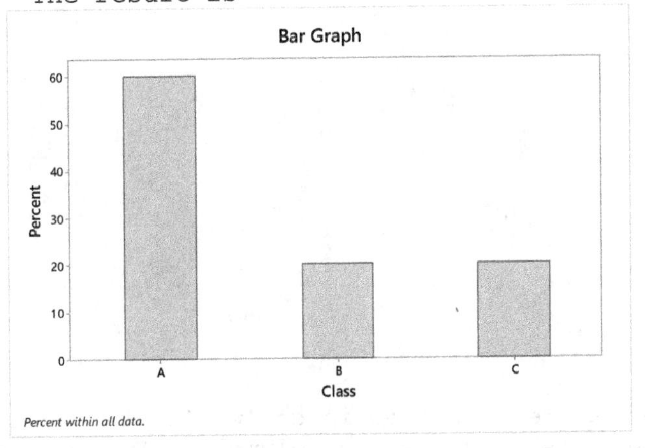

2.23 (a)-(b)

The classes are presented in column 1. The frequency distribution of the classes is presented in column 2. Dividing each frequency by the total number of observations, which is 10, results in each class's relative frequency. The relative frequency distribution is presented in column 3.

Class	Frequency	Relative Frequency
A	4	0.40
B	3	0.30
C	1	0.10
D	2	0.20
	10	1.00

(c) We multiply each of the relative frequencies by 360 degrees to obtain the portion of the pie represented by each class. The result using Minitab is

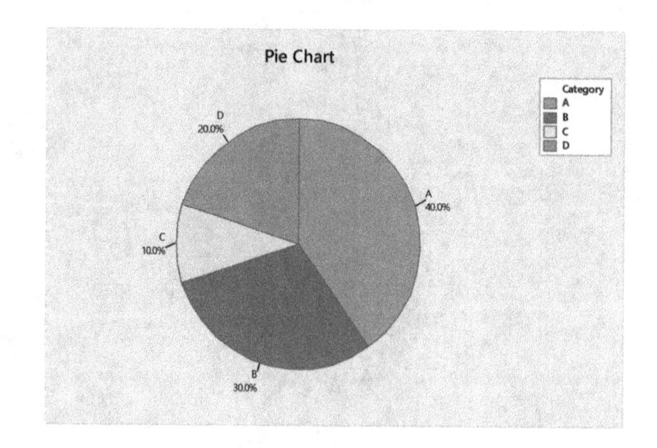

(d) We use the bar chart to show the relative frequency with which each class occurs. The result is

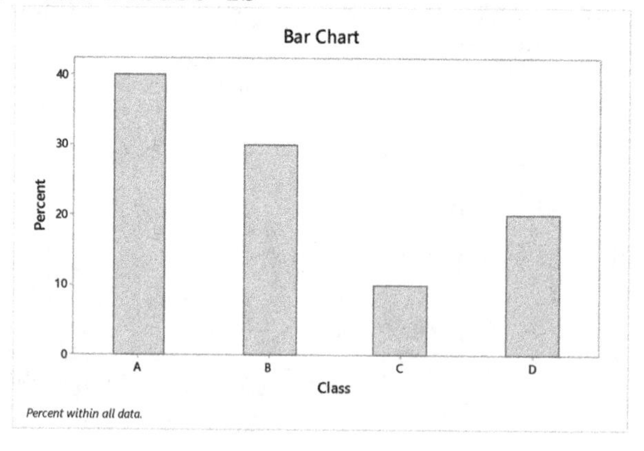

2.25 (a)-(b)

The classes are presented in column 1. The frequency distribution of the classes is presented in column 2. Dividing each frequency by the total number of observations, which is 20, results in each class's relative frequency. The relative frequency distribution is presented in column 3.

Class	Frequency	Relative Frequency
A	1	0.05
B	3	0.15
C	7	0.35
D	7	0.35
E	2	0.10
	20	1.00

(c) We multiply each of the relative frequencies by 360 degrees to obtain the portion of the pie represented by each class. The result using Minitab is

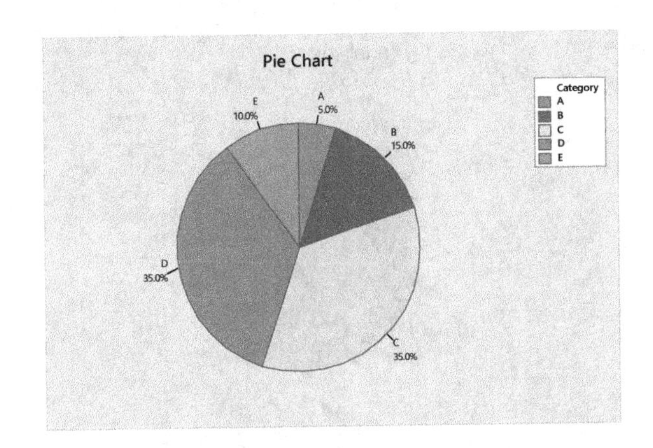

(d) We use the bar chart to show the relative frequency with which each class occurs. The result is

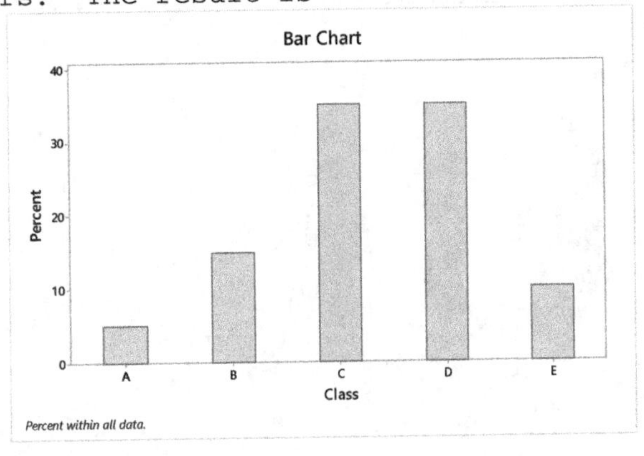

2.27 (a)-(b)

The classes are the NCAA wrestling champions and are presented in column 1. The frequency distribution of the champions is presented in column 2. Dividing each frequency by the total number of champions, which is 25, results in each class's relative frequency. The relative frequency distribution is presented in column 3.

Champion	Frequency	Relative Frequency
Iowa	12	0.48
Penn State	3	0.12
Minnesota	3	0.12
Oklahoma St.	7	0.28
	25	1.00

(b) We multiply each of the relative frequencies by 360 degrees to obtain the portion of the pie represented by each team. The result is

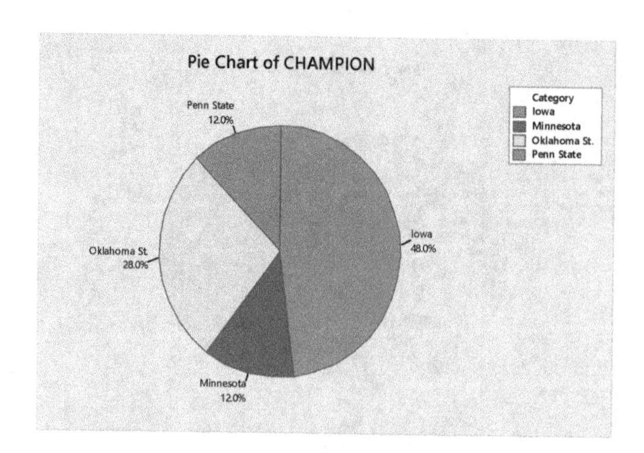

(c) We use the bar chart to show the relative frequency with which each TEAM occurs. The result is

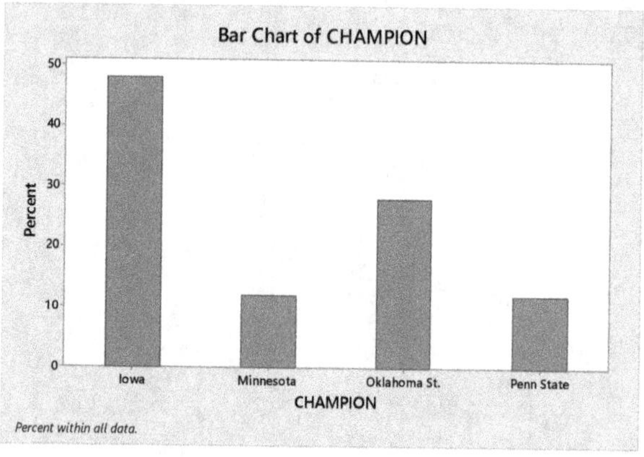

2.29 (a)-(b)

The classes are the class levels and are presented in column 1. The frequency distribution of the class levels is presented in column 2. Dividing each frequency by the total number of students in the introductory statistics class, which is 40, results in each class's relative frequency. The relative frequency distribution is presented in column 3.

Class Level	Frequency	Relative Frequency
Fr	6	0.150
So	15	0.375
Jr	12	0.300
Sr	7	0.175
	40	1.000

(c) We multiply each of the relative frequencies by 360 degrees to obtain the portion of the pie represented by each class level. The result is

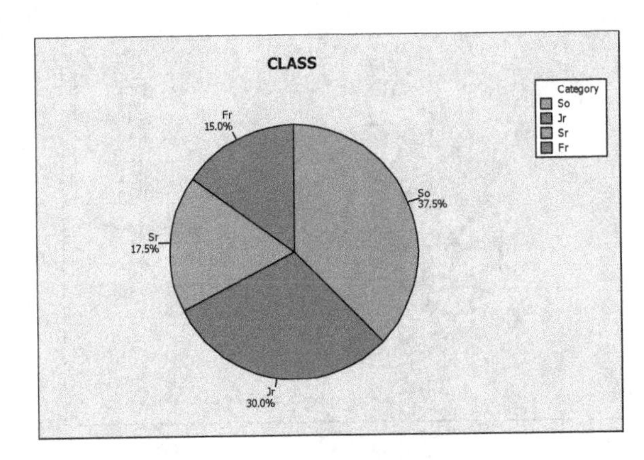

(d) We use the bar chart to show the relative frequency with which each CLASS level occurs. The result is

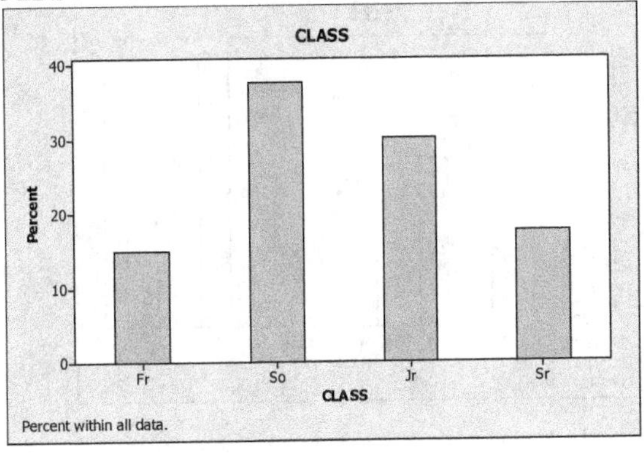

2.31 (a)-(b)

The classes are the days and are presented in column 1. The frequency distribution of the days is presented in column 2. Dividing each frequency by the total number road rage incidents, which is 69, results in each class's relative frequency. The relative frequency distribution is presented in column 3.

Class Level	Frequency	Relative Frequency
Su	5	0.0725
M	5	0.0725
Tu	11	0.1594
W	12	0.1739
Th	11	0.1594
F	18	0.2609
Sa	7	0.1014
	69	1.0000

(c) We multiply each of the relative frequencies by 360 degrees to obtain the portion of the pie represented by each day. The result is

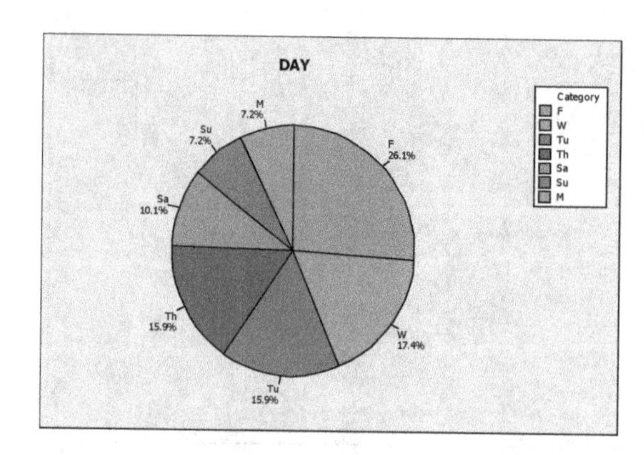

(d) We use the bar chart to show the relative frequency with which each DAY occurs. The result is

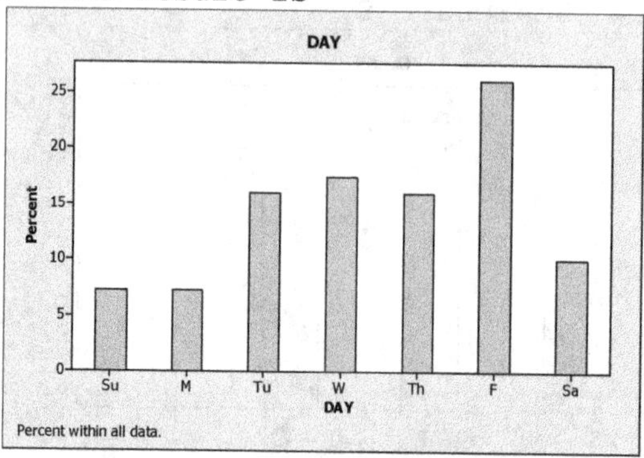

2.33 (a) We find the relative frequencies by dividing each of the frequencies by the total sample size of 509.

Color	Frequency	Relative Frequency
Brown	152	0.2986
Yellow	114	0.2240
Red	106	0.2083
Orange	51	0.1002
Green	43	0.0845
Blue	43	0.0845
	509	1.0000

(b) We multiply each of the relative frequencies by 360 degrees to obtain the portion of the pie represented by each color of M&M. The result is

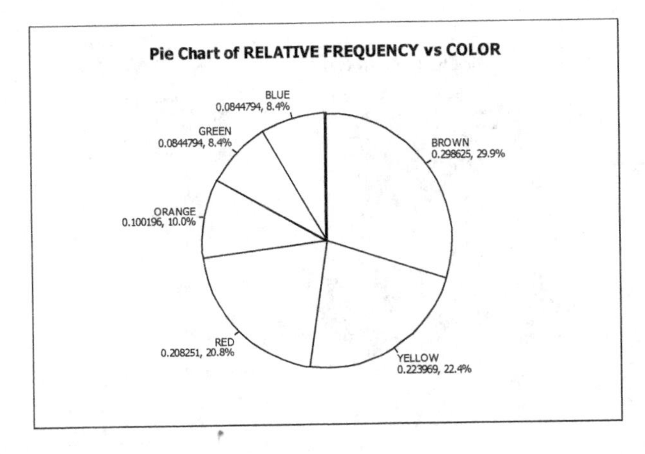

(c) We use the bar chart to show the relative frequency with which each color occurs. The result is

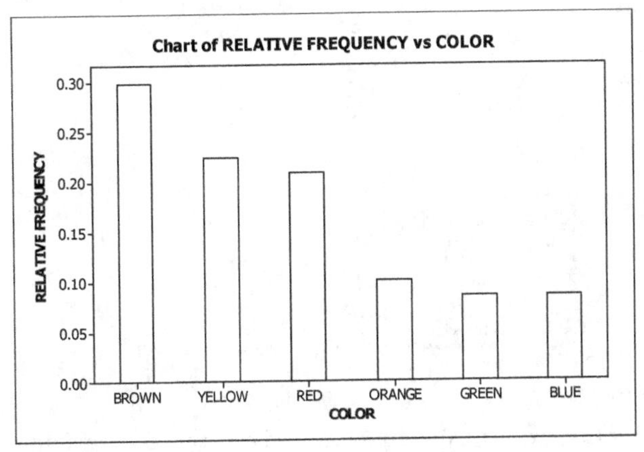

2.35 (a) We find the relative frequencies by dividing each of the frequencies by the total sample size of 137,925.

Rank	Frequency	Relative Frequency
Professor	32,511	0.2357
Associate professor	28,572	0.2072
Assistant professor	59,277	0.4298
Instructor	14,289	0.1036
Other	3,276	0.0238
	137,925	1.0000

(b) We multiply each of the relative frequencies by 360 degrees to obtain the portion of the pie represented by each rank. The result is

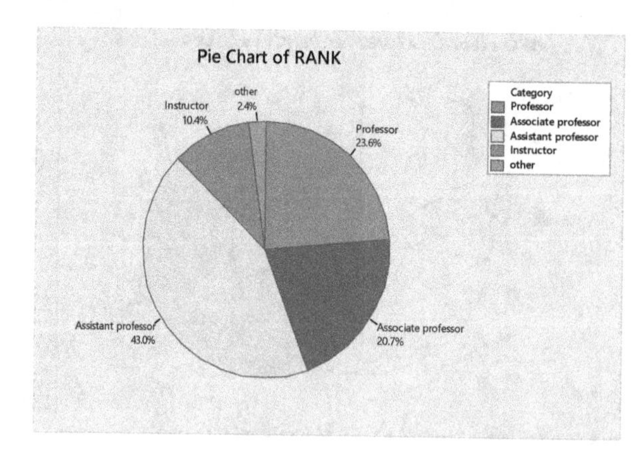

(c) We use the bar chart to show the relative frequency with which each rank occurs. The result is

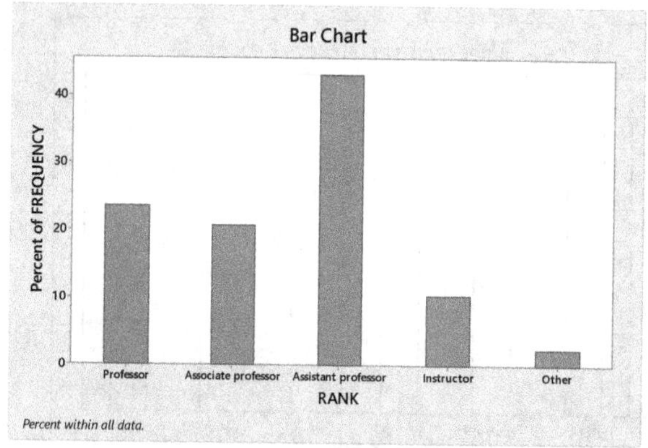

2.37 (a) We first find the relative frequencies by dividing each of the frequencies by the total sample size of 200.

Color	Frequency	Relative Frequency
Red	88	0.44
Black	102	0.51
Green	10	0.05
	200	1.00

(b) We multiply each of the relative frequencies by 360 degrees to obtain the portion of the pie represented by each color. The result is

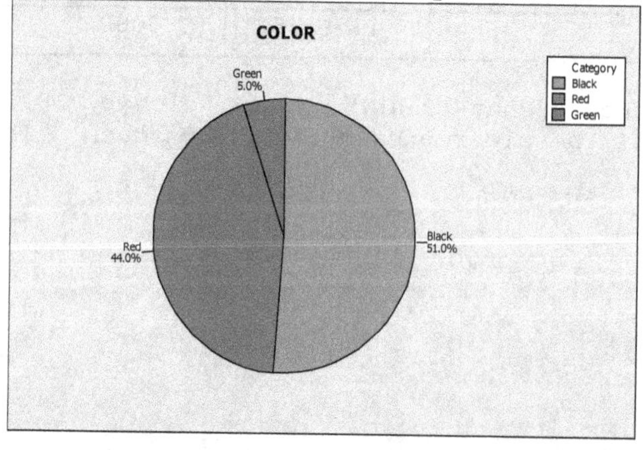

(c) We use the bar chart to show the relative frequency with which each color occurs. The result is

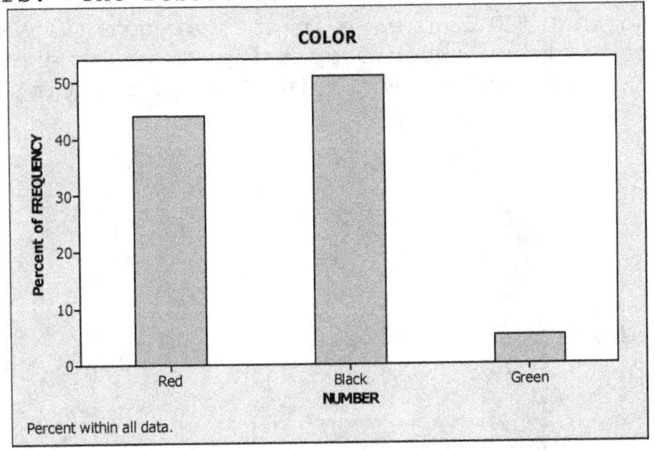

2.39 (a) Using Minitab, retrieve the data from the WeissStats Resource Site. Column 1 contains the type of the vehicle. From the tool bar, select **Stat ▶ Tables ▶ Tally Individual Variables**, double-click on TYPE in the first box so that TYPE appears in the **Variables** box, put a check mark next to Counts and Percents under Display, and click **OK**. The result is

TYPE	Count	Percent
Bus	21	2.80
Car	667	88.93
Truck	62	8.27
N=	750	

(b) The relative frequencies were calculated in part(a) by putting a check mark next to Percents. 2.8% of the vehicles were busses, 88.93% of the vehicles were cars, and 8.27% of the vehicles were trucks.

(c) Using Minitab, select **Graph ▶ Pie Chart**, check Chart counts of unique values, double-click on TYPE in the first box so that TYPE appears in the Categorical Variables box. Click Pie Options, check decreasing volume, click OK. Click Labels, enter TYPE in for the title, click Slice Labels, check Category Name, Percent, and Draw a line from label to slice, Click OK twice. The result is

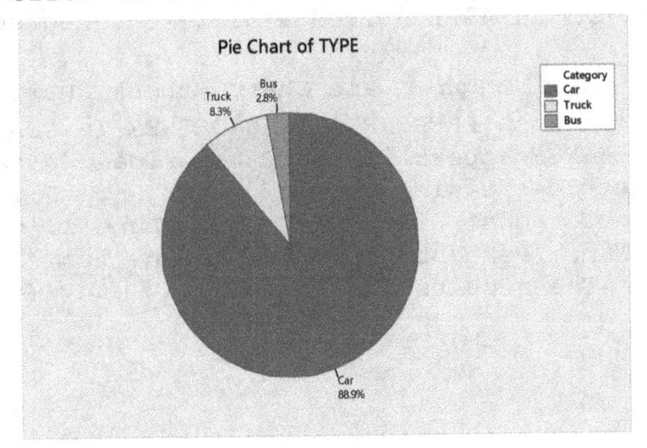

(d) Using Minitab, select **Graph ▶ Bar Chart**, select Counts of unique values, select Simple option, click OK. Double-click on TYPE in the first box so that TYPE appears in the Categorical Variables box. Select Chart Options, check decreasing Y, check show Y as a percent, click OK. Select Labels, enter in TYPE as the title. Click OK twice. The result is

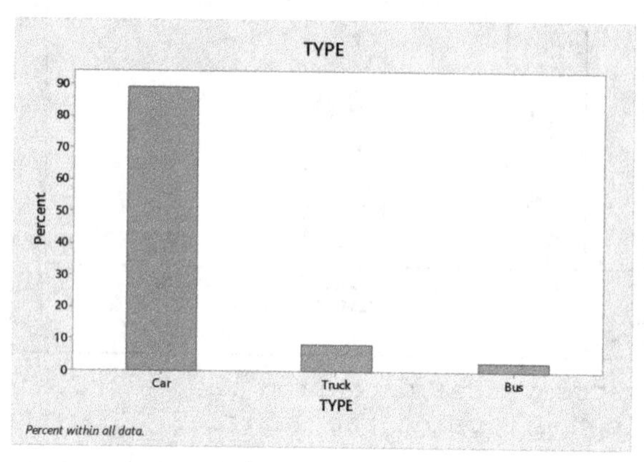

2.41 (a) Using Minitab, retrieve the data from the WeissStats Resource Site. Column 2 contains the preference for how the members want to receive the ballots and column 3 contains the highest degree obtained by the members. From the tool bar, select **Stat ▶ Tables ▶ Tally Individual Variables**, double-click on PREFERENCE and DEGREE in the first box so that both PREFERENCE and DEGREE appear in the **Variables** box, put a check mark next to Counts and Percents under Display, and click **OK**. The results are

PREFERENCE	Count	Percent
Both	112	19.79
Email	239	42.23
Mail	86	15.19
N/A	129	22.79
N=	566	

DEGREE	Count	Percent
MA	167	29.51
Other	11	1.94
PhD	388	68.55
N=	566	

(b) The relative frequencies were calculated in part(a) by putting a check mark next to Percents. For the PREFERENCE variable; 19.79% of the members prefer to receive the ballot by both e-mail and mail, 42.23% prefer e-mail, 15.19% prefer mail, and 22.79% didn't list a preference. For the Degree variable; 29.51% obtained a Master's degree, 68.55% obtained a PhD, and 1.94% received a different degree.

(c) Using Minitab, select **Graph ▶ Pie Chart**, check Chart counts of unique values, double-click on PREFERENCE and DEGREE in the first box so that PREFERENCE and DEGREE appear in the Categorical Variables box. Click Pie Options, check decreasing volume, click OK. Click Multiple Graphs, check On the Same Graphs, Click OK. Click Labels, click Slice Labels, check Category Name, Percent, and Draw a line from label to slice, Click OK twice. The results are

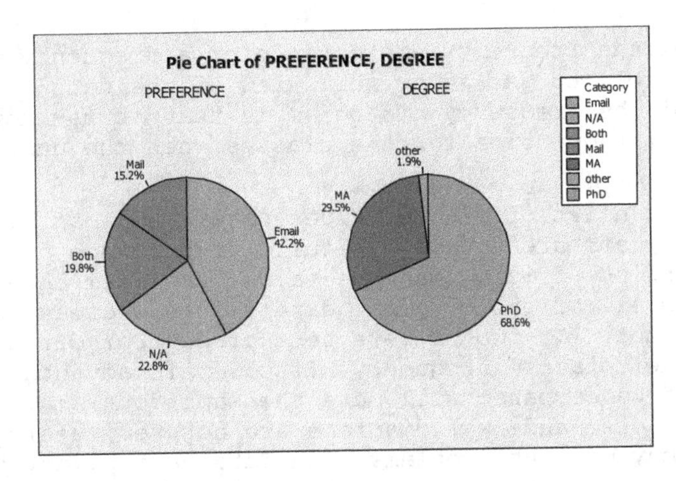

(d) Using Minitab, select **Graph** ▶ **Bar Chart**, select Counts of unique values, select Simple option, click OK. Double-click on PREFERENCE and DEGREE in the first box so that PREFERENCE and DEGREE appear in the Categorical Variables box. Select Chart Options, check decreasing Y, check show Y as a percent, click OK. Click OK twice. The results are

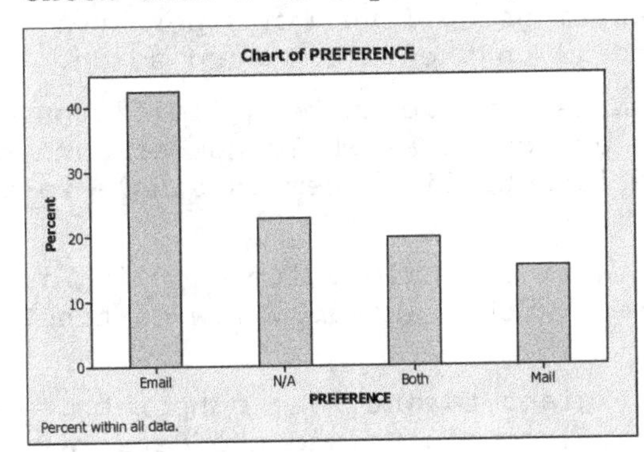

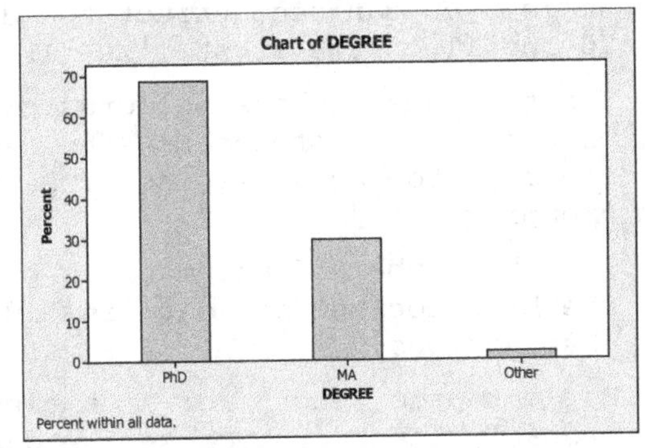

Exercises 2.3

2.43 For class limits, marks, cutpoints and midpoints to make sense, data must be numerical. They do not make sense for qualitative data classes because such data are nonnumerical.

2.45 In the first method for depicting classes called cutpoint grouping, we used the notation **a - under b** to mean values that are greater than or equal to **a** and up to, but not including **b**, such as 30 - under 40 to mean a range of values greater than or equal to 30, but strictly less than 40. In the alternate method called limit grouping, we used the notation **a-b** to indicate a class that extends from **a** to **b**, including both. For example, 30-39 is a class that includes both 30 and 39. The alternate method is especially appropriate when all of the data values are integers. If the data include values like 39.7 or 39.93, the first method is more advantageous since the cutpoints remain integers; whereas, in the alternate method, the upper limits for each class would have to be expressed in decimal form such as 39.9 or 39.99.

2.47 For limit grouping, we find the class mark, which is the average of the lower and upper class limit. For cutpoint grouping, we find the class midpoint, which is the average of the two cutpoints.

2.49 An advantage of the frequency histogram over a frequency distribution is that it is possible to get an overall view of the data more easily. A disadvantage of the frequency histogram is that it may not be possible to determine exact frequencies for the classes when the number of observations is large.

2.51 If the classes consist of single values, stem-and-leaf diagrams and frequency histograms are equally useful. If only one diagram is needed and the classes consist of more than one value, the stem-and-leaf diagram allows one to retrieve all of the original data values whereas the frequency histogram does not. If two or more sets of data of different sizes are to be compared, the relative frequency histogram is advantageous because all of the diagrams to be compared will have the same total relative frequency of 1.00. Finally, stem-and-leaf diagrams are not very useful with very large data sets and may present problems with data having many digits in each number.

2.53 You can reconstruct the stem-and-leaf diagram using two lines per stem. For example, instead of listing all of the values from 10 to 19 on a '1' stem, you can make two '1' stems. On the first, you record the values from 10 to 14 and on the second, the values from 15 to 19. If there are still two few stems, you can reconstruct the diagram using five lines per stem, recording 10 and 11 on the first line, 12 and 13 on the second, and so on.

2.55 For the ages of householders, given as a whole number, limit grouping is probably the best because the data are given as whole numbers and there are probably too many distinct observations to list them as single-value grouping.

2.57 For the number of automobiles per family, single-value grouping is probably the best because the data is discrete with relatively few distinct observations.

2.59 For carapace length for a sample of giant tarantulas, cutpoint grouping is probably the best because the data is continuous and the data was recorded to the nearest hundredth of a millimeter.

2.61 (a) Since the data values range from 1 to 4, we construct a table with classes based on a single value. The resulting table follows.

Single Value Class	Frequency
1	2
2	2
3	5
4	1
	10

(b) To get the relative frequencies, divide each frequency by the sample size of 10.

Single Value Class	Relative Frequency
1	0.20
2	0.20
3	0.50
4	0.10
	1.00

(c) The frequency histogram in Figure (a) is constructed using the frequency distribution presented in part (a) of this exercise. Column 1 demonstrates that the data are grouped using classes based on a single value. These single values in column 1 are used to label the horizontal axis of the frequency histogram. When classes are based on a single value, the middle of each histogram bar is placed directly over the single numerical value represented by the class. Also, the height of each bar in the frequency histogram matches the respective frequency in column 2.

Figure (a)

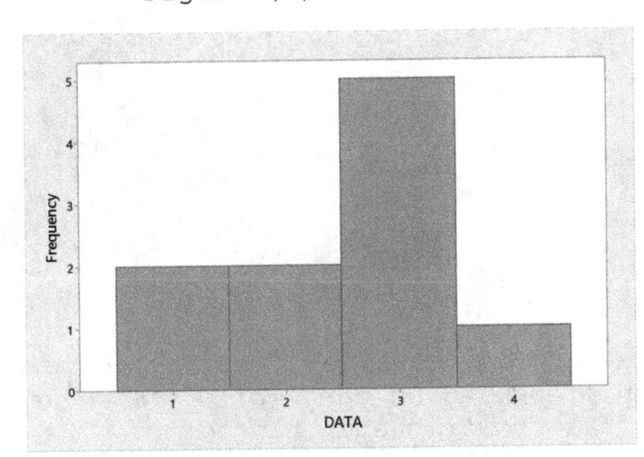

Figure (b)

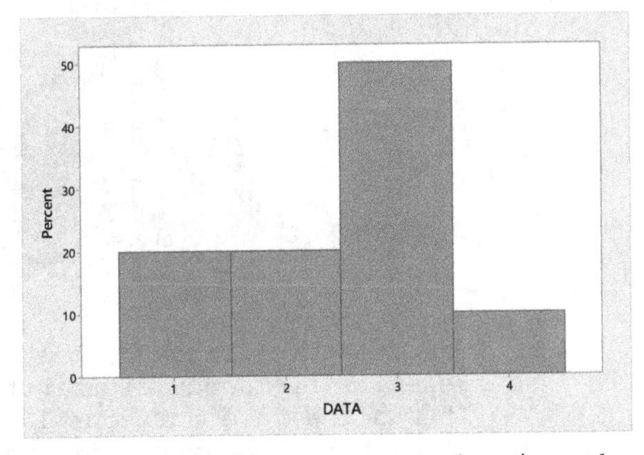

(d) The relative-frequency histogram in Figure (b) is constructed using the relative-frequency distribution presented in part (b) of this exercise. It has the same horizontal axis as the frequency histogram. The middle of each histogram bar is placed directly over the single numerical value represented by the class. Also, the height of each bar in the relative-frequency histogram matches the respective relative frequency in column 2.

2.63 (a) Since the data values range from 0 to 4, we construct a table with classes based on a single value. The resulting table follows.

Single Value Class	Frequency
0	1
1	1
2	4
3	8
4	6
	20

(b) To get the relative frequencies, divide each frequency by the sample size of 20.

Single Value Class	Relative Frequency
0	0.05
1	0.05
2	0.20
3	0.40
4	0.30
	1.00

(c) The frequency histogram in Figure (a) is constructed using the frequency distribution presented in part (a) of this exercise. Column 1 demonstrates that the data are grouped using classes based on a single value. These single values in column 1 are used to label the horizontal axis of the frequency histogram. When classes are based on a single value, the middle of each histogram bar is placed directly over the single numerical value represented by the class. Also, the height of each bar in the frequency histogram matches the respective frequency in column 2.

Figure (a)

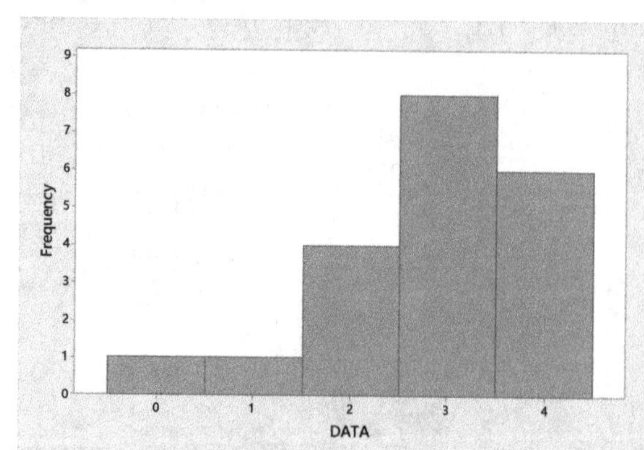

Figure (b)

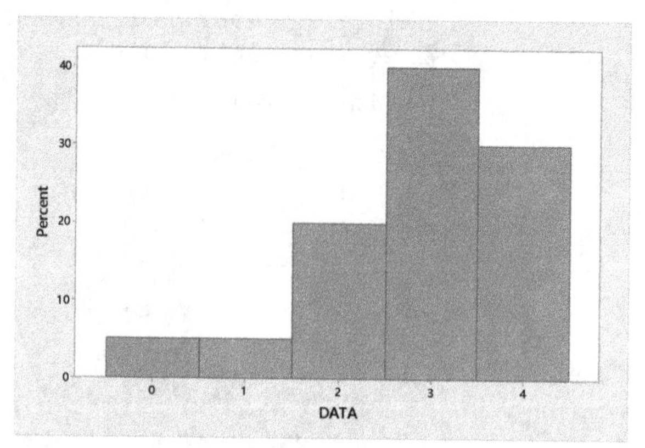

(d) The relative-frequency histogram in Figure (b) is constructed using the relative-frequency distribution presented in part (b) of this exercise. It has the same horizontal axis as the frequency histogram. The middle of each histogram bar is placed directly over the single numerical value represented by the class. Also, the height of each bar in the relative-frequency histogram matches the respective relative frequency in column 2.

2.65 (a) The first class to construct is 0-4. The width of all the classes is 5, so the next class would be 5-9. The classes are presented in column 1. The last class to construct is 25-29, since the largest single data value is 26. The tallied results are presented in column 2, which lists the frequencies.

Limit Grouping Classes	Frequency
0-4	5
5-9	5
10-14	1
15-19	4
20-24	2
25-29	3
	20

(b) Dividing each frequency by the total number of observations, which is 20, results in each class's relative frequency. The relative frequencies for all classes are presented in column 2. The resulting table follows.

Limit Grouping Classes	Relative Frequency
0-4	0.25
5-9	0.25
10-14	0.05
15-19	0.20
20-24	0.10
25-29	0.15
	1.00

(c) The frequency histogram in Figure (a) is constructed using the frequency distribution presented in part (a) of this exercise. The lower class limits of column 1 are used to label the horizontal axis of the frequency histogram. The height of each bar in the frequency histogram matches the respective frequency in column 2.

(d) The relative-frequency histogram in Figure (b) is constructed using the relative-frequency distribution presented in part (b) of this exercise. It has the same horizontal axis as the frequency histogram. The height of each bar in the relative-frequency histogram matches the respective relative frequency in column 2.

Figure (a)

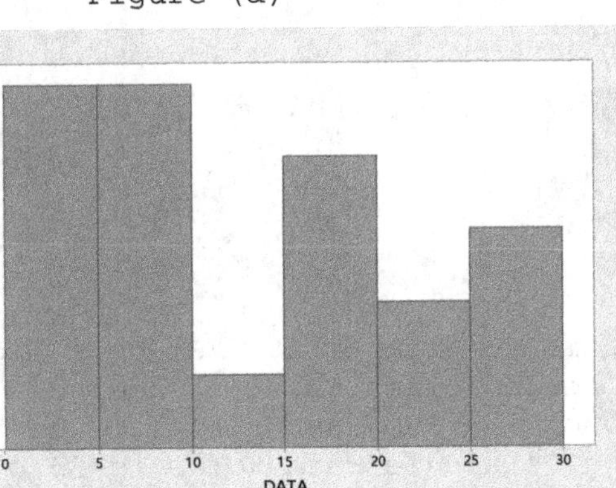

Figure (b)

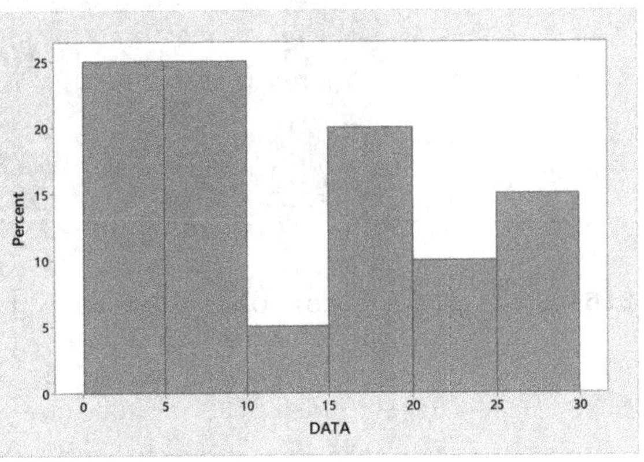

2.67 (a) The first class to construct is 50-59. The width of all the classes is 10, so the next class would be 60-69. The classes are presented in column 1. The last class to construct is 90-99, since the largest single data value is 98. The tallied results are presented in column 2, which lists the frequencies.

Limit Grouping Classes	Frequency
50-59	6
60-69	3
70-79	6
80-89	7
90-99	3
	25

(b) Dividing each frequency by the total number of observations, which is 25, results in each class's relative frequency. The relative frequencies for all classes are presented in column 2. The resulting table follows.

Limit Grouping Classes	Relative Frequency
50-59	0.24
60-69	0.12
70-79	0.24
80-89	0.28
90-99	0.12
	1.00

(c) The frequency histogram in Figure (a) is constructed using the frequency distribution presented in part (a) of this exercise. The lower class limits of column 1 are used to label the horizontal axis of the frequency histogram. The height of each bar in the frequency histogram matches the respective frequency in column 2.

(d) The relative-frequency histogram in Figure (b) is constructed using the relative-frequency distribution presented in part (b) of this exercise. It has the same horizontal axis as the frequency histogram. The height of each bar in the relative-frequency histogram matches the respective relative frequency in column 2.

Figure (a)

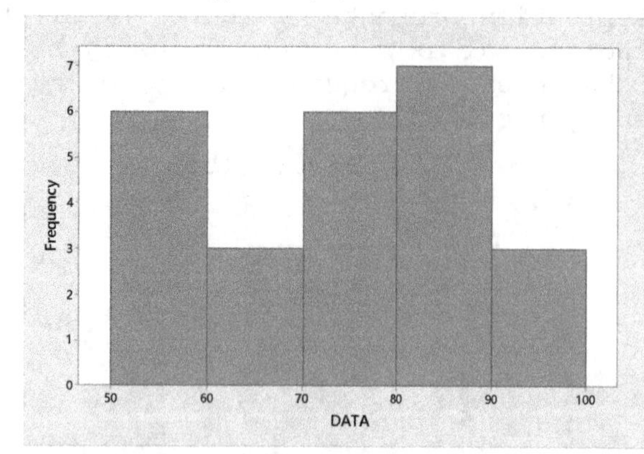

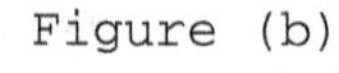
Figure (b)

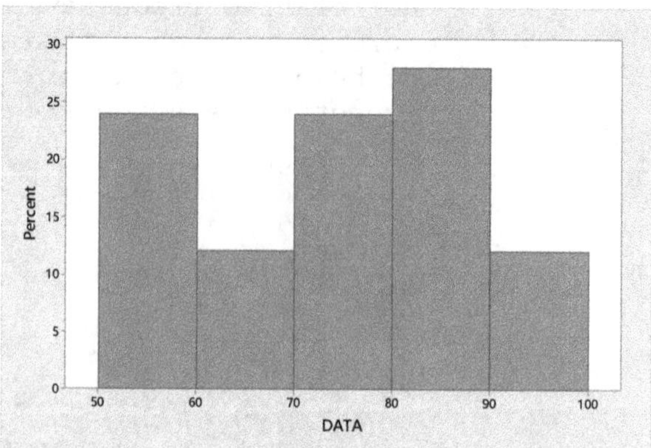

2.69 (a) The first class to construct is 40 - under 46. Since all classes are to be of equal width 6, the second class is 46 - under 52. All of the classes are presented in column 1. The last class to construct is 64 - under 70, since the largest single data value is 65.4. The results of the tallying are presented in column 2, which lists the frequencies.

Cutpoint Grouping Classes	Frequency
40 - under 46	3
46 - under 52	6
52 - under 58	10
58 - under 64	0
64 - under 70	1
	20

(b) Dividing each frequency by the total number of observations, which is 20, results in each class's relative frequency. The relative frequencies for all classes are presented in column 2.

Cutpoint Grouping	Relative Frequency
40 - under 46	0.15
46 - under 52	0.30
52 - under 58	0.50
58 - under 64	0.00
64 - under 70	0.05
	1.00

(c) The frequency histogram in Figure (a) is constructed using the frequency distribution obtained in part (a) of this exercise. The cutpoints are used to label the horizontal axis. Also, the height of each bar in the frequency histogram matches the respective frequency in column 2.

(d) The relative-frequency histogram in Figure (b) is constructed using the relative-frequency distribution obtained in part (b) of this exercise. It has the same horizontal axis as the frequency histogram. The height of each bar in the relative-frequency histogram matches the respective relative frequency in column 2.

Figure (a)

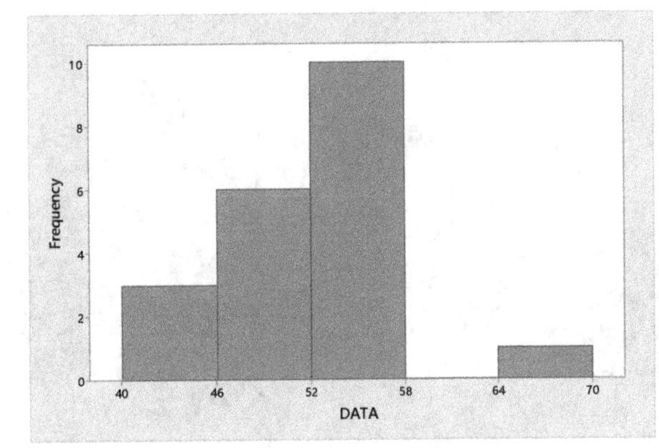

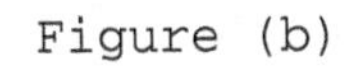

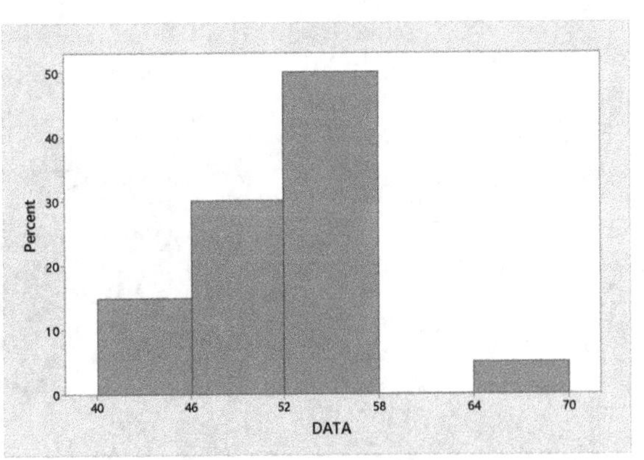

2.71 (a) The first class' cutpoint is 25 with a width of 3. Therefore, the first class to construct is 25 - under 28. Since all classes are to be of equal width 3, the second class is 28 - under 31. All of the classes are presented in column 1. The last class to construct is 43 - under 46, since the largest single data value is 43.01. The results of the tallying are presented in column 2, which lists the frequencies.

Cutpoint Grouping Classes	Frequency
25 - under 28	1
28 - under 31	2
31 - under 34	5
34 - under 37	7
37 - under 40	5
40 - under 43	4
43 - under 46	1
	25

(b) Dividing each frequency by the total number of observations, which is 25, results in each class's relative frequency. The relative frequencies for all classes are presented in column 2.

Cutpoint Grouping Classes	Relative Frequency
25 - under 28	0.04
28 - under 31	0.08
31 - under 34	0.20
34 - under 37	0.28
37 - under 40	0.20
40 - under 43	0.16
43 - under 46	0.04
	1.00

(c) The frequency histogram in Figure (a) is constructed using the frequency distribution obtained in part (a) of this exercise. The cutpoints are used to label the horizontal axis. Also, the height of each bar in the frequency histogram matches the respective frequency in column 2.

(d) The relative-frequency histogram in Figure (b) is constructed using the relative-frequency distribution obtained in part (b) of this exercise. It has the same horizontal axis as the frequency histogram. The height of each bar in the relative-frequency histogram matches the respective relative frequency in column 2.

Figure (a)

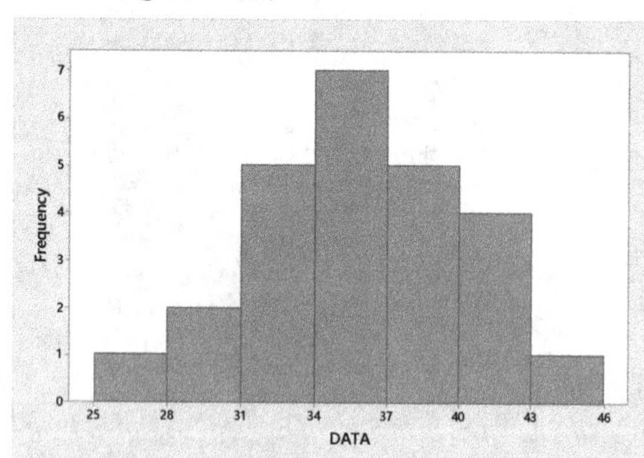

Figure (b)

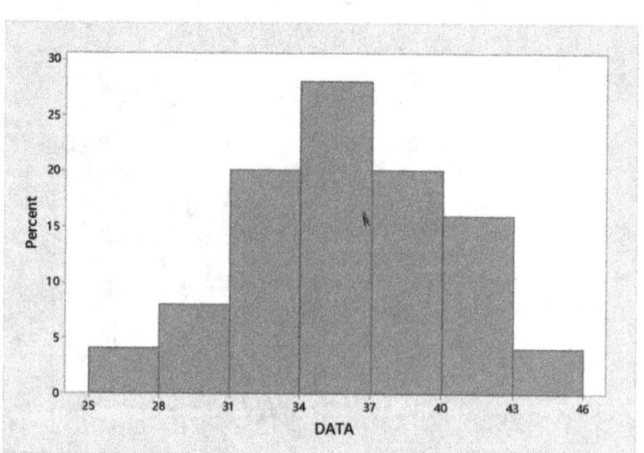

2.73 The horizontal axis of this dotplot displays a range of possible values. To complete the dotplot, we go through the data set and record data value by placing a dot over the appropriate value on the horizontal axis.

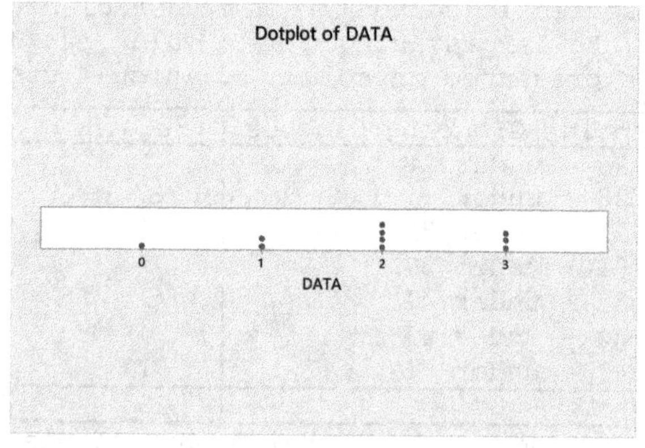

2.75 The horizontal axis of this dotplot displays a range of possible values. To complete the dotplot, we go through the data set and record data value by placing a dot over the appropriate value on the horizontal axis.

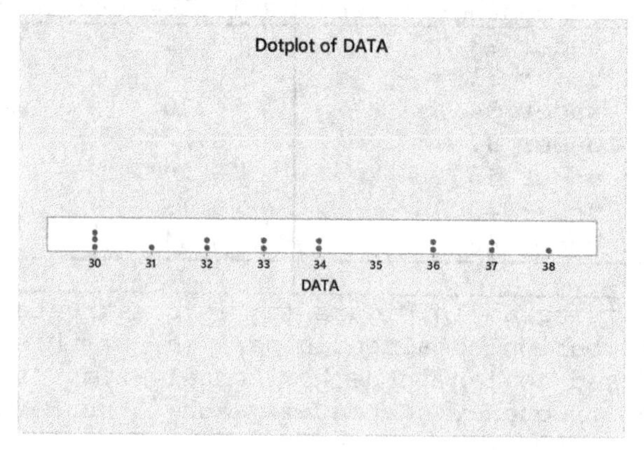

2.77 Each data value consists of 2 digit numbers ranging from 20 to 62. The last digit becomes the leaf and the remaining digits are the stems, so we have stems of 2 to 6. The resulting stem-and-leaf diagram is

```
2| 01
3| 2278
4| 13
5| 5
6| 2
```

2.79 Each data value consists of 2 digit numbers ranging from 22 to 46. The last digit becomes the leaf and the remaining digits are the stems, so we have stems of 2 to 4. Splitting the stems into two lines per stem, the resulting stem-and-leaf diagram is

```
2| 224
2| 5577789
3| 12234
3| 67
4| 0
4| 56
```

2.81 (a) Since the data values range from 1 to 7, we construct a table with classes based on a single value. The resulting table follows.

Number of Persons	Frequency
1	7
2	13
3	9
4	5
5	4
6	1
7	1
	40

(b) To get the relative frequencies, divide each frequency by the sample size of 40.

Number of Persons	Relative Frequency
1	0.175
2	0.325
3	0.225
4	0.125
5	0.100
6	0.025
7	0.025
	1.000

(c) The frequency histogram in Figure (a) is constructed using the frequency distribution presented in part (a) of this exercise. Column 1 demonstrates that the data are grouped using classes based on a single value. These single values in column 1 are used to label the horizontal axis of the frequency histogram. Suitable candidates for vertical axis units in the frequency histogram are the integers within the range 0 through 13, since these are representative of the magnitude and spread of the frequencies presented in column 2. When classes are based on a single value, the middle of each histogram bar is placed directly over the single numerical value represented by the class. Also, the height of each bar in the frequency histogram matches the respective frequency in column 2.

Figure (a)

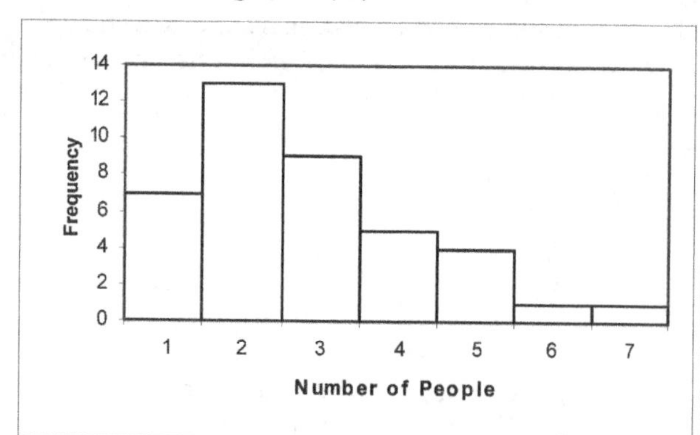

Figure (b)

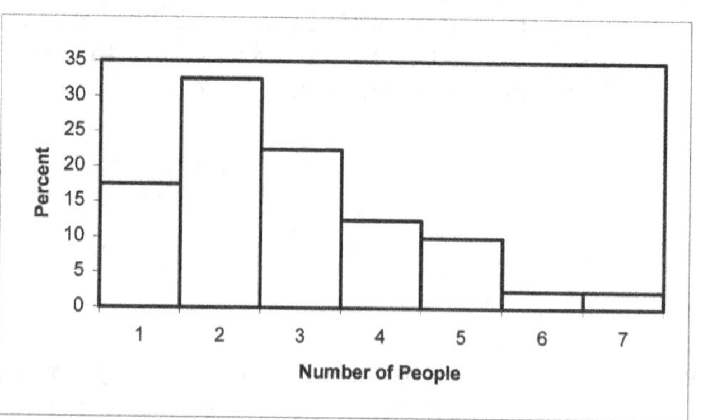

(d) The relative-frequency histogram in Figure (b) is constructed using the relative-frequency distribution presented in part (b) of this exercise. It has the same horizontal axis as the frequency histogram. We notice that the relative frequencies presented in column 2 range in size from 0.025 to 0.325. Thus, suitable candidates for vertical-axis units in the relative-frequency histogram are increments of 0.05 (or 5%), from zero to 0.35 (or 35%). The middle of each histogram bar is placed directly over the single numerical value represented by the class. Also, the height of each bar in the relative-frequency histogram matches the respective relative frequency in column 2.

2.83 (a) Since the data values range from 0 to 5, we construct a table with classes based on a single value. The resulting table follows.

Number of Computers	Frequency
0	5
1	22
2	11
3	2
4	4
5	1
	45

(b) To get the relative frequencies, divide each frequency by the sample size of 45. The sum of the relative frequency column is 0.999 due to rounding

Number of Computers	Relative Frequency
0	0.111
1	0.489
2	0.244
3	0.044
4	0.089
5	0.022
	0.999

(c) The frequency histogram in Figure (a) is constructed using the frequency distribution presented in part (a) of this exercise. Column 1 demonstrates that the data are grouped using classes based on a single value. These single values in column 1 are used to label the horizontal axis of the frequency histogram. When classes are based on a single value, the middle of each histogram bar is placed directly over the single numerical value represented by the class. Also, the height of each bar in the frequency histogram matches the respective frequency in column 2.

Figure (a)

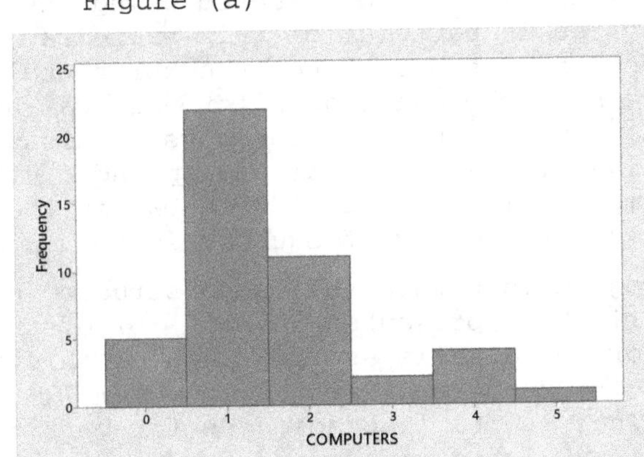

Figure (b)

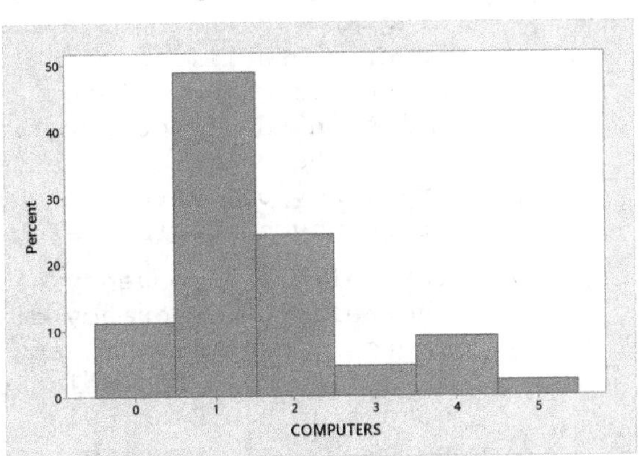

(d) The relative-frequency histogram in Figure (b) is constructed using the relative-frequency distribution presented in part (b) of this exercise. It has the same horizontal axis as the frequency histogram. The middle of each histogram bar is placed directly over the single numerical value represented by the class. Also, the height of each bar in the relative-frequency histogram matches the respective relative frequencies in column 2.

2.85 (a) The first class to construct is 40-44. Since all classes are to be of equal width, and the second class begins with 45, we know that the width of all classes is 45 - 40 = 5. All of the classes are presented in column 1. The last class to construct is 60-64, since the largest single data value is 61. Having established the classes, we tally the age figures into their respective classes. These results are presented in column 2, which lists the frequencies.

Age	Frequency
40-44	4
45-49	3
50-54	4
55-59	8
60-64	2
	21

(b) Dividing each frequency by the total number of observations, which is 21, results in each class's relative frequency. The relative frequencies for all classes are presented in column 2. The resulting table follows.

Age	Relative Frequency
40-44	0.190
45-49	0.143
50-54	0.190
55-59	0.381
60-64	0.095
	1.000

(c) The frequency histogram in Figure (a) is constructed using the frequency distribution presented in part (a) of this exercise. The lower class limits of column 1 are used to label the horizontal axis of the frequency histogram. Suitable candidates for vertical-axis units in the frequency histogram are the even integers 2 through 8, since these are representative of the magnitude and spread of the frequency presented in column 2. The height of each bar in the frequency histogram matches the respective frequency in column 2.

(d) The relative-frequency histogram in Figure (b) is constructed using the relative-frequency distribution presented in part (b) of this exercise. It has the same horizontal axis as the frequency histogram. We notice that the relative frequencies presented in column 2 vary in size from 0.095 to 0.381. Thus, suitable candidates for vertical axis units in the relative-frequency histogram are increments of 0.10 (or 10%), from zero to 0.40 (or 40%). The height of each bar in the relative-frequency histogram matches the respective relative frequency in column 2.

Figure (a) Figure (b)

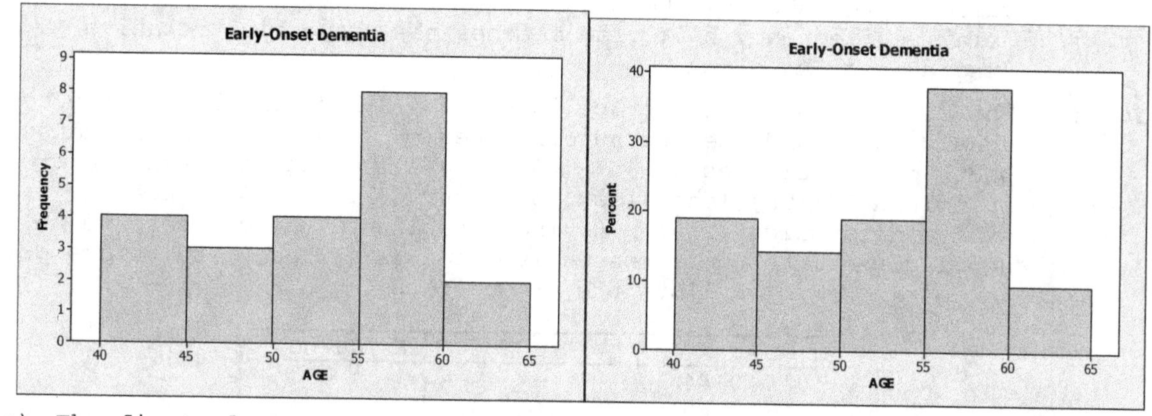

2.87 (a) The first class to construct is 12-17. Since all classes are to be of equal width, and the second class begins with 18, we know that the width of all classes is 18 - 12 = 6. All of the classes are presented in column 1. The last class to construct is 60-65, since the largest single data value is 61. Having established the classes, we tally the anxiety questionnaire score figures into their respective classes. These results are presented in column 2, which lists the frequencies.

Anxiety	Frequency
12-17	2
18-23	3
24-29	6
30-35	5
36-41	10
42-47	4
48-53	0
54-59	0
60-65	1
	31

(b) Dividing each frequency by the total number of observations, which is 31, results in each class's relative frequency. The relative frequencies for all classes are presented in column 2. The resulting table follows.

Anxiety	Relative Frequency
12-17	0.065
18-23	0.097
24-29	0.194
30-35	0.161
36-41	0.323
42-47	0.129
48-53	0.000
54-59	0.000
60-65	0.032
	1.000

(c) The frequency histogram in Figure (a) is constructed using the frequency distribution presented in part (a) of this exercise. The lower class limits of column 1 are used to label the horizontal axis of the frequency histogram. Suitable candidates for vertical-axis units in the frequency histogram are the even integers 0 through 10, since these are representative of the magnitude and spread of the frequency presented in column 2. The height of each bar in the frequency histogram matches the respective frequency in column 2.

(d) The relative-frequency histogram in Figure (b) is constructed using the relative-frequency distribution presented in part (b) of this exercise. It has the same horizontal axis as the frequency histogram. We notice that the relative frequencies presented in column 2 vary in size from 0.000 to 0.323. Thus, suitable candidates for vertical axis units in the relative-frequency histogram are increments of 0.05 (or 5%), from zero to 0.35 (or 35%). The height of each bar in the relative-frequency histogram matches the respective relative frequency in column 2.

Figure (a) Figure (b)

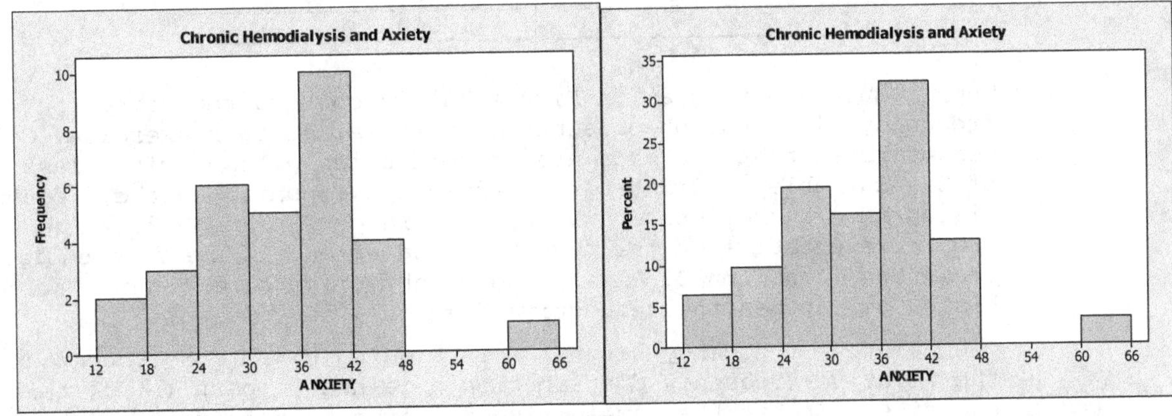

2.89 (a) The first class to construct is 52 - under 54. Since all classes are to be of equal width 2, the second class is 54 - under 56. All of the classes are presented in column 1. The last class to construct is 74 - under 76, since the largest single data value is 75.3. Having established the classes, we tally the cheetah speeds into their respective classes. These results are presented in column 2, which lists the frequencies.

Speed	Frequency
52 - under 54	2
54 - under 56	5
56 - under 58	6
58 - under 60	8
60 - under 62	7
62 - under 64	3
64 - under 66	2
66 - under 68	1
68 - under 70	0
70 - under 72	0
72 - under 74	0
74 - under 76	1
	35

(b) Dividing each frequency by the total number of observations, which is 35, results in each class's relative frequency. The relative frequencies for all classes are presented in column 2

Speed	Relative Frequency
52 - under 54	0.057
54 - under 56	0.143
56 - under 58	0.171
58 - under 60	0.229
60 - under 62	0.200
62 - under 64	0.086
64 - under 66	0.057
66 - under 68	0.029
68 - under 70	0.000
70 - under 72	0.000
72 - under 74	0.000
74 - under 76	0.029
	1.000

(c) The frequency histogram in Figure (a) is constructed using the frequency distribution obtained in part (a) of this exercise Column 1 demonstrates that the data are grouped using classes with class widths of 2. Suitable candidates for vertical axis units in the frequency histogram are the integers within the range 0 through 8, since these are representative of the magnitude and spread of the frequencies presented in column 2. Also, the height of each bar in the frequency histogram matches the respective frequency in column 2.

(d) The relative-frequency histogram in Figure (b) is constructed using the relative-frequency distribution obtained in part (b) of this exercise. It has the same horizontal axis as the frequency histogram. We notice that the relative frequencies presented in column 3 range in size from 0.000 to 0.229. Thus, suitable candidates for vertical axis units in the relative-frequency histogram are increments of 0.05 (5%), from zero to 0.25 (25%). The height of each bar in the relative-frequency histogram matches the respective relative frequency in column 2.

Figure (a)

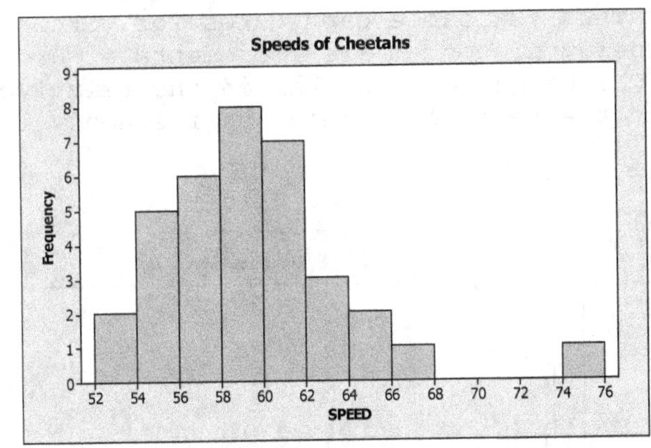

Figure (b)

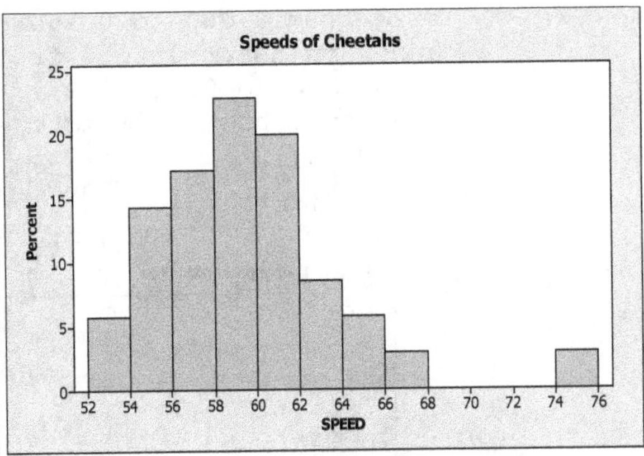

2.91 (a) The first class to construct is 0 - under 1. Since all classes are to be of equal width 1, the second class is 1 - under 2. All of the classes are presented in column 1. The last class to construct is 7 - under 8, since the largest single data value is 7.6. Having established the classes, we tally the fuel tank capacities into their respective classes. These results are presented in column 2, which lists the frequencies.

Oxygen Distribution	Frequency
0 - under 1	1
1 - under 2	10
2 - under 3	5
3 - under 4	4
4 - under 5	0
5 - under 6	0
6 - under 7	1
7 - under 8	1
	22

(b) Dividing each frequency by the total number of observations, which is 22, results in each class's relative frequency. The relative frequencies for all classes are presented in column 2

Oxygen Distribution	Relative Frequency
0 - under 1	0.045
1 - under 2	0.455
2 - under 3	0.227
3 - under 4	0.182
4 - under 5	0.000
5 - under 6	0.000
6 - under 7	0.045
7 - under 8	0.045
	1.000

(c) The frequency histogram in Figure (a) is constructed using the frequency distribution obtained in part (a) of this exercise Column 1 demonstrates that the data are grouped using classes with class widths of 2. Suitable candidates for vertical axis units in the frequency histogram are the integers within the range 0 through 10, since these are representative of the magnitude and spread of the frequencies presented in column 2. Also, the height of each bar in the frequency histogram matches the respective frequency in column 2.

(d) The relative-frequency histogram in Figure (b) is constructed using the relative-frequency distribution obtained in part (b) of this exercise. It has the same horizontal axis as the frequency histogram.

We notice that the relative frequencies presented in column 3 range in size from 0.000 to 0.455. Thus, suitable candidates for vertical axis units in the relative-frequency histogram are increments of 0.05 (5%), from zero to 0.50 (50%). The height of each bar in the relative-frequency histogram matches the respective relative frequency in column 2.

Figure (a)

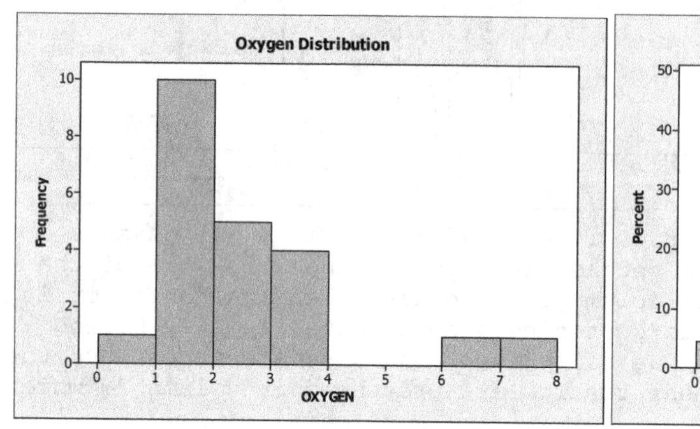

Figure (b)

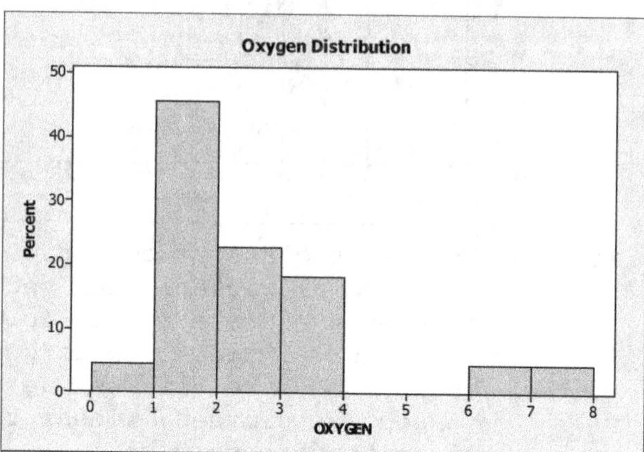

2.93 The horizontal axis of this dotplot displays of range of possible ages of the passenger cars. To complete the dotplot, we go through the data set and record each age by placing a dot over the appropriate value on the horizontal axis.

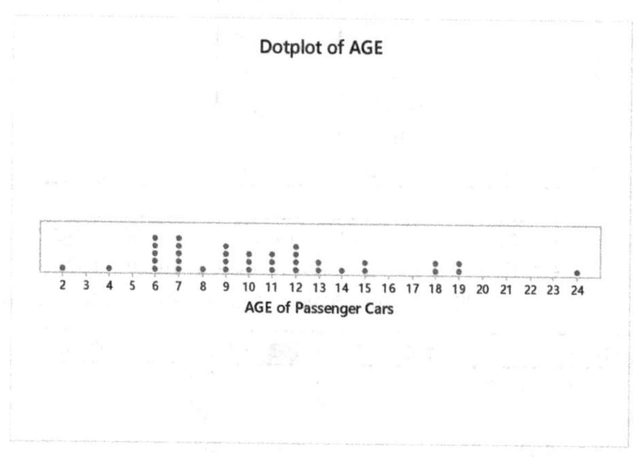

2.95 (a) The data values range from 7 to 18, so the scale must accommodate those values. We stack dots above each value on two different lines using the same scale for each line. The result is

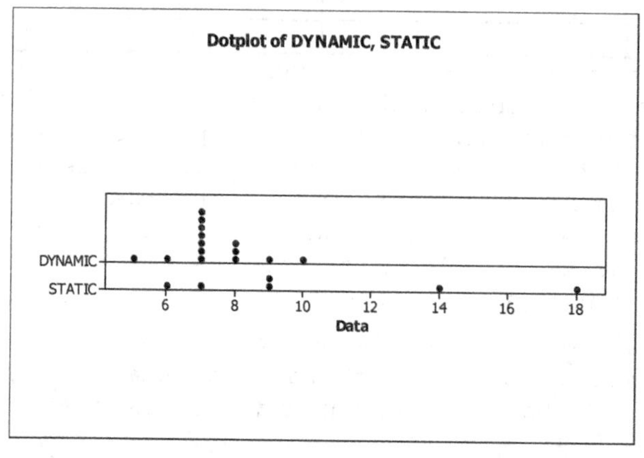

(b) The Dynamic system does seem to reduce acute postoperative days in the hospital on the average. The Dynamic data are centered at about 7 days, whereas the Static data are centered at about 11 days and are much more spread out than the Dynamic data.

2.97 Since each data value consists of 2 digits, each beginning with 1, 2, 3, or 4, we will construct the stem-and-leaf diagram with these four values as the stems. The result is

```
1| 238
2| 1678899
3| 34459
4| 04
```

2.99 (a) Since each data value lies between 2 and 93, we will construct the stem-and-leaf diagram with one line per stem. The result is

```
0| 2234799
1| 11145566689
2| 023479
3| 004555
4| 19
5| 5
6| 9
7| 9
8|
9| 3
```

(b) Using two lines per stem, the same data result in the following diagram:

```
0| 2234
0| 799
1| 1114
1| 5566689
2| 0234
2| 79
3| 004
3| 555
4| 1
4| 9
5|
5| 5
6|
6| 9
7|
7| 9
8|
8|
9| 3
```

(c) The stem with one line per stem is more useful. One gets the same impression regarding the shape of the distribution, but the two lines per stem version has numerous lines with no data, making it take up more space than necessary to interpret the data and giving it too many lines.

2.101 (a) Since we have two digit numbers, the last digit becomes the leaf and the first digit becomes the stem. For this data, we have stems of 6, 7, and 8. Splitting the data into five lines per stem, we put the leaves 0-1 in the first stem, 2-3 in the second stem, 4-5 in the third stem, 6-7 in the fourth stem, and 8-9 in the fifth stem. The result is

```
6| 8
7| 11111111
7| 22222222233333
7| 4444555555
7| 66666777777
7| 8
8| 0
```

(b) Using one or two lines per stem would have given us too few lines.

2.103 The graph indicates that:

(a) 20% of the patients have cholesterol levels between 205 and 209, inclusive.

(b) 20% are between 215 and 219; and 5% are between 220 and 224. Thus, 25% (i.e., 20% + 5%) have cholesterol levels of 215 or higher.

(c) 35% of the patients have cholesterol levels between 210 and 214, inclusive. With 20 patients in total, the number having cholesterol levels between 210 and 214 is 7 (i.e., 0.35 x 20).

2.105 The graph indicates that:

(a) 1 of these 14 patients was under 45 years old at the onset of symptoms.

(b) 3 of these 14 patients were at least 65 years old at the onset of symptoms.

(c) 8 of these 14 patients were between 50 and 64 years old, inclusive, at the onset of symptoms.

2.107 (a) Using Minitab, there is not a direct way to get a grouped frequency distribution. However, you can use an option in creating your histogram that will report the frequencies in each of the classes, essentially creating a grouped frequency distribution. Retrieve the data from the WeissStats Resource Site. Column 2 contains the length of the Beatles songs, in seconds. From the tool bar, select **Graph ▶ Histogram,** choose **Simple** and click **OK**. Double click on LENGTH to enter LENGTH in the **Graph variables** box. Click **Labels,** click the **Data Labels** tab, then check **Use y-value labels.** Click **OK** twice. The result is

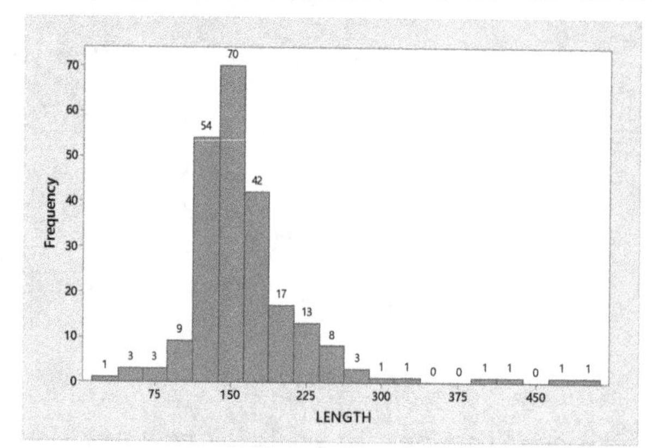

Above each bar is a label for each of the frequencies. Also, the labeling on the horizontal axis is the midpoint for class, where each class has a width of 25. To change the width of the bars, right Click on the bars and choose **Edit Bars,** click **Binning,** click **Midpoint/Cutpoint positions:,** type 0 50 100 150 200 250 300 350 400 450 500. The result is

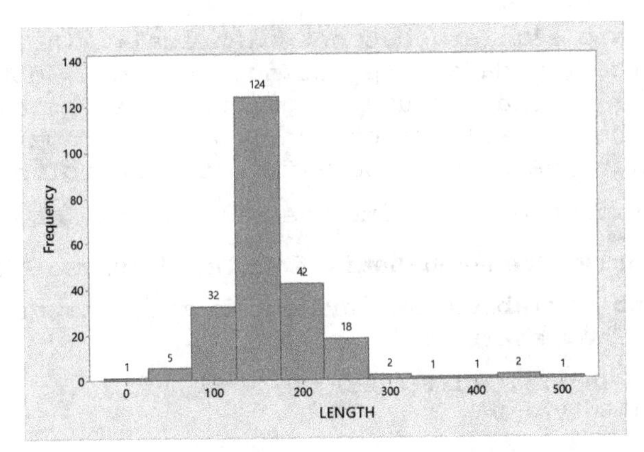

To get the relative-frequency distribution, follow the same steps, but also Click the **Scale** button, click the **Y-scale Type,** check **Percent,** then click **OK** twice. The result is

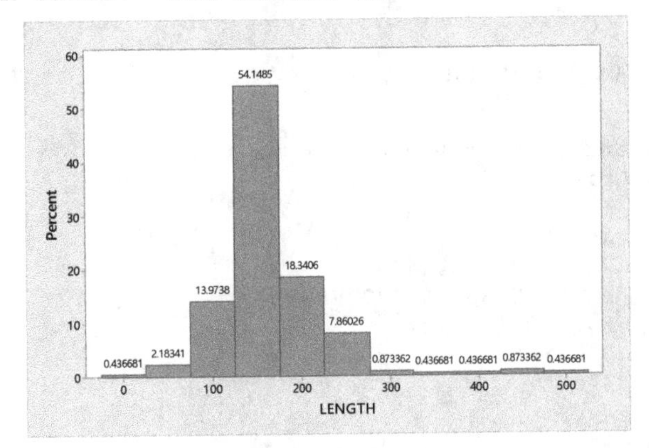

Above each bar is the percentage for that class, essentially creating a relative-frequency distribution. You could also transfer these results into a table.

(b) A frequency histogram and a relative frequency distribution were created in part (a).

(c) To obtain the dotplot, select **Graph ▶ Dotplot,** select **Simple** in the **One Y** row, and click **OK.** Double click on LENGTH to enter LEGNTH in the **Graph variables** box and click **OK.** The result is

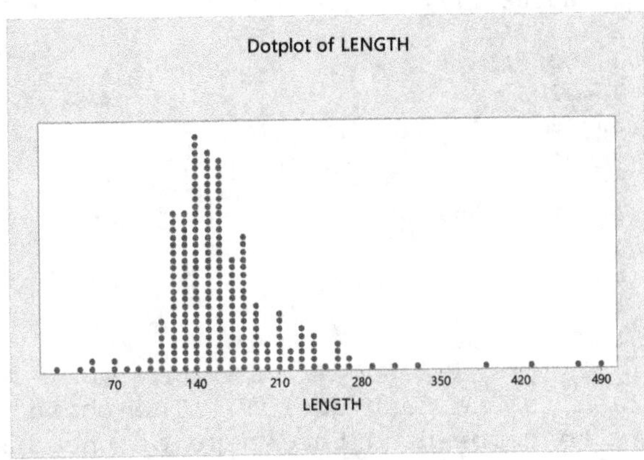

(d) The graphs are similar, but not identical. This is because the dotplot preserves the raw data by plotting individual dots and the histogram looses the raw data because it groups observations into grouped classes. The overall impression, however, remains the same. They both are generally the same shape with outliers to the right.

2.109 (a) After entering the data from the WeissStats Resource Site, in Minitab, select **Graph ▶ Stem-and-Leaf,** double click on PERCENT to enter PERCE T in the **Graph variables** box and enter a 10 in the **Increment** box, and click **OK.** The result is

```
Stem-and-leaf of PERCENT  N  = 51
Leaf Unit = 1.0

 2     1  79
(32)   2  01122333444455555566666777778999
 17    3  0001112223456668
 1     4  9
```

(b) Repeat part (a), but this time enter a 5 in the **Increment** box. The result is

```
Stem-and-leaf of PERCENT  N  = 51
Leaf Unit = 1.0

 2     1  79
 14    2  011223334444
(20)   2  55555566666777778999
 17    3  00011122234
 6     3  56668
 1     4
 1     4  9
```

(c) Repeat part (a) again, but this time enter a 2 in the **Increment** box. The result is

```
Stem-and-leaf of PERCENT  N  = 51
Leaf Unit = 1.0

 1     1  7
 2     1  9
 5     2  011
 10    2  22333
 20    2  4444555555
(10)   2  6666677777
 21    2  8999
 17    3  000111
 11    3  2223
 7     3  45
 5     3  666
 2     3  8
 1     4
 1     4
 1     4
 1     4
 1     4  9
```

(d) The second graph is the most useful. The third one has more classes than necessary to comprehend the shape of the distribution and has a number of empty stems. Typically, we like to have five to fifteen classes and the first and second diagrams satisfy that condition, but the second one provides a better idea of the shape of the distribution.

2.111 (a) The classes are presented in column 1. With the classes established, we then tally the exam scores into their respective classes. These results are presented in column 2, which lists the frequencies. Dividing each frequency by the total number of exam scores, which is 20, results in each class's relative frequency. The relative frequencies for all classes are presented in column 3. The class mark of each class is the average of the lower and upper limits. The class marks for all classes are presented in column 4.

Score	Frequency	Relative Frequency	Class Mark
30-39	2	0.10	34.5
40-49	0	0.00	44.5
50-59	0	0.00	54.5
60-69	3	0.15	64.5
70-79	3	0.15	74.5
80-89	8	0.40	84.5
90-100	4	0.20	95.0
	20	1.00	

(b) The first six classes have width 10; the seventh class had width 11.

(c) Answers will vary, but one choice is to keep the first six classes the same and make the next two classes 90-99 and 100-109. Another possibility is 31-40, 41-50, …, 91-100.

2.113 Answers will vary, but by following the steps we first decide on the approximate number of classes. Since there are 37 observations, we should have 7-14 classes. This exercise states we should have approximately eight classes. Step 2 says that we calculate an approximate class width as (278.8 − 129.2)/8 = 18.7. A convenient class width close to 18.7 would be a class width of 20. Step 3 says that we choose a number for the lower cutpoint which is less than or equal to our minimum observation of 129.2. Let's choose 120. Beginning with a lower cutpoint of 120 and width of 20, we have a first class of 120 - under 140, a second class of 140 - under 160, a third class of 160 - under 180, a fourth class of 180 - under 200, a fifth class of 200 - under 220, a sixth class of 220 - under 240, a seventh class of 240 - under 260, and an eighth class of 260 - under 280. This would be our last class since the largest observation is 278.8.

2.115 Consider columns 1 and 2 of the energy-consumption data given in Exercise 2.56 part (b). Compute the class mark for each class presented in column 1. Pair each class mark with its corresponding relative frequency found in column 2. Construct a horizontal axis, where the units are in terms of class marks and a vertical axis where the units are in terms of relative frequencies. For each class mark on the horizontal axis, plot a point whose height is equal to the relative frequency of the class. Then join the points with connecting lines. The result is a relative-frequency polygon.

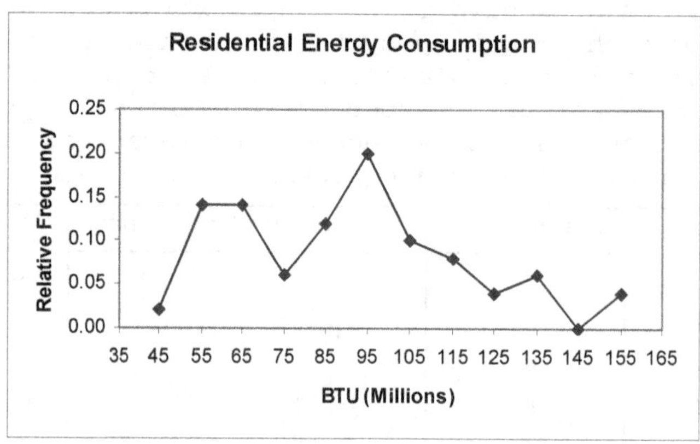

2.117 In single value grouping the horizontal axis would be labeled with the value of each class.

2.119 (a) Consider parts (a) and (b) of the Cheetah speed data given in Exercise 2.61. The classes are now reworked to present just the lower cutpoint of each class. The frequencies are reworked to sum the frequencies of all classes representing values less than the specified lower cutpoint. These successive sums are the cumulative frequencies. The relative frequencies are reworked to sum the relative frequencies of all classes representing values less than the specified cutpoints. These successive sums are the cumulative relative frequencies. (Note: The cumulative relative frequencies can also be found by dividing the each cumulative frequency by the total number of data values.)

Less than	Cumulative Frequency	Cumulative Relative Frequency
52	0	0.000
54	2	0.057
56	7	0.200
58	13	0.371
60	21	0.600
62	28	0.800
64	31	0.886
66	33	0.943
68	34	0.971
70	34	0.971
72	34	0.971
74	34	0.971
76	35	1.000

(b) Pair each cutpoint with its corresponding cumulative relative frequency found in column 3. Construct a horizontal axis, where the units are in terms of the cutpoints and a vertical axis where the units are in terms of cumulative relative frequencies. For each cutpoint on the horizontal axis, plot a point whose height is equal to the cumulative relative frequency. Then join the points with connecting lines. The result, presented in the following figure, is an *ogive* using cumulative relative frequencies. (Note: A similar procedure could be followed using cumulative frequencies.)

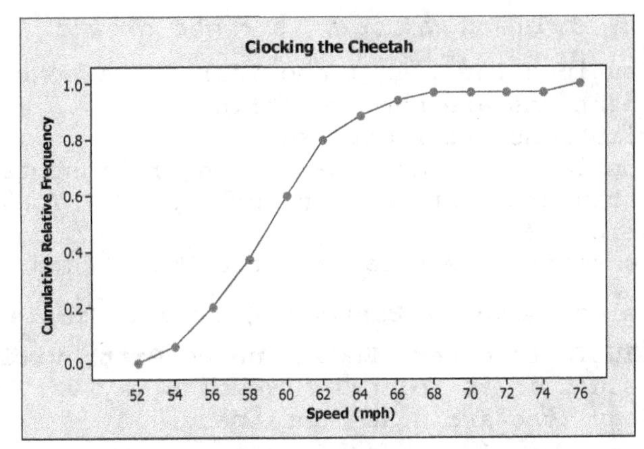

2.121 Minitab used truncation. Note that there was a data point of 5.8 in the sample. It would have been plotted with a stem of 0 and a leaf of 6 if it had been rounded. Instead Minitab plotted the observation with a stem of 0 and a leaf of 5.

Section 2.4

2.123 Sample data are the values of a variable for a sample of the population.

2.125 A sample distribution is the distribution of sample data.

2.127 A distribution of a variable is the same as a population distribution.

2.129 A large simple random sample from a bell-shaped distribution would be expected to have roughly a bell-shaped distribution since more sample values should be obtained, on average, from the middle of the distribution.

2.131 Three distribution shapes that are symmetric are bell-shaped, triangular, and Uniform (or rectangular), shown in that order below. It should be noted that there are others as well.

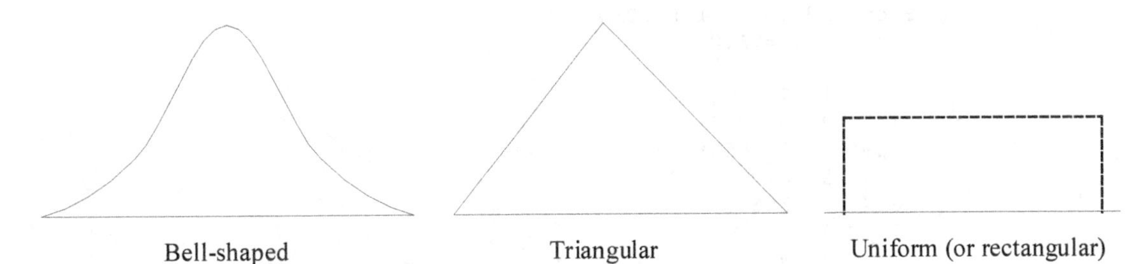

2.133 (a) The shape of the distribution is unimodal.
(b) The shape of the distribution is symmetric.

2.135 (a) The shape of the distribution is unimodal.
(b) The shape of the distribution is not symmetric.
(c) The shape of the distribution is left-skewed.

2.137 (a) The shape of the distribution is unimodal.
(b) The shape of the distribution is not symmetric.
(c) The shape of the distribution is left-skewed.

2.139 (a) The shape of the distribution is multimodal.
(b) The shape of the distribution is not symmetric.

2.141 Except for the one data value between 74 and 76, this distribution is close to bell-shaped. That one value makes the distribution slightly right skewed.

2.143 The distribution of depths of the burrows is left skewed.

2.145 The distribution of PCB concentration is symmetric.

2.147 The distribution of cholesterol levels appears to be slightly left skewed.

2.149 The distribution of length of stay is right skewed.

2.151 (a) The distributions for Year 1 and Year 2 are both unimodal.
(b) Both distributions are not symmetric.
(c) Both distributions are right skewed.
(d) The distribution for Year 1 has a longer right tail indicating more variation than the distribution for Year 2. They are also not centered in the same place.

2.153 (a) After entering the data from the WeissStats Resource Site, in Minitab, select **Graph ▶ Histogram,** select **Simple** and click **OK.** Double click on LENGTH to enter LENGTH in the **Graph variables** box and click **OK**. Our result is as follows. Results may vary depending on the type of technology used and graph obtained.

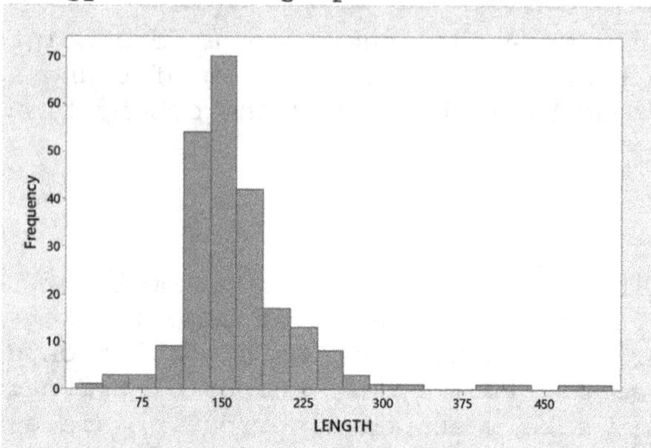

The distribution of LENGTH is unimodal and not symmetric.

(b) The distribution is right skewed.

2.155 (a) In Exercise 2.109, we used Minitab to obtain a stem-and-leaf diagram using 2 lines per stem. That diagram is shown below.

```
Stem-and-leaf of PERCENT  N  = 51
Leaf Unit = 1.0

  2    1  79
 14    2  011223334444
(20)   2  55555566666777778999
 17    3  00011122234
  6    3  56668
  1    4
  1    4  9
```

The overall shape of this distribution is unimodal and roughly symmetric.

2.157 (a) After entering the data from the WeissStats Resource Site, in Minitab, select **Graph ▶ Histogram,** select **Simple** and click **OK.** Double click on LENGTH to enter LENGTH in the **Graph variables** box and click **OK**. Our result is as follows. Results may vary depending on the type of technology used and graph obtained. The distribution of LENGTH is approximately unimodal and symmetric.

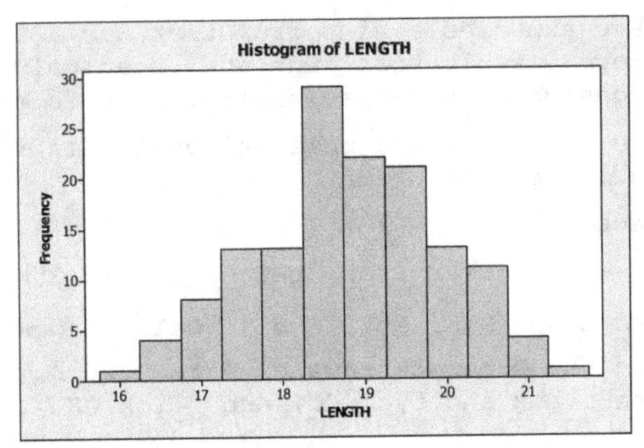

2.159 The precise answers to this exercise will vary from class to class or individual to individual. Thus your results will likely differ from our results shown below.

(a) We obtained 50 random digits from a table of random numbers. The digits were

4 5 4 6 8 9 9 7 7 2 2 2 9 3 0 3 4 0 0 8 8 4 4 5 3

9 2 4 8 9 6 3 0 1 1 0 9 2 8 1 3 9 2 5 8 1 8 9 2 2

(b) Since each digit is equally likely in the random number table, we expect that the distribution would look roughly uniform.

(c) Using single value classes, the frequency distribution is given by the following table. The relative frequency histogram is shown below.

Value	Frequency	Relative-Frequency
0	5	.10
1	4	.08
2	8	.16
3	5	.10
4	6	.12
5	3	.06
6	2	.04
7	2	.04
8	7	.14
9	8	.16

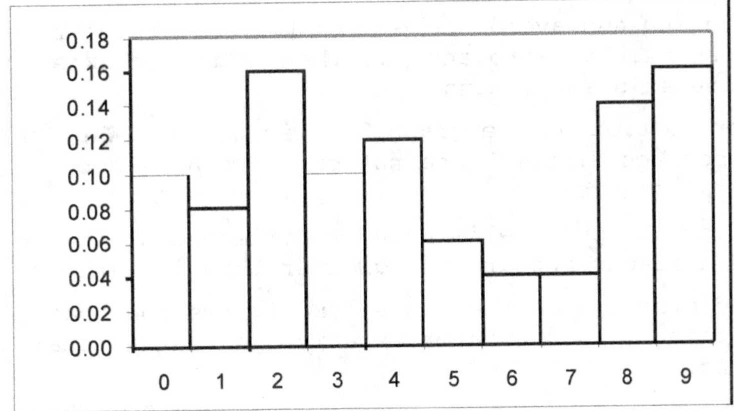

We did not expect to see this much variation.

(d) We would have expected a histogram that was a little more 'even', more like a uniform distribution, but the relatively small sample size can result in considerable variation from what is expected.

(e) We should be able to get a more uniformly distributed set of data if we choose a larger set of data.

(f) Class project.

2.161 (a) Your result will differ from, but be similar to, the one below which was obtained using Minitab. Choose **Calc ▶ Random Data ▶ Normal...**, type 3000 in the **Generate rows of data** text box, click in the **Store in column(s)** text box and type C1, and click **OK**.

(b) Then choose **Graph ▶ Histogram**, choose the **Simple** version, click **OK**, enter C1 in the **Graph variables** text box, click on the **Scale** button and then on the **Y-Scale Type** tab. Check the **Percent** box and click **OK** twice.

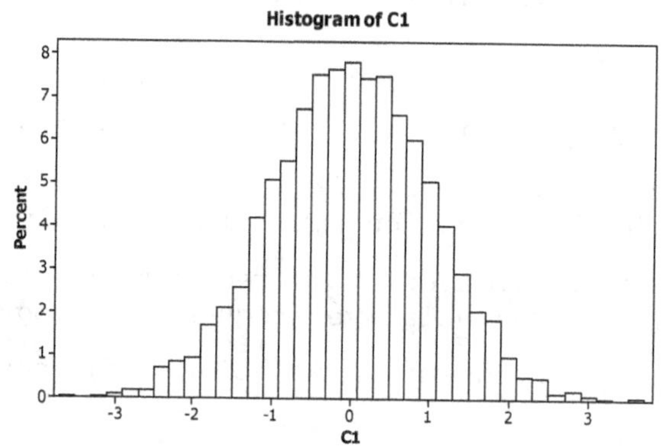

(c) The histogram in part (b) has the bell, unimodal and symmetric distribution. The sample of 3000 is representative of the population from which the sample was taken.

Section 2.5

2.163 (a) A truncated graph is one for which the vertical axis starts at a value other than its natural starting point, usually zero.

(b) A legitimate motivation for truncating the axis of a graph is to place the emphasis on the ups and downs of the distribution rather than on the actual height of the graph.

(c) To truncate a graph and avoid the possibility of misinterpretation, one should start the axis at zero and put slashes in the axis to indicate that part of the axis is missing.

2.165 (a) A large lower portion of the graph is eliminated. When this is done, differences between district and national averages appear greater than in the original figure.

(b) Even more of the graph is eliminated. Differences between district and national averages appear even greater than in part (a).

(c) The truncated graphs give the misleading impression that, in 2013, the district average is much greater relative to the national average than it actually is.

2.167 (a) The problem with the bar chart is that it is truncated. That is, the vertical axis, which should start at $0 (in trillions), starts with $7.6 (in trillions) instead. The part of the graph from $0 (in trillions) to $7.6 (in trillions) has been cut off. This truncation causes the bars to be out of correct proportion and hence creates the misleading impression that the money supply is changing more than it actually is.

(b) A version of the bar chart with an untruncated and unmodified vertical axis is presented in Figure (a). Notice that the vertical axis starts at $0.00 (in trillions). Increments are in trillion dollars. In contrast to the original bar chart, this one illustrates that the changes in money supply from week to week are not that different. However, the "ups" and "downs" are not as easy to spot as in the original, truncated bar chart.

(c) A version of the bar chart in which the vertical axis is modified in an acceptable manner is presented in Figure (b). Notice that the special symbol "//" is used near the base of the vertical axis to signify that the vertical axis has been modified. Thus, with this version of the bar chart, not only are the "ups" and "downs" easy to spot but the reader is also aptly warned by the slashes that part of the vertical axis between $0.00 (in trillions) and $7.6 (in trillions) has been removed.

Figure (a)

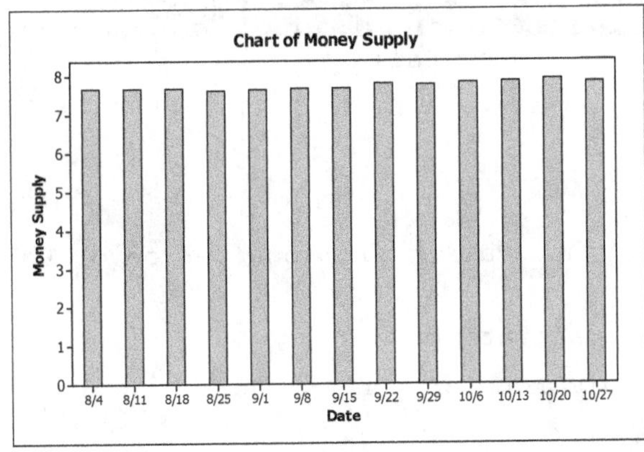

Figure (b)

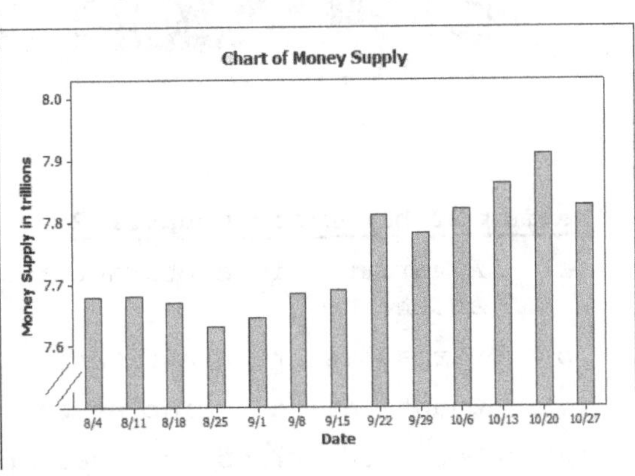

2.169 (b) Without the vertical scale, it would appear that the happiness score dropped about 25% between the ages of 15 and 20.

(c) The actual drop was from a happiness score of about 5.5 to a happiness score of 5.25 between the ages of 15 and 20. This is a drop of 0.25/5.5 = 0.045 or about 4.5%.

(d) Without the vertical scale, it would look like the happiness scale has dropped about 50% from 5.5 to 5.0 where actually it really is closer to a 20% drop. The peak at 74 looks like a bigger significant increase without the vertical scale as well. Because the scale on the horizontal axis is not evenly spaced, the time between events is misleading.

(e) The graph could be made less potentially misleading by either making the vertical scale range from zero to 6.0 and by making the scale on the vertical axis the same width.

2.171 (a) The brochure shows a "new" ball with twice the radius of the "old" ball. The intent is to give the impression that the "new" ball lasts roughly twice as long as the "old" ball. However, if the "new" ball has twice the radius of the "old" ball, the "new" ball will have eight times the volume of the "old" ball (since the volume of a sphere is proportional to the cube of its radius, or the radius $2^3 = 8$).

Thus, the scaling is improper because it gives the impression that the "new" ball lasts eight times as long as the "old" ball rather than merely two times as long.

Old Ball

New Ball

(b) One possible way for the manufacturer to illustrate the fact that the "new" ball lasts twice as long as the "old" ball is to present pictures of two balls, side by side, each of the same magnitude as the picture of the "old" ball and to label this set of two balls "new ball". This will illustrate the point that a purchaser will be getting twice as much for the money.

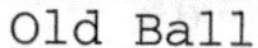

Old Ball

New Ball

<u>Review Problems For Chapter 2</u>

1. (a) A variable is a characteristic that varies from one person or thing to another.

(b) Variables are quantitative or qualitative.

(c) Quantitative variables can be discrete or continuous.

2. (d) Data are values of a variable.

(e) The data type is determined by the type of variable being observed.

3. A frequency distribution of qualitative data is a table that lists the distinct values of data and their frequencies. It is useful to organize the data and make it easier to understand. A relative-frequency distribution of qualitative data is a table that lists the distinct values of data and a ratio of the class frequency to the total number of observations, which is called the relative frequency.

4. For both quantitative and qualitative data, the frequency and relative-frequency distributions list the values of the distinct classes and their frequencies and relative frequencies. For single value grouping of quantitative data, it is the same as the distinct classes for qualitative data. For class limit and cutpoint grouping in quantitative data, we create groups that form distinct classes similar to qualitative data.

5. The two main types of graphical displays for qualitative data are the bar chart and the pie chart.

6. The bars do not abut in a bar chart because there is not any continuity between the categories. Also, this differentiates them from histograms.

7. Answers will vary. One advantage of pie charts is that it shows the proportion of each class to the total. One advantage of bar charts is that it emphasizes each individual class in relation to each other. One disadvantage of pie charts is that if there are too many classes, the chart becomes confusing. Also, if a class is really small relative to the total, it is hard to see the class in a pie chart.

8. Single value grouping is appropriate when the data are discrete with relatively few distinct observations.

9. (a) The second class would have lower and upper limits of 9 and 14. The class mark of this class would be the average of these limits and equal 11.5.

(b) The third class would have lower and upper limits of 15 and 20.

(c) The fourth class would have lower and upper limits of 21 and 26. This class would contain an observation of 23.

10. (a) If the width of the class is 5, then the class limits will be four whole numbers apart. Also, the average of the two class limits is the class mark of 8. Therefore, the upper and lower class limits must be two whole numbers above and below the number 8. The lower and upper class limits of the first class are 6 and 10.

(b) The second class will have lower and upper class limits of 11 and 15. Therefore, the class mark will be the average of these two limits and equal 13.

(c) The third class will have lower and upper class limits of 16 and 20.

(d) The fourth class has limits of 21 and 25, the fifth class has limits of 26 and 30. Therefore, the fifth class would contain an observation of 28.

11. (a) The common class width is the distance between consecutive cutpoints, which is 15 - 5 = 10.

(b) The midpoint of the second class is halfway between the cutpoints 15 and 25, and is therefore 20.

(c) The sequence of the lower cutpoints is 5, 15, 25, 35, 45, ... Therefore, the lower and upper cutpoints of the third class are 25 and 35.

(d) Since the third class has lower and upper cutpoints of 25 and 35, an observation of 32.4 would belong to this class.

12. (a) The midpoint is halfway between the cutpoints. Since the class width is 8, 10 is halfway between 6 and 14.

(b) The class width is also the distance between consecutive midpoints. Therefore, the second midpoint is at 10 + 8 = 18.

(c) The sequence of cutpoints is 6, 14, 22, 30, 38, ... Therefore the lower and upper cutpoints of the third class are 22 and 30.

(d) An observation of 22 would go into the third class since that class contains data greater than or equal to 22 and strictly less than 30.

13. (a) If lower class limits are used to label the horizontal axis, the bars are placed between the lower class limit of one class and the lower class limit of the next class.

(b) If lower cutpoints are used to label the horizontal axis, the bars are placed between the lower cutpoint of one class and the lower cutpoint of the next class.

(c) If class marks are used to label the horizontal axis, the bars are placed directly above and centered over the class mark for that class.

(d) If midpoints are used to label the horizontal axis, the bars are placed directly above and centered over the midpoint for that class.

14. (a) Bell-shaped (b) Triangular (c) Reverse J shape (d) Uniform

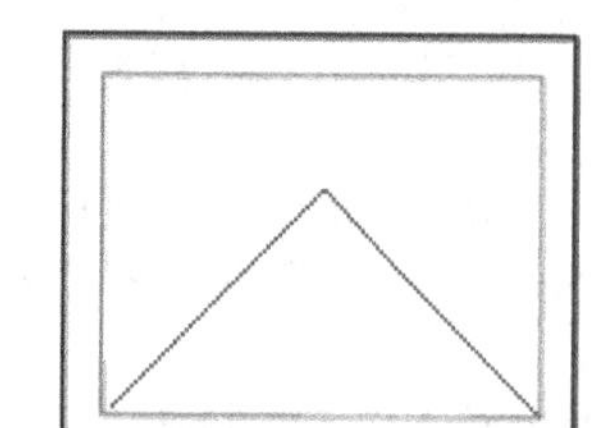

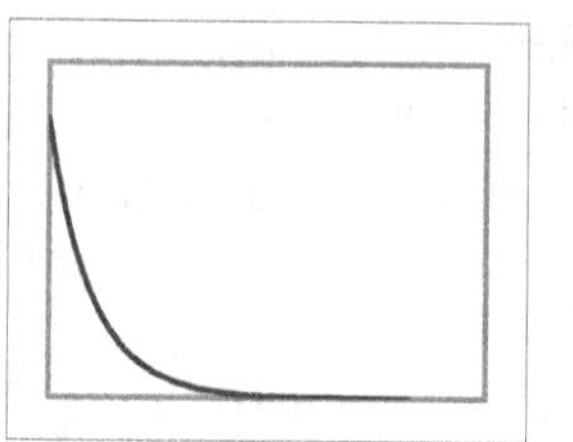

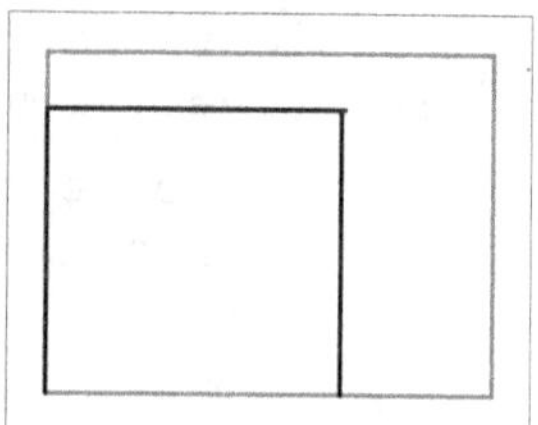

15. Answers will vary but here is one possibility.

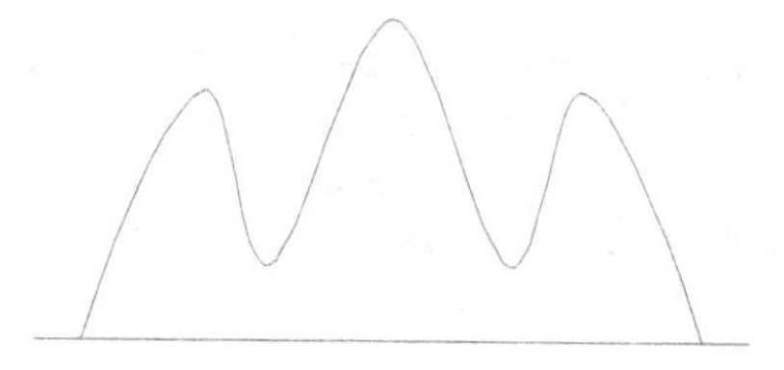

16. (a) The distribution of the large simple random sample will reflect the distribution of the population, so it would be left-skewed as well.

(b) No. The randomness in the samples will almost certainly produce different sets of observations resulting in shapes that are not identical.

(c) Yes. We would expect both of the simple random samples to reflect the shape of the population and be left-skewed.

17. (a) The first column ranks the hydroelectric plants. Thus, it consists of *quantitative, discrete* data.

(b) The fourth column provides measurements of capacity. Thus, it consists of *quantitative, continuous* data.

(c) The third column provides nonnumerical information. Thus, it consists of *qualitative* data.

18. (a) A single value frequency histogram for the prices of DVD players would be identical to the dotplot in example 2.16 because the classes would be 197 through 224 and the height for each bar in the frequency histogram would reflect the number of dots above each observation in the dotplot.

(b) No. The frequency histogram with cutpoint or class limit grouping would combine some of the single values together which would change the frequencies and the heights of the bars corresponding to those classes.

19. (a) The first class to construct is 40-44. Since all classes are to be of equal width, and the second class begins with 45, we know that the width of all classes is 45 - 40 = 5. All of the classes are presented in column 1 of the grouped-data table in the following figure. The last class to construct does not go beyond 65-69, since the largest single data value is 69. The sum of the relative frequency column is 0.999 due to rounding.

Age at Inauguration	Frequency	Relative Frequency	Class Mark
40-44	2	0.045	42
45-49	7	0.159	47
50-54	13	0.295	52
55-59	12	0.273	57
60-64	7	0.159	62
65-69	3	0.068	67
	44	0.999	

(b) By averaging the lower and upper limits for each class, we arrive at the class mark for each class. The class marks for all classes are presented in column 4.

(c) Having established the classes, we tally the ages into their respective classes. These results are presented in column 2, which lists the frequencies. Dividing each frequency by the total number of observations, which is 44, results in each class's relative frequency. The relative frequencies for all classes are presented in column 3.

(d) The frequency histogram presented below is constructed using the frequency distribution presented above; i.e., columns 1 and 2. Notice that the lower cutpoints of column 1 are used to label the horizontal axis of the frequency histogram. Suitable candidates for vertical-axis units in the frequency histogram are the even integers within the range 2 through 14, since these are representative of the magnitude and spread of the frequencies presented in column 2. The height of each bar in the frequency histogram matches the respective frequency in column 2.

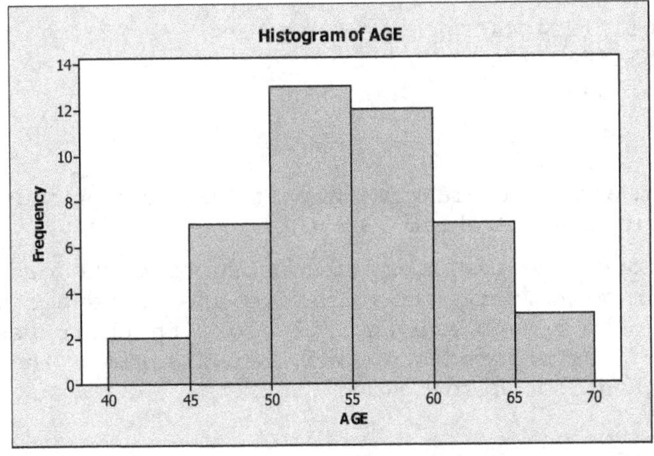

(e) The shape of the inauguration ages is unimodal and symmetric.

20. The horizontal axis of this dotplot displays a range of possible ages for the 44 Presidents of the United States. To complete the dotplot, we go through the data set and record each age by placing a dot over the appropriate value on the horizontal axis.

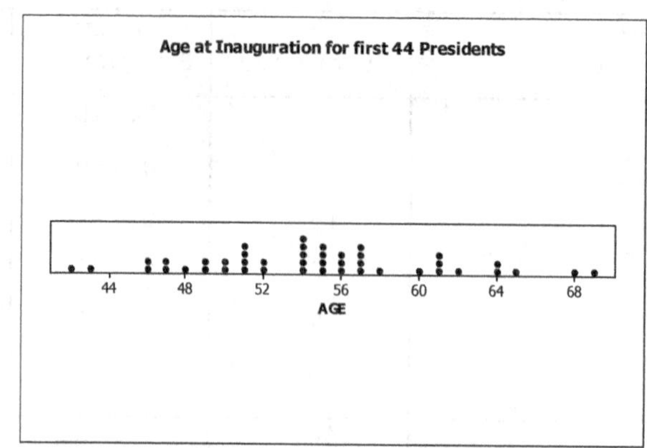

21. (a) Using *one* line per stem in constructing the ordered stem-and-leaf diagram means vertically listing the numbers comprising the stems *once*. The leaves are then placed with their respective stems in order. The ordered stem-and-leaf diagram using one line per stem is presented in the following figure.

```
4| 236677899
5| 00111122444445555566677778
6| 0111244589
```

(b) Using *two* lines per stem in constructing the ordered stem-and-leaf diagram means vertically listing the numbers comprising the stems *twice*. In turn, if the leaf is one of the digits 0 through 4, it is ordered and placed with the first of the two stem lines. If the leaf is one of the digits 5 through 9, it is ordered and placed with the second of the two stem lines. The ordered stem-and-leaf diagram using two lines per stem is presented in the following figure.

```
4| 23
4| 6677899
5| 0011112244444
5| 555566677778
6| 0111244
6| 589
```

(c) The stem-and-leaf diagram in part (b) corresponds to the frequency distribution of Problem 19.

22. (a) Using *one* line per stem in constructing the ordered stem-and-leaf diagram means vertically listing the numbers comprising the stems *once*. The leaves are then placed with their respective stems in order. The ordered stem-and-leaf diagram using one line per stem is presented in the following figure.

```
3| 4778
4| 0467
5| 446677778
6| 0
```

(b) Using *two* lines per stem in constructing the ordered stem-and-leaf diagram means vertically listing the numbers comprising the stems *twice*. In turn, if the leaf is one of the digits 0 through 4, it is ordered and placed with the first of the two stem lines. If the leaf is one of the digits 5 through 9, it is ordered and placed with the second of the two stem lines. The ordered stem-and-leaf diagram using two lines per stem is presented in the following figure.

```
3| 4
3| 778
4| 04
4| 67
5| 44
5| 6677778
6| 0
```

(c) The second graph is the most useful. Typically, we like to have five to fifteen classes and the second one provides a better idea of the shape of the distribution.

23. (a) The frequency and relative-frequency distribution presented below is constructed using classes based on a single value. Since each data value is one of the integers 0 through 6, inclusive, the classes will be 0 through 6, inclusive. These are presented in column 1. Having established the classes, we tally the number of busy tellers into their respective classes. These results are presented in column 2, which lists the frequencies. Dividing each frequency by the total number of observations, which is 25, results in each class's relative frequency. The relative frequencies for all classes are presented in column 3.

Number Busy	Frequency	Relative Frequency
0	1	0.04
1	2	0.08
2	2	0.08
3	4	0.16
4	5	0.20
5	7	0.28
6	4	0.16
	25	1.00

(b) The following relative-frequency histogram is constructed using the relative-frequency distribution presented in part (a); i.e., columns 1 and 3. Column 1 demonstrates that the data are grouped using classes based on a single value. These single values in column 1 are used to label the horizontal axis of the relative-frequency histogram. We notice that the relative frequencies presented in column 3 range in size from 0.04 to 0.28 (4% to 28%). Thus, suitable candidates for vertical axis units in the relative-frequency histogram are increments of 0.05, starting with zero and ending at 0.30. The middle of each histogram bar is placed directly over the single numerical value represented by the class. Also, the height of each bar in the relative-frequency histogram matches the respective relative frequency in column 3.

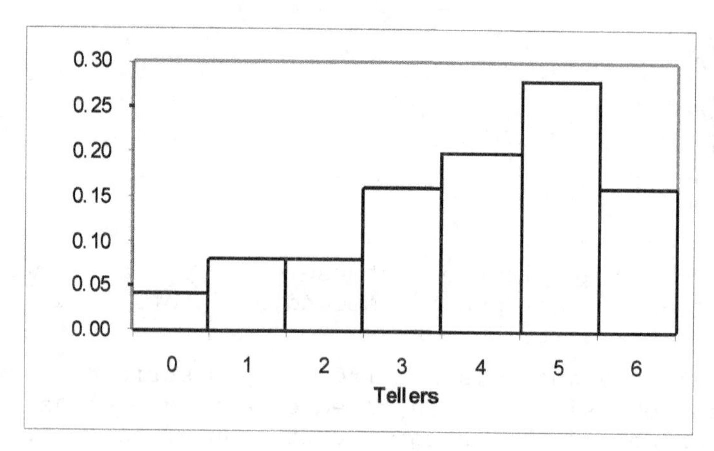

(c) The distribution is unimodal.

(d) The distribution is left skewed.

(e)

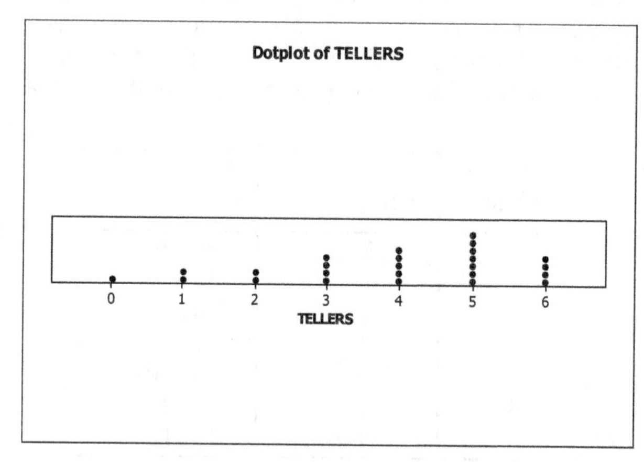

(f) Since both the histogram and the dotplot are based on single value grouping, they both convey exactly the same information.

24. (a) The classes will begin with the class 60 - under 65. The second class will be 65 - under 70. The classes will continue like this until the last class, which will be 90 - under 95, since the largest observation is 93.1. The classes can be found in column 1 of the frequency distribution in part (c).

(b) The midpoints of the classes are the averages of the lower and upper cutpoint for each class. For example, the first midpoint will be 62.5, the second midpoint will be 67.5. The midpoints will continue like this until the last midpoint, which will be 92.5. The midpoints can be found in column 4 of the frequency distribution in part (c).

(c) The first and second columns of the following table provide the frequency distribution. The first and third columns of the following table provide the relative-frequency distribution for percentages of on-time arrivals for the airlines.

Percent On-Time	Frequency	Relative Frequency	Midpoint
60 - under 65	1	0.0625	62.5
65 - under 70	5	0.3125	67.5
70 - under 75	5	0.3125	72.5
75 - under 80	3	0.1875	77.5
80 - under 85	0	0.0000	82.5
85 - under 90	1	0.0625	87.5
90 - under 95	1	0.0625	92.5
	16	1.0000	

(d) The frequency histogram is constructed using columns 1 and 2 from the frequency distribution from part (c). The cutpoints are used to label the horizontal axis. The vertical axis is labeled with the frequencies that range from 0 to 5.

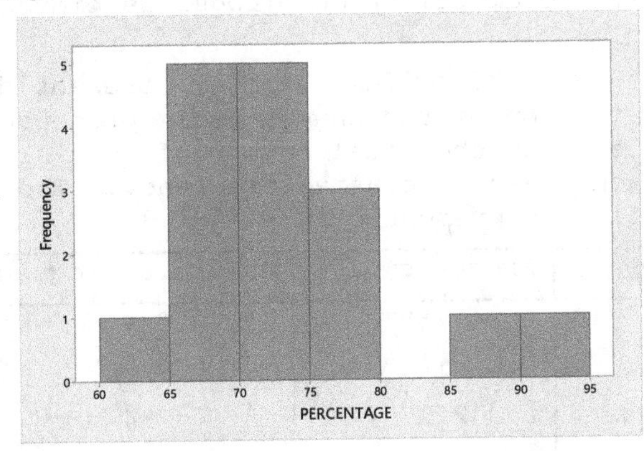

(e) After rounding each observation to the nearest whole number, the stem-and-leaf diagram with two lines per stem for the rounded percentages is

```
6| 2
6| 669
7| 0011334
7| 678
8|
8| 8
9| 3
```

(f) After obtaining the greatest integer in each observation, the stem-and-leaf diagram with two lines per stem is

```
6| 1
6| 56999
7| 01233
7| 677
8|
8| 7
9| 3
```

(g) The stem-and-leaf diagram in part (f) corresponds to the frequency distribution in part (d) because the observations weren't rounded in constructing the frequency distribution.

25. (a) The dotplot of the ages of the oldest player on each major league baseball team is

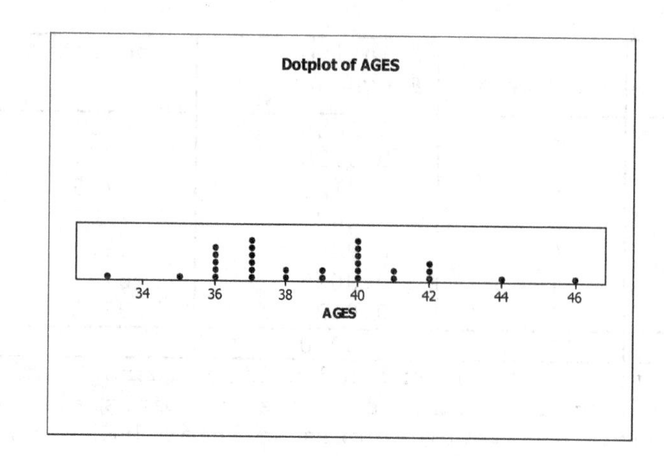

(b) The overall shape of the distribution of ages is bimodal and roughly symmetric.

26. (a) The classes are the types of evidence and are presented in column 1. The frequency distribution of the champions is presented in column 2. Dividing each frequency by the total number of submissions, which is 3436, results in each class's relative frequency. The relative frequency distribution is presented in column 3.

Evidence Type	Frequency	Relative Frequency
Firearm	289	0.084
Magazine	161	0.047
Live Cartridge	2727	0.794
Spent Cartridge	259	0.075
	3436	1.000

(b) We multiply each of the relative frequencies by 360 degrees to obtain the portion of the pie represented by each evidence type. The result is

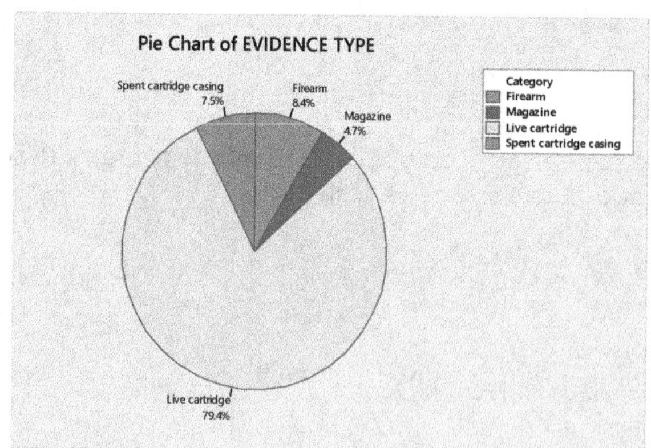

(c) We use the bar chart to show the relative frequency with which each EVIDENCE TYPE occurs. The result is

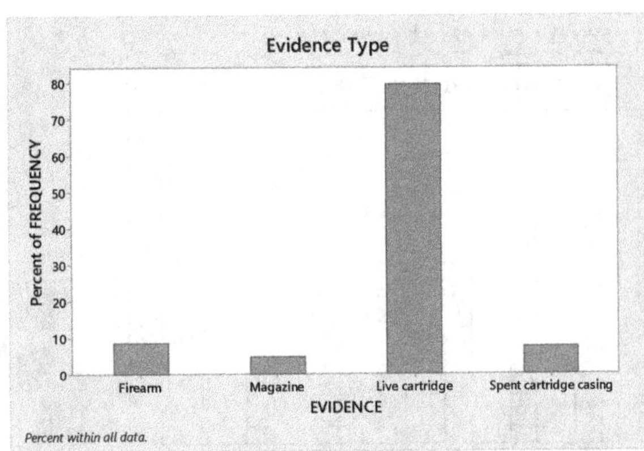

(d) The most common type of evidence type submitted to crime labs is a live cartridge which was the type submitted 79.4% of the time.

27. (a) The population consists of the states in the United States. The variable is the division.

(b) The frequency and relative frequency distribution for Region is shown below.

Region	Frequency	Relative Frequency
East North Central	5	0.10
East South Central	4	0.08
Middle Atlantic	3	0.06
Mountain	8	0.16
New England	6	0.12
Pacific	5	0.10
South Atlantic	8	0.16
West North Central	7	0.14
West South Central	4	0.08
	50	1.000

(c) The pie chart for Region is shown below.

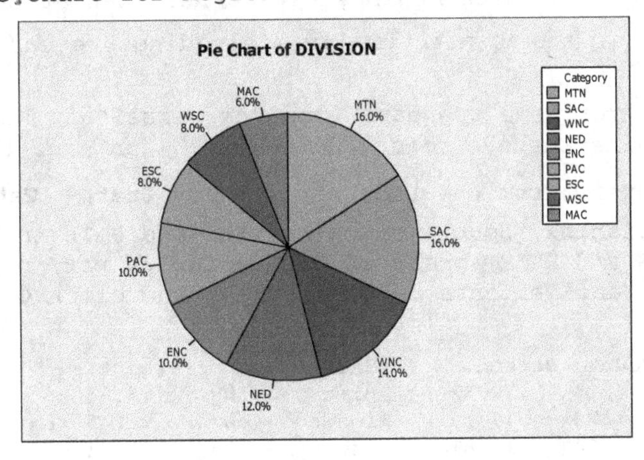

(d) The bar chart for Region is shown below.

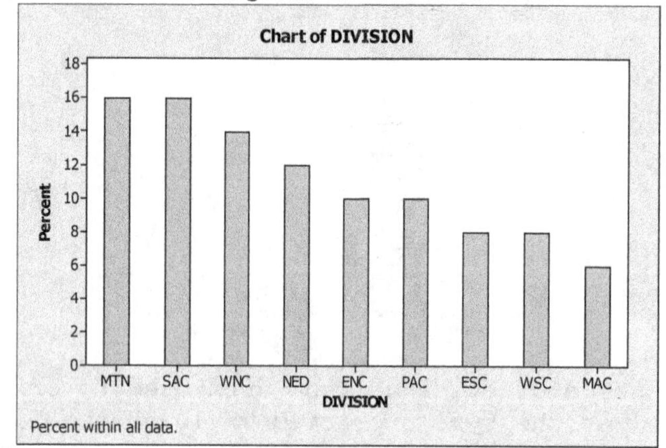

(e) The distribution of Region seems to be almost uniformly distributed between the categories. The smallest region is the Middle Atlantic at a frequency of 3, the largest region is the Mountain and South Atlantic at a frequency of 8.

28. (a) The break in the third bar is to emphasize that the bar as shown is not as tall as it should be.

(b) The bar for the space available in coal mines is at a height of about 30 billion tonnes. To accurately represent the space available in saline aquifers (10,000 billion tonnes), the third bar would have to be over 300 times as high as the first bar and more than 10 times as high as the second bar. If the first two bars were kept at the sizes shown, there wouldn't be enough room on the page for the third bar to be shown at its correct height. If a reasonable height were chosen for the third bar, the first bar wouldn't be visible. The only apparent solution is to present the third bar as a broken bar.

29. (a) Covering up the numbers on the vertical axis totally obscures the percentages.

(b) Having followed the directions in part (a), we might conclude that the percentage of women in the labor force for 2000 is about three and one-half times that for 1960.

(c) Not covering up the vertical axis, we find that the percentage of women in the labor force for 2000 is about 1.8 times that for 1960.

(d) The graph is potentially misleading because it is truncated. Notice that vertical axis units begin at 30 rather than at zero.

(e) To make the graph less potentially misleading, we can start it at zero instead of 30.

30. (a) Using Minitab, retrieve the data from the WeissStats Resource Site. Column 2 contains the eye color and column 3 contains the hair color for the students. From the tool bar, select **Stat ▶ Tables ▶ Tally Individual Variables**, double-click on EYES and HAIR in the first box so that both EYES and HAIR appear in the **Variables** box, put a check mark next to Counts and Percents under Display, and click **OK**. The results are

EYES	Count	Percent
Blue	215	36.32
Brown	220	37.16
Green	64	10.81
Hazel	93	15.71
N=	592	

HAIR	Count	Percent
Black	108	18.24
Blonde	127	21.45
Brown	286	48.31
Red	71	11.99
N=	592	

(b) Using Minitab, select **Graph** ▶ **Pie Chart**, check Chart counts of unique values, double-click on EYES and HAIR in the first box so that EYES and HAIR appear in the Categorical Variables box. Click Pie Options, check decreasing volume, click OK. Click Multiple Graphs, check On the Same Graphs, Click OK. Click Labels, click Slice Labels, check Category Name, Percent, and Draw a line from label to slice, Click OK twice. The results are

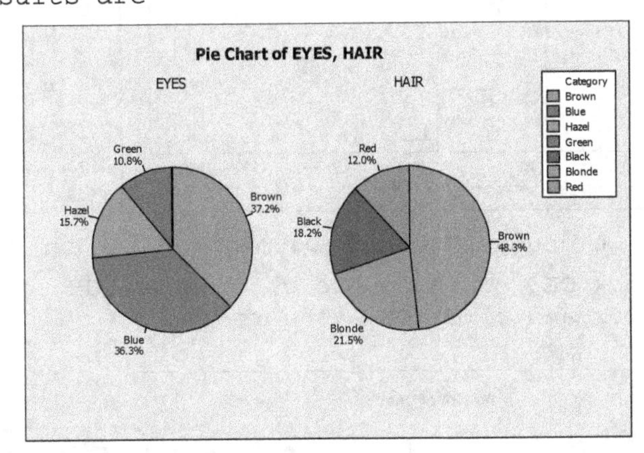

(c) Using Minitab, select **Graph** ▶ **Bar Chart**, select Counts of unique values, select Simple option, click OK. Double-click on EYES and HAIR in the first box so that EYES and HAIR appear in the Categorical Variables box. Select Chart Options, check decreasing Y, check show Y as a percent, click OK. Click OK twice. The results are

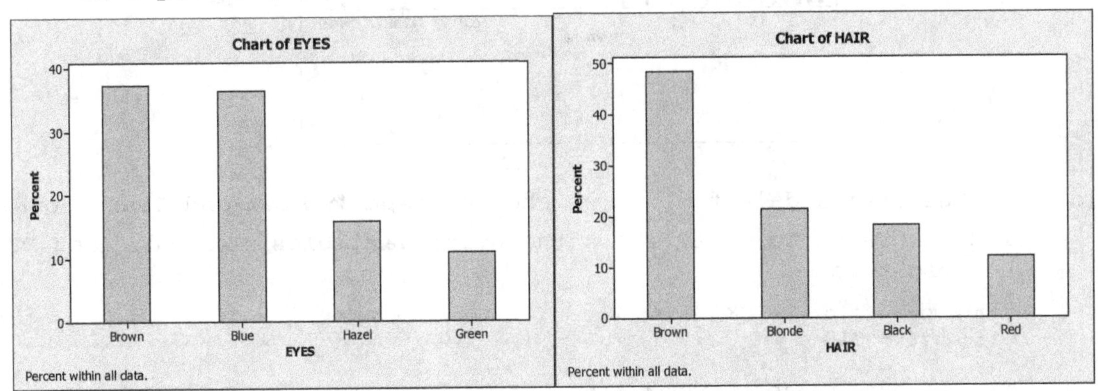

31. (a) The population consists of the states of the U.S. and the variable under consideration is the value of the exports of each state.

(b) Using Minitab, we enter the data from the WeissStats Resource Site, choose **Graph** ▶ **Histogram**, click on **Simple** and click **OK**. Then double click on VALUE to enter it in the **Graph variables** box and click **OK**. The result is

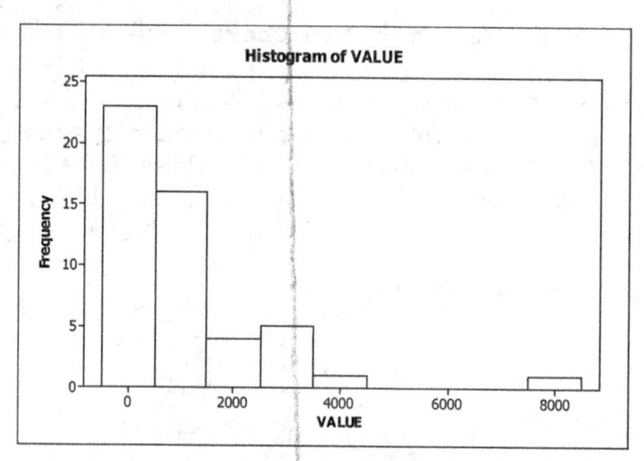

(c) For the dotplot, we choose **Graph ▶ Dotplot**, click on **Simple** from the **One Y** row and click **OK**. Then double click on VALUE to enter it in the **Graph variables** box and click **OK**. The result is

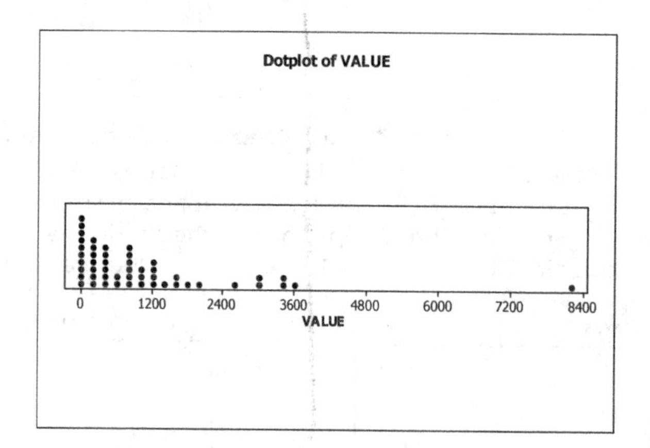

(d) For the stem-and-leaf plot, we choose **Graph ▶ Stem-and-Leaf**, double click on VALUE to enter it in the **Graph variables** box and click **OK**. The result is

```
Stem-and-leaf of VALUE  N  = 50
Leaf Unit = 100

 23    0  00000000001112222344444
(10)   0  5677888899
 17    1  012224
 11    1  5679
 7     2
 7     2  69
 5     3  034
 2     3  6
 1     4
 1     4
 1     5
 1     5
 1     6
 1     6
 1     7
 1     7
 1     8  2
```

(e) The overall shape of the distribution is unimodal and not symmetric.

(f) The distribution is right skewed.

32. (a) The population consists of countries of the world, and the variable under consideration is the expected life in years for people in those countries.

(b) Using Minitab, we enter the data from the WeissStats Resource Site, choose **Graph ▶ Histogram**, click on **Simple** and click **OK**. Then double click on YEARS to enter it in the **Graph variables** box and click **OK**. The result is

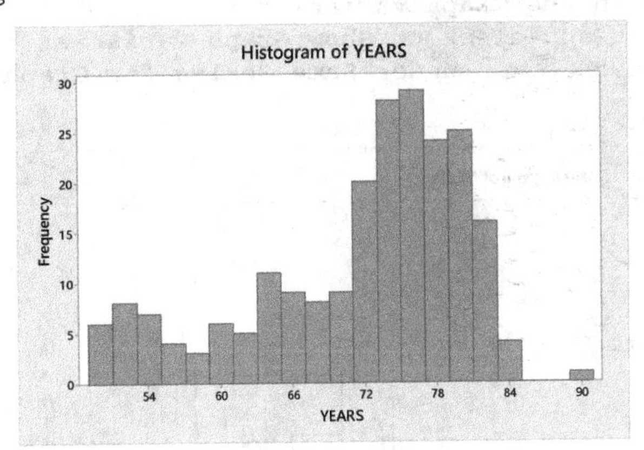

(c) For the dotplot, we choose **Graph ▶ Dotplot**, click on **Simple** from the **One Y** row and click **OK**. Then double click on YEARS to enter it in the **Graph variables** box and click **OK**. The result is

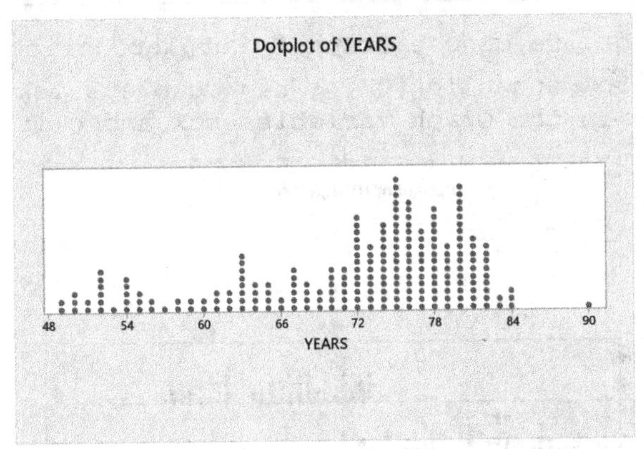

(d) For the stem-and-leaf plot, we choose **Graph ▶ Stem-and-Leaf**, double click on YEARS to enter it in the **Graph variables** box and click **OK**. The result is

```
Stem-and-leaf of YEARS  N  = 223
Leaf Unit = 1.0

  3    4  999
 21    5  000112222223344444
 30    5  556677899
 50    6  00001112233333333444
 73    6  55556666667778888899999
(51)   7  000111111111112222222222333333333334444444444444444
 99    7  555555555555566666666666666667777777777788888888888899999999999999
 33    8  00000000000011111111111122223444
  1    8  9
```

(e) The overall shape of the distribution is unimodal and not symmetric.

(f) This distribution is classified as left skewed.

33. (a) The population consists of cities in the U.S., and the variables under consideration are their annual average maximum and minimum temperatures.

(b) Using Minitab, we enter the data from the WeissStats Resource Site, choose **Graph ▶ Histogram**, click on **Simple** and click **OK**. Double click on HIGH to enter it in the **Graph variables** box, and double click on LOW to enter it in the **Graph variables** box. Now click on the **Multiple graphs** button and click to **Show Graph Variables on separate graphs** and also check both boxes under **Same Scales for Graphs**, and click **OK** twice. The result is

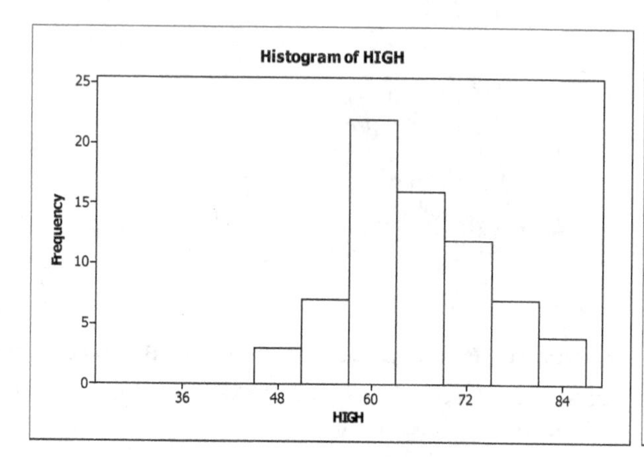

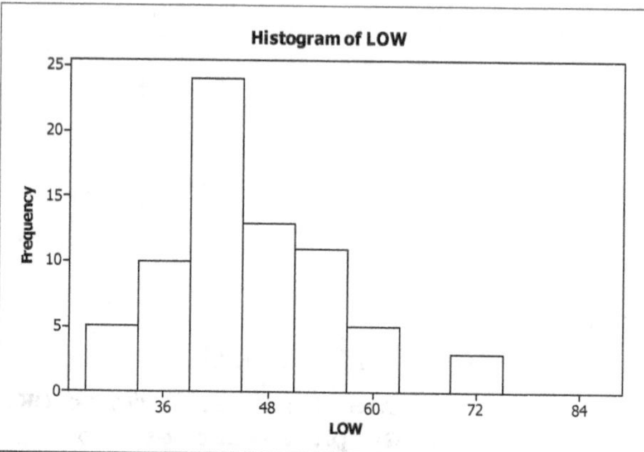

(c) For the dotplot, we choose **Graph ▶ Dotplot**, click on **Simple** from the **Multiple Y's** row and click **OK**. Then double click on HIGH and then LOW to enter them in the **Graph variables** box and click **OK**. The result is

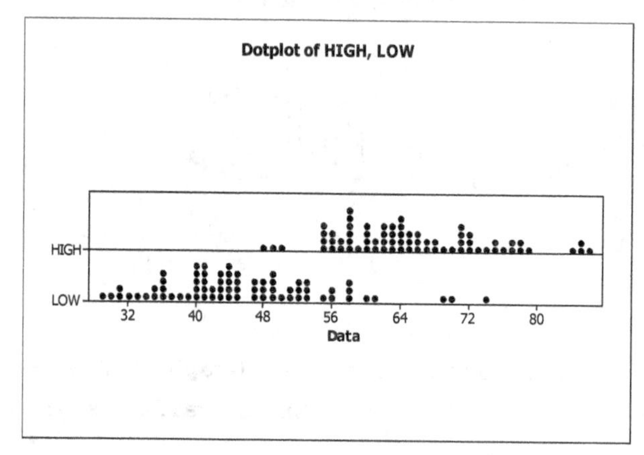

(d) For the stem-and-leaf diagram, we choose **Graph ▶ Stem-and-Leaf**, double click on HIGH and then on LOW to enter then in the **Graph variables** box and click **OK**. The result is

```
Stem-and-leaf of HIGH  N  = 71
Leaf Unit = 1.0
```

```
   3    4  789
   6    5  444
  21    5  555677777888999
 (17)   6  00001122222333444
  33    6  55556677779
  22    7  00001122234
  11    7  5577889
   4    8  444
   1    8  5

Stem-and-leaf of LOW  N  = 71
Leaf Unit = 1.0

   1    2  9
   7    3  001234
  19    3  555556779999
 (20)   4  00011111223333344444
  32    4  5567777888899
  19    5  11122223
  11    5  5667889
   4    6  1
   3    6  9
   2    7  04
```

(e) Both variables have distributions that are unimodal. HIGH is close to symmetric and LOW is not symmetric.

(f) LOW is slightly right skewed.

CHAPTER 3 SOLUTIONS

Exercises 3.1

3.1 The purpose of a measure of center is to indicate where the center or most typical value of a data set lies.

3.3 Of the mean, median, and mode, only the mode is appropriate for use with qualitative data.

3.5 (a) The mean is the sum of the values (45) divided by n (9). The result is 5. The median is the middle value in the ordered list and is also 5.

(b) The mean is the sum of the values (135) divided by n (9). The result is 15. The median is the middle value in the ordered list and is 5, as previously. The median is more typical of most of the data than is the mean and works better here.

(c) The mean does not have the property of being resistant to the influence of extreme observations.

3.7 The median is more appropriate as a measure of central tendency than the mean because, unlike the mean, the median is not affected strongly by the relatively few homes that have extremely large or small areas.

3.9 (a) The mean is calculated first by summing all 3 data values presented, which is 9, and then dividing this sum by 3. Thus the mean is 3.

(b) The median requires ordering the data from the smallest to the largest values: 0, 4, 5. Since the number of data values is odd, the median will be the middle data value in the ordered list. The middle value is the second ordered observation. Thus, the median is 4.

(c) The mode is the most frequently occurring data value or values. In these data, no value occurs more than once, so there is no mode.

3.11 (a) The mean is calculated first by summing all 4 data values presented, which is 11, and then dividing this sum by 4. Thus the mean is 2.75.

(b) The median requires ordering the data from the smallest to the largest values: 1, 2, 4, 4. Since the number of data values is even, the median will be the mean of the two middle data values in the ordered list. The two middle values are the second and third ordered observation. Thus, the median is 3.

(c) The mode is the most frequently occurring data value or values. In these data, the mode is 4 because it occurs twice.

3.13 (a) The mean is calculated first by summing all 5 data values presented, which is 25, and then dividing this sum by 5. Thus the mean is 5.

(b) The median requires ordering the data from the smallest to the largest values: 1, 3, 4, 8, 9. Since the number of data values is odd, the median will be the middle data value in the ordered list. The middle value is the third ordered observation. Thus, the median is 4.

(c) The mode is the most frequently occurring data value or values. In these data, no value occurs more than once, so there is no mode.

3.15 (a) $\sum$ represents the word 'summation'. It indicates that one should sum the values of the quantity that follows it.

(b) n represents the number data values

(c) $\bar{x}$ represents the sample mean

3.17 (a) n = number of data values = 5

(b) $\sum x_i = 1 + 7 + 4 + 5 + 10 = 27$

(c) $\bar{x} = \sum x_i/n = 27/5 = 5.4$

3.19 The mean number of values is calculated first by summing all seven values presented, which results in 51 days, and then dividing this by 7. Carrying this result to one more decimal place than the original data, the mean is 7.3 days.

Calculating the median requires ordering the data from the smallest to the largest values: 5 5 6 **6** 7 11 11. Since the number of data values is odd, the median will be the observation exactly in the middle of this ordering, i.e., in the fourth ordered position. Thus, the median number of days is 6.

The mode is the most frequently occurring data value or values. For these data, there are three modes because 5, 6, and 11 each occur twice.

3.21 The mean number of tornado touchdowns is calculated by first summing all 12 data values presented, which results in 941, and then dividing this sum by 12. Thus, the mean is 78.4. This figure is already rounded to one more decimal place than the original data.

Calculating the median requires ordering the data from the smallest to the largest values: 2, 3, 47, 57, 62, **68, 86**, 97, 98, 99, 118, 204. Since the number of data values is even, the median will be the mean of the two middle data values in the ordered list. The two middle values are the sixth and seventh ordered observations, or 68 and 86. Thus, the median age is found as (68 + 86)/2 = 77.0.

The mode is the most frequently occurring data value or values. In these data, no value occurs more than once, so there is no mode.

3.23 The mean wealth is calculated first by summing all 10 data values presented, which results in 347.1, and then dividing this sum by 10. Thus, the mean is $34.71 billion. This figure is already rounded to one more decimal place than the original data.

Calculating the median requires ordering the data from the smallest to the largest values: 25.0, 26.1, 26.3, 26.8, 27.9, 31.0, 31.0, 41.0, 46.0, 66.0. Since the number of data values is even, the median will be the mean of the two middle data values in the ordered list. The two middle values are the fifth and sixth ordered observations. Thus, the median wealth is (27.9 + 31.0)/2 = $29.45 billion dollars.

The mode is the most frequently occurring data value or values. In these data, $31.0 billion dollars is the mode because it occurs twice.

3.25 The mean is calculated first by summing all 12 data values presented, which results in 175, and then dividing this sum by 12. Thus, the mean is 14.6 songs.

Calculating the median requires ordering the data from the smallest to the largest values. Since the number of data values is even, the median will be the mean of the two middle data values in the ordered list. The two middle values are the sixth and seventh ordered observations. Thus, the median number of songs is (14 + 14)/2 = 14.0.

The mode is the most frequently occurring data value or values. In these data, the mode is 14 songs since 14 occurs 6 times.

3.27 (a) The mean number of cremation burials is calculated first by summing all 17 data values presented, which results in 4977, and then dividing this sum by 17. Thus, the mean is 292.8. This figure is already rounded to one more decimal place than the original data.

Calculating the median requires ordering the data from the smallest to the largest values: 21, 34, 35, 46, 46, 48, 51, 64, **83**, 86, 119, 258,

265, 385, 429, 523, 2484. Since the number of data values is odd, the median will be the middle data value in the ordered list. The middle value is the ninth ordered observation. Thus, the median is 83 cremation burials.

The mode is the most frequently occurring data value or values. In these data, the mode is 46 since 46 occurs two times, more than any other value.

(b) The median works best. The mean is highly affected by the largest value and the mode is well to the left of the distribution's center.

3.29 (a) The mean number of accidents is calculated first by summing all 7 data values presented, which results in 619, and then dividing this sum by 7. Thus, the mean is 88.4. This figure is already rounded to one more decimal place than the original data.

Calculating the median requires ordering the data from the smallest to the largest values: 69, 76, 85, **88**, 98, 100, 103. Since the number of data values is odd, the median will be the middle data value in the ordered list. The middle value is the fourth ordered observation. Thus the median is 88.

The mode is the most frequently occurring data value or values. In these data, there is no mode since no number occurs more than once.

(b) The mean number of accidents is calculated first by summing all 7 data values presented, which results in 492, and then dividing this sum by 7. Thus, the mean is 70.3. This figure is already rounded to one more decimal place than the original data.

Calculating the median requires ordering the data from the smallest to the largest values: 53, 56, 58, **59**, 70, 94, 102. Since the number of data values is even, the median will be the middle data value in the ordered list. The middle value is the fourth ordered observation. Thus the median is 59.

(c) For built-up roads, the modal day is Friday since Friday would appear 103 times in the list, more than any other day. For non-built-up roads the modal day is Sunday since Sunday would appear 102 times in the list, more than any other day.

(d) Perhaps on Sunday, motorcycles are being used more for pleasure and sightseeing, possibly on unfamiliar roads. This explanation seems reasonable since Saturday is the day with the second most number of accidents on non-built-up roads. On the weekdays, which include the modal day Friday, motorcycles are perhaps being used more for transportation to and from work. This traffic is more likely to be on built-up roads.

3.31 (a) n = number of data values = 10

(b) $\sum x_i = 23.3$

(c) $\bar{x} = \sum x_i/n = 23.3/10 = 2.33$, which is given to two decimal places.

3.33 (a) n = number of data values = 9

(b) $\sum x_i = 607$

(c) $\bar{x} = \sum x_i/n = 607/9 = 67.4444$, which, rounded to one decimal place, is 67.4 years.

3.35 (a) The mode is defined as the data value or values that occur most frequently. In this exercise, the data values are the NCAA wrestling champion universities. During the sampled period, Iowa was champion twelve times. Since Iowa is the data value that occurs most frequently, it is the mode.

(b) Again, the data values are the champions. There is no way to compute a mean or median for such nonnumerical data. Thus, it would not be appropriate to use either the mean or median here. The mode is the only measure of center that can be used for qualitative data.

3.37 (a) The mode is defined as the data value or values that occur most frequently. In this exercise, the data values are the law schools of the justices of the U.S Supreme Court. For this sample, Harvard was the law school for five of the justices. Since Harvard is the data value that occurs most frequently, it is the mode.

(b) Again, the data values are the law schools. There is no way to compute a mean or median for such nonnumerical data. Thus, it would not be appropriate to use either the mean or median here. The mode is the only measure of center that can be used for qualitative data.

3.39 (a) The mode is defined as the data value or values that occur most frequently. In this exercise, the data values are the political view of the freshmen. During the year, Moderate was the most frequently occurring political view for the freshmen with a frequency of 237. Since moderate is the data value that occurs most frequently, it is the mode.

(b) Again, the data values are the three different political views. There is no way to compute a mean or median for such nonnumerical data. Thus, it would not be appropriate to use either the mean or median here. The mode is the only measure of center that can be used for qualitative data.

3.41 (a) The mode is defined as the data value or values that occur most frequently. In this exercise, the data values are the colors that the Roulette ball landed on. During 200 trials, black was the most frequently occurring color with a frequency of 102. Since black is the data value that occurs most frequently, it is the mode.

(b) Again, the data values are the three different colors. There is no way to compute a mean or median for such nonnumerical data. Thus, it would not be appropriate to use either the mean or median here. The mode is the only measure of center that can be used for qualitative data.

3.43 Using Minitab, retrieve the data from the WeissStats Resource Site. Since the data are qualitative, the only appropriate measure of center is the mode. Select **Stat ▶ Tables ▶ Tally Individual Variables**, then enter TYPE in the **Variables** box and click **OK**. The result is

TYPE	Count
NPC	2919
IOC	889
SLC	1119
FGH	221
NLT	129
NFP	451
HUI	19
N=	5747

The mode is the most occurring value. We see that NPC (Nongovernment not-for-profit community hospitals) occurred 2919 times, more than any other value. Therefore, NPC is the mode.

3.45 Using Minitab, retrieve the data from the WeissStats Resource Site. Since both of the variables are qualitative, the only appropriate measure of center for preference and highest degree obtained is the mode. Select **Stat ▶ Tables ▶ Tally Individual Variables**, then enter PREFERENCE and DEGREE in the **Variables** box and click **OK**. The result is

PREFERENCE	Count	DEGREE	Count
Both	112	MA	167
Email	239	Other	11
Mail	86	PhD	388
N/A	129	N=	566
N=	566		

The mode is the most occurring value. For PREFERENCE, we see that email occurred 239 times, more than any other value. Therefore, email is the mode for PREFERENCE. For DEGREE, we see that PhD occurred 388 times, more than any other value. Therefore, PhD is the mode for DEGREE.

3.47 Using Minitab, retrieve the data from the WeissStats Resource Site. Since the data are quantitative, any of the three measures of center could be used. To obtain the mode, select **Stat ▶ Tables ▶ Tally Individual Variables**, then enter LENGTH in the **Variables** box and click **OK**. The mode is the most occurring value, which, in this case, is 158 (occurring 9 times).

To obtain the mean and median, select **Stat ▶ Basic Statistics ▶ Display Descriptive Statistics**, then enter LENGTH in the **Variables** box and click **OK**. The result is

Variable	N	N*	Mean	SE Mean	StDev	Minimum	Q1	Median	Q3	Maximum
LENGTH	229	0	163.75	3.80	57.57	23.00	132.50	153.00	181.50	493.00

From this display, we see that the mean is 163.75 seconds and the median is 153.00 seconds. In this case, the median is the most useful statistic since it tells us that half of the songs are longer than 153.00 seconds. The mean is heavily influenced by a few songs that are longer than the others. The median is more representative of the data as a whole.

3.49 Using Minitab, retrieve the data from the WeissStats Resource Site. Since the data are quantitative, any of the three measures of center could be used. To obtain the mode, select **Stat ▶ Tables ▶ Tally Individual Variables**, then enter PERCENT in the **Variables** box and click **OK**. The result is

PERCENT	Count		
17	1	35	1
19	1	36	3
20	1	38	1
21	2	49	1
22	2	N=	51
23	3		
24	4		
25	6		
26	5		
27	5		
28	1		
29	3		
30	3		
31	3		
32	3		
33	1		
34	1		

The mode is the most occurring value, which is 25 (occurring 6 times).

To obtain the mean and median, select **Stat ▶ Basic Statistics ▶ Display Descriptive Statistics**, then enter PERCENT in the **Variables** box and click **OK**. The result is

Variable	N	N*	Mean	SE Mean	StDev	Minimum	Q1	Median	Q3	Maximum
PERCENT	51	0	27.706	0.788	5.626	17.000	24.000	27.000	31.000	49.000

From this display, we see that the mean is 27.706% and the median is 27.000%. In this case, the median is the most useful statistic since it tells us that half of the states had a bachelor's completion percentage of 27% or greater. The mean is not a true national crime rate since all states are counted equally in computing this mean, whereas states with greater populations should have more effect on the mean than states with smaller populations.

3.51 (a) In Minitab, open the worksheet from the WeissStats Resource Site. This will enter the data values in columns C1 and C2 under the names INTEGRATED and STANDARD. Select **Stat ▶ Basic Statistics ▶ Display Descriptive Statistics**, enter INTEGRATED and STANDARD in the **Variables** text box and click **OK**. The output appears in the Session Window as

Variable	N	N*	Mean	SE Mean	StDev	Minimum	Q1	Median	Q3
INTEGRATED	227	0	24.916	0.300	4.513	11.000	22.000	25.000	28.000
STANDARD	192	0	23.000	0.521	7.221	6.000	18.000	23.000	28.750

Variable	Maximum
INTEGRATED	35.000
STANDARD	40.000

From the output, we see that the mean score on client satisfaction for the integrated treatment is 24.916 and the median score is 25.000. The mean score on client satisfaction for the standard treatment is 23.000 and the median score is 23.000.

(b) It would appear from this data that the integrated treatment group had a slightly higher mean and median score than the standard treatment group. However, we can also see from the data that the standard treatment group had a higher maximum score than the integrated group.

3.53 (a) The mean of the ratings is 2.5. This is derived by summing the 14 individual ratings, which yields 35, and dividing by 14.

(b) The median of the data is the average of the two observations appearing in the middle positions after ordering the data from the smallest to the largest values: 1, 1, 2, 2, 2, 2, **2, 2**, 3, 3, 3, 4, 4, 4. The median is the average of the seventh and eighth ordered observations, or (2 + 2)/2 = 2.

(c) In this case, the median provides a better descriptive summary of the data than does the mean. Notice that six of the fourteen values are 2s and these are more in line with the median 2 than with the mean 2.5. Furthermore, the data are not actually measurements, but reflect the opinions of the participants, each of whom must decide what a '2' or a '3' really means.

3.55 The expression $(\sum x_i)^2$ represents the square of the sum of the data, whereas $\sum x_i^2$ represents the sum of the squares of the data. To show that these two quantities are generally unequal, consider three observations on x as presented in column 1 of the following table.

Also, consider each observation's square, as presented in column 2.

x	x^2
1	1
3	9
5	25
9	35

From the bottom row, we see that $(\sum x_i)^2 = 9^2 = 81$ and $\sum x_i^2 = 35$. Thus, $(\sum x_i)^2 \neq \sum x_i^2$.

Exercises 3.2

3.57 The purpose of a measure of variation is to show the amount of variation or spread in a data set. Any value of central tendency does not adequately characterize the elements of a data set. Measures of central tendency merely describe the center of the observations, but do not show how observations differ from each other. A measure of variation is intended to capture the degree to which observations differ among themselves.

3.59 When we use the standard deviation as a measure of variation, the reference point is the mean since all of the differences used in computing the standard deviation are between the data values and the mean.

3.61 (a) The mean of this data set is 5. Thus

$$\sum (x_i-\bar{x})^2 = (-4)^2 + (-3)^2 + (-2)^2 + (-1)^2 + 0^2 + 1^2 + 2^2 + 3^2 + 4^2 = 60$$

Dividing by $n - 1 = 8$, $s^2 = 60/8 = 7.5$. The standard deviation, s, is the square root of 7.5, or 2.739.

(b) The mean of this data set is 15. Thus

$$\sum (x_i-\bar{x})^2 = (-14)^2 + (-13)^2 + (-12)^2 + (-11)^2 + 10^2 + (-9)^2 + (-8)^2 + (-7)^2 + 84^2 = 7980$$

Dividing by $n - 1 = 8$, $s^2 = 7980/8 = 997.5$. The standard deviation, s, is the square root of 997.5, or 31.583.

(c) Changing the 9 to 99 greatly increases the standard deviation, illustrating that it lacks the property of resistance to the influence of extreme data values.

3.63 (a) The range is 54 - 9 = 45.

(b) The defining formula for *s* is $s = \sqrt{\dfrac{\sum(x_i - \bar{x})^2}{n-1}}$.

The first three columns of the following table present the calculations that are needed to compute *s* by the defining formula:

x	$x-\bar{x}$	$(x-\bar{x})^2$	x^2
21	-15	225	441
54	18	324	2916
9	-27	729	81
45	9	81	2025
51	15	225	2601
180	0	1584	8064

With $n = 5$ and using the bottom figure of column 1, we find that $\bar{x} = \sum x_i/n = 180/5 = 36$.

This value is needed to construct the differences presented in column 2. Column 3 squares these differences. The figure presented at the bottom of column 3 is the computation needed for the numerator of the defining formula for *s*. Thus,

$$s = \sqrt{\frac{1584}{5-1}} = 19.900$$

(c) The computing formula for s is

$$s = \sqrt{\frac{\sum x_i^2 - (\sum x_i)^2 / n}{n-1}}$$

The computing formula for s with figures used from the bottom of Columns 1 and 4 of the previous table yields

$$s = \sqrt{\frac{8064 - 180^2 / 5}{5-1}} = 19.900$$

(d) The computing formula was a time-saver since only one column of intermediate numbers needed to be calculated, whereas the defining formula required three additional columns.

3.65 (a) The range is 5 - 0 = 5.

(b) The defining formula for *s* is $s = \sqrt{\frac{\sum (x_i - \bar{x})^2}{n-1}}$.

The following table presents the calculations that are needed to compute *s* by the defining formula:

X	x-x̄	(x-x̄)²
4	1.0	1.0
0	-3.0	9.0
5	2.0	4.0
9	0.0	14.0

With n = 3 and using the bottom figure of column 1, we find that $\bar{x} = \sum x_i / n = 9/3 = 3.0$. This value is needed to construct the differences presented in column 2. Column 3 squares these differences. The figure presented at the bottom of column 3 is the computation needed for the numerator of the defining formula for *s*. Thus,

$$s = \sqrt{\frac{14.0}{3-1}} = 2.6$$

3.67 (a) The range is 4 - 1 = 3.

(b) The defining formula for *s* is $s = \sqrt{\frac{\sum (x_i - \bar{x})^2}{n-1}}$.

The following table presents the calculations that are needed to compute *s* by the defining formula:

X	x-x̄	(x-x̄)²
1	-1.75	3.06
2	-0.75	0.56
4	1.25	1.56
4	1.25	1.56
11	0.00	6.74

With n = 4 and using the bottom figure of column 1, we find that $\bar{x} = \sum x_i/n$ = 11/4 = 2.75. This value is needed to construct the differences presented in column 2. Column 3 squares these differences. The figure presented at the bottom of column 3 is the computation needed for the numerator of the defining formula for *s*. Thus,

$$s = \sqrt{\frac{6.74}{4-1}} = 1.5$$

3.69 (a) The range is 9 - 1 = 8.

(b) The defining formula for *s* is $s = \sqrt{\frac{\sum (x_i - \bar{x})^2}{n-1}}$.

The following table presents the calculations that are needed to compute *s* by the defining formula:

X	x-x̄	(x-x̄)²
1	-4.0	16.0
9	4.0	16.0
8	3.0	9.0
4	-1.0	1.0
3	-2.0	4.0
25	0.00	46.0

With n = 5 and using the bottom figure of column 1, we find that $\bar{x} = \sum x_i/n$ = 25/5 = 5.0. This value is needed to construct the differences presented in column 2. Column 3 squares these differences. The figure presented at the bottom of column 3 is the computation needed for the numerator of the defining formula for *s*. Thus,

$$s = \sqrt{\frac{46.00}{5-1}} = 3.4$$

3.71 The range is 11 - 5 = 6.

The defining formula for *s* is $s = \sqrt{\frac{\sum (x_i - \bar{x})^2}{n-1}}$.

The following table presents the calculations that are needed to compute *s* by the defining formula:

x	x-x̄	(x-x̄)²
6	-1.3	1.69
7	-0.3	0.09
11	3.7	13.69
6	-1.3	1.69
5	-2.3	5.29
5	-2.3	5.29
11	3.7	13.69
51	-0.1	41.43

With n = 7 and using the bottom figure of column 1, we find that $\bar{x} = \sum x_i/n = 51/7 = 7.3$. This value is needed to construct the differences presented in column 2. Column 3 squares these differences. The figure presented at the bottom of column 3 is the computation needed for the numerator of the defining formula for *s*. Thus,

$$s = \sqrt{\frac{41.43}{7-1}} = 2.63$$

3.73 The range of a data set is defined to be the difference between the largest and smallest data values in the data set. The largest number of tornado touchdowns is 204. The smallest number is 2. The range is 204 - 2 = 202 tornadoes. The defining formula for *s* is $s = \sqrt{\frac{\sum (x_i - \bar{x})^2}{n-1}}$.

The first three columns of the following table present the calculations that are needed to compute *s* by the defining formula:

x	x-x̄	(x-x̄)²	x²
3	-75.4	5685.16	9
2	-76.4	5836.96	4
47	-31.4	985.96	2204
118	39.6	1568.16	13924
204	125.6	15775.36	41616
97	18.6	345.96	9409
68	-10.4	108.16	4624
86	7.6	57.76	7396
62	-16.4	268.96	3844
57	-21.4	457.96	3249
98	19.6	384.16	9604
99	20.6	424.36	9801
941	0.2	31898.92	105689

$$s = \sqrt{\frac{31898.92}{12-1}} = 53.85$$

The computing formula for *s* is

$$s = \sqrt{\frac{\sum x_i^2 - (\sum x_i)^2 / n}{n-1}}$$

Columns 1 and 4 of the previous table present the calculations needed to compute *s* by the computing formula. Using the figures presented at the bottom of each of these columns and substituting into the formula itself, we get:

$$s = \sqrt{\frac{105689 - 941^2 / 12}{12-1}} = 53.85$$

This is, of course, the same result that we obtained by the defining formula.

3.75 The range of a data set is defined to be the difference between the largest and smallest data values in the data set. The highest value is 66 billion and the lowest value is 25 billion. The range is 66 - 25 = 41 billion.

The defining formula for *s* is $s = \sqrt{\frac{\sum (x_i - \bar{x})^2}{n-1}}$.

The following table presents the calculations that are needed to compute *s* by the defining formula:

x	x-x̄	(x-x̄)^2
66.0	31.29	979.06
46.0	11.29	127.46
41.0	6.29	39.56
31.0	-3.71	13.76
31.0	-3.71	13.76
27.9	-6.81	46.38
26.8	-7.91	62.57
26.3	-8.41	70.73
26.1	-8.61	74.13
25.0	-9.71	94.28
347.1	0.00	1521.71

With n = 10 and using the bottom figure of column 1, we find that $\bar{x} = \sum x_i / n = 347.1/10 = 34.71$. This value is needed to construct the differences in column 2. Column 3 squares the differences presented in column 2. The figure presented at the bottom of column 3 is the computation needed for the numerator of the defining formula *s*. Thus,

$$s = \sqrt{\frac{1521.71}{10-1}} = 13.00$$

3.77 The range of a data set is defined to be the difference between the largest and smallest data values in the data set. The highest overall value is 30 and the lowest overall value is 6. The range is 30 - 6 = 24.

The defining formula for s is $s=\sqrt{\frac{\sum(x_i-\bar{x})^2}{n-1}}$.

The following table presents the calculations that are needed to compute s by the defining formula:

x	x-x̄	(x-x̄)²
14	-0.58	0.3364
14	-0.58	0.3364
13	-1.58	2.4964
14	-0.58	0.3364
14	-0.58	0.3364
14	-0.58	0.3364
14	-0.58	0.3364
13	-1.58	2.4964
30	15.42	237.7764
6	-8.58	73.6164
17	2.42	5.8564
12	-2.58	6.6564
175	0.00	330.9168

With n = 12 and using the bottom figure of column 1, we find that $\bar{x}$ = $\sum x_i/n$ = 175/12 = 14.58. This value is needed to construct the differences in column 2. Column 3 squares the differences presented in column 2. The figure presented at the bottom of column 3 is the computation needed for the numerator of the defining formula s. Thus,

$$s=\sqrt{\frac{330.9168}{12-1}}=5.5$$

3.79 (a) We will use the computing formula to find s. The following table presents the calculations that are needed to compute s by the computing formula:

x	x^2
83	6889
46	2116
64	4096
385	148225
46	2116
21	441
48	2304
86	7396
523	273529
429	184041

35	1225
51	2601
34	1156
258	66564
265	70225
119	14161
2484	6170256
4977	6957341

The computing formula for *s* is

$$s = \sqrt{\frac{\sum x_i^2 - (\sum x_i)^2 / n}{n-1}}$$

The previous table presents the calculations needed to compute *s* by the computing formula. Using the figures presented at the bottom of the columns and substituting into the formula itself, we get:

$$s = \sqrt{\frac{6957341 - 4977^2 / 17}{17-1}} = 586.3$$

(b) No. The standard deviation is based on deviations from the mean. However, these data are right skewed and contain a very large outlier (2484) that greatly increases the value of the mean. The deviation of 2484 from the mean accounts for about 87% of the total of the squared deviations from the mean, so the influence of that data value on both the mean and the standard deviation is very great.

3.81 (a) We would guess that the non-built up roads would have the greater variation due to more variation in the recreational use of those roads at various times of the week.

(b) The range for built-up roads is 103 - 69 = 34, and the range for non-built-up roads is 102 - 53 = 49.

For built-up roads, n = 17, $\sum x_i = 619$, $\sum x_i^2 = 55719$, and

$$s = \sqrt{\frac{55719 - 619^2 / 7}{7-1}} = 12.79$$

For non-built-up-roads, n = 7, $\sum x_i = 492$, $\sum x_i^2 = 36930$, and

$$s = \sqrt{\frac{36930 - 492^2 / 7}{7-1}} = 19.79$$

Thus both measures of variation confirm that our guess is correct. The range and standard deviation are greater for non-built-up roads than for built-up roads.

3.83 The data in this set are qualitative (the hospital type for U.S. registered hospitals). Being nonnumerical, it is not possible to compute a range or standard deviation.

3.85 The data in this set are qualitative (ballot preference and highest degree obtained). Being nonnumerical, it is not possible to compute a range or standard deviation.

3.87 Using Minitab, retrieve the data from the WeissStats Resource Site. Select **Stat ▶ Basic Statistics ▶ Display Descriptive Statistics**, enter LENGTH in the **Variables** text box. Now click on the **Statistics** button and check the range to add it to the list of statistics to be computed. Click **OK** twice. The output appears in the Session Window as

Variable	N	N*	Mean	SE Mean	StDev	Minimum	Q1	Median	Q3
LENGTH	229	0	163.75	3.80	57.57	23.00	132.50	153.00	181.50

Variable	Maximum	Range
LENGTH	493.00	470.00

The range (the difference between the maximum and the minimum data values) is 470.00 and the standard deviation is 57.57. Roughly speaking, on the average, the lengths differ from the overall mean length by 57.57 minutes. For these data, which are right skewed and contain a few very large values, the standard deviation may not be a very good measure of variation due to the large influence of those large values. Typically, the range will be about 4 to 6 times the standard deviation. In this case, it is about 8 times the standard deviation, indicating that there may be outliers or highly influential data values.

3.89 Using Minitab, retrieve the data from the WeissStats Resource Site. Select **Stat ▶ Basic Statistics ▶ Display Descriptive Statistics**, enter PERCENT in the **Variables** text box. Now click on the **Statistics** button and check the range to add it to the list of statistics to be computed. Click **OK** twice. The output appears in the Session Window as

Variable	N	N*	Mean	SE Mean	StDev	Minimum	Q1	Median	Q3
PERCENT	51	0	27.706	0.788	5.626	17.000	24.000	27.000	31.000

Maximum	Range
49.000	32.000

The range (the difference between the maximum and the minimum data values) is 32% and the standard deviation is 5.626%. Roughly speaking, on the average, the completion percetages differ from the overall mean completion percentage by 5.626%. One should keep in mind, however, that the overall mean computed by Minitab assumes that the completion percentages for each state count equally. In fact, the true overall mean completion percentage would take into account that there are different numbers of students in each state.

3.91 (a) Using Minitab, retrieve the data from the WeissStats Resource Site. Select **Stat ▶ Basic Statistics ▶ Display Descriptive Statistics**, enter INTEGRATED and STANDARD in the **Variables** text box. Now click on the **Statistics** button and check the range to add it to the list of statistics to be computed. Click **OK** twice. The output appears in the Session Window as

Variable	N	N*	Mean	SE Mean	StDev	Minimum	Q1	Median	Q3
INTEGRATED	227	0	24.916	0.300	4.513	11.000	22.000	25.000	28.000
STANDARD	192	0	23.000	0.521	7.221	6.000	18.000	23.000	28.750

Variable	Maximum	Range
INTEGRATED	35.000	24.000
STANDARD	40.000	34.000

The range of the questionnaire scores for the integrated treatment is 24. The range of the questionnaire scores for the standard treatment is 34. The standard deviation of the questionnaire scores for the integrated treatment is 4.513 and 7.221 for the standard treatment.

(b) Both the range and the standard deviation are greater for the standard treatment. This indicates that the standard treatment had more variation in its observations than the integrated treatments observations.

3.93 (a) We will compute *s* for each data set using the computing formula. The computing formula requires the following column manipulations.

Data Set I

x	x^2
0	0
0	0
10	100
12	144
14	196
14	196
14	196
15	225
15	225
15	225
16	256
17	289
23	529
24	576
189	3,157

Data Set II

x	x^2
10	100
12	144
14	196
14	196
14	196
15	225
15	225
15	225
16	256
17	289
142	2,052

For each data set, the calculations for *s* are presented in column 2 of the following table.

Data Set	s	Range
I	$s=\sqrt{\dfrac{3157-189^2/14}{14-1}}=6.8$	24
II	$s=\sqrt{\dfrac{2052-142^2/10}{10-1}}=0.20.$	7

(b) The range for Data Set I is 24 - 0 = 24. The range for Data Set II is 17 - ′10 = 7. These are recorded in column 3 of the previous table.

(c) Outliers increase the variation in a data set; in other words, removing the outliers from a data set results in a decrease in the variation.

3.95 (a) We create a table with columns headed by x, f, xf, $(x-\bar{x})$, $(x-\bar{x})^2$, and $(x-\bar{x})^2f$ to estimate the sample mean and standard deviation of the days-to-maturity data.

Class Mark x	Frequency f	xf	$x-\bar{x}$	$(x-\bar{x})^2$	$f(x-\bar{x})^2$
34.5	3	103.5	-33.5	1122.25	3366.75
44.5	1	44.5	-23.5	552.25	552.25
54.5	8	436.0	-13.5	182.25	1458.00
64.5	10	645.0	-3.5	12.25	122.50
74.5	7	521.5	6.5	42.25	295.75
84.5	7	591.5	16.5	272.25	1905.75
94.5	4	378.0	26.5	702.25	2809.00
	40	2720.0			10510.00

The mean $\bar{x}$ = 2720/40 =68. This is subtracted from each of the class marks in column 1 to get the entries in column 4. The grouped data formula yields the following estimate of the standard deviation:

$$s = \sqrt{\frac{\sum (x-\bar{x})^2 f}{n-1}} = \sqrt{\frac{10510}{40-1}} = 16.4$$

(b) There is a small discrepancy between the estimated mean and estimated standard deviation and the actual values. These discrepancies occur because in the grouped data formulas, every actual data value in a given class is replaced by the class mark even though most values in the class are not equal to the class mark. For example, every data value in the first class is represented by the class mark 34.5, but the three data values are 36, 38, and 39, all of them larger than the class mark. This introduces a small error in each difference between an observation and the mean. It is unlikely that all of these small errors will cancel each other out causing both the mean and the standard deviation to exhibit small errors.

Exercises 3.3

3.97 The empirical rule improves on the estimates given by Chebyshev's rule.

3.99 (a) If k = 4, using Chebyshev's rule, at least $1 - 1/k^2 = 1 - 1/16 = 15/16$, or 93.75%, of the observations lie within 4 standard deviations to either side of the mean.

(b) If k = 2.5, using Chebyshev's rule, at least $1 - 1/k^2 = 1 - 1/6.25 =$ 0.84, or 84%, of the observations lie within 2.5 standard deviations to either side of the mean.

3.101 If k = 5, using Chebyshev's rule, at least $1 - 1/k^2 = 1 - 1/25 = 24/25$, or 96%, of the observations lie within 5 standard deviations to either side of the mean.

3.103 (a) The mean of the data set is 83.0 and the standard deviation is 7.85. Locate the mean on the horizontal axis and measure intervals equal in length to the standard deviation. The dotplot of the data set is as follows.

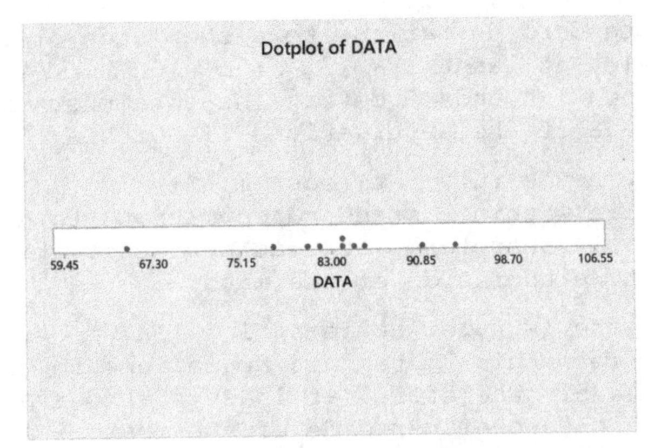

(b) Two standard deviations to either side of the mean would be 83.00 - 2(7.85) to 83.00 + 2(7.85) or 67.3 to 98.70. 9 of the 10 observations are within this range, which is 90% of the observations. If $k = 2$, using Chebyshev's rule, at least $1 - 1/k^2 = 1 - 1/4 = 3/4$, or 75%, of the observations are estimated to lie within 2 standard deviations to either side of the mean. The actual percentage is higher than the Chebyshev's estimated percentage.

(c) Three standard deviations to either side of the mean would be 83.00 - 3(7.85) to 83.00 + 3(7.85) or 59.45 to 106.55. All of the 10 observations are within three standard deviations to either side of the mean. This is 100% of the observations. If $k = 3$, using Chebyshev's rule, at least $1 - 1/k^2 = 1 - 1/9 = 8/9$, or 88.9%, of the observations is estimated to lie within 3 standard deviations to either side of the mean. The actual percentage is higher than the Chebyshev's estimated percentage.

3.105 (a) The shape of the distribution is bell-shaped. Therefore application of the empirical rule is appropriate.

(b) The shape of the distribution is very right-skewed. Therefore application of the empirical rule is not appropriate.

(c) The shape of the distribution is left skewed. Therefore, application of the empirical rule is not appropriate.

3.107 (a) Chebyshev's rule estimates that, for any distribution, at least 75% of the observations will be within two standard deviations to either side of the mean. The empirical rule states that in a roughly bell-shaped distribution, 95% of the observations will be within two standard deviations to either side of the mean. The empirical rule improved on the estimated percentages of Chebyshev's rule.

(b) Chebyshev's rule estimates that, for any distribution, at least 89% of the observations will be within three standard deviations to either side of the mean. The empirical rule states that in a roughly bell-shaped distribution, 99.7% of the observations will be within three standard deviations to either side of the mean. The empirical rule improved on the estimated percentages of Chebyshev's rule.

3.109 (a) 89% corresponds to $k = 3$ for Chebyshev's rule. At least 89% of the observations lie between 25 - 3(5) = <u>10</u> and 25 + 3(5) = <u>40</u>.

(b) Since 15 and 35 are two standard deviations to either side of the mean, Chebyshev's rule estimates that at least <u>75%</u> of the observations lie within 15 and 35.

3.111 10 and 50 are five standard deviations to either side of the mean. If $k = 5$, using Chebyshev's rule, at least $1 - 1/k^2 = 1 - 1/25 = 24/25$, or 96%, of the observations lie within 5 standard deviations to either side of the mean. Therefore, at least 96% of the observations lie within 10 and 50.

3.113 If $k = 2$, using Chebyshev's rule, at least $1 - 1/k^2 = 1 - 1/4 = 3/4$, or 75%, of the observations lie within 2 standard deviations to either side of the mean. 75% of 80 is 60. Therefore, at least 60 observations lie within two standard deviations to either side of the mean.

3.115 If $k = 3$, using Chebyshev's rule, at least $1 - 1/k^2 = 1 - 1/9 = 8/9$, or 89%, of the observations lie within 3 standard deviations to either side of the mean. 89% of 50 is 44.5. Therefore, at least 45 observations lie within three standard deviations to either side of the mean.

3.117 68 and 132 are two standard deviations to either side of the mean. If $k = 2$, using Chebyshev's rule, at least $1 - 1/k^2 = 1 - 1/4 = 3/4$, or 75%, of the observations lie within 2 standard deviations to either side of the mean. Furthermore, 75% of 60 is 45. Therefore, at least 45 of the observations lie within 68 and 132.

3.119 23 and 47 are three standard deviations to either side of the mean. If $k = 3$, using Chebyshev's rule, at least $1 - 1/k^2 = 1 - 1/9 = 8/9$, or 89%, of the observations lie within 3 standard deviations to either side of the mean. Furthermore, 89% of 150 is 133.5. Therefore, at least 134 of the observations lie within 23 and 47.

3.121 (a) The empirical rule states that 68% of the observations will be within one standard deviation to either side of the mean. Therefore, approximately 68% of the observations lie between $25 - 5 = \underline{20}$ and $25 + 5 = \underline{30}$.

(b) The empirical rule states that 95% of the observations will be within two standard deviations to either side of the mean. Therefore, approximately 95% of the observations lie between $25 - 2(5) = \underline{15}$ and $25 + 2(5) = \underline{35}$.

(c) The empirical rule states that 99.7% of the observations will be within three standard deviations to either side of the mean. Therefore, approximately 99.7% of the observations lie between $25 - 3(5) = \underline{10}$ and $25 + 3(5) = \underline{40}$.

3.123 22 and 38 are two standard deviations to either side of the mean. Using the empirical rule, 95% of the observations lie within 2 standard deviations to either side of the mean.

3.125 Using the empirical rule, 95% of the observations lie within 2 standard deviations to either side of the mean. 95% of 80 is 76. Therefore, approximately 76 of the observations are estimated to be within 2 standard deviations to either side of the mean.

3.127 Using the empirical rule, 68% of the observations lie within 1 standard deviation to either side of the mean. 68% of 50 is 34. Therefore, approximately 34 of the observations are estimated to be within 1 standard deviation to either side of the mean.

3.129 52 and 148 are three standard deviations to either side of the mean. Using the empirical rule, 99.7% of the observations lie within 3 standard deviations to either side of the mean. 99.7% of 250 is 249.25. Therefore, approximately 249 of the observations are estimated to be between 52 and 148.

3.131 31 and 39 are one standard deviation to either side of the mean. Using the empirical rule, 68% of the observations lie within 1 standard deviation to either side of the mean. 68% of 150 is 102. Therefore, approximately 102 of the observations are estimated to be between 31 and 39.

3.133 (a) Chebyshev's rule with k = 2 says that at least 75% of the observations lie within two standard deviations to either side of the mean. The sample mean and sample standard deviation are 11.9 and 6.5, respectively. Two standard deviations to either side of the mean would be 11.9 - 2(6.5) = -1.1 and 11.9 + 2(6.5) = 24.9. Thirty-one out of the thirty-three observations are within this range. Therefore, actually 93.9% of the observations lie within two standard deviations to either side of the mean. This is greater than the Chebyshev's estimate.

(b) Chebyshev's rule with k = 3 says that at least 89% of the observations lie within three standard deviations to either side of the mean The sample mean and sample standard deviation are 11.9 and 6.5, respectively. Three standard deviations to either side of the mean would be 11.9 - 3(6.5) = -7.6 and 11.9 + 3(6.5) = 31.4. All of the thirty observations are within this range. Therefore, actually 100% of the observations lie within three standard deviations to either side of the mean. This is greater than the Chebyshev's estimate.

(c) The actual percentage of observations within 2 and within 3 standard deviations to either side of the mean were greater that each of the Chebyshev's estimates. Chebychev's estimates are not necessarily precise.

3.135 (a) Based on the fact that the sample mean is 14.7 mg and the sample standard deviation is 3.1 mg, we constructed the following figure. Note, for instance, that two standard deviations to the left is 14.7 - 2(3.1) = 8.5.

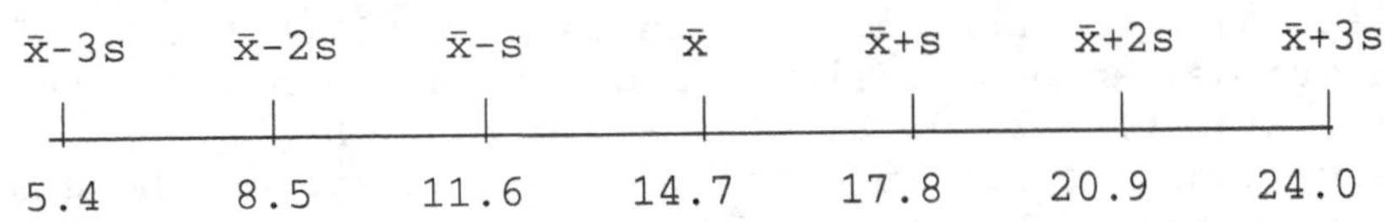

(b) Chebyshev's rule with k = 2 says that at least 75% of the observations lie within two standard deviations to either side of the mean. 75% of 45 is 33.75. Therefore, at least 34 people in this sample of females under the age of 51 have intakes between 8.5 and 20.9 mg of iron during a 24-hour period.

(c) Chebyshev's rule with k = 3 says that at least 89% of the observations lie within three standard deviations to either side of the mean. 89% of 45 is 40.05. Therefore, at least 41 people in this sample of females under the age of 51 have intakes between 5.4 and 24.0 mg of iron during a 24-hour period.

3.137 (a) Based on the fact that the sample mean is 98.1°F and the sample standard deviation is 0.65°F, we constructed the following figure. Note, for instance, that two standard deviations to the left is 98.1 - 2(0.65) = 96.8.

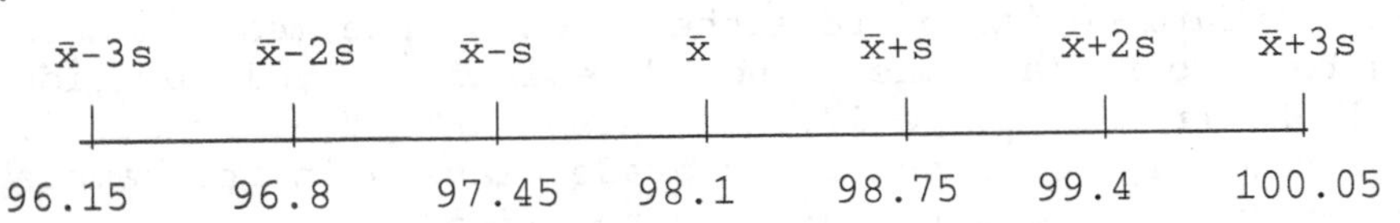

(b) Chebyshev's rule with $k = 2$ says that at least 75% of the observations lie within two standard deviations to either side of the mean. 75% of 93 is 69.75. Therefore, at least 70 people in this sample of healthy humans have body temperatures between 96.8°F and 99.4°F.

(c) Chebyshev's rule with $k = 3$ says that at least 89% of the observations lie within three standard deviations to either side of the mean. 89% of 93 is 82.77. Therefore, at least 83 people in this sample of healthy humans have body temperatures between 96.15°F and 100.05°F.

3.139 (a) The data set is approximately bell-shaped. It is reasonable to use the empirical rule.

(b) The empirical rule estimates that 68% of the observations will lie within one standard deviation to either side of the mean, 95% will lie within two standard deviations to either side of the mean, and 99.7% will lie within three standard deviations to either side of the mean.

(c) Based on the fact that the sample mean is 45.3 kg and the sample standard deviation is 4.16 kg, we constructed the following figure.

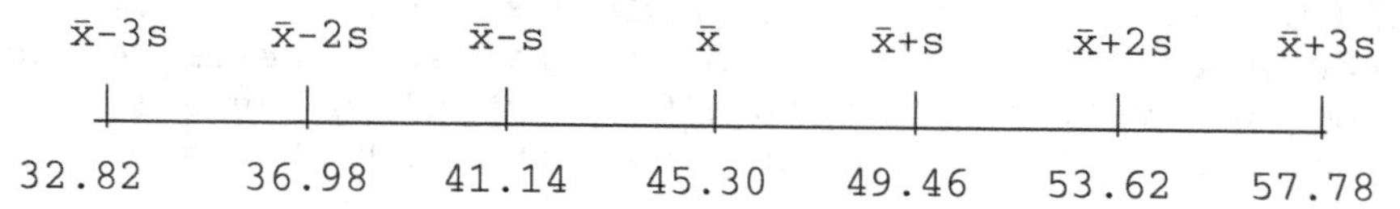

One standard deviation to either side of the mean would be the range of 45.3 - 4.16 = 41.14 to 45.3 + 4.16 = 49.46 kg. Forty-four of the observations lie within this range. Therefore, the exact percentage of observations within one standard deviation to either side of the mean is 44/60 * 100 = 73.3%. Two standard deviations to either side of the mean would be the range of 45.3 - 2(4.16) = 36.98 to 45.3 + 2(4.16) = 53.62 kg. Fifty-seven of the observations lie within this range. Therefore, the exact percentage of observations within two standard deviations to either side of the mean is 57/60 * 100 = 95%. Three standard deviations to either side of the mean would be the range of 45.3 - 3(4.16) = 32.82 to 45.3 + 3(4.16) = 57.78. All of the observations lie within this range. Therefore, the exact percentage of observations that lie within three standard deviations to either side of the mean is 100%.

(d) The actual percentages calculated in part (c) are very close to the estimated empirical percentages in part (b). Due to the fact that the shape was close to bell-shaped, it is expected that the empirical rule will give good estimates.

3.141 (a) Based on the fact that the sample mean is 15.83 and the sample standard deviation is 1.74, we constructed the following figure.

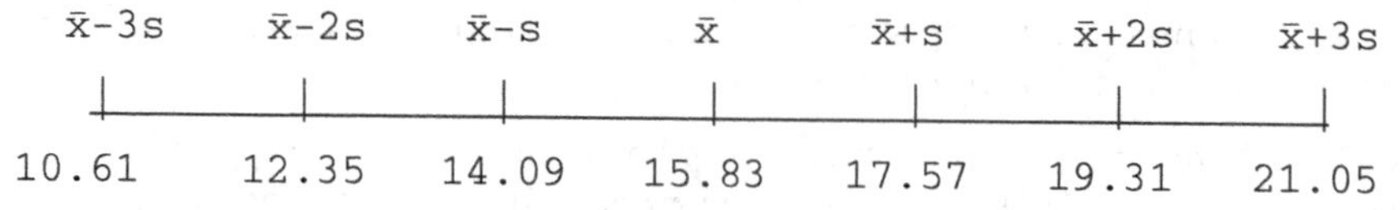

(b) According to Property 1 of the empirical rule, approximately 68% of the fifth grade classes in the sample have student-to-faculty ratios within one standard deviation to either side of the mean. According to the figure in part (a), one standard deviation to either side of the mean would be from 14.09 to 17.57. Also, 68% of 81 is 55.08. Therefore, approximately 55 of the fifth grade classes in the sample have student-to-faculty ratios between 14.09 and 17.57.

(c) According to Property 2 of the empirical rule, approximately 95% of the fifth grade classes in the sample have student-to-faculty ratios within two standard deviations to either side of the mean. According to the figure in part (a), two standard deviations to either side of the mean would be from 12.35 to 19.31. Also, 95% of 81 is 76.95. Therefore, approximately 77 of the fifth grade classes in the sample have student-to-faculty ratios between 12.35 and 19.31.

(d) According to Property 3 of the empirical rule, approximately 99.7% of the fifth grade classes in the sample have student-to-faculty ratios within three standard deviations to either side of the mean. According to the figure in part (a), three standard deviations to either side of the mean would be from 10.61 to 21.05. Also, 99.7% of 81 is 80.757. Therefore, approximately all of fifth grade classes in the sample have student-to-faculty ratios between 10.61 and 21.05.

3.143 (a) Based on the fact that the sample mean is 1.40 kg and the sample standard deviation is 0.11 kg, we constructed the following figure.

$\bar{x}-3s$	$\bar{x}-2s$	$\bar{x}-s$	$\bar{x}$	$\bar{x}+s$	$\bar{x}+2s$	$\bar{x}+3s$
1.07	1.18	1.29	1.40	1.51	1.62	1.73

(b) According to Property 1 of the empirical rule, approximately 68% of the Swedish men in the sample have brain weights within one standard deviation to either side of the mean. According to the figure in part (a), one standard deviation to either side of the mean would be from 1.29 to 1.51 kg. Also, 68% of 225 is 153. Therefore, approximately 153 of the Swedish men in the sample have brain weights between 1.29 and 1.51 kg.

(c) According to Property 2 of the empirical rule, approximately 95% of the Swedish men in the sample have brain weights within two standard deviations to either side of the mean. According to the figure in part (a), two standard deviations to either side of the mean would be from 1.18 to 1.62 kg. Also, 95% of 225 is 213.75. Therefore, approximately 214 of the Swedish men in the sample have brain weights between 1.18 and 1.62 kg.

(d) According to Property 3 of the empirical rule, approximately 99.7% of the Swedish men in the sample have brain weights within three standard deviations to either side of the mean. According to the figure in part (a), three standard deviations to either side of the mean would be from 1.07 to 1.73 kg. Also, 99.7% of 225 is 224.325. Therefore, approximately 224 of the Swedish men in the sample have brain weights between 1.07 and 1.73 kg.

3.145 We are trying to solve for k in Chebyshev's rule such that it estimates at least 99% of the observations will lie within k standard deviations to either side of the mean. Setting Chebyshev's rule equal to 0.99 and solving for k, we have

$$1-1/k^2=0.99$$
$$0.01=1/k^2$$
$$k^2=1/0.01=100$$
$$k=10$$

Therefore, Chebyshev's rule estimates that at least 99% of the observations for any data set will lie within 10 standard deviations to either side of the mean.

3.147 (a) If you added $2m^2-1$ zeros, one $-m$ and one m, you would get a sum of zero. Therefore, the sample mean would be 0. Using the defining formula to calculate the sample standard deviation, we would get the following.

x	$x-\bar{x}$	$(x-\bar{x})^2$
0	0	0
0	0	0
0	0	0
etc.	etc.	etc.
$-m$	$-m$	m^2
m	m	m^2
0	0.00	$2m^2$

The defining formula for s is $s=\sqrt{\frac{\sum(x-\bar{x})^2}{n-1}}$. There are $2m^2 - 1 + 2 = 2m^2 + 1$ observations in the sample. Plugging this in for n in the defining formula, we have $s=\sqrt{\frac{\sum(x-\bar{x})^2}{n-1}}=\sqrt{\frac{2m^2}{(2m^2+1)-1}}=\sqrt{\frac{2m^2}{2m^2}}=1$. The sample mean is 0 and the sample standard deviation is 1.

(b) Since the sample standard deviation is 1 and the sample mean is 0, the number 5 would be 5 standard deviations from the mean. Therefore, the observation m would be m standard deviations from the mean.

(c) If $m \geq 4$, then m and $-m$ would be at least four standard deviations from the mean. The remaining observations of the data set are all equal to zero. These remaining observations are all equal to the sample mean of 0 and therefore all within three standard deviations from the mean. All but two observations are within three standard deviations of the mean. Since there are $2m^2 + 1$ observations and $2m^2 - 1$ zeros, this would be $\left(\frac{2m^2-1}{2m^2+1}\right)*100$ percent of observations that lie within three standard deviations to either side of the mean.

Exercises 3.4

3.149 The median and the interquartile range have the advantage over the mean and standard deviation that they are not sensitive to extreme values; they are said to be resistant.

3.151 No. An extreme observation may be an outlier, but it may also be an indication of skewness.

3.153 (a) The interquartile range is a descriptive measure of variation.

(b) It measures the spread of the middle two quarters (50%) of the data.

3.155 The adjacent values are just the minimum and maximum observations when there are no potential outliers.

3.157 Roughly, when arranged in increasing order, the middle 50% of a data set are found between $\underline{Q_1}$ and $\underline{Q_3}$.

3.159 (a) Following Procedure 3.1, Step 1: First arrange the 4 data values in increasing order and note the two middle values:

1 **2** **3** 4

Step 2: The median (second quartile) is the average of the two middle values, 2 and 3, and thus equals (2 + 3)/2 = 2.5.

Step 3 and 4: The first quartile Q_1 is the median of the lower half of the ordered data set 1 2 and thus is the average of the first and second values, 1 and 2. $Q_1 = (1 + 2)/2 = 1.5$

Step 5: The third quartile Q_3 is the median of the upper half of the ordered data set 3 4 and thus is the average of the first and second values, 3 and 4. $Q_3 = (3 + 4)/2 = 3.5$

Step 6: The quartiles are $Q_1 = 1.5$, $Q_2 = 2.5$, and $Q_3 = 3.5$.

(b) The interquartile range is $Q_3 - Q_1 = 3.5 - 1.5 = 2.0$.

(c) The five-number summary consists of the Minimum, Q_1, Median, Q_3, and the Maximum. Respectively, these are 1, 1.5, 2.5, 3.5, and 4.

3.161 (a) Following Procedure 3.1, Step 1: First arrange the 5 data values in increasing order:

1 2 3 4 5

Step 2: Since this data set has an odd number of values (5), the median is the middle value (3rd) in the ordered list. The median, or second quartile, is 3.

Step 3 and 4: The first quartile Q_1 is the median of the lower half of the ordered data set (including the median) 1 **2** 3 and thus is the second value. $Q_1 = 2$

Step 5: The third quartile Q_3 is the median of the upper half of the ordered data set (including the median) 3 **4** 5 and thus is the second value. $Q_3 = 4$

Step 6: The quartiles are $Q_1 = 2$, $Q_2 = 3$, and $Q_3 = 4$.

(b) The interquartile range is $Q_3 - Q_1 = 4 - 2 = 2$.

(c) The five-number summary consists of the Minimum, Q_1, Median, Q_3, and the Maximum. Respectively, these are 1, 2, 3, 4, and 5.

3.163 (a) Following Procedure 3.1, Step 1: First arrange the 6 data values in increasing order and note the two middle values:

1 2 **3** **4** 5 6

Step 2: The median (second quartile) is the average of the two middle values, 3 and 4, and thus equals (3 + 4)/2 = 3.5.

Step 3 and 4: The first quartile Q_1 is the median of the lower half of the ordered data set 1 **2** 3 and thus is the second value. $Q_1 = 2$

Step 5: The third quartile Q_3 is the median of the upper half of the ordered data set 4 **5** 6 and thus is second values. $Q_3 = 5$

Step 6: The quartiles are $Q_1 = 2$, $Q_2 = 3.5$, and $Q_3 = 5$.

(b) The interquartile range is $Q_3 - Q_1 = 5 - 2 = 3$.

(c) The five-number summary consists of the Minimum, Q_1, Median, Q_3, and the Maximum. Respectively, these are 1, 2, 3.5, 5, and 6.

3.165 (a) Following Procedure 3.1, Step 1: First arrange the 7 data values in increasing order:

1 2 3 4 5 6 7

Step 2: Since this data set has an odd number of values (7), the median is the middle value (4th) in the ordered list. The median, or second quartile, is 4.

Step 3 and 4: The first quartile Q_1 is the median of the lower half of the ordered data set (including the median) 1 **2** **3** 4 and thus is the mean of the second and third values. $Q_1 = (2 + 3)/2 = 2.5$

Stpe 5: The third quartile Q_3 is the median of the upper half of the ordered data set (including the median) 4 **5** **6** 7 and thus is the mean of the second and third values. $Q_3 = (5 + 6)/2 = 5.5$

Step 6: The quartiles are $Q_1 = 2.5$, $Q_2 = 4$, and $Q_3 = 5.5$.

(b) The interquartile range is $Q_3 - Q_1 = 5.5 - 2.5 = 3$.

(c) The five-number summary consists of the Minimum, Q_1, Median, Q_3, and the Maximum. Respectively, these are 1, 2.5, 4, 5.5, and 7.

3.167 (a) Following Procedure 3.1, Step 1: First arrange the 20 data values in increasing order and note the two middle values:

45 48 64 70 73 74 74 78 78 **79** **79** 80 80 80 80 80 80 81 82 82

Step 2: The median (second quartile) is the average of the two middle values, 79 and 79, and thus equals $(79 + 79)/2 = 79$.

Step 3 and 4: The first quartile Q_1 is the median of the lower half of the ordered data set 45 48 64 70 **73** **74** 74 78 78 79 and thus is the average of the fifth and sixth values, 73 and 74. $Q_1 = (73 + 74)/2 = 73.5$

Step 5: The third quartile Q_3 is the median of the upper half of the ordered data set 79 80 80 80 **80** **80** 80 81 82 82 and thus is the average of the fifth and sixth values, 80 and 80. $Q_3 = (80 + 80)/2 = 80.0$

Step 6: The quartiles are $Q_1 = 73.5$, $Q_2 = 79$, and $Q_3 = 80$.

Thus 25% of the data values are less than or equal to 73.5, the next 25% are between 73.5 and the median 79, the next 25% are between 79 and 80, and the upper 25% are greater than or equal to 80.

(b) The interquartile range is $Q_3 - Q_1 = 80.0 - 73.5 = 6.5$.

(c) The five-number summary consists of the Minimum, Q_1, Median, Q_3, and the Maximum. Respectively, these are 45, 73.5, 79.0, 80.0, and 82. In addition to the explanation of the quartiles in part (a), all of the values are greater than or equal to the Minimum, and all of the values are less than or equal to the Maximum.

(d) To identify any potential outliers, we first find the lower and upper limits. These are found by subtracting 1.5 x IQR from Q_1 and adding 1.5 x IQR to Q_3. The lower limit is $73.5 - 1.5(6.5) = 63.75$ and the upper limit is $80.0 + 1.5(6.5) = 89.75$. Potential outliers consist of any values that are less than 63.75 or greater than 89.75. We see that 45 and 48 are both less than 63.75, so these two values are potential outliers.

(e) We will construct a modified boxplot so as to show the potential outliers. The boxplot consists of a rectangle with its ends at the first and third quartiles, a line through it at the median, and two "whiskers" that extend from the quartiles outward to the adjacent values. The adjacent values are the most extreme data values that do not lie outside the lower and upper limits. For these data, the adjacent values are 64 and 82. The potential outliers, 45 and 48, are plotted individually. The modified boxplot, prepared using Minitab, is

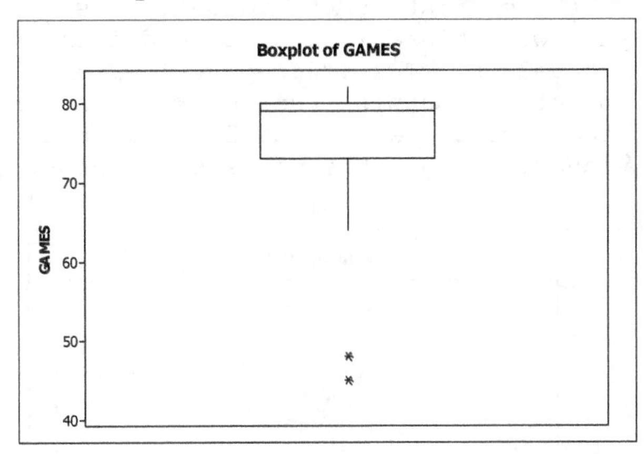

We see that the data is very left skewed, even without considering the two potential outliers. This is because, in most seasons, Gretzky played almost every game, but in a few seasons, he played considerably fewer than the maximum number of games, probably due to injuries.

3.169 (a) Following Procedure 3.1, Step 1: First arrange the data in increasing order:

1 1 3 3 4 4 5 6 6 7 **7** 9 9 10 12 12 13 15 18 23 55

Step 2: Since this data set has an odd number of values (21), the median is the middle value (11th) in the ordered list. The median, or second quartile, is 7.

Step 3 and 4: The first quartile Q_1 is the median of the lower half of the ordered data set (including the median) 1 1 3 3 4 **4** 5 6 6 7 7. Since this data set has an odd number of values (11), Q_1 is the sixth value, so $Q_1 = 4$.

Step 5: The third quartile Q_3 is the median of the upper half of the ordered data set (including the median) 7 9 9 10 12 **12** 13 15 18 23 55. Since this data set has an odd number of values (11), Q_3 is the sixth value, so $Q_3 = 12$.

Step 6: The quartiles are $Q_1 = 4$, $Q_2 = 7$, and $Q_3 = 12$.

Thus 25% of the data values are less than or equal to 4, the next 25% are between 4 and the median 7, the next 25% are between 7 and 12, and the upper 25% are greater than or equal to 12.

(b) The interquartile range is $Q_3 - Q_1 = 12 - 4 = 8$.

(c) The five-number summary consists of the Minimum, Q_1, Median, Q_3, and the Maximum. Respectively, these are 1, 4, 7, 12, and 55. In addition to the explanation of the quartiles in part (a), all of the values are greater than or equal to the Minimum, and all of the values are less than or equal to the Maximum.

(d) To identify any potential outliers, we first find the lower and upper limits. These are found by subtracting 1.5 x IQR from Q_1 and adding 1.5 x IQR to Q_3. Thus the lower limit is 4 - 1.5(8) = -8 and the upper limit is 12 + 1.5(8) = 24. Potential outliers consist of any values that are less than -8 or greater than 24. We see that 55 is the only potential outlier.

(e) We will construct a modified boxplot, which consists of a rectangle with its ends at the first and third quartiles, a line through it at the median, and two "whiskers" that extend from the quartiles outward to the adjacent values. The adjacent values are the most extreme data values that do not lie outside the lower and upper limits. For these data, the adjacent values are 1 and 23. The potential outlier, 55, is plotted individually. Thus the modified boxplot, prepared using Minitab, is

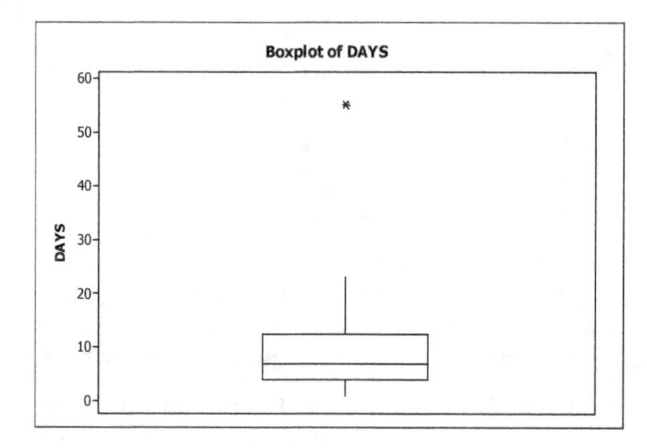

We see that the data are slightly right skewed if the potential outlier is not considered. The potential outlier is far from the other values, making it important to investigate the observation further.

3.171 (a) Following Procedure 3.1, Step 1: First arrange the data in increasing order:

57 66 88 96 **116** **147** 147 154 154 175

Step 2: Since this data set has an even number of values (10), the median is the mean of the middle two values (5th and 6th) in the ordered list. Thus the median, or second quartile, is (116 + 147)/2 = 131.5.

Step 3 and 4: The first quartile Q_1 is the median of the lower half of the ordered data set 57 66 **88** 96 116. Since this data set has an odd number of values (5), Q_1 is the middle value (3rd) in the ordered list = 88.

Step 5: The third quartile Q_3 is the median of the upper half of the ordered data set 147 147 **154** 154 175. Since this data set has an odd number of values (5), Q_3 is the middle value in the ordered list = 154.

Step 6: The quartiles are Q_1 = 88, Q_2 = 131.5, and Q_3 = 154.

Thus 25% of the data values are less than or equal to 88, the next 25% are between 88 and the median 131.5, the next 25% are between 131.5 and 154, and the upper 25% are greater than or equal to 154.

(b) The interquartile range is $Q_3 - Q_1$ = 154 - 88 = 66.

(c) The five-number summary consists of the Minimum, Q_1, Median, Q_3, and the Maximum. Respectively, these are 57, 88, 131.5, 154, and 175.

In addition to the explanation of the quartiles in part (a), all of the values are greater than or equal to the Minimum, and all of the values are less than or equal to the Maximum.

(d) To identify any potential outliers, we first find the lower and upper limits. These are found by subtracting 1.5 x IQR from Q_1 and adding 1.5 x IQR to Q_3. Thus the lower limit is 88 − 1.5(66) = −11 and the upper limit is 154 + 1.5(66) = 253. Potential outliers consist of any values that are less than −11 or greater than 253. We see that there are no potential outliers.

(e) We will construct a boxplot, which consists of a rectangle with its ends at the first and third quartiles, a line through it at the median, and two "whiskers" that extend from the quartiles outward to the minimum and maximum values. Following is the boxplot,

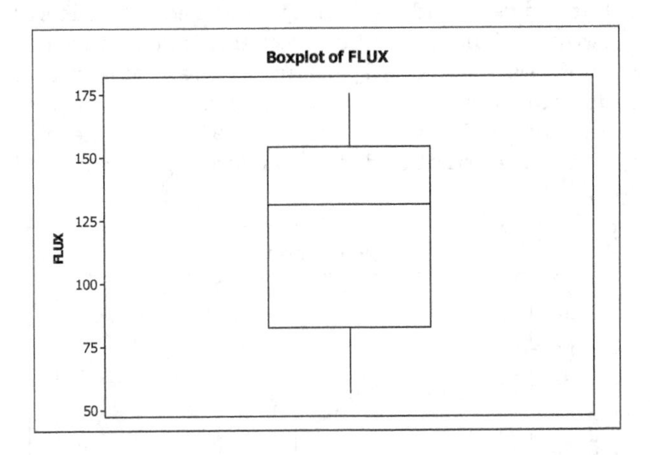

We see that the data are roughly symmetrical with no potential outliers.

3.173 (a) Following Procedure 3.1, Step 1: First arrange the data in increasing order:

21 70 125 195 389 649 656 664 682 1006 1300 1403 1433 **1800** 1982 2205 2515 3027 3634 4200 5299 5947 7886 8543 9310 11189 17341

Step 2: Since this data set has an odd number of values (27), the median is the middle value (14th) in the ordered list. The median, or second quartile, is 1800.

Step 3 and 4: The first quartile Q_1 is the median of the lower half of the ordered data set (including the median) 21 70 125 195 389 649 **656** **664** 682 1006 1300 1403 1433 1800. Since this data set has an even number of values (14), Q_1 is the average of the seventh and eighth values, so Q_1 = (656 + 664)/2 = 660.0.

Step 5: The third quartile Q_3 is the median of the upper half of the ordered data set (including the median) 1800 1982 2205 2515 3027 3634 **4200** **5299** 5947 7886 8543 9310 11189 17341. Since this data set has an even number of values (14), Q_3 is the average of the seventh and eighth values, so Q_3 = (4200 + 5299)/2 = 4749.5.

Step 6: The quartiles are Q_1 = 660.0, Q_2 = 1800.0, and Q_3 = 4749.5.

Thus 25% of the data values are less than or equal to 660.0, the next 25% are between 660.0 and the median 1800, the next 25% are between 1800 and 1719.5, and the upper 25% are greater than or equal to 4749.5.

(b) The interquartile range is $Q_3 - Q_1$ = 4749.5 - 660.0 = 4089.5

(c) The five-number summary consists of the Minimum, Q_1, Median, Q_3, and the Maximum. Respectively, these are 21, 660, 1800, 4749.5, and 17341. In addition to the explanation of the quartiles in part (a), all of the values are greater than or equal to the Minimum, and all of the values are less than or equal to the Maximum.

(d) To identify any potential outliers, we first find the lower and upper limits. These are found by subtracting 1.5 x IQR from Q_1 and adding 1.5 x IQR to Q_3. Thus the lower limit is 660.0 - 1.5(4089.5) = -5474.25 and the upper limit is 4749.5 + 1.5(4089.5) = 10883.75. Potential outliers consist of any values that are less than -5474.25 or greater than 10883.75. We see that 11189 and 17341 are potential outliers.

(e) We will construct a modified boxplot that consists of a rectangle with its ends at the first and third quartiles, a line through it at the median, and two "whiskers" that extend from the quartiles outward to the adjacent values. The adjacent values are the most extreme data values that do not lie outside the lower and upper limits. For these data, the adjacent values are 21 and 9310. The potential outliers, 11189 and 17341, are plotted individually. Thus the boxplot, prepared using Minitab, is

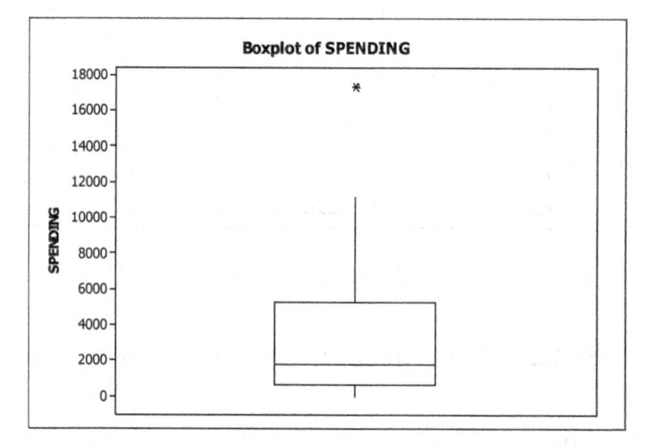

We have shown you the modified boxplot from Minitab to illustrate that software may produce a slightly different boxplot than we expected from our own computations. Minitab uses a different definition of the quartiles, which leads to a larger value of the IQR (4643). This, in turn, results in upper and lower limits that are farther from the median than ours are. Consequently, 11189 shows up as an adjacent value, not an outlier, in Minitab's modified boxplot. The overall picture of the data being right skewed with outlier(s) remains much the same.

3.175 (a) Following Procedure 3.1, Step 1: First arrange the data in increasing order:

26.09	26.16	26.26	26.38	26.50	26.54	26.55	26.61	26.64	26.67
26.70	26.70	26.71	26.79	26.83	26.87	26.97	27.04	27.04	27.05
27.12	27.12	27.14	27.27	27.45	27.52	28.32	28.44	28.78	28.79
28.84	28.90	28.97	29.14	29.17	29.19	29.23	29.32	29.33	29.33
29.40									

Step 2: Since this data set has an odd number of values (41), the median is the middle value (21st) in the ordered list. The median, or second quartile, is 27.12.

Step 3 and 4: The first quartile Q_1 is the median of the lower half of the ordered data set (including the median) 26.09 26.16 26.26 26.38 26.50 26.54 26.55 26.61 26.64 26.67 **26.70** 26.70 26.71 26.79 26.83 26.87 26.97 27.04 27.04 27.05 27.12 . Since this data set has an odd number of values (21), Q_1 is the middle value, so Q_1 = 26.70.

Step 5: The third quartile Q_3 is the median of the upper half of the ordered data set (including the median) 27.12 27.12 27.14 27.27 27.45 27.52 28.32 28.44 28.78 28.79 **28.84** 28.90 28.97 29.14 29.17 29.19 29.23 29.32 29.33 29.33 29.40. Since this data set has an odd number of values (21), Q_3 is the middle value, so Q_3 = 28.84.

Step 6: The quartiles are Q_1 = 26.70, Q_2 = 27.12, and Q_3 = 28.84.

Thus 25% of the data values are less than or equal to 26.70, the next 25% are between 26.70 and the median 27.12, the next 25% are between 27.12 and 28.84, and the upper 25% are greater than or equal to 28.84.

(b) The interquartile range is $Q_3 - Q_1$ = 28.84 - 26.70 = 2.14.

(c) The five-number summary consists of the Minimum, Q_1, Median, Q_3, and the Maximum. Respectively, these are 26.09, 26.70, 27.12, 28.84, and 29.40. In addition to the explanation of the quartiles in part (a), all of the values are greater than or equal to the Minimum, and all of the values are less than or equal to the Maximum.

(d) To identify any potential outliers, we first find the lower and upper limits. These are found by subtracting 1.5 x IQR from Q_1 and adding 1.5 x IQR to Q_3. Thus the lower limit is 26.70 - 1.5(2.14) = 23.49, and the upper limit is 28.84 + 1.5(2.14) = 29.94. Potential outliers consist of any values that are less than 23.49 or greater than 29.94. We see that there are no potential outliers.

(e) We will construct a boxplot, which consists of a rectangle with its ends at the first and third quartiles, a line through it at the median, and two "whiskers" that extend from the quartiles outward to the minimum and maximum values. Thus the boxplot, prepared using Minitab, is

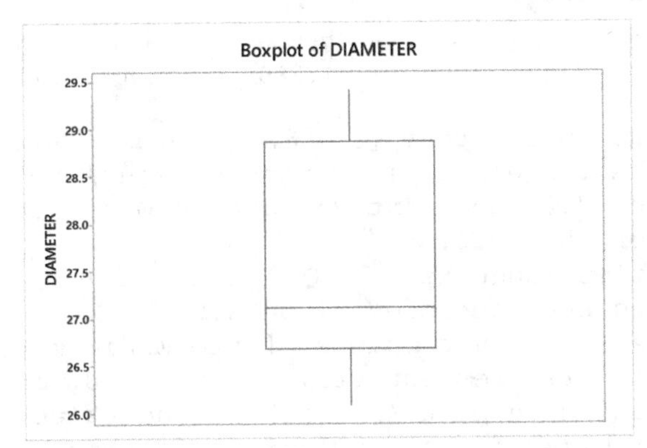

We see that the data are roughly symmetric with a slight right skew. Also, there are no potential outliers.

3.177 (a) Following Procedure 3.1, Step 1: First, arrange the data in increasing order

6 7 8 8 8 8 9 **9** 9 10 10 10 10 10 10

Step 2: Since there is an odd number of data values (15), the median is the middle (8th) value. Thus the median is 9.

Step 3 and 4: The first quartile Q_1 is the median of the lower half of the ordered data set (including the median) 6 7 8 **8 8** 8 9 9. Since this data set has an even number of values (8), Q_1 is the average of the fourth and fifth values, so $Q_1 = (8 + 8)/2 = 8$.

Step 5: The third quartile Q_3 is the median of the upper half of the ordered data set (including the median) 9 9 10 **10 10** 10 10 10. Since this data set has an even number of values (8), Q_3 is the average of the fourth and fifth values, so $Q_3 = (10 + 10)/2 = 10$.

Step 6: The quartiles are $Q_1 = 8$, $Q_2 = 9$, and $Q_3 = 10$.

(b) The quartiles are not particularly useful for this set of data due to the small range of the data and the large numbers of identical values at 8, 9, and 10. For example, Q_3 and the maximum are equal as a result of the highest six values all being 10.

3.179 The center of the distribution of weight loss, as measured by the median, is very similar for the two groups. However, group 1 has less variation in weight loss, both in the middle 50% and overall. Group 2 has a larger range than group 1. The distribution of weight loss for both groups is approximately symmetric.

3.181 The median hemoglobin level is much lower in the HB SS patients than in the other two types of patients (which were roughly the same). The least variation in hemoglobin levels occurred in the HB SC patients, while the greatest variation occurred in the HB ST patients All three distributions were roughly symmetric about their medians.

3.183 (a) Using Minitab, retrieve the data from the WeissStats Resource Site.

Then choose **Stat ▶ Basic Statistics ▶ Display Descriptive Statistics**, enter PUPS in the **Variables** text box, and click **OK**. The output appears in the Session Window as

Variable	N	N*	Mean	SE Mean	StDev	Minimum	Q1	Median	Q3	Maximum
PUPS	80	0	7.125	0.211	1.885	3.000	6.000	7.000	8.000	12.000

We see from the output that $Q_1 = 6.0$, the median = 7.0, and $Q_3 = 8.0$. Thus 25% of the values are less than or equal to 6.0, another 25% are between 6.0 and 7.0, the next 25% are between 7.0 and 8.0, and the top 25% are greater than 8.0.

(b) The interquartile range is $Q_3 - Q_1 = 8.0 - 6.0 = 2.0$. The middle 50% of the data values span an interval of 2.0.

(c) The five-number summary consists of the Minimum, Q_1, median, Q_3, and the Maximum. These are, respectively, 3, 6, 7, 8, and 12. In addition to the interpretation in part (a), all of the values are greater than or equal to the minimum, 3, and all are less than or equal to the maximum, 12.

(d) To identify any potential outliers, we first find the lower and upper limits. These are found by subtracting 1.5 x IQR from Q_1 and adding 1.5 x IQR to Q_3. Thus the lower limit is $6 - 1.5(2) = 3$, and the upper limit is $8 + 1.5(2) = 11$. Potential outliers consist of any values that are less than 3 or greater than 11. We see that 12 is a potential outlier. The maximum, 128, is greater than the upper limit, and it occurs once, so there is one potential outlier.

(e) To obtain a boxplot, we choose **Graph ▶ Boxplot,** select **Simple** in the **One Y** row and click **OK**. Enter PUPS in the **Graph Variables** text box and click **OK**. The result is

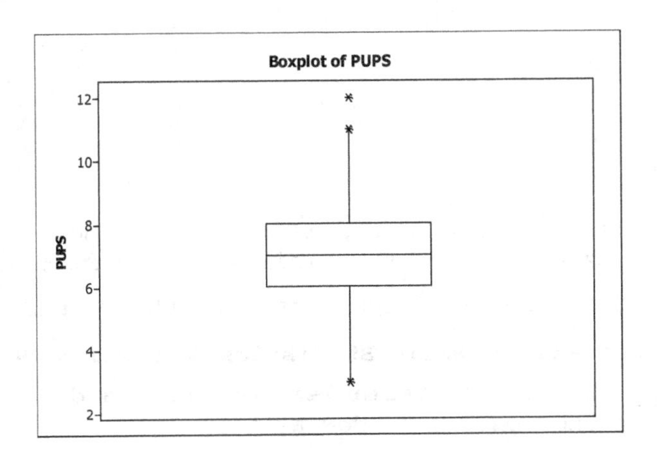

With the exception of the one potential outlier, the distribution of the data is very symmetric, approximately bell-shaped.

3.185 (a) Using Minitab, retrieve the data from the WeissStats Resource Site. Then choose **Stat ▶ Basic Statistics ▶ Display Descriptive Statistics**, enter PERCENT in the **Variables** text box, and click **OK**. The output appears in the Session Window as

```
Variable   N  N*    Mean  SE Mean  StDev  Minimum      Q1  Median      Q3
PERCENT   51   0  86.902    0.474  3.384   80.000  84.000  87.000  90.000

Variable  Maximum
PERCENT    92.000
```

We see from the output that Q_1 = 84.0, the median = 87.0, and Q_3 = 90.0. Thus 25% of the values are less than or equal to 84.0, another 25% are between 84.0 and 87.0, the next 25% are between 87.0 and 90.0, and the top 25% are greater than 90.0.

(b) The interquartile range is $Q_3 - Q_1 = 90.0 - 84.0 = 6.0$. The middle 50% of the data values span an interval of 6.0.

(c) The five-number summary consists of the Minimum, Q_1, median, Q_3, and the Maximum. These are, respectively, 80.0, 84.0, 87.0, 90.0, and 92.0. In addition to the interpretation in part (a), all of the values are greater than or equal to the minimum, 80.0, and all are less than or equal to the maximum, 92.0.

(d) To identify any potential outliers, we first find the lower and upper limits. These are found by subtracting 1.5 x IQR from Q_1 and adding 1.5 x IQR to Q_3. Thus the lower limit is $84.0 - 1.5(6.0) = 75.0$, and the upper limit is $90.0 + 1.5(6.0) = 99$. Potential outliers consist of any values that are less than 75 or greater than 99. We see that there are no potential outliers.

(e) To obtain a boxplot, we choose **Graph ▶ Boxplot,** select **Simple** in the **One Y** row and click **OK**. Enter PERCENT in the **Graph Variables** text box and click **OK**. The result is

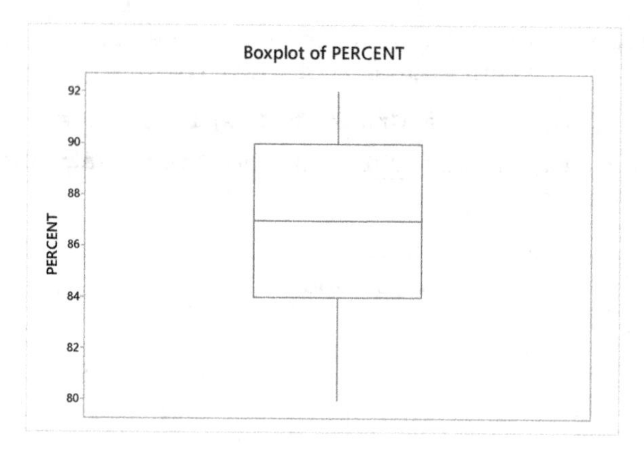

The distribution of the data is roughly symmetrical. The whiskers on the graph extend out to the minimum and maximum values.

3.187 (a) Using Minitab, retrieve the data from the WeissStats Resource Site.

Then choose **Stat ▶ Basic Statistics ▶ Display Descriptive Statistics**, enter BODYTEMP in the **Variables** text box, and click **OK**. The output appears in the Session Window as

```
Variable   N  N*    Mean  SE Mean  StDev  Minimum      Q1  Median      Q3
BODYTEMP  93   0  98.124   0.0671  0.647   96.700  97.650  98.200  98.600

Variable  Maximum
BODYTEMP   99.400
```

We see from the output that Q_1 = 97.65, the median = 98.20, and Q_3 = 98.60. Thus 25% of the values are less than or equal to 97.65, another 25% are between 97.65 and 98.20, the next 25% are between 98.20 and 98.60, and the top 25% are greater than 98.60.

(b) The interquartile range is $Q_3 - Q_1$ = 98.60 − 97.65 = 0.95. The middle 50% of the data values span an interval of 0.95.

(c) The five-number summary consists of the Minimum, Q_1, median, Q_3, and the Maximum. These are, respectively, 96.70, 97.65, 98.20, 98.60, and 99.40. In addition to the interpretation in part (a), all of the values are greater than or equal to the minimum, 96.70, and all are less than or equal to the maximum, 99.40.

(d) To identify any potential outliers, we first find the lower and upper limits. These are found by subtracting 1.5 x IQR from Q_1 and adding 1.5 x IQR to Q_3. Thus the lower limit is 97.65 − 1.5(.095) = 96.225, and the upper limit is 98.60 + 1.5(0.95) = 100.025. Potential outliers consist of any values that are less than 96.225 or greater than 100.025. We see that there are no potential outliers.

(e) To obtain a boxplot, we choose **Graph ▶ Boxplot,** select **Simple** in the **One Y** row and click **OK**. Enter RATE in the **Graph Variables** text box and click **OK**. The result is

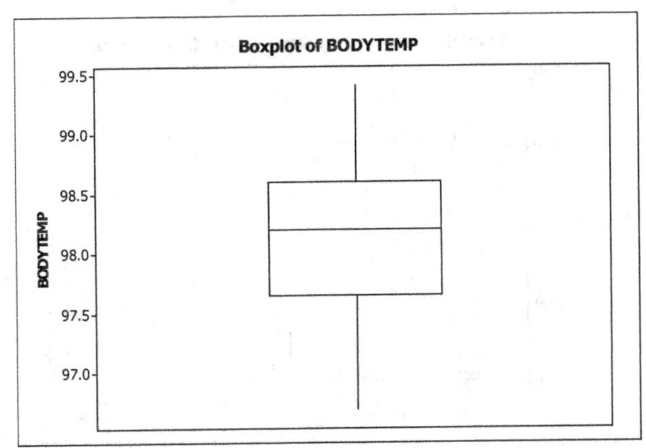

We see that the distribution of the data is roughly symmetrical with no outliers.

3.189 (a) Using Minitab, retrieve the data from the WeissStats Resource Site. To obtain a boxplot choose **Graph ▶ Boxplot,** select **Simple** in the **Multiple Y's** row and click **OK**. Enter ITALIANS and ETRUSCANS in the **Graph Variables** text box and click **OK**. The result is

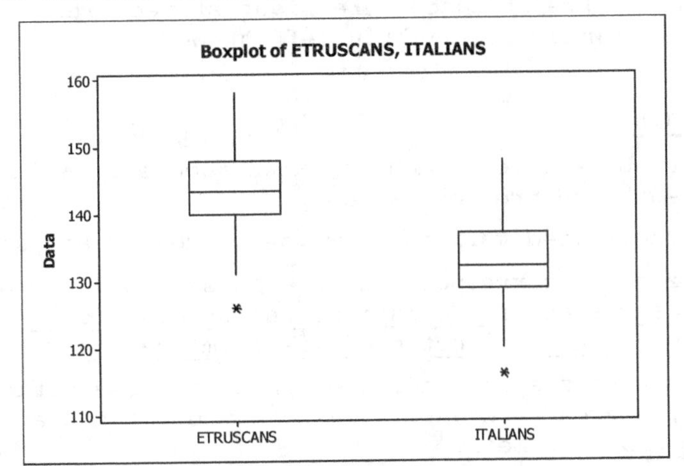

(b) The overall shape of the distributions for the Italian and Etruscan head breadths are very similar. They are both symmetric with a single lower potential outlier and similarly spread out quartiles. However, the distribution for the Etruscans is greater than the Italians. The median, quartiles, maximum, etc. are all larger for the Etruscans than the Italians.

3.191 (a) Using Minitab, retrieve the data from the WeissStats Resource Site. Then choose **Graph ▶ Boxplot,** select **Simple** in the **Multiple Y's** row and click **OK**. Enter STOMACH, BRONCHUS, COLON, OVARY, and BREAST in the **Graph Variables** text box and click **OK**. The result is

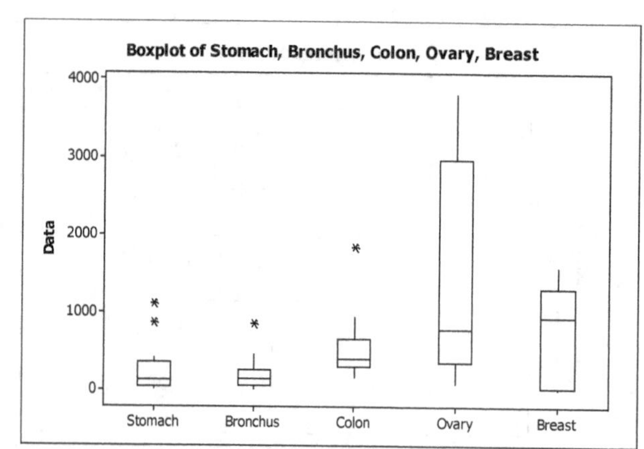

(b) The median survival times were greatest for breast cancer patients, followed by those with cancer of the ovary, colon, bronchus, and stomach. The variation in survival times for colon, bronchus, and stomach cancer patients, as measured by the interquartile range, were considerably smaller than the variation experienced by the ovary and breast cancer patients. The first three also contained outliers in the data, whereas the last two did not. All of the data distributions except for breast cancer are right skewed, whereas the breast cancer survival times are slightly left skewed.

Exercises 3.5

3.193 The ultimate objective of dealing with sample data in inferential studies is to describe the entire population.

3.195 (a) A standardized variable always has mean 0 and standard deviation 1.

(b) The z-score corresponding to an observed value of a variable tells you how many standard deviations the observation is from the mean and, by its sign, what direction it is from the mean.

(c) A positive z-score indicates that the observation is greater than the mean, whereas a negative z-score indicates that the observation is less than the mean.

3.197 The number 413.1 is a parameter since it is the mean of the entire population of lengths of holes at Augusta National Golf Club.

3.199 If we consider the heights as a population:

(a) $\mu = \dfrac{\sum x_i}{N} = \dfrac{375}{5} = 75$ inches

(b) $\sigma = \sqrt{\dfrac{\sum (x_i - \bar{x})^2}{N}} = \sqrt{\dfrac{156}{5}} = 5.6$ inches

3.201 (a) The population mean is $\mu = \dfrac{\sum x_i}{N} = \dfrac{9}{3} = 3$.

(b) We will use the computing formula to find the standard deviation. For that, we need $\sum x_i^2 = 4^2 + 0^2 + 5^2 = 41$. The standard deviation is then

$$\sigma = \sqrt{\frac{\sum x_i^2}{N} - \mu^2} = \sqrt{\frac{41}{3} - 3^2} = 2.2 .$$

3.203 (a) The population mean is $\mu = \frac{\sum x_i}{N} = \frac{11}{4} = 2.75$.

(b) We will use the computing formula to find the standard deviation. For that, we need $\sum x_i^2 = 1^2 + 2^2 + 4^2 + 4^2 = 37$. The standard deviation is then $\sigma = \sqrt{\frac{\sum x_i^2}{N} - \mu^2} = \sqrt{\frac{37}{4} - 2.75^2} = 1.3$.

3.205 (a) The population mean is $\mu = \frac{\sum x_i}{N} = \frac{25}{5} = 5$.

(b) We will use the computing formula to find the standard deviation. For that, we need $\sum x_i^2 = 1^2 + 9^2 + 8^2 + 4^2 + 3^2 = 171$. The standard deviation is then $\sigma = \sqrt{\frac{\sum x_i^2}{N} - \mu^2} = \sqrt{\frac{171}{5} - 5^2} = 3.0$.

3.207 (a) The population consists of all people living in the United States. The variable of interest is the age of each person.

(b) The median is the mean of the two middle ages when they are listed in increasing order as 7, 9, **29, 45**, 51, 54. Thus the median is (29 + 45)/2 = 37.0. This is a statistic since it is derived from a sample. We write our results as M = 37.0.

(c) This median is a parameter since it is derived from the entire population. Thus we write η = 37.2.

3.209 (a) The population mean is $\mu = \frac{\sum x_i}{N} = \frac{970}{10} = 97.0$.

(b) We will use the computing formula to find the standard deviation. For that we need $\sum x_i^2 = 85^2 + 100^2 + \ldots + 115^2 = 95800$. The standard deviation is then $\sigma = \sqrt{\frac{\sum x_i^2}{N} - \mu^2} = \sqrt{\frac{95800}{10} - 97.0^2} = 13.1$.

(c) The median is the mean of the two middle observations in the following ordered list of 10 observations: 80 80 85 90 **90 100** 105 110 115 115. The mean of the two middle observations is (90 + 100)/2 = 95.0, so η = 95.0.

(d) The mode is the value(s) that occurs more often than any other value, so the modes are 80, 90, and 115.

(e) The first quartile is the median of the first five values in the ordered list of part (c), so it is the third value. Thus Q_1 = 85. The third quartile is the median of the last five values in the ordered list of part (c), so Q_3 = 110. The IQR is the difference between Q_1 and Q_3. Thus IQR = $Q_3 - Q_1$ = 110 − 85 = 25.

3.211 (a) For Orlando, the population mean is $\mu = \frac{\sum x_i}{N} = \frac{2327}{5} = 465.4$. For Cincinnati, the population mean is $\mu = \frac{\sum x_i}{N} = \frac{1329}{5} = 265.8$.

(b) The range of values for Cincinnati is about 400, while it is only a little under 200 for Orlando, so Orlando is likely to have the smaller standard deviation.

(c) We will use the computing formula to find the standard deviation. For Orlando, we need $\sum x_i^2 = 583^2 + 460^2 + ... + 485^2 = 1,106,059$. The standard deviation is then $\sigma = \sqrt{\frac{\sum x_i^2}{N} - \mu^2} = \sqrt{\frac{1106059}{5} - 465.4^2} = 67.93$. For Cincinnati, we need $\sum x_i^2 = 77^2 + 105^2 ... + 436^2 = 492,835$. The standard deviation is then $\sigma = \sqrt{\frac{\sum x_i^2}{N} - \mu^2} = \sqrt{\frac{492835}{5} - 265.8^2} = 167.08$.

(d) Yes. We expected the standard deviation to be smaller for Orlando than for Cincinnati, and that turned out to be the case.

3.213 (a) The standardized version of x is z = (x - 32.9)/17.9.

(b) The mean and standard deviation of z are 0 and 1 respectively.

(c) For x = 81.3, the z score is (81.3 - 32.9)/17.9 = 2.70

For x = 20.8, the z score is (20.8 - 32.9)/17.9 = -0.68

(d) The value 81.3 is 2.70 standard deviations above the mean 32.9.

The value 20.8 is 0.68 standard deviations below the mean 32.9.

(e)

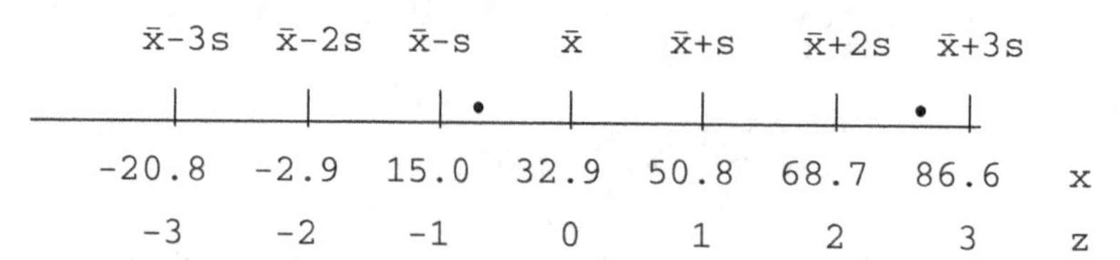

3.215 (a) The standardized version of x is z = (x - 6.71)/0.67.

(b) For x = 5.2, the z score is (5.2 - 6.71)/0.67 = -2.25

For x = 8.1, the z score is (8.1 - 6.71)/0.67 = 2.07

The value 5.2 is 2.25 standard deviations below the mean 6.71.

The value 8.1 is 2.07 standard deviations above the mean 6.71.

3.217 (a) z = (21.4 - 25)/1.15 = -3.13.

(b) Yes. If the gas mileages have a distribution that is roughly bell-shaped, we would expect more than 99% of cars to get mileage within three standard deviations from the mean of 25 mpg. Your car's mpg is more than three standard deviations below the mean. Even if the distribution is not bell-shaped, Chebychev's Rule says that at least 89% of the mileages must be within three standard deviations.

3.219 The formulas used here to compute s and σ are

$$s = \sqrt{\frac{\sum (x_i - \bar{x})^2}{n-1}} \quad \text{and} \quad \sigma = \sqrt{\frac{\sum (x_i - \mu)^2}{N}}$$

For each data set, the relevant calculations are:

Data Set	Number of Observations	$\Sigma(x_i\text{-mean})^2$	s	σ
1	4	14.00	2.16	1.87
2	7	24.86	2.04	1.88
3	10	36.40	2.01	1.91

(a) The sample standard deviations are found in the fourth column of the previous table.

(b) The population standard deviations are found in the fifth column of the previous table.

(c) Comparing s and σ for a given data set, the two measures will tend to be closer together if the data set is large.

3.221 (a) Using Minitab, retrieve the data from the WeissStats Resource Site.

Then choose **Stat ▶ Basic Statistics ▶ Display Descriptive Statistics**, enter WEIGHT in the **Variables** text box, and click **OK**. The output appears in the Session Window as

```
Variable   N  N*    Mean  SE Mean  StDev  Minimum      Q1  Median      Q3
WEIGHT    21   0  214.05     5.86  26.88   176.00  190.00  211.00  237.50

Maximum
270.00
```

The sample mean is calculated the same as the population mean, so the population mean is 214.05 pounds.

(b) Unfortunately, this procedure produces a sample standard deviation instead of a population standard deviation, so you than have to convert s to σ using the formula derived in Exercise 3.220. The sample standard deviation is computed using Minitab in part (a) to be 26.88.

In Exercise 3.220, the conversion formula is $\sigma = s\sqrt{\dfrac{m-1}{m}}$. Thus the population standard deviation would be $\sigma = 26.88\sqrt{\dfrac{21-1}{21}} = 26.23$ pounds.

3.223 z = (205500 − 220258)/5237 = −2.82. Applying Chebychev's rule to that z-score with k = 82.82, we conclude that at least $100(1 - 1/2.82^2)\% = 87.4\%$ of the sale prices lie within 2.82 standard deviations to either side of the mean of $220,258. Therefore, the $205,500 price of the home you are contemplating buying is lower than at least 87.4% of the prices of comparable homes. It appears that this home is bargain.

3.225 (a) In the population of lowland copperhead snakes, for x = 850, the z score is (850 − 812.07)/330.24 = 0.115. In the population of tiger snakes, for x = 850, the z score is (850 − 743.65)/336.36 = 0.316.

(b) If the distributions of the two types of snakes are approximately the same shape, it would be reasonable to use z-scores to compare the relative standings of the weights of the two snakes.

(c) The value 850 is 0.115 standard deviations above the mean for lowland copperhead snakes. The value 850 is 0.316 standard deviations above the mean for tiger snakes. The tiger snake of 850g is larger relative to other tiger snakes than the copperhead of 850g is to other copperheads.

REVIEW PROBLEMS FOR CHAPTER 3

1. (a) Descriptive measures are numbers used to describe data sets.

   (b) Measures of center indicate where the center or most typical value of a data set lies.

   (c) Measures of variation indicate how much variation or spread the data set has.

2. The two most commonly used measures of center for quantitative data are the mean and the median. The mean uses all of the data, but can be influenced by the presence of a few outliers. The median is computed from the one or two center values in an ordered list of the data. It does not make use of all of the data, but it has the advantage that it is not influenced by the presence of a few outliers.

3. The only measure of center, among those we discussed, that is appropriate for qualitative data is the mode.

4. (a) standard deviation

   (b) interquartile range

5. (a) $\bar{x}$ (b) s (c) μ (d) σ

6. (a) This is not necessarily true.

   (b) This is necessarily true.

7. Almost all of the observations in any data set lie within <u>3</u> standard deviations to either side of the mean.

8. (a) If $k = 6$, using Chebyshev's rule, at least $1 - 1/k^2 = 1 - 1/36 = 35/36$, or 97.2%, of the observations lie within 6 standard deviations to either side of the mean.

   (b) If $k = 1.5$, using Chebyshev's rule, at least $1 - 1/k^2 = 1 - 1/2.25 = 0.556$, or 55.6%, of the observations lie within 1.5 standard deviations to either side of the mean.

9. 60 and 100 are two standard deviations to either side of the mean. Chebyshev's rule states that at least 75% of the observations for any data set lie within two standard deviations to either side of the mean. 75% of 87 is 65.25. Therefore, at least 66 observations in the data set will be between 60 and 100.

10. 33 and 57 are one standard deviation to either side of the mean. The empirical rule states that approximately 68% of the observations in a data set that is bell shaped will lie within one standard deviation to either side of the mean. Therefore, approximately 68% of the observations in this data set will lie between 33 and 57.

11. 17 and 33 are two standard deviations to either side of the mean. The empirical rule states that approximately 95% of the observations in a data set that is bell shaped will lie within two standard deviations to either side of the mean. 95% of 152 is 144.4. Therefore, approximately 144 of the observations in this data set will lie between 17 and 33.

12. (a) The components of the five-number summary are the minimum, Q_1, median, Q_3, and the maximum.

    (b) The median is a measure of the center. The interquartile range, which is found as $Q_3 - Q_1$, and the range, which is the difference between the maximum and the minimum, are measures of variation.

    (c) The boxplot is based on the five-number summary.

13. (a) An outlier is an observation that falls well outside the overall pattern of the data.

(b) First compute the interquartile range IQR = $Q_3 - Q_1$. Then compute two limits as $Q_1 - 1.5IQR$ and $Q_3 + 1.5IQR$. Observations that are lower than the lower limit or higher than the upper limit are potential outliers and require further study.

14. (a) A z-score for an observation x is obtained by subtracting the mean of the data set from x and then dividing the result by the standard deviation; i.e., $z = (x - \mu)/\sigma$ for population data or $z = (x - \bar{x})/s$ for sample data.

(b) A z-score indicates how many standard deviations an observation is below or above the mean of the data set.

(c) An observation with a z-score of 2.9 is likely to be greater than all or almost all of the other data values.

15. (a) The mean is calculated as $\bar{x} = \Sigma x_i/n$. For the sample of 20 guests, $\bar{x}$ = 47/20 = 2.35 alcoholic drinks.

The median is found by ordering the observations from lowest to highest and finding the average of the two observations in the middle. The list is

0 0 1 1 1 1 1 2 2 **2** **2** 2 3 3 4 4 4 4 5 5

The median is the average of the 10th and 11th values, and is therefore (2 + 2)/2 = 2.

The mode is the value that occurs the most times. In this data set, 1 and 2 both occur 5 times, and so the modes are 1 and 2.

(b) If the purpose is to help in estimating the cost of the party, the mean is the best since it takes into account the drinking practices of all of the guests. If the purpose is to describe the average guest, the median is more typical of the list of values. The mode is less useful for this data since there are two modes.

16. Many more marriages are characterized as being of short duration than other durations. Since the median is ordinarily preferred for data sets that have exceptional (very large or small) values, the median is more appropriate than the mean as a measure of central tendency for data on the duration of marriages.

17. Causes of death are qualitative. There is no way to compute a mean or median for such data. The mode is the only measure of central tendency that can be used for qualitative data.

18. The mean is $\bar{x} = \frac{\sum x_i}{n} = \frac{305.3}{10} = 30.53$; since n is even (10), the median is the mean of the two middle observations in the ordered list of the data values: 17.4 21.0 17.4 31.5 **32.0** **33.0** 33.0 34.5 37.5 38.0. Thus the median is (32 + 33)/2 = 32.5. The mode is the most frequently occurring value. Since 33.0 occurs twice and no other value occurs more than once, 33.0 is the mode.

19. (a) $\bar{x} = \frac{\sum x_i}{n} = \frac{457}{10} = 45.7$ kilograms

(b) Range = 54 - 37 = 17 kilograms

(c) $s=\sqrt{\dfrac{\sum x^2-(\sum x)^2/n}{n-1}}=\sqrt{\dfrac{21109-457^2/10}{9}}=5.0$ kg

20. (a)

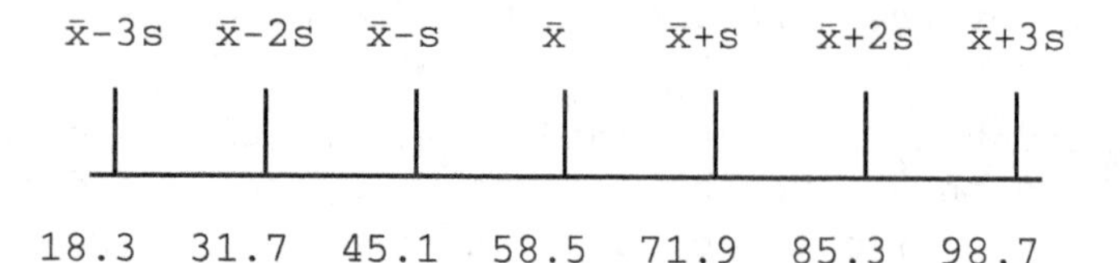

(b) Almost all of the ages of the 36 millionaires are between 18.3 and 98.7 years old.

21. (a) 15.1 and 36.3 are two standard deviations to either side of the mean. Chebyshev's rule states that at least 75% of the observations in any data set will lie within two standard deviations to either side of the mean. 75% of 10,215 is 7,661.25. Therefore, at least 7,662 women in the sample will have a BMI between 15.1 and 36.3.

(b) Chebyshev's rule states that at least 89% of the observations in any data set will lie within three standard deviations to either side of the mean. Therefore, at least 89% of the women in the sample have BMIs between 25.7 - 3(5.3) = 9.8 and 25.7 + 3(5.3) = 41.6.

22. (a) The Empirical rule states that for a roughly bell shaped distribution, approximately 95% of the observations will lie within two standard deviations to either side of the mean. Therefore, at least 95% of the mobile homes in the sample have prices between 63.3 - 2(7.9) = $47.5 thousand and 63.3 + 2(7.9) = $79.1 thousand.

(b) 39.6 and 87.0 are three standard deviations to either side of the mean. The empirical rule states that approximately 99.7% of the observations in a data set that is bell-shaped will lie within three standard deviations to either side of the mean. Therefore, approximately 99.7% of the mobile homes in the sample have prices between $39.6 thousand and $87.0 thousand.

(c) 55.4 and 71.2 are one standard deviation to either side of the mean. The empirical rule states that approximately 68% of the observations in a data set that is bell-shaped will lie within one standard deviation to either side of the mean. 68% of 250 is 170. Therefore, approximately 170 mobile homes in the sample will have prices between $55.4 thousand and $71.2 thousand.

23. (a) Following Procedure 3.1, Step 1: The data set is already ordered.

Step 2: The second quartile is the median of the entire data set. The number of pieces of data is 36, and so the position of the median is (36 + 1)/2 = 18.5, halfway between the eighteenth and nineteenth data values. Thus the median of the entire data set is (59 + 60)/2 = 59.5. That is, Q_2 = 59.5.

Step 3 and 4: The first quartile is the median of the lower half of the ordered list. Since there is a total of 36 ages, the first 18 are in the lower half and the median of these is the average of the middle two, the 9th and 10th values. Thus Q_1 = (48 + 48)/2 = 48.

Step 5: The third quartile is the median of the upper half of the ordered list. Since there is a total of 36 ages, the last 18 are in the upper half and the median of these is the average of the middle two, the 9th and 10th values. Thus Q_3 = (68 + 69)/2 = 68.5.

Step 6: Q_1 = 48, Q_2 = 59.5, and Q_3 = 68.5.

Interpreting our results, we conclude that 25% of the ages are less than 48 years; 25% of the ages are between 48 and 59.5 years; 25% of the ages are between 59.5 and 68.5 years; and 25% of the ages are greater than 68.5 years.

(b) The IQR = Q_3 - Q_1 = 68.5 - 48 = 20.5. Thus, the middle 50% of the ages has a range of 20.5 years.

(c) Min = 31, Q_1 = 48, Q_2 = 59.5, Q_3 = 68.5, Max = 79

(d) The limits are given by:

Lower limit = Q_1 - 1.5(IQR) = 48.0 - 1.5(20.5) = 17.25

Upper limit = Q_3 + 1.5(IQR) = 68.5 + 1.5(20.5) = 99.25

(e) There are no values below 17.25 or above 99.25, so there are no potential outliers.

(f) A boxplot is constructed easily using the information in part (a).

(i) low data value = 31 years

(ii) high data value = 79 years

(iii) Q_1 = 48.0 years

(iv) Q_3 = 68.5 years

(v) median = Q_2 = 59.5 years.

These values are used to construct the boxplot as follows:

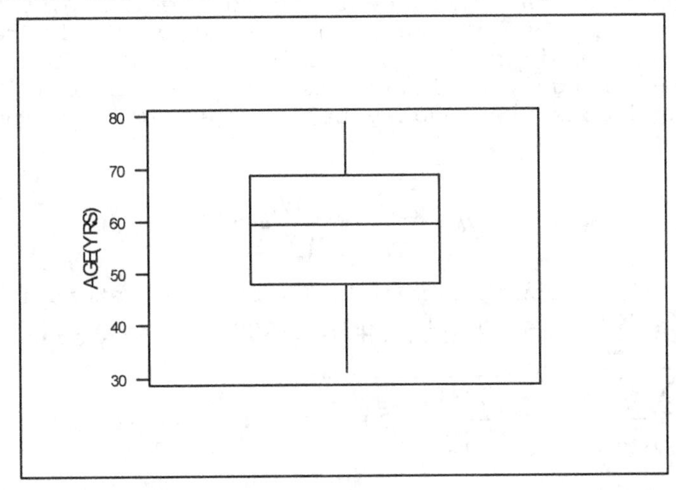

24. (a) Following Procedure 3.1, Step 1: The order list of the data is 0.7 1.0 1.1 1.1 1.2 1.5 1.8 1.8 1.8 1.8 **1.9** **2.0** 2.0 2.0 2.3 2.7 3.3 3.4 3.6 3.8 6.7 7.6.

Step 2: Since n is even (22), the median is mean of the two middle values, the 11th and 12th. Thus the median is (1.9 + 2.0)/2 = 1.95.

Step 3 and 4: The first quartile is the median of the first 11 values in the list, so Q_1 = the 6th value = 1.5.

Step 5: The third quartile is the median of the last 11 values in the list, so Q_3 = the 6th value from the upper end = 3.3.

Step 6: Thus, the five-number summary is Minimum = 0.7, Q_1 = 1.5, Median = 1.95, Q_3 = 3.3, and Maximum = 7.6.

(b) The IQR = 3.3 - 1.95 = 1.35. The lower limit = Q1 - 1.5 x IQR = 1.5 - 1.5(1.35) = -0.525, while the upper limit = Q3 + 1.5 x IQR = 3.3 + 1.5(1.35) = 5.325. There are no values below the lower limit, but 6.7 and 7.6 lie above the upper limit and are therefore potential outliers.

(c) The boxplot consists of a rectangle with its ends at 1.5 and 3.3 and a line across the middle at 1.95. Whiskers extend from the ends of the box out to the adjacent values (the most extreme values on each end of the box that are not beyond the limits). Thus the lower whisker extends to 0.7, and the upper whisker extends to 3.8. The two potential outliers are plotted individually. The result is

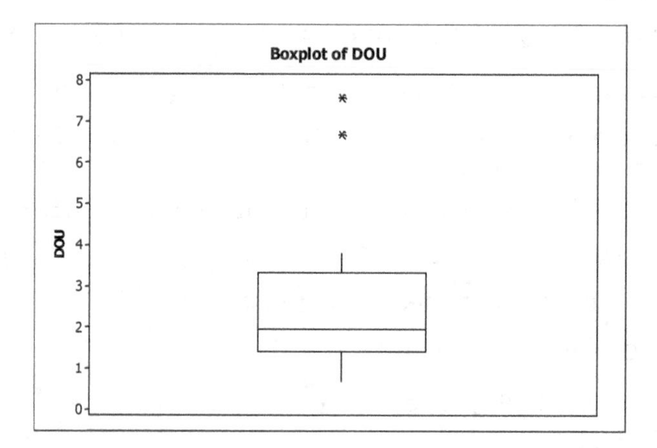

25. The median distance traveled for golf balls hit off the wooden tee is about 227 yards while the median distance traveled when hit off the Stinger tee is about 240 yards. In fact, the maximum distance with the regular wooden tee is shorter than the minimum distance with the Stinger tee. Both distributions have roughly a similar amount of variation (Stinger perhaps slightly more) and both are roughly bell-shaped (Stinger perhaps slightly right-skewed).

26. (a) The population mean is $\mu = \frac{\sum x_i}{N} = \frac{659}{12} = 54.9$ minutes.

(b) We will use the computing formula to find the standard deviation. For that we need $\sum x_i^2 = 53^2 + 58^2 + ... + 16^2 = 39311$. The standard deviation is then $\sigma = \sqrt{\frac{\sum x_i^2}{N} - \mu^2} = \sqrt{\frac{39311}{12} - 54.9^2} = 16.1$ minutes.

27. (a) $\mu = \frac{\sum x_i}{n} = \frac{183.1}{9} = 20.34$ thousand students.

(b) Squaring each data value and then summing, we find that $\sum x_i^2 = 4094.39$. Then the population standard deviation is

$$\sigma = \sqrt{\frac{\sum x_i^2}{N} - \mu^2} = \sqrt{\frac{4094.39}{9} - 20.34^2} = 6.41 \text{ thousand students.}$$

(c) $z = (x - \mu)/\sigma = (x - 20.34)/6.41$.

(d) The mean of z is 0, and the standard deviation of z is 1. All standardized variables have mean 0 and standard deviation 1.

(e) Converting each x value (dotplot on the left) to its corresponding z-score results in the dotplot on the right.

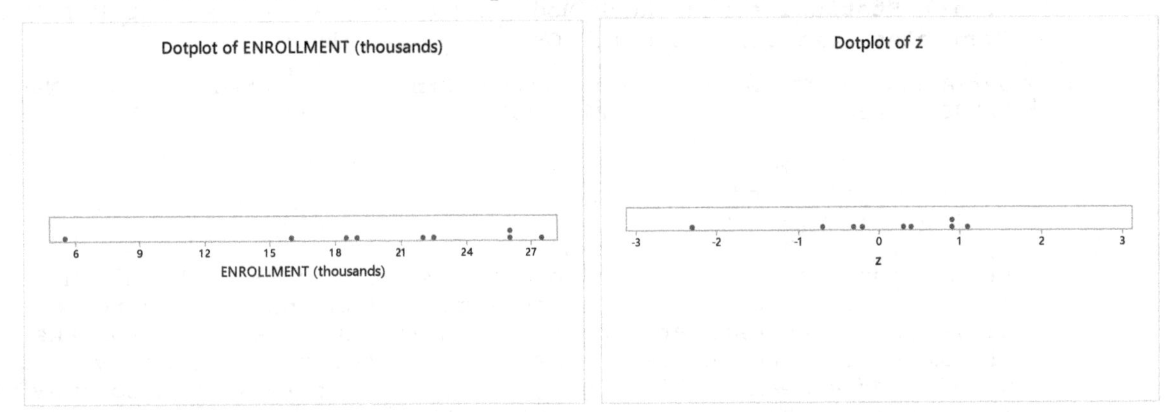

The relative positions of the points in the two plots are the same, but the scales are different.

(f)

x	$z = (x - \mu)/\sigma$
27.7	(27.7 - 20.34)/6.41 = 1.15
18.6	(18.6 - 20.34)/6.41 = -0.27

The enrollment at Los Angeles is 1.15 standard deviations above the mean of 27.7 thousand students and the enrollment at Riverside is 0.27 standard deviations below the mean.

28. (a) The mean price given here is a sample mean. This is because it is the mean price per gallon for the sample of 800 gasoline service stations.

(b) The symbol used to designate the mean of $3.547 is $\bar{x}$.

(c) The mean price given here is a statistic. It is a descriptive measure.

29. Using Minitab, retrieve the data from the WeissStats Resource Site. Then choose **Stat ▶ Tables ▶ Tally Individual Variables**. Enter REGION and DIVISION in the **Variables** text box and click **OK**. The result is the table below.

REGION	Count
Midwest	12
Northeast	9
South	16
West	13
N=	50

DIVISION	Count
East North Central	5
East South Central	4
Middle Atlantic	3
Mountain	8
New England	6
Pacific	5
South Atlantic	8
West North Central	7
West South Central	4
N=	50

(a) We see from the first column of the table that the South region occurs 16 times, more than any other region, so South is the mode of the regions.

(b) We see from the second column of the table that the Mountain and the South Atlantic divisions each occur 8 times, more than any other region, so Mountain and South Atlantic are both modal divisions.

30. (a) Using Minitab, retrieve the data from the WeissStats Resource Site.

Then choose **Stat ▶ Basic Statistics ▶ Display Descriptive Statistics**. Click **Statistics** and check **Mode** and click **OK.** Enter VALUE in the **Variables** text box and click **OK**. The result is

```
Variable   N  N*  Mean  SE Mean  StDev  Minimum   Q1  Median    Q3  Maximum
VALUE     50   0  1563      281   1983        4  234     995  2118    11302

                   N for
Variable  Mode      Mode
VALUE        *         0
```

We see that the mean is 1563 and the Median is 995. Examining the data, we find that no value occurs more than once, so there is no mode. If we wanted to estimate the total agricultural exports for the nation, the mean would be the best measure of center to use since we could just multiply the mean by 50 to get the total. If we wanted to know what value half of the states are below, then the median is the best measure of center to use. Another reason for using the median is that there are several outliers that lie above the median, including one that is very large (11301.7 for California). Since the mean is quite sensitive to outliers such as this one, the median is the better choice for describing these data.

(b) From the table, Range = 11302 - 4 = 11298, and the sample standard deviation s = 1983.

(c) The five-number summary consists of the minimum, Q_1, the median, Q_3, and the maximum. These are, respectively, 4, 234, 995, 2118, and 11302. The interquartile range IQR = $Q_3 - Q_1$ = 2118 - 234= 1884.

(d) We first find the lower limit as Q_1 - 1.5 x IQR = 234 - 1.5(1884) = -2592. The upper limit is Q_3 + 1.5 x IQR = 2118 + 1.5(1884) = 4944. Any data values below the lower limit or above the upper limit are potential outliers. Thus 5198.6, 5246.8, and 11301.7 are potential outliers.

(e) To obtain a boxplot of the data, choose **Graph ▶ Boxplot,** select **Simple** from the **One Y** row and click **OK**. Enter VALUE in the **Variables** text box and click **OK**. The result is

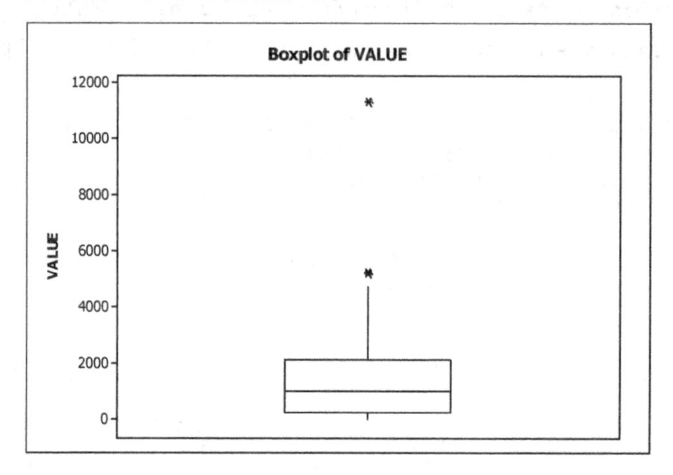

The data distribution is quite right skewed with a median of 995. The plot shows potential outliers on the high side of the data.

31. (a) Using Minitab, retrieve the data from the WeissStats Resource Site. Then choose **Stat ▶ Basic Statistics ▶ Display Descriptive Statistics.** Enter 'LIFE EXP' in the **Variables** text box and click **OK.** The result is

```
Variable          N   N*    Mean  SE Mean  StDev  Minimum      Q1   Median
LIFE EXPECTANCY  193   0  69.968    0.708  9.839   47.800  64.400  73.300

  Q3   Maximum
76.700   83.400
```

We see that the mean is 69.968 and the median is 73.300. Finding the mode is easiest if we choose **Stat ▶ Tables ▶ Tally Individual Variables** and enter 'LIFE EXP' in the **Variables** text box. Examining the resulting table (not shown here), we find that seventeen values each occur three times, more than any other values, so there are seventeen modes. The mean is not appropriate for these data since each value is a mean life expectancy based on population sizes that are different for different countries. The median is the best measure of center to use since it tells us that half of the listed countries have life expectancies below this number. Another reason for using the median is that the distribution is left skewed with several potential outliers at the lower end, as will be seen in parts (d) and (e).

(b) From the table above, Range = 83.4 - 47.8 = 35.6, and the sample standard deviation s = 9.839. [Note: The sample standard deviation is based on a calculation that uses the sample mean. Since the sample mean being used does not account for the fact that the different countries have different populations, the true sample standard deviation may differ somewhat from this value.]

(c) The five-number summary consists of the minimum, Q_1, the median, Q_3, and the maximum. These are, respectively, 47.8, 64.4, 73.3, 76.7, 83.4. The interquartile range $IQR = Q_3 - Q_1 = 76.7 - 64.4 = 12.3$.

(d) We first find the lower limit as $Q_1 - 1.5 \times IQR = 64.4 - 1.5(12.3) = 45.95$. The upper limit is $Q_3 + 1.5 \times IQR = 76.7 + 1.5(12.3) = 95.15$. Any data values below the lower limit or above the upper limit are potential outliers. Thus there are no potential outliers.

(e) To obtain a boxplot of the data, choose **Graph ▶ Boxplot,** select **Simple** from the **One Y** row and click **OK.** Enter 'LIFE EXP' in the **Variables** text box and click **OK.** The result is

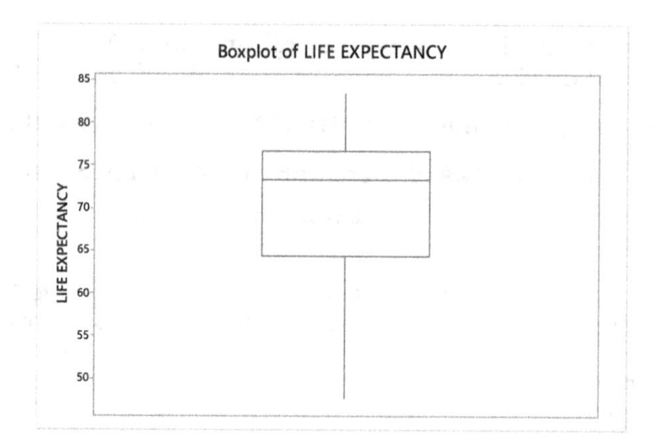

We see that the data are slightly left-skewed with the median at about 73.

32. (a) Using Minitab, retrieve the data from the WeissStats Resource Site. Then choose **Stat ▶ Basic Statistics ▶ Display Descriptive Statistics.** Enter HIGH and LOW in the **Variables** text box and click **OK**. The result is

Variable	N	N*	Mean	SE Mean	StDev	Minimum	Q1	Median	Q3	Maximum
HIGH	71	0	65.33	1.04	8.72	47.60	58.40	64.00	71.10	85.50
LOW	71	0	45.52	1.11	9.34	29.30	39.80	44.20	51.40	74.20

We see that the mean HIGH temperature is 65.33 and the median HIGH is 64.00. The mean LOW temperature is 45.52 and the median LOW is 44.20.

Finding the mode is easiest if we choose **Stat ▶ Tables ▶ Tally Individual Variables** and enter HIGH and LOW in the **Variables** text box. Examining the resulting table (not shown here), we find for HIGH that 54.5, 55.9, 57.6, 59.8, 62.6, 63.6, 65.1, 67.4, 67.8, and 70.6 all occur twice, more than any other values, so there are ten modes. For LOW, the values 35.8, 39.8, 41.2, 43.2, 44.8, 47.4, 48.6, and 52.5 all occur twice, more than any other values, so there are eight modes. The median is the best measure of center to use since it tells us that half of the listed cities have minimums (or maximums) below this number. Another reason for using the median is that the distribution for LOW has several potential outliers at the upper end, as will be seen in parts (d) and (e).

(b) From the table above, for HIGH, Range = 85.5 - 47.6 = 37.9, and the sample standard deviation s = 8.72. For LOW, Range = 74.20 - 29.30 = 44.90 and the sample standard deviation s = 9.34.

(c) The five-number summary consists of the minimum, Q_1, the median, Q_3, and the maximum. For HIGH, these are, respectively, 47.60, 58.40, 64.00, 71.10, and 85.50. The interquartile range for HIGH is IQR = Q_3 - Q_1 = 71.10 - 58.40= 12.70. For LOW, the five-number summary is 29.30, 39.80, 44.20, 51.40, and 74.20. IQR = Q_3 - Q_1 = 51.40 - 39.80 = 11.60.

(d) For HIGH, we first find the lower limit as Q_1 - 1.5 x IQR = 58.40 - 1.5(12.70) = 39.35. The upper limit is Q_3 + 1.5 x IQR = 71.10 + 1.5(12.70) = 90.15. Any data values below the lower limit or above the upper limit are potential outliers. Examining the data, we find that there are no potential outliers for HIGH.

For LOW, we first find the lower limit as Q_1 - 1.5 x IQR = 39.80 - 1.5(11.60) = 22.40. The upper limit is Q_3 + 1.5 x IQR = 51.40 + 1.5(11.60) = 68.80. Any data values below the lower limit or above the upper limit are potential outliers. Examining the data, we see that

69.1, 70.2, and 74.2 are potential outliers. If you are looking for someplace that is warm year around, these values are for Miami, Honolulu, and San Juan.

(e) To obtain a boxplot of the data, choose **Graph ▶ Boxplot,** select **Simple** from the **Multiple Y's** row and click **OK**. Enter HIGH and LOW in the **Variables** text box and click **OK**. The result is

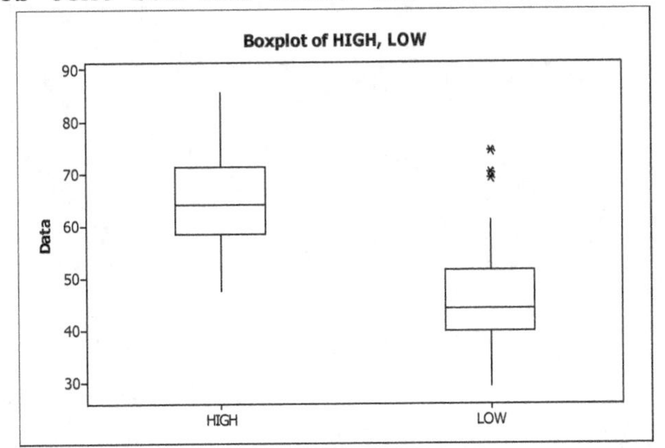

The distribution of HIGH temperatures is quite symmetrical, while the distribution of LOW temperatures is also quite symmetrical except for three outliers. The three cities with high LOW temperatures are all affected by warm southern ocean waters that keep the temperature quite moderate year around.

33. **(a)** To obtain a boxplot of the height data, choose **Graph ▶ Boxplot,** select **Simple** from the **Multiple Y's** row and click **OK**. Enter 'VEGETARIAN' and 'OMNIVORE' in the **Variables** text box and click **OK**. The result is

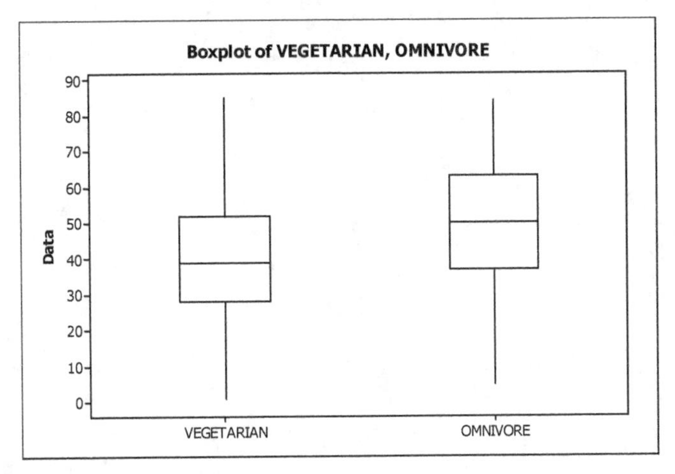

(b) The range from minimum to maximum seems to be similar for the vegetarian and omnivore daily protein intake. Also, the IQR seem to be similar in value for the two groups. The biggest difference is that the quartiles (including the median) for the omnivore daily protein intake is greater than that for the vegetarians. There were not any outliers in either group.

CHAPTER 4 SOLUTIONS

Exercises 4.1

4.1 An experiment is an action the result of which cannot be predicted with certainty. An event is a specified result that may or may not occur when the experiment is performed.

4.3 There is no difference.

4.5 An experiment has 20 possible outcomes, all equally likely. An event can occur in five ways. The probability that the event occurs is 5/20 = 0.25.

4.7 The frequentist interpretation of probability is that the probability of an event is the proportion of times the event occurs in a large number of repetitions of the experiment.

4.9 (a) In a large number of Texas hold'em poker hands, a pocket pair will be dealt approximately 5.9% of the time.

(b) If three balanced dimes are tossed a large number of times, it will come up all three heads approximately 12.5% of the time.

4.11 The following could not possibly be a probability:

(b) 3.5: A probability cannot exceed 1.

4.13 (a) G,L,S,A G,L,S,T G,L,A,T G,S,A,T L,S,A,T

(b) Two samples include the governor, attorney general, and treasurer. Therefore, the probability is f/N = 2/5.

(c) Three samples include the governor and treasurer. Therefore, the probability is f/N = 3/5.

(d) Four samples include the governor. Thus, the probability is f/N = 4/5.

4.15 (a) There are 3 chips that are red. Therefore, the probability is f/N = 3/12 = 1/4.

(b) There are 3 + 4 = 7 chips that are red or white. Therefore, the probability is f/N = 7/12.

(c) There are 3 + 5 = 8 chips that are not white. Therefore, the probability is f/N = 8/12 = 2/3.

4.17 (a) There was a total of 70,686,784 votes cast. Therefore, the probability that a randomly selected voter voted for Vladimir Putin is f/n = 45,513,001/70,686,784 = 0.644.

(b) There were 4,448,959 + 2,755,642 = 7,204,601 votes cast for Zhirinovsky or Mironov. Therefore, the probability that a randomly selected voter voted for Zhirinovsky or Mironov is f/n = 7,204,601/70,686,784 = 0.102.

(c) The number of votes not for Putin is 70,686,784 - 45,513,001 = 25,173,783. Thus, the probability that a randomly selected voter did not vote for Putin is 25,173,783/70,686,784 = 0.356.

4.19 The total number of units (in thousands) is N = 132,418.

(a) The probability that the unit has 4 rooms is f/N = 23636/132418 = 0.178 (to three decimal places).

(b) The probability that the unit has more than 4 rooms is f/N = (30440 + 27779 + 17868 + 19257)/132418 = 95344/132418 = 0.720.

(c) The probability that the unit has 1 or 2 rooms is f/N = (601 + 1404)/ 132418 = 2005/132418 = 0.015.

(d) The probability that the unit has fewer than one room is f/N = 0/132418 = 0.000.

(e) The probability that the unit has one or more rooms is f/N = 132418/132418 = 1.000.

4.21 The total number of graduate students is 49,088.

(a) The probability that his occupation is service is f/N = 9274/49088 = 0.189.

(b) The probability that his occupation is administrative is f/N = (2197 + 6450)/49088 = 8647/49088 = 0.176.

(b) The probability that his occupation is manufacturing is f/N = (2197 + 2166 + 1640 + 5721)/49088 = 11724/49088 = 0.239

(c) The number whose occupation is not manufacturing is 49088 - 11724 = 37364. The probability that his occupation is not manufacturing is f/N = 37364/49088 = 0.761.

4.23 (a) The total number of graduate science students in doctorate-granting institutions is 349.3 thousand. Of these, 45.6 thousand are in psychology. Thus, the probability that a randomly selected graduated student in science is in the field of psychology is f/N = 45.6/349.3 = 0.131.

(b) There are 38.7 + 97.2 = 135.9 thousand graduate students in physical or social science. Thus, the probability that a randomly selected graduated student in science is in physical or social science is f/N = 135.9/349.3 = 0.389.

(c) There are 46.1 thousand graduate students in computer science and 349.3 - 46.1 = 303.2 thousand graduate students not in computer science. Thus, the probability that a randomly selected graduated student in science is not in computer science is f/N = 303.2/349.3 = 0.868.

4.25 (a) There are five ways [(1,5), (2,4), (3,3), (4,2), (5,1)] among the 36 possibilities that sum to 6. Thus, f/N = 5/36 = 0.139.

(b) There are 18 ways among the 36 possibilities that provide a sum that is even. Thus, f/N = 18/36 = 0.500.

(c) There are six ways among the 36 possibilities that sum to 7 and two ways among the 36 possibilities that sum to 11. Thus, f/N = (6 + 2)/36 = 8/36 = 0.222.

(d) There is one way among the 36 possibilities that sums to 2, two ways in which the sum is 3, and one way in which the sum is 12. Thus, f/N = (1 + 2 + 1)/36 = 4/36 = 0.111.

4.27 (a) The event in part (e), that the housing unit has one or more rooms, is a certainty. The event in part (d), that the housing unit has less than one room, is impossible.

(b) The event that is a certainty has a probability of 1. The event that is impossible has a probability of 0.

4.29 Answers will vary

4.31 In a large sample of horse races, approximately 66.7% of the favorites will finish in the money. Therefore, 66.7% of 500 is 333.5. Approximately 334 of the favorites will finish in the money.

4.33 When rolling two balanced dice and observing the sum, there are 36 possible equally likely outcomes, not 11. The probability that the sum is 12 equals 1/36, not 1/11.

4.35 (a) Answers will vary. For my experiment, here are the outcomes from tossing a coin five times and keeping track of the running totals and running proportions.

Toss	Outcome	Number of Heads	Proportion of heads
1	H	1	1.00
2	H	2	1.00
3	T	2	0.67
4	T	2	0.50
5	T	2	0.40

(b) Based on my five tosses, I would estimate that the probability of a head when this coin is tossed is 0.40 because that was my running total for the proportion of heads after tossing the coin five times.

(c)

Toss	Outcome	Number of Heads	Proportion of heads
1	H	1	1.00
2	H	2	1.00
3	T	2	0.67
4	T	2	0.50
5	T	2	0.40
6	H	3	0.50
7	T	3	0.43
8	H	4	0.50
9	T	4	0.44
10	H	5	0.50

Based on my ten tosses, I would estimate that the probability of a head when this coin is tossed is 0.50 because that was my running total for the proportion of heads after tossing the coin ten times.

(d)

Toss	Outcome	Number of Heads	Proportion of heads
1	H	1	1.00
2	H	2	1.00
3	T	2	0.67
4	T	2	0.50
5	T	2	0.40
6	H	3	0.50
7	T	3	0.43
8	H	4	0.50
9	T	4	0.44
10	H	5	0.50
11	T	5	0.45
12	T	5	0.42
13	T	5	0.38

14	T	5	0.36
15	H	6	0.40
16	T	6	0.38
17	H	7	0.41
18	T	7	0.39
19	H	8	0.42
20	H	9	0.45

Based on my twenty tosses, I would estimate that the probability of a head when this coin is tossed is 0.45 because that was my running total for the proportion of heads after tossing the coin twenty times.

(d) For my answers in part (b)-(d), the frequency interpretation cannot be used as the definition of probability because the number of trials was small. The frequency interpretation requires a large number of trials before it can be used as the definition of probability.

4.37 If p is the probability that a randomly selected adult woman believes that a "cyber affair" is cheating, then p = 0.75. The odds against selecting such a woman are in the ratio of 1-p to p, or 0.25 to 0.75. This is normally expressed in integers as 1 to 3.

4.39 If p is the probability that a randomly selected person age 18-34 years has cursed at their computer, then p = 0.46. The odds against selecting such a person are in the ratio of 1-p to p, or 0.54 to 0.46. This is normally expressed in integers and reduced as 27 to 23.

Exercises 4.2

4.41 Venn diagrams are useful for portraying events and relationships between events.

4.43 The shaded areas correspond to the desired events in the diagrams below.

(a) A&B

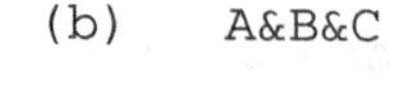

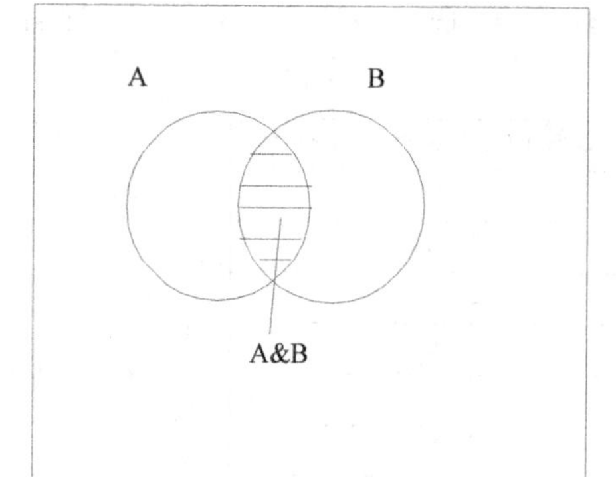

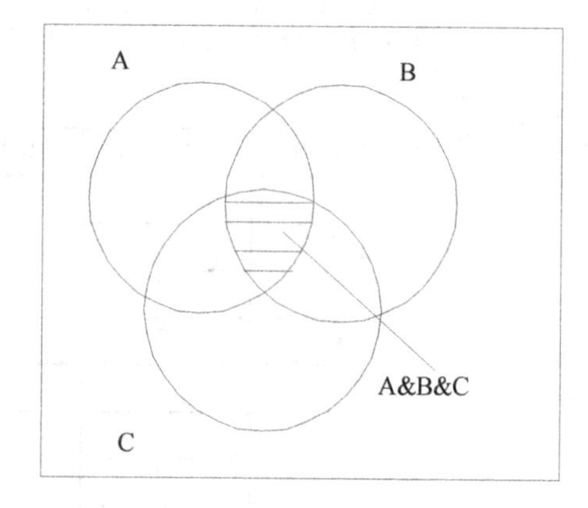

4.45 The shaded areas correspond to the desired events in the diagrams below.

(a) (A & (not B))

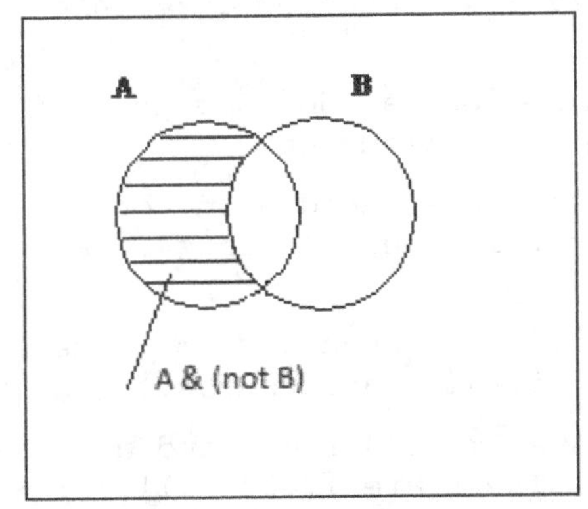

(b) ((A or B) & (not (A & B)))

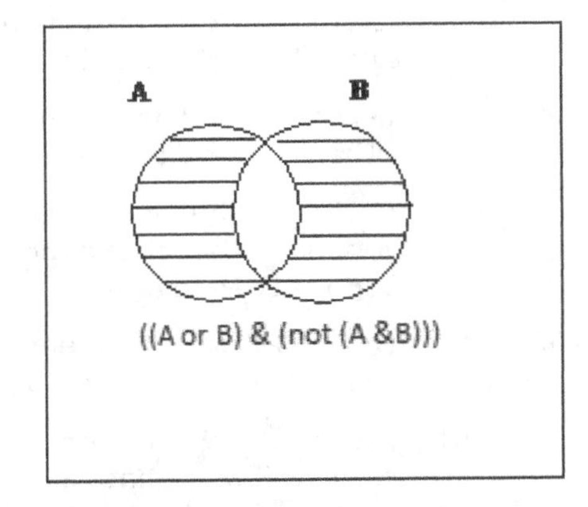

4.47 (a) At least six would be 6, 7, 8, 9, and 10.

(b) At most 3 would be 1, 2, and 3.

(c) Between 2 and 5, inclusive, would be 2, 3, 4, and 5.

4.49 Three events are mutually exclusive if no two of them can occur at the same time, i.e., no pair of the events has any outcomes in common.

4.51 False. C could have outcomes in common with A or with B even though A and B have no outcomes in common. For example, if a die is tossed and A = even number, B = odd number, C = divisible by 3, then A and B are mutually exclusive, but A and C have the outcome 6 in common while B and C have the number 3 in common.

4.53 A = {2, 4, 6}; B = {4, 5, 6}; C = {1, 2}; D = {3}

4.55 A contains the outcomes JM, JS, JH, JB, WM, WS, WH, WJ

B contains the outcomes HM, HS, HJ, HW

C contains the outcomes MW, SW, HW, JW

D contains the outcomes MS, MH, SM, SH, HM, HS

4.57 (a) (not A) = {1, 3, 5} = the event the die comes up odd.

(b) (A & B) = {4, 6} = the event the die comes up four or six.

(c) (B or C) = {1, 2, 4, 5, 6} = the event the die does *not* come up three.

4.59 (a) (not A) = (MS, MH, MJ, MW, SM, SH, SJ, SW, HM, HS, HJ, HW) = the event that a female is appointed chairperson

(b) (B & D) = (HM, HS) = the event that Holly is appointed chairperson and a woman is appointed secretary

(c) (B or C) = (HM, HS, HJ, HW, MW, SW, JW) = the event that Holly is appointed chairperson or Will is appointed secretary.

4.61 (a) (not C) = the event that a state has a diabetes prevalence percentage less than 6% or is 13% or more. There are 1 + 1 = 2 states in (not C).

(b) (A&B) = the event that a state has a diabetes prevalence percentage at least 8% and is less than 7%. There are no states in this event.

(c) (C or D) = the event that a state has a diabetes prevalence percentage that is less than 13%. There are 49 states in this event.

(d) (C&B) = the event that a state has a diabetes prevalence percentage of at least 6% and less than 7%. There are 5 states in this event.

4.63 (a) (not A) = the event that the World Series was decided in at least 5 games. There are 24 + 24 + 36 = 84 World Series in this event.

(b) (A&B) = the event that the World Series was decided in 4 games. There are 21 World Series in this event.

(c) (A or C) = the event that the World Series was decided in 4 or 7 games. There are 21 + 36 = 57 World Series in this event.

(d) (A&C) = the event that the World Series was decided in both 4 and 7 games. There are no World Series in this event. This is an impossible event.

4.65 (a) (not A) is the event that the unit has more than 4 rooms. There are 30,440 + 27,779 + 17,868 + 19,257 = 95,344 (thousand) such units.

(b) (A&B) is the event that the unit has at most 4 rooms and at least 2 rooms, i.e., has 2 or 3 or 4 rooms. There are 1,404 + 11,433 + 23,636 = 36,473 (thousand) such units.

(c) The event (C or D) is the event that the unit has between 5 and 7 rooms inclusive or more than seven rooms, i.e., has 5 or more rooms. This is the same as the event (not A) and therefore there are 95,344 (thousand) such units.

4.67 (a) Events A and B are not mutually exclusive. They have a four and a six in common.

(b) Events B and C are mutually exclusive. They have no outcomes in common.

(c) Events A, C, and D are not mutually exclusive. The outcome two is common to A and C.

(d) Among A, B, C, and D, there are three mutually exclusive events. These are B, C, and D. Among B, C, and D, there are no outcomes in common. There are not, however, four mutually exclusive events. The outcome two is common to A and C, four is common to A and B, and six is common to A and B.

4.69 The groups that are mutually exclusive are A and C; A and D; C and D; and A, C, and D.

4.71 (a) The first number in each sample space outcome is the number of dots that are face up after rolling the die. The possible outcomes for the first number are one through six. The second number in each sample space outcome is the total number of heads that are observed from a coin toss out of the number of trials that the die first showed. Therefore, the second number can be the numbers zero up to and including the number that the die first showed. The sample space is (1,0), (1,1), (2,0), (2,1), (2,2), (3,0), (3,1), (3,2), (3,3), (4,0), (4,1), (4,2), (4,3), (4,4), (5,0), (5,1), (5,2), (5,3), (5,4), (5,5), (6,0), (6,1), (6,2), (6,3), (6,4), (6,5), (6,6).

(b) The event that the total number of heads is even would consist of outcomes from the sample space that have an even second number (count zero as even). The event is (1,0), (2,0), (2,2), (3,0), (3,2), (4,0), (4,2), (4,4), (5,0), (5,2), (5,4), (6,0), (6,2), (6,4), (6,6).

4.73 (a) If events A and (not B) are mutually exclusive, this means that A and (not B) never overlap. Thus, if A occurs (not B) does not occur. This means that B does occur.

(b) If B occurs whenever A occurs, this means that (not B) never occurs when A occurs. This implies that A and (not B) are mutually exclusive.

4.75 To say that A, B, and C do not all occur simultaneously does not necessarily imply that A, B, and C are mutually exclusive. In the following diagram, C does not touch A or B. That is, we have a situation where A, B, and C do not occur simultaneously. However, these three events clearly are not mutually exclusive.

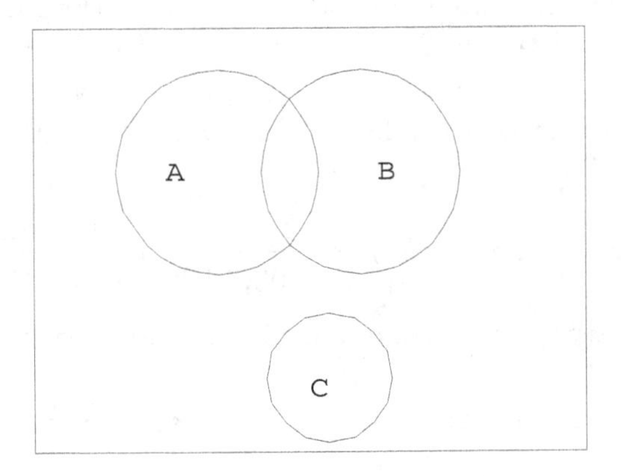

Exercises 4.3

4.77 There are 26 cards that are red. Therefore, the probability is f/N = 26/52 = 1/2. P(R) = 1/2.

4.79 Since A and B are mutually exclusive, the Special Addition Rule applies. P(A or B) = P(A) + P(B) = 0.25 + 0.40 = 0.65.

4.81 Using the Complementation rule, P(E) = 1 - P(not E), we have 0.35 = 1 - P(not E). Solving this equation results in P(not E) = 0.65.

4.83 Since C and D are not mutually exclusive, the General Addition Rule applies. P(C or D) = P(C) + P(D) - P(C&D) = 0.35 + 0.40 - 0.30 = 0.45.

4.85 (a) A and B are not mutually exclusive since P(A&B) = 1/10.

(b) Apply the General Addition Rule, which states P(A or B) = P(A) + P(B) - P(A&B). Plugging in the known values, we have 1/2 = 1/3 + P(B) - 1/10. Solve for P(B).
P(B) = 1/2 + 1/10 - 1/3 = 4/15.

4.87 (a) P(S) = f/N = (11 + 30 + 37)/100 = 0.78.

(b) S = (A or B or C)

(c) P(A) = 11/100 = 0.11; P(B) = 30/100 = 0.30;

P(C) = 37/100 = 0.37.

(d) P(S) = P(A or B or C) = P(A) + P(B) + P(C) = 0.11 + 0.30 + 0.37 = 0.78. This result is identical to the one found in part (a).

4.89 (a) The percentage who smoked within the last 30 days is 9.8 + 7.8 + 5.3 + 2.8 + 0.7 + 0.3 = 26.7. Therefore, the probability that a randomly selected twelfth grader smoked is 0.267.

(b) The percentage who smoked at least one cigarette per day within the last 30 days is 7.8 + 5.3 + 2.8 + 0.7 + 0.3 = 16.9. Therefore, the probability that a randomly selected twelfth grader smoked at least one cigarette per day is 0.169.

(d) The percentage who smoked between 6 and 34 cigarettes per day, inclusive, within the last 30 days is 5.3 + 2.8 + 0.7 = 8.8. Therefore, the probability that a randomly selected twelfth grader smoked between 6 and 34 cigarettes per day is 0.088.

4.91 (a) P(Income under $50,000) = P(A or B) = P(A) + P(B) = 0.171 + 0.240 = 0.411.

(b) P(Income of $25,000 or above) = P (B or C or D or E) = P(B) + P(C) + P(D) +P(E) = 0.240 + 0.350 + 0.141 + 0.098 = 0.829.

(c) P(Income between $25,000 and $149,999, inclusive) = P(B or C or D) = P(B) + P(C) + P(D) = 0.240 + 0.350 + 0.141 = 0.731.

(d) Of households with Internet access, 41.1% have incomes under $50,000, 82.9% have incomes of $25,000 or above, and 73.1% have incomes between $25,000 and $149,999, inclusive.

4.93 (a) P(senator is at least 50) = $\frac{30}{100}+\frac{37}{100}+\frac{20}{100}+\frac{2}{100}=0.89$.

This is accomplished more easily using the complementation rule:

P(senator is at least 50) = 1 - P(senator is under 50) = $1-\frac{11}{100}=0.89$.

(b) P(senator is under 70) = $\frac{11}{100}+\frac{30}{100}+\frac{37}{100}=\frac{78}{100}=0.78$.

This is accomplished more easily using the complementation rule:

P(senator is under 70) = 1 - P(senator is at least 70) =

$1-\left(\frac{20}{100}+\frac{2}{100}\right)=1-\frac{22}{100}=\frac{78}{100}=0.78$.

4.95 (a) The probability that the individual obtained a loan balance between $10,001 and $100,000, inclusive, is 0.292 + 0.165 + 0.058 + 0.023 = 0.538. We are asked the probability of this OR the event of at most $75,000. The only loan balance of this second event that is not already included in the event between $10,001 and $100,000, inclusive, is $1 - $10,000. Adding this to our previous sum, we have 0.538 + 0.431 = 0.969. The event that the loan balance is between $10,001 and $100,000, inclusive, or at most $75,000 is 0.969.

(b) Let A = the event that the loan balance is between $10,001 and $100,000, inclusive. Let B = the event that the loan balance is at most $75,000. P(A) = 0.538 from part (a). P(B) = 0.431 + 0.292 + 0.165 + 0.058 = 0.946. P(A&B) = the event that the loan balance is between $10,001 and $75,000, inclusive. P(A&B) = 0.292 + 0.165 + 0.058 = 0.515. Using the General Addition Rule, P(A or B) = P(A) + P(B) - P(A&B) = 0.538 + 0.946 - 0.515 = 0.969

(c) Answers will vary.

4.97 (a) $P(A)=\frac{6}{36}=0.167$ $P(B)=\frac{2}{36}=0.056$ $P(C)=\frac{1}{36}=0.028$ $P(D)=\frac{2}{36}=0.056$

$P(E)=\frac{1}{36}=0.028$ $P(F)=\frac{5}{36}=0.139$ $P(G)=\frac{6}{36}=0.167$

(b) P(7 or 11)= P(A) + P(B) = 0.167 + 0.056 = 0.223.

(c) P(2 or 3 or 12) = P(C) or P(D) or P(E).

= 0.028 + 0.056 + 0.028 = 0.112

(d) P(8 or doubles): Using Figure 4.1: $\frac{10}{36} = 0.278$

(e) P(8) + P(doubles) - P(8 & doubles) = 0.139 + 0.167 - 0.028 = 0.278.

4.99 P(Public school or college) = P(Public school) + P(College) - P(Public school & college) = 0.853 + 0.279 - 0.201 = 0.931. 93.1% of students attend either public school or college.

4.101 Let H = the event that the household gets the *Herald*, E = the event that the household gets the *Examiner*, and T = the event that the household gets the *Times*. The event that a household gets at least one of the three major newspapers can be written (H or E or T).

The general addition rule for three events is

P(H or E or T) = P(H) + P(E) + P(T) - P(H&E) - P(H&T) - P(E&T) + P(H&E&T).

Substituting in the values given in the problem we have

P(H or E or T) = 0.334 + 0.346 + 0.470 - 0.104 - 0.119 - 0.151 + 0.048 = 0.824.

Therefore, the probability that a household gets at least one of the three major newspapers is 0.824.

Exercises 4.4

4.103 The total number of observations of bivariate data can be obtained from the frequencies in a contingency table by summing the counts in the cells, summing the row totals, or summing the column totals.

4.105 (a) Data obtained by observing values of one variable of a population are called univariate data.

(b) Data obtained by observing values of two variables of a population are called bivariate data.

4.107 (a) The missing entries are filled in the following complete contingency table.

	C_1	C_2	**Total**
R_1	3	7	10
R_2	8	7	15
Total	11	14	25

(b) $P(C_1)$ = 11/25 = 0.44, $P(R_2)$ = 15/25 = 0.60, $P(C_1\&R_2)$ = 8/25 = 0.32

(c) The following table is the joint probability distribution.

	C_1	C_2	**Total**
R_1	0.12	0.28	0.25
R_2	0.32	0.28	0.60
Total	0.44	0.56	1.00

4.109 (a) The missing entries are filled in the following complete contingency table.

	C_1	C_2	C_3	**Total**
R_1	12	4	6	22
R_2	5	10	13	28
Total	17	14	19	50

(b) $P(C_1) = 17/50 = 0.34$, $P(R_2) = 28/50 = 0.56$, $P(C_1\&R_2) = 5/50 = 0.10$

(c) The following table is the joint probability distribution.

	C_1	C_2	C_3	**Total**
R_1	0.24	0.08	0.12	0.44
R_2	0.10	0.20	0.26	0.56
Total	0.34	0.28	0.38	1.00

4.111 (a) This contingency table has 12 cells.

(b) The total number of players on the New England Patriots as of September 26, 2013 is 62.

(c) There are 18 rookies.

(d) There are 41 players who weigh between 200 and 300 pounds.

(e) There are 11 rookies who weigh between 200 and 300 pounds.

4.113 (a) The missing entries are filled in the following completed contingency table.

Marital status	**Pay grade** Enlisted G_1	Officer G_2	Warrant G_3	**Total**
Single w/o children M_1	118,116	13,915	75	132,106
Single with children M_2	14,784	1,482	99	16,365
Joint service marriage M_3	14,722	2,681	67	17,470
Civilian marriage M_4	124,976	32,031	1,423	158,430
Total	272,598	50,109	1,664	324,371

(b) The number of officers is 50,109.

(c) The number that are enlisted and have a civilian marriage are 124,976.

(b) The number that are either enlisted or have a civilian marriage are 272,598 + 32,031 + 1,423 = 306,052.

(c) The number that are neither enlisted nor have a civilian marriage are 324,371 - 306,052 = 18,319.

(d) The number that do not have a joint service marriage is 324,371 - 17,470 = 306,901.

4.115 (a) Thirty-two teachers offered field trips.

(b) Twenty-three of the teachers have master's degrees.

(c) Fourteen of the bachelor's degree teachers offered field trips.

(d) D_1 is the event that one of these teachers has a bachelor's degree.

(D_2 and F_2) is the event that one of these teachers has a master's degree and does not offer field trips.

(e) $P(D_1) = 28/51 = 0.549$. $P(D_2 \text{ and } F_2) = 5/51 = 0.098$.

4.117 (a) Y_3 is the event that the player has 6-10 years of experience. W_2 is the event that the player weighs between 200 and 300 pounds. $W_1\&Y_2$ is the event that the player has between 1 and 5 years of experience and weighs under 200 pounds.

(b) $P(Y_3) = 12/62 = 0.194$; $P(W_2) = 41/62 = 0.661$; $P(W_1\&Y_2) = 5/62 = 0.081$. Thus 19.4% of the players have 6-10 years of experience; 66.1% of the players weigh between 200 and 300 pounds; and 8.1% of the players have 1-5 years of experience and weigh less than 200 pounds.

(c) We divide each cell frequency by the total 62 to obtain the following table. Due to rounding, the total does not equal 1.000

		Rookie	1-5	6-10	Over 10	Total
		Y_1	Y_2	Y_3	Y_4	
Under 200	W_1	0.048	0.081	0.000	0.000	0.129
200-300	W_2	0.177	0.339	0.113	0.032	0.661
Over 300	W_3	0.065	0.065	0.081	0.000	0.211
Total		0.290	0.485	0.194	0.032	1.001

(b) The sum of each row and column of joint probabilities equals the marginal probability to within 0.001.

4.119 (a) M_2 is the event the person obtained is single with children; G_3 is the event the person obtained is a warrant officer; and $M_1\&G_1$ is the event that the person obtained is single without children and enlisted.

(b) $P(M_2) = 16365/324371 = 0.050$; $P(G_3) = 1664/324371 = 0.005$; $P(M_1\&G_1) = 118116/324371 = 0.364$.

(c) Percentage Distribution. Due to rounding, the total does not equal 100.

		Enlisted	Officer	Warrant	Total
		G_1	G_2	G_3	
Single w/o children	M_1	36.4	4.3	0.0	40.7
Single with children	M_2	4.6	0.5	0.0	5.1
Joint service marriage	M_3	4.5	0.8	0.0	5.3
Civilian marriage	M_4	38.5	9.9	0.4	48.8
Total		84.0	15.5	0.4	99.9

4.121 The categories for each of the characteristics in the cross classification are selected so that no member of the population (or sample) can belong to more than one of the categories (cells).

4.123 (a) A member of the population (or sample) that belongs to category R_1 must also belong to exactly one of the categories $C_1, C_2, \ldots, C_n$.

(b) The events (R_1 & C_1), (R_1 & C_2), ..., (R_1 & C_n) are mutually exclusive since no member of the population (or sample) can belong to more than one of the categories $C_1, C_2, \ldots, C_n$.

(c) We can conclude that this equation holds because of the special addition rule that applies to mutually exclusive events.

Exercises 4.5

4.125 The conditional probability of tossing a head on the second toss given that a head occurred on the first toss is the same as the unconditional probability of tossing a head on the second toss without knowing the result of the first toss.

4.127 Apply the Conditional Probability Rule, which states $P(D|C)=\dfrac{P(C\,\&\,D)}{P(C)}$.

Plugging in the given values, we have $P(D|C)=\dfrac{0.2}{0.5}=\dfrac{2}{5}=0.40$.

4.129 Apply the Conditional Probability Rule, which states $P(B|A)=\dfrac{P(A\,\&\,B)}{P(A)}$.

Plugging in the given values, we have $P(B|A)=\dfrac{3/25}{2/7}=\dfrac{21}{50}=0.42$.

4.131 Apply the Conditional Probability Rule, which states $P(C_1|R_2)=\dfrac{P(C_1\,\&\,R_2)}{P(R_2)}$.

From the contingency table calculate, $P(C_1\&R_2)$ = 20/50, $P(R_2)$ = 32/50, and $P(C_1)$ = 35/50. Plugging in the given values, we have

$P(C_1|R_2)=\dfrac{20/50}{32/50}=5/8=0.625$. Reapplying the rule for $P(R_2|C_1)$, we have

$P(R_2|C_1)=\dfrac{20/50}{35/50}=4/7=0.571$.

4.133 Apply the Conditional Probability Rule, which states $P(C_1|R_2)=\dfrac{P(C_1\,\&\,R_2)}{P(R_2)}$.

From the contingency table calculate, $P(C_1\&R_2)$ = 0.10, $P(R_2)$ = 0.30, and $P(C_1)$ = 0.60. Plugging in the given values, we have $P(C_1|R_2)=\dfrac{0.10}{0.30}=0.333$.

Reapplying the rule for $P(R_2|C_1)$, we have $P(R_2|C_1)=\dfrac{0.10}{0.60}=0.167$.

4.135 (a) P(B) = 4/52 = 0.077; the probability of randomly selecting a king from an ordinary deck of 52 playing cards is 0.077.

(b) P(B|A) = 4/12 = 0.333; given that the selection is made from among (the 12) face cards, the probability of selecting a king is 0.333.

(c) P(B|C) = 1/13 = 0.077; given that the selection is made from among (the 13) hearts, the probability of selecting a king is 0.077.

(d) P(B|(not A)) = 0/40 = 0; given that the selection is made from among (the 40) non-face cards, the probability of selecting a king is 0.

(e) P(A) = 12/52 = 0.231; the probability of selecting a face card from an ordinary deck of 52 playing cards is 0.231.

(f) P(A|B) = 4/4 = 1; given that the selection is made from among (the four) kings, the probability of selecting a face card is 1.

(g) P(A|C) = 3/13 = 0.231; since the selection is made from among (the 13) hearts, the probability of selecting a face card is 0.231.

(h) P(A|(not B)) = 8/48 = 0.167; given that the selection is made from among (the 48) non-kings, the probability of selecting a face card is 0.167.

4.137 (a) P(exactly 4 rooms) = 23636/132418 = 0.178

(b) P(exactly 4 rooms | at least 2 rooms)

= 23636/(132418 - 601) = 23636/131817 = 0.179

(c) P(at most 4 rooms | at least 2 rooms)

= (1404 + 11433 + 23636)/131817 = 36473/131817 = 0.277

(d) (i) 17.8% of the units have exactly 4 rooms;

(ii) Of those with at least 2 rooms, 17.9% have exactly 4 rooms;

(iii) Of those with at least 2 rooms, 27.7% have at most 4 rooms.

4.139 (a) P(Rookie) = 18/62 = 0.290

(b) P(Weighs under 200 pounds) = 8/62 = 0.129

(c) P(Rookie | Weighs under 200 Pounds) = 3/8 = 0.375

(d) P(Weighs under 200 pounds | Rookie) = 3/18 = 0.167

(e) Of the players, 29.0% are rookies and 12.9% weigh under 200 pounds; 37.5% of those under 200 pounds are rookies; and 16.7% of the rookies weigh under 200 pounds.

4.141 (a) $P(V_2)$ = 60/160 = 0.375. Of all ads Indian adolescents, 37.5% were blind.

(b) $P(V_2\&S_3)$ = 17/160 = 0.106. Of all the Indian adolescents, 10.6% were blind and had a low self concept.

(c) $P(S_3|V_2)$ = 17/60 = 0.283. Of those Indian adolescents that were blind, 28.3% had a low self concept.

(d) $P(S_3|V_2) = P(V_2\&S_3)/P(V_2)$ = 0.106/0.375 = 0.283

4.143 (a) $P(S_1)$ = 0.564

(b) $P(E_4)$ = 0.391

(c) $P(S_1 \& E_4)$ = 0.274

(d) $P(S_1|E_4) = P(S_1 \& E_4)/P(E_4)$ = 0.274/0.391 = 0.701

(e) $P(E_4|S_1) = P(S_1 \& E_4)/P(S_1)$ = 0.274/0.564 = 0.486

(f) 56.4% of U.S. adults own a cell phone. 39.1% of U.S. adults are college graduates. 27.4% of U.S. adults are college graduates that own a cell phone. Of those U.S. adults that are college graduates, 70.1% own a cell phone. Of those U.S. adults that own a cell phone, 48.6% are college graduates.

4.145 (a) $P(T_1) = 0.441$

(b) $P(\text{not } A_2) = 1 - P(\text{not } A_2) = 1 - 0.314 = 0.686$

(c) $P(A_4 \& T_5) = 0.022$

(d) $P(T_4|A_1) = P(T_4 \& A_1)/P(A_1) = 0.070/0.526 = 0.133$

(e) $P(A_1|T_4) = P(A_1 \& T_4)/P(T_4) = 0.070/0.122 = 0.574$

Of U.S. adults, for 44.1%, it has been less than half a year since their last visit to a dentist. 69.9% of U.S. adults are not between 45 and 64 years old. 2.2% are over 75 and have not seen a dentist for 5 or more years. Of those 18-44 years old, for 13.3%, it has been between 2 and 5 years since they last have seen a dentist. Of those for whom it has been between 2 and five years since their last dental visit, 57.4% are between 18 and 44 years old.

4.147 Let: R = property crime that is committed in a rural area;

B = property crime that is a burglary.

We are given: $P(R) = 0.051$ and $P(B \& R) = 0.016$. The percentage of property crimes committed in rural areas that are burglaries is

$P(B|R) = P(B \& R)/ P(R) = 0.016/0.051 = 0.314$ or 31.4%.

4.149 Let: I = The U.S. physician is in internal medicine;

A = The U.S. physician is under age 55.

We are given: $P(I) = 0.136$ and $P(I \& A) = 0.0867$. The percentage of U.S. physicians who specialize in internal medicine that are under age 55 is

$P(A|I) = P(A \& I)/ P(I) = 0.0867/0.136 = 0.6375$ or 63.75%.

4.151 (a) The sample space for this experiment is given in the following list. Let B stand for boy and G stand for girl

BB BG GB GG

P(both children are boys | the first child is a boy)

= P(the first and second child are boys) / P(the first child is a boy)

= (1/4) / (2/4) = 1/2 = 0.50

(b) P(both children are boys | at least one child is a boy)

= P(the first and second child are boys) / P(at least one child is a boy)

= (1/4) / (3/4) = 1/3 = 0.333

4.153 (a) Event B is positively correlated with Event A if the probability of B increases when it becomes known that A has occurred. Event B is negatively correlated with Event A if the probability of B decreases when it becomes known that A has occurred. Event B is independent of Event A if the probability of B does not change when it becomes known that A has occurred.

(b) We need to show that $P(A|B) > P(A)$ if and only if $P(B|A) > P(B)$. Formula 4.4 implies that $P(A \& B) = P(A|B)P(B) = P(A \& B) = P(B|A)P(A)$.

$$P(A|B) = \frac{P(A \& B)}{P(B)} = \frac{P(B|A)P(A)}{P(B)} > \frac{P(B)P(A)}{P(B)} = P(A)$$ where the ">" sign occurs because $P(B|A) > P(B)$. This proves that $P(A|B) > P(A)$ <u>if</u>

P(B|A) > P(B). By interchanging A & B in the proof, we can show that P(B|A) > P(B) if P(A|B) > P(A). This is the proof of the only if part of the statement.

(c) We need to show that P(A|B) < P(A) if and only if P(B|A) < P(B).

$$P(A \mid B) = \frac{P(A \& B)}{P(B)} = \frac{P(B \mid A)P(A)}{P(B)} < \frac{P(B)P(A)}{P(B)} = P(A)$$ where the "<" sign occurs because P(B|A) < P(B). This proves that P(A|B) < P(A) if P(B|A) < P(B). By interchanging A & B in the above proof, we can show that P(B|A) < P(B) if P(A|B) < P(A). This is the proof of the only if part of the statement.

(d) We need to show that P(A|B) = P(A) if and only if P(B|A) = P(B).

$$P(A \mid B) = \frac{P(A \& B)}{P(B)} = \frac{P(B \mid A)P(A)}{P(B)} = \frac{P(B)P(A)}{P(B)} = P(A)$$ where the third "=" sign occurs because P(B|A) = P(B). This proves that P(A|B) = P(A) if P(B|A) = P(B). By interchanging A & B in the above proof, we can show that P(B|A) = P(B) if P(A|B) = P(A). This is the proof of the only if part of the statement.

Exercises 4.6

4.155 (a) General Multiplication rule: P(A&B) = P(A)P(B|A)

Conditional probability rule: P(B|A) = P(A&B)/P(A)

(b) The general probability can be obtained from the conditional probability rule by multiplying both sides of the equation by P(A).

(c) We can use the general multiplication rule when we know the marginal and conditional probabilities to find a joint probability. We can use the conditional probability rule when we know the joint and marginal probabilities to find a conditional probability.

4.157 The General Multiplication Rule states for two events A and B, P(A&B) = P(A)*P(B|A). Plugging in the given values, we have P(A&B) = 0.6*0.4 = 0.24.

4.159 The General Multiplication Rule states for two events C and D, P(C&D) = P(C)*P(D|C). Plugging in the given values, we have P(C&D) = 1/4*4/5 = 1/5 = 0.20.

4.161 The Special Multiplication Rule (for Two Independent events) states that if C and D are independent events, then P(C&D) = P(C)*P(D). Plugging in the given values, we have P(C&D) = 0.7*0.6 = 0.42.

4.163 The Special Multiplication Rule (for Two Independent events) states that if A and B are independent events, then P(A&B) = P(A)*P(B). Plugging in the given values, we have P(A&B) = 5/8*4/7 = 5/14 = 0.357.

4.165 A and B are not independent because P(B|A) does not equal P(B).

4.167 C and D are independent because P(D|C) equals P(D).

4.169 It is not possible to tell if A and B are independent. We need to know either P(A) or P(B|A).

4.171 It is not possible to tell if C and D are independent. We need to know either P(C) or P(D|C).

4.173 C and D are independent because P(C&D) equals P(C)*P(D).

4.175 A and B are not independent because P(A&B) does not equal P(A)*P(B).

4.177 The Special Multiplication Rule states that if events A, B, and C are independent, then P(A&B&C) = P(A)*P(B)*P(C). Plugging in the given values, we have P(A&B&C) = 0.8*0.5*0.3 = 0.12.

4.179 We have P(Holiday Depression | Woman) = 0.44 and P(Woman) = 0.52. The probability that a randomly selected U.S. adult is a woman who suffers from holiday depression is P(Woman & Holiday Depression) = P(Woman)P(Holdiay Depression | Woman) = (0.52)(0.44) = 0.2288. Thus, about 23% of U.S. adults are women who suffer from holiday depression.

4.181 (a) P(H) = 1/6 = 0.167.

(b) After selecting a 3, there are five numbers remaining. With two numbers exceeding 4, P(K|H) = 2/5 = 0.4.

(c) P(H & K) = P(H)·P(K|H) = (0.167)(0.4) = 0.067.

(d) The probability that both numbers picked are less than 3 is (2/6)(1/5) = 2/30 = 0.067.

(e) The probability that both numbers picked are greater than 3 is (3/6)(2/5) = 6/30 = 0.2.

4.183 (a) P(First is Jr & Second is Sr) = P(First is Jr)P(Second is Sr | First is Jr) = (12/40)(7/39) = 84/1560 = 0.054.

(b) P(First is So & Second is So) = P(First is So)P(Second is So | First is So) = (15/40)(14/39) = 210/1560 = 0.135.

(c)

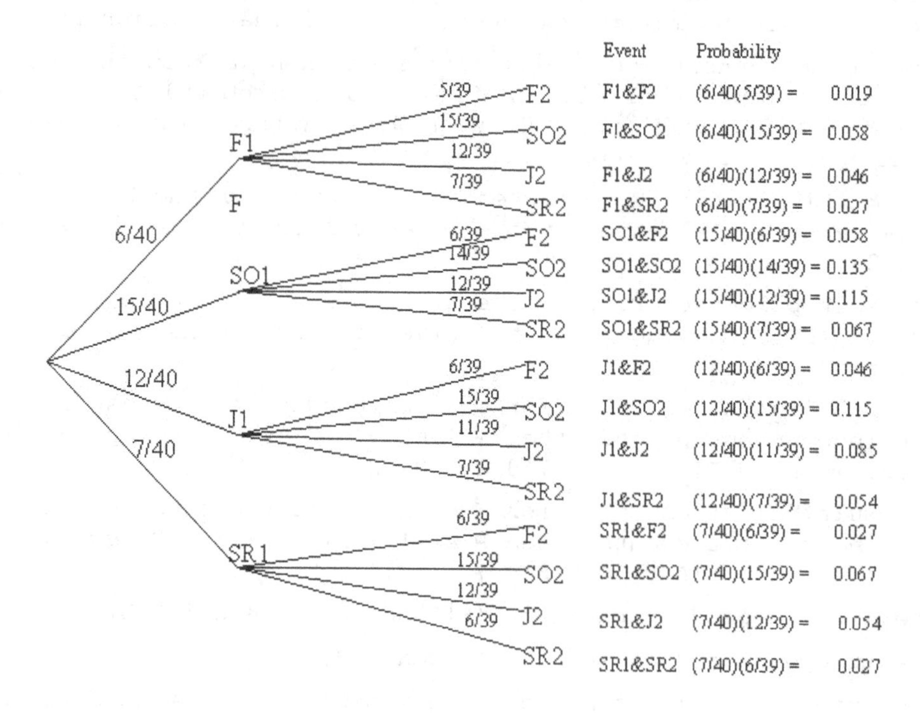

(d) Using the tree diagram, P(One is F and one is SO) = P(F1 & SO2) + P(SO1 & F2) = 0.058 + 0.058 = 0.116

4.185 (a) $P(R_3)$ = 40379/98993 = 0.408

(b) $P(R_3|G_1)$ = (25888/70000) = 0.370

(c) No. The probability of R_3 changes when it becomes know that G_1 has occurred. To be independent, the probability must not change.

(d) Since $P(R_2) = 21732/98993 = 0.220$ and $P(R_2|G_2) = 5400/28993 = 0.186$, the events of being female and being associate professor are not independent. To be independent, the probability must not change.

4.187 (a) $P(P_1) = 0.471$; $P(C_2) = 0.187$; $P(P_1 \& C_2) = 0.097$

(b) $P(P_1)\ P(C_2) = (0.471)(0.187) = 0.088$. Since this is not equal to $P(P_1 \& C_2)$, these events are not independent.

4.189 (a) $P(A) = 4/8 = 1/2$; $P(B) = 4/8 =1/2$; $P(C) = 3/8$.

(b) $P(B|A) =2/4 = 1/2$

(c) Yes. $P(B|A)$ is the same as $P(B)$.

(d) $P(C|A) = 1/4$

(e) No. $P(C|A)$ and $P(C)$ are not equal.

4.191 Let A_i = draw an ace on selection i, where i = 1,2.

(a) If the first card is replaced before the second card is drawn:

$P(A_1 \& A_2) = (4/52)(4/52) = 0.006$.

(b) If the first card is not replaced before the second card is drawn:

$P(A_1 \& A_2) = (4/52)(3/51) = 0.005$.

4.193 Let:

$\overline{F}_i$ = non-failure of "criticality 1" item i, where i = 1, 2, …, 748

F_i = failure of "criticality 1" item i, where i = 1, 2, …, 748

Also, $P(\overline{F}_i) = 0.9999$ and $P(F_i) = 0.0001$

(a) The probability that none of the "criticality 1" items would fail is

$$P(\overline{F}_1)\cdot P(\overline{F}_2)\cdot \ldots \cdot P(\overline{F}_{748}) = (0.9999)^{748} = 0.928$$

(b) The probability that at least one "criticality 1" item would fail is

$1 - 0.928 = 0.072$.

(c) There was a 7.2% chance that at least one "criticality 1" item would fail. In other words, on the average at least one "criticality 1" item will fail in 7.2 out of every 100 such missions.

4.195 (a) The age of a driver at fault in one fatal accident is independent of the age of a driver at fault in any other fatal accident. Therefore, P(16-24 & 25-34 & 34-64) = P(16-24)P(25-34)P(34-64) = (0.255)(0.238)(0.393) = 0.024.

(b) Let A be the event that a driver at fault is between 16 and 24 years old and let B be the event that a driver at fault is 65 or older. Having two drivers at fault between 16 and 24 years old and one being 65 or older can occur as A&A&B, A&B&A, or B&A&A. These three outcomes are mutually exclusive, so we can find the probabilities of each one and then add the three results. P(A&A&B) = P(A)P(A)P(B) = (0.255)(0.255)(0.114) = 0.0074. The other two products have the same factors, but written in a different order. Therefore all three probabilities are the same, so the probability of having two of the drivers being between 16 and 24 and one being 65 or older is 0.0074 + 0.0074 + 0.0074 = 0.0222.

4.197 (a) P(Nut defective or bolt defective) = P(Nut defective) + P(Bolt defective) − P(Bolt defective & Nut defective) = P(Nut defective) + P(Bolt defective) − P(Nut defective)P(Bolt defective) = 0.02 + 0.03 − (0.02)(0.03) = 0.0494. Thus 4.94% of the nut-bolt units will be defective.

(b) We are assuming that the events that a nut is defective and a bolt is defective are independent. This seems reasonable since the machines that make the nuts are not the same ones that make the bolts. We are also assuming that a defective nut or bolt is not discarded before it is paired up with a bolt or nut, respectively.

4.199 Let F be the event that the daughter does not win the first grant. Let S be the event that the daughter does not win the second grant. The probability that the daughter does not win the first grant is P(F) = 3/5. The probability that the daughter does not win the second grant is P(S) = 4/5. Assuming that winning either grant is independent of each other, the probability that the daughter wins neither grant is P(F&S) = 3/5*4/5 = 12/25 = 0.48. The event that the daughter wins at least one grant is the complement of (F&S), or (not(F&S)). Applying the complementation rule P(not(F&S)) = 1 − P(F&S) = 1 − 12/25 = 13/25 = 0.52.

4.201 P(F1 & F2 & M3) = P(F1)P(F2|F1)P(M3|F1 & F2) = (23/40)(22/39)(17/38) = 0.145

4.203 (a) If two events A and B are mutually exclusive, they cannot both happen, so P(A & B) = 0.

(b) If neither A nor B is impossible, then P(A) > 0 and P(B) > 0. Since A and B are independent, P(A & B) = P(A) P(B), but this cannot be zero since neither A nor B has a probability of 0.

(c) Answers will vary. If A and B are two events such that P(A) = 0.4, P(B) = 0.3, and P(A & B) = 0.2, then A and B are not mutually exclusive since they can happen simultaneously, and they are not independent since P(A & B) ≠ P(A) P(B).

4.205 (a) We have

$$P(A) = P(B) = P(C) = \frac{18}{36} = \frac{1}{2}$$

$$P(A\& B) = P(A\& C) = P(B\& C) = \frac{9}{36} = \frac{1}{4}$$

$$P(A\& B\& C) = \frac{9}{36} = \frac{1}{4}$$

Consequently,

$$P(A\&B) = P(A)P(B)$$
$$P(A\&C) = P(A)P(C)$$
$$P(B\&C) = P(B)P(C)$$

However, $P(A\&B\&C) = \frac{1}{4} \neq \frac{1}{2}\cdot\frac{1}{2}\cdot\frac{1}{2} = P(A)\cdot P(B)\cdot P(C)$.

Thus, A, B, and C are not independent.

(b) We have

$P(D) = \frac{18}{36} = \frac{1}{2}$ $P(E) = \frac{18}{36} = \frac{1}{2}$ $P(F) = \frac{4}{36} = \frac{1}{9}$

$P(D \& E) = \frac{6}{36} = \frac{1}{6}$ $P(D \& F) = \frac{3}{36} = \frac{1}{12}$ $P(E \& F) = \frac{2}{36} = \frac{1}{18}$

$P(D \& E \& F) = \frac{1}{36}$

Consequently, $P(D \& E \& F) = \frac{1}{36} = \frac{1}{2}\cdot\frac{1}{2}\cdot\frac{1}{9} = P(D)P(E)P(F)$

However, since $P(D \& E) = \frac{1}{6} \neq \frac{1}{2}\cdot\frac{1}{2} = P(D)P(E)$,

the events D, E, and F are *not* independent.

Exercises 4.7

4.207 Four events are exhaustive if at least one of them must occur.

4.209 Exhaustive events are not necessarily mutually exclusive. Consider an example in which a die is rolled. Let A be the event that the number showing is even, let B be the event that the number is odd, let C be the event that the number is a 1, and let D be the event that the number is 4 or greater. These four events contain all of the possible outcomes from 1 to 6 and are thus exhaustive. However, B and C both contain 1, A and D both contain 4 and 6, and B and D both contain 5. Therefore the events are not mutually exclusive.

Mutually exclusive events are not necessarily exhaustive. Consider an example in which two dice are tossed. Let A be the event the total is 2, let B be the event that the total is between 3 and 5 inclusive, let C be the event that the total is between 7 and 9 inclusive, and let D be the event that the total is 12. These four events are mutually exclusive, but if the total is 6, 10, or 11, none of these events will occur. Thus, these events are not exhaustive.

4.211 (a) The Rule of Total Probability states $P(B) = P(B|A_1)P(A_1) + P(B|A_2)P(A_2)$. Plugging in the given values, we have $P(B) = 0.8*0.4 + 0.7*0.6 = 0.74$.

(b) Baye's Rule states $P(A_1 | B) = \frac{P(A_1)P(B|A_1)}{P(A_1)P(B|A_1) + P(A_2)P(B|A_2)}$. Plugging in the values calculated in part (a), we have $P(A_1|B) = (0.8*0.4)/0.74 = 0.432$.

4.213 (a) The Rule of Total Probability states
$P(B) = P(B|A_1)P(A_1) + P(B|A_2)P(A_2) + P(B|A_3)P(A_3)$.

Plugging in the given values, we have
$P(B) = 0.4*0.5 + 0.8*0.2 + 0.6*0.3 = 0.54$.

(b) Baye's Rule states $P(A_1 | B) = \frac{P(A_1)P(B|A_1)}{P(A_1)P(B|A_1) + P(A_2)P(B|A_2) + P(A_3)P(B|A_3)}$.

Plugging in the values calculated in part (a), we have
$P(A_1|B) = (0.4*0.5)/0.54 = 0.370$.

4.215 (a) $P(R_3)$

(b) $P(S|R_3)$

(c) $P(R_3|S)$

4.217 (a) P(believe in aliens) = 0.54 · 0.48 + 0.33 · 0.52 = 0.431, so 43.1% of U.S. adults believe in aliens.

(b) P(believe in aliens|woman) = 0.33; so 33% of women believe in aliens.

(c) P(woman|believe in aliens) = (0.33)(0.52)/.431 = 0.398; so 39.8% of adults who believe in aliens are women.

4.219 (a) Let A be the event that the selected person reads an astrology report eve1ry day, let L be the event that the person has less than high school education, let H be the event that the person is a high school graduate, and let B be the event that the person has Baccalaureate or higher. P(A) = P(A|L)P(L) + P(A|H)P(H) + P(A|B)P(B) = (0.09)(0.074) + (0.07)(0.533) + (0.04)(0.393) = 0.0597

(b) P(H|A) = P(H & A)/P(A) = P(A|H)P(H)/P(A) = (0.09)(0.074)/0.0597 = 0.1116. From this, P((not H)|A) = 1 - P(H|A) = 1 - 0.1116 = 0.8884.

(c) P(B|A) = P(B & A)/P(A) = P(A|B)P(B)/P(A) = (0.04)(0.393)/0.0597 = 0.2634.

4.221 (a) Let A1 to A4 represent the four age categories in the order shown in the table and let F by the event that a selected person is obese or overweight. P(F) = P(A1)P(F|A1) + P(A2)P(F|A2) + P(A3)P(F|A3) + P(A4)P(F|A4) = (0.425)(0.411) + (0.285)(0.579) + (0.164)(0.682) + (0.126)(0.553) = 0.5212. Thus 52.12% of Utah adults are overweight or obese.

(b) 57.9% of Utah adults between 35 and 49 years old are overweight or obese.

(c) P(A2|D) = P(A2 & D)/P(D) = P(D|A2)P(A2)/P(D) = (0.579)(0.285)/0.5212 = 0.3166. Thus 31.66% of overweight or obese Utah adults are between 35 and 49 years old, inclusive.

(d) 52.12% of Utah adults are overweight or obese. 57.9% of Utah adults between 35 and 49 years old are overweight or obese. 31.66% of overweight or obese Utah adults are between 35 and 49 years old, inclusive.

4.223 Let: M = sell more than projected P(M) = 0.10 P(R|M) = 0.70
C = sell close to projected P(C) = 0.30 P(R|C) = 0.50
L = sell less than projected P(L) = 0.60 P(R|L) = 0.20
R = revised for a second edition

(a) P(R) = P(M)·P(R|M) + P(C)·P(R|C) + P(L)·P(R|L)

= (0.10 · 0.70) + (0.30 · 0.50) + (0.60 · 0.20) = 0.34 or 34%

(b)

$$P(L \mid R) = \frac{P(L \,\&\, R)}{P(R)} = \frac{P(L)P(R \mid L)}{P(R)} = \frac{(0.60)(0.20)}{0.34} = 0.353 \text{ or } 35.3\%$$

4.225 (a) Of those women who have precancerous or cancerous cells in the cervix, 51% will test positive. Also, of those individuals who do not have precancerous or cancerous cells in the cervix, 98% will test negative.

(b) Let: P= test is positive

not P= test is negative

D= person has cancer

not D= person does not have cancer

$P(P|D) = 0.51$

$P(\text{not } P|\text{not } D) = 0.98$

$P(D) = 0.000079$

$$P(P) = P(D) \cdot P(P \mid D) + P(not\ D) \cdot P(P \mid not\ D)$$
$$= 0.000079(0.51) + 0.999921(0.02) = 0.0200$$

Two of the items on the right-hand side of the equation above are not provided in the statement of the problem. These are P(not D) and P(P|not D). Because D and not D are complements, we know that P(not D) = 1 - P(D) = 1 - 0.000079 = 0.999921. Also, P(P|not D) and P(not P|not D) are complements. (Although not so readily apparent, this can be demonstrated with the assistance of a tree diagram.) Thus, P(P|not D) = 1 - P(not P|not D) = 1 - 0.98 = 0.02.

(c)

$$P(D \mid P) = \frac{P(D) \cdot P(P \mid D)}{P(D) \cdot P(P \mid D) + P(not\ D) \cdot P(P \mid not\ D)}$$
$$= \frac{0.000079(0.51)}{0.02} = 0.0020$$

(d) In terms of a percentage, the result from part (b) means that 2% of the population of women getting a pap test will test positive. Interpreting part (c), of those that test positive, only 0.2% have cervical cancer.

4.227 (a) One card is red on both sides. Let the outcomes for that card be R1 and R2. The other card is mixed with red on one side and black on the other. Let the outcomes for that card be MR and MB. If you lift your hand and see a red card, you have the R1, R2, or MR card. If you have the R1 card, the other side is red. If you have the R2, card, the other side is red. If you have the MR card, the other side is black. Since getting the R1, R2, or MR card is equally likely, the probability that the other side is red is 2/3.

(b) The explanation is included in part (a). If you know that the card has at least one red side, there is a 2/3 chance that the other side is red. Two out of the three red cards have a second red side.

Exercises 4.8

4.229 Counting rules are techniques for determining the number of ways something can happen without directly listing all of the possibilities. They are important because sometimes the number of possibilities is so large that a direct listing is impractical.

4.231 (a) A permutation of r objects from a collection of m objects is any *ordered* rearrangement of r of the m objects.

(b) A combination of r objects from a collection of m objects is any *unordered* set of r of the m objects.

(c) Permutations are ordered rearrangements while combinations are sets in

4.233 Applying the Basic Counting Rule (BCR), there are $4 \cdot 5 \cdot 2 = 40$ possibilities altogether for the three actions.

4.235 $0! = 1$, $5! = 5 \cdot 4 \cdot 3 \cdot 2 \cdot 1 = 120$, $6! = 6 \cdot 5 \cdot 4 \cdot 3 \cdot 2 \cdot 1 = 720$

4.237 (a) ${}_4P_3 = 4 \cdot 3 \cdot 2 = 24$

(b) ${}_{15}P_4 = 15 \cdot 14 \cdot 13 \cdot 12 = 32{,}760$

(c) ${}_6P_2 = 6 \cdot 5 = 30$

(d) ${}_{10}P_0 = 1$

(e) ${}_8P_8 = 8 \cdot 7 \cdot 6 \cdot 5 \cdot 4 \cdot 3 \cdot 2 \cdot 1 = 40{,}320$

4.239 (a)

$${}_4C_3 = \frac{4!}{3!(4-3)!} = \frac{4!}{3!1!} = \frac{4}{1!} = \frac{4}{1} = 4$$

(b)

$${}_{15}C_4 = \frac{15!}{4!(15-4)!} = \frac{15!}{4!11!} = \frac{15 \cdot 14 \cdot 13 \cdot 12 \cdot 11!}{4!11!} = \frac{15 \cdot 14 \cdot 13 \cdot 12}{4 \cdot 3 \cdot 2 \cdot 1} = 1365$$

(c)

$${}_6C_2 = \frac{6!}{2!(6-2)!} = \frac{6!}{2!4!} = \frac{6 \cdot 5 \cdot 4!}{2!4!} = \frac{6 \cdot 5}{2 \cdot 1} = 15$$

(d)

$${}_{10}C_0 = \frac{10!}{0!(10-0)!} = \frac{10!}{0!10!} = 1$$

(e)

$${}_8C_8 = \frac{8!}{8!(8-8)!} = \frac{8!}{8!0!} = 1$$

4.241 The number of possible permutations of m objects among themselves is $\underline{m!}$.

4.243 There are

$${}_{70}C_5 = \frac{70!}{5!(70-5)!} = \frac{70!}{5!65!} = \frac{70 \cdot 69 \cdot 68 \cdot 67 \cdot 66 \cdot 65!}{5!65!} = \frac{70 \cdot 69 \cdot 68 \cdot 67 \cdot 66}{5 \cdot 4 \cdot 3 \cdot 2 \cdot 1} = 12{,}103{,}014$$

possible samples of size 5 from a population of size 70.

4.245 (a)

MODEL	ELEVATION	OUTCOME
S	A	SA
	B	SB
	C	SC
P	A	PA
	B	PB
	C	PC
V	A	VA
	B	VB
	C	VC
M	A	MA
	B	MB
	C	MC
N	A	NA
	B	NB
	C	NC

(b) There are 15 choices for the selection of a home, including both model and elevation.

(c) There are 5 ways to choose the model and for each of these, there are 3 ways to choose the elevation. By the BCR, there are (5)(3) = 15 ways to choose both model and elevation.

4.247 (a) Let m_i equal the number of ways of choosing the ith digit. Then $m_1 = m_2 = m_3 = m_4 = 10$ and $m_5 = 8$. Thus the number of possible 5-digit zip codes is (10)(10)(10)(10)(8) = 80,000.

(b) For the plus four zip codes, let $m_6 = m_7 = m_8 = 10$ and $m_9 = 9$. Thus the number of plus four zips codes is (10)(10)(10)(9) = 9,000.

(c) The total possible number of zip codes is (10)(10)(10)(10)(8)(10)(10)(10)(9) = 720,000,000

4.249 (a) There are (9)(10)(10) = 900 possible different area codes.

(b) There are (9)(10)(10)(10)(10)(10)(10) = 9,000,000 possible local telephone numbers in any given area code.

(c) There are (900)(9,000,000) = 8,100,000,000 possible phone numbers in the United States.

4.251 There are 30·29·28·27 = 657,720 ways to make the four investments.

4.253 Don Mattingly would have $_5P_3$ = (5)(4)(3) = 60 possible assignments for the three outfield positions.

4.255 There are $_7P_7$ = 7·6·5·4·3·2·1 = 5040 ways to assign the seven salespeople to seven different territories.

4.257 The number of possibilities is

$$_{18}C_3 = \frac{18!}{3!(18-3)!} = \frac{18!}{3!15!} = \frac{18\cdot 17\cdot 16\cdot 15!}{3!15!} = \frac{18\cdot 17\cdot 16}{3\cdot 2\cdot 1} = 816$$

4.259 The order of the people in each pair does not matter. The number of handshakes is the same as the number of ways that two people can be chosen from 10 or

$$_{10}C_2 = \frac{10!}{2!(10-2)!} = \frac{10!}{2!8!} = \frac{10\cdot 9\cdot 8!}{2!8!} = \frac{10\cdot 9}{2\cdot 1} = 45$$

4.261 (a) Since the order of the cards is not important, there are $_{52}C_5$ = 52!/5!·47! = 52·51·50·49·48 / 5·4·3·2·1 = 2,598,960 different five-card hands.

(b) There are $_4C_3$ = 4!/3!1! = 4 ways of getting three kings and $_4C_2$ = 4!/2!2! = 4·3/2·1 = 6 ways of getting two queens. Altogether, there are 4·6 = 24 ways of getting three kings and two queens.

(c) There are 13 ways to pick the three-card denomination and then 12 ways to pick the two-card denomination. Then as in part (b), there are 24 ways to pick the three cards of one denomination and two cards of the other denomination. Altogether, there are 13·12·24 = 3744 different full houses possible.

(d) From parts (a) and (c), P(full house) = 3744/2,598,960 = 0.00144

(e) The answers are the same in both exercises.

4.263 There are 6^4 = 1296 possible outcomes when you toss four balanced dice. The number of ways that the four outcomes result in four different numbers is $_6P_4 = 6\cdot 5\cdot 4\cdot 3 = 360$. The probability of tossing four balanced dice resulting in four different numbers is 360/1296 = 0.278.

4.265 (a) The probability that exactly one of the TVs selected is defective is

$$\frac{_6C_1 \cdot {_{94}C_4}}{_{100}C_5} = \frac{\frac{6!}{1!(6-1)!}\cdot\frac{94!}{4!(94-4)!}}{75,287,520} = \frac{6\cdot(3,049,501)}{75,287,520} = 0.243$$

(b) The probability that at most one of the TVs selected is defective is the sum of the probabilities for zero and one defectives. The probability of zero defectives is

$$\frac{_6C_0 \cdot {_{94}C_5}}{_{100}C_5} = \frac{\frac{6!}{0!(6-0)!}\cdot\frac{94!}{5!(94-5)!}}{75,287,520} = \frac{54,891,018}{75,287,520} = 0.729$$

The probability of one defective is 0.243, as calculated in part (a). Thus, the probability that at most one of the TVs selected is defective is 0.243 + 0.729 = 0.972.

(c) The probability that at least one of the TVs selected is defective is

1 - probability of zero defectives = 1 - 0.729 = 0.271.

4.267 (a) The probability that you win the jackpot is

$$\frac{_5C_5 \cdot {_{70}C_0}}{_{75}C_5} \bullet \frac{_1C_1 \cdot {_{14}C_0}}{_{15}C_1} = \frac{\frac{5!}{5!(5-5)!}\cdot\frac{70!}{0!(70-0)!}}{\frac{75!}{5!(75-5)!}} \bullet \frac{\frac{1!}{1!(1-1)!}\cdot\frac{14!}{0!(14-0)!}}{\frac{15!}{1!(15-1)!}}$$

$$= \frac{1\cdot 1}{17259390} \bullet \frac{1\cdot 1}{15} = \frac{1}{258,890,850} = 0.000000003863 = 3.86\times 10^{-9}$$

(b) Winning $5,000 means that you matched 4 of the white balls and the Mega ball. The probability of winning $5,000 is

$$\frac{{}_5C_4 \cdot {}_{70}C_1}{{}_{75}C_5} \bullet \frac{{}_1C_1 \cdot {}_{14}C_0}{{}_{15}C_1} = \frac{\frac{5!}{4!(5-4)!} \cdot \frac{70!}{1!(70-1)!}}{\frac{75!}{5!(75-5)!}} \bullet \frac{\frac{1!}{1!(1-1)!} \cdot \frac{14!}{0!(14-0)!}}{\frac{15!}{1!(15-1)!}}$$

$$= \frac{5 \cdot 70}{17259390} \bullet \frac{1 \cdot 1}{15} = \frac{350}{258,890,850} = 0.000001352 = 1.35 \times 10^{-6}$$

(c) Winning at least $5,000 means you win $5,000, $1,000,000, or the jackpot. P($5,000) = 350/258,890,850 from part (b). P(jackpot) = 1/258,890,850 from part (a). P($1,000,000) is the probability that you matched 5 of the white balls and mismatched the Mega Ball. The probability of winning $1,000,000 is

$$\frac{{}_5C_5 \cdot {}_{70}C_0}{{}_{75}C_5} \bullet \frac{{}_1C_0 \cdot {}_{14}C_1}{{}_{15}C_1} = \frac{\frac{5!}{5!(5-5)!} \cdot \frac{70!}{0!(70-0)!}}{\frac{75!}{5!(75-5)!}} \bullet \frac{\frac{1!}{1!(1-1)!} \cdot \frac{14!}{1!(14-1)!}}{\frac{15!}{1!(15-1)!}}$$

$$= \frac{1 \cdot 1}{17259390} \bullet \frac{1 \cdot 14}{15} = \frac{14}{258,890,850} = 0.000000005408 = 5.41 \times 10^{-9}$$

Therefore, the Probability of winning at least $5,000 is (1 + 14 + 350)/258,890,850 = 365/258,890,850 = 0.000001410 = 1.41×10^{-6}.

(b) There are three ways to not win a prize: two white balls with no Mega Ball, one White Ball with no Mega Ball, and no white balls with no Mega Ball. The probability of each of these are calculated separately in the following.

$$\frac{{}_5C_2 \cdot {}_{70}C_3}{{}_{75}C_5} \bullet \frac{{}_1C_0 \cdot {}_{14}C_1}{{}_{15}C_1} = \frac{10 \cdot 54740}{17259390} \bullet \frac{1 \cdot 14}{15} = \frac{7,663,600}{258,890,850} = 0.0296$$

$$\frac{{}_5C_1 \cdot {}_{70}C_4}{{}_{75}C_5} \bullet \frac{{}_1C_0 \cdot {}_{14}C_1}{{}_{15}C_1} = \frac{5 \cdot 916895}{17259390} \bullet \frac{1 \cdot 14}{15} = \frac{64,182,650}{258,890,850} = 0.2479$$

$$\frac{{}_5C_0 \cdot {}_{70}C_5}{{}_{75}C_5} \bullet \frac{{}_1C_0 \cdot {}_{14}C_1}{{}_{15}C_1} = \frac{1 \cdot 12103014}{17259390} \bullet \frac{1 \cdot 14}{15} = \frac{169,442,196}{258,890,850} = 0.6545$$

Using the complementation rule, the probability that you win a prize would be 1 - (0.0296 + 0.2479 + 0.6545) = 0.0680.

4.269 Barack Obama received about 50% of the popular vote in Florida. Ten Floridians who voted are selected at random.

(a) The probability that exactly five voted for Obama is

$$\frac{{}_{10}C_5}{2^{10}} = \frac{252}{1024} = 0.246$$

(b) The required calculations for determining the approximate probability that at least eight voted for Obama are

$$\frac{{}_{10}C_8}{2^{10}}=\frac{45}{1024}=0.0439 \qquad \frac{{}_{10}C_9}{2^{10}}=\frac{10}{1024}=0.0097 \qquad \frac{{}_{10}C_{10}}{2^{10}}=\frac{1}{1024}=0.0010$$

The approximate probability that at least voted for Obama is the sum of the three items above, or 0.0546.

(c) The answers are exact for sampling with replacement. However, if you are selecting the voters you really are selecting without replacement. Since the population size is large relative to the sample size, the answers are excellent approximations when the sampling is without replacement.

4.271 (a) Given 365 days in a year and N students, the probability that at least two students have the same birthday is

$$1-\left[\frac{365!}{(365-N)!}/365^N\right]$$

(b) If the values for N = 2, 3, ..., 70 are successively substituted into the formula, we end up with a table giving the probability that at least two of the students in the class have the same birthday, for N = 2, 3, ..., 70.

N	Probability	N	Probability	N	Probability
2	0.003	25	0.569	48	0.961
3	0.008	26	0.598	49	0.966
4	0.016	27	0.627	50	0.970
5	0.027	28	0.654	51	0.974
6	0.040	29	0.681	52	0.978
7	0.056	30	0.706	53	0.981
8	0.074	31	0.730	54	0.984
9	0.095	32	0.753	55	0.986
10	0.117	33	0.775	56	0.988
11	0.141	34	0.795	57	0.990
12	0.167	35	0.814	58	0.992
13	0.194	36	0.832	59	0.993
14	0.223	37	0.849	60	0.994
15	0.253	38	0.864	61	9.995
16	0.284	39	0.878	62	0.996
17	0.315	40	0.891	63	0.997
18	0.347	41	0.903	64	0.997
19	0.379	42	0.914	65	0.998
20	0.411	43	0.924	66	0.998
21	0.444	44	0.933	67	0.998
22	0.476	45	0.941	68	0.999
23	0.507	46	0.948	69	0.999
24	0.538	47	0.955	70	0.999

Review Problems for Chapter 4

1. Probability theory enables us to control and evaluate the likelihood that a statistical inference is correct, and it provides the mathematical basis for statistical inference.

2. (a) The equal-likelihood model is used for computing probabilities when an experiment has N possible outcomes, all equally likely.

   (b) If an experiment has N equally likely outcomes and an event can occur in f ways, then the probability of the event is f/N.

3. In the frequentist interpretation of probability, the probability of an event is the relative frequency of the event in a large number of experiments.

4. The numbers (b) -0.047 and (c) 3.5 cannot be probabilities because probabilities must always be between 0 and 1.

5. The Venn diagram is a common graphical technique for portraying events and relationships among events.

6. Two or more events are mutually exclusive if no two of the events have any outcomes in common, i.e., no two of the events can occur at the same time.

7. (a) P(E)

   (b) P(E) = 0.436

8. (a) False. For any two events, the probability that one or the other occurs equals the sum of the two individual probabilities minus the probability that both events occur.

   (b) True. Either the event occurs or it does not. Therefore, the probability that the event occurs plus the probability that it does not occur equals 1. Thus the probability that it occurs is 1 minus the probability that it does not occur.

9. Frequently, it is quicker to compute the probability of the complement of an event than it is to compute the probability of the event itself. The complement rule allows one to use the probability of the complement to obtain the probability of the event itself.

10. (a) Data obtained by observing values of one variable of a population are called <u>univariate</u> data.

    (b) Data obtained by observing values of two variables of a population are called <u>bivariate</u> data.

    (c) A frequency distribution for bivariate data is called a <u>contingency table</u>.

11. The sum of the joint probabilities in a row or column of a joint probability distribution equals the <u>marginal</u> probability of that row or column.

12. (a) P(B|A)

    (b) A

13. Conditional probabilities can sometimes be computed directly as f/N or by using the formula P(B|A) = P(A&B)/P(A).

14. The joint probability of two independent events is the product of their marginal probabilities.

15. If two or more events have the property that at least one of them must occur when the experiment is performed, then the events are said to be <u>exhaustive</u>.

16. If r actions are to be performed in a definite order and there are m_1 possibilities for the first action, m_2 possibilities for the second action, ..., and m_r possibilities for the rth action, then there are $m_1 m_2 m_3 \ldots m_r$ possibilities altogether for the r actions.

17. Since A ,B, and C are mutually exclusive, the Special Addition Rule applies. P(A or B or C) = P(A) + P(B) +P(C) = 0.2 + 0.6 + 0.1 = 0.9.

18. Using the Complementation rule, P(E) = 1 - P(not E), we have P(E) = 1 - 0.4 = 0.6.

19. Since A and B are not mutually exclusive, the General Addition Rule applies. P(A or B) = P(A) + P(B) - P(A&B) = 0.2 + 0.6 - 0.1 = 0.7.

20. Apply the Conditional Probability Rule, which states $P(A|B) = \frac{P(A \& B)}{P(B)}$.

Plugging in the given values, we have $P(A|B) = \frac{0.1}{0.6} = \frac{1}{6} = 0.167$.

$P(B|A) = \frac{P(A \& B)}{P(A)} = \frac{0.1}{0.2} = \frac{1}{2} = 0.50$.

21. (a) A and B are not mutually exclusive because the P(A&B) is not equal to zero.

(b) A and B are independent because P(A&B) = P(A)P(B).

22. The General Multiplication Rule states for two events A and B, P(A&B) = P(A)*P(B|A). Plugging in the given values, we have P(A&B) = 0.4*0.7 = 0.28.

23. The Special Multiplication Rule (for Two Independent events) states that if A and B are independent events, then P(A&B) = P(A)*P(B). Plugging in the given values, we have P(A&B) = 0.3*0.6 = 0.18.

24. (a) The Rule of Total Probability states
$P(B) = P(B|A_1)P(A_1) + P(B|A_2)P(A_2) + P(B|A_3)P(A_3)$.

Plugging in the given values, we have
P(B) = 0.7*0.2 + 0.8*0.5 + 0.4*0.3 = 0.66.

Baye's Rule states $P(A_2 | B) = \frac{P(A_2)P(B | A_2)}{P(A_1)P(B | A_1) + P(A_2)P(B | A_2) + P(A_3)P(B | A_3)}$.

The denominator is P(B) and plugging in the given values into the numerator, we have $P(A_2 | B) = \frac{0.5 \cdot 0.8}{0.66} = 0.606$.

(b) P(B) is calculated in part (a) to be 0.66.

25. There are 3·4·2·3 = 72 possibilities altogether for the four actions.

26. 0! = 1, 1! = 1, 3! = 3·2·1 = 6, and 4! = 4·3·2·1 = 24

27. There are ${}_{10}C_3$ = 120 possible samples of size 3 from a population of size 10.

28. (a)

ABC	ACB	BAC	BCA	CAB	CBA
ABD	ADB	BAD	BDA	DAB	DBA
ACD	ADC	CAD	CDA	DAC	DCA
BCD	BDC	CBD	CDB	DBC	DCB

(b) ABC ABD ACD BCD

(c) $_4P_3 = 24$ $_4C_3 = 4$

(d) $_4P_3 = 4!/(4-3)! = 24$ $_4C_3 = 4!/3!1! = 4$

29. (a) The probability that a randomly selected house television is located in a bedroom is 0.24 + 0.19 = 0.43.

(b) Applying the complementation rule, the probability that the television is not located in a bedroom is 1 – 0.43 = 0.57.

(c) The probability that a randomly selected house television is located in the kitchen, dining room, basement, or garage is 0.05 + 0.05 = 0.10.

30. (a) $P(A) = \dfrac{26268}{142450} = 0.184$

(b) $P(D \text{ or } E \text{ or } F) = P(D) + P(E) + P(F) = \dfrac{14554 + 11087 + 30926}{142450} = \dfrac{56567}{142450} = 0.397$

(c)

Event	Probability
A	0.184
B	0.160
C	0.131
D	0.102
E	0.078
F	0.217
G	0.128

31. (a) (not J) is the event that the return selected shows an adjusted gross income of at least $100,000. There are 18,227 thousand such returns.

(b) (H & I) is the event that the return selected shows an adjusted gross income of between $20,000 and $50,000. There are 44,251 thousand such returns.

(c) (H or K) is the event that the return selected shows an adjusted gross income of at least $20,000. There are 93,404 thousand such returns.

(d) (H & K) is the event that the return selected shows an adjusted gross income of between $50,000 and $100,000. There are 30,926 thousand such returns.

32. (a) Events H and I are not mutually exclusive. Both have events C, D, and E in common.

(b) Events I and K are mutually exclusive. They have no events in common.

(c) Events H and (not J) are mutually exclusive. H = (C or D or E or F) while (not J) = G. H and (not J) have no events in common.

(d) Events H, (not J), and K are not mutually exclusive. While H and (not J) have no events in common, a part of K (i.e., event F) is common to event H, and the other part of K (i.e., event G) is common to event (not J).

33. (a)

$$P(H) = \frac{18610 + 14554 + 11087 + 30926}{142450} = \frac{75177}{142450} = 0.528$$

$$P(I) = 1 - \frac{30926 + 18227}{142450} = 1 - \frac{49153}{142450} = 0.655$$

$$P(J) = 1 - \frac{18227}{142450} = 0.872$$

$$P(K) = \frac{30926 + 18227}{142450} = \frac{49153}{142450} = 0.345$$

(b) H = (C or D or E or F)

I = (A or B or C or D or E)

J = (A or B or C or D or E or F)

K = (F or G)

(c) P(H) = P(C) + P(D) + P(E) + P(F) = 0.131 + 0.102 + 0.078 + 0.217 = 0.528

P(I) = P(A) + P(B) + P(C) + P(D) + P(E) = 0.184 + 0.160 + 0.131 + 0.102 + 0.078 = 0.655

P(J) = P(A) + P(B) + P(C) + P(D) + P(E) + P(F) = 0.184 + 0.160 + 0.131 + 0.102 + 0.078 + 0.217 = 0.872

P(K) = P(F) + P(G) = 0.217 + 0.128 = 0.345

34. (a)

$$P(not\ J) = \frac{18227}{142450} = 0.128$$

$$P(H\ \&\ I) = \frac{44251}{142450} = 0.311$$

$$P(H\ or\ K) = \frac{93404}{142450} = 0.656$$

$$P(H\ \&\ K) = \frac{30926}{142450} = 0.217$$

(b) P(J) = 1 - P(not J) = 1 - 0.128 = 0.872

(c) P(H or K) = P(H) + P(K) - P(H & K) = 0.528 + 0.345 - 0.217 = 0.656

(d) The answer in (c) agrees with that in (a).

35. (a) The contingency table has six cells.

(b) There are 15,980 thousand students in high school.

(c) There are 66,947 thousand students in public schools.

(d) There are 6,059 thousand students in private colleges.

36. (a) (i) L_3 = the student selected is in college;

(ii) T_1 = the student selected attends a public school;

(iii) (T_1 & L_3) = the student selected attends a public college.

(b)

$$P(L_3) = \frac{21968}{78182} = 0.281$$

$$P(T_1) = \frac{66947}{78182} = 0.856$$

$$P(T_1 \ \& \ L_3) = \frac{15909}{78180} = 0.203$$

The interpretation of each result above as a percentage is as follows: 28.1% of students attend college, 85.6% attend public schools, and 20.3% attend public colleges.

(c)

	Public T_1	Private T_2	$P(L_i)$
Elementary L_1	0.463	0.052	0.515
High School L_2	0.190	0.014	0.204
College L_3	0.203	0.077	0.281
$P(T_j)$	0.856	0.144	1.000

(d) $P(T_1 \text{ or } L_3) = (36161 + 14877 + 15909 + 6059)/78182 = 0.934$

(e) $P(T_1 \text{ or } L_3) = P(T_1) + P(L_3) - P(T_1 \ \& \ L_3) = 0.856 + 0.281 - 0.203 = 0.934$.

(f) The answers are the same in parts (d) and (e).

37. (a) $P(L_3 \mid T_1) = \dfrac{15909}{66947} = 0.238$

23.8% of students attending public schools are in college.

(b) $P(L_3 \mid T_1) = \dfrac{P(L_3 \& T_1)}{P(T_1)} = \dfrac{0.203}{0.856} = 0.237$

(c) The answers are very close in parts (a) and (b). They are slightly different because in part (b) we did some intermediate rounding. Therefore, the discrepancy is due to roundoff error.

38. (a) $P(T_2) = \dfrac{11235}{78182} = 0.144; P(T_2 \mid L_2) = \dfrac{1103}{15980} = 0.069$

(b) No. For L_2 and T_2 to be independent, $P(T_2|L_2)$ must equal $P(T_2)$. From part (a), however, we see that this is not the case. In terms of percentages, each of these probabilities, respectively, means that 6.9% of high school students attend private schools, whereas 14.4% of all students attend private schools.

(c) Events L_2 and T_2 are not mutually exclusive. From Table 4.17, we see that both events can occur simultaneously. There are 1,103 thousand students who attend private high schools.

39. Let MA1, MP1, and MS1 denote, respectively, the events that the first student selected received a master of arts, a master of public administration, and a master of science; and let MA2, MP2, and MS2 denote, respectively, the events that the second student selected received a master of arts, a master of public administration, and a master of science.

(a) $P(MA1 \ \& \ MS2) = \frac{3}{50} \cdot \frac{19}{49} = 0.023$

(b) $P(MP1 \ \& \ MP2) = \frac{28}{50} \cdot \frac{27}{49} = 0.309$

(c)

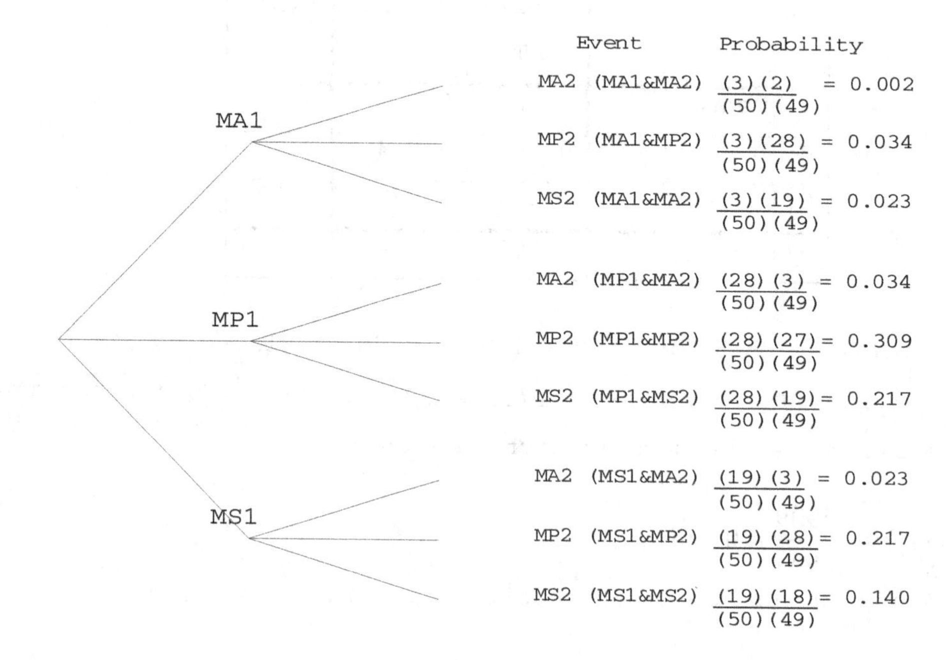

(d) P(MA1 & MA2) + P(MP1 & MP2) + P(MS1 & MS2)

= 0.002 + 0.309 + 0.140 = 0.451.

40. Let F = event that pairs raised their offspring to the Fledgling stage, and D = event that pairs divorced. The given information is

P(D) = 0.63, P(D|F) = 0.81, P(D|not F) = 0.43.

(a) We are trying to find P(Offspring died) = P(not F). The total probability rule states that

P(D) = P(D|F)P(F) + P(D|not F)P(not F) or

P(D) = P(D|F)P(F) + P(D|not F)[1 - P(F)]. Substituting known values,

0.63 = (0.81)P(F) + (0.43)[1 - P(F)] or

0.20 = 0.38 P(F). Thus

P(F) = 0.20/0.38 = 0.526. Therefore, P(not F) = 1 - 0.526 = 0.474 or 47.4%.

(b) We can use the multiplication rule to find

P(D & not F) = P(D|not F)P(not F) = (0.43)(0.474) = 0.204 or 20.4%

Thus 20.4% divorced and had offspring that died.

(c) P(not F|D) = P(D & not F)/P(D) = (0.204)/(0.63) = 0.324

Thus, among those that divorced, 32.4% had offspring that died.

41. (a) Let C_i be the event that the ith man selected is color blind.

P(none color blind) = P{(not C_1)& (not C_2)& (not C_3)& (not C_4)} =

P(not C_1)P(not C_2)P(not C_3)P(not C_4) = 0.91^4 = 0.686

(b) P{(not C_1) & (not C_2) & (not C_3) & (C_4)} =

P(not C_1)P(not C_2)P(not C_3)P(C_4) = 0.91· 0.91· 0.91· 0.09 = 0.068

(c) There are four sequences like the one in part (b) in which the fourth man is color blind. The others have the first man color blind and the others not; the second color blind and the others not; and the third color blind and the others not. Each sequence has the same probability as the one in part (b). Thus, P(exactly one man is color blind) is 4 $(0.91)^3 \cdot (0.09)$ = 0.271.

42. Let: A1 = 18-24 years old

A2 = 25-34 years old

A3 = 35-44 years old

A4 = 45-54 years old

A5 = 55-64 years old

A6 = 65+ years old

B = Driver has a smartphone

(a) P(B|A1) = 0.79

(b)

$$P(B) = \sum P(B \mid A_i)\, P(A_i)$$
$$= (0.79)(0.125) + (0.81)(0.180) + \ldots + (0.18)(0.189) = 0.554$$

(c)

P(A1|B) = P(A1 & B)/P(B) = P(B|A1)P(A1)/P(B) = (0.79)(0.125)/0.554 = 0.178

(d) Thus 79% of U.S. adults who are aged 18-24 own a smartphone; 55.4% of all U.S. adults own a smartphone; Of those U.S. adults that own a smartphone, 17.8% are aged 18-24.

43. (a) $_{12}C_2 = \dfrac{12!}{2!10!} = \dfrac{12 \cdot 11}{2 \cdot 1} = 66$

(b) $_{12}P_3 = \dfrac{12!}{(12-3)!} = 12 \cdot 11 \cdot 10 = 1320$

(c) (i) $_8C_2 = \dfrac{8!}{2!6!} = \dfrac{8 \cdot 7}{2 \cdot 1} = 28$

(ii) $${}_8P_3 = \frac{12!}{(8-3)!} = 8 \cdot 7 \cdot 6 = 336$$

44. (a) $${}_{52}C_{13} = \frac{52!}{13!\,39!} = 635{,}013{,}559{,}600$$

(b) $$\frac{{}_4C_2\;{}_{48}C_{11}}{{}_{52}C_{13}} = \frac{\frac{4!}{2!(4-2)!} \cdot \frac{48!}{11!(48-11)!}}{\frac{52!}{13!(52-13)!}} = 0.213$$

(c) With four choices for the eight-card suit, three choices for the four-card suit, and two choices for the one-card suit, the probability of being dealt an 8-4-1 distribution is

$$4 \cdot 3 \cdot 2 \cdot \frac{{}_{13}C_8\;{}_{13}C_4\;{}_{13}C_1\;{}_{13}C_0}{{}_{52}C_{13}} = 24 \cdot \frac{\frac{13!}{8!(13-8)!} \cdot \frac{13!}{4!(13-4)!} \cdot \frac{13!}{1!(13-1)!} \cdot \frac{13!}{0!(13-13)!}}{635{,}013{,}559{,}600}$$

$$= 24 \cdot \frac{1287 \cdot 715 \cdot 13 \cdot 1}{635{,}013{,}559{,}600} = 24 \cdot \frac{11{,}962{,}665}{635{,}013{,}559{,}600} = 0.00045$$

(d) Initially, two suits from among the four are to be selected, from which five cards from each suit are drawn. From the remaining two suits, one is to be selected from which two cards are drawn. Finally, only one suit remains from which to draw the final card. The probability of being dealt a 5-5-2-1 distribution is

$${}_4C_2 \cdot {}_2C_1 \cdot {}_1C_1 \cdot \frac{{}_{13}C_5\;{}_{13}C_5\;{}_{13}C_2\;{}_{13}C_1}{{}_{52}C_{13}} = 12 \cdot \frac{\frac{13!}{5!(13-5)!} \cdot \frac{13!}{5!(13-5)!} \cdot \frac{13!}{2!(13-2)!} \cdot \frac{13!}{1!(13-1)!}}{635{,}013{,}559{,}600}$$

$$= 12 \cdot \frac{1287 \cdot 1287 \cdot 78 \cdot 13}{635{,}013{,}559{,}600} = 12 \cdot \frac{1{,}679{,}558{,}166}{635{,}013{,}559{,}600} = 0.032$$

(e) The probability of being dealt a hand void in a specified suit is

$$\frac{{}_{39}C_{13}\;{}_{13}C_0}{{}_{52}C_{13}} = \frac{\frac{39!}{13!(39-13)!} \cdot \frac{138!}{0!(39-0)!}}{\frac{52!}{13!(52-13)!}} = 0.013$$

45. $${}_{64}C_8 = \frac{64!}{8!56!} = 4{,}426{,}165{,}368$$

46. (a) We would need to assume that people don't own a DVD player unless they also have a TV set. This is a reasonable assumption since the DVD is almost useless without a TV set.

(b) $P(DVD) = P(DVD \& TV) = P(TV)P(DVD|TV) = (0.986)(0.852) = 0.840$ or 84.0%.

(c) We would need to know how many (or the percentage of) households have a DVD player, but don't have a TV.

CHAPTER 5 SOLUTIONS

Exercises 5.1

5.1 (a) probability

(b) probability

5.3 The notation {X=3} denotes an event that occurs when X=3, whereas P(X=3) denotes the probability that the event {X=3} will occur. Another way of thinking about the difference is that the first is an event and the second is a number between zero and one.

5.5 This table will resemble the probability distribution of the random variable.

5.7 (a)

x	Probability
1	0.30
2	0.40
3	0.30

(b) $(X = 2)$ is the event that X takes on the value 2, $(X \leq 2)$ is the event that X takes on a value of at most 2, and $(X > 2)$ is the event that X takes on a value greater than 2.

(c) $P(X = 2) = 0.40$, $P(X \leq 2) = 0.70$, $P(X > 2) = 0.30$
The probability that X takes on a value of 2 is 0.40. The probability that X takes on a value of at most 2 is 0.70. The probability that X takes on a value more than 2 is 0.30.

(d)

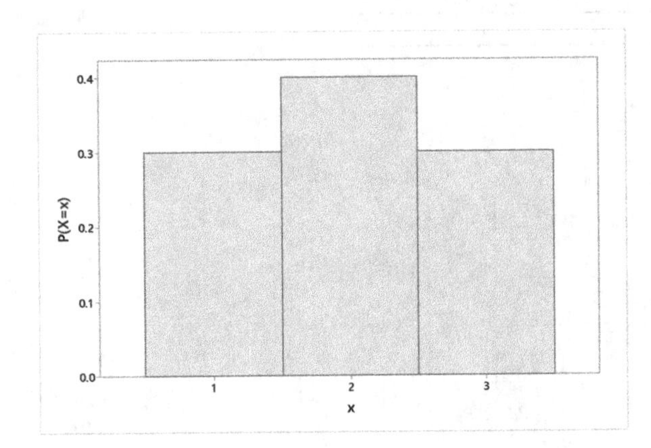

5.9 (a)

y	Probability
0	0.36
1	0.28
4	0.16
6	0.20

(b) $(Y = 3)$ is the event that Y takes on the value 3, $(Y < 3)$ is the event that Y takes on a value less than 3, and $(Y \geq 3)$ is the event that Y takes on a value of at least 3.

(c) $P(Y = 3) = 0$, $P(Y < 3) = 0.64$, $P(Y \geq 3) = 0.36$

The probability that Y takes on the value of 3 is 0. The probability that Y takes on a value less than 3 is 0.64. The probability that Y takes on a value of at least 3 is 0.36

(d)

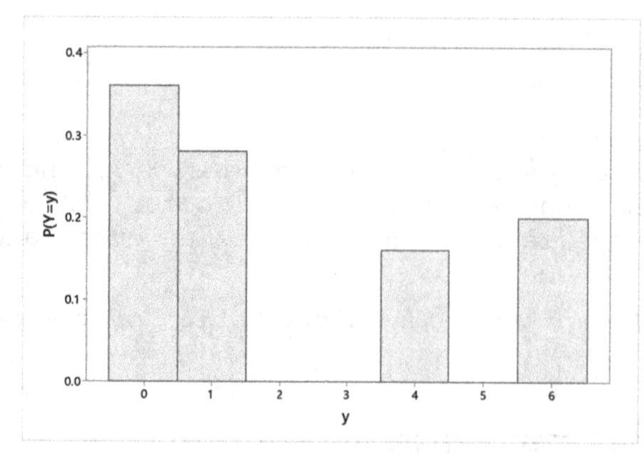

5.11 (a) X = 2, 4, 5, 6, 7, 8

(b) {X = 7}

(c) The total number of shuttle missions is 135. P(X = 4) = 3/135 = 0.022, so 2.2% of the shuttle crews consist of exactly four people.

(d)

Size of Crew	Probability
2	0.030
3	0.000
4	0.022
5	0.267
6	0.207
7	0.467
8	0.007

(e)

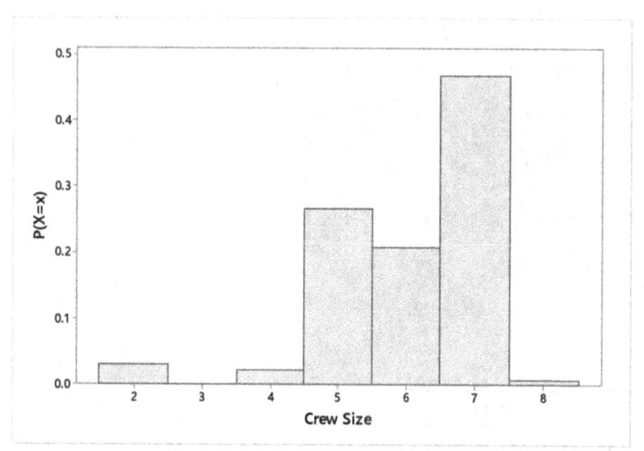

5.13 (a) $\{Y \geq 1\}$ (b) $\{Y = 3\}$ (c) $\{2 \leq Y \leq 4\}$

(d) $P(Y \geq 1\} = P(Y = 1) + P(Y = 2) + P(Y = 3) + P(Y = 4) + P(Y = 5) + P(Y = 6) + P(Y = 7) + P(Y = 8)$

$= 0.296 + 0.266 + 0.093 + 0.049 + 0.056 + 0.037 + 0.012 + 0.006 = 0.815$

(e) $P(Y = 3) = 0.093$

(f) $P(2 \leq y \leq 4) = P(Y = 2) + P(Y = 3) + P(Y = 4) = 0.266 + 0.093 + 0.049 = 0.408$

5.15 (a) Y = 2,3,4,5,6,7,8,9,10,11,12

(b) {Y = 7}

(c) $P(Y = 7) = 6/36 = 1/6 = 0.167$

(d)

Y	P(Y)
2	1/36
3	2/36
4	3/36
5	4/36
6	5/36
7	6/36
8	5/36
9	4/36
10	3/36
11	2/36
12	1/36

(e)

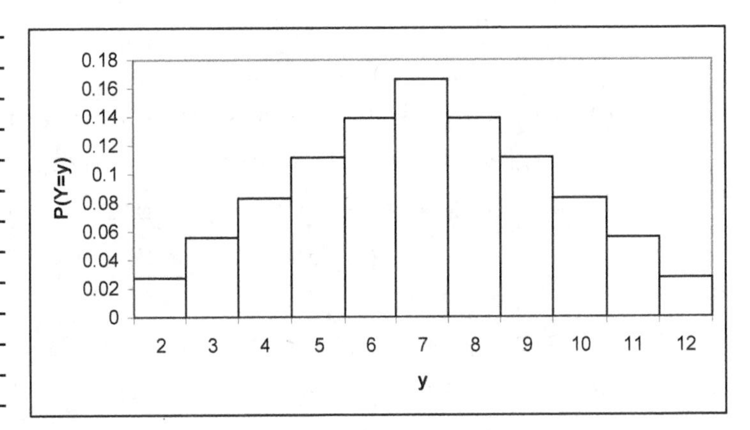

5.17 (a) The area of the square is $6^2 = 36$ square feet. The area of the bull's eye is $\pi(1)^2 = 3.1416$ square feet. The area of the ring is the area of a circle with radius 2 less the area of the bull's eye, or $\pi(2)^2 - \pi(1)^2 = 12.5664 - 3.1416 = 9.4248$. The remaining area of the square outside the ring has area $36 - 12.5664 = 23.4336$. Thus, $P(S = 10) = 3.1416/36 = 0.087$; $P(S = 5) = 9.4248/36 = 0.262$; and $P(S = 0) = 23.3226/36 = 0.651$. These are shown in the table below.

S	P(S = s)
0	0.651
5	0.262
10	0.087

Assuming that all points in the six foot square are equally likely, the archer has almost a 2/3 chance of scoring 0 points and about a 1/12 chance of scoring 10 points.

(b) $P(S = 5) = 0.262$

$P(S > 0) = P(S = 5 \text{ or } 10) = 0.262 + 0.087 = 0.349$

$P(S \leq 7) = P(S = 0 \text{ or } 5) = 0.651 + 0.262 = 0.913$

$P(5 < S \leq 15) = P(S = 10) = 0.087$

$P(S < 15) = P(S = 0 \text{ or } 5 \text{ or } 10) = 1$; This event always occurs.

$P(S < 0) = 0.000$; This event cannot occur.

5.19 (a) The probabilities are only an estimate of the probability distribution, because the information was gathered from a sample.

(b) Let X = the number of young per litter for female Florida black bears. The probability that X is 2 or 3 is $P(X = 2) + P(X = 3) = 0.606 + 0.152 = 0.758$.

5.21 $P(Z \leq 1.96) + P(Z > 1.96) = 1$.

Since $P(Z > 1.96) = 0.025$,

$P(Z \leq 1.96) + 0.025 = 1$, or

$P(Z \leq 1.96) = 1 - 0.025 = 0.975$.

5.23 (a) $P(X \le c) + P(X > c) = 1.$

Since $P(X > c) = \alpha$,

$P(X \le c) + \alpha = 1$, or

$P(X \le c) = 1 - \alpha.$

(b) $P(Y < -c) + P(-c \le Y \le c) + P(Y > c) = 1.$

Since $P(Y < -c) = P(Y > c) = \alpha/2$,

$\alpha/2 + P(-c \le Y \le c) + \alpha/2 = 1$, or

$P(-c \le Y \le c) = 1 - \alpha/2 - \alpha/2 = 1 - \alpha.$

(c) $P(T < -c) + P(-c \le T \le c) + P(T > c) = 1.$

Since $P(-c \le T \le c) = 1 - \alpha$,

$P(T < -c) + (1 - \alpha) + P(T > c) = 1$, or

$P(T < -c) + P(T > c) = 1 - 1 + \alpha = \alpha.$

Since $P(T < -c) = P(T > c)$,

$2 \cdot [P(T > c)] = \alpha$ or $P(T > c) = \alpha/2.$

Exercises 5.2

5.25 The mean of a discrete random variable generalizes the concept of a population mean.

5.27 The required calculations are

x	P(X=x)	xP(X=x)	x^2	X^2 P(X=x)
1	0.3	0.3	1.0	0.3
2	0.4	0.8	4.0	1.6
3	0.3	0.9	9.0	2.7
		2.0		4.6

(a) $\mu_x = \sum xP(X=x) = 2.0$

(b) $\sigma = \sqrt{\sum x^2 P(X=x) - \mu_x^2} = \sqrt{4.6 - 2.0^2} = 0.8$

5.29 The required calculations are

x	P(X=x)	xP(X=x)	x^2	X^2 P(X=x)
0	0.36	0.00	0.0	0.00
1	0.28	0.28	1.0	0.28
4	0.16	0.64	16.0	2.56
6	0.20	1.20	36.0	7.20
		2.12		10.04

(a) $\mu_x = \sum xP(X=x) = 2.12$

(b) $\sigma = \sqrt{\sum x^2 P(X=x) - \mu_x^2} = \sqrt{10.04 - 2.12^2} = 2.35$

5.31 The required calculations are

x	P(X=x)	xP(X=x)	x²	X² P(X=x)
2	0.030	0.060	4.000	0.120
4	0.022	0.088	16.000	0.352
5	0.267	1.335	25.000	6.675
6	0.207	1.242	36.000	7.452
7	0.467	3.269	49.000	22.883
8	0.007	0.056	64.000	0.448
		6.050		37.930

(a) μ_x = $\sum$xP(X=x) = 6.050. The average number of persons in a shuttle crew is about 6.05.

(b) $\sigma = \sqrt{\sum x^2 P(X=x) - \mu_x^2} = \sqrt{37.930 - 6.05^2} = 1.15$

(c)

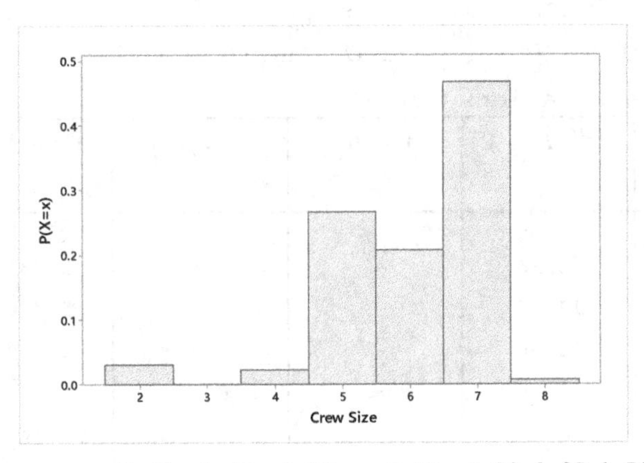

2.60	3.75	4.90	6.05	7.20	8.35	9.50
μ-3σ	μ-2σ	u-σ	μ	μ+σ	μ+2σ	μ+3σ

5.33 The required calculations are

y	P(Y=y)	yP(Y=y)	y²	y²P(Y=y)
0	0.185	0.000	0	0.000
1	0.296	0.296	1	0.296
2	0.266	0.532	4	1.064
3	0.093	0.279	9	0.837
4	0.049	0.196	16	0.784
5	0.056	0.280	25	1.400
6	0.037	0.222	36	1.332
7	0.012	0.084	49	0.588
8	0.006	0.048	64	0.384
	1	1.937		6.685

μ_y = $\sum$yP(Y=y) = 1.937. The mean number of major hurricanes per year is 1.934.

(b) $\sigma=\sqrt{\sum y^2 P(Y=y)-\mu_y^2}=\sqrt{6.685-1.937^2}=1.713$. Roughly speaking, the number of major hurricane in a given year is about 1.7 away from the mean of 1.937.

(c)

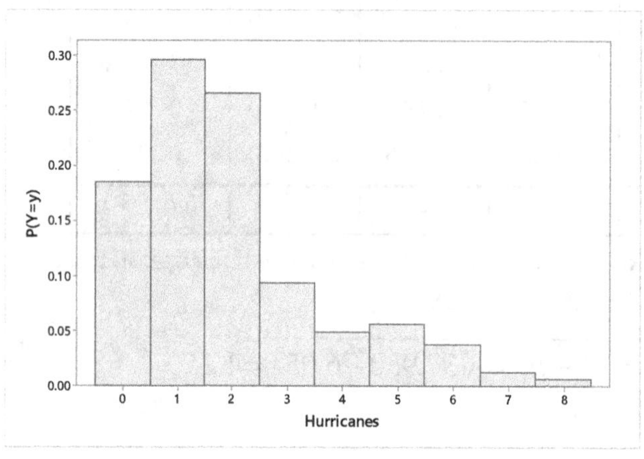

-3.20	-1.49	0.22	1.94	3.65	5.36	7.08
$\mu-3\sigma$	$\mu-2\sigma$	$u-\sigma$	μ	$\mu+\sigma$	$\mu+2\sigma$	$\mu+3\sigma$

5.35 The required calculations are

y	$P(Y=y)$	$yP(Y=y)$	$y-\mu_y$	$(y-\mu_y)^2$	$(y-\mu_y)^2 P(Y=y)$	y^2	$y^2P(Y=y)$
2	1/36	2/36	-5	25	25/36	4	4/36
3	2/36	6/36	-4	16	32/36	9	18/36
4	3/36	12/36	-3	9	27/36	16	48/36
5	4/36	20/36	-2	4	16/36	25	100/36
6	5/36	30/36	-1	1	5/36	36	180/36
7	6/36	42/36	0	0	0/36	49	294/36
8	5/36	40/36	1	1	5/36	64	320/36
9	4/36	36/36	2	4	16/36	81	324/36
10	3/36	30/36	3	9	27/36	100	300/36
11	2/36	22/36	4	16	32/36	121	242/36
12	1/36	12/36	5	25	25/36	144	144/36
		252/36=7			210/36		1,974/36

(a) $\mu_y = \sum$ yP(Y=y) = 252/36 = 7. When rolling two balanced dice, the mean of their sum is 7.

(b) $\sigma=\sqrt{\sum(y-\mu_y)^2 P(Y=y)}=\sqrt{210/36}=2.415$

$$\sigma=\sqrt{\sum y^2 P(Y=y)-\mu_y^2}=\sqrt{\frac{1974}{36}-7^2}=2.415$$

(c)

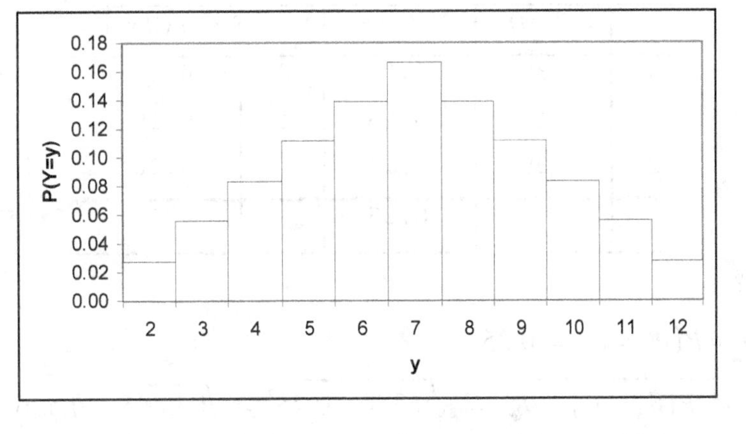

-0.2	2.2	4.6	7	9.4	11.8	14.2
$\mu-3\sigma$	$\mu-2\sigma$	$\mu-\sigma$	μ	$\mu+\sigma$	$\mu+2\sigma$	$\mu+3\sigma$

5.37 (a) The necessary calculations are

s	P(S=s)	sP(S=s)	s^2	s^2P(S=s)
0	0.651	0.000	0	0.000
5	0.262	1.310	25	6.550
10	0.087	0.870	100	8.700
		2.180		15.250

On average, the archer will score 2.180 points per arrow shot.

(b) $\sigma = \sqrt{\sum s^2 P(S=s) - \mu_s^2} = \sqrt{15.250 - 2.180^2} = 3.24$. Roughly speaking, on average, the number of points on a shot will differ from the mean by 3.24 points.

5.39 (a)

$$P(X=1) = \frac{18}{38} = 0.474$$

$$P(X=-1) = \frac{20}{38} = 0.526; \quad 0.474 + 0.526 = 1.000$$

(b) $\mu = \sum x \cdot P(X=x) = (1)(0.474) + (-1)(0.526) = -0.052$

(c) On the average, you will lose 5.2¢ per play.

(d) If you bet $1 on red 100 times, you can expect to lose 100 · 0.052 = $5.20. If you bet $1 on red 1000 times, you can expect to lose $52.00.

(e) Roulette is not a profitable game for a person to play. Parts (c) and (d) demonstrate that, no matter how much you play, you can expect to lose. Also, part (a) shows that a higher probability is associated with losing rather than winning.

5.41 (a) $\mu = \sum x \cdot P(X=x) = (0)(0.95) + (10)(0.045) + (50)(0.004) + 100(0.0009) + (200)(0.0001) = 0.76$. The expected annual claim amount per homeowner is 0.76 thousand dollars, or $760.

(b) They want an average net profit of $50 per policy, so they need to charge $50 more than the expected claim amount of $760. Therefore, they should charge $810.

5.43

w	P(W=w)	w·P(w)	w^2	w^2·P(W=w)
0	0.80	0.00	0	0.00
1	0.15	0.15	1	0.15
2	0.05	0.10	4	0.20
		0.25		0.35

(a)

$$\mu_w = \sum wP(W = w) = 0.25$$

$$\sigma_w = \sqrt{w^2\ P(W = w) - \mu_w^2} = \sqrt{0.350 - 0.25^2} = \sqrt{0.2875} = 0.536$$

(b) On the average, there are 0.25 breakdowns per day.

(c) Assuming 250 workdays per year, the number of breakdowns expected per year is (0.25)(250) = 62.5.

5.45 (a) The necessary calculations are

y	P(Y=y)	yP(Y=y)	y^2	y^2P(Y=y)
0	0.424	0.000	0	0.000
1	0.161	0.161	1	0.161
2	0.134	0.268	4	0.536
3	0.111	0.333	9	0.999
4	0.093	0.372	16	1.488
5	0.077	0.385	25	1.925
		1.519		5.109

The mean of Y, the mean number of customers waiting is 1.519. On the average, there are about 1.5 customers waiting at any given time.

(b) In a large number of independent observations, 1.519 customers will be waiting, on average.

(c) We will use Minitab to simulate these observations. Enter the values of Y in column C1 and their probabilities in column C2. Then select **Calc ▶ Random data ▶ Discrete,** enter 500 in the **Generate Rows of data** text box, enter Customers in the **Store in Column(s)** text box, then enter C1 in the **Values in** text box and C2 in the **Probabilities in** text box and click **OK**. This will produce 500 random values in a column named Customeers. Now select **Stat ▶ Basic Statistics ▶ Display Descriptive Statistics,** enter Customers in the **Variables** text box and click **OK**. The result appears in the Sessions Window. Our result is

```
Variable     N  N*    Mean  SE Mean   StDev      Minimum           Q1  Median
Customers  500   0  1.5800   0.0767  1.7151  0.000000000  0.000000000  1.0000

Variable        Q3  Maximum
Customers   3.0000   5.0000
```

(d) The mean for our observations was 1.58, a little bit higher than the exact mean of 1.519. Your data and mean will likely be different.

(e) Part (d) illustrates that simulations yield sample means that are close to the population mean.

5.47 (a)

		y			
		0	1	2	P(X=x)
x	0	0.6400	0.1200	0.0400	0.8000
	1	0.1200	0.0225	0.0075	0.1500
	2	0.0400	0.0075	0.0025	0.0500
	P(Y=y)	0.8000	0.4500	0.0500	1.0000

(b)

x + y	P(x+y)	(x+y)·P(x+y)	$(x + y)^2$	$(x+y)^2 \cdot P(x+y)$
0	0.6400	0.000	0	0.000
1	0.2400	0.240	1	0.240
2	0.1025	0.205	4	0.410
3	0.0150	0.045	9	0.135
4	0.0025	0.010	16	0.040
		0.500		0.825

Columns 1 and 2 of the preceding table comprise the probability distribution of the random variable X + Y. We have used P(x+y) for P(X+Y=x+y) in the table headings.

Thus the table to be completed is

u	0	1	2	3	4
P(X + Y) = u	0.6400	0.2400	0.1025	0.0150	0.0025

(c) Column 3 of the table above provides the calculations for the mean: $\mu_{x+y} = 0.500$. Columns 4 and 5 provide the calculations for the variance:

$$\sigma_{x+y}^2 = \sum (x+y)^2 P(X+Y=x+y) - \mu_{x+y}^2 = 0.825 - (0.50)^2 = 0.5750$$

(d) Using the result from Exercise 5.43 and the information about X and Y in this part of the exercise, we have

$\mu_w = \mu_x = \mu_y = 0.25$; $\sigma_w^2 = \sigma_x^2 = \sigma_y^2 = 0.2875$.

Now, $\mu_{x+y} = \mu_x + \mu_y = 0.25 + 0.25 = 0.50$

and $\sigma_{x+y}^2 = \sigma_x^2 + \sigma_y^2 = 0.2875 + 0.2875 = 0.5750$

Notice that these two results match those obtained in part (c).

(e) The mean of the sum of two random variables equals the sum of their means. The variance of the sum of two *independent* random variables equals the sum of their variances.

Exercises 5.3

5.49 Each repetition of an experiment is called a trial.

5.51 The binomial coefficient equals the number of outcomes that contain exactly x successes in n Bernoulli trials.

5.53 The binomial distribution is the probability distribution for the number of successes in a sequence of Bernoulli trials.

5.55 (1) Randomly selected pieces of identical rope are tested by subjecting each to a 1000 pound force. Each rope either breaks or it doesn't. (2) People are selected at random by ticket numbers at a large convention and their gender is noted. Each is either male or female.

5.57 $3! = 3 \cdot 2 \cdot 1 = 6$; $7! = 7 \cdot 6 \cdot 5 \cdot 4 \cdot 3 \cdot 2 \cdot 1 = 5{,}040$;

$8! = 8 \cdot 7 \cdot 6 \cdot 5 \cdot 4 \cdot 3 \cdot 2 \cdot 1 = 40{,}320$; $9! = 9 \cdot 8 \cdot 7 \cdot 6 \cdot 5 \cdot 4 \cdot 3 \cdot 2 \cdot 1 = 362{,}880$

5.59 (a) $\binom{5}{2} = \frac{5!}{2!(5-2)!} = \frac{5 \cdot 4 \cdot 3 \cdot 2 \cdot 1}{2 \cdot 1 \cdot 3 \cdot 2 \cdot 1} = 10$

(b) $\binom{7}{4} = \frac{7!}{4!(7-4)!} = \frac{7 \cdot 6 \cdot 5 \cdot 4 \cdot 3 \cdot 2 \cdot 1}{4 \cdot 3 \cdot 2 \cdot 1 \cdot 3 \cdot 2 \cdot 1} = 35$

(c) $\binom{10}{3} = \frac{10!}{3!(10-3)!} = \frac{10 \cdot 9 \cdot 8 \cdot 7 \cdot 6 \cdot 5 \cdot 4 \cdot 3 \cdot 2 \cdot 1}{3 \cdot 2 \cdot 1 \cdot 7 \cdot 6 \cdot 5 \cdot 4 \cdot 3 \cdot 2 \cdot 1} = 120$

(d) $\binom{12}{5} = \frac{12!}{5!(12-5)!} = \frac{12 \cdot 11 \cdot 10 \cdot 9 \cdot 8 \cdot 7 \cdot 6 \cdot 5 \cdot 4 \cdot 3 \cdot 2 \cdot 1}{5 \cdot 4 \cdot 3 \cdot 2 \cdot 1 \cdot 7 \cdot 6 \cdot 5 \cdot 4 \cdot 3 \cdot 2 \cdot 1} = 792$

5.61 (a) $\binom{4}{1} = \frac{4!}{1!(4-1)!} = \frac{4 \cdot 3 \cdot 2 \cdot 1}{1 \cdot 3 \cdot 2 \cdot 1} = 4$

(b) $\binom{6}{2} = \frac{6!}{2!(6-2)!} = \frac{6 \cdot 5 \cdot 4 \cdot 3 \cdot 2 \cdot 1}{2 \cdot 1 \cdot 4 \cdot 3 \cdot 2 \cdot 1} = 15$

(c) $\binom{8}{3} = \frac{8!}{3!(8-3)!} = \frac{8 \cdot 7 \cdot 6 \cdot 5 \cdot 4 \cdot 3 \cdot 2 \cdot 1}{3 \cdot 2 \cdot 1 \cdot 5 \cdot 4 \cdot 3 \cdot 2 \cdot 1} = 56$

(d) $\binom{9}{6} = \frac{9!}{6!(9-6)!} = \frac{9 \cdot 8 \cdot 7 \cdot 6 \cdot 5 \cdot 4 \cdot 3 \cdot 2 \cdot 1}{6 \cdot 5 \cdot 4 \cdot 3 \cdot 2 \cdot 1 \cdot 3 \cdot 2 \cdot 1} = 84$

5.63 (a) Since the graph is symmetric, $p = 0.5$.

(b) Since the graph is right skewed, p is less than 0.5.

5.65 (a) Each trial consists of observing whether the child is cured of the pinworm infestation by pyrantel pamoate. Each is either cured or not cured. The results are independent assuming that the children are not in the same families. Since a success s is that the child is cured, the success probability p is 0.90.

(b)

Outcome	Probability
sss	(0.90)(0.90)(0.90)=0.729
ssf	(0.90)(0.90)(0.10)=0.081
sfs	(0.90)(0.10)(0.90)=0.081
sff	(0.90)(0.10)(0.10)=0.009
fss	(0.10)(0.90)(0.90)=0.081
fsf	(0.10)(0.90)(0.10)=0.009
ffs	(0.10)(0.10)(0.90)=0.009
fff	(0.10)(0.10)(0.10)=0.001

(c)

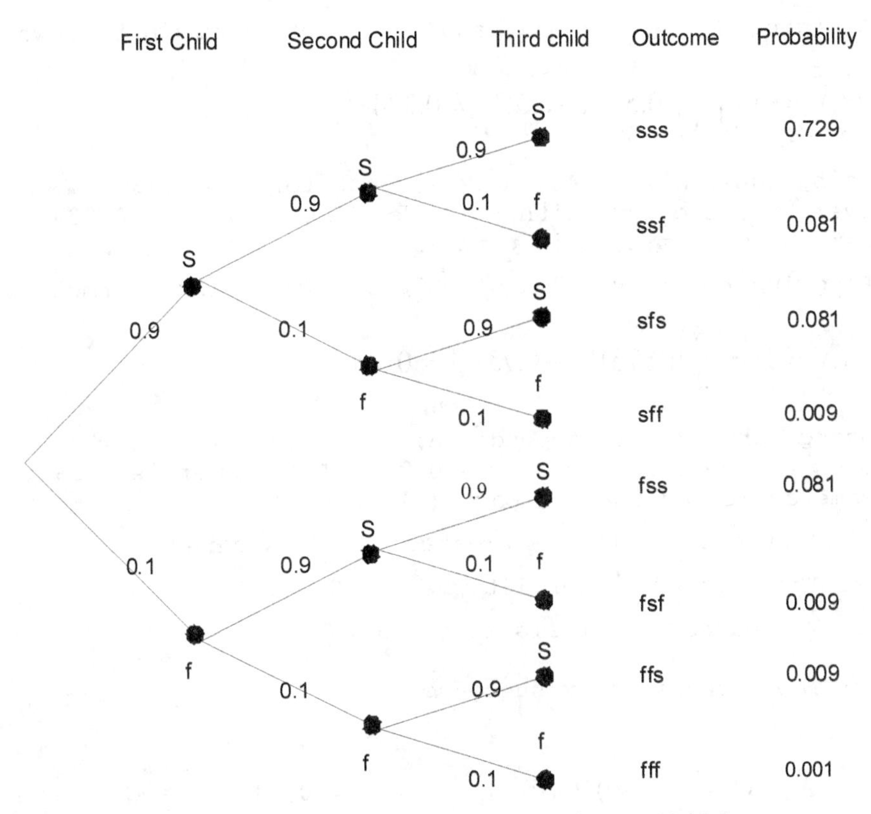

(d) The outcomes in which exactly two of the three children are cured are ssf, sfs, and fss.

(e) Each outcome in part (d) has the probability 0.081. This probability is the same for each outcome because each probability is obtained by multiplying two success probabilities of 0.90 and one failure probability of 0.10.

(f) P(exactly two children are cured) = P(ssf) + P(sfs) + P(fss)

= 0.081 + 0.081 + 0.081 = 0.243

(g) P(exactly one child is cured) = P(sff) + P(fsf) + P(ffs)

= 0.009 + 0.009 + 0.009 = 0.027

P(exactly three children are cured) = P(sss) = 0.729, and

P(exactly zero children are cured) = P(fff) = 0.001.

Thus the complete probability distribution of X, the number of children out of three that are cured is

X	0	1	2	3
P(X=x)	0.001	0.027	0.243	0.729

5.67 (a) Plugging n = 4, p = 0.3, and x = 2 into formula 5.1, we have

$$P(X=2)=\binom{4}{2}(0.3)^2(1-0.3)^{4-2}=0.265.$$

(b) Using Table XII in Appendix A, look for n = 4, x = 2, and then navigate over to the column with p = 0.3. The answer is 0.265. This is the same answer as that in part (a).

5.69 (a) Plugging n = 6, p = 0.5, and x = 4 into formula 5.1, we have

$$P(X=4)=\binom{6}{4}(0.5)^4(1-0.5)^{6-4}=0.234.$$

(b) Using Table XII in Appendix A, look for n = 6, x = 4, and then navigate over to the column with p = 0.5. The answer is 0.234. This is the same answer as that in part (a).

5.71 (a) Plugging n = 5, p = 3/4 = 0.75, and x = 4 into formula 5.1, we have

$$P(X=4)=\binom{5}{4}(0.75)^4(1-0.75)^{5-4}=0.396.$$

(b) Using Table XII in Appendix A, look for n = 5, x = 4, and then navigate over to the column with p = 0.75. The answer is 0.396. This is the same answer as that in part (a).

5.73 Step 1: A success is that a treated child is cured.

Step 2: The success probability is p = 0.90.

Step 3: The number of trials is n = 3.

Step 4: The formula for x successes is

$P(X=x)=\binom{3}{x}(.90)^x(.10)^{3-x}$. For x = 0, 1, 2, and 3, the probabilities are

$$P(X=0)=\binom{3}{0}(.90)^0(.10)^3=\frac{3!}{0!\,3!}(.90)^0(.10)^3=0.001$$

$$P(X=1)=\binom{3}{1}(.90)^1(.10)^2=\frac{3!}{1!\,2!}(.90)^1(.10)^2=0.027$$

$$P(X=2)=\binom{3}{2}(.90)^2(.10)^1=\frac{3!}{2!\,1!}(.90)^2(.10)^1=0.243$$

$$P(X=3)=\binom{3}{3}(.90)^3(.10)^0=\frac{3!}{0!\,3!}(.90)^3(.10)^0=0.729$$

5.75 Step 1: A success is that the coin will come up heads.

Step 2: The success probability is p = 0.5.

Step 3: The number of trials is n = 10.

Step 4: The formula for y successes is

$$P(Y=y)=\binom{10}{y}(.50)^y(.50)^{10-y}.$$

We want the probability of exactly half of 10, or 5 successes

$$P(Y=5)=\binom{10}{5}(.50)^5(.50)^{10-5}=0.246$$

5.77 The calculations required to answer all parts of this exercise are

$$P(0)=\binom{5}{0}\cdot(.67)^0(.33)^5=0.004 \quad P(3)=\binom{5}{3}\cdot(.67)^3(.33)^2=0.328$$

$$P(1)=\binom{5}{1}\cdot(.67)^1(.33)^4=0.040 \quad P(4)=\binom{5}{4}\cdot(.67)^4(.33)^1=0.332$$

$$P(2)=\binom{5}{2}\cdot(.67)^2(.33)^3=0.161 \quad P(5)=\binom{5}{5}\cdot(.67)^5(.33)^0=0.135$$

(a) P(2) = 0.161

(b) P(4) = 0.332

(c) $P(X \geq 4) = P(4) + P(5) = 0.332 + 0.135 = 0.467$ (actually 0.468 if rounding is done after the addition).

(d) $P(2 \leq X \leq 4) = P(2) + P(3) + P(4) = 0.161 + 0.328 + 0.332 = 0.821$

(e)

x	P(X=x)
0	0.004
1	0.040
2	0.161
3	0.328
4	0.332
5	0.135

(f) Left skewed

(g)

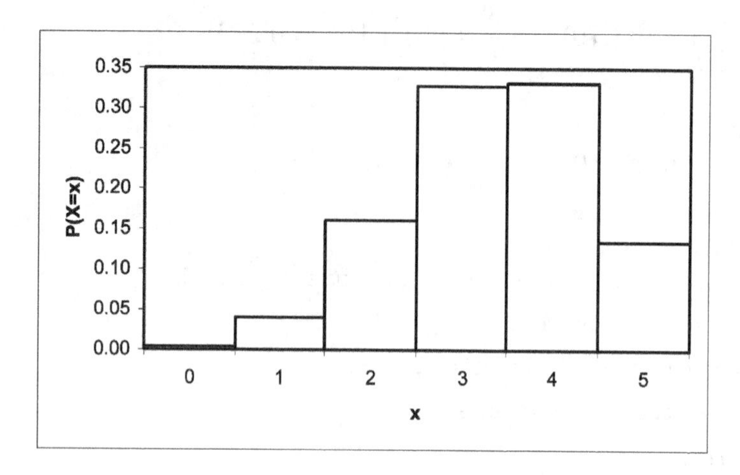

(h)

X	P(X=x)	xP(X=x)	x^2	x^2P(X=x)
0	0.004	0.000	0	0.000
1	0.040	0.040	1	0.040
2	0.161	0.322	4	0.644
3	0.328	0.984	9	2.952
4	0.332	1.328	16	5.312
5	0.135	0.675	25	3.375
		3.349		12.323

$\mu = 3.349$

$\sigma^2 = 12.323 - 3.349^2 = 1.107;\ \sigma = \sqrt{1.107} = 1.052$

(i) $\mu = np = 5(0.67) = 3.35$

$\sigma^2 = np(1-p) = 5(0.67)(0.33) = 1.1055$

$\sigma = \sqrt{1.1055} = 1.051$

(j) Out of any five races, the average number of favorites that finish in the money is 3.35.

5.79 (a) n = 8; p = 0.40

$$P(\text{Exactly } 3) = \binom{8}{3}(0.4)^3(0.6)^5 = 0.0241$$

$$P(3 \text{ or more}) = 1 - P(0 \text{ or } 1 \text{ or } 2) = 1 - \binom{8}{0}(0.4)^0(0.6)^8 - \binom{8}{1}(0.4)^1(0.6)^7 - \binom{8}{2}(0.4)^2(0.6)^6$$

$$= 1 - 0.0168 - 0.0896 - 0.2090 = 0.6846$$

$$P(\text{At most } 3) = P(0) + P(1) + P(2) + P(3)$$

$$= \binom{8}{0}(0.4)^0(0.6)^8 + \binom{8}{1}(0.4)^1(0.6)^7 + \binom{8}{2}(0.4)^2(0.6)^6 + \binom{8}{3}(0.4)^3(0.6)^5$$

$$= 0.0168 + 0.0896 + 0.2090 + 0.2787 = 0.5941$$

(b)

$$P(2 \le Y \le 4) = P(2) + P(3) + P(4)$$

$$= \binom{8}{2}(0.4)^2(0.6)^6 + \binom{8}{3}(0.4)^3(0.6)^5 + \binom{8}{4}(0.4)^4(0.6)^4 = 0.2092 + 0.2787 + 0.2322 = 0.7201$$

(If rounding is done after the addition, you will get 0.7200.)

(c) The mean of Y is $\mu = np = 8(0.4) = 3.2$. On average, of eight traffic fatalities, 3.2 will involve an intoxicated or alcohol-impaired driver or non-occupant.

(d) $\sigma^2 = np(1 - p) = 8(0.4)(0.6) = 19.2$, so $\sigma = 4.3418$.

5.81 (a) $n = 9$, $p = 0.35$, so

$$P(\text{Exactly } 5) = \binom{9}{5}(0.35)^5(0.65)^4 = 0.1181$$

$$P(\text{At least } 5) = \binom{9}{5}(0.35)^5(0.65)^4 + \binom{9}{6}(0.35)^6(0.65)^3 + \binom{9}{7}(0.35)^7(0.65)^2$$

$$+ \binom{9}{8}(0.35)^8(0.65)^1 + \binom{9}{9}(0.35)^9(0.65)^1$$

$$= 0.1181 + 0.0424 + 0.0098 + 0.0013 + 0.0001 = 0.1717$$

$$P(\text{At most } 5) = \binom{9}{0}(0.35)^0(0.65)^9 + \binom{9}{1}(0.35)^1(0.65)^8 + \binom{9}{2}(0.35)^2(0.65)^7$$

$$+ \binom{9}{3}(0.35)^3(0.65)^6 + \binom{9}{4}(0.35)^4(0.65)^5 + \binom{9}{5}(0.35)^5(0.65)^4$$

$$= 0.0207 + 0.1004 + 0.2162 + 0.2716 + 0.2194 + 0.1181 = 0.9464$$

$$P(\text{At least 1 correct}) = 1 - \text{P(0 Correct)} = 1 - \binom{9}{0}(0.35)^0(0.65)^9 = 1 - 0.0207 = 0.9793$$

(b) $$\text{P(At most 1 correct)} = \text{P(0)} + \text{P(1)} = \binom{9}{0}(0.35)^0(0.65)^9 + \binom{9}{1}(0.35)^1(0.65)^8$$

$$= 0.0207 + 0.1004 = 0.1211$$

(c) $$P(6 \text{ or } 7 \text{ or } 8) = \binom{9}{6}(0.35)^6(0.65)^3 + \binom{9}{7}(0.35)^7(0.65)^2 + \binom{9}{8}(0.35)^8(0.65)^1$$

$$= 0.0424 + 0.0098 + 0.0013 = 0.0535$$

(d) The formula is $P(X = x) = \binom{9}{x}(0.35)^x(0.65)^{9-x}$ for x = 0, 1, 2, ..., 9

Applying this formula for each value of x results in the table below.

x	P(X=x)
0	0.0207
1	0.1004
2	0.2162
3	0.2716
4	0.2194
5	0.1181
6	0.0424
7	0.0098
8	0.0013
9	0.0001

(e) The sampling was actually done without replacement, so the trials are not independent and the success probability changes very slightly from trial to trial. The exact probability distribution is called a hypergeometric distribution.

5.83 (a) n = 7 and p = 0.09, so the formula for the probability distribution is

$$P(X=x)=\binom{7}{x}(0.09)^x(0.91)^{7-x} \text{ for x } = 0,\ 1,\ 2, \ldots 6,\ 7$$

Applying this formula for each value of x results in the table below

x	P(X=x)
0	0.517
1	0.358
2	0.106
3	0.017
4	0.002
5	0.000
6	0.000
7	0.000

(b) The mean μ = np = 7(0.09) = 0.63. Thus, on the average, we would expect 0.63 youths out of seven children selected will be a PVGU.

(c) Yes. If, in fact, 9% of children in grades 3-8 are PVGUs, then finding that 3 or more out of 7 were PVGUs would mean that an event with probability 0.017 + 0.002 + 0 = 0.019 had just occurred. This is a fairly rare event if p is really 0.09. Rather than believe that we had just observed a rare event, we would be inclined to conclude that our assumption (p = 0.09) was incorrect and that p has increased from the 9% rate of 2011.

(d) No. If, in fact, 9% of children in grades 3-8 are PVGUs, then finding that two or more out of seven were PVGUs would mean that an event with probability 0.106 + 0.019 = 0.125 had just occurred. This is a fairly common event if p is really 0.09, so we would not be inclined to take 2 out of 7 as evidence that the PVGU rate has increased from the 9% rate in 2011.

5.85 (a) n = 4, p = 18/38 = 0.4737

$$P(2)=\binom{4}{2}(0.4737)^2(0.5263)^2 = 0.3729$$

(b) $P(\text{At least } 1) = 1 - P(0) = 1 - \binom{4}{0}(0.4737)^0(0.5263)^4 = 1 - 0.0767 = 0.9233$

5.87 (a) P(both are male) = (17/40)(17/40) = 289/1600 = 0.180625 if sampling is done with replacement.

(b) P(both are male) = (17/40)(16/39) = 272/1560 = 0.174359 if sampling is done without replacement.

(c) The probability without replacement is slightly smaller than the probability with replacement. There is about a 3% difference in the probabilities.

(d) P(both are male) = (170/400)(170/400) = 28900/160000 = 0.180625 if sampling is done with replacement. P(both are male) = (170/400)(169/399) = 28730/159600 = 0.180013 if sampling is done without replacement. The probability without replacement is just barely smaller than the probability with replacement. There is only about a 0.3% difference in the probabilities.

(e) In the second case, there is less difference between sampling without and with replacement. This is because removing the first male from the population makes less difference in the probability of the second person being a male when there are 400 people in the class than when there are only 40 people in the class. In other words, 169/399 is closer to 170/400 than 16/39 is to 17/40.

5.89 (a) p = 0.029; $P(X = x) = 0.029(0.971)^{x-1}$

This formula gives the probability that a person will first win a prize in week x.

(b) $P(3) = 0.029(0.971)^2 = 0.0273$

$P(X \leq 3) = P(1) + P(2) + P(3) = 0.0290 + 0.0282 + 0.0273 = 0.0845$

$P(X \geq 3) = 1 - P(X \leq 2) = 1 - [0.0290 + 0.0282] = 1 - 0.0572$

$= 0.9428$

(c) On the average, it will take $\mu = 1/p = 1/0.029 = 34.5$ weeks until you win a prize.

Exercises 5.4

5.91 Two uses of Poisson distributions would be (1) to model the frequency with which a specified event occurs during a particular period of time, and (2) to approximate binomial distribution probabilities when n is at least 100 and np is no more than 10.

5.93 The mean of a Poisson random variable is that same as its parameter λ.

5.95 The parameter for a Poisson distribution is $\lambda = np$.

5.97 (a) Using Formula 5.3 and $\lambda = 5$,

$$P(X = 5) = e^{-5} \cdot \frac{5^5}{5!} = 0.175$$

$$P(X < 2) = P(0) + P(1) = e^{-5} \cdot \frac{5^0}{0!} + e^{-5} \cdot \frac{5^1}{1!} = 0.00674 + 0.0337 = 0.040$$

$$P(X \geq 3) = 1 - P(X < 3) = 1 - P(0) - P(1) - P(2) = 1 - e^{-5} \cdot \frac{5^0}{0!} - e^{-5} \cdot \frac{5^1}{1!} - e^{-5} \cdot \frac{5^2}{2!}$$
$$= 1 - 0.00674 - 0.0337 - 0.0842 = 0.875$$

(b) Using Formula 5.4, the mean and standard deviation of a Poisson random variable with parameter λ are $\mu = \lambda$ and $\sigma = \sqrt{\lambda}$. Therefore, the mean is 5 and the standard deviation is $\sqrt{5} = 2.2$.

5.99 (a) Using Formula 5.3 and $\lambda = 4.7$,

$$P(X = 3) = e^{-4.7} \cdot \frac{4.7^3}{3!} = 0.157$$

$$P(5 \leq X \leq 7) = P(5) + P(6) + P(7) = e^{-4.7} \cdot \frac{4.7^5}{5!} + e^{-4.7} \cdot \frac{4.7^6}{6!} + e^{-4.7} \cdot \frac{4.7^7}{7!}$$
$$= 0.174 + 0.136 + 0.0914 = 0.401$$

$$P(X > 2) = 1 - P(X \leq 2) = 1 - P(0) - P(1) - P(2) = 1 - e^{-4.7} \cdot \frac{4.7^0}{0!} - e^{-4.7} \cdot \frac{4.7^1}{1!} - e^{-4.7} \cdot \frac{4.7^2}{2!}$$
$$= 1 - 0.00910 - 0.0427 - 0.100 = 0.848$$

(b) Using Formula 5.4, the mean and standard deviation of a Poisson random variable with parameter λ are $\mu = \lambda$ and $\sigma = \sqrt{\lambda}$. Therefore, the mean is 4.7 and the standard deviation is $\sqrt{4.7} = 2.2$.

5.101 (a) $P(Y = 4) = e^{-3.87} \cdot \dfrac{3.87^4}{4!} = 0.1949$

(b)

$$P(\text{At most } 1) = P(Y = 0 \text{ or } 1) = e^{-3.87} \cdot \frac{3.87^0}{0!} + e^{-3.87} \cdot \frac{3.87^1}{1!} = 0.0209 + 0.0807 = 0.1016$$

(c)

$$P(2 \leq Y \leq 5) = e^{-3.87} \cdot \frac{3.87^2}{2!} + e^{-3.87} \cdot \frac{3.87^3}{3!} + e^{-3.87} \cdot \frac{3.87^4}{4!} + e^{-3.87} \cdot \frac{3.87^5}{5!}$$
$$= 0.1561 + 0.2015 + 0.1949 + 1509 = 0.7034$$

(d)

y	P(Y=y)
0	0.021
1	0.081
2	0.156
3	0.201
4	0.195
5	0.151
6	0.097
7	0.054
8	0.026
9	0.011
10	0.004
11	0.002
12	0.000

(e)

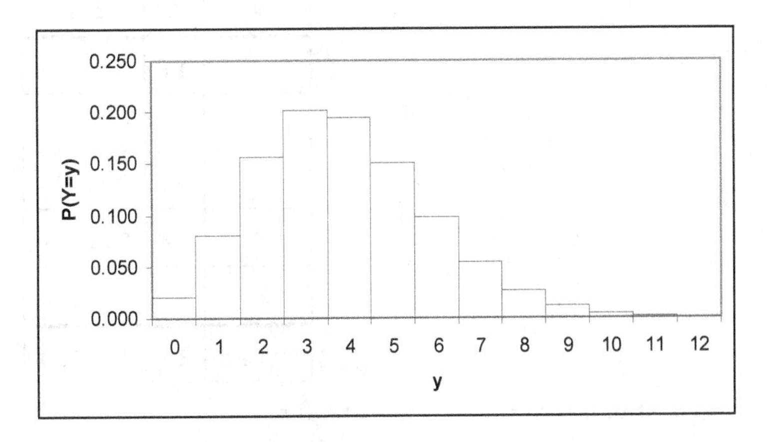

(f) On average, the number of alpha particles reaching the screen during an 8-minute interval is $\mu = \lambda = 3.87$.

5.103 (a) $P(X=0) = e^{-0.7} \cdot \dfrac{0.7^0}{0!} = 0.4966$

(b) $$\text{P(At most 2)} = P(X=0 \text{ or } 1 \text{ or } 2) = e^{-0.7} \cdot \frac{0.7^0}{0!} + e^{-0.7} \cdot \frac{0.7^1}{1!} + e^{-0.7} \cdot \frac{0.7^2}{2!}$$
$$= 0.4966 + 0.3476 + 0.1217 = 0.9659$$

(c) $$\text{P}(1 \le \text{X} \le 3) = P(X=1 \text{ or } 2 \text{ or } 3) = e^{-0.7} \cdot \frac{0.7^1}{1!} + e^{-0.7} \cdot \frac{0.7^2}{2!} + e^{-0.7} \cdot \frac{0.7^3}{3!}$$
$$= 0.3476 + 0.1217 + 0.0284 = 0.4977$$

(d) The mean number of wars that begin during a given year is the mean of X, $\mu = \lambda = 0.7$.

(e) $\sigma_x = \sqrt{\lambda} = \sqrt{0.7} = 0.8367$

5.105 (a) The mean number of cherries per pie is 42/35 = 1.2

(b)

x	f	f/n
0	11	0.3143
1	12	0.3429
2	8	0.2286
3	2	0.0571
4	2	0.0571
Total	35	1.0000

(c) Using λ = 1.2, we obtain

x	P(X=x)
0	0.301
1	0.361
2	0.217
3	0.087
4	0.026
5	0.006
6	0.001
7	0.000

(d) All of the actual relative frequencies are quite close to the Poisson probabilities using a mean of 1.2. They would be even closer if the counts for 3 and 4 cherries were 3 and 1 instead of 2 and 2, respectively. We conclude that the Poisson distribution with a mean of 1.2 does a very good job of modeling the distribution of the number of cherries in the pies.

5.107 We use $\lambda = \mu = np = 500(1/200) = 2.5$. First, we note that $n \geq 100$ and $np \leq 10$. Then

$$\text{P(At most 3)} = P(X = 0 \text{ or } 1 \text{ or } 2 \text{ or } 3) = e^{-2.5} \cdot \frac{2.5^0}{0!} + e^{-2.5} \cdot \frac{2.5^1}{1!}$$

$$+ e^{-2.5} \cdot \frac{2.5^2}{2!} + e^{-2.5} \cdot \frac{2.5^3}{3!}$$

$$= 0.082 + 0.205 + 0.257 + 0.214 = 0.758$$

5.109 (a) We would expect $\mu = np = (10000)(1/1500) = 6.667$ to have Fragile X Syndrome.

(b) First, we note that $n \geq 100$ and $np \leq 10$. Then we use the Poisson probability formula to obtain the probabilities for x = 0 through 10. The required calculations are

$$P(0)=\frac{e^{-6.67}6.67^0}{0!}=0.0013 \qquad P(1)=\frac{e^{-6.67}6.67^1}{1!}=0.0085$$

$$P(2)=\frac{e^{-6.67}6.67^2}{2!}=0.0282 \qquad P(3)=\frac{e^{-6.67}6.67^3}{3!}=0.0627$$

$$P(4)=\frac{e^{-6.67}6.67^4}{4!}=0.1046 \qquad P(5)=\frac{e^{-6.67}6.67^5}{5!}=0.1395$$

$$P(6)=\frac{e^{-6.67}6.67^6}{6!}=0.1551 \qquad P(7)=\frac{e^{-6.67}6.67^7}{7!}=0.1478$$

$$P(8)=\frac{e^{-6.67}6.67^8}{8!}=0.1232 \qquad P(9)=\frac{e^{-6.67}6.67^9}{9!}=0.0913$$

$$P(10)=\frac{e^{-6.67}6.67^{10}}{10!}=0.0609$$

The probabilities that more than 7 of the males have Fragile X Syndrome and that at most 10 of the males have Fragile X Syndrome are (from the preceding calculations)

$$\begin{aligned}P(X>7)&=1-P(X\le 7)=1-(P(0)+P(1)+\cdots+P(7))\\&=1-0.0013-0.0085-0.0282-0.0627-0.1046-0.1395-0.1551-0.1478\\&=1-0.6477=0.3523\end{aligned}$$

$$P(X\le 10)=0.0013+0.0085+\cdots+0.0609=0.9231$$

5.111 $\mu=np=(100{,}000{,}000)\cdot\frac{1}{30{,}000{,}000}=\frac{10}{3}=3.33$

(a) $$P(3\le X\le 5)=\frac{e^{-3.33}3.33^3}{3!}+\frac{e^{-3.33}3.33^4}{4!}+\frac{e^{-3.33}3.33^5}{5!}=0.5258$$

(b) Assuming that n lobsters must be hatched, the mean will be n/30,000,000.

Then $P(X\ge 1)=1-P(X=0)=1-\frac{e^{-n/30000000}(n/30000000)^0}{0!}=1-e^{-(n/30000000)}\ge 0.9$

This implies that $e^{-(n/30000000)}\le 0.1$. Taking the natural log of both sides of this inequality, we find that $-n/30000000\le \ln(0.1)$. Multiplying both sides of this inequality by -30,000,000 and reversing the direction of the inequality, we have $n\ge(-30000000)\ln(0.1)=69{,}077{,}552.79\ \textit{or}\ 69{,}077{,}553$.

5.113 When n is large and the success probability p is near 1, 1 - p will be near zero. If we let q = 1 - p, then we can approximate the probabilities of the number of failures Y = n - X by letting λ = nq. The probability of x successes in n trials is the same as the probability of n - x failures in n trials. Thus

$$P(X=x)=P(Y=n-x)=e^{-\lambda}\frac{\lambda^{n-x}}{(n-x)!}\ \textit{where}\ \lambda=nq\ .$$

Review Problems for Chapter 5

1. (a) random variable

   (b) finite (or countably infinite)

2. A probability distribution of a discrete distribution of a discrete random variable gives us a listing of the possible values of the random variable and their probabilities; or a formula for the probabilities.

3. Probability histogram

4. 1

5. (a) P(X = 2) = 0.386

   (b) 38.6%

   (c) 50(0.386) = 19.3 or about 19; 500(0.386) = 193

6. 3.6

7. X is more likely to take a value close to its mean because it has a smaller standard deviation and therefore less variation.

8. 0! = 1 3! = 3·2·1 = 6 4! = 4·3·2·1 = 24

   7! = 7·6·5·4·3·2·1 = 5040

9.

$$(a)\ \binom{8}{3}=\frac{8!}{3!5!}=56 \qquad (b)\ \binom{8}{5}=\frac{8!}{5!3!}=56 \qquad (c)\ \binom{6}{6}=\frac{6!}{6!6!}=1$$

$$(d)\ \binom{10}{2}=\frac{10!}{2!8!}=45 \qquad (e)\ \binom{40}{4}=\frac{40!}{4!36!}=91390 \qquad (f)\ \binom{100}{0}=\frac{100!}{0!100!}=1$$

10. Each trial must have the same two possible outcomes (success and failure), the trials must be independent, and they must have a probability of success p that remains constant for all trials.

11. The binomial distribution is a probability distribution for the number of successes in a sequence of n Bernoulli trials.

12. $$\binom{10}{3} = \frac{10\,!}{3\,!\,7\,!} = \frac{10(9)\ (8)}{3(2)\ (1)} = 120$$

13. (a) The success probability is p = 0.493.

    (b) The outcomes and their probabilities are shown in the table.

Outcome	Probability
WWW	0.120
WWL	0.123
WLW	0.123
WLL	0.127
LWW	0.123
LWL	0.127
LLW	0.127
LLL	0.130

See the following tree diagram for the details of each probability.

(c)

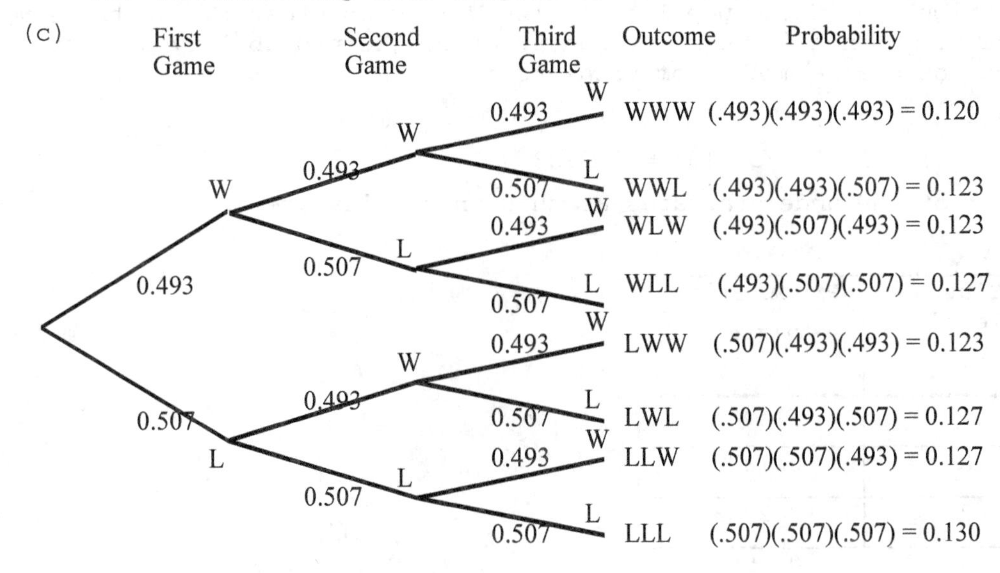

(d) The outcomes in which the player wins two out of the three times are WWL, WLW, and LWW.

(e) Each outcome in part (d) has probability 0.123. The probabilities are equal for each outcome because each probability is obtained by multiplying two WIN probabilities of 0.493 and one LOSS probability of 0.507.

(f) P(2 wins) = P(WWL) + P(WLW) + P(LWW) = 0.123 + 0.123 + 0.123
= 0.369

(g) P(0 wins) = P(LLL) = 0.130

P(1 win) = P(WLL) + P(LWL) + P(LLW) = 0.127 + 0.127 + 0.127 = 0.381

P(3 wins) = P(WWW) = 0.120

Y	P(Y=y)
0	0.120
1	0.381
2	0.369
3	0.130

(h) Binomial distribution with parameters n = 3 and p = 0.493.

14. (a) p < 0.5 since the distribution is left-skewed.

(b) p = 0.5 since the distribution is symmetric.

15. Definition 5.4 for the mean and definition 5.5 for the standard deviation are applied to the binomial and Poisson distribution. For example, in Definition 5.4, the formula for the binomial probability function is substituted for P(X = x) to get the mean.

16. (a) Binomial distribution

(b) Hypergeometric distribution

(c) The hypergeometric distribution may be approximated by the binomial distribution if the population size N is much greater than the sample size n. When n is no more than 5% of N, the probability of a success does not change much from trial to trial.

17. (a) X = 1, 2, 3, 4 (b) {X = 3}

(c) P(X = 3) = 17,302/59,183 = 0.2923.

29.2% of the undergraduates at this university are juniors.

(d)

Class level x	Probability P(X=x)
1	0.1631
2	0.1878
3	0.2923
4	0.3568

(e)

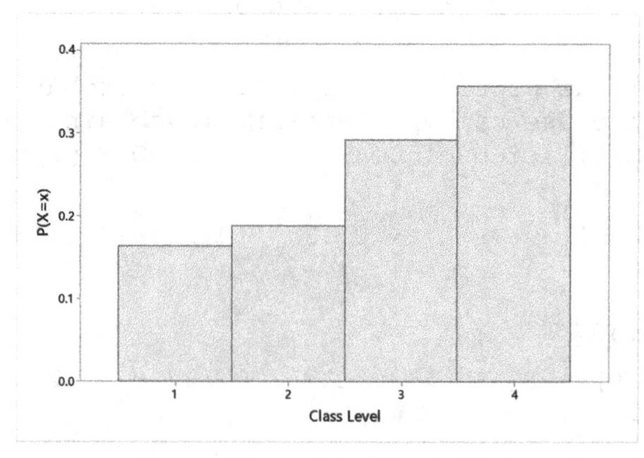

18. (a) {Y = 4}

(b) {Y $\geq$ 4}

(c) {2 $\leq$ Y $\leq$ 4}

(d) {Y $\geq$ 1}

(e) P(Y = 4) = 0.174

(f) P(Y $\geq$ 4) = P(4) + P(5) + P(6) = 0.174 + 0.105 + 0.043 = 0.322

(g) P(2 $\leq$ Y $\leq$ 4) = P(2) + P(3) + P(4) = 0.232 + 0.240 + 0.174 = 0.646

(h) P(Y $\geq$ 1) = 1 - P(Y $\leq$ 0) = 1 - 0.052 = 0.948

19. The required calculations are

y	P(Y=y)	yP(Y=y)	y^2	y^2P(Y=y)
0	0.052	0.000	0	0.000
1	0.154	0.154	1	0.154
2	0.232	0.464	4	0.928
3	0.240	0.720	9	2.160
4	0.174	0.696	16	2.784
5	0.105	0.525	25	2.625
6	0.043	0.258	36	1.548
		2.817		10.199

(a) (a) $\mu_Y = \sum yP(Y=y) = 2.817$.

(b) On the average, the number of busy lines is about 2.817.

(c)

$$\sigma_y = \sqrt{\sum (Y-\mu_y)^2 P(Y=y)} = \sqrt{2.2635} = 1.504, \text{ or}$$
$$\sigma_y = \sqrt{\sum Y^2 P(Y=y) - \mu_y^2} = \sqrt{10.199 - 2.817^2} = 1.504$$

(d)

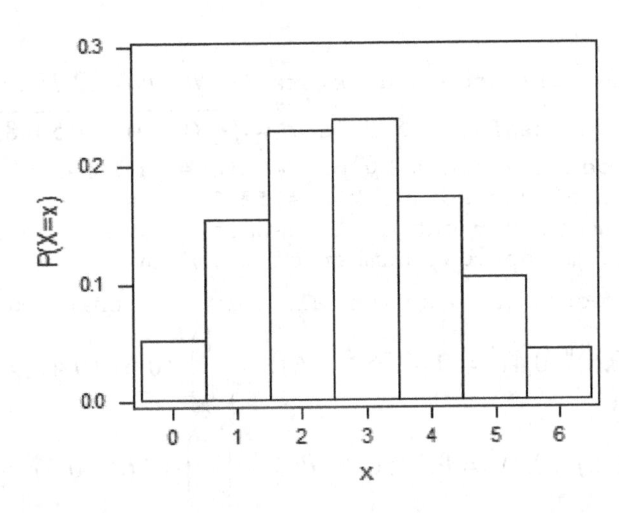

-1.7	-0.2	1.3	2.8	4.3	5.8	7.3
$\mu - 3\sigma$	$\mu - 2\sigma$	$\mu - \sigma$	μ	$\mu + \sigma$	$\mu + 2\sigma$	$\mu - 3\sigma$

20. Using the Binomial formula with n = 3 and p = 0.493, the probability distribution for Y, the number of times out of three that the player wins is given by the following table.

x	P(X=x)
0	$\binom{3}{0}0.493^0 0.507^3 = 0.130$
1	$\binom{3}{1}0.493^1 0.507^2 = 0.380$
2	$\binom{3}{2}0.493^2 0.507^1 = 0.370$
3	$\binom{3}{3}0.493^3 0.507^0 = 0.120$

21. (a) Let X be a random variable equal to the number of penalty kicks by professional soccer players in a sample of 15 that are successful. X is a Binomial random variable with n = 15 and p = 0.85. The probability that all of the kicks in the sample are successful is $P(X=15)=\binom{15}{15}0.85^{15}0.15^0 = 0.087$.

(b) The Probability that at least 13 are successful is

$$P(X \ge 13) = P(X=13)+P(X=14)+P(X=15)$$

$$=\binom{15}{13}0.85^{13}0.15^2+\binom{15}{14}0.85^{14}0.15^1+\binom{15}{15}0.85^{15}0.15^0 = 0.286+0.231+0.087=0.604$$

(c) The mean number of successful kicks is $\mu = np = 12.75$. The standard deviation of successful kicks is $\sigma = \sqrt{np(1-p)} = \sqrt{15(0.85)(0.15)} = 1.38$. In a sample of 15 penalty kicks by professional soccer players, the expected average number of successful kicks is 12.75. And, roughly speaking, on average, the number out of 15 penalty kicks that will be successful will differ from the mean number of 12.75 by 1.38.

22. The calculations required to answer all parts of this exercise are

$$P(0)=\binom{4}{0}\cdot(0.6)^0(0.4)^4 = 0.0256 \qquad P(1)=\binom{4}{1}\cdot(0.6)^1(0.4)^3 = 0.1536$$

$$P(2)=\binom{4}{2}\cdot(0.6)^2(0.4)^2 = 0.3456 \qquad P(3)=\binom{4}{3}\cdot(0.6)^3(0.4)^1 = 0.3456$$

$$P(4)=\binom{4}{4}\cdot(0.6)^4(0.4)^0 = 0.1296$$

(a) $P(3) = 0.3456$, $P(X \ge 3) = P(3) + P(4) = 0.3456 + 0.1296 = 0.4752$, $P(X \le 3) = P(0) + P(1) + P(2) + P(3) = 0.0256 + 0.1536 + 0.3456 + 0.3456 = 0.8704$

(b)

Number with Pets x	Probability P(X=x)
0	0.0256
1	0.1536
2	0.3456
3	0.3456
4	0.1296

(c) Left skewed, since p > 0.5.

23. (a)

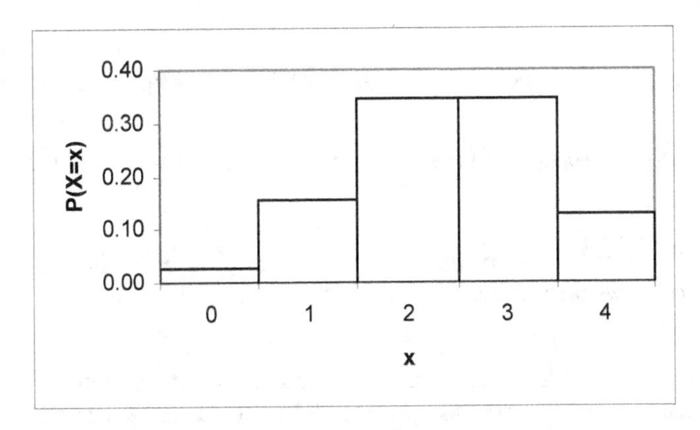

(b) The distribution is only approximate for several reasons: the actual distribution is hypergeometric, based on sampling without replacement; and the success probability p = 0.60 is probably based on a sample.

(c) Yes, because the sample size most likely does not exceed 5% of the population side.

24. (a) $P(X=2)=e^{-1.75}\dfrac{1.75^2}{2!}=0.266$

(b)

$$P(4\le X\le 6)=e^{-1.75}\frac{1.75^4}{4!}+e^{-1.75}\frac{1.75^5}{5!}+e^{-1.75}\frac{1.75^6}{6!}=0.068+0.024+0.007$$

$=0.099$

(c) $P(X\ge 1)=1-P(0)=1-e^{-1.75}\dfrac{1.75^0}{0!}=1-0.174=0.826$

25.

(a)

x	P(X=x)
0	0.174
1	0.304
2	0.266
3	0.155
4	0.068
5	0.024
6	0.007
7	0.002
8	0.000

(b)

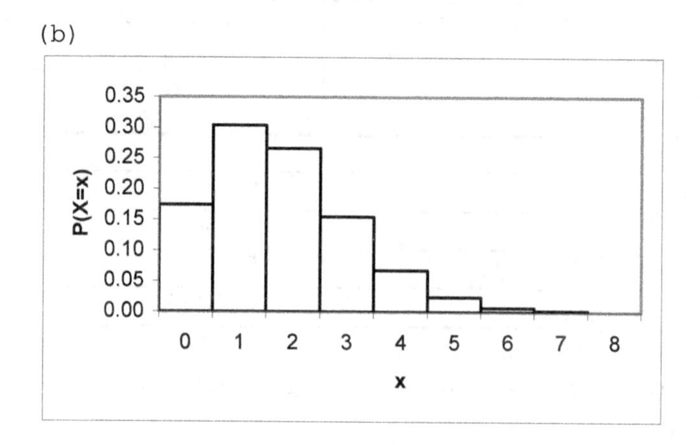

(c) The distribution is right skewed. Yes, this is typical of Poisson distributions.

(d) $\mu = \lambda = 1.75$; on the average, there are 1.75 calls to a wrong number per minute.

(e) $\sigma = \sqrt{\lambda} = \sqrt{1.75} = 1.323$. On average, during a random minute, the number of calls to a wrong number differ from the mean of 1.75 by 1.323.

26. (a) We would expect about 3 meteoroids to be alien matter from outside our solar system. Assuming meteoroids to be independent of each other, the number out of 300 consisting of such matter would have a binomial distribution with n = 300 and $\mu = 0.01$. The mean, or expected number, is np = 300(0.01) = 3.

(b) Since n is at least 100 and np is less than 10, we can approximate the binomial distribution with the Poisson distribution having parameter λ = 3. Then,

$$P(2 \le Y \le 4) = e^{-3} \cdot \frac{3^2}{2!} + e^{-3} \cdot \frac{3^3}{3!} + e^{-3} \cdot \frac{3^4}{4!} = 0.2240 + 0.2240 + 0.1680 = 0.6160 .$$

(c) $P(Y \ge 1) = 1 - P(Y = 0) = 1 - e^{-3} \cdot \frac{3^0}{0!} = 1 - 0.0498 = 0.9502$.

27. (a) n = 100 and p = 0.015.

(b) λ = np = 1.5

(c) The binomial probability function is

$$P(Y = y) = \binom{100}{y} (0.015)^y (0.985)^{100-y} \text{ for } y = 0, 1, 2, \ldots, 100$$

Starting at y = 0 and evaluating this function until it is zero to four decimal places, we obtain the following table.

y	P(Y = y)
0	0.2206
1	0.3360
2	0.2532
3	0.1260
4	0.0465
5	0.0136
6	0.0033
7	0.0007
8	0.0001
9	0.0000

(d) The Poisson probability function is

$$P(Y=y)=e^{-1.5}\cdot\frac{1.5^y}{y!} \text{ for } y = 0, 1, 2, \ldots$$

Starting at 7 = 0 and evaluating this function until it is zero to four decimal places, we obtain the following table.

Y	P(Y = y)
0	0.2231
1	0.3347
2	0.2510
3	0.1255
4	0.0471
5	0.0141
6	0.0035
7	0.0008
8	0.0001
9	0.0000

(e) All of the probabilities for y = 0 to 9 agree to two decimal places and are usually at most a couple of thousandths apart. The agreement is very good.

28.

Event	Binomial Probability	Poisson Probability
(a) Y = 3	0.1260	0.1255
(b) $2 \le Y \le 5$	0.4393	0.4377
(c) Y < 4	0.9358	0.9343
(d) Y > 2	0.1902	0.1912

Each pair of probabilities agree to within 0.0016 or less.

CHAPTER 6 SOLUTIONS

Exercises 6.1

6.1 A density curve is a smooth curve that identifies the shape of a distribution for a variable.

6.3 The percentage of all possible observations of a variable that lie within a specified range is approximately equal to the corresponding area under its density curve.

6.5 The percentage of all possible observations of the variable that lie to the right of 4 equals the area under its density curve to the right of 4, expressed as a percentage.

6.7 (a) Since the area to the right of 15 is 0.324, 32.4% of all possible observations of the variable exceed 15.

(b) Since the total area under the density curve is one and the area to the right of 15 is 0.324, the area to the left of 15 must be 1 - 0.324 = 0.676. Therefore, 67.6% of all possible observations of the variable are at most 15.

6.9 Since the area under the curve that lies between 15 and 20 is 0.414 and the total area under the density curve is one, the area not between 15 and 20 is 1 - 0.414 = 0.586. The area not between 15 and 20 would be the area to the left of 15 or the area to the right of 20. Therefore, 58.6% of all possible observations of the variable are either less than 15 or greater than 20.

6.11 (a) Since 28.4% of all possible observations of the variable are less than 11, the area under the density curve that lies to the left of 11 is 0.284.

(b) Since the total area under the density curve is one and the area to the left of 11 is 0.284, the area to the right of 11 must be 1 - 0.284 = 0.716.

6.13 No, because the total area to the right and left of 65 should be 1, not 0.9.

6.15 The histogram will be roughly bell-shaped.

6.17 Their distributions are identical. The mean and standard deviation completely determine the shape of a normal distribution. Thus if two normally distributed variables have the same mean and standard deviation, they also have the same distribution.

6.19 (a) True. Both normal curves have the same spread. Spread is represented by σ. For each normal curve, $\sigma = 3$.

(b) False. The parameter μ affects where the normal curve is centered. Since this parameter is different for each normal curve, each normal curve is centered at a different place.

6.21 True. The value of the parameter μ has no effect on the spread of a normal curve. The parameter μ affects only where the normal curve is centered. The spread of the normal curve is determined by the parameter σ.

6.23

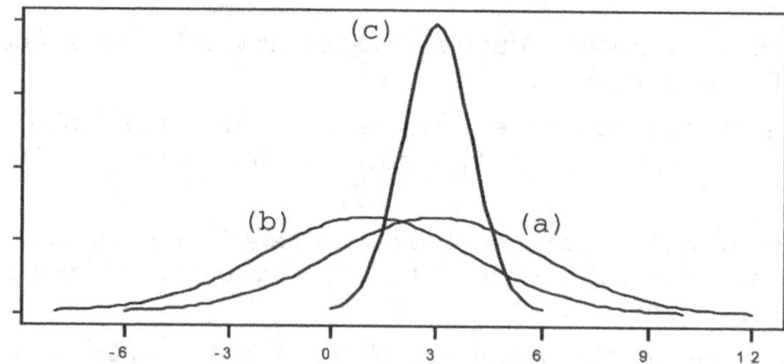

6.25 The percentage of all possible observations of a normally distributed variable that lie between 2 and 3 is the same as the area under the associated normal curve between 2 and 3. If the variable is only approximately normally distributed, the percentage of all possible observations between 2 and 3 is only approximately the area under the associated normal curve between 2 and 3.

6.27 If the area under a particular normal curve to the left of 105 is 0.6227, then 62.27% of all possible observations of the variable lie to the left of 105.

6.29 (a) The following is a graph of the density curve with the equation $y = 2x$ for $0 < x < 1$.

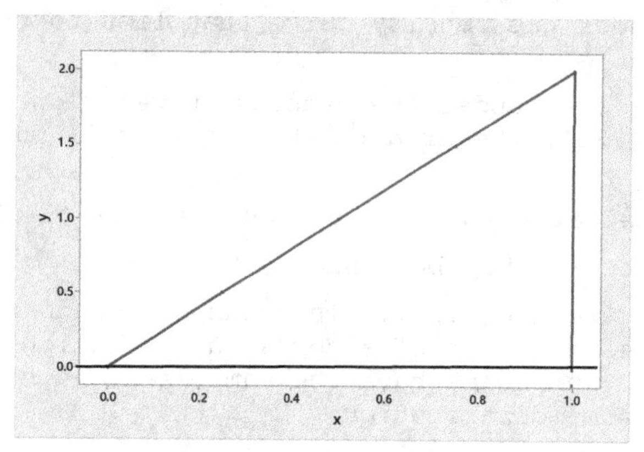

(b) The area to the left of any number x between 0 and 1 would be a triangle with a base of x and a height of $2x$. The area of a triangle is $1/2 \cdot base \cdot height = 1/2 \cdot x \cdot 2x = x^2$.

(c) The area between 1/2 and 3/4 can be calculated by finding the area to the left of 1/2 and subtracting that from the area to the left of 3/4. Using part (b), the area to the left of 1/2 is $(1/2)^2 = 1/4$ and the area to the left of 3/4 is $(3/4)^2 = 9/16$. Therefore, the area between 1/2 and 3/4 is $9/16 - 1/4 = 5/16$.

(d) The area to the left of 1/4, using part (b), is $(1/4)^2 = 1/16$. Therefore, the area to the right of 1/4 is one minus the area to the left, or $1 - 1/16 = 15/16$.

6.31 (a) The following is a graph of the density curve with the equation $y = 1/30$ for $0 < x < 30$.

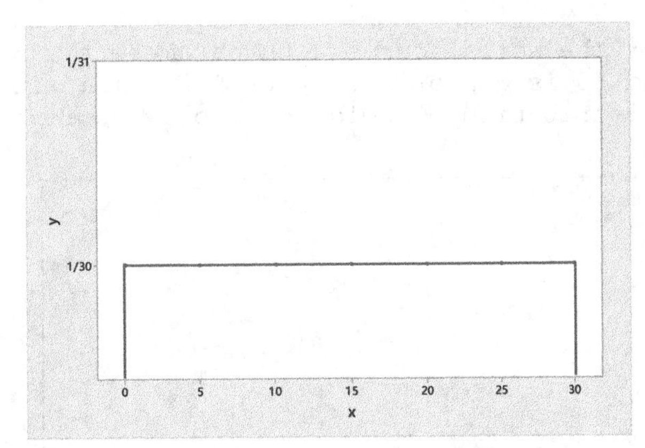

(b) The area to the left of any number x between 0 and 30 would be a rectangle with a base of x and a height of $1/30$. The area of a rectangle is $base \cdot height = x \cdot 1/30 = x/30$.

(c) The probability that John waits for the train less than 5 minutes is the area under the density curve to the left of 5. Following part (b), the area to the left of 5 is $5/30 = 1/6$. Therefore, John waits for the train less than 5 minutes about 16.7% of the time.

(d) The probability that John waits for the train between 10 and 15 minutes is the area under the density curve between 10 and 15. This area can be calculated by finding the area to the left of 10 and subtracting that from the area to the left of 15. Using part (b), the area to the left of 10 is 10/30 = 1/3 and the area to the left of 15 is 15/30 = 1/2. The area between 10 and 15 is 1/2 - 1/3 = 1/6. Therefore, John waits for the train between 10 and 15 minutes about 16.7% of the time.

(e) The probability that John waits for the train at least 20 minutes is the area under the density curve to the right of 20. The area to the left of 20, using part (b), is 20/30 = 2/3. The area to the right of 20 is one minus the area to the left, or 1 - 2/3 = 1/3. Therefore, John waits for the train at least 20 minutes about 33.3% of the time.

6.33 (a) The following is a graph of the density curve with the equation

$y = 1 - x/2$ for $0 < x < 2$.

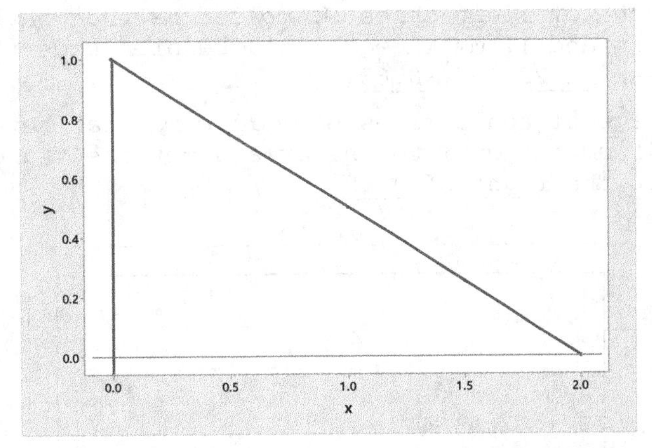

(b) The probability that the losses exceed \$1.5 million is the area to the right of 1.5. The area to the left of 1.5, using the given information for area to the left, is $1.5 - 1.5^2/4 = 0.9375$. The area to the right of 1.5 is one minus the area to the left, or 1 - 0.9375 = 0.0625. Therefore, the losses exceed \$1.5 million about 6.25% of the time.

6.35 (a) The percentage of female students who are between 60 and 65 inches tall is 100(0.0450 + 0.0757 + 0.1170 + 0.1480 + 0.1713) = 55.70%.

(b) The area under the normal curve with parameters $\mu = 64.4$ and $\sigma = 2.4$ between 60 and 65 is approximately 0.5570. This is only an estimate because the distribution of heights is only approximately normally distributed.

6.37 (a)

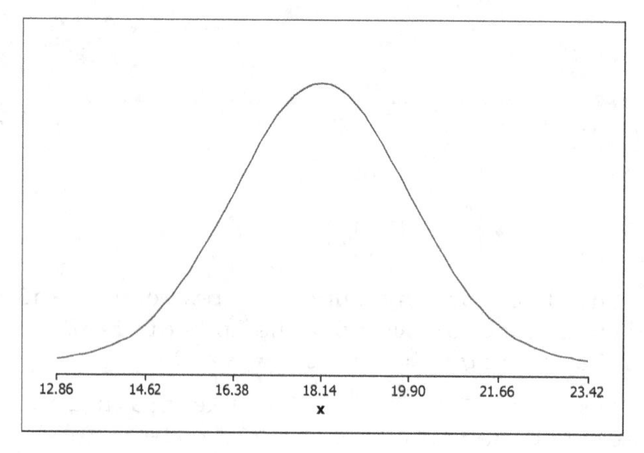

(b) $z = (x - 18.14)/1.76$

(c) z has a standard normal distribution ($\mu = 0$ and $\sigma = 1$).

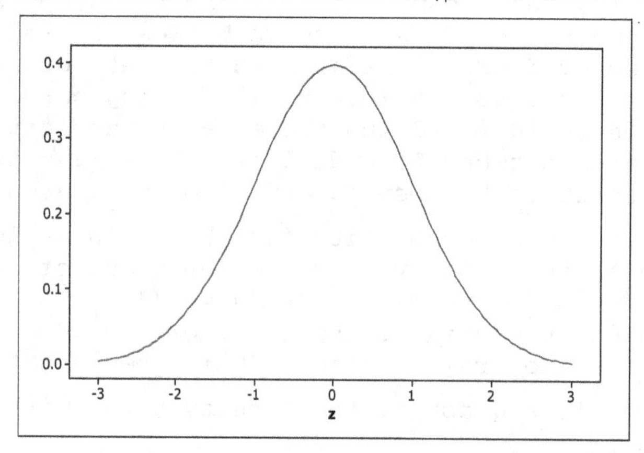

(d) The percentage of adult males *G. mollicoma* that have carapace lengths between 16 mm and 17 mm is equal to the area under the standard normal curve between <u>-1.22</u> and <u>-0.65</u>.

(e) The percentage of adult males *G. mollicoma* that have carapace lengths exceeding 19 mm is equal to the area under the standard normal curve that lies to the <u>right</u> of <u>0.49</u>.

6.39 (a)

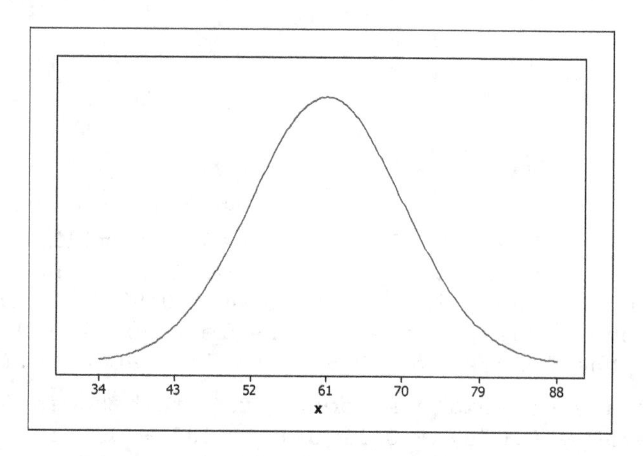

(b) z = (x - 61)/9

(c) z has a standard normal distribution (μ = 0 and σ = 1).

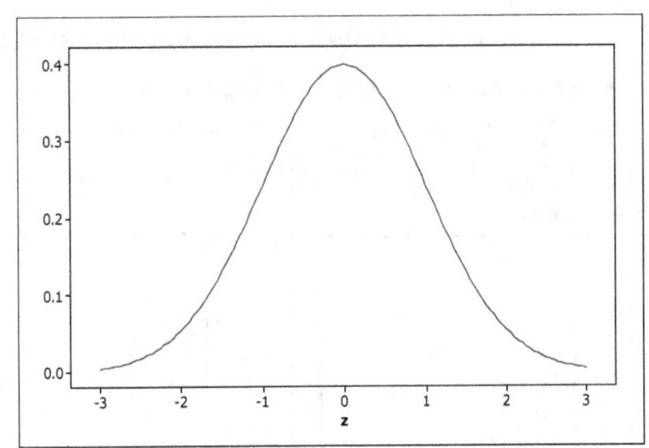

(d) The percentage of finishers with times between 50 and 70 minutes is equal to the area under the standard normal curve between -1.22 and 1.00.

(e) The percentage of finishers with times less than 75 minutes is equal to the area under the standard normal curve that lies to the left of 1.56.

6.41 (a)

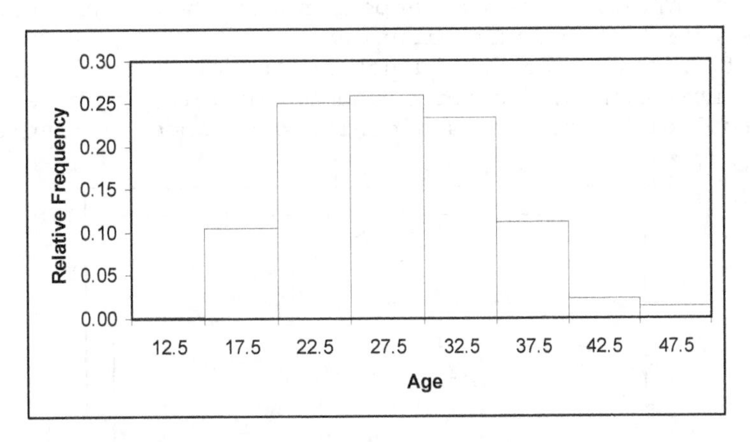

(b) Yes. The distribution appears to be slightly right skewed, but could be approximated quite well by a normal distribution.

6.43 (a)

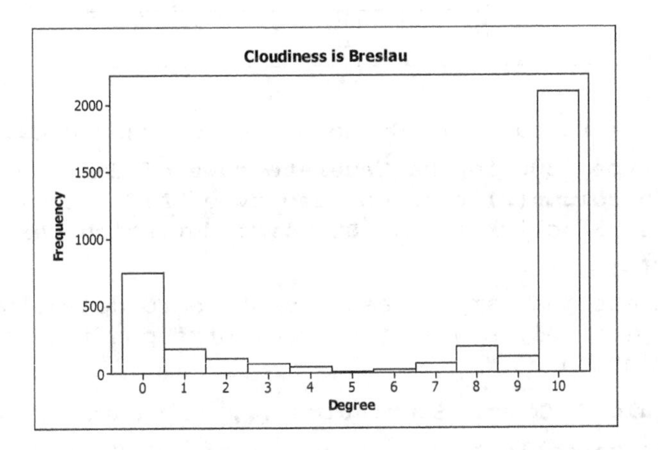

(b) The data is definitely not normal. The distribution instead looks bimodal with the most frequent values of 0 and 10 occurring at the far left and far right of the distribution. The rest of the distribution looks evenly distributed or uniform.

6.45 (a) Using Minitab, we obtain the data from the WeissStats Resource Site and choose **Graph ▶ Histogram,** select **Simple** from the first row, and click **OK**. Then enter VERBAL and MATH in the Graphs variables text box and click OK. The resulting graphs follow.

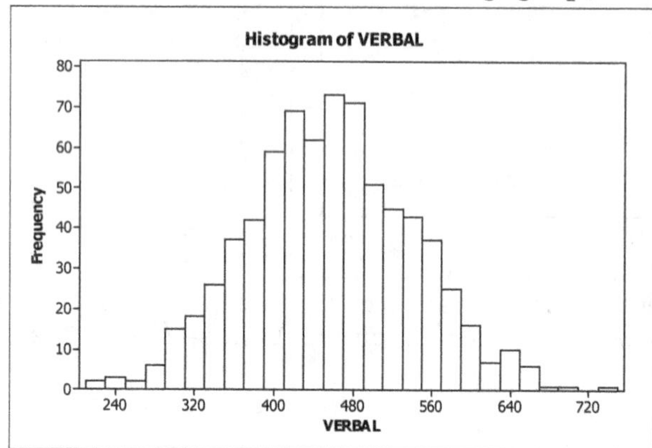

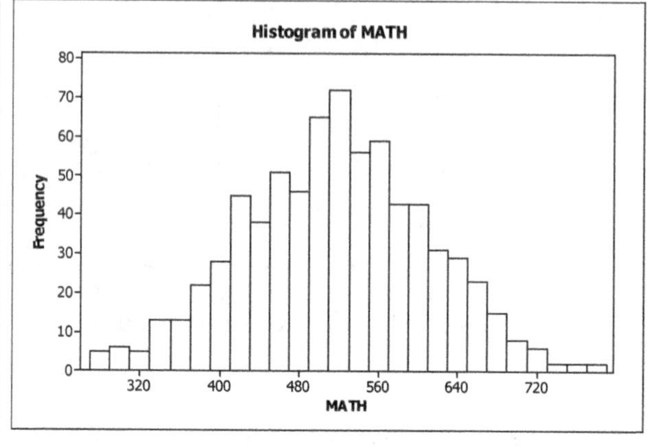

(a) Yes. The SAT verbal scores appearing in the left hand histogram have roughly a bell shaped distribution.

(b) Yes. The SAT verbal scores appearing in the right hand histogram have roughly a bell shaped distribution.

6.47 The number of chips per bag could not be exactly normally distributed because that number is a discrete random variable, whereas any variable having a normal distribution is a continuous random variable.

6.49 (a)

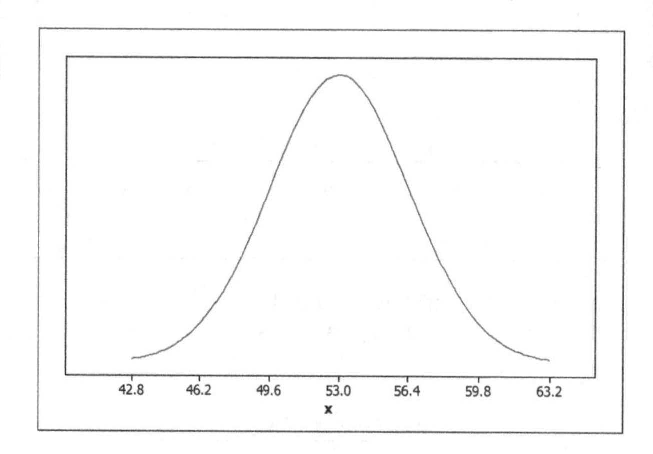

(b) Following the procedure in Example 6.4, we choose **Calc ▶ Random Data ▶ Normal...**, type 1500 in the **Generate rows of data** text box, click in the **Store in column(s)** text box and type DAYS, click in the **Mean** text box and type 53, click in the **Standard deviation** text box and type 3.4, and click **OK**.

(c) We would expect the sample mean and standard deviation to be near 53 and 3.4 respectively since these the corresponding values for the entire population.

(d) We choose **Calc ▶ Column Statistics...,** click on the **Mean** button, click in the **Input variable** text box and select DAYS, and click **OK**.

Then repeat the process, selecting the **Standard deviation** button. The results are shown in the Session Window. The results we obtained were

Mean of DAYS = 52.9267 and Standard deviation of DAYS = 3.19167. Your results will vary from ours.

(e) We would expect the histogram for the sample to be bell-shaped since it should be similar to the normal distribution of the population.

(f) Using Minitab, we obtain the histogram below for the sample. Your histogram will likely be similar to ours, but not identical.

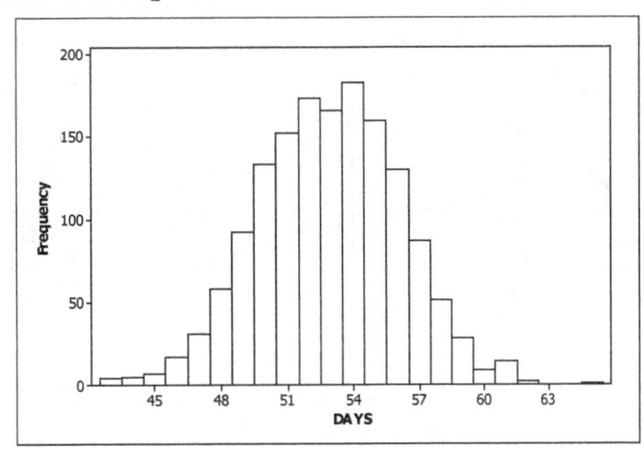

Exercises 6.2

6.51 The total area under the curve is 1, and the standard normal curve is symmetric about 0. Therefore, the area to the left of 0 is 0.5, and the area to the right of 0 is 0.5.

6.53 The area under the standard normal curve to the right of 0.43 is 1 - the area to the left of 0.43. The area to the left of 0.43 is 0.6664. Therefore, the area to the right of 0.43 is 1 - 0.6664 = 0.3336.

6.55 The area to the left of z = 3.00 is 0.9987 and the area to the left of -3.00 is 0.0013. Therefore the area between -3.00 and 3.00 is 0.9987 - 0.0013 = 0.9974 (99.74%).

6.56 The standard normal curve is the associated normal curve for a standardized normally distributed variable. The standardized variable is labeled with the letter z, hence the name "z-curve."

6.57 (a) Locate the row (tenths digit) and column (hundredths digit) of the specified z-score. The corresponding table entry is the area under the standard normal curve that lies to the left of the z-score.

(b) The area that lies under the standard normal curve to the right of a specified z-score is 1 - (area to the left of the z-score).

(c) The area that lies under the standard normal score between two specified z-scores, say *a* and *b*, where a < b, is found by subtracting the area to the left of *a* from the area to the left of *b*.

6.59 (a)

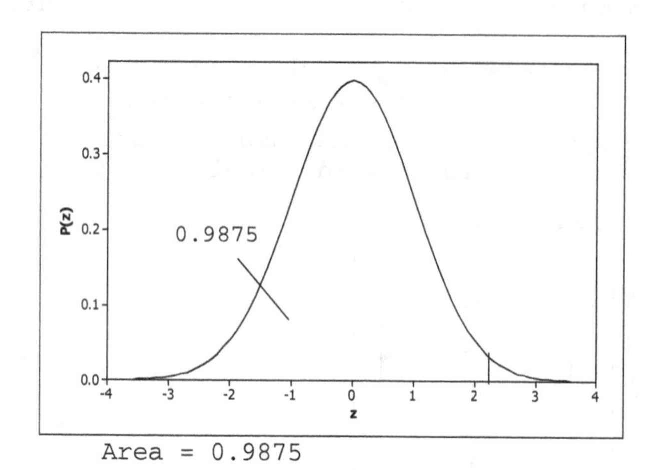

Area = 0.9875

(b)

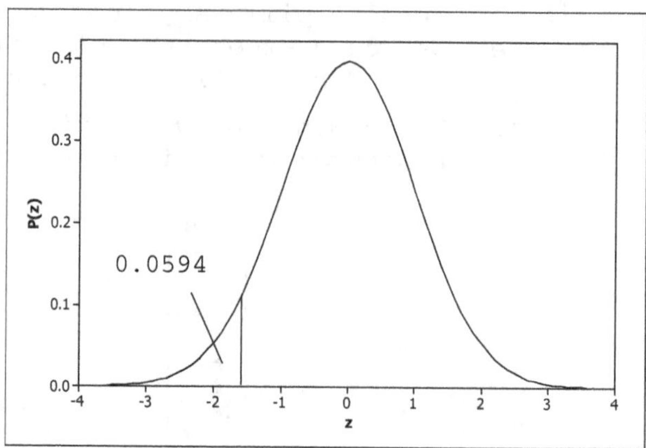

Area = 0.0594

(c)

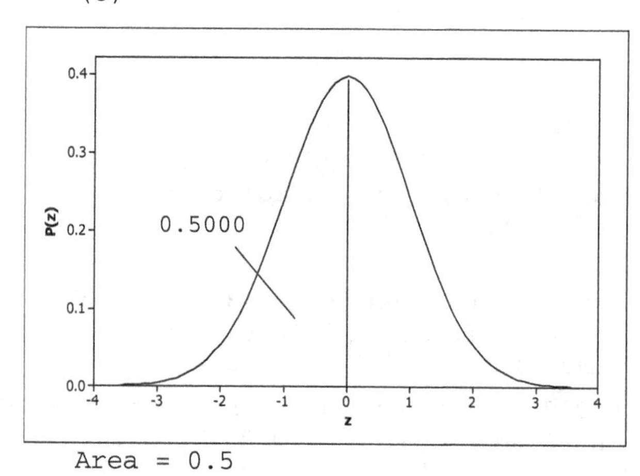

Area = 0.5

(d)

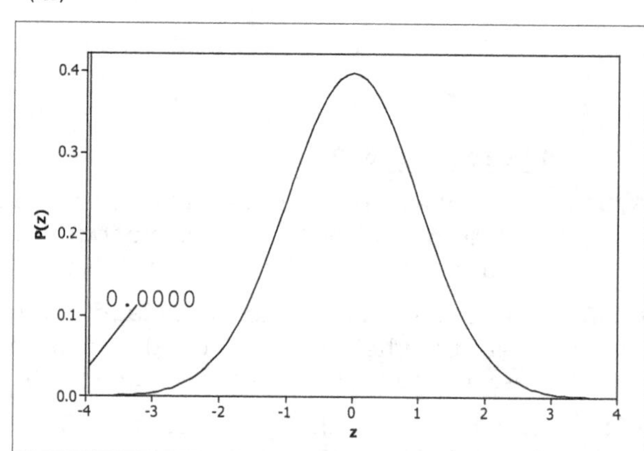

Area = 0.0000 (to 4 dp)

6.61 (a)

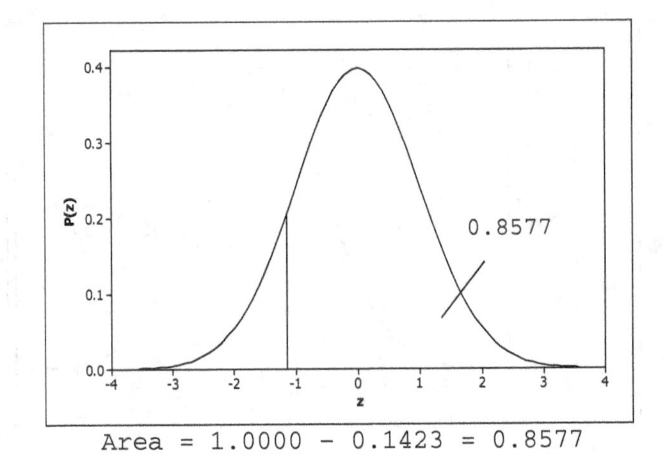

Area = 1.0000 − 0.1423 = 0.8577

(b)

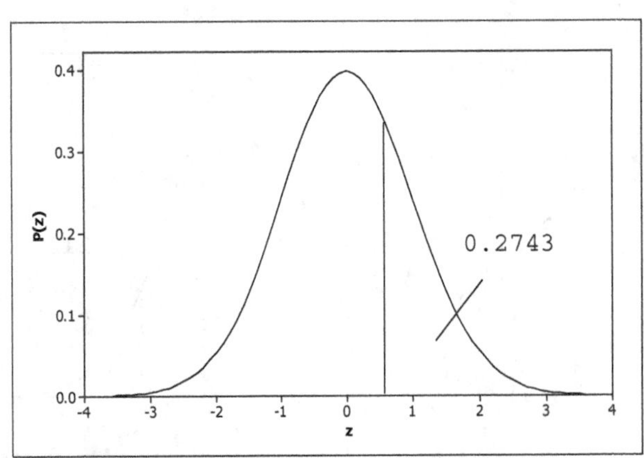

Area = 1.0000 − 0.7257 = 0.2743

(c)

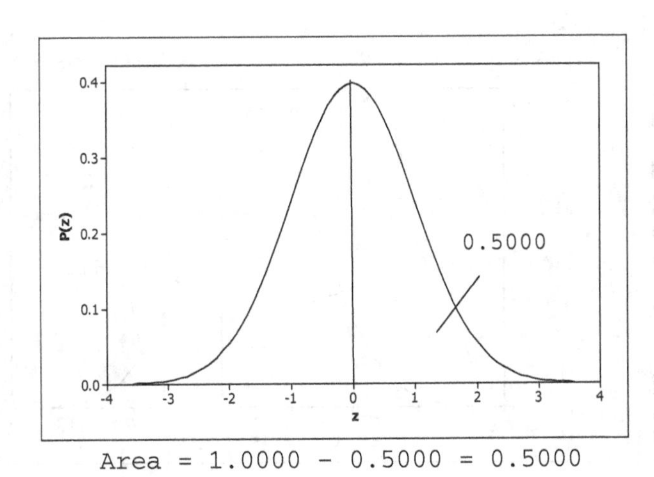

Area = 1.0000 − 0.5000 = 0.5000

(d)

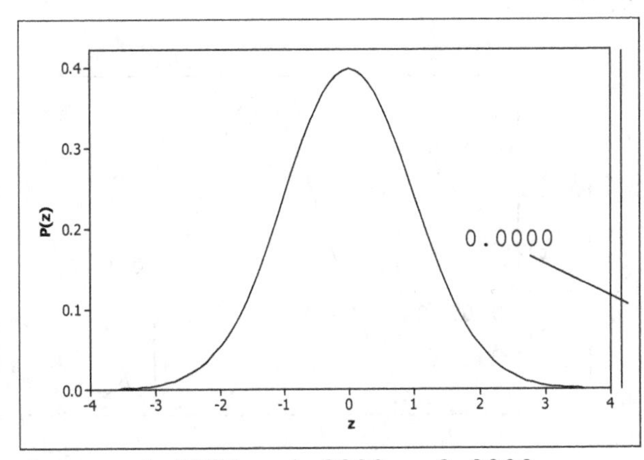

Area = 1.0000 − 1.0000 = 0.0000

6.63 (a)

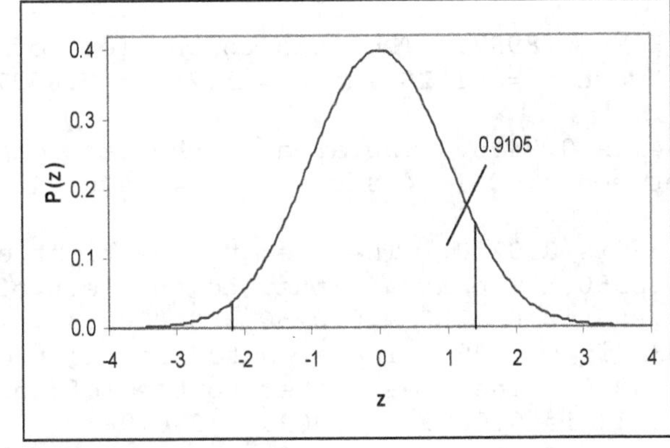

Area = 0.9251 − 0.0146 = 0.9105

(b)

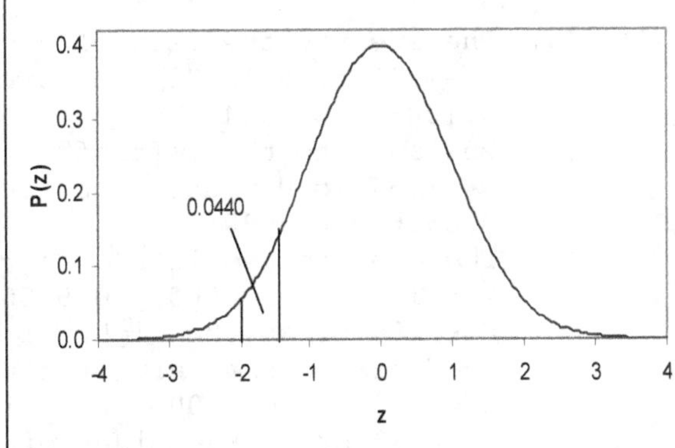

Area = 0.0668 − 0.0228 = 0.0440

(c)

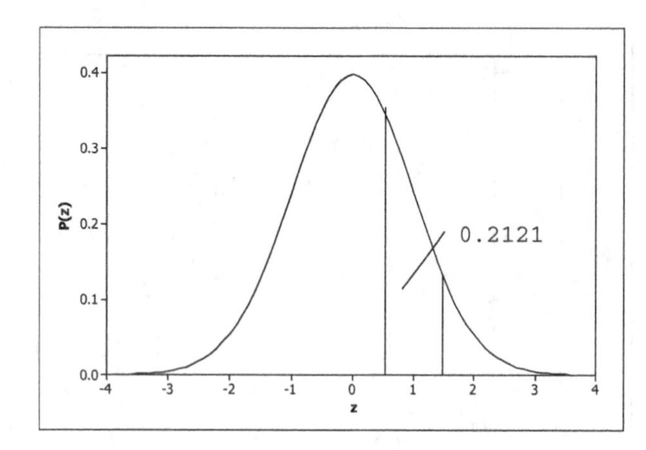

Area = 0.9345 - 0.7224 = 0.2121

(d)

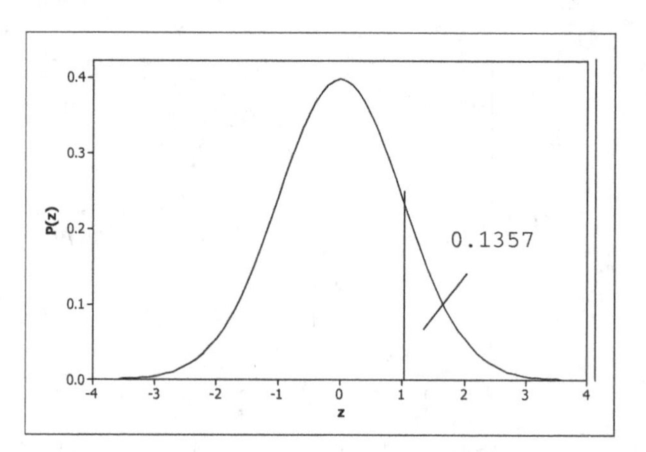

Area = 1.0000 - 0.8643 = 0.1357

6.65 (a)

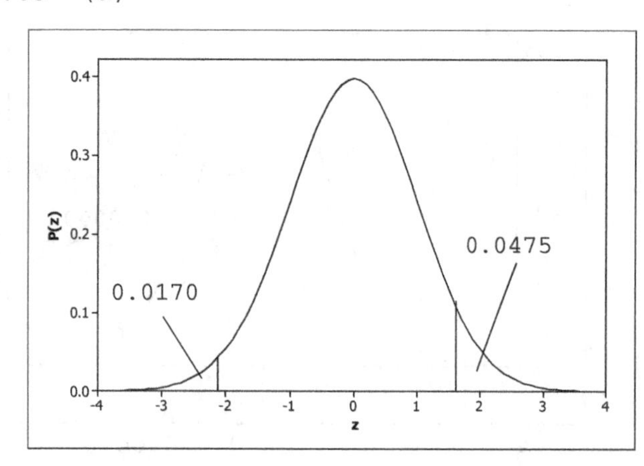

Area = 0.0170 + (1 - 0.9525) = 0.0645

(b)

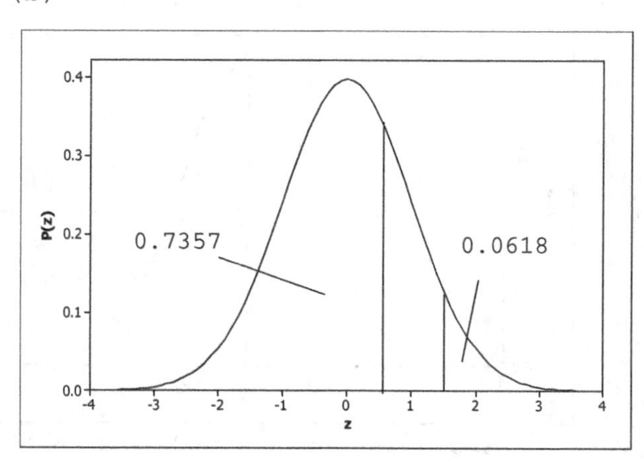

Area = 0.7357 + (1 - 0.9382) = 0.7975

6.67 (a) The area to the left of z = 1.28 is 0.8997. The area to the left of z = -1.28 is 0.1003. The area between z = -1.28 and z = 1.28 is 0.8997 - 0.1003 = 0.7994.

(b) The area to the left of z = 1.64 is 0.9495. The area to the left of z = -1.64 is 0.0505. The area between z = -1.64 and z = 1.64 is 0.9495 - 0.0505 = 0.8990.

(c) The area to the left of z = -1.96 is 0.0250. The area to the right of z = 1.96 is 1.0000 - 0.9750 = 0.0250. The area either to the left of z = -1.96 or to the right of z = 1.96 is 0.0250 + 0.0250 = 0.0500.

(d) The area to the left of z = -2.33 is 0.0099. The area to the right of z = 2.33 is 1.0000 - 0.9901 = 0.0099. The area either to the left of z = -2.33 or to the right of z = 2.33 is 0.0099 + 0.0099 = 0.0198.

6.69 (a)

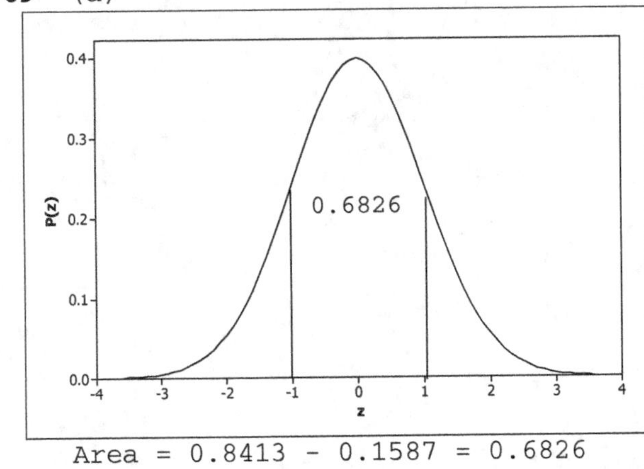

Area = 0.8413 - 0.1587 = 0.6826

(b)

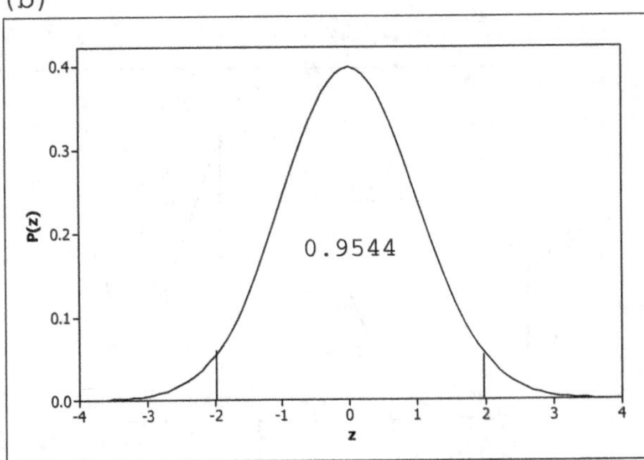

Area = 0.9772 - 0.0228 = 0.9544

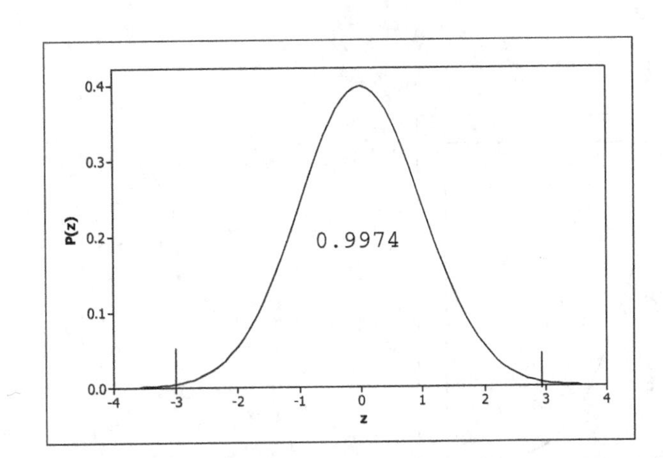

(c)

Area = 0.9987 - 0.0013 = 0.9974

6.71

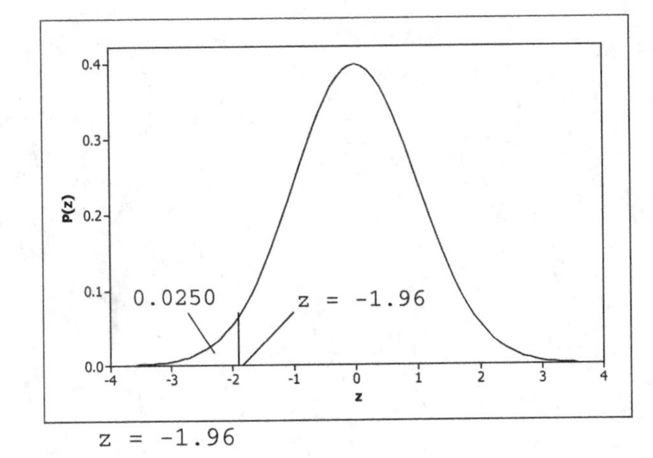

z = -1.96

6.73

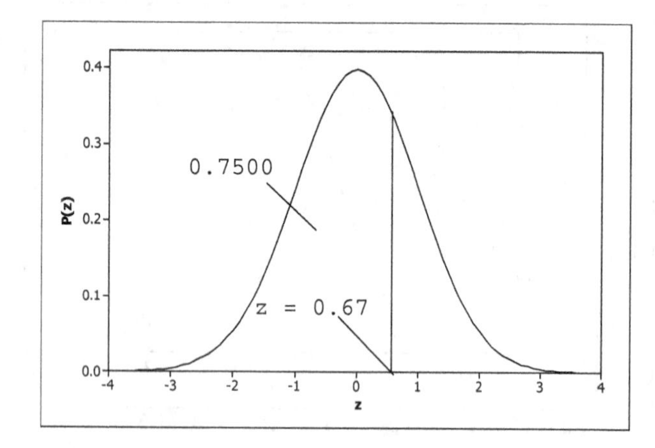

z = 0.67

6.75

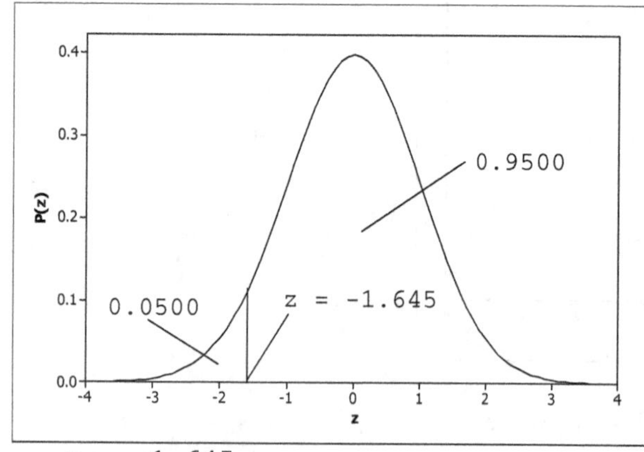

z = -1.645

Note: The area 0.0500 was exactly between the area of two z-scores on the table. Therefore, the two z-scores of -1.64 and -1.65 were averaged.

6.77

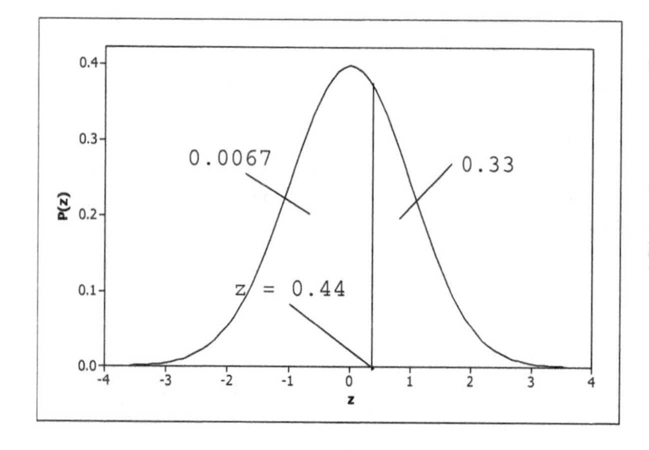

z= 0.44

6.79 (a)

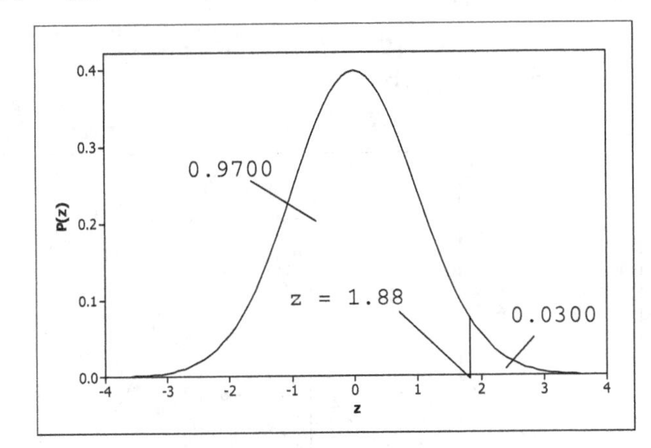

$z = 1.88$

(b)

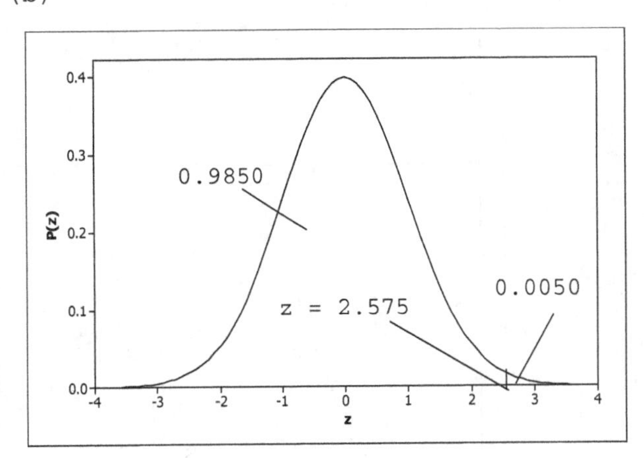

$z = 2.575$

6.81

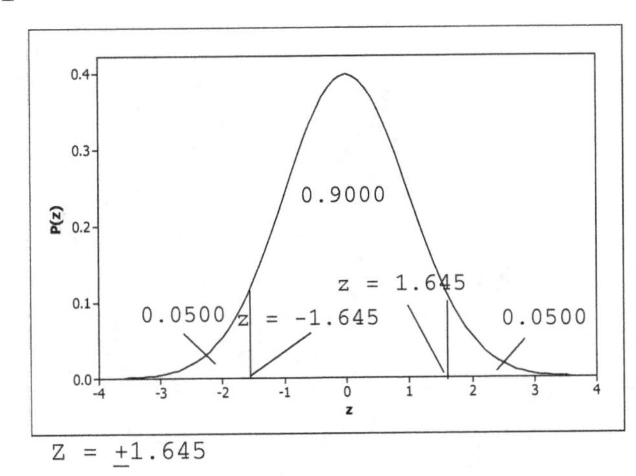

$Z = \pm 1.645$

6.83

$z_{0.10}$	$z_{0.05}$	$z_{0.025}$	$z_{0.01}$	$z_{0.005}$
1.28	1.645	1.96	2.33	2.575

6.85 (a) z_α (b) $-z_\alpha$ (c) $\pm z_{\alpha/2}$

(d)

For part (a):

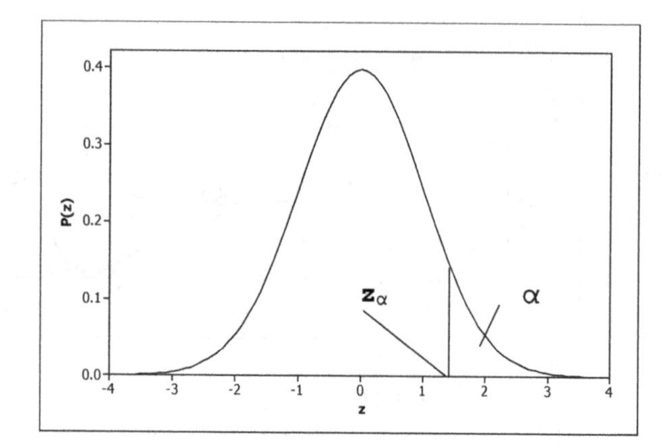

For part (b):

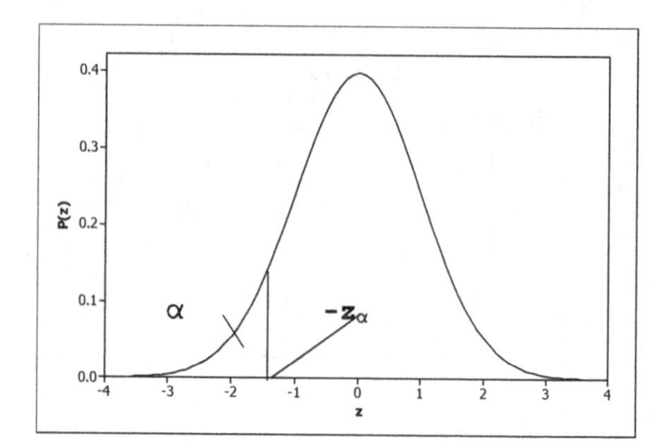

For part (c):

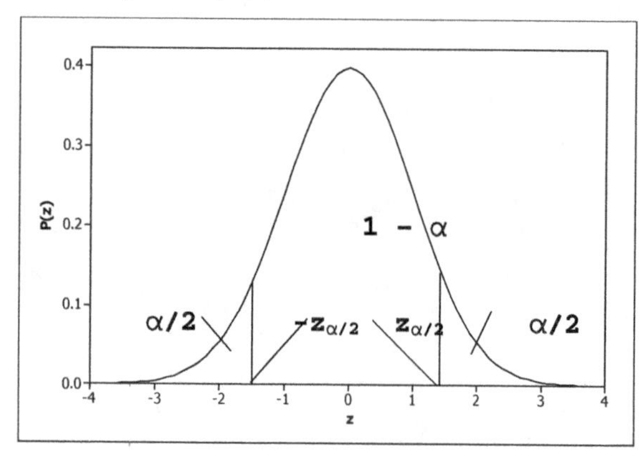

Exercises 6.3

6.87 The x values delimiting the interval within two standard deviations either side of the mean are $\mu - 2\sigma$ and $\mu + 2\sigma$. When these are standardized using $z = (x - \mu)/\sigma$, the former becomes $z = [(\mu - 2\sigma) - \mu]/\sigma = -2$ and the latter becomes $z = [(\mu + 2\sigma) - \mu]/\sigma = +2$.

6.89 (a) For the variable values 1 and 7, the z-values are

$$z = \frac{1-6}{2} = -2.50 \text{ and } z = \frac{7-6}{2} = 0.50$$

The area to the left of $z = -2.50$ is 0.0062 and the area to the left of $z = 0.50$ is 0.6915. Therefore the area between $z = -2.50$ and $z = 0.50$ is $0.6915 - 0.0062 = 0.6853$. Thus the percentage of all possible values of the variable that lie between 1 and 7 is 68.53%.

(b) For a variable value of 5, the z value is

$$z = \frac{5-6}{2} = -0.50$$

The area to the left of $z = -0.50$ is 0.3085. The area to right of $z = -0.50$ is $1 - 0.3085 = 0.6915$. Thus the percentage of values of the variable exceeding 5 is 69.15%.

(c) For a variable value of 4, the z value is

$$z = \frac{4-6}{2} = -1.00$$

The area to the left of z = -1.00 is 0.1587. Thus the percentage of values of the variable that are less than 4 is 15.87%.

6.91 (a) For the variable values 6 and 7, the z-values are

$$z = \frac{6-10}{3} = -1.33 \text{ and } z = \frac{7-10}{3} = -1.00$$

The area to the left of z = -1.33 is 0.0918 and the area to the left of z = -1.00 is 0.1587. Therefore the area between z = -1.33 and z = -1.00 is 0.1587 - 0.0918 = 0.0669. Thus the percentage of all possible values of the variable that lie between 6 and 7 is 6.69%.

(b) For a variable value of 10, the z value is

$$z = \frac{10-10}{3} = 0.00$$

The area to the left of z = 0.00 is 0.5000. The area to right of z = 0.00 is 1 - 0.5000 = 0.5000. Thus the percentage of values of the variable that are at least 10 is 50.00%.

(c) For a variable value of 17.5, the z value is

$$z = \frac{17.5-10}{3} = 2.50$$

The area to the left of z = 2.50 is 0.9938. Thus the percentage of values of the variable that are at most 17.5 is 99.38%.

6.93 (a) Using Table II, we find that an area of 0.2500 lies to left of z = -0.67, an area of .5000 lies to the left of z = 0.00 and an area of 0.7500 lies to the left of 0.67. We convert each of these z-values to x-values using $x = \mu + z\sigma$. Thus 25% of the values of the variable lie to the left of the first quartile, x = 6 + (-0.67)(2) = 4.66. Half (50%) of the values of the variable lie to the left of the second quartile (median), x = 6 + (0.00)(2) = 6. Finally, 75% of the values of the variable lie to the left of the third quartile, x = 6 + (0.67)(2) = 7.34.

(b) Using Table II, we find that an area of 0.8500 lies to the left of z = 1.04. We convert this z-value to an x-value using $x = \mu + z\sigma$. Thus 85% of the values of the variable lie to the left of the 85th percentile, x = 6 + (1.04)(2) = 8.08.

(c) If 65% of the possible values exceed the value, then 35% of the possible values are less than the value. Using Table II, we find that an area of 0.3500 lies to the left of z = -0.39. We convert this z-value to an x-value using $x = \mu + z\sigma$. Thus 65% of the values of the variable exceed x = 6 + (-0.39)(2) = 5.22.

(d) The two values that divide the area into a middle area of 0.95 and two outside areas of 0.025 have 2.5% of the curve and 97.5% of the curve to their left. Using Table II, we find that an area of 0.0250 lies to the left of z = -1.96 and an area of 0.9750 lies to the left of z = 1.96. We convert these z-values to x-values using $x = \mu + z\sigma$. Thus 95% of the variable lie between x = 6 + (-1.96)(2) = 2.08 and x = 6 + (1.96)(2) = 9.92.

6.95 (a) Using Table II, we find that an area of 0.2500 lies to left of $z = -0.67$, an area of .5000 lies to the left of $z = 0.00$ and an area of 0.7500 lies to the left of 0.67. We convert each of these z-values to x-values using $x = \mu + z\sigma$. Thus 25% of the values of the variable lie to the left of the first quartile, $x = 10 + (-0.67)(3) = 7.99$. Half (50%) of the values of the variable lie to the left of the second quartile (median), $x = 10 + (0.00)(3) = 10$. Finally, 75% of the values of the variable lie to the left of the third quartile, $x = 10 + (0.67)(3) = 12.01$.

(b) The seventh decile is the same as the 70th percentile. Using Table II, we find that an area of 0.7000 lies to the left of $z = 0.52$. We convert this z-value to an x-value using $x = \mu + z\sigma$. Thus 70% of the values of the variable lie to the left of the seventh decile, $x = 10 + (0.52)(3) = 11.56$.

(c) If 35% of the possible values exceed the value, then 65% of the possible values are less than the value. Using Table II, we find that an area of 0.6500 lies to the left of $z = 0.39$. We convert this z-value to an x-value using $x = \mu + z\sigma$. Thus 35% of the values of the variable exceed $x = 10 + (0.39)(3) = 11.17$.

(d) The two values that divide the area into a middle area of 0.99 and two outside areas of 0.005 have 0.5% of the curve and 99.5% of the curve to their left. Using Table II, we find that an area of 0.0050 lies to the left of $z = -2.575$ and an area of 0.9950 lies to the left of $z = 2.575$. We convert these z-values to x-values using $x = \mu + z\sigma$. Thus 99% of the variable lie between $x = 10 + (-2.575)(3) = 2.275$ and $x = 10 + (2.575)(3) = 17.725$.

6.97 (a) For carapace lengths 16 mm and 17 mm, the z-values are

$$z = \frac{16-18.14}{1.76} = -1.22 \text{ and } z = \frac{17-18.14}{1.76} = -0.65$$

The area to the left of $z = -1.22$ is 0.1112 and the area to the left of $z = -0.65$ is 0.2578. Therefore the area between $z = -1.22$ and $z = -0.65$ is $0.2578 - 0.1112 = 0.1466$. Thus the percentage of carapace lengths that are between 16 mm and 17 mm is 14.66%.

(b) For a carapace length of 19 mm, the z value is

$$z = \frac{19-18.14}{1.76} = 0.49$$

The area to the left of $z = 0.49$ is 0.6879. The area to right of 0.49 is $1 - 0.6879 = 0.3121$. Thus the percentage of adult *G. mollicoma* with carapace lengths exceeding 19 mm is 31.21%.

(c) Using Table II, we find that an area of 0.2500 lies to left of $z = -0.67$, an area of .5000 lies to the left of $z = 0.00$ and an area of 0.7500 lies to the left of 0.67. We convert each of these z-values to x-values using $x = \mu + z\sigma$. Thus 25% of the carapace lengths of adult *G. mollicoma* lie to the left of the first quartile, $x = 18.14 + (-0.67)(1.76) = 16.96$ mm. Half (50%) of the carapace lengths of adult *G. mollicoma* lie to the left of the second quartile (median), $x = 18.14 + (0.00)(1.76) = 18.14$ mm. Finally. 75% of the carapace lengths of adult *G. mollicoma* lie to the left of the third quartile, $x = 18.14 + (0.67)(1.76) = 19.32$ mm.

(d) Using Table II, we find that an area of 0.9500 lies to the left of $z = 1.645$. We convert this z-value to an x-value using $x = \mu + z\sigma$. Thus, the 95th percentile of the carapace lengths of adult *G. mollicoma* is $x = 18.14 + (1.645)(1.76) = 21.04$ mm.

6.99 (a) For finishers with times of 50 and 70 minutes, the z-values are

$$z = \frac{50-61}{9} = -1.22 \text{ and } z = \frac{70-61}{9} = 1.00$$

The area to the left of $z = -1.225$ is 0.1112 and the area to the left of $z = 1.00$ is 0.8413. Therefore the area between $z = -1.22$ and $z = 1.00$ is $0.8413 - 0.1112 = 0.7301$. Thus the percentage of finishers with times between 50 minutes and 70 minutes in the New York City 10 km run is 73.01%.

(b) For a finishing time of 75 minutes, the z value is

$$z = \frac{75-61}{9} = 1.56$$

The area to the left of $z = 1.56$ is 0.9406. Thus the percentage of finishers with times less than 765 minutes is 94.06%.

(c) Using Table II, we find that an area of 0.4000 lies to left of $z = -0.25$. We convert this z-value to an x-value using $x = \mu + z\sigma$. Thus 40% of the finishers had times less than the 40th percentile, $x = 61 + (-0.25)(9) = 58.75$ minutes.

(d) The eighth decile is the same as the 80th percentile. Using Table II, we find that an area of 0.8000 lies to the left of $z = 0.84$. We convert this z-value to an x-value using $x = \mu + z\sigma$. Thus, 80% of the finishing times were less than $x = 61 + (0.84)(9) = 68.56$ minutes.

6.101 (a) For umbilical cord PH levels that are 7.00 and 7.50, the z values are

$$z = \frac{7-7.32}{0.1} = -3.20 \quad \textit{and} \quad z = \frac{7.50-7.32}{0.1} = 1.80$$

The area to the left of $z = -3.20$ is 0.0007 and the area to the left of $z = 1.80$ is 0.9641. The area between $z = -3.20$ and $z = 1.80$ is $0.9641 - 0.0007 = 0.9634$. Therefore 96.34% of umbilical cord PH levels for preterm infants are between 7.00 and 7.50.

(b) For umbilical cord PH levels that are 7.40, the z value is

$$z = \frac{7.40-7.32}{0.1} = 0.80.$$ The area to the left of $z = 0.80$ is 0.7881.

Thus the area to the right of $z = 0.80$ is $1 - 0.7881 = 0.2119$, implying that 21.19% of umbilical cord PH levels for preterm infants are over 7.40.

6.103

Part	Standard deviations to either side of the mean	Area under normal curve	Percent
(a)	1	0.3413 x 2 = 0.6826	68
(b)	2	0.4772 x 2 = 0.9544	95
(c)	3	0.4987 x 2 = 0.9974	99.7

6.105 (a) Approximately 68% of Swedish men have brain weights between <u>1.29</u> and <u>1.51</u> kg.

(b) Approximately 95% of Swedish men have brain weights between <u>1.18</u> and <u>1.62</u> kg.

(c) Approximately 99.7% of Swedish men have brain weights between <u>1.07</u> and <u>1.73</u> kg.

(d)

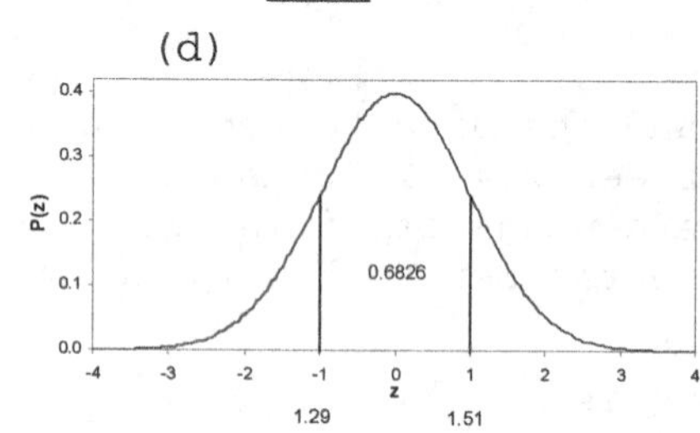

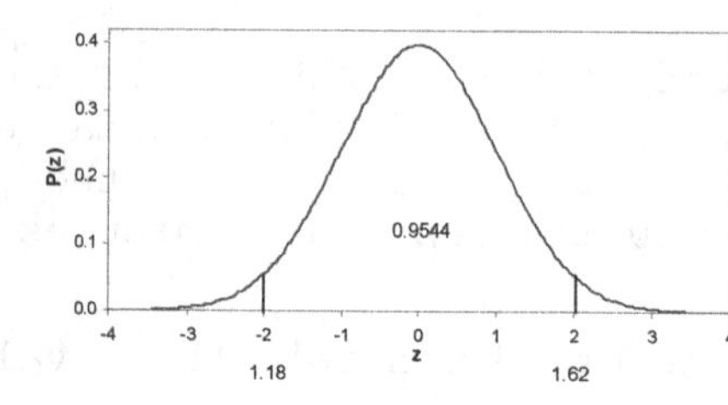

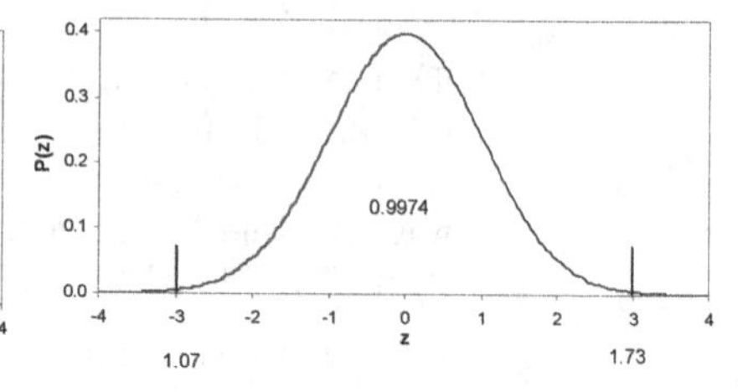

6.107 (a) From Table 6.1, exactly 11.70% of the heights of female students are between 62 and 63. For heights of 62 and 63 inches, the z values are

$$z = \frac{62-64.4}{2.4} = -1.00 \text{ and } z = \frac{63-64.4}{2.4} = -0.58$$

The area to the left of z = -1.00 is 0.1587 and the area to the left of z = -0.58 is 0.2810. Thus, the area between z = -1.00 and z = -0.58 is 0.2810 - 0.1587 = 0.1223 (12.23%). The two percentages are quite close.

(b) From Table 6.1, exactly = 0.1575 + 0.1100 + 0.0735 + 0.0374 + 0.0199 = 0.3983 (39.83%) of the heights of female students are between 65 and 70 inches. For heights of 65 and 70 inches, the z values are

$$z = \frac{65-64.4}{2.4} = 0.25 \text{ and } z = \frac{70-64.4}{2.4} = 2.33$$

The area to the left of z = 0.25 is 0.5987 and the area to the left of z = 2.33 is 0.9901. Thus, the area between z = 0.25 and a = 2.33 is 0.9901 - 0.5987 = 0.3914 (39.14%). The two percentages are quite close.

6.109 (a) All of the times, 100%, should be at least 0 years because negative times do not make sense in the context of this problem.

(b) $$P(X \geq 0) = 1 - P(X \leq 0) = 1 - P(Z \leq \frac{0-0.18}{0.624}) = 1 - P(Z \leq -0.29) = 1 - 0.3859 = 0.6141$$

61.41% of times are at least 0 years.

(c) If x was normally distributed, the probability that the time is at least 0 is approximated to be 61.41%. However, really all of the times should be at least 0 because negative values do not make sense for the variable length of time since taking a math course. Therefore, the variable x is not approximately normally distributed.

6.111 (a) $P(Y > 5) = P(Z > \frac{5-4.66}{0.75}) = P(Z > 0.45) = 1 - 0.6736 = 0.3264$. We interpret this to mean that about 32.64% of the nests of the booted eagle of western Europe are more than 5 km to the nearest marshland.

(b) $P(3 \le Y \le 6) = P(\frac{3-4.66}{0.75} \le Z \le \frac{6-4.66}{0.75}) = P(-2.21 \le Z \le 1.79)$

$= 0.9633 - 0.0136 = 0.9497$

We interpret this to mean that about 95% of the nests of the booted eagle of western Europe are between 3 and 6 km from the nearest marshland.

6.113 $P(\mu - z_{\alpha/2}\,\sigma < x < \mu + z_{\alpha/2}) = P(-z_{\alpha/2}\,\sigma < x - \mu < z_{\alpha/2}\,\sigma)$

$= P(-z_{\alpha/2} < (x - \mu)/\sigma < z_{\alpha/2})$

$= P(-z_{\alpha/2} < z < z_{\alpha/2}) = (1 - \alpha/2) - \alpha/2$

$= 1 - \alpha$

6.115 The *k*th percentile can be found by first finding the z-score with area of *k*/100 to its left or 1-*k*/100 to its right. Then, multiplying this z-score by the standard deviation and adding it to the mean. Therefore, the *k*th percentile is $P_k = \mu + z_{(1-k/100)}\,\sigma$.

Exercises 6.4

6.117 Decisions about whether a variable is normally distributed often are important in subsequent analyses such as percentile calculations or in determining the type of statistical inference procedure to be used.

6.119 If the normal probability plot is roughly linear except for a small number of points that lie well outside the overall pattern of the plot, it is possible that some or all of those points are outliers. If there is sufficient reason to remove the potential outliers from the sample and doing so results in a plot that is linear without outliers, the analysis can often be carried out with the remaining observations.

6.121 The variable is approximately normally distributed because the normal probability plot is approximately linear.

6.123 The variable is not normally distributed because the normal probability plot is not linear and is instead curved.

6.125 The variable is not normally distributed because the normal probability plot is not linear and is instead curved.

6.127 (a)

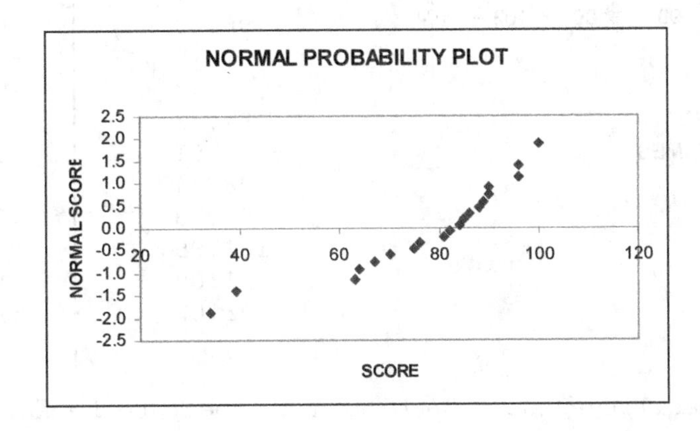

Exam Score X	Normal Score y
34	-1.87
39	-1.40
63	-1.13
64	-0.92
67	-0.74
70	-0.59
75	-0.45
76	-0.31
81	-0.19
82	-0.06
84	0.06
85	0.19
86	0.31
88	0.45
89	0.59
90	0.74
90.01	0.92
96	1.13
96.01	1.40
100	1.87

(b) Based on the probability plot, there appear to be two outliers in the sample: 34 and 39.

(c) Based on the probability plot, the sample does not appear to come from a normally distributed population.

6.129 (a)

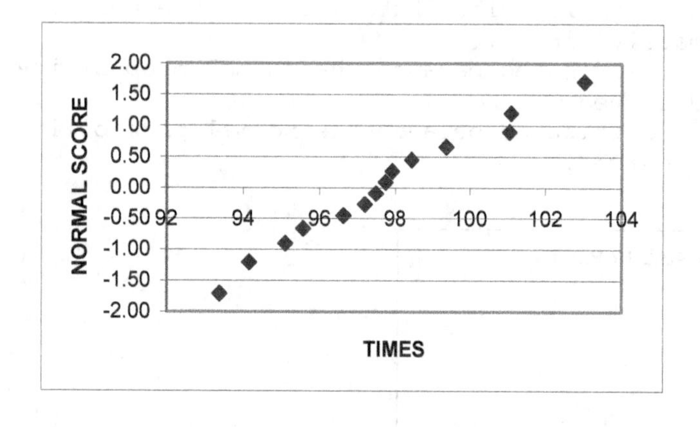

Times X	Normal Score y
93.37	-1.71
94.15	-1.20
95.10	-0.90
95.57	-0.66
96.63	-0.45
97.19	-0.27
97.47	-0.09
97.73	0.09
97.91	0.27
98.44	0.45
99.38	0.66
101.05	0.90
101.09	1.20
103.02	1.71

(b) Based on the probability plot, there do not appear to be any outliers in the sample.

(c) Based on the probability plot, the sample appears to be from an approximately normally distributed population.

6.131 (a) We enter the data in Excel in a column which we named "x" at the top. Highlight the name and the data with your mouse and then choose **Charts and Plots.** Select **Normal Probability Plot** from the drop down **Function type** box and enter x in both the **Quantitative variable** and **Label variable** boxes. Click **OK**. The resulting plot is

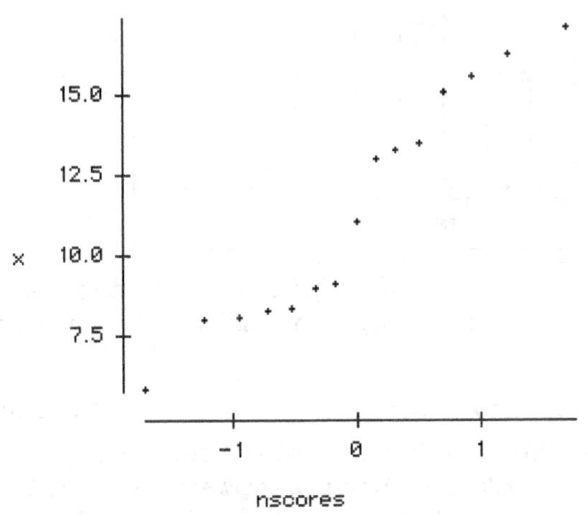

(b) From the plot, there do not appear to be any outliers.

(c) From the plot, normality seems to be a reasonable assumption.

6.133 (a) We enter the data in Excel in a column which we named "DOU" at the top. Highlight the name and the data with your mouse and then choose **Charts and Plots.** Select **Normal Probability Plot** from the drop down **Function type** box and enter DOU in both the **Quantitative variable** and **Label variable** boxes. Click **OK**. The resulting plot is

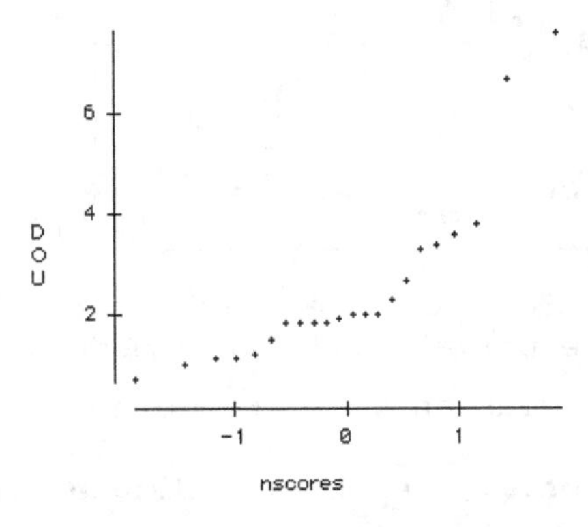

(b) From the plot, it appears that there may be two outliers, 7.6 and 6.7.

(c) From the plot, the data do not appear to be normally distributed. However, if the two outliers were excluded from the data, the remaining data may be normally distributed.

6.135 (a) Using Minitab and assuming that the data are in a column named 'TEMP', choose **Graph ▶ Histogram...**, select the **Simple** plot and click **OK**. Specify TEMP in the **Graph Variables** text box and click **OK**. The resulting plot is

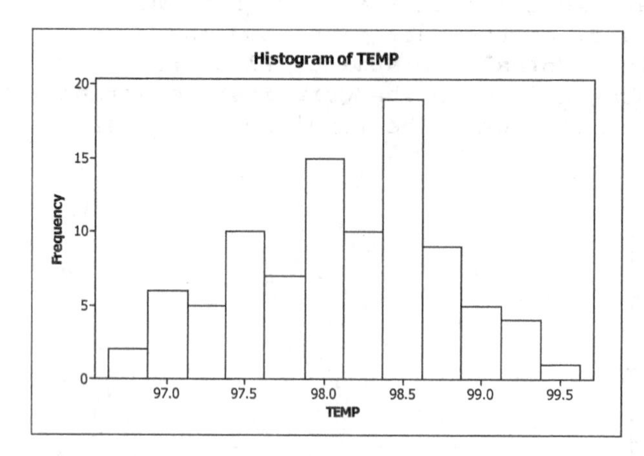

Aside from the unusual 'saw tooth' pattern of the left side of the graph, an assumption of normality for this data appears to be reasonable.

(b) Choose **Graph ▶ Probability Plot...**, select the **Single** plot and click **OK**. Specify TEMP in the Graph Variables text box. Click on the **Distribution** button, click the **Data Display** tab, select the **Symbols only** option button from the **Data Display** list and click **OK**. Now click the **Scale** button, click on the **Y-Scale Type** tab, select the **Score** option button from the **Y-Scale Type** list, and click **OK** twice. The resulting plot is

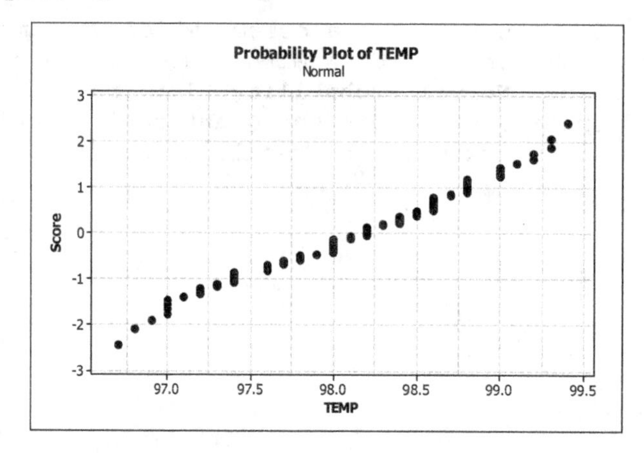

(b) There are no apparent outliers and the graph is roughly linear, so it seems reasonable to assume normality for these data.

(c) We arrived at the same conclusion from both graphs.

6.137 Choose **Graph ▶ Probability Plot...**, select the **Single** plot and click **OK**. Specify CHIPS in the **Graph Variables** text box. Click on the **Distribution** button, click the **Data Display** tab, select the **Symbols only** option button from the **Data Display** list and click **OK**. Now click the **Scale** button, click on the **Y-Scale Type** tab, select the **Score** option button from the **Y-Scale Type** list, and click **OK** twice. The resulting plot is

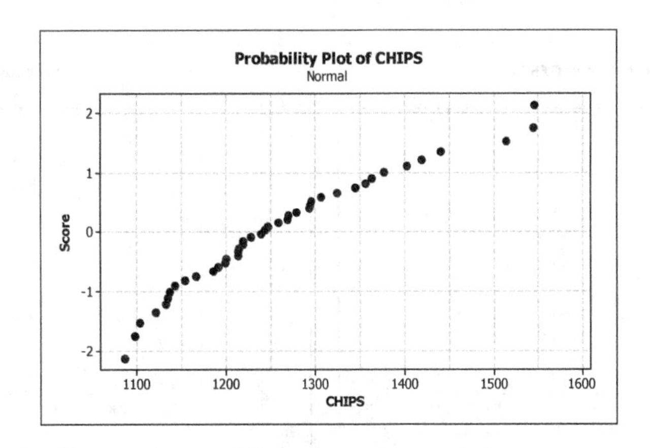

Yes. There do not appear to be any outliers in these data. Since the plot follows a fairly straight line, normality of the chips data seems to be a reasonable assumption.

6.139 (a) Using Minitab, we will generate four columns of fifty observations each by choosing **Calc ▶ Random data ▶ Normal...**, entering 50 in the **Generate rows of data** text box, clicking in the **Store in column(s)** text box and typing C1-C4, clicking in the **Mean** text box and entering 266, clicking in the **Standard deviation** text box and entering 16, and clicking **OK**. Now click in the worksheet column title row and name the four columns GEST1, GEST2, GEST3, and GEST4.

Next we select **Graph ▶ Probability Plot**, select the **Single** version and click **OK**, then enter GEST1, GEST2, GEST3, and GEST4 in the **Graph Variables** text box. Click **OK**. The resulting four graphs are shown as follows.

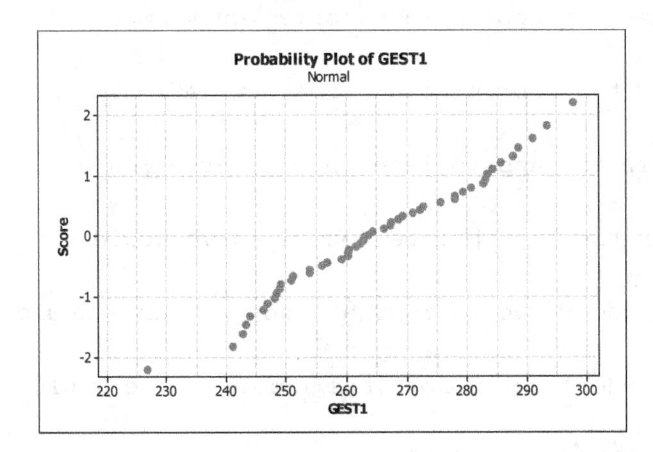

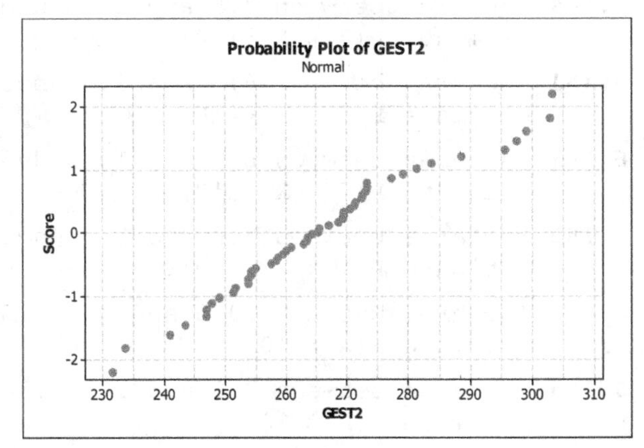

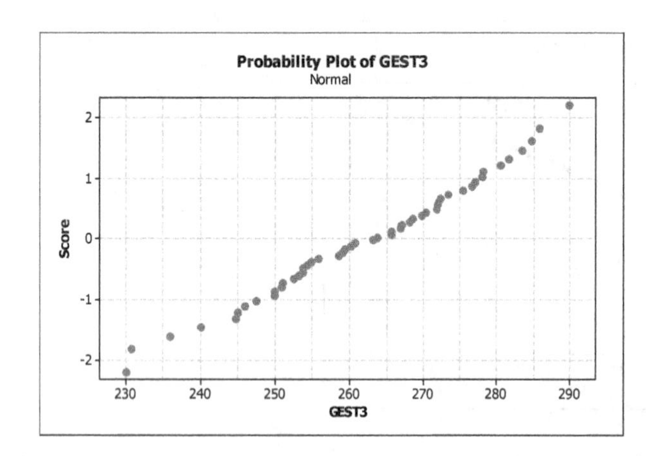

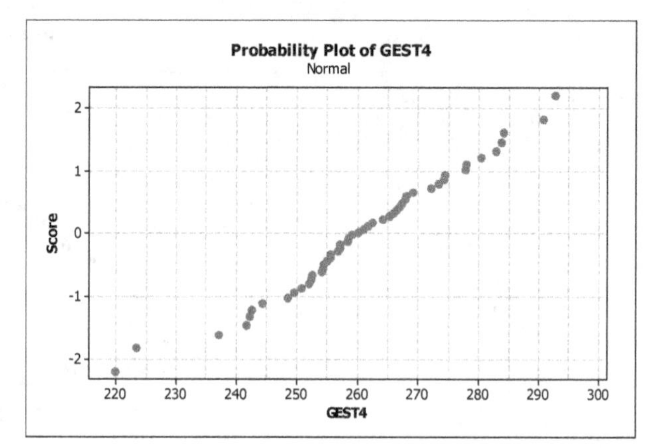

(c) Yes. Since the data were generated from a normal distribution, we would expect the four plots to be roughly linear, as they are. Your simulation will likely result in data and graphs that differ from ours.

Exercises 6.5

6.141 It is not practical to use the binomial probability formula when the number of trials, n, is large.

6.143 The area between 7.5 and 8.5 under the appropriate normal curve would estimate $P(X = 8)$.

6.145 The area between -0.5 and 6.5 under the appropriate normal curve would estimate $P(X < 7)$.

6.147 The area between -0.5 and 7.5 under the appropriate normal curve would estimate $P(X \le 7)$.

6.149 The area between 7.5 and 10.5 under the appropriate normal curve would estimate $P(7 < X \le 10)$.

6.151 The area between 6.5 and 9.5 under the appropriate normal curve would estimate $P(7 \le X < 10)$.

6.153 The area between 6.5 and 10.5 under the appropriate normal curve would estimate $P(7 \le X \le 10)$.

6.155 The area between 7.5 and 9.5 under the appropriate normal curve would estimate $P(7 < X < 10)$.

6.157 The area between 10.5 and $n + 0.5$ under the appropriate normal curve would estimate $P(X > 10)$.

6.159 The area between 9.5 and $n + 0.5$ under the appropriate normal curve would estimate $P(X \ge 10)$.

6.161 (a) (i) $P(x = 4 \text{ or } 5) = 0.2051 + 0.2461 = 0.4512$

(ii) $P(3 \le x \le 7) = 0.1172 + 0.2051 + 0.2461 + 0.2051 + 0.1172 = 0.8907$

(b) (i) $P(x = 4 \text{ or } 5)$:

Step 1: $n = 10$; $p = 0.5$

Step 2: $np = 5$; $n(1 - p) = 5$. Since both np and $n(1 - p)$ are at least 5, the normal approximation can be used.

Step 3: $\mu_x = np = 10(0.5) = 5$

$\sigma_x = \sqrt{np(1-p)} = \sqrt{10(0.5)(0.5)} = 1.58$

Step 4: Making the continuity correction, we find the area under the normal curve with parameters μ = 5 and σ = 1.58 that lies between x = 3.5 and x = 5.5. The z-values for these x values are

$$z = \frac{3.5-5}{1.58} = -0.95 \text{ and } z = \frac{5.5-5}{1.58} = 0.32$$

The area to the left of z = -0.95 is 0.1711 and the area to the left of z = 0.32 is 0.6255. Therefore, the approximate probability that x equals 4 or 5 is 0.6255 - 0.1711 = 0.4544. This is very close the actual probability of 0.4512.

(ii) $P(3 \le x \le 7)$:

Steps 1, 2 and 3 are the same as the previous problem.

Step 4: Making the continuity correction, we find the area under the normal curve with parameters μ = 5 and σ = 1.58 that lies between x = 2.5 and x = 7.5. The z-values for these x values are

$$z = \frac{2.5-5}{1.58} = -1.58 \text{ and } z = \frac{7.5-5}{1.58} = 1.58$$

The area to the left of z = -1.58 is 0.0571 and the area to the left of z = 1.58 is 0.9429. Therefore, the approximate probability that x lies between 3 and 7, inclusive, is 0.9429 - 0.0571 = 0.8858. This is very close the actual probability of 0.8907.

6.163 Since both np and n(1 - p) are at least 5, we would use the normal curve with

$$\mu = np = 25(0.5) = 12.5 \text{ and } \sigma = \sqrt{np(1-p)} = \sqrt{25(0.5)(1-0.5)} = \sqrt{6.25} = 2.5.$$

6.165 (a) The mean is $\mu = np = 100(0.13) = 13$ and $\sigma = \sqrt{np(1-p)} = \sqrt{100(0.13)(0.87)} = 3.363$. Note that np = 13 and n(1 - p) = 87, so it is reasonable to use the normal approximation for binomial probabilities. If X represents the number of college women that have been stalked, we want P(X < 10). With the continuity correction, this becomes the area between -0.5 and 9.5. For X = -0.5, we have z = (-0.5 - 13)/3.363 = -4.01, for X = 9.5, we have z = (9.5 - 13)/3.363 = -1.04. The area to the left of z = -4.01 is approximately 0.0000 and the area to the left of -1.04 is 0.1492. Therefore P(X < 10) = 0.1492 - 0.0000 = 0.1492 (approximately).

(b) If X represents the number of college women that have been stalked, we want $P(15 \le X \le 20)$. With the continuity correction, this becomes P(14.5 < X < 20.5). For X = 14.5, we have z = (14.5 - 13)/3.363 = 0.45. For X = 20.5, we have z = (20.5 - 13)/3.363 = 2.23. The area to the left of z = 0.45 is 0.6736. The area to the left of z = 2.23 is 0.9871. The area between z = 0.45 and z = 2.23 is 0.9871 - 0.6736 = 0.3135. Therefore $P(15 \le X \le 20)$ = 0.3135 (approximately).

(c) If X represents the number of college women that have been stalked, we want $P(X \geq 15)$. With the continuity correction, this becomes the area between 14.5 and 100.5. For X = 14.5, we have z = (14.5 - 13)/3.363 = 0.45. For X = 100.5, we have z = (100.5 - 13)/3.363 = 26.02. The area to the left of z = 0.45 is 0.6736 and the area to the left of z = 5.20 is approximately 1.000. The area between 0.45 and 26.02 is 1.0000 - 0.6736 = 0.3264 = 0.3264. Therefore $P(X \geq 15) = 0.3264$ (approximately).

6.167 For parts (a), (b),and (c), steps 1-3 are as follows:

Step 1: n = 100; p = 0.304

Step 2: np = 30.4; n(1 - p) = 69.6. Since both np and n(1 - p) are at least 5, the normal approximation can be used.

Step 3:

$$\mu_x = np = 100(0.304) = 30.4$$

$$\sigma_x = \sqrt{np(1-p)} = \sqrt{100(0.304)(0.696)} = 4.60$$

(a) P(X = 32)

Step 4: For x = 31.5 and x = 32.5, the z-scores are

$$z = \frac{31.5-30.4}{4.60} = 0.24 \text{ and } z = \frac{32.5-30.4}{4.60} = 0.46$$

The area to the left of z = 0.24 is 0.5948 and the area to the left of z = 0.46 is 0.6772. Therefore, the approximate probability that X = 32 is 0.6772 - 0.5948 = 0.0824.

(b) $P(30 \leq X \leq 35)$

Step 4: For X = 29.5 and X = 35.5, the z-scores are

$$z = \frac{29.5-30.4}{4.60} = -0.20 \text{ and } z = \frac{35.5-30.4}{4.60} = 1.11$$

The area to the left of z = -0.20 is 0.4207 and the area to the left of z = 1.11 is 0.8665. Therefore, the approximate probability that X is between 30 and 35, inclusive, is 0.8665 - 0.4207 = 0.4458.

(c) $P(X \geq 25)$

Step 4: For X = 24.5 and X = *n* + 0.5 = 100.5, the z-scores are

$$z = \frac{24.5-30.4}{4.60} = -1.28 \quad \textit{and} \quad z = \frac{100.5-30.4}{4.60} = 15.24$$

The area to the left of z = -1.28 is 0.1003 and the area to the left of z = 15.24 is 1.0000. Therefore, the approximate probability that X is at least 25 is 1.0000 - 0.1003 = 0.8997.

6.169 For parts (a), (b), (c), and (d), steps 1-3 are as follows:

Step 1: n = 42; p = 0.16

Step 2: np = 6.72; n(1 - p) = 35.28. Since both np and n(1 - p) are at least 5, the normal approximation can be used.

Step 3:

$$\mu_x = np = 42(0.16) = 6.72$$
$$\sigma_x = \sqrt{np(1-p)} = \sqrt{42(0.16)(0.84)} = 2.38$$

(a) P(x = 5):

Step 4: For x = 4.5 and x = 5.5, the z-scores are

$$z = \frac{4.5 - 6.72}{2.38} = -0.93 \text{ and } z = \frac{5.5 - 6.72}{2.38} = -0.51.$$

The area to the left of z = -0.93 is 0.1762 and the area to the left of z = -0.51 is 0.3050. Therefore, the approximate probability that X is five is 0.3050 - 0.1762 = 0.1288.

(b) P(9 ≤ x ≤ 12):

Step 4: For x = 8.5 and x = 12.5, the z-scores are

$$z = \frac{8.5 - 6.72}{2.38} = 0.75 \text{ and } z = \frac{12.5 - 6.72}{2.38} = 2.43$$

The area to the left of z = 0.75 is 0.7734 and the area to the left of z = 2.43 is 0.9925. The probability that X is between 9 and 12, inclusive, is 0.9925 - 0.7734 = 0.2191.

(c) P(x ≥ 1):

Step 4: For x = 0.5 and x = *n* + 0.5 = 42.5, the z-scores are

$$z = \frac{0.5 - 6.72}{2.38} = -2.61 \quad \textit{and} \quad z = \frac{42.5 - 6.72}{2.38} = 15.03$$

The area to the left of z = -2.61 is 0.0045 and the area to the left of z = 15.03 is 1.0000. Therefore, the approximate probability that X is at least one is 1.0000 - 0.0045 = 0.9955.

(d) P(x ≤ 2):

Step 4: For x = 2.5 and x = -0.5, the z-scores are

$$z = \frac{2.5 - 6.72}{2.38} = -1.77 \quad \textit{and} \quad z = \frac{-0.5 - 6.72}{2.38} = -3.03$$

The area to the left of z = -1.77 is 0.0384 and the area to the left of -3.03 is 0.0012. Therefore, the approximate probability that X is at most two is 0.0384 - 0.0012 = 0.0372. (answer also 0.0363 depending on rounding of the standard deviation)

(e) In Step 2, we found that np = 6.72. This meets the minimum requirement of at least 5, but it is close to 5. The larger the values of np and n(1-p), the better the normal approximation to the binomial distribution. Since 6.72 is a smaller number, the normal approximation is acceptable, but not the most accurate approximation.

6.171 For parts (a), (b), and (c), steps 1-3 are as follows:

Step 1: n = 250; p = 0.34

Step 2: np = 85; n(1 - p) = 165. Since both n(1 -p) and np are at least 5, the normal approximation can be used.

Step 3:

$$\mu_x = np = 250(0.34) = 85$$

$$\sigma_x = \sqrt{np(1-p)} = \sqrt{250(0.34)(0.66)} = 7.49$$

(a) Thirty four percent of 250 is 85. P(X = 85):

Step 4: For x = 84.5 and x = 85.5, the z scores are

$$z = \frac{84.5-85}{7.49} = -0.07 \text{ and } z = \frac{85.5-85}{7.49} = 0.07.$$

The area to the left of z = -0.07 is 0.4721 and the area to the left of z = 0.07 is 0.5279. Therefore, the approximate probability that X is 85 is 0.5279 - 0.4721 = 0.0558.

(b) Step 4: We need to find P(X > 85). For x = 85.5 and x = 250.5, the z-scores are

$$z = \frac{85.5-85}{7.49} = 0.07 \quad \textit{and} \quad z = \frac{250.5-85}{7.49} = 22.10.$$

The area to the left of z = 0.07 is 0.5279 and the area to the left of z = 22.10 is 1.0000. Therefore, the approximate probability that X exceeds 85 is 1.0000 - 0.5279 = 0.4721.

(c) Step 4: P(X < 85). For x =84.5 and x = -0.5, the z-scores are

$$z = \frac{84.5-85}{7.49} = -0.07 \quad \textit{and} \quad z = \frac{-0.5-85}{7.49} = -11.42.$$

The area to the left of z = -0.07 is 0.4721 and the area to the left of z = -11.42 is 0.0000. Therefore, the approximate probability that X is fewer than 85 is 0.4721 - 0.0000 = 0.4721.

6.173 (a) 100 bets are made. The gambler is ahead if she has won more than 50 bets. So, we are concerned with finding P(x > 50) or P(x ≥ 51).

Step 1: n = 100; p = 18/38 = 0.47368

Step 2: np = 47.368; n(1 - p) = 52.632. Since both np and n(1-p) are at least 5, the normal approximation can be used.

Step 3:

$$\mu_x = np = 100(0.47368) = 47.368$$

$$\sigma_x = \sqrt{np(1-p)} = \sqrt{100(0.47368)(0.52632)} = 4.993$$

Step 4: For x = 50.5 and x = 100.5, the z-scores are

$$z = \frac{50.5-47.368}{4.993} = 0.63 \quad \textit{and} \quad z = \frac{100.5-47.368}{4.993} = 10.64$$

The area to the left of z = 0.63 is 0.7357 and the area to the left of z = 10.64 is 1.0000. Therefore, the approximate probability that X is at least 51 is 1.0000 - 0.7357 = 0.2643.

(b) 1000 bets are made. The gambler is ahead if she has won more than 500 bets. So, we are concerned with finding P(x > 500) or P(x ≥ 501).

Step 1: n = 1000; p = 18/38 = 0.47368

Step 2: np = 473.68; n(1 - p) = 526.32. Since both np and n(1-p) are at least 5, the normal approximation can be used.

Step 3:

$$\mu_x = np = 1000(0.47368) = 473.68$$

$$\sigma_x = \sqrt{np(1-p)} = \sqrt{1000(0.47368)(0.52632)} = 15.789$$

Step 4: For x = 500.5 and x = 1000.5, the z-scores are

$$z = \frac{500.5 - 473.68}{15.789} = 1.70 \quad and \quad z = \frac{1000.5 - 473.68}{15.789} = 33.37$$

The area to the left of z = 1.70 is 0.9554 and the area to the left of z = 33.37 is 1.0000. Therefore, the approximate probability that X is at least 501 is 1.0000 - 0.9554 = 0.0446.

(c) 5000 bets are made. The gambler is ahead if she has won more than 2500 bets. So, we are concerned with finding P(x > 2500) or P(x ≥ 2501).

Step 1: n = 5000; p = 18/38 = 0.47368

Step 2: np = 2368.4; n(1 - p) = 2631.6. Since both np and n(1-p) are at least 5, the normal approximation can be used.

Step 3:

$$\mu_x = np = 5000(0.47368) = 2368.4$$

$$\sigma_x = \sqrt{np(1-p)} = \sqrt{300(0.47368)(0.52632)} = 35.306$$

Step 4: For x = 2500.5 and x = 5000.5, the z-scores are

$$z = \frac{2500.5 - 2368.4}{35.306} = 3.74 \quad and \quad z = \frac{5000.5 - 2368.4}{35.306} = 74.55$$

The area to the left of z = 3.74 is 0.9999 and the area to the left of 74.55 is 1.0000. Therefore, the approximate probability that X is at least 2501 is 1.0000 - 0.9999 = 0.0001.

6.175 (a) The expected number of males with Fragile X Syndrome is np = 10000(1/1500) = 6.67.

(b) To use the normal approximation to the binomial distribution, we first need the standard deviation, which is

$$\sigma_x = \sqrt{np(1-p)} = \sqrt{10000(\frac{1}{1500})(\frac{1499}{1500})} = 2.581$$

To find the approximate probability that more than 7 males have Fragile X syndrome, we want P(X > 7) or P(X ≥ 8) or, approximating, we use P(X > 7.5). For x = 7.5 and x = 10000.5, the z-scores are

$$z = \frac{7.5 - 6.67}{2.581} = 0.32 \quad and \quad z = \frac{1000.5 - 6.67}{2.581} = 3872.1$$

The area to the left of z = 0.32 is 0.6255 and the area to the left of z = 3872.1 is 1.0000. Thus, the probability that more than 7 males have Fragile X Syndrome is 1.0000 - 0.6255 = 3745.

To find the approximate probability that at most 10 males have Fragile X syndrome, we want P(X ≤ 10) or, approximating, we use P(X < 10.5). For x = 10.5 and x = -0.5, the z=scores are

$$z = \frac{10.5 - 6.67}{2.581} = 1.48 \quad and \quad z = \frac{-0.5 - 6.67}{2.581} = -2.78$$

The area to the left of z = 1.48 is 0.9306 and the area to the left of -2.78 is 0.0027. Thus, the probability that at most 10 males have Fragile X Syndrome is 0.9306 - 0.0027 = 0.9279.

(c) Although both np and n(1 - p) are at least 5, a binomial distribution with p = 1/1500 is highly skewed to the right whereas the normal distribution being used to approximate it is a symmetric distribution. The Poisson distribution used in Exercise 5.109 reflects the skewness of the binomial distribution and thus we would expect it to provide the better approximation.

Review Problems for Chapter 6

1. A density curve is a smooth curve that identifies the shape of a distribution for a variable. They are used to approximate percentages of all possible observations that lie within a range by finding corresponding area under its curve.

2. The percentage of all possible observations of a variable that lie between 25 and 50 equals the area under its density curve between 25 and 50, expressed as a percentage.

3. (a) Since the area that lies to the left of 60 is 0.364, 36.4% of all possible observations of the variable are less than 60.

(b) Since the area that lies to the left of 60 is 0.364, the area that lies to the right of 60 is 1.000 - 0.364 = 0.636. Thus, 63.6% of all possible observations of the variable are at least 60.

4. The area between 5 and 6 is 0.728. Therefore, the area outside of 5 and 6 (i.e. the area to the left of 5 or to the right of 6) is 1.000 - 0.728 = 0.272. Thus, 27.2% of the possible observations are either less than 5 or greater than 6.

5. (a) The following is a graph of the density curve with the equation

$y = 1 - x/2$ for $0 < x < 2$.

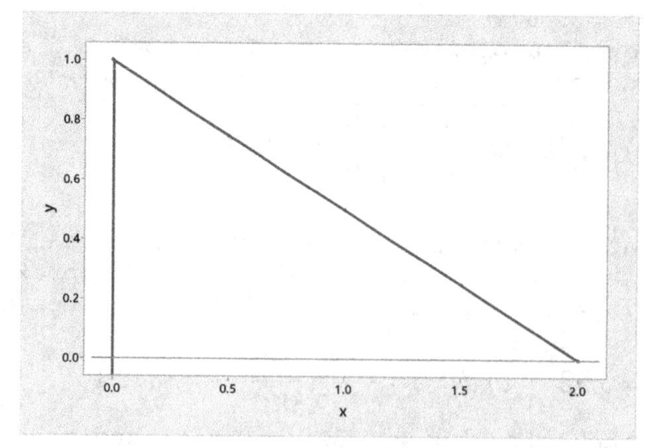

(b) The area to the right of any number x between 0 and 2 would be a triangle with a base of $2 - x$ and a height of $1 - x/2$. The area of a triangle is $1/2 \cdot base \cdot height = 1/2 \cdot (2 - x) \cdot (1 - x/2) = 1 - x + x^2/4$. This is the area the right of x. The area to the left of x would be $1 - (1 - x + x^2/4) = x - x^2/4$.

(c) The area between 1/2 and 1 can be calculated by finding the area to the left of 1/2 and subtracting that from the area to the left of 1. Using part (b), the area to the left of 1/2 is $1/2 - (1/2)^2/4 = 7/16$ and the area to the left of 1 is $1 - (1)^2/4 = 3/4$. Therefore, the area between 1/2 and 1 is $3/4 - 7/16 = 5/16$. 31.25% of the possible observations of the variable are between 1/2 and 1.

(d) The probability that the observation of the variable is at least 1.5 is the area to the right of 1.5. The area to the left of 1.5, using part (b), is $1.5 - 1.5^2/4 = 0.9375$. The area to the right of 1.5 is one minus the area to the left, or $1 - 0.9375 = 0.0625$. Therefore, 6.25% of the possible observations of the variable are at least 1.5.

6. Two primary reasons for studying the normal distribution are that:

(a) It is often appropriate to use the normal distribution as the distribution of a population or random variable.

(b) The normal distribution is frequently employed in inferential statistics.

7. (a) A variable is normally distributed if its distribution has the shape of a normal curve.

(b) A population is normally distributed if a variable of the population is normally distributed and it is the only variable under consideration.

(c) The parameters for a normal curve are the mean μ and the standard deviation σ.

8. (a) False. There are many types of distributions that could have the same mean and standard deviation.

(b) True. The mean and standard deviation completely determine a normal distribution, so if two normal distributions have the same mean and standard deviation, then those two distributions are identical.

9. The percentages for a normally distributed variable and areas under the corresponding normal curve (expressed as a percentage) are identical.

10. The distribution of the standardized version of a normally distributed variable is the standard normal distribution, that is, a normal distribution with a mean of 0 and standard deviation of 1.

11. (a) True. The mean completely determines the center of the distribution.

(b) True. The standard deviation completely determines the spread of the distribution.

12. (a) The (second) curve with $\sigma = 6.2$ has the largest spread.

(b) The first and second curves are centered at $\mu = 1.5$.

(c) The first and third curves have the same spread because $\sigma = 3$ for both.

(d) The third curve is centered farthest to the left because it has the smallest value of $\mu = -2.7$.

(e) The fourth curve is the standard normal curve because $\mu = 0$ and $\sigma = 1$.

13. Key Fact 6.4.

14. (a) The table entry corresponding to the specified z-score is the area to the left of that z-score.

(b) The area to the right of a specified z-score is found by subtracting the table entry from 1.

(c) The area between two specified z-scores is found by subtracting the table entry for the smaller z-score from the table entry for the larger z-score.

15. (a) Find the table entry that is closest to the specified area. The z-score determined by locating the corresponding marginal values is the z-score that has the specified area to its left. If the specified area is exactly between two entries, then the two z-scores should be averaged.

(b) Subtract the specified area from 1. Find the entry in the table that is closest to the result of the subtraction, averaging if the area is exactly between two table entries. The z-score determined by locating the corresponding marginal values is the z-score that has the specified area to its right.

16. The value z_α is the z-score that has area α to its right under the standard normal curve.

17. The empirical rule for variables states that for a normally distributed variable: 68% of all possible observations lie within one standard deviation to either side of the mean; 95% of all possible observations lie within two standard deviations to either side of the mean; 99.7% of all possible observations lie within three standard deviation to either side of the mean.

18. The normal scores for a sample of observations are the observations we would expect to get for a sample of the same size for a variable having the standard normal distribution.

19. If we observe the values of a normally distributed variable for a sample, then a normal probability plot should be roughly <u>linear</u>.

20.

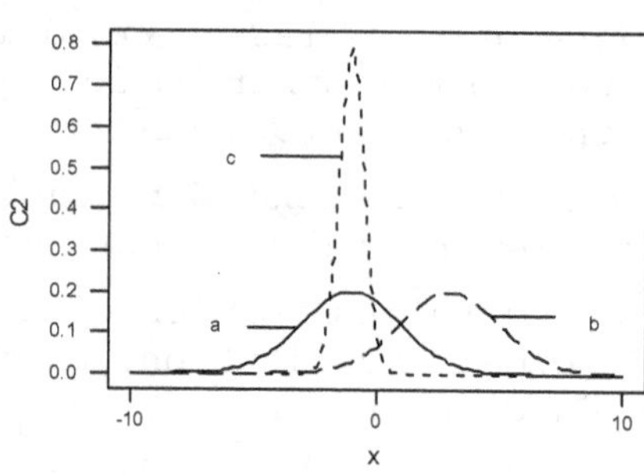

21. (a) The area to the right of 1.05 is 1.0000 - 0.8531 = 0.1469.
(b) The area to the left of -1.05 is 0.1469 (by symmetry).
(c) The area between -1.05 and 1.05 is 0.8531 - 0.1469 = 0.7062.

22. (a) Area = 0.0013

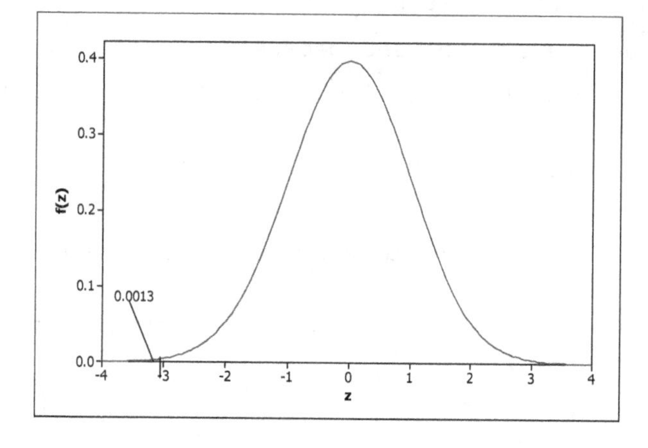

(b) Area = 1 - 0.7291 = 0.2709

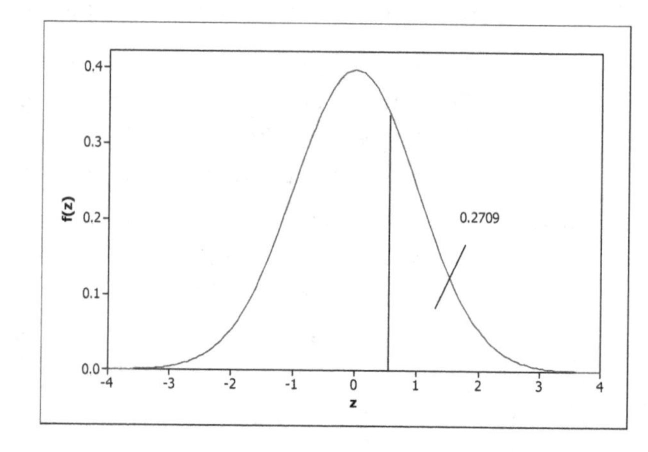

(c) Area = 0.9970 - 0.8665 = 0.1305

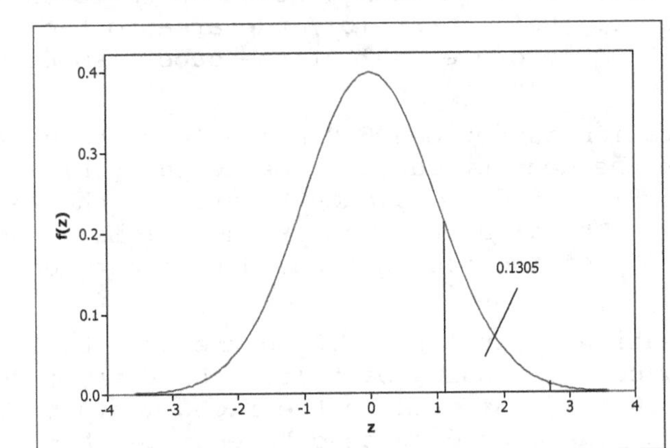

(d) Area = 1.000 - 0.0197 = 0.9803

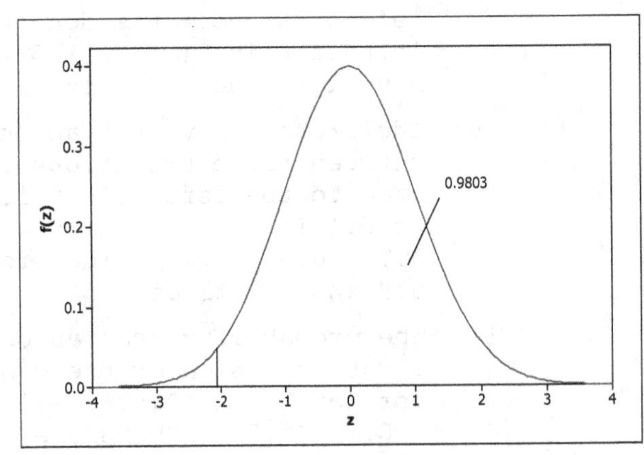

(e) Area = 0.0668 - 0.0000 = 0.0668

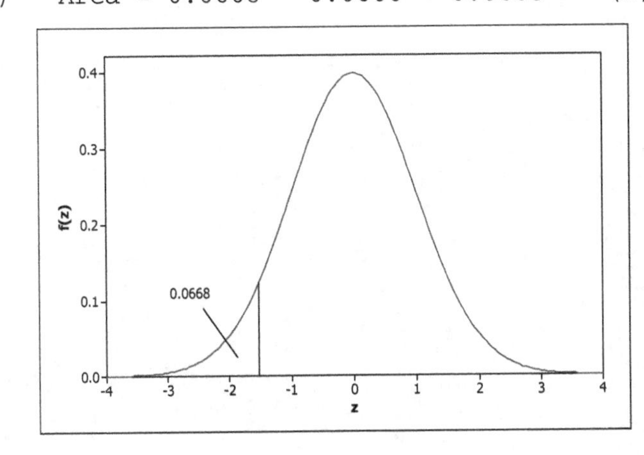

(f) Area = 0.8413 + (1 - 0.9987) = 0.8426

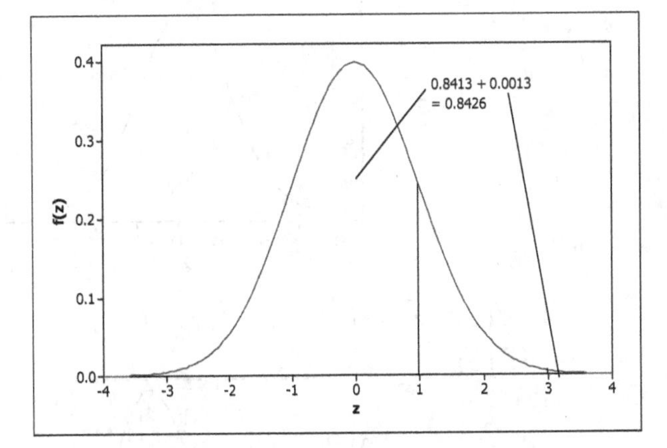

23. (a) z = -0.52

(b) z will have 0.9 to its left: z = 1.28

(c) $z_{0.025} = 1.96$; $z_{0.05} = 1.645$; $z_{0.01} = 2.33$; $z_{0.005} = 2.575$

(d) -2.575 and +2.575

24. (a) The following is a graph of the density curve with the equation $y = 2$ for $5.75 < x < 6.25$.

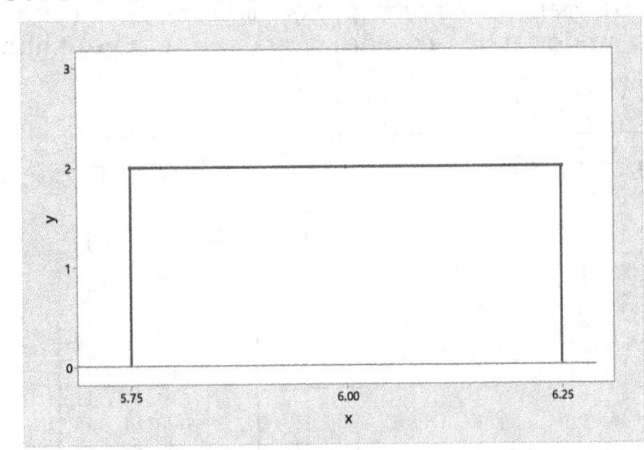

(b) The area to the left of any number x between 5.75 and 6.25 would be a rectangle with a base of $x - 5.75$ and a height of 2. The area of a rectangle is *base·height* $= (x - 5.75)\cdot 2 = 2x - 11.5$.

(c) The probability that an observation is less than 6 is the area to the left of 6 under the density curve. Using part (b), the area to the left of 6 is 2(6) - 11.5 = 0.5. Therefore, 50% of the cups dispensed have less than 6 fl oz.

(d) The probability that an observation is between 5.9 and 6.1 is the area between those two points under the density curve. Using part (b), the area to the left of 5.9 is 2(5.9) - 11.5 = 0.3 and the area to the left of 6.1 is 2(6.1) - 11.5 = 0.7. The area between 5.9 and 6.1 is then 0.7 - 0.3 = 0.4. Therefore, 40% of the cups dispensed have between 5.9 and 6.1 fl oz.

(e) The probability that an observation is at least 5.8 is the area to the right of 5.8 under the density curve. Using part (b), the area to the left of 5.8 is 2(5.8) - 11.5 = 0.1. The area to the right would be 1 - 0.1 = 0.9. Therefore, 90% of the cups dispensed have at least 5.8 fl oz.

25. (a)

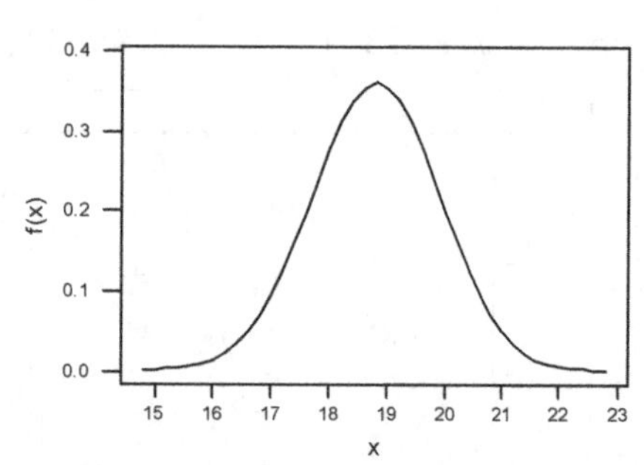

(b) $z = (x - 18.8)/1.1$

(c)

(d) $P(17 \le x \le 20) = 0.8115$

(e) The percentage of men who have forearm lengths less than 16 inches equals the area under the standard normal curve that lies to the left of -2.55.

26. (a)

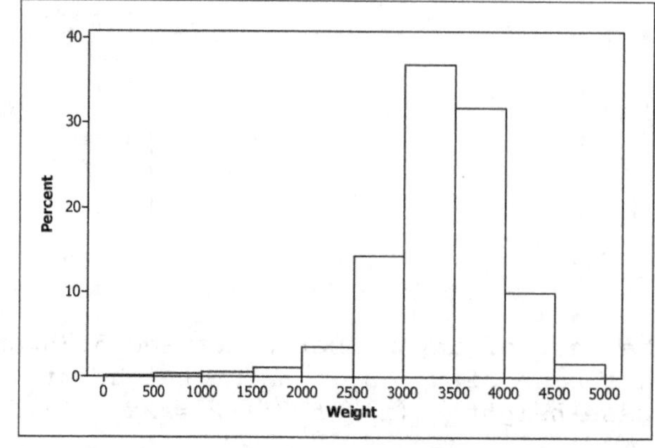

(b) No. The histogram indicates that the data are left skewed.

27. (a) The z-score for x = 304 is (304 − 283.3)/83.3 = 0.25. The area to the left of z = 0.25 is 0.5987, so the percentage of patients whose amount of blood loss during surgery is less than 304 ml is 59.87%.

(b) The z-score for x = 221 is (221 − 283.3)/83.3 = −0.75 and for x = 429, z = (429 − 283.3)/83.3 = 1.75. The area to the left of z = −0.75 is 0.2266 and the area to the left of z = 1.75 is 0.9599. Therefore, the percentage of patients whose amount of blood loss during surgery is between 221 and 429 ml is 0.9599 − 0.2266 = 0.7333, or 73.33%.

(c) The z-score for x = 450 is (450 − 283.3)/83.3 = 2.00. The area to the left of z = 2.00 is 0.9772, so the percentage of patients whose amount of blood loss during surgery is above 450 ml is 1.0000 − 0.9772 = 0.0228, or 2.28%.

28. (a) An area of 0.2500 lies to the left of z = −0.67; an area of 0.5000 lies to the left of z = 0.00; and an area of 0.7500 lies to the left of z = 0.67. To convert these z-scores to verbal GRE scores (x), we use $x = \mu + z\sigma$. Thus Q_1 = 150 + (−0.67)(8.75) = 144.14; the median (Q_2) = 150 + (0.00)(8.75) = 150; and Q_3 = 150 + (0.67)(8.75) = 155.86. 25% of the verbal GRE scores will be less than 144.14; 50% of the scores will be less than 150; and 75% of the scores will be less than 155.86.

(b) An area of 0.9900 lies to the left of z = 2.33. To convert this z-scores to a verbal GRE score (x), we use $x = \mu + z\sigma$. Thus the 99th percentile is 150 + (2.33)(8.75) = 170.39. 99% of the verbal GRE scores will be less than 170.39.

29. (a) Approximately 68% of students who took the verbal portion of the GRE scored between 141.25 and 158.75.

(b) Approximately 95% of students who took the verbal portion of the GRE scored between 132.50 and 167.50.

(c) Approximately 99.7% of students who took the verbal portion of the GRE scored between 123.75 and 176.25.

30. (a) Order the lengths from smallest to largest. Obtain the normal scores for 12 observations from Table III. The result is the table below. The normal probability plot is to the left of the table.

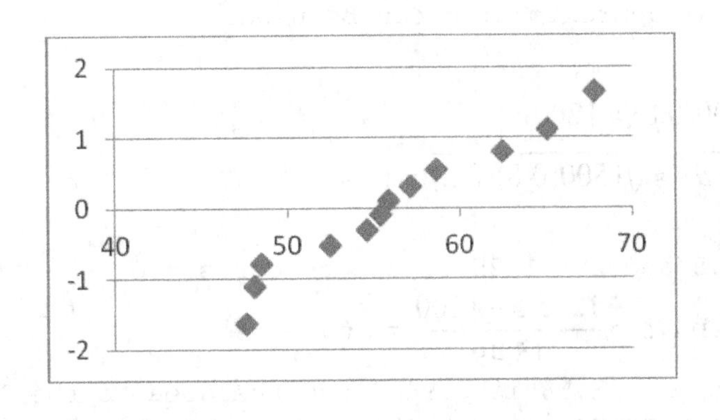

Length	Normal score
47.6	-1.64
48.1	-1.11
48.5	-0.79
52.4	-0.53
54.6	-0.31
55.4	-0.10
55.8	0.10
57.1	0.31
58.6	0.53
62.5	0.79
65.1	1.11
67.9	1.64

(b) There do not appear to be any outliers.
(c) The plot does not appear linear, so it does not seem reasonable to assume that the variable is normally distributed.

31. (a) Using Minitab, enter the data in a column named EMPLOYEES. Select **Graph ▶ Probability Plot**, select **Single** and click **OK**. Enter EMPLOYEES in the **Graph variables,** click on the **Distributions** button and the **Data Display** tab. Check the **Symbols only** option and click **OK**. Then click on the **Scale** button and the **Y-Scale Type** tab and select the **Score** option. Click OK twice. The resulting graph is

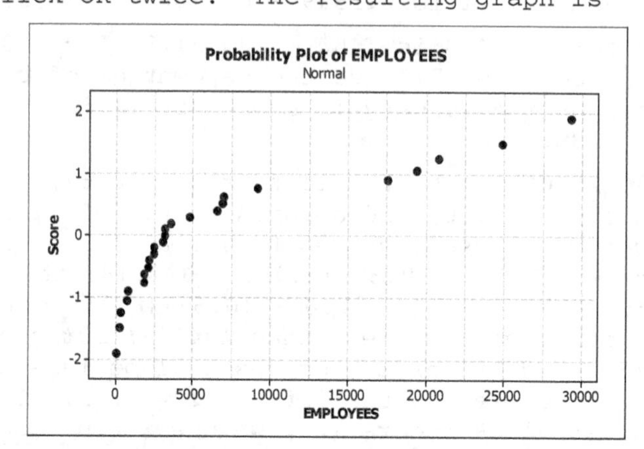

(b) There are five observations that are considerably greater than the other 20. If the distribution were otherwise normally distributed, these might be possible outliers.

(c) The graph is not close to being roughly linear. The data are very right skewed and hence not normally distributed, so the five possible outliers may be just part of the pattern of right skewness.

32. For parts (a), (b), and (c), steps 1-3 are as follows:

Step 1: n = 1500; p = 0.80

Step 2: np = 1200; n(1 - p) = 300. Since both np and n(1 - p) are at least 5, the normal approximation can be used.

Step 3:

$$\mu_x = np = 1500(0.8) = 1200$$
$$\sigma_x = \sqrt{np(1-p)} = \sqrt{1500(0.8)(0.2)} = 15.49$$

(a) P(x = 1225):
Step 4: For x = 1224.5 and x = 1225.5, the z-scores are

$$z = \frac{1224.5 - 1200}{15.49} = 1.58 \text{ and } z = \frac{1225.5 - 1200}{15.49} = 1.65$$

The area to the left of z = 1.58 is 0.9429 and the area to the left of z = 1.65 is 0.9505. The probability that the vaccine will be effective in exactly 1225 cases is approximately = 0.9505 - 0.9429 = 0.0076.

(b) $P(x \geq 1175)$:

Step 4: For x = 1174.5 and x = 1500.5, the z-scores are

$$z = \frac{1174.5 - 1200}{15.49} = -1.65 \quad \textit{and} \quad z = \frac{1500.5 - 1200}{15.49} = 19.40$$

The area to the left of z = -1.65 is 0.0495 and the area to the left of 19.40 is 1.0000. The probability that the vaccine will be effective in at least 1175 cases is approximately 1.0000 - 0.0495 = 0.9505.

(c) $P(1150 \leq x \leq 1250)$:

Step 4: For x = 1149.5 and x = 1250.5, the z-scores are

$$z = \frac{1149.5 - 1200}{15.49} = -3.26 \text{ and } z = \frac{1250.5 - 1200}{15.49} = 3.26.$$

The area to the left of z = -3.26 is 0.0006 and the area to the left of z = 3.26 is 0.9994. The probability that the vaccine will be effective in 11509 to 1250 cases, inclusive, is approximately = 0.9994 - 0.0006 = 0.9988.

CHAPTER 7 SOLUTIONS

Exercises 7.1

7.1 Sampling is often preferable to conducting a census because it is quicker, less costly, and sometimes it is the only practical way to get information.

7.3 (a) $\mu = \Sigma x/N = 6/3 = 2$

(b) for n = 1

Sample	$\bar{x}$
1	1.0
2	2.0
3	3.0

for n = 2

Sample	$\bar{x}$
1,2	1.2
1,3	2.0
2,3	2.5

for n = 3

Sample	$\bar{x}$
1,2,3	3.0

See the dotplots in part (c).

(c)

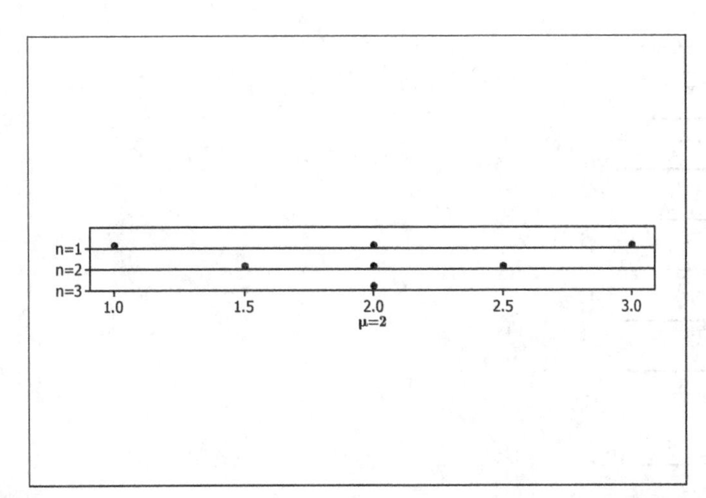

The sample means cluster more closely around the population mean as the sample size increases. Sampling error is smaller for large samples than for small samples.

(d) For n = 1, one out of the three sample means equaled the population mean. Thus, the probability that the sample mean equals the population mean for n = 1 is 1/3.

For $n = 2$, one out of the three sample means equaled the population mean. Thus, the probability that the sample mean equals the population mean for $n = 2$ is 1/3.

For $n = 3$, the one sample mean equals the population mean, Thus, the probability that the sample mean equals the population mean for $n = 3$ is 1.0.

(e) If the absolute value of the difference between the sample mean and the population mean is at most 0.5, that means that we are looking for sample means within the range 1.5 to 2.5, inclusive.

For $n = 1$, one out of the three sample means is in this range. Thus, the probability that the sampling error will be 0.5 or less for $n = 1$ is 1/3.

For $n = 2$, all of the three sample means are in this range. Thus, the probability that the sampling error will be 0.5 or less for $n = 2$ is 1.0.

For $n = 3$, the one sample mean is in this range. Thus, the probability that the sampling error will be 0.5 or less for $n = 3$ is 1.0.

7.5 (a) $\mu = \Sigma x/N = 10/4 = 2.5$

(b) for $n = 1$

Sample	$\bar{x}$
1	1.0
2	2.0
3	3.0
4	4.0

for $n = 2$

Sample	$\bar{x}$
1,2	1.5
1,3	2.0
1,4	2.5
2,3	2.5
2,4	3.0
3,4	3.5

for $n = 3$

Sample	$\bar{x}$
1,2,3	2.0
1,2,4	2.3
1,3,4	2.7
2,3,4	3.0

for $n = 4$

Sample	$\bar{x}$
1,2,3,4	2.5

See the dotplots in part (c).

(c)

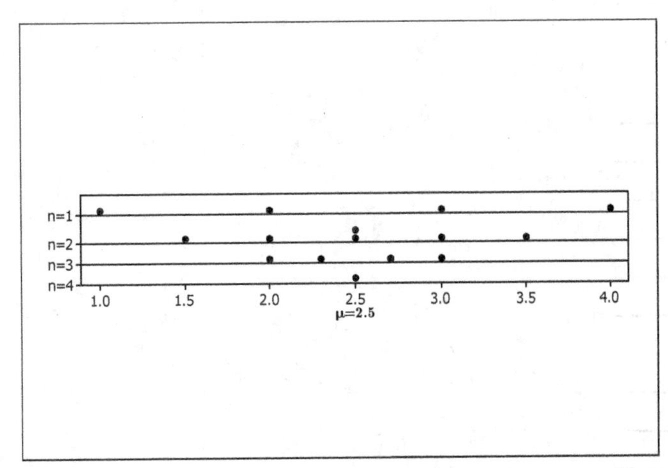

The sample means cluster more closely around the population mean as the sample size increases. Sampling error is smaller for large samples than for small samples.

(d) For $n = 1$, none of the four sample means equaled the population mean. Thus, the probability that the sample mean equals the population mean for $n = 1$ is 0.

For $n = 2$, two out of the six sample means equaled the population mean. Thus, the probability that the sample mean equals the population mean for $n = 2$ is $2/6 = 1/3$.

For $n = 3$, none of the four sample means equaled the population mean. Thus, the probability that the sample mean equals the population mean for $n = 3$ is 0.

For $n = 4$, the one sample mean equals the population mean, Thus, the probability that the sample mean equals the population mean for $n = 4$ is 1.

(e) If the absolute value of the difference between the sample mean and the population mean is at most 0.5, that means that we are looking for sample means within the range 2.0 to 3.0, inclusive.

For $n = 1$, two out of the four sample means is in this range. Thus, the probability that the sampling error will be 0.5 or less for $n = 1$ is $2/4 = 1/2$.

For $n = 2$, four of the six sample means are in this range. Thus, the probability that the sampling error will be 0.5 or less for $n = 2$ is $4/6 = 2/3$.

For $n = 3$, all four of the sample means are in this range. Thus, the probability that the sampling error will be 0.5 or less for $n = 3$ is 1.

For $n = 4$, the one sample mean is in this range. Thus, the probability that the sampling error will be 0.5 or less for $n = 4$ is 1.0.

7.7 (a) $\mu = \Sigma x/N = 15/5 = 3.0$

(b) for n = 1

Sample	$\bar{x}$
1	1.0
2	2.0
3	3.0
4	4.0
5	5.0

for n = 2

Sample	$\bar{x}$
1,2	1.5
1,3	2.0
1,4	2.5
1,5	3.0
2,3	2.5
2,4	3.0
2,5	3.5
3,4	3.5
3,5	4.0
4,5	4.5

for n = 3

Sample	$\bar{x}$
1,2,3	2.0
1,2,4	2.3
1,2,5	2.7
1,3,4	2.7
-1,3,5	3.0
1,4,5	3.3
2,3,4	3.0
2,3,5	3.3
2,4,5	3.7
3,4,5	4.0

for n = 4

Sample	$\bar{x}$
1,2,3,4	2.50
1,2,3,5	2.75
1,2,4,5	3.00
1,3,4,5	3.25
2,3,4,5	3.50

for n = 5

Sample	$\bar{x}$
1,2,3,4,5	3.0

See the dotplots in part (c).

(c)

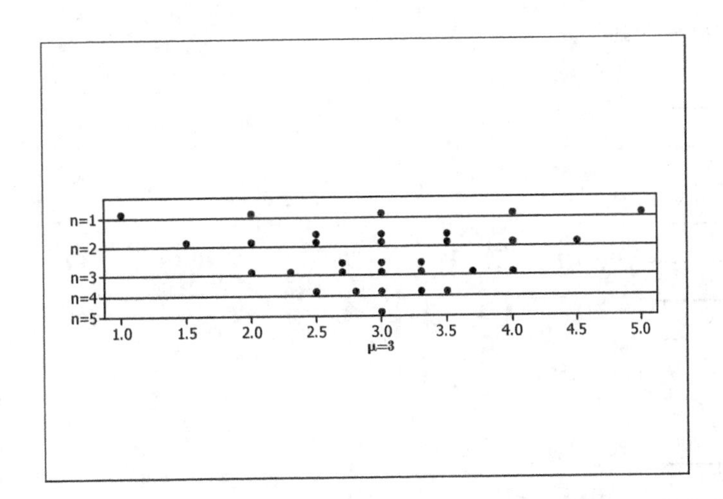

The sample means cluster more closely around the population mean as the sample size increases. Sampling error is smaller for large samples than for small samples.

(d) For n = 1, one of the five sample means equaled the population mean. Thus, the probability that the sample mean equals the population mean for n = 1 is 1/5.

For n = 2, two out of the ten sample means equaled the population mean. Thus, the probability that the sample mean equals the population mean for n = 2 is 2/10 = 1/5.

For n = 3, two out of the ten sample means equaled the population mean. Thus, the probability that the sample mean equals the population mean for n = 3 is 2/10 = 1/5.

For n = 4, one out of the five sample means equaled the population mean. Thus, the probability that the sample mean equals the population mean for n = 4 is 1/5.

For n = 5, the one sample mean equals the population mean, Thus, the probability that the sample mean equals the population mean for n = 5 is 1.

(e) If the absolute value of the difference between the sample mean and the population mean is at most 0.5, that means that we are looking for sample means within the range 2.5 to 3.5, inclusive.

For n = 1, one of the five sample means is in this range. Thus, the probability that the sampling error will be 0.5 or less for n = 1 is 1/5.

For n = 2, six of the ten sample means are in this range. Thus, the probability that the sampling error will be 0.5 or less for n = 2 is 6/10 = 3/5.

For n = 3, six of the ten sample means are in this range. Thus, the probability that the sampling error will be 0.5 or less for n = 3 is 6/10 = 3/5.

For n = 4, all of the five sample means are in this range. Thus, the probability that the sampling error will be 0.5 or less for n = 4 is 1.

For n = 5, the one sample mean is in this range. Thus, the probability that the sampling error will be 0.5 or less for n = 5 is 1.0.

7.9 (a) $\mu = \Sigma x/N = 21/6 = 3.5$

(b) for $n = 1$

Sample	$\bar{x}$
1	1.0
2	2.0
3	3.0
4	4.0
5	5.0
6	6.0

for $n = 2$

Sample	$\bar{x}$
1,2	1.5
1,3	2.0
1,4	2.5
1,5	3.0
1,6	3.5
2,3	2.5
2,4	3.0
2,5	3.5
2,6	4.0
3,4	3.5
3,5	4.0
3,6	4.5
4,5	4.5
4,6	5.0
5,6	5.5

for $n = 3$

Sample	$\bar{x}$
1,2,3	2.0
1,2,4	2.3
1,2,5	2.7
1,2,6	3.0
1,3,4	2.7
1,3,5	3.0
1,3,6	3.3
1,4,5	3.3
1,4,6	3.7
1,5,6	4.0
2,3,4	3.0
2,3,5	3.3
2,3,6	3.7
2,4,5	3.7

2,4,6	4.0
2,5,6	4.3
3,4,5	4.0
3,4,6	4.3
3,5,6	4.7
4,5,6	5.0

for n = 4

Sample	$\bar{x}$
1,2,3,4	2.50
1,2,3,5	2.75
1,2,3,6	3.00
1,2,4,5	3.00
1,2,4,6	3.25
1,2,5,6	3.50
1,3,4,5	3.25
1,3,4,6	3.50
1,3,5,6	3.75
1,4,5,6	4.00
2,3,4,5	3.50
2,3,4,6	3.75
2,3,5,6	4.00
2,4,5,6	4.25
3,4,5,6	4.50

for n = 5

Sample	$\bar{x}$
1,2,3,4,5	3.0
1,2,3,4,6	3.2
1,2,3,5,6	3.4
1,2,4,5,6	3.6
1,3,4,5,6	3.8
2,3,4,5,6	4.0

for n = 6

Sample	$\bar{x}$
1,2,3,4,5,6	3.5

See the dotplots in part (c)

(c)

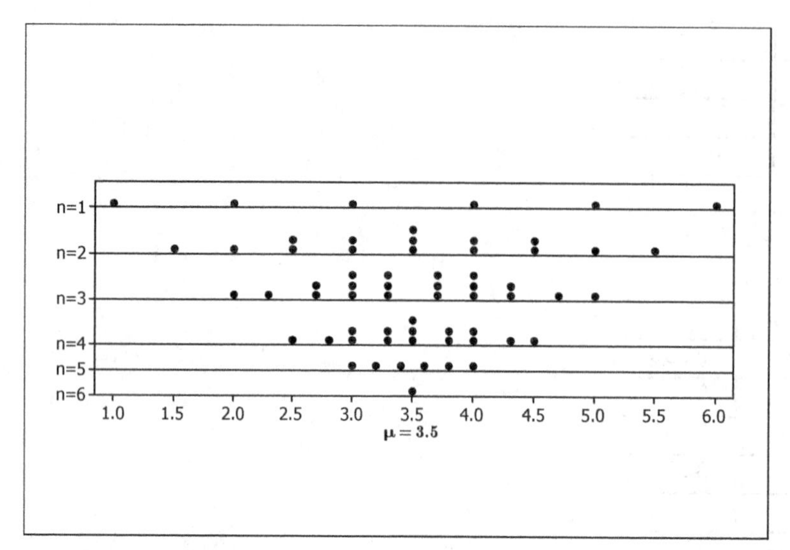

The sample means cluster more closely around the population mean as the sample size increases. Sampling error is smaller for large samples than for small samples.

(d) For n = 1, none of the six sample means equaled the population mean. Thus, the probability that the sample mean equals the population mean for n = 1 is 0.

For n = 2, three out of the fifteen sample means equaled the population mean. Thus, the probability that the sample mean equals the population mean for n = 2 is 3/15 = 1/5.

For n = 3, none of the twenty sample means equaled the population mean. Thus, the probability that the sample mean equals the population mean for n = 3 is 0.

For n = 4, three out of the fifteen sample means equaled the population mean. Thus, the probability that the sample mean equals the population mean for n = 4 is 3/15 = 1/5.

For n = 5, none of the six sample means equaled the population mean. Thus, the probability that the sample mean equals the population mean for n = 5 is 0.

For n = 6, the one sample mean equals the population mean, Thus, the probability that the sample mean equals the population mean for n = 6 is 1.

(e) If the absolute value of the difference between the sample mean and the population mean is at most 0.5, that means that we are looking for sample means within the range 3.0 to 4.0, inclusive.

For n = 1, two of the six sample means is in this range. Thus, the probability that the sampling error will be 0.5 or less for n = 1 is 2/6 = 1/3.

For n = 2, seven of the fifteen sample means are in this range. Thus, the probability that the sampling error will be 0.5 or less for n = 2 is 7/15.

For n = 3, twelve of the twenty sample means are in this range. Thus, the probability that the sampling error will be 0.5 or less for n = 3 is 12/20 = 3/5.

For n = 4, eleven of the fifteen sample means are in this range. Thus, the probability that the sampling error will be 0.5 or less for n = 4 is 11/15.

For n = 5, all of the six sample means are in this range. Thus, the probability that the sampling error will be 0.5 or less for n = 5 is 1.

For n = 6, the one sample mean is in this range. Thus, the probability that the sampling error will be 0.5 or less for n = 6 is 1.0.

7.11 (a) $\mu = \Sigma x/N = 393/5 = 78.6$ inches.

(b)

Sample	Heights	$\bar{x}$
B,W	83, 76	79.5
B,J	83, 80	81.5
B,C	83, 74	78.5
B,H	83, 80	81.5
W,J	76, 80	78.0
W,C	76, 74	75.0
W,H	76, 80	78.0
J,C	80, 74	77.0
J,H	80, 80	80.0
C,H	74, 80	77.0

(c)

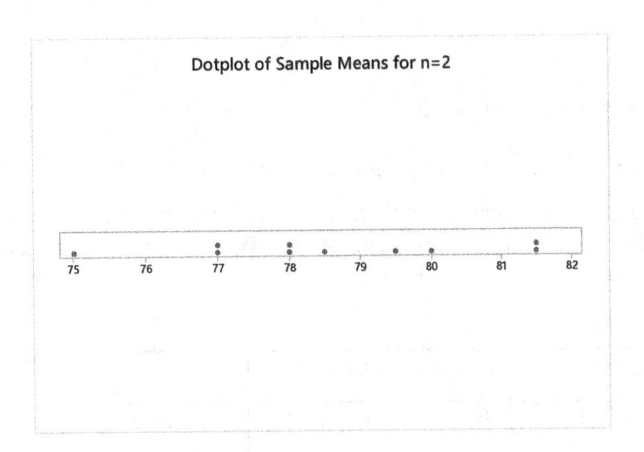

(d) $P(\bar{x} = \mu) = P(\bar{x} = 78.6) = 0.0$

(e) $P(78.6 - 1 \leq \bar{x} \leq 78.6 + 1) = P(77.6 \leq \bar{x} \leq 79.6)$
$= 4/10 = 0.4$

If we take a random sample of two heights, there is a 40% chance that the mean of the sample selected will be within one inch of the population mean.

7.13 (b)

Sample	Heights	$\bar{x}$
B,W,J	83,76,80	79.67
B,W,C	83,76,74	77.67
B,W,H	83,76,80	79.67
B,J,C	83,80,74	79.00
B,J,H	83,80,80	81.00
B,C,H	83,74,80	79.00
W,J,C	76,80,74	76.67
W,J,H	76,80,80	78.67
W,C,H	76,74,80	76.67
J,C,H	80,74,80	78.00

(c)

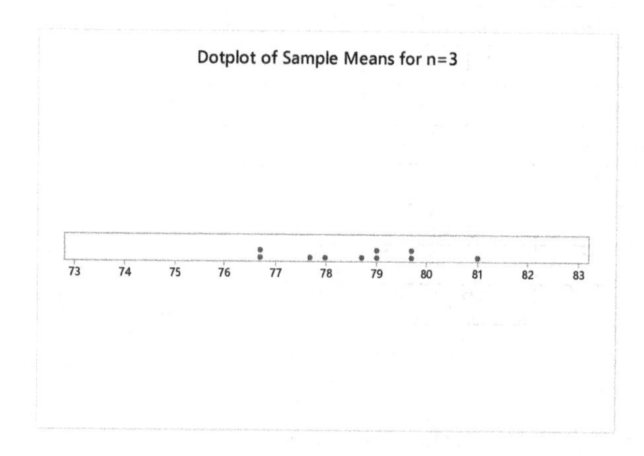

(d) $P(\bar{x} = \mu) = P(\bar{x} = 78.6) = 0.0$

(e) $P(78.6 - 1 \le \bar{x} \le 78.6 + 1) = P(77.6 \le \bar{x} \le 79.6) = 5/10 = 0.5$

If we take a random sample of three heights, there is a 50% chance that the mean of the sample selected will be within one inch of the population mean.

7.15 (b)

Sample	Heights	$\bar{x}$
B,W,J,C,H	83,76,80,74,80	78.60

(c)

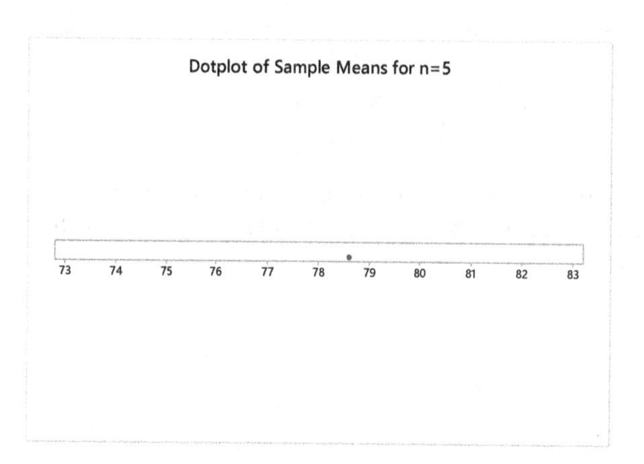

(d) $P(\bar{x} = \mu) = P(\bar{x} = 78.6) = 1.0$

(e) $P(78.6 - 1 \le \bar{x} \le 78.6 + 1) = P(77.6 \le \bar{x} \le 79.6) = 1.0$

If we take a random sample of five heights, there is a 100% chance that the mean of the sample selected will be within one inch of the population mean.

7.17 (a) $\mu = \Sigma x/N = 279/6 = 46.5$ billion

(b)

Sample	Wealth	$\bar{x}$
G,B	72, 59	65.5
G,E	72, 41	56.5
G,C	72, 36	54.0
G,D	72, 36	54.0
G,W	72, 35	53.5
B,E	59, 41	50.0
B,C	59, 36	47.5
B,D	59, 36	47.5
B,W	59, 35	47.0
E,C	41, 36	38.5
E,D	41, 36	38.5
E,W	41, 35	38.0
C,D	36, 36	36.0
C,W	36, 35	35.5
D,W	36, 35	35.5

(c)

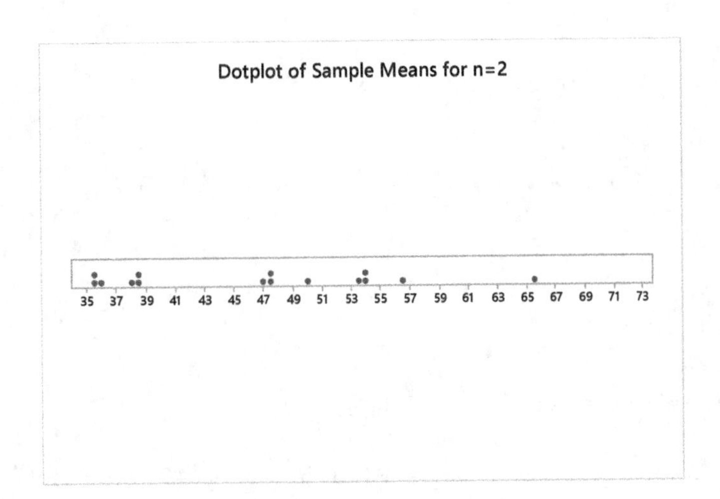

(d) $P(\bar{x} = \mu) = P(\bar{x} = 46.5) = 0/15 = 0$

(e) $P(46.5 - 3 \le \bar{x} \le 46.5 + 3) = P(43.5 \le \bar{x} \le 49.5) = 3/15 = 0.20$

If we take a random sample of two of these six rich Americans, there is a 20% chance that their mean wealth will be within three billion of the population mean wealth.

7.19 (b)

Sample	Wealth	$\bar{x}$
G,B,E	72, 59, 41	57.3
G,B,C	72, 59, 36	55.7
G,B,D	72, 59, 36	55.7
G,B,W	72, 59, 35	55.3
G,E,C	72, 41, 36	49.7
G,E,D	72, 41, 36	49.7
G,E,W	72, 41, 35	49.3
G,C,D	72, 36, 36	48.0
G,C,W	72, 36, 35	47.7
G,D,W	72, 36, 35	47.7
B,E,C	59, 41, 36	45.3
B,E,D	59, 41, 36	45.3
B,E,W	59, 41, 35	45.0
B,C,D	59, 36, 36	43.7
B,C,W	59, 36, 35	43.3
B,D,W	59, 36, 35	43.3
E,C,D	41, 36, 36	37.7
E,C,W	41, 36, 35	37.3
E,D,W	41, 36, 35	37.3
C,D,W	36, 36, 35	35.7

(c)

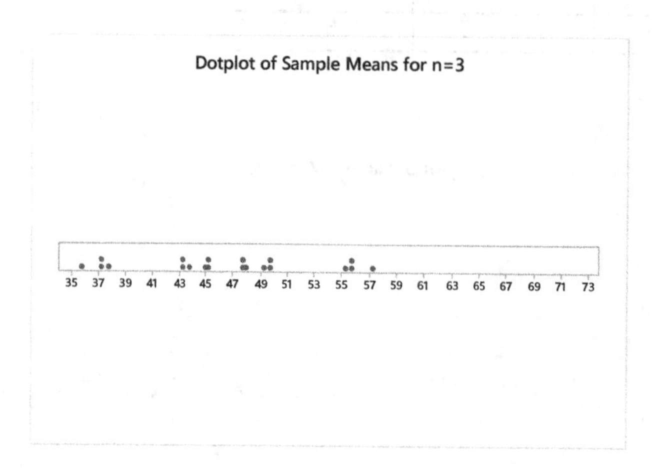

(d) $P(\bar{x} = \mu) = P(\bar{x} = 46.5) = 0/20 = 0.0$

(e) $P(46.5 - 3 \leq \bar{x} \leq 46.5 + 3) = P(43.5 \leq \bar{x} \leq 49.5)$

$= 8/20 = 0.40$

If we take a random sample of three wealthy Americans, there is a 40% chance that their mean wealth will be within three billion of the population mean wealth.

7.21 (b)

Sample	Wealth	$\bar{x}$
G,B,E,C,D	72,59,41,36,36	48.8
G,B,E,C,W	72,59,41,36,35	48.6
G,B,E,D,W	72,59,41,36,35	48.6
G,B,C,D,W	72,59,36,36,35	47.6
G,E,C,D,W	72,41,36,36,35	44.0
B,E,C,D,W	59,41,36,36,35	41.4

(c)

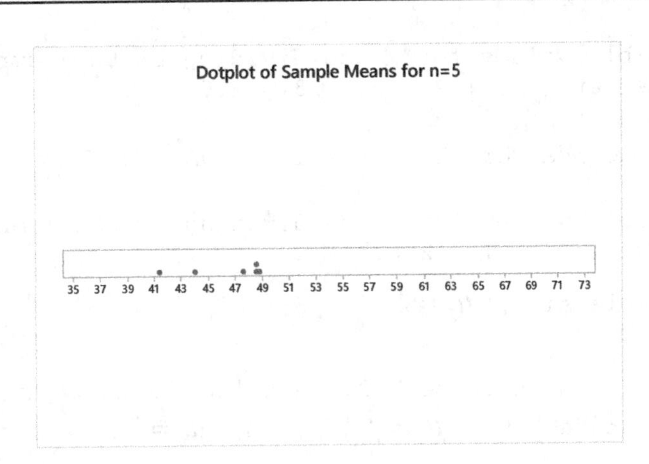

(d) $P(\bar{x} = \mu) = P(\bar{x} = 46.5) = 0/6 = 0.000$

(e) $P(46.5 - 3 \leq \bar{x} \leq 46.5 + 3) = P(43.5 \leq \bar{x} \leq 49.5)$
$= 5/6 = 0.833$

If we take a random sample of five wealthy Americans, there is a 83.3% chance that their mean wealth will be within three billion of the population mean wealth.

7.23 Increasing the sample size tends to reduce the sampling error.

7.25 (a) If a sample of size n = 1 is taken from a population of size N, there are N possible samples.

(b) Since each sample mean is based upon a single observation, the possible values of the sample mean and the population values are the same.

(c) There is no difference between taking a random sample of size n = 1 from a population and selecting a member at random from the population.

Exercises 7.2

7.27 Obtaining the mean and standard deviation of $\bar{x}$ is a first step in approximating the sampling distribution of the mean by a normal distribution because the normal distribution is completely determined by its mean and standard deviation.

7.29 Yes. The spread of the distribution of sample means becomes smaller as the sample size increases. Since that spread is measured by the standard deviation of all possible sample means, the standard deviation also gets smaller.

7.31 Standard error of the sample mean. The standard deviation of $\bar{x}$ determines the amount of sampling error to be expected when a population mean is estimated by a sample mean.

7.33 (a) For n = 1, the mean of the variable $\bar{x}$, $\mu_{\bar{x}}$, is found by summing up all of the possible values for $\bar{x}$ and dividing by the number of possible samples. We get (1 + 2 + 3)/3 = 2.0.

For n = 2, the mean of the variable $\bar{x}$, $\mu_{\bar{x}}$, is found by summing up all of the possible values for $\bar{x}$ and dividing by the number of possible samples. We get (1.5 + 2 + 2.5)/3 = 2.0.

For n = 3, the mean of the variable $\bar{x}$, $\mu_{\bar{x}}$, is found by summing up all of the possible values for $\bar{x}$ and dividing by the number of possible samples. We get (2)/1 = 2.0.

For each sample size, $\mu_{\bar{x}} = 2$.

(b) From part (a), in Exercise 7.3, we calculated $\mu = \Sigma x/N = 6/3 = 2$. Formula 7.1 states that $\mu_{\bar{x}} = \mu$. Thus, $\mu_{\bar{x}} = 2$.

7.35 (a) For n = 1, the mean of the variable $\bar{x}$, $\mu_{\bar{x}}$, is found by summing up all of the possible values for $\bar{x}$ and dividing by the number of possible samples. We get (1 + 2 + 3 + 4)/4 = 2.5.

For n = 2, the mean of the variable $\bar{x}$, $\mu_{\bar{x}}$, is found by summing up all of the possible values for $\bar{x}$ and dividing by the number of possible samples. We get (1.5 + 2 + 2.5 + 2.5 + 3 + 3.5)/6 = 2.5.

For n = 3, the mean of the variable $\bar{x}$, $\mu_{\bar{x}}$, is found by summing up all of the possible values for $\bar{x}$ and dividing by the number of possible samples. We get (2 + 2.3 + 2.7 + 3)/4 = 2.5.

For n = 4, the mean of the variable $\bar{x}$, $\mu_{\bar{x}}$, is found by summing up all of the possible values for $\bar{x}$ and dividing by the number of possible samples. We get (2.5)/1 = 2.5.

For each sample size, $\mu_{\bar{x}} = 2.5$.

(b) From part (a), in Exercise 7.5, we calculated $\mu = \Sigma x/N = 10/4 = 2.5$. Formula 7.1 states that $\mu_{\bar{x}} = \mu$. Thus, $\mu_{\bar{x}} = 2.5$.

7.37 (a) For n = 1, the mean of the variable $\bar{x}$, $\mu_{\bar{x}}$, is found by summing up all of the possible values for $\bar{x}$ and dividing by the number of possible samples. We get (1 + 2 + 3 + 4 + 5)/5 = 3.0.

For n = 2, the mean of the variable $\bar{x}$, $\mu_{\bar{x}}$, is found by summing up all of the possible values for $\bar{x}$ and dividing by the number of possible samples. We get (1.5 + 2 + 2.5 + 3 + 2.5 + 3 + 3.5 + 3.5 + 4 + 4.5)/10 = 3.0.

For n = 3, the mean of the variable $\bar{x}$, $\mu_{\bar{x}}$, is found by summing up all of the possible values for $\bar{x}$ and dividing by the number of possible samples. We get (2 + 2.3 + 2.7 + 2.7 + 3 + 3.3 + 3 + 3.3 + 3.7 + 4)/10 = 3.0.

For n = 4, the mean of the variable $\bar{x}$, $\mu_{\bar{x}}$, is found by summing up all of the possible values for $\bar{x}$ and dividing by the number of possible samples. We get (2.5 + 2.75 + 3 + 3.25 + 3.5)/5 = 3.0.

For n = 5, the mean of the variable $\bar{x}$, $\mu_{\bar{x}}$, is found by summing up all of the possible values for $\bar{x}$ and dividing by the number of possible samples. We get (3)/1 = 3.0. For each sample size, $\mu_{\bar{x}} = 3.0$.

(b) From part (a), in Exercise 7.7, we calculated μ = Σx/N = 15/5 = 3. Formula 7.1 states that $\mu_{\bar{x}} = \mu$. Thus, $\mu_{\bar{x}} = 3$.

7.39 (a) For n = 1, the mean of the variable $\bar{x}$, $\mu_{\bar{x}}$, is found by summing up all of the possible values for $\bar{x}$ and dividing by the number of possible samples. We get (1 + 2 + 3 + 4 + 5 + 6)/6 = 3.5.

For n = 2, the mean of the variable $\bar{x}$, $\mu_{\bar{x}}$, is found by summing up all of the possible values for $\bar{x}$ and dividing by the number of possible samples. We get (1.5 + 2 + 2.5 + 3 + 3.5 + 2.5 + 3 + 3.5 + 4 + 3.5 + 4 + 4.5 + 4.5 + 5 + 5.5)/15 = 3.5.

For n = 3, the mean of the variable $\bar{x}$, $\mu_{\bar{x}}$, is found by summing up all of the possible values for $\bar{x}$ and dividing by the number of possible samples. We get (2 + 2.3 + 2.7 + 3 + 2.7 + 3 + 3.3 + 3.3 + 3.7 + 4 + 3 + 3.3 + 3.7 + 3.7 + 4 + 4.3 + 4 + 4.3 + 4.7 + 5)/20 = 3.5.

For n = 4, the mean of the variable $\bar{x}$, $\mu_{\bar{x}}$, is found by summing up all of the possible values for $\bar{x}$ and dividing by the number of possible samples. We get (2.5 + 2.75 + 3 + 3 + 3.25 + 3.5 + 3.25 + 3.5 + 3.75 + 4 + 3.5 + 3.75 + 4 + 4.25 + 4.5)/15 = 3.5.

For n = 5, the mean of the variable $\bar{x}$, $\mu_{\bar{x}}$, is found by summing up all of the possible values for $\bar{x}$ and dividing by the number of possible samples. We get (3 + 3.2 + 3.4 + 3.6 + 3.8 + 4)/6 = 3.5.

For each sample size, $\mu_{\bar{x}} = 3.5$.

(b) From part (a), in Exercise 7.9, we calculated $\mu = \Sigma x/N = 21/6 = 3.5$. Formula 7.1 states that $\mu_{\bar{x}} = \mu$. Thus, $\mu_{\bar{x}} = 3.5$.

7.41 (a) $\mu_{\bar{x}} = \frac{\sum x}{N} = \frac{393}{5} = 78.6$

(b)

$$\mu_{\bar{x}} = \frac{\sum \bar{x}}{N} = \frac{79.5+81.5+78.5+81.5+78.0+75.0+78.0+77.0+80.0+77.0}{10}$$
$$= \frac{786}{10} = 78.6$$

(c) $\mu_{\bar{x}} = \mu = 78.6$

7.43 (b)

$$\mu_{\bar{x}} = \frac{\sum \bar{x}}{N} = \frac{79.7+77.7+79.7+79.0+81.0+79.0+76.7+78.7+76.7+78.0}{10}$$
$$= \frac{786}{10} = 78.6$$

(c) $\mu_{\bar{x}} = \mu = 78.6$

7.45 (b)

$$\mu_{\bar{x}} = \frac{\sum \bar{x}}{N} = \frac{78.6}{1} = 78.6$$

(c) $\mu_{\bar{x}} = \mu = 78.6$

7.47 (a) The population consists of all babies born in 1991. The variable is the birth weight of the baby.

(b) $\mu_{\bar{x}} = \mu = 3369$ grams; $\sigma_{\bar{x}} = \sigma/\sqrt{n} = 581/\sqrt{200} = 41.08$ grams

(c) $\mu_{\bar{x}} = \mu = 3369$ grams; $\sigma_{\bar{x}} = \sigma/\sqrt{n} = 10/\sqrt{400} = 29.05$ grams

7.49 (a) $\mu_{\bar{x}} = \mu = \$65{,}100$; $\sigma_{\bar{x}} = \sigma/\sqrt{n} = 7200/\sqrt{50} = \$1018.23$

Thus, for samples of size 50, the mean and standard deviation of all possible sample means are respectively, $65,100 and $1018.23.

(b) $\mu_{\bar{x}} = \mu = \$65{,}100$; $\sigma_{\bar{x}} = \sigma/\sqrt{n} = 7200/\sqrt{100} = \$720.00$

Thus, for samples of size 100, the mean and standard deviation of all possible sample means are respectively, $65,100 and $720.

7.51 (a) On the average, we would expect the sample mean of the four times to be the same as the population mean of 437 days.

(b) The standard deviation of the sample mean is $\sigma_{\bar{x}} = \sigma / \sqrt{n} = 399/\sqrt{4} = 199.5$ days, so 99.74% of the time we would expect the sample mean to fall within 3 standard deviations of 437 days, i.e., between 437 - 3(199.5) = -161.5 and 437 + 3(199.5) = 1035.5. Since the lower limit is clearly below zero and the time between earthquakes cannot be negative, we would expect the sample mean to fall between 0 and 1035.5 days 99.74% of the time.

7.53 (a) Enter the data into cells A1 to A487 of Excel. In cell B1, enter the expression =STDEVP(A1:A487) to obtain the population standard deviation of the data in B1. The result is 94.334 Note: Using procedures in Excel will yield a SAMPLE standard deviation of 94.431, which you could then convert to a POPULATION standard deviation by multiplying by the square root of 486/487 to obtain the same result.

(b) In Excel, in cells C1 to C487, enter the numbers 1 to 487. In cell D1, enter the expression =SQRT((487-C1)/(487-1))*94.334/SQRT(C1). In Cell E1, enter the expression =94.334/SQRT(C1). Copy cells D1 and E1 to the clipboard and then paste them into cells D2 through D487. This will produce your table in three columns. To save space, we show the table below using six columns and every tenth value of n after n = 30.

n	**Eq. 7.1**	**Eq. 7.2**	**n**	**Eq. 7.1**	**Eq. 7.2**
1	94.33	94.33	120	7.48	8.61
2	66.64	66.70	130	7.09	8.27
3	54.35	54.46	140	6.74	7.97
4	47.02	47.17	150	6.41	7.70
5	42.01	42.19	160	6.12	7.46
6	38.31	38.51	170	5.84	7.24
7	35.43	35.65	180	5.59	7.03
8	33.11	33.35	190	5.35	6.84
9	31.18	31.44	200	5.13	6.67
10	29.55	29.83	210	4.91	6.51
11	28.15	28.44	220	4.71	6.36
12	26.92	27.23	230	4.52	6.22
13	25.84	26.16	240	4.34	6.09
14	24.87	25.21	250	4.17	5.97
15	24.00	24.36	260	4.00	5.85
16	23.22	23.58	270	3.84	5.74
17	22.50	22.88	280	3.68	5.64
18	21.84	22.23	290	3.53	5.54
19	21.24	21.64	300	3.38	5.45
20	20.68	21.09	310	3.23	5.36
21	20.16	20.59	320	3.09	5.27
22	19.67	20.11	330	2.95	5.19
23	19.22	19.67	340	2.81	5.12
24	18.79	19.26	350	2.68	5.04
25	18.40	18.87	360	2.54	4.97
26	18.02	18.50	370	2.41	4.90
27	17.66	18.15	380	2.27	4.84
28	17.33	17.83	390	2.13	4.78
29	17.01	17.52	400	2.00	4.72
30	16.70	17.22	410	1.85	4.66
40	14.30	14.92	420	1.71	4.60
50	12.65	13.34	430	1.56	4.55

60	11.42	12.18	440	1.40	4.50
70	10.44	11.28	450	1.23	4.45
80	9.65	10.55	460	1.04	4.40
90	8.99	9.94	470	0.81	4.35
100	8.42	9.43	480	0.52	4.31
110	7.92	8.99	487	0.00	4.27

The results obtained with Equation 7.2 are reasonably accurate when n is 5% of N or less (n < 25). As n gets larger, the discrepancy between the results of the two equations continues to grow.

7.55 (a) We start with Equation 7.1 and substitute 1 for the sample size n.

$\sigma_{\bar{x}} = \sqrt{\frac{N-n}{N-1}}\frac{\sigma}{\sqrt{n}} = \sqrt{\frac{N-1}{N-1}}\frac{\sigma}{\sqrt{n}} = \frac{\sigma}{\sqrt{n}}$. The result on the right side of this expression is the same as Equation 7.2.

(b) The two equations should be the same when n = 1 because it makes no difference whether you are sampling with (Equation 7.2) or without (Equation 7.1) replacement when there is no second observation to be made.

(c) If a sample of size N is drawn from a population of size N without replacement, all observations will have been drawn and there can be only one result for the sample mean and that is the population mean. There can be no variation from the population mean and therefore the standard deviation of the distribution of the sample mean must be zero.

(d) $\sigma_{\bar{x}} = \sqrt{\frac{N-n}{N-1}}\frac{\sigma}{\sqrt{n}} = \sqrt{\frac{N-N}{N-1}}\frac{\sigma}{\sqrt{n}} = \sqrt{\frac{0}{N-1}}\frac{\sigma}{\sqrt{n}} = 0 \bullet \frac{\sigma}{\sqrt{n}} = 0$.

7.57 (a) When n = 1, $\sqrt{\frac{N-n}{N-1}} = \sqrt{\frac{N-1}{N-1}} = 1$. When n > 1, N - n will be smaller than N - 1, and the square root will be less than 1.

If $n \le 0.05N$, then N - n > N - 0.05N = 0.95N and

$\sqrt{\frac{N-n}{N-1}} \geq \sqrt{\frac{0.95N}{N-1}} = \sqrt{\frac{0.95}{1-1/N}} \geq \sqrt{\frac{0.95}{1}} = 0.975 > 0.97$. In the third expression, we divided the numerator and denominator by N. In the fourth expression, we used the fact that 1 - 1/N is smaller than 1 and therefore the ratio under the square root sign in the third expression is larger than 0.95/1. These two results demonstrate that when

$n \le 0.05N$, $0.97 \leq \sqrt{\frac{N-n}{N-1}} \leq 1$.

(b) Since Equation 7.1 is the same as Equation 7.2 multiplied by the factor in part (a), our result in part (a) shows that whenever $n \le 0.05N$, Equation 7.1 will yield a result that is at least as large as 97% of that yielded by Equation 7.2 and never any larger than that yielded by Equation 7.2. There will never be more than a 3% discrepancy between Equation 7.1 and Equation 7.2 when $n \le 0.05N$.

(c) Equation 7.2 is used when the population is infinite or can be considered infinite due to sampling **with** replacement. If sampling is from a finite population **without** replacement, we must account for that fact by multiplying the result of Equation 7.2 by this factor to get the result of Equation 7.1, which should be used in this situation.

This multiplication 'corrects' the result of Equation 7.2 to get the correct result from Equation 7.1 when the population is finite. Thus, this factor is called the finite population correction factor.

7.59 (a) Theoretically, the mean of possible sample means is 266 and the standard deviation is $16/\sqrt{9} = 5.33$.

We have Minitab take 2000 random samples of size n = 9 from a normally distributed population with mean 266 and standard deviation 16 by choosing **Calc ▶ Random Data ▶ Normal...**, entering 2000 in the **Generate rows of data** text box, entering C1-C9 in the **Store in column(s)** text box, entering 266 in the **Mean** text box and 16 in the **Standard deviation** text box, and clicking **OK**. These commands tell Minitab to place a total of 18000 observations from the population into columns C1-C9, with 2000 observations in each column. Then, our first random sample of size n = 9 is the first row of columns C1-C9, our second random sample of size n = 9 is the second row of columns C1-C9, and so on.

(c) We compute the sample mean of each of the 2000 samples by choosing **Calc ▶ Row statistics...**,clicking on the **Mean** button, entering C1-C9 in the **Input variables** text box and XBAR in the **Store result in** text box, and clicking **OK**. This command instructs Minitab to compute the means of the 2000 rows of C1-C9 and to place those means in a column named XBAR. Thus, the 2000 -values are now in XBAR. (We will not print these values.)

(d) We would expect the mean of the 2000 sample means to be roughly 266 and the standard deviation of the sample means to be about 5.33 since we are taking a sample of 2000 means from a theoretical sampling distribution with mean and standard deviation given in part (a).

(e) The mean is obtained by choosing **Calc ▶ Column statistics...**, clicking on the **Mean** button, selecting XBAR in the **Input variable** text box, and clicking **OK**. The result is

Mean of XBAR = 266.03

Similarly, the standard deviation is obtained by choosing **Calc ▶ Column statistics...**, clicking on the **Standard deviation** button, selecting XBAR in the **Input variable** text box, and clicking **OK**. The result is

Standard deviation of XBAR = 5.2753

Thus, the mean of our 2000 means was 266.03 and the standard deviation was 5.2753.

(b) The answers in part (e) differ from the theoretical values given in part (d) as a result of random error or sampling variability.

Exercises 7.3

7.61 (a) The sampling distribution of the mean is approximately normally distributed with mean

μ = 100 and standard deviation $\sigma_{\bar{x}} = 28/\sqrt{49} = 4$.

(b) No assumptions were made about the distribution of the population.

(c) Part (a) cannot be answered if the sample size is n = 16. Since the distribution of the population is not specified, we need a sample size of at least 30 to apply Key Fact 7.6.

7.63 (a) The probability distribution of $\bar{x}$ is normal with mean μ and standard deviation $\sigma_{\bar{x}} = \sigma/\sqrt{n}$.

(b) The answer to part (a) does not depend on how large the sample size is because the population being sampled is normally distributed.

(c) The mean of $\bar{x}$ is μ; its standard deviation is $\sigma_{\bar{x}} = \sigma/\sqrt{n}$.

(d) No. The mean and standard deviation of the sampling distribution of the sample mean are always as given in part (c) regardless of the distribution of the variable under consideration.

7.65 (a) All four graphs are centered at the same place because the population mean of $\bar{x}$ is μ and because normal curves are centered at their μ-parameter.

(b) Since $\sigma_{\bar{x}} = \sigma/\sqrt{n}$, we see that $\sigma_{\bar{x}}$ decreases as n increases. This results in a diminishing of the spread because the spread of a distribution is determined by its σ-parameter. As a consequence, we see that the larger the sample size, the greater the likelihood for small sampling error.

(c) The graphs in Figure 7.6(a) are bell-shaped because, for normally distributed populations, the random variable $\bar{x}$ is always normally distributed regardless of the sample size.

(d) The graphs in Figures 7.6(b) and 7.6(c) become bell-shaped as the sample size increases because of the central limit theorem; the probability distribution of $\bar{x}$ tends to a normal distribution as the sample size increases.

7.67 (a) Because the weights themselves are normally distributed, the sampling distribution for means of samples of size 3 will also be normal and will have mean $\mu_{\bar{x}} = \mu = 1.40\ kg$ and standard deviation $\sigma_{\bar{x}} = 0.11/\sqrt{3} = 0.064$. Thus, the distribution of the possible sample means for samples of three brain weights will be normal with mean 1.40 kg and standard deviation 0.064 kg.

(b) Because the weights themselves are normally distributed, the sampling distribution for means of samples of size 12 will also be normal and will have mean $\mu_{\bar{x}} = \mu = 1.40\ kg$ and standard deviation $\sigma_{\bar{x}} = 0.11/\sqrt{12} = 0.032$. Thus, the distribution of the possible sample means for samples of twelve brain weights will be normal with mean 1.40 kg and standard deviation 0.032 kg.

(c) To facilitate the comparison of the three graphs, we have overlaid them on one set of axes.

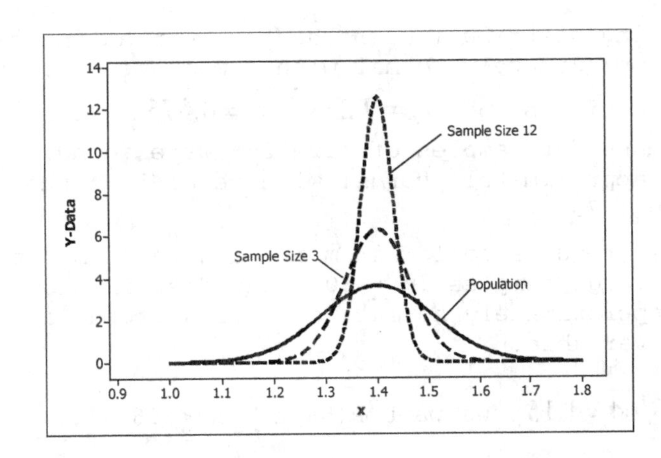

(d) $\sigma_{\bar{x}} = 0.11/\sqrt{3} = 0.064$; we want $P(1.30 \le \bar{x} \le 1.50)$.

z-score computations: Area less than z:

$\bar{x} = 1.30 \rightarrow z = \dfrac{1.30 - 1.40}{0.064} = -1.56$ 0.0594

$\bar{x} = 1.50 \rightarrow z = \dfrac{1.50 - 1.40}{0.064} = 1.56$ 0.9406

Total area = 0.9406 - 0.0594 = 0.8812

Note: If you round the standard deviation of the sample means differently, you may reach a slightly different answer.

Thus, 88.12% of samples of three Swedish men will have mean brain weights within 0.1 kg of the population mean brain weight of 1.40 kg.

(e) $\sigma_{\bar{x}} = 0.11/\sqrt{12} = 0.032$; we want $P(1.30 \le \bar{x} \le 1.50)$.

z-score computations: Area less than z:

$\bar{x} = 1.30 \rightarrow z = \dfrac{1.30 - 1.40}{0.032} = -3.13$ 0.0009

$\bar{x} = 1.50 \rightarrow z = \dfrac{1.50 - 1.40}{0.032} = 3.13$ 0.9991

Total area = 0.9991 - 0.0009 = 0.9982

Note: If you round the standard deviation of the sample means differently, you may reach a slightly different answer.

Thus, 99.82% of samples of twelve Swedish men will have mean brain weights within 0.1 kg of the population mean brain weight of 1.40 kg.

7.69 (a) The sampling distribution of the sample mean for samples of size 64 will be approximately normal with a mean of $55.4 thousand and a standard deviation of $\sigma_{\bar{x}} = 9.2/\sqrt{64} = 1.15$. Thus, if all possible sample means for samples of size 64 were found, their distribution would be approximately normal with mean $55.4 thousand and standard deviation $1,150.

(b) The sampling distribution of the sample mean for samples of size 256 will be approximately normal with a mean of \$55.4 thousand and a standard deviation of $\sigma_{\bar{x}} = 9.2/\sqrt{256} = 0.575$. Thus, if all possible sample means for samples of size 256 were found, their distribution would be approximately normal with mean \$55.4 thousand and standard deviation \$575.

(c) No. These results follow from the central limit theorem, which implies that for large samples ($n \geq 30$), the distribution of the sample mean will be approximately normal regardless of the distribution of the original variable.

(d) $\sigma_{\bar{x}} = 9.2/\sqrt{64} = 1.15$; we want $P(54.4 \leq \bar{x} \leq 56.4)$.

z-score computations:	Area less than z:
$\bar{x} = 54.4 \rightarrow z = \frac{54.4 - 55.4}{1.15} = -0.87$	0.1922
$\bar{x} = 56.4 \rightarrow z = \frac{56.4 - 55.4}{1.15} = 0.87$	0.8078

Total area = 0.8078 - 0.1922 = 0.6156

Thus, 61.56% of samples of size 64 will have a mean that is within \$1000 of the population mean, i.e., 61.56% of the samples will have a sampling error less than \$1000 for samples of size 64.

(e) $\sigma_{\bar{x}} = 9.2/\sqrt{256} = 0.575$; we want $P(54.4 \leq \bar{x} \leq 56.4)$.

z-score computations:	Area less than z:
$\bar{x} = 54.4 \rightarrow z = \frac{54.4 - 55.4}{0.575} = -1.74$	0.0409
$\bar{x} = 56.4 \rightarrow z = \frac{56.4 - 55.4}{0.575} = 1.74$	0.9591

Total area = 0.9591 - 0.0409 = 0.9182

Thus, 91.82% of samples of size 256 will have a mean that is within \$1000 of the population mean, i.e., 91.82% of the samples will have a sampling error less than \$1000 for samples of size 256.

7.71 (a) The sampling distribution of the sample mean for n = 80 will be approximately normal with some mean μ and standard deviation $\sigma_{\bar{x}} = 8.3/\sqrt{80} = 0.928$.

(b) No. The sample size of 80 is large enough so that the Central Limit Theorem applies and the sampling distribution will be approximately normal.

(c) We want $P(\mu - 2 \leq \bar{x} \leq \mu + 2)$

z-score computations:	Area less than z:
$\bar{x} = \mu - 2 \rightarrow z = \frac{(\mu - 2) - \mu}{0.928} = -2.16$	0.0154
$\bar{x} = \mu + 2 \rightarrow z = \frac{(\mu + 2) - \mu}{0.928} = 2.16$	0.9846

Total area = 0.9846 - 0.0154 = 0.9692

Thus, the probability is approximately 0.9692 that the sampling error made in estimating the population mean length of stay on the intervention ward by the mean length of stay of a sample of 80 patients will be at most 2 days.

7.73 $\sigma_{\bar{x}} = 1150/\sqrt{500} = 51.43; P(|\bar{x} - \mu| \leq 100)$

z-score computations: Area less than z:

$\bar{x} = \mu - 100 \rightarrow z = \dfrac{(\mu - 100) - \mu}{51.43} = -1.94$ 0.0262

$\bar{x} = \mu + 100 \rightarrow z = \dfrac{(\mu + 100) - \mu}{51.43} = 1.94$ 0.9738

Required Area = 0.9738 - 0.0262 = 0.9476

Thus, there is a 0.9476 probability that the mean tariff rate of a sample of 500 randomly selected ethanol shipments used to estimate the mean tariff rate of all ethanol shipments will have a sampling error less than $100.

7.75 In solving this problem, we have to assume that the distribution of calcium intakes of adults with incomes below the poverty level is approximately normally distributed. Then, the distribution of the sample means will be approximately normally distributed. The distribution of the sample means have a mean 1000 mg and standard deviation of $\sigma_{\bar{x}} = \sigma/\sqrt{n} = 188/\sqrt{18} = 44.312.$

We want $P(\bar{x} \leq 947.4)$:

z-score computation: Area less than z:

$\bar{x} = 947.4 \rightarrow z = \dfrac{947.4 - 1000}{44.312} = -1.19$ 0.1170

Total area = 0.1170

11.70% of all samples of size 18 of adults with incomes below the poverty level will have mean calcium intakes of at most 947.4 mg.

7.77 In solving this problem, we have to assume that the distribution of post-work hear rate for casting workers is approximately normally distributed. Then, the distribution of the sample means will be approximately normally distributed. The distribution of the sample means have a mean 72 bpm and standard deviation of $\sigma_{\bar{x}} = \sigma/\sqrt{n} = 11.2/\sqrt{29} = 2.080.$ We want $P(\bar{x} > 78.3)$:

z-score computation: Area less than z:

$\bar{x} = 78.3 \rightarrow z = \dfrac{78.3 - 72}{2.080} = 3.03$ 0.9988

Total area = 1 - 0.9988 = 0.0012.

The probability is 0.0012 that samples of 29 casting workers will have a mean post=work hear rate exceeding 78.3 bpm.

7.79 (a) 68.26 (b) 95.44 (c) 99.74 (d) 100(1 - α)

7.81 (a) No. The original population is quite skewed. A sample size of 4 is quite small, so we would expect that a histogram of the means of such a sample would still exhibit some degree of skewness.

(b) We'll use Excel to carry out the simulation. First enter the values 1 through 7 in cells A1 to A7, then enter the frequencies 19.4 through 1.6 from the table in Example 7.9 into cells C1 through C7.

In C8, enter the expression =Sum(C1:C7). Then in cell B1, enter the expression =C1/C8, copy this cell to the clipboard and then paste it into cells B2 through B7. Columns A and B now contain the probability distribution of x, the household size. Now from the Tools menu, select **Data Analysis ▶ Random Number Generation.** Enter 4 in the **Number of Variables** text box, enter 1000 in the **Number of Random Numbers** text box, choose **Discrete** in the **Distribution** box, enter the range A1:B7 in the **Value and Probability Input Range** text box, and enter A11 in the **Output Range** box. Click on **OK.** You will get 4 columns of 100 numbers between 1 and 7 in columns A, B, C, and D, starting in row 11 and ending in row 1010. In cell E11, enter the expression =AVERAGE(A11:D11). Copy this cell to the clipboard and then paste it into the cells E12 through E1010. Type the label **Mean** in cell E10. Use the mouse to highlight cells E10 though E1010. Now from the menu, select **Charts and Plots.** In the **Function Type** box, select **Histogram,** Click on **Mean** in the **Names and Columns** box and enter it in the **Quantitative Variables** box. Then click **OK.** A histogram of the sampling distribution of $\overline{x}$ for our samples of size 4 is shown below. Yours is very likely to be different. Our histogram is slightly skewed to the right.

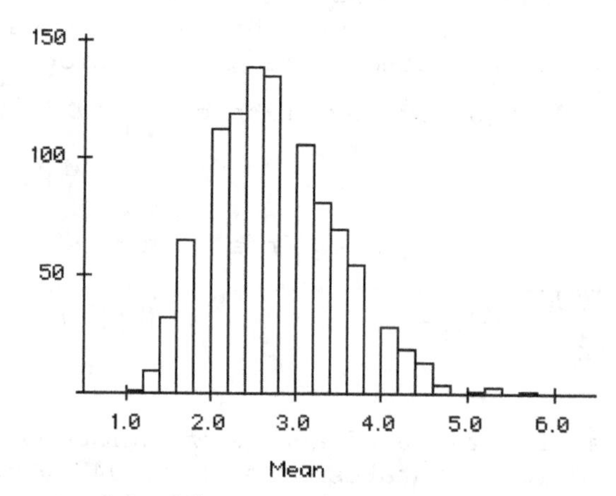

(c) We would expect this histogram to be closer to bell-shaped than the one in part (a), but 10 is still a relatively small sample size, so there may still be a trace of skewness. Using the same procedure as in part (b), but with 10 variables, our result below is more symmetrical, but we definitely did not expect the one tall spike. Note that the scale is different from the first histogram. These means are more closely concentrated.

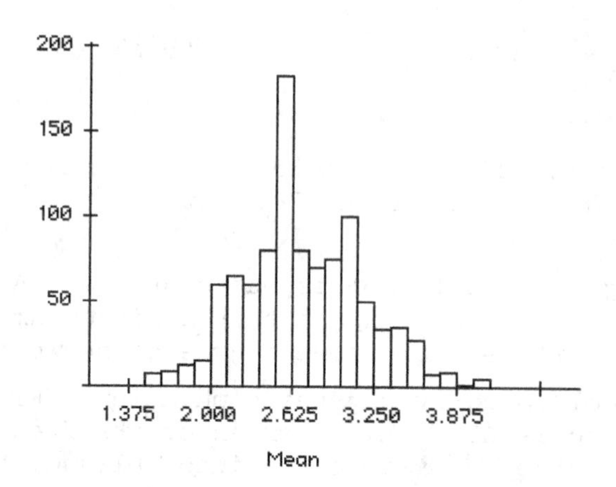

(d) Since 100 is definitely a large sample size, we would expect the histogram of the sample means to be bell-shaped. Using the same procedure as in part (b), but with 100 variables, our result below is close to bell-shaped and again is more concentrated than the first two.

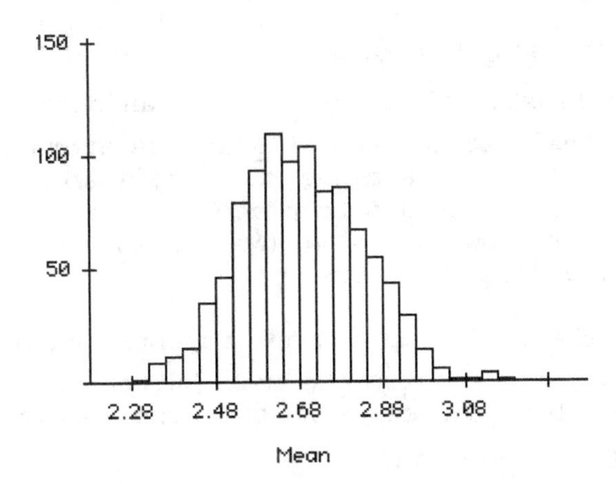

7.83 (a) We will describe a procedure using Minitab to sketch the exponential curve for this population, although the sketch could be done using any spreadsheet program as well. We begin by selecting typical values of x that will eventually be substituted into the exponential function itself to trace out values for y. Values of x selected are the integers 0 through 15. Name two columns in the Data window X and Y and enter the numbers 0, 1, 2, ...,15 in the X column. Then choose **Calc ▶ Probability Distributions ▶ Exponential...,** click on the **Probability density** button, click in the **Mean** text box and type 8.7. Then enter **X** in the **Input column** text box and **Y** in the **Optional storage** text box. Click **OK**. Now choose **Graph ▶ Scatterplot...,** select the **With connect line** version and click **OK**. Select Y for row 1 of the **Y** column and X for row 1 of the **X** column. To plot a continuous curve, click on the **Data view** button and check only the **Connect line** box. Click **OK** twice. The result is shown below. The exponential distribution is reverse J-shaped.

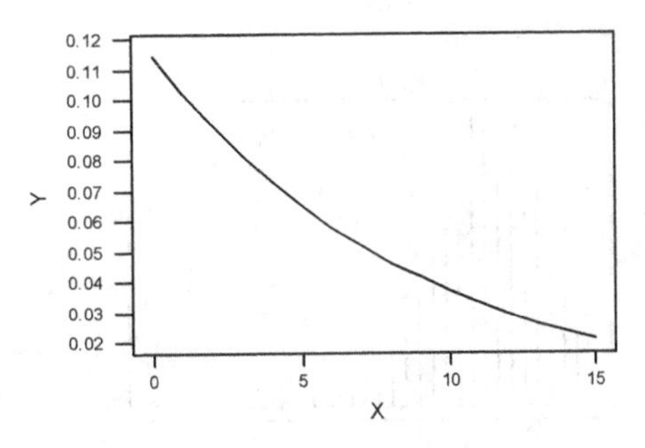

(b) To have Minitab take 1000 random samples of size n = 4 from an exponentially distributed population with mean μ = 8.7, we choose **Calc ▶ Random Data ▶ Exponential...,** enter 1000 in the **Generate rows of data** text box, enter C1-C4 in the **Store in column(s)** text box, enter 8.7 in the **Mean** text box, and click **OK**. These commands tell

Minitab to place a total of 4000 observations from the population into columns C1-C4, with 1000 observations in each column. Then, our first random sample of size n = 4 is the first row of columns C1-C4, our second random sample of size n = 4 is the second row of columns C1-C4, and so on.

(c) We compute the sample mean of each of the 1000 samples by choosing **Calc ▶ Row statistics...**, clicking on the **Mean** button, entering C1-C4 in the **Input variables** text box and XBAR in the **Store result in** text box, and clicking **OK**. This command instructs Minitab to compute the means of the 1000 rows of C1-C4 and to place those means in XBAR. Thus, the 1000 -values are now in XBAR. (We will not print these values.) Your results will differ from ours.

(d) The mean of the 1000 sample means is obtained by choosing **Calc ▶ Column statistics...**, clicking on the **Mean** button, selecting XBAR in the **Input variable** text box, and clicking **OK**. The result is

Mean of XBAR = 8.4743

Similarly, the standard deviation of the 1000 sample means is obtained by choosing **Calc ▶ Column statistics...**, clicking on the **Standard deviation** button, selecting XBAR in the **Input variable** text box, and clicking **OK**. The result is

Standard deviation of XBAR = 4.2498

(e) Theoretically, for all possible sample means for samples of size 4, the mean is the same as the population mean, 8.7, and the standard deviation is $\sigma/\sqrt{n} = 8.7/\sqrt{4} = 4.35$. Both of the simulated values obtained in part (d) are close to these values.

(f) To get a histogram of the 1000 sample means stored in XBAR, choose **Graph ▶ Histogram...**, selecting the **Simple** version and click **OK**. Enter XBAR in the **Graph variables** text box, and click **OK**. The result is shown below at the right. The histogram is not bell-shaped; given the right-skewness of the population, we would expect for samples of only size four, that the distribution of means would be somewhat skewed to the right.

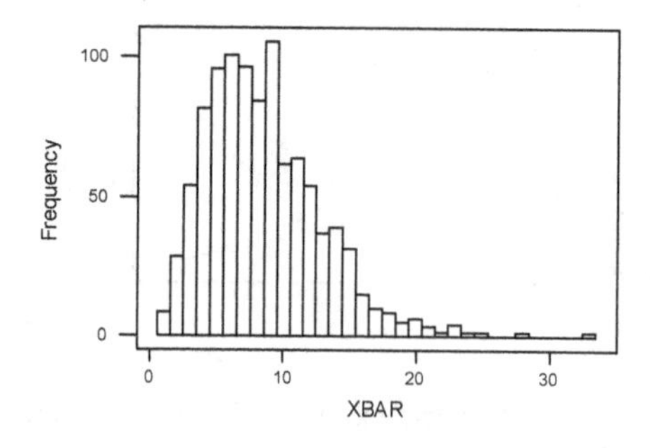

(g) Repeating the process for samples of size 40 (use columns C1-C40 for the samples), we find that

Mean of XBAR = 8.7367 and Standard deviation of XBAR = 1.4204

In theory, the mean of all possible samples of size 40 is 8.7 and the standard deviation is $\sigma/\sqrt{n} = 8.7/\sqrt{40} = 1.38$. The simulated values are very close to these.

A histogram of the means of samples of size 40 is shown following, obtained as was done previously. The simulated distribution of sample means is close to bell-shaped as we would expect for samples over size 30.

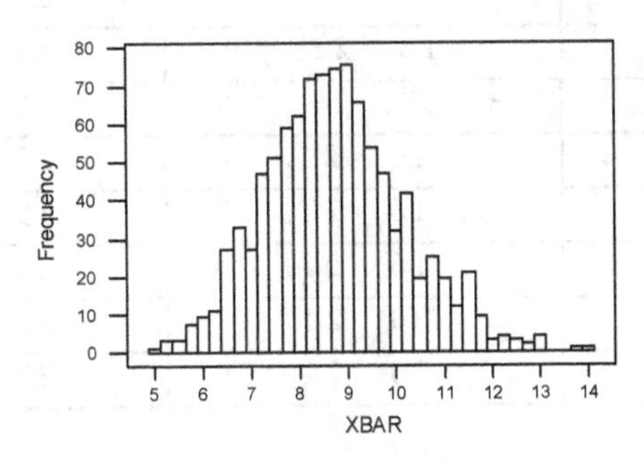

Review Problems for Chapter 7

1. Sampling errors are errors resulting from using a sample to estimate a population characteristic.

2. The sampling distribution of a statistic is the set of all possible observations of the statistic for a sample of a given size. It is important to know the sampling distribution in order to answer questions about the accuracy of estimating a population parameter using the sample statistic.

3. Two other terms are 'sampling distribution of the sample mean' and 'the distribution of the variable $\bar{x}$.'

4. The set of possible sample means exhibits less and less variability as the sample size increases, that is, the set becomes more and more clustered about the population mean. This means that as the sample size increases, there is a greater chance that the value of the sample mean from any sample is close to the value of the population mean.

5. (a) μ = \$108/6 = \$18 (thousands)

(b)

Sample	Salaries	
A,B,C,D	8, 12, 16, 20	14
A,B,C,E	8, 12, 16, 24	15
A,B,C,F	8, 12, 16, 28	16
A,B,D,E	8, 12, 20, 24	16
A,B,D,F	8, 12, 20, 28	17
A,B,E,F	8, 12, 24, 28	18
A,C,D,E	8, 16, 20, 24	17
A,C,D,F	8, 16, 20, 28	18
A,C,E,F	8, 16, 24, 28	19
A,D,E,F	8, 20, 24, 28	20
B,C,D,E	12, 16, 20, 24	18
B,C,D,F	12, 16, 20, 28	19
B,C,E,F	12, 16, 24, 28	20
B,D,E,F	12, 20, 24, 28	21
C,D,E,F	16, 20, 24, 28	22

(c)

```
                    •
          •    •    •    •    •
•    •    •    •    •    •    •    •    •
+----+----+----+----+----+----+----+----+
14   15   16   17   18   19   20   21   22
```

(d) $P(|\bar{x}-\mu| \le 1) = P(\bar{x}=17)+P(\bar{x}=18)+P(\bar{x}=19)$
$= 2/15+3/15+2/15 = 7/15 = 0.4667$

6. (a) $\mu_{\bar{x}} = \dfrac{\sum \bar{x}}{N} = \dfrac{270}{15} = 18.0$. The mean of the means of all of the possible samples of size 4 is $18.0 thousand.

(b) Yes. The mean of the sampling distribution of $\bar{x}$ is always the same as the mean of the population, which is $18 thousand in this case.

7. (a) For a normally distributed population, the random variable $\bar{x}$ is normally distributed, regardless of the sample size. Also, we know that $\mu_{\bar{x}} = \mu$. Consequently, since the normal curve for a normally distributed population or random variable is centered at its μ-parameter, all three curves are centered at the same place.

(b) Curve B corresponds to the larger sample size. Since $\sigma_{\bar{x}} = \sigma/\sqrt{n}$, the larger the sample size, the smaller the value of σ and, hence, the smaller the spread of the normal curve for $\bar{x}$. Thus, Curve B, which has the smaller spread, corresponds to the larger sample size.

(c) The spread of each curve is different because $\sigma_{\bar{x}} = \sigma / \sqrt{n}$ and the spread of a normal curve is determined by $\sigma_{\bar{x}}$, Thus, different sample sizes result in normal curves with different spreads.

(d) Curve B corresponds to the sample size that will tend to produce less sampling error. The smaller the value of $\sigma_{\bar{x}}$, the smaller the sampling error tends to be.

(e) When x is normally distributed, $\bar{x}$ always has a normal distribution as well.

8. (a) The sampling error results from using the sample mean income tax, $\bar{x}$, of the 308,946 tax returns as an estimate of the mean income tax, μ, of all 2010 tax returns.

(b) The sampling error is \$11,266 - \$11,354 = -\$88.

(c) No, not necessarily. However, increasing the sample size from 308,946 to 400,000 would increase the likelihood for a smaller sampling error.

(d) Increase the sample size.

9. (a) The population consists of all new car sales in the U.S. in the given year. The variable under consideration is the amount spent for a new car.

(b) $\mu_{\bar{x}} = \$30,803; \sigma_{\bar{x}} = \$10200 / \sqrt{50} = \$1442.50$

(c) $\mu_{\bar{x}} = \$30,803; \sigma_{\bar{x}} = \$10200 / \sqrt{100} = \$1020.00$

(d) The value of $\sigma_{\bar{x}}$ will be smaller than \$1020 because $\sigma_{\bar{x}} = \sigma / \sqrt{n}$. Thus, the larger the sample size, the smaller the value of $\sigma_{\bar{x}}$.

10. (a) False. By the central limit theorem, the random variable $\bar{x}$ is approximately normally distributed. Furthermore, $\mu_{\bar{x}} = \mu = 45$, and $\sigma_{\bar{X}} = \sigma / \sqrt{n} = 7 / \sqrt{196} = 0.5.$ Thus, $P(31 \leq \bar{x} \leq 59)$ equals the area under the normal curve with parameters $\mu_{\bar{x}} = 45$ and $\sigma_{\bar{x}} = 0.5$ that lies between 31 and 59. Applying the usual techniques, we find that area to be 1.0000 to four decimal places. Hence, there is almost a 100% chance that the mean of the sample will be between 31 and 59.

(b) This is not possible to tell, since we do not know the distribution of the population.

(c) True. Referring to part (a), we see that $P(44 \leq \bar{x} \leq 46)$ equals the area under the normal curve with parameters $\mu_{\bar{x}} = 45$ and $\sigma_{\bar{x}} = 0.5$ that lies between 44 and 46. Applying the usual techniques, we find that area to be 0.9544. Hence, there is approximately a 95.44% chance that the mean of the sample will be between 44 and 46.

11. (a) False. Since the population is normally distributed, so is the random variable $\bar{x}$. Furthermore, $\mu_{\bar{x}} = \mu = 45$ and $\sigma_{\bar{X}} = \sigma / \sqrt{n} = 7 / \sqrt{196} = 0.5.$

Thus, $P(31 \leq \bar{x} \leq 59)$ equals the area under the normal curve with parameters $\mu_{\bar{x}} = 45$ and $\sigma_{\bar{x}} = 0.5$ that lies between 31 and 59.

Applying the usual techniques, we find that area to be 1.0000 to four decimal places. Hence, there is almost a 100% chance that the mean of the sample will be between 31 and 59.

(b) True. Since the population is normally distributed, percentages for the population are equal to areas under the normal curve with parameters $\mu = 45$ and $\sigma = 7$. Applying the usual techniques, we find that the area under that normal curve between 31 and 59 is 0.9544.

(c) True. From part (a), we see that the random variable $\bar{x}$ is normally distributed with $\mu_{\bar{x}} = 45$ and $\sigma_{\bar{x}} = 0.5$. Applying the usual techniques, we find that area to be 0.9544. Hence, we find that there is a 95.44% chance that the mean of the sample will be between 44 and 46.

12. (a) $\mu = 8.5g \text{ and } \sigma = 0.3g$

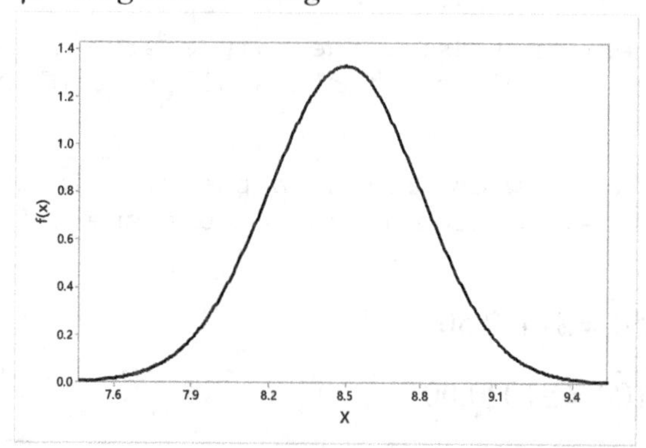

(b) normal distribution with $\mu_{\bar{x}} = 8.5g \text{ and } \sigma_{\bar{x}} = 0.3/\sqrt{4} = 0.15g$

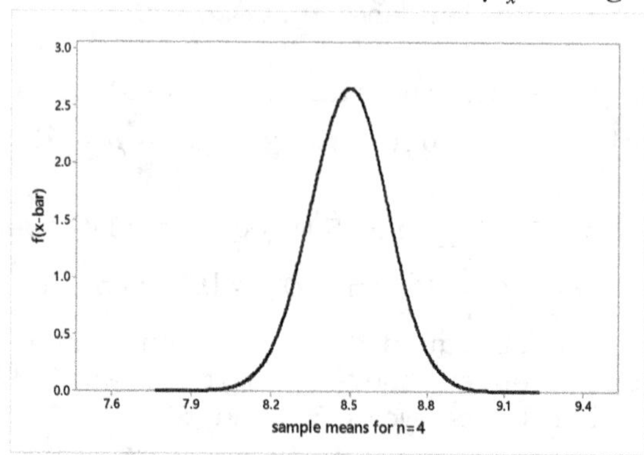

(c) normal distribution with $\mu_{\bar{x}} = 8.5g$ $\text{and } \sigma_{\bar{x}} = 0.3/\sqrt{9} = 0.10g$

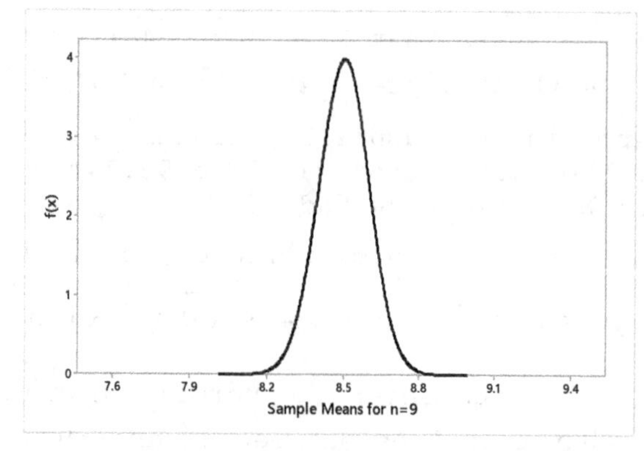

13. $n = 4 : \mu_{\bar{x}} = 8.5 \text{ and } \sigma_{\bar{x}} = 0.3/\sqrt{4} = 0.15$

(a) $P(8.275 \le \bar{x} \le 8.725)$:

z-score computations: Area less than z:

$\bar{x} = 8.275 \rightarrow z = \frac{8.275 - 8.5}{0.15} = -1.50$ 0.0668

$\bar{x} = 8.725 \rightarrow z = \frac{8.725 - 8.5}{0.15} = 1.50$ 0.9332

Total area = 0.9332 - 0.0668 = 0.8664, or 86.64%.

(b) The probability is 0.8664 that a sample of size four will have a mean within 0.225g of the population mean of 8.5g.

(c) There is an 86.64% chance that the sampling error when using the mean weights, $\bar{x}$, of the four possums will be less than 0.225g.

(d) $n = 9 : \mu_{\bar{x}} = 8.5 \text{ and } \sigma_{\bar{x}} = 0.3/\sqrt{9} = 0.10$

z-score computations: Area less than z:

$\bar{x} = 8.275 \rightarrow z = \frac{8.275 - 8.5}{0.10} = -2.25$ 0.0122

$\bar{x} = 8.725 \rightarrow z = \frac{8.725 - 8.5}{0.1} = 2.25$ 0.9878

Total area = 0.9878 - 0.0122 = 0.9756 or 97.56%

The probability is 0.9756 that a sample of size 9 will have a mean within 0.225g of the population mean of 8.5g.

There is an 97.56% chance that the sampling error when using the mean weights, $\bar{x}$, of the nine possums will be less than 0.225g

14. (a) Since 60 is a fairly large sample size, the distribution of the sample mean will be approximately normal with a mean of 4.60 mmol/l and a standard deviation of $\sigma_{\bar{x}} = \sigma/\sqrt{n} = 0.16/\sqrt{60} = 0.0207$.

(b) Since 120 is a large sample size, the distribution of the sample mean will be approximately normal with a mean of 4.60 mmol/l and a standard deviation of $\sigma_{\bar{x}} = \sigma/\sqrt{n} = 0.16/\sqrt{120} = 0.0146$.

(c) No. The Central Limit Theorem ensures that when n is large, the distribution of the sample mean will be approximately normal regardless of whether the blood glucose levels themselves are normally distributed.

15. (a) $n = 500 : \mu_{\bar{x}} = \mu \text{ and } \sigma_{\bar{x}} = 50{,}900/\sqrt{500} = 2276.32$

$P(\mu - 2{,}000 \le \bar{x} \le \mu + 2{,}000)$:

z-score computations: Area less than z:

$\bar{x} = \mu - 2000 \rightarrow z = \frac{(\mu - 2000) - \mu}{2276.32} = -0.88$ 0.1894

$\bar{x} = \mu + 2000 \rightarrow z = \frac{(\mu + 2000) - \mu}{2276.32} = 0.88$ 0.8106

Total area = 0.8106 - 0.1894 = 0.6212

(b) To answer part (a), it is not necessary to assume that the population is normally distributed because the sample size is large and,

therefore, $\bar{x}$ is approximately normally distributed, regardless of the distribution of the population of life-insurance amounts.

If the sample size were 20 instead of 500, it would be necessary to assume normality because the sample size is small.

(c) $n = 5000 : \mu_{\bar{x}} = \mu$ *and* $\sigma_{\bar{x}} = 50{,}900 / \sqrt{5000} = 719.83$

z-score computation: Area less than z:

$\bar{x} = \mu - 2000 \rightarrow \frac{(\mu - 2000) - \mu}{719.83} = -2.78$ 0.0027

$\bar{x} = \mu + 2000 \rightarrow \frac{(\mu + 2000) - \mu}{719.83} = 2.78$ 0.9973

Total area = 0.9973 - 0.0027 = 0.9946

16. (a) $P(\bar{x} \leq 4.5)$:

z-score computation: Area less than z:

$\bar{x} = 4.5 \rightarrow z = \frac{4.5 - 5}{0.5} = -1.00$ 0.1587

Total area = 0.1587

If the paint lasts 4.5 years, one would not consider this to be substantial evidence against the manufacturer's claim that the paint will last an average of five years.

Assuming the manufacturer's claim is correct, the probability is 0.1587 that the paint will last 4.5 years or less on a (randomly selected) house painted with the paint. In other words, there is a (fairly high) 15.87% chance that the paint would last 4.5 years or less, even if the manufacturer's claim is correct.

(b) $P(\bar{x} \leq 4.5)$:

$n = 10 : \mu_{\bar{x}} = 5$ *and* $\sigma_{\bar{x}} = 0.5 / \sqrt{10} = 0.158$

z-score computation: Area less than z:

$\bar{x} = 4.5 \rightarrow z = \frac{4.5 - 5}{0.158} = -3.16$ 0.0008

Total area = 0.0008

For 10 houses, if the paint lasts an average of 4.5 years, I would consider this to be substantial evidence against the manufacturer's claim that the paint will last an average of five years.

Assuming the manufacturer's claim is correct, the probability is 0.0008 that the paint will last an average of 4.5 years or less for 10 (randomly selected) houses painted with the paint. In other words, there is less than a 0.1% chance that that would occur, if the manufacturer's claim is correct.

(c) $P(\bar{x} \leq 4.9)$:

z-score computation: Area less than z:

$\bar{x} = 4.9 \rightarrow z = \dfrac{4.9-5}{0.158} = -0.63$ 0.2643

Total area = 0.2643

For 10 houses, if the paint lasts an average of 4.9 years, I would not consider this to be substantial evidence against the manufacturer's claim that the paint will last an average of five years.
Assuming the manufacturer's claim is correct, the probability is 0.2643 that the paint will last an average of 4.9 years or less for 10 (randomly selected) houses painted with the paint. In other words, there is a (fairly high) 26.43% chance that that would occur, even if the manufacturer's claim is correct.

17. (a) The sample size is large enough that the distribution of the sample means will be approximately normal with mean 6.83 and standard deviation of $\sigma_{\bar{x}} = \sigma / \sqrt{n} = 4.28 / \sqrt{100} = 0.428$.

We want $P(\bar{x} > 7.5)$:

z-score computation: Area less than z:

$\bar{x} = 7.5 \rightarrow z = \dfrac{7.5-6.83}{0.428} = 1.57$ 0.9418

Total area = 1 - 0.9418 = 0.0582.

5.82% of all samples of size 100 days during the decade in question will have a mean degree of cloudiness exceeding 7.5.

(b) For samples of size 5, it is not reasonable to use a normal distribution to obtain the percentage required in part (a). From the frequency distribution, we can see that the distribution of degree of cloudiness is definitely not normally distributed. Therefore a sample of at least 30 is required before we can assume that the distribution of the sample means is approximately normally distributed.

18. (a) We have Minitab take 1000 random samples of size n = 4 from a normally distributed population with mean 49.34 and standard deviation 10.14 by choosing **Calc** ▶ **Random Data** ▶ **Normal...**, entering 1000 in the **Generate rows of data** text box, entering C1-C4 in the **Store in column(s)** text box, entering 49.34 in the **Mean** text box and 10.14 in the **Standard deviation** text box, and clicking **OK**.

(b) We compute the sample mean of each of the 1000 samples by choosing **Calc** ▶ **Row statistics...**,clicking on the **Mean** button, typing C1-C4 in the **Input variables** text box and XBAR in the **Store result in** text box, and clicking **OK**. (We will not print these values.)

(c) The mean is obtained by choosing **Calc ▶ Column statistics...**, clicking on the **Mean** button, selecting XBAR in the **Input variable** text box, and clicking **OK**. Our result is

```
Mean of XBAR = 49.1875
```

Similarly, the standard deviation is obtained by choosing **Calc ▶ Column statistics...**, clicking on the **Standard deviation** button, selecting XBAR in the **Input variable** text box, and clicking **OK**. Our result is

```
Standard deviation of XBAR = 5.18840
```

To get a histogram of the 1000 sample means stored in XBAR, choose **Graph ▶ Histogram...**, select the **Simple** version and click **OK**. Enter XBAR in the **Graph variables** text box, and click **OK**. Our result follows. Individual results will vary.

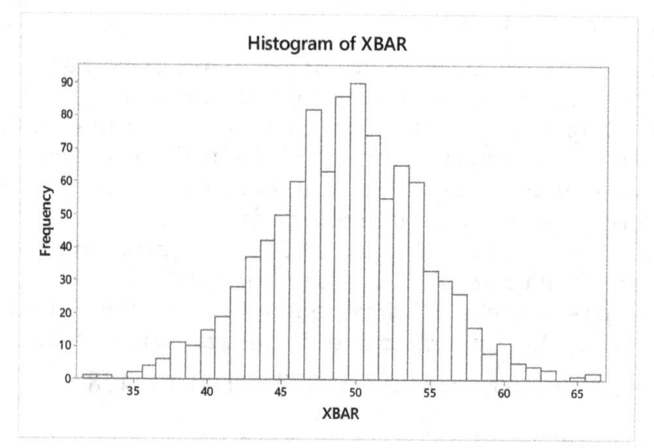

(d) Theoretically, the distribution of all possible sample means for samples of size four from this normal population should have mean 49.34, standard deviation $\sigma_{\bar{x}} = \sigma / \sqrt{n} = 10.14 / \sqrt{4} = 5.07$, and a normal distribution.

(e) The histogram is close to bell-shaped, is centered near 49.34, and most of the data lies within three standard deviations (3 x 5.07 = 15.21) of 49.34, i.e. between 34.13 and 64.55. The mean of the 1000 sample means is 49.19, very close to 49.34, and the standard deviation of the sample means is 5.19, very close to 5.07.

19. (a) A uniform distribution between 0 and 1:

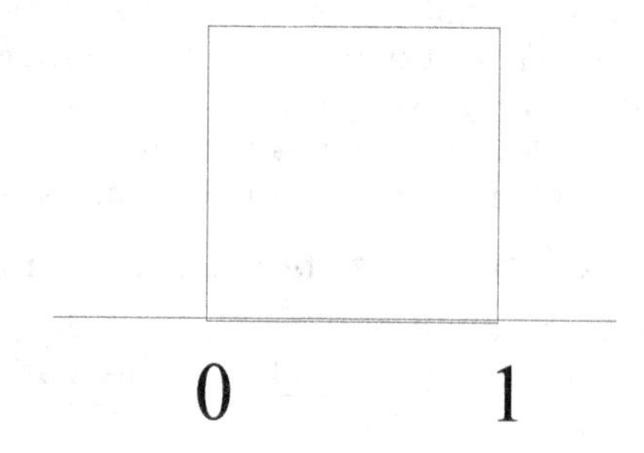

(b) We have Minitab take 2000 random samples of size n = 2 from a uniformly distributed population between 0 and 1 by choosing **Calc ▶ Random Data ▶ Uniform...**, entering 2000 in the **Generate rows of data** text box, entering C1-C2 in the **Store in column(s)** text box, entering 0.0 in the **Lower endpoint** text box and 1.0 in the **Upper endpoint** text box, and clicking **OK**.

(c) We compute the sample mean of each of the 1000 samples by choosing **Calc ▶ Row statistics...**,clicking on the **Mean** button, typing C1-C2 in the **Input variables** text box and XBAR in the **Store result in** text box, and clicking **OK**. (We will not print these values.)

(d) The mean of the sample means is obtained by choosing **Calc ▶ Column statistics...**, clicking on the **Mean** button, selecting XBAR in the **Input variable** text box, and clicking **OK**. Our result is

Mean of XBAR = 0.50454

Similarly, the standard deviation is obtained by choosing **Calc ▶ Column statistics...**, clicking on the **Standard deviation** button, selecting XBAR in the **Input variable** text box, and clicking **OK**. Our result is

Standard deviation of XBAR = 0.20545

(e) Theoretically, the distribution of all possible sample means for samples of size two from this uniform normal population should have mean

(0 + 1)/2 = .5 and standard deviation $\sigma/\sqrt{n} = ((1-0)/\sqrt{12})/\sqrt{2} = 0.2041$. Minitab simulation results are very close to these theoretical values.

(f) To get a histogram of the 1000 sample means stored in XBAR, choose **Graph ▶ Histogram...**, select the **Simple** version and click **OK**. Enter XBAR in the **Graph variables** text box, and click **OK**. The result is shown following.

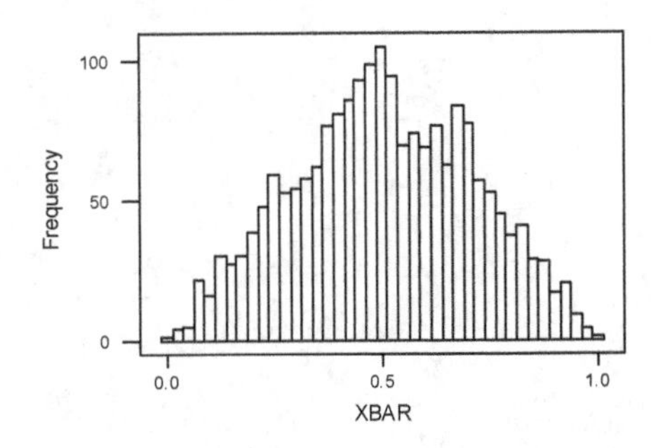

The histogram is more triangle-shaped than bell-shaped. Since the population is far from normal and the sample size is only two, we would not expect the sampling distribution to be bell-shaped. In fact, it can be shown using advanced methods that the sum of two uniformly distributed variables has a triangular distribution.

(g) Repeating the process, but using C1-C35 for the data in each sample, we obtained

Mean of XBAR = 0.49932

Standard deviation of XBAR = 0.049240

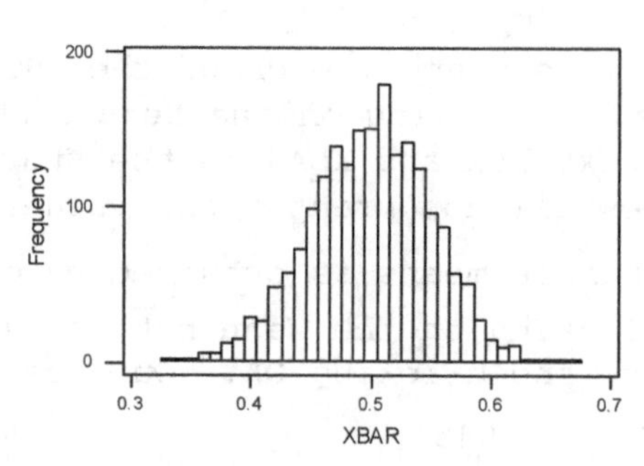

The theoretical distribution of the means has mean 0.5 and standard deviation $\sigma/\sqrt{n} = ((1-0)/\sqrt{12})/\sqrt{35} = 0.0488.$

The Minitab simulation values are very close to these. The histogram for the means of the samples of size 35 is shown. It is much more bell-shaped than for samples of size two, as we would expect since n > 30.

CHAPTER 8 SOLUTIONS

Exercises 8.1

8.1 The value of a statistic that is used to estimate a parameter is called a point estimate of the parameter.

8.3 Margin of error indicates how accurate our estimate ($\bar{x}$) is an estimate for the value of the unknown parameter (μ).

8.5 Approximately 95%, or 950, of the confidence intervals will contain the value of the unknown parameter.

8.7 $\bar{x} = 230/10 = 23.0$

8.9 $s=\sqrt{\dfrac{\sum x^2-\left(\sum x\right)^2/n}{n-1}}=\sqrt{\dfrac{6140-(230)^2/10}{9}}=\sqrt{94.44}=9.72$

8.11 (a) The approximate confidence interval will be

$$\bar{x}-2\sigma/\sqrt{n} \text{ to } \bar{x}+2\sigma/\sqrt{n}$$
$$20-2(3)/\sqrt{36} \text{ to } 20+2(3)/\sqrt{36}$$
$$19 \text{ to } 21$$

(b) The margin of error is 1. We can be approximately 95% confident that the unknown parameter, μ, will be within 1 of our sample mean, $\bar{x} = 20$.

(c) The endpoints of a confidence interval are found by point estimate ± margin of error. For the interval in part (a), we have 20 ± 1.

8.13 (a) The approximate confidence interval will be

$$\bar{x}-2\sigma/\sqrt{n} \text{ to } \bar{x}+2\sigma/\sqrt{n}$$
$$30-2(4)/\sqrt{25} \text{ to } 30+2(4)/\sqrt{25}$$
$$28.4 \text{ to } 31.6$$

(b) The margin of error is 1.6. We can be approximately 95% confident that the unknown parameter, μ, will be within 1.6 of our sample mean, $\bar{x} = 30$.

(c) The endpoints of a confidence interval are found by point estimate ± margin of error. For the interval in part (a), we have 30 ± 1.6.

8.15 (a) The approximate confidence interval will be

$$\bar{x}-2\sigma/\sqrt{n} \text{ to } \bar{x}+2\sigma/\sqrt{n}$$
$$50-2(5)/\sqrt{16} \text{ to } 50+2(5)/\sqrt{16}$$
$$47.5 \text{ to } 52.5$$

(b) The margin of error is 2.5. We can be approximately 95% confident that the unknown parameter, μ, will be within 2.5 of our sample mean, $\bar{x} = 50$.

(c) The endpoints of a confidence interval are found by point estimate ± margin of error. For the interval in part (a), we have 50 ± 2.5.

8.17 (a) $\bar{x}$ = \$526538/20 = \$26326.90

(b) Since some sampling error is expected, it is unlikely that the sample mean will exactly equal the population mean μ.

8.19 (a) The approximate confidence interval will be

$$\bar{x} - 2\sigma/\sqrt{n} \text{ to } \bar{x} + 2\sigma/\sqrt{n}$$

$$26326.90 - 2(8100)/\sqrt{20} \text{ to } 26326.90 + 2(8100)/\sqrt{20}$$

$$\$22{,}704.47 \text{ to } \$29{,}949.33$$

(b) Since we know that approximately 95% of all samples of 20 wedding costs have the property that the interval from $\bar{x}$ - \$3622.43 to $\bar{x}$ + \$3622.43 contains μ, we can be 95% confident that the interval from \$22707.47 to \$29952.33 contains μ.

(c) We can't be certain that the population mean lies in the interval, but we are 95% confident that it does.

8.21 (a) n = 35; a point estimate for the mean fuel tank capacity is

$\bar{x}$ = 664.9/35 = 19.00 gallons. This number is called a point estimate because it consists of a single value.

(b) The approximate confidence interval will be

$$\bar{x} - 2\sigma/\sqrt{n} \text{ to } \bar{x} + 2\sigma/\sqrt{n}$$

$$19.00 - 2(3.50)/\sqrt{35} \text{ to } 19.00 + 2(3.50)/\sqrt{35}$$

$$17.82 \text{ to } 20.18$$

(c) We could see if a histogram looked bell-shaped or if a normal probability plot produced a relatively straight line.

(d) It is not necessary that the fuel tank capacities be exactly normally distributed because for samples of size 35 by the Central Limit Theorem, the distribution of the sample means will be approximately normally distributed, regardless of the population distribution. Therefore, the confidence interval will be approximately correct.

8.23 (a) Using Minitab to retrieve the data from the WeissStats Resource Site, we obtained the following normal probability plot by choosing **Stat ▶ Basic Statistics ▶ Normality Test,** entering LENGTHS in the **Variable** text box, and clicking **OK**.

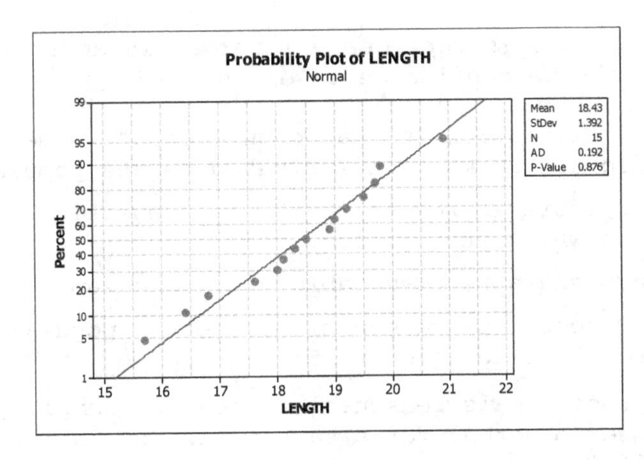

(b) Yes. There do not appear to be any outliers, and the data points fall very close to a straight line.

(c) The sample mean is 18.43 mm. Since $P(\bar{x}-2\sigma/\sqrt{n}<\mu<\bar{x}+2\sigma/\sqrt{n})=0.9544$, we can be approximately 95% confident that the mean μ is somewhere between $\bar{x}-2\sigma/\sqrt{n}$ and $\bar{x}+2\sigma/\sqrt{n}$. Since n = 15, $\bar{x}$ = 18.43, and σ = 1.76, the 95% confidence interval is

$$\bar{x}-2\sigma/\sqrt{n} \ \ to \ \ \bar{x}+2\sigma/\sqrt{n}$$

$$18.43-2(1.76)/\sqrt{15} \ to \ 18.43+2(1.76)/\sqrt{15}$$

$$17.52 \ to \ 19.34$$

We can be approximately 95% confident that the interval 17.52 mm to 19.34 mm contains the value of the mean μ.

(d) Yes. 18.14 lies between 17.52 and 19.34, so the interval we obtained in part (c) does contain μ. It is possible that our confidence interval will not contain μ. This should happen only about 5% of the time.

8.25 Since $P(\bar{x}-3\sigma/\sqrt{n}<\mu<\bar{x}+3\sigma/\sqrt{n})=0.9974$, we can be 99.7% confident that the mean μ is somewhere between $\bar{x}-3\sigma/\sqrt{n}$ and $\bar{x}+3\sigma/\sqrt{n}$. Since n = 36, $\bar{x}$ = 63.28, and σ = 7.2, the 99.7% confidence interval is (in thousands)

$$\bar{x}-3\sigma/\sqrt{n} \ \ to \ \ \bar{x}+3\sigma/\sqrt{n}$$

$$63.28-3(7.2)/\sqrt{36} \ to \ 63.28+3(7.2)/\sqrt{36}$$

$$59.68 \ to \ 66.88$$

We can be 99.7% confident that the interval $59,680 to $66,880 contains the value of the mean μ.

Exercises 8.2

8.27 (a) Confidence level = 0.90; α = 0.10

(b) Confidence level = 0.99; α = 0.01

8.29 (a) By saying that a 1-α confidence interval is exact, we mean that the true confidence level is equal to 1-α.

(b) By saying that a 1-α confidence interval is approximately correct, we mean that the true confidence level is only approximately equal to 1-α.

8.31 When we use the abbreviation "normal population," we mean that the variable under consideration is normally distributed on the population of interest.

8.33 A statistical procedure is robust if it is insensitive to departures from the assumptions on which it is based.

8.35 The z-interval procedure is reasonable since the sample size is very large.

8.37 The z-interval procedure is reasonable since the population is roughly normal and the sample size is over 15.

8.39 The z-interval procedure is reasonable since the sampling distribution of $\bar{x}$ will be very close to normal for a sample size of 250 even though the population itself is far from normal.

8.41 A 95% confidence interval will give a more precise (shorter interval) estimate of μ than will a 99% confidence interval.

8.43 The margin of error is the standard error multiplied by $z_{\alpha/2}$.

8.45 Increasing the sample size while keeping the same confidence level will decrease the margin of error and hence increase the accuracy of estimating a population mean by a sample mean.

8.47 Increasing the confidence level while keeping the same sample size will increase the margin of error and hence decrease the accuracy of estimating a population mean by a sample mean.

8.49 (a) The length of the confidence interval is twice the margin of error; i.e., 2 x 3.4 = 6.8.

(b) The confidence interval is 52.8 $\pm$ 3.4 = (49.4, 56.2).

(c)

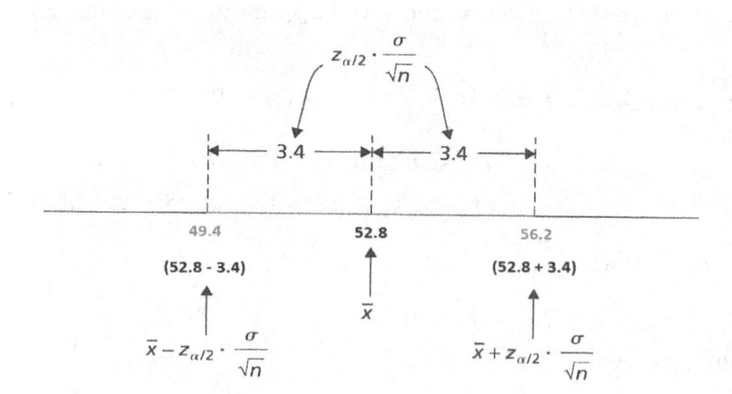

8.51 (a) The margin of error is 1/2 the length of the confidence interval; i.e., (1/2) x 20 = 10.

(b) The confidence interval is 60 $\pm$ 10 = (50, 70)

(c)

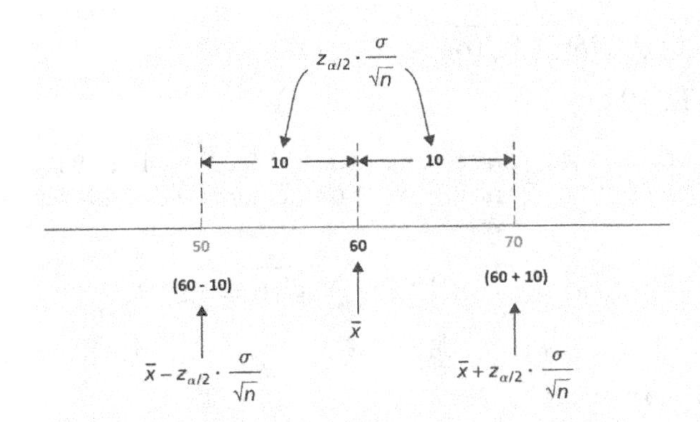

8.53 True. The length of the confidence interval is twice the margin of error.

8.55 False. One must also know the sample mean.

8.57 False. One must also know n and σ.

8.59 True. $E = z_{\alpha/2}(\sigma/\sqrt{n})$.

8.61 (a) We want a whole number because the number of observations to be taken cannot be fractional. It must be an integer because we cannot sample a partial member of a population.

(b) The original number computed is the smallest value of n that will provide the required margin of error. If we were to round down, the actual sample size would be slightly too small.

8.63 (a) The approximate 95% confidence interval for μ is

$$\bar{x} - z_{\alpha/2}\sigma/\sqrt{n} \quad to \quad \bar{x} + z_{\alpha/2}\sigma/\sqrt{n}$$

$$20 - 1.96(3)/\sqrt{36} \; to \; 20 + 1.96(3)/\sqrt{36}$$

$$19.02 \; to \; 20.98$$

(b) The length of the confidence interval is 20.98 - 19.03 = 1.96. The margin of error is half the length of the interval and equal to 1.96/2 = 0.98.

(c) Formula 8.1 states $E = z_{\alpha/2}\sigma/\sqrt{n} = 1.96(3)/\sqrt{36} = 0.98$.

8.65 (a) The 90% confidence interval for μ is

$$\bar{x} - z_{\alpha/2}\sigma/\sqrt{n} \quad to \quad \bar{x} + z_{\alpha/2}\sigma/\sqrt{n}$$

$$30 - 1.645(4)/\sqrt{25} \; to \; 30 + 1.645(4)/\sqrt{25}$$

$$28.684 \; to \; 31.316$$

(b) The length of the confidence interval is 31.316 - 28.684 = 2.632. The margin of error is half the length of the interval and equal to 2.632/2 = 1.316.

(c) Formula 8.1 states $E = z_{\alpha/2}\sigma/\sqrt{n} = 1.645(4)/\sqrt{25} = 1.316$.

8.67 (a) The 99% confidence interval for μ is

$$\bar{x} - z_{\alpha/2}\sigma/\sqrt{n} \quad to \quad \bar{x} + z_{\alpha/2}\sigma/\sqrt{n}$$
$$50 - 2.575(5)/\sqrt{16} \; to \; 50 + 2.575(5)/\sqrt{16}$$
$$46.781 \; to \; 53.219$$

(b) The length of the confidence interval is 53.219 - 46.781 = 6.438. The margin of error is half the length of the interval and equal to 6.438/2 = 3.219.

(c) Formula 8.1 states $E = z_{\alpha/2}\sigma/\sqrt{n} = 2.575(5)/\sqrt{16} = 3.219$.

8.69 The sample mean is $113.97 million / 18 = $6.332 million. The 95% confidence interval for μ is

$$\bar{x} - z_{\alpha/2}\sigma/\sqrt{n} \quad to \quad \bar{x} + z_{\alpha/2}\sigma/\sqrt{n}$$
$$6.332 - 1.96(2.04)/\sqrt{18} \; to \; 6.332 + 1.96(2.04)/\sqrt{18}$$
$$\$5.389 \; to \; \$7.274 \text{ (million)}$$

We can be 95% confident that the interval from $5.389 million to $7.274 million contains the population mean venture capital investment in the fiber optics business sector.

8.71 The sample mean is 6.31 / 12 = 0.526 ppm. The 95% confidence interval for μ is

$$\bar{x} - z_{\alpha/2}\sigma/\sqrt{n} \quad to \quad \bar{x} + z_{\alpha/2}\sigma/\sqrt{n}$$
$$0.526 - 2.575(0.37)/\sqrt{12} \quad to \quad 0.526 + 2.575(0.37)/\sqrt{12}$$
$$0.251 \; to \; 0.801 \text{ ppm}$$

We can be 99% confident that the interval from 0.251 to 0.801 ppm contains the population mean cadmium level in *Boletus pinicola* mushrooms.

8.73 n = 32; $\bar{x}$ = 33.4; σ = 42

Step 1: α = 0.05; $z_{\alpha/2} = z_{0.025} = 1.96$

Step 2:

$$\bar{x} - z_{\alpha/2}\sigma/\sqrt{n} \text{ to } \bar{x} + z_{\alpha/2}\sigma/\sqrt{n}$$
$$33.4 - 1.96(42)/\sqrt{32} \quad \text{to} \quad 33.4 + 1.96(42)/\sqrt{32}$$
$$18.8 \text{ to } 48.0$$

We can be 95% confident that the mean duration of imprisonment, μ, of all East German political prisoners with chronic PTSD is somewhere between 18.8 and 48.0 months.

8.75 n = 18; $\bar{x}$ = 6.33; σ = 2.04

(a) Step 1: α = 0.01; $z_{\alpha/2}$ = $z_{0.005}$ = 2.575

Step 2:

$$\bar{x} - z_{\alpha/2}\sigma/\sqrt{n} \text{ to } \bar{x} + z_{\alpha/2}\sigma/\sqrt{n}$$

$$6.33 - 2.575(2.04)/\sqrt{18} \quad \text{to} \quad 6.33 + 2.575(2.04)/\sqrt{18}$$

$$5.09 \text{ to } 7.57$$

(b) The confidence interval in part (a) is longer than the one in Exercise 8.69 because we have changed the confidence level from 95% in Exercise 8.69 to 99% in this exercise. Notice that increasing the confidence level from 95% to 99% increases the $z_{\alpha/2}$-value from 1.96 to 2.575. The larger z-value, in turn, results in a longer interval. In order to achieve a higher level of confidence that the interval contains the population mean, we need a longer interval.

(c)

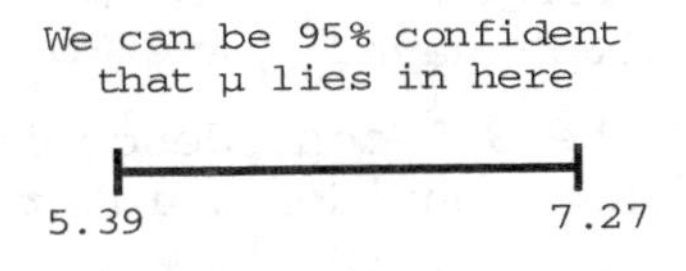

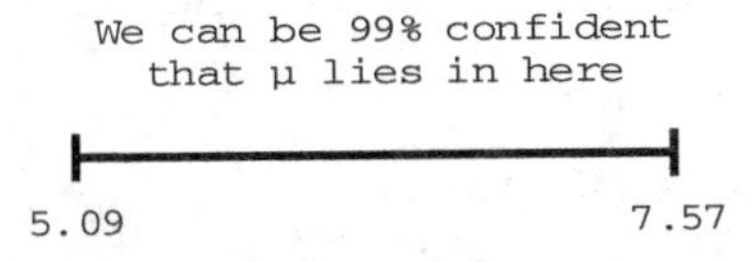

(d) The 95% confidence interval is shorter and therefore provides a more accurate estimate of μ.

8.77 n = 120; $\bar{x}$ = 102.72 days; σ = 32 days

(a) Step 1: α = 0.05; $z_{\alpha/2}$ = $z_{0.025}$ = 1.96

Step 2:

$$\bar{x} - z_{\alpha/2}\sigma/\sqrt{n} \text{ to } \bar{x} + z_{\alpha/2}\sigma/\sqrt{n}$$

$$102.72 - 1.96(32)/\sqrt{120} \quad \text{to} \quad 102.72 + 1.96(32)/\sqrt{120}$$

$$96.994 \quad \text{to} \quad 108.446$$

(b) Step 1: α = 0.10; $z_{\alpha/2}$ = $z_{0.05}$ = 1.645

Step 2:

$$\bar{x} - z_{\alpha/2}\sigma/\sqrt{n} \text{ to } \bar{x} + z_{\alpha/2}\sigma/\sqrt{n}$$

$$102.72 - 1.645(32)/\sqrt{120} \quad \text{to} \quad 102.72 + 1.645(32)/\sqrt{120}$$

$$97.915 \quad \text{to} \quad 107.525$$

(c)

We can be 90% confident
that μ lies within

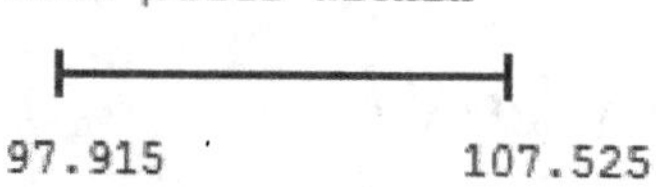

97.915 107.525

We can be 95% confident that
μ lies within

96.994 108.446

(d) The 90% interval is narrower and therefore provides a more accurate estimate of μ.

8.79 (a) The margin of error for the 95% confidence interval is (108.446 - 96.994)/2 = 5.726 days.

(b) The margin of error for the 90% confidence interval is (107.525 - 97.915)/2 = 4.805 days.

(c) The margin of error for the 90% confidence interval is smaller than the margin of error for the 95% confidence interval.

(d) This is illustrating the principle that decreasing the confidence level, while keeping the sample size the same, decreases the margin of error.

8.81 n = 30; $\bar{x}$ = 105.43 days; σ = 32 days

(a) Step 1: $\alpha = 0.05$; $z_{\alpha/2} = z_{0.025} = 1.96$

Step 2:

$$\bar{x} - z_{\alpha/2}\sigma/\sqrt{n} \text{ to } \bar{x} + z_{\alpha/2}\sigma/\sqrt{n}$$

$$105.43 - 1.96(32)/\sqrt{30} \text{ to } 105.43 + 1.96(32)/\sqrt{30}$$

$$93.979 \text{ to } 116.881$$

(b)

Ex 8.77

We can be 95% confident
that μ lies within

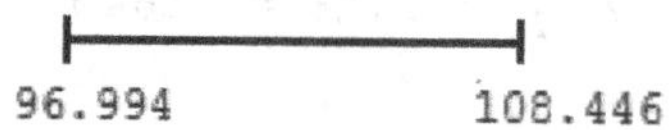

96.994 108.446

Ex 8.81

We can be 95% confident that
μ lies within

93.979 116.881

(c) The margin of error for this exercise is (116.881 - 93.979)/2 = 11.451 days while the margin of error for the 95% confidence interval in

Exercise 8.77 is 5.726 days. The margin of error for this exercise is larger than the margin of error for the 95% confidence interval in Exercise 8.77.

(d) This is illustrating the principle that decreasing the sample size, while keeping the confidence level fixed, increases the margin of error.

8.83 n = 36; $\bar{x}$ = \$63.28 thousand; σ = \$7.2 thousand

(a) Step 1: α = 0.05; $z_{\alpha/2}$ = $z_{0.025}$ = 1.96

Step 2:

$$\bar{x} - z_{\alpha/2}\sigma/\sqrt{n} \text{ to } \bar{x} + z_{\alpha/2}\sigma/\sqrt{n}$$

$$63.28 - 1.96(7.2)/\sqrt{36} \quad \text{to} \quad 63.28 + 1.96(7.2)/\sqrt{36}$$

$$60.928 \quad \text{to} \quad 65.632$$

(b) The interval in Example 8.2 is 60.88 to 65.68. Example 8.2 used 2 (From Property 2 of the empirical rule) as an approximation of $z_{\alpha/2}$. This exercise used 1.96. Example 8.2 gave an approximate 95% confidence interval.

8.85 n = 10; $\bar{x}$ = 34.2 cm; σ = 2.1 cm

(a) α = 0.10; $z_{\alpha/2}$ = $z_{0.05}$ = 1.645, so the 90% confidence interval for μ is

$$\bar{x} - 1.645(2.1)/\sqrt{10} \quad to \quad \bar{x} + 1.645(2.1)/\sqrt{10}$$

$$34.2 - 1.645(2.1)/\sqrt{10} \; to \; 34.2 + 1.645(2.1)/\sqrt{10}$$

$$33.1 \; to \; 35.3 \text{ cm}$$

(b) The margin of error is $E = 1.645\sigma/\sqrt{n} = 1.645(2.1)/\sqrt{10} = 1.1$ cm.

(c) We are 90% confident that our error in estimating μ by $\bar{x}$ is at most 1.1 cm.

(d) α = 0.05; $z_{\alpha/2}$ = $z_{0.025}$ = 1.96, so $n = \left(\frac{z_{\alpha/2} \cdot \sigma}{E}\right)^2 = \left(\frac{1.96 \cdot 2.1}{0.5}\right)^2 = 67.76 \; or \; 68.$

8.87 (a) E = 0.5(\$7.247 million− \$5.389 million) = 0.943 million

(b) $E = 1.96\sigma/\sqrt{n} = 1.96(2.04)/\sqrt{18} = \$0.942$ million. Note: there is a small difference in the answers do to rounding.

8.89 (a) E = (48.0 - 18.8)/2 = 14.6 months

(b) We are 95% confident that the maximum error made in using $\bar{x}$ to estimate μ is 14.6 months.

(c) The margin of error of the estimate is specified to be E = 12.0 months. Also, for a 99% confidence interval, $z_{\alpha/2}$ = $z_{0.005}$ = 2.575.

$$n = \left[\frac{z_{\alpha/2}\sigma}{E}\right]^2 = \left[\frac{2.575(42)}{12}\right]^2 = 81.2 \rightarrow 82$$

(d)

$$\bar{x} - z_{\alpha/2}\sigma/\sqrt{n} \text{ to } \bar{x} + z_{\alpha/2}\sigma/\sqrt{n}$$

$$36.2 - 2.575(12)/\sqrt{82} \text{ to } 36.2 + 2.575(12)/\sqrt{82}$$

$$24.3 \text{ to } 48.1$$

8.91 Using Minitab, retrieve the data from the WeissStats Resource Site. Click on the **Graphs** button and check the box for **Histogram** then click **OK**. Enter BRIGHTNESS in the **Variable** text box and click **OK.** To get the probability plot and stem-and leaf diagram, choose **Stat, Basic Statistics,** then **Normality Test.** Enter BRIGHTNESS in the **Variable** text box and click **OK**. The resulting graphs are as follows.

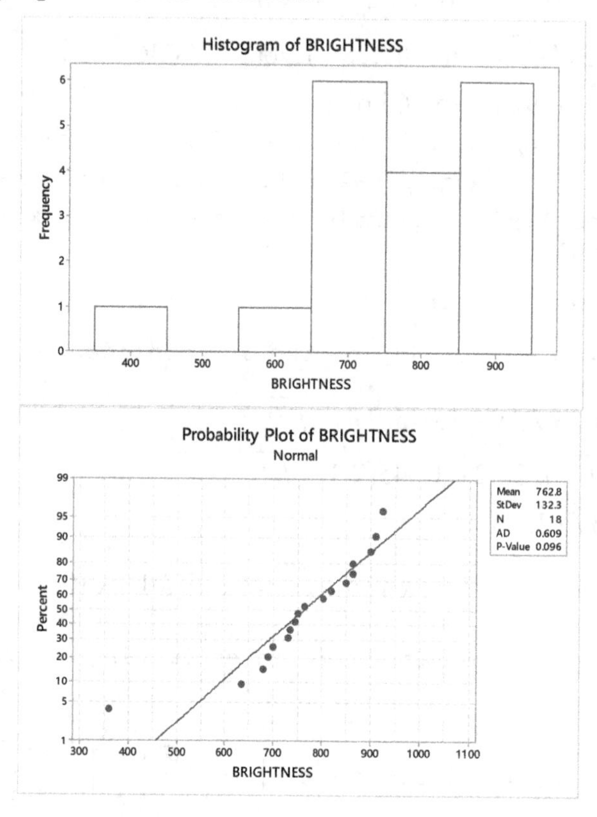

From the graphs, we determine that the variable is not normally distributed and contains an outlier. Also the sample size is small. Therefore it is not reasonable to apply the z-interval procedure.

8.93 (a) Using Minitab, retrieve the data from the WeissStats Resource Site.

Then choose **Stat ▶ Basic statistics ▶ 1-Sample z.** Enter TIME in the **Samples in Columns** text box and enter 30 in the **Standard deviation** text box. Click on the **Graphs** button and check the boxes for **Histogram of data** and **Boxplot of data**, then click **OK** twice. The confidence interval appears in the Sessions Window s

Variable	N	Mean	StDev	SE Mean	95% CI
TIME	20	289.950	30.741	6.708	(276.802, 303.098)

The histogram and boxplot are shown following.

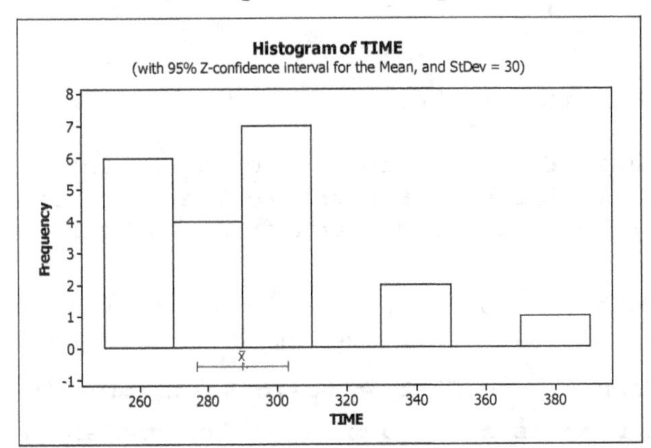

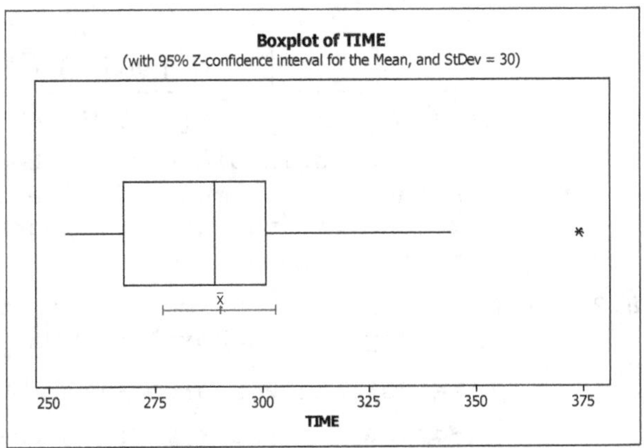

To get the probability plot and stem-and leaf diagram, choose **Stat ▶ Basic statistics ▶ Normality test**, enter TIME in the **Variable** text box and click **OK**. Then choose **Graph ▶ Stem-and-Leaf**, and enter TIME in the **Graph Variables** text box and click **OK**. The results are

```
Stem-and-leaf of TIME  N  = 20
Leaf Unit = 1.0

 2   25  45
 6   26  0779
 9   27  235
 10  28  7
 10  29  01257
 5   30  22
 3   31
 3   32
 3   33  3
 2   34  4

  HI 374
```

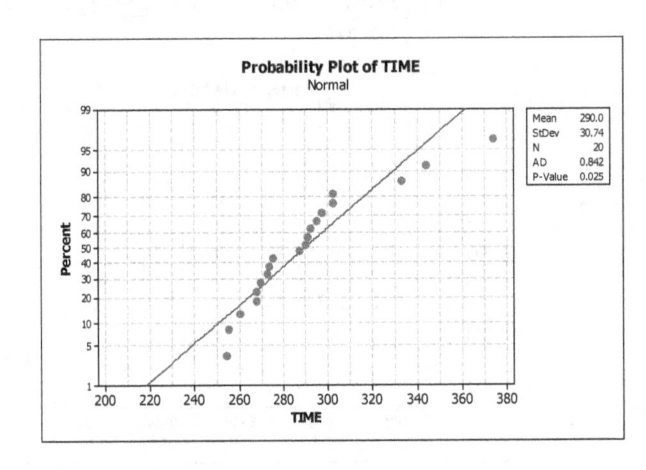

(c) The value 374 is a potential outlier. Remove it from the data and repeat the process in part (a). The result is

Variable	N	Mean	StDev	SE Mean	95% CI
TIME	19	285.526	24.174	6.882	(272.037, 299.016)

(d) The graphs (not shown here) that were produced with the confidence interval now show that 344 is an outlier in the sample of 19 observations. The data are also skewed to the right both before and after deleting the first outlier. The relatively small sample size makes it unwise to use the z-interval procedure with these data.

8.95 $n = 900$; $\alpha = 0.05$; $z_{\alpha/2} = z_{0.025} = 1.96$; $\sigma = 12.1$

$$E = 1.96(12.1)/\sqrt{900} = 0.791$$

8.97 (a) The margin of error of the estimate is specified to be E = 2 years.

$$n = \left[\frac{z_{\alpha/2}\sigma}{E}\right]^2 = \left[\frac{1.96(13.36)}{2.0}\right]^2 = 171.4 \rightarrow 172$$

(b) We used s in place of σ because σ was unknown. We can do this because the sample of size 36 is large enough to provide an estimate of σ and the variation is not likely to change much from one year to the next.

8.99 (a) We will get the confidence interval for part (c) along with the histogram and boxplot. Using Minitab, retrieve the data from the WeissStats Resource Site. Then choose **Stat ▶ Basic statistics ▶ 1-Sample z.** Enter TEMP in the **Samples in Columns** text box and enter 0.63 in the **Standard deviation** text box. Click on the Options button and enter 99.0 in the **Confidence level** text box and click **OK**. Click on the **Graphs** button and check the boxes for **Histogram of data** and **Boxplot of data**, then click **OK** twice. To get the probability plot and stem-and leaf diagram, choose **Stat ▶ Basic statistics ▶ Normality test**, enter TEMP in the **Variable** text box and click **OK**. Then choose **Graph ▶ Stem-and-Leaf**, and enter TEMP in the **Graph Variables** text box and click **OK**. The results are

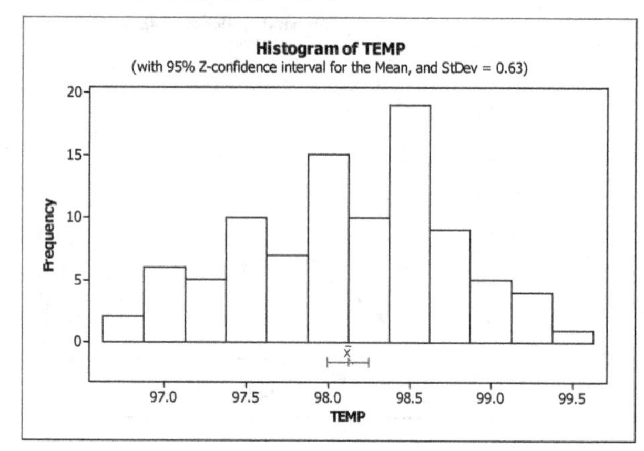

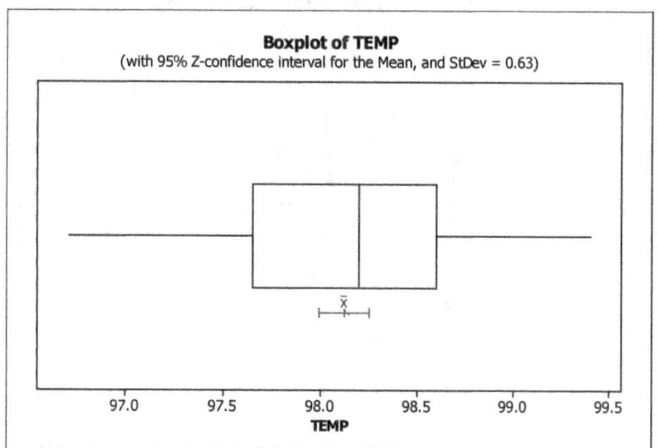


```
Stem-and-leaf of TEMP  N  = 93
Leaf Unit = 0.10

  1    96  7
  3    96  89
  8    97  00001
 13    97  22233
 19    97  444444
 26    97  6666777
 31    97  88889
 45    98  00000000000111
(10)   98  2222222233
 38    98  4444445555
 28    98  66666666677
 17    98  8888888
 10    99  00001
  5    99  2233
  1    99  4
```

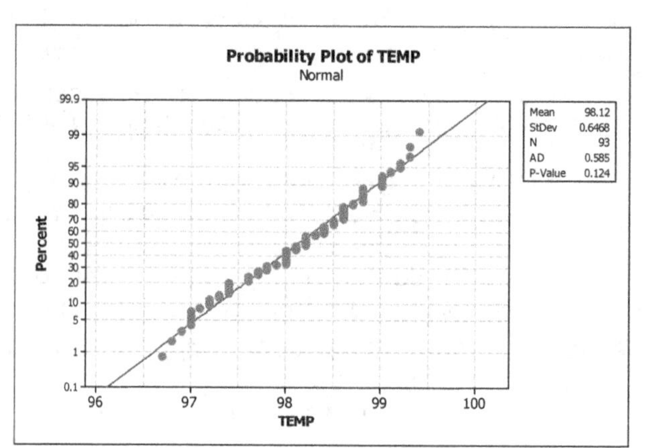

(b) Yes. There are no outliers, the sample size is large (93), and the distribution of the data appears to be approximately normal.

(c) The output from part (a) in the Sessions Window is

```
Variable   N     Mean   StDev  SE Mean        99% CI
TEMP      93  98.1237  0.6468   0.0653  (97.9554, 98.2919)
```

We can be 99% confident that the interval (97.9554, 98.2919) contains the mean body temperature μ of healthy humans. The output is surprising since the confidence interval does not inclue the generally accepted normal temperature of 98.6° F.

8.101 (a) We will get the confidence interval for part (c) along with the histogram and boxplot. Using Minitab, retrieve the data from the WeissStats Resource Site. Then choose **Stat ▶ Basic statistics ▶ 1-Sample z.** Enter SPEED in the **Samples in Columns** text box and enter 3.2 in the **Standard deviation** text box. Click on the **Options** button and enter 95.0 in the **Confidence level** text box and click **OK**. Click on the **Graphs** button and check the boxes for **Histogram of data** and **Boxplot of data**, then click **OK** twice. To get the probability plot and stem-and leaf diagram, choose **Stat ▶ Basic statistics ▶ Normality test**, enter SPEED in the **Variable** text box and click **OK**. Then choose **Graph ▶ Stem-and-Leaf**, and enter SPEED in the **Graph Variables** text box and click **OK**. The results are

```
Variable   N     Mean   StDev  SE Mean        95% CI
SPEED     35  59.5257  4.2739   0.5409  (58.4656, 60.5859)
```

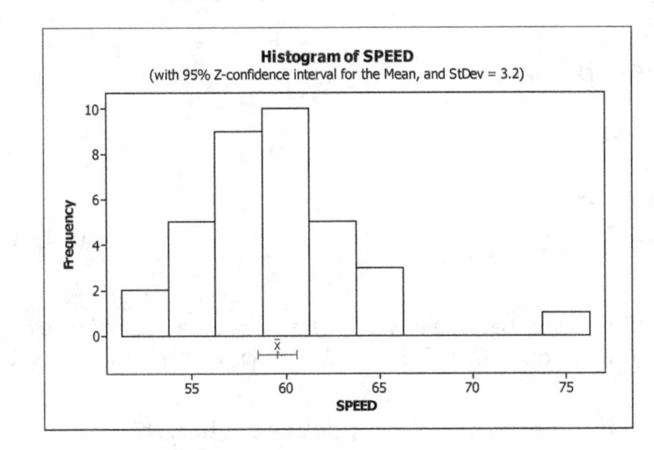

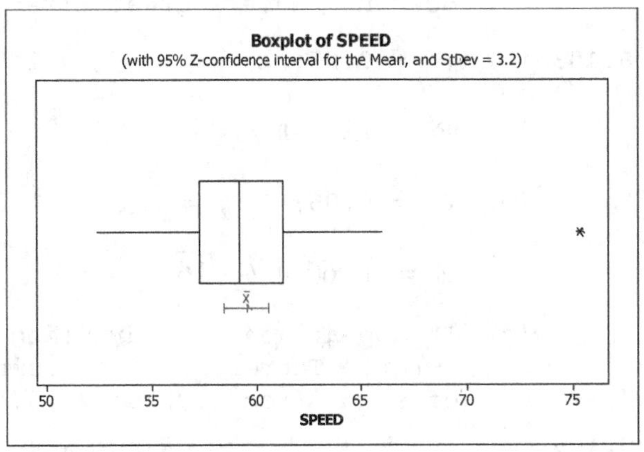

```
Stem-and-leaf of SPEED  N  = 35
Leaf Unit = 0.10

  2   52  46
  2   53
  4   54  78
  7   55  459
  8   56  5
 13   57  35688
 16   58  137
 (5)  59  02678
 14   60  12679
  9   61  36
  7   62  36
  5   63  4
  4   64
  4   65  02
  2   66  0
HI 753
```

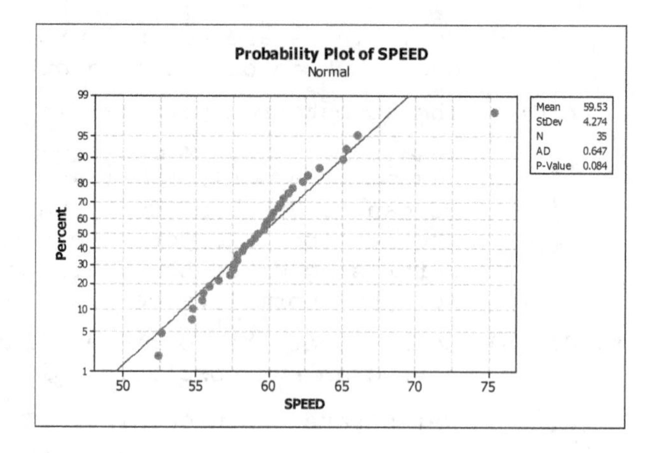

(c) After removing 75.3 from the data, the 95% confidence interval is

```
Variable   N     Mean    StDev  SE Mean        95% CI
SPEED     34  59.0618  3.3253   0.5488  (57.9861, 60.1374)
```

(d) The outlier had an effect on the mean of about 0.46 mph and, therefore, about the same effect on the endpoints of the confidence interval. The movement of the interval is about one-fourth the length of the interval and is significant in that sense. On the other hand, the change in the mean represents less than a 1% error and is probably not very important, although this is a matter of judgment. The sample size of 35 appears to be large enough that one outlier of the size observed does not have a great effect.

8.103 (a) $\alpha = 0.05$; $z_{\alpha/2} = z_{0.025} = 1.96$; $\sigma = 10$

$$E = 1.96(10) / \sqrt{4} = 9.80$$

(b) $\alpha = 0.05$; $z_{\alpha/2} = z_{0.025} = 1.96$; $\sigma = 10$

$$E = 1.96(10) / \sqrt{16} = 4.90$$

(c) It appears that quadrupling the sample size will halve the margin of error. Therefore, increasing n from 16 to 64 will decrease the margin of error from 4.90 to 2.45.

8.105 We can be 100(1 - α)% confident that the mean μ is greater than the lower confidence bound and 100(1 - α)% confident that the mean μ is less than the upper confidence bound.

8.107 (a) The sample mean is 6.31 / 12 = 0.526 ppm. The 99% lower confidence bound for μ is

$$\bar{x} - z_{\alpha}\sigma / \sqrt{n}$$

$$0.526 - 2.33(0.37) / \sqrt{12}$$

$$0.277 \text{ ppm}$$

We can be 99% confident that the population mean cadmium level in *Boletus pinicola* mushrooms is greater than 0.277 pp

(b) This lower confidence bound is greater than the lower confidence limit of 0.251 found in Exercise 8.71. This is because the z-value used for a one-sided 99% confidence bound is 2.33, whereas, the z-value used for a two-sided 99% confidence interval is 2.575.

Exercises 8.3

8.109 (a) The standardized version of $\bar{x}$ is

$$z = (\bar{x} - \mu)/(\sigma/\sqrt{n}) = (108 - 100)/(16/\sqrt{4}) = 1.00 .$$

(b) The studentized version of $\bar{x}$ is

$$t = (\bar{x} - \mu)/(s/\sqrt{n}) = (108 - 100)/(12/\sqrt{4}) = 1.333 .$$

8.111 (a) Standard normal

(b) t-distribution with 11 degrees of freedom

8.113 The variation in the possible values of the standardized version of $\bar{x}$ is due only to the variation in $\bar{x}$ while the variation in the studentized version results not only from the variation in $\bar{x}$, but also from the variation in the sample standard deviation.

8.115 For df = 6:

(a) $t_{0.10} = 1.440$ (b) $t_{0.025} = 2.447$ (c) $t_{0.01} = 3.143$

8.117 (a) $t_{0.10} = 1.323$ (b) $t_{0.01} = 2.518$

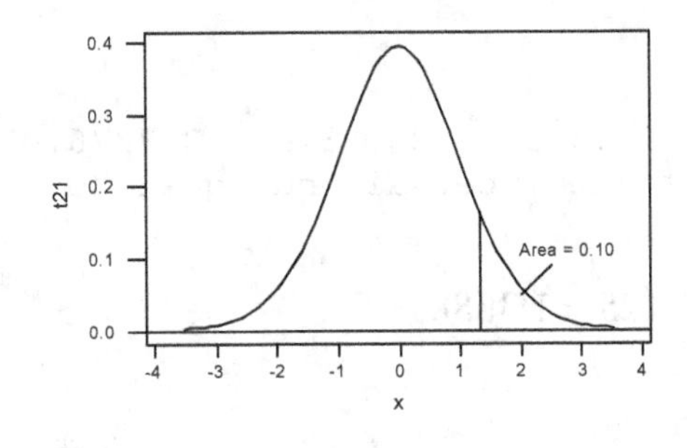

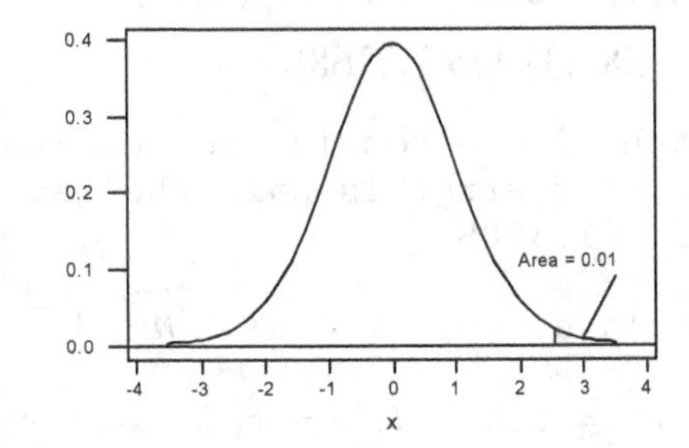

(c) $-t_{0.025} = -2.080$ (d) $\pm t_{0.05} = \pm 1.721$

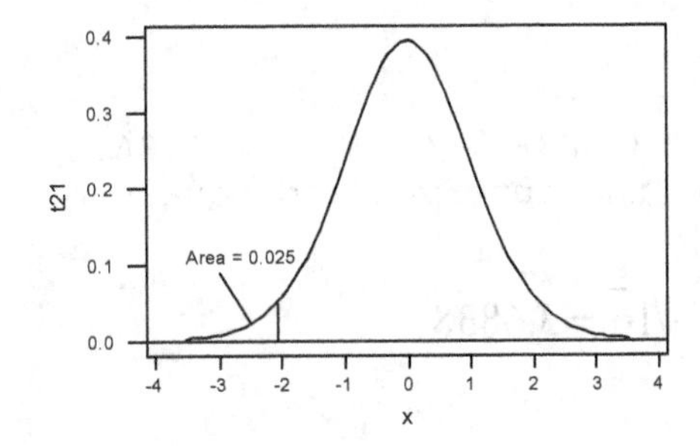

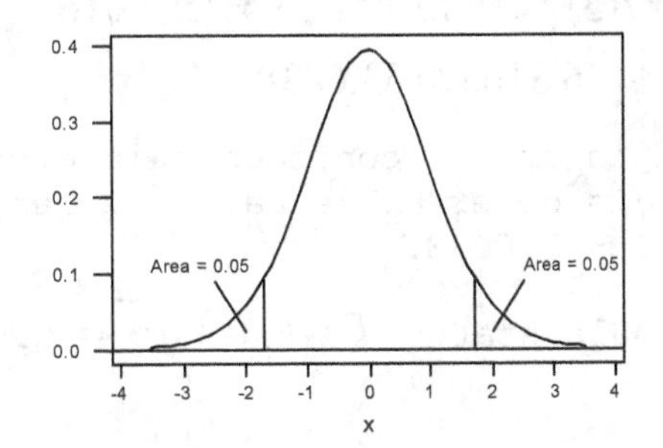

8.119 It is reasonable to use the t-interval procedure since the sample size is large and for large degrees of freedom (99), the t-distribution is very similar to the standard normal distribution. Another way of expressing this is that the sampling distribution of $\bar{x}$ is approximately normal when n is large, so the standardized and studentized versions of $\bar{x}$ are essentially the same.

8.121 $E = t_{\alpha/2} s / \sqrt{n}$

8.123 (a) df = 35 α = 0.05; $t_{\alpha/2}$ = $t_{0.025}$ = 2.030

The 95% confidence interval for μ is

$$\bar{x} - t_{\alpha/2} s / \sqrt{n} \quad to \quad \bar{x} + t_{\alpha/2} s / \sqrt{n}$$

$$20 - 2.030(3)/\sqrt{36} \ to \ 20 + 2.030(3)/\sqrt{36}$$

$$18.985 \ to \ 21.015$$

(b) The length of the confidence interval is 21.015 - 18.985 = 2.030. The margin of error is half the length of the interval and equal to 2.030/2 = 1.015.

(c) The formula states $E = t_{\alpha/2} s / \sqrt{n} = 2.030(3)/\sqrt{36} = 1.015$.

8.125 (a) df = 24 α = 0.10; $t_{\alpha/2}$ = $t_{0.05}$ = 1.711

The 90% confidence interval for μ is

$$\bar{x} - t_{\alpha/2} s / \sqrt{n} \quad to \quad \bar{x} + t_{\alpha/2} s / \sqrt{n}$$

$$30 - 1.711(4)/\sqrt{25} \ to \ 30 + 1.711(4)/\sqrt{25}$$

$$28.6312 \ to \ 31.3688$$

(b) The length of the confidence interval is 31.3688 - 28.6312 = 2.7376. The margin of error is half the length of the interval and equal to 2.7376/2 = 1.3688.

(c) The formula states $E = t_{\alpha/2} s / \sqrt{n} = 1.711(4)/\sqrt{25} = 1.3688$.

8.127 (a) df = 15 α = 0.01; $t_{\alpha/2}$ = $t_{0.005}$ = 2.947

The 99% confidence interval for μ is

$$\bar{x} - t_{\alpha/2} s / \sqrt{n} \quad to \quad \bar{x} + t_{\alpha/2} s / \sqrt{n}$$

$$50 - 2.947(5)/\sqrt{16} \ to \ 50 + 2.947(5)/\sqrt{16}$$

$$46.3163 \ to \ 53.6838$$

(b) The length of the confidence interval is 53.6838 - 46.3163 = 7.3675. The margin of error is half the length of the interval and equal to 7.3675/2 = 3.6838.

(c) The formula states $E = t_{\alpha/2} s / \sqrt{n} = 2.947(5)/\sqrt{16} = 3.6838$.

8.129 n = 30; df = 29; $t_{\alpha/2}$ = $t_{0.05}$ = 1.699; $\bar{x}$ = 27.97 minutes; s = 10.04 minutes

$$\bar{x} - t_{\alpha/2} s / \sqrt{n} \text{ to } \bar{x} + t_{\alpha/2} s / \sqrt{n}$$

$$27.97 - 1.699(10.04) / \sqrt{30} \quad \text{to} \quad 27.97 + 1.699(10.04) / \sqrt{30}$$

$$24.86 \text{ to } 31.08$$

We are 90% confident that the interval 24.86 to 31.08 minutes contains the mean commute times for all working adults in the Washington, D.C. area.

8.131 n = 10; df = 9; $t_{\alpha/2}$ = $t_{0.025}$ = 2.262; $\bar{x}$ = 2.33 hours; s = 2.002 hours

(a)

$$\bar{x} - t_{\alpha/2} s / \sqrt{n} \text{ to } \bar{x} + t_{\alpha/2} s / \sqrt{n}$$

$$2.33 - 2.262(2.002) / \sqrt{10} \quad \text{to} \quad 2.33 + 2.262(2.002) / \sqrt{10}$$

$$0.90 \text{ to } 3.76$$

(b) We are 95% confident that the interval 0.90 to 3.76 hours contains the mean of additional sleep obtained by all patients using laevohysocyamine hydrobromide. Since this interval was entirely above zero, we could conclude that the treatment was effective.

8.133 n = 42; df = 41; $t_{\alpha/2}$ = $t_{0.025}$ = 2.020; $\bar{x}$ = 1261.6 chips; s = 117.6

(a)

$$\bar{x} - t_{\alpha/2} s / \sqrt{n} \text{ to } \bar{x} + t_{\alpha/2} s / \sqrt{n}$$

$$1261.6 - 2.020(117.6) / \sqrt{42} \quad \text{to} \quad 1261.6 + 2.020(117.6) / \sqrt{42}$$

$$1224.94 \text{ to } 1298.26$$

(b) We are 95% confident that the interval 1224.94 to 1298.26 chips contains the mean number of chips per bag for all 18oz bags of Chips Ahoy! cookies. Since this interval was entirely above 1000, we could conclude that the average 18oz bag of Chips Ahoy! cookies contains at least 1000 chips.

8.135 We used Minitab to obtain a boxplot and a normal probability plot of the data, which are shown as follows. The plots show that the value 1643 is a potential outlier and the distribution is skewed right. Since the sample size is only 10, it is not reasonable to use the t-interval procedure to obtain a confidence interval for the population mean.

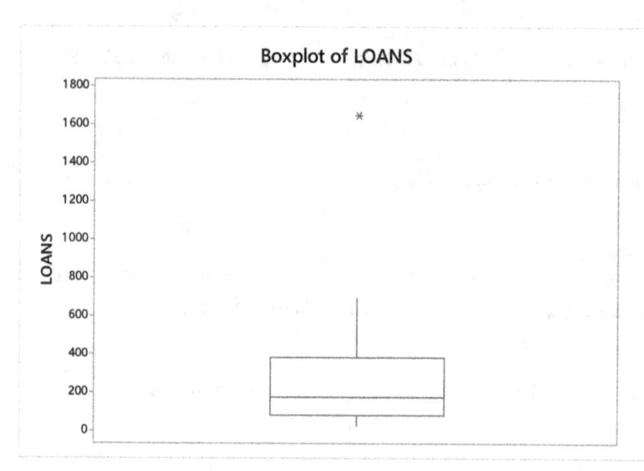

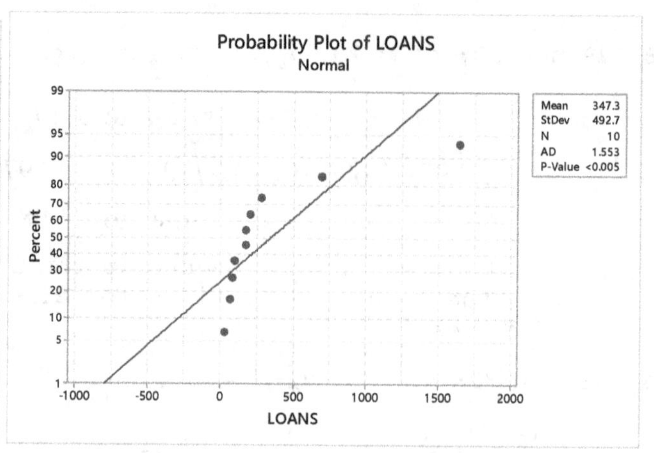

8.137 We used Minitab to obtain a boxplot and a normal probability plot of the data, which are shown below. The plots show no outliers and the distribution is close to symmetrical. Since the sample size is 20, it is reasonable to use the t-interval procedure to obtain a confidence interval for the population mean.

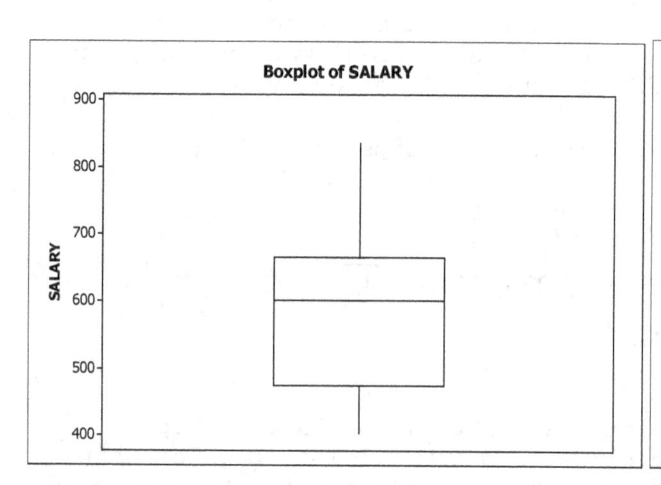

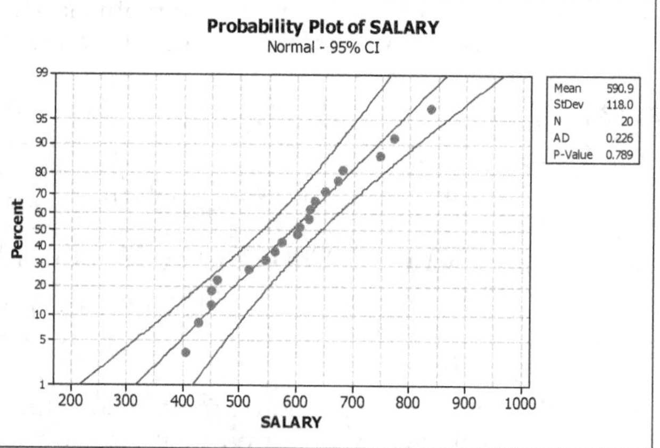

8.139 (a) We will get the confidence interval for part (c) along with the histogram and boxplot. Using Minitab, retrieve the data from the WeissStats Resource Site. Then choose **Stat ▶ Basic statistics ▶ 1-Sample t.** Enter DEPTH in the **Samples in Columns** text box. Click on the **Options** button and enter 90.0 in the **Confidence level** text box and click **OK**. Click on the **Graphs** button and check the boxes for **Histogram of data** and **Boxplot of data**, then click **OK** twice. To get the probability plot and stem-and leaf diagram, choose **Stat ▶ Basic statistics ▶ Normality test**, enter DEPTH in the **Variable** text box and click **OK**. Then choose **Graph ▶ Stem-and-Leaf**, and enter DEPTH in the **Graph Variables** text box and click **OK**. The results are

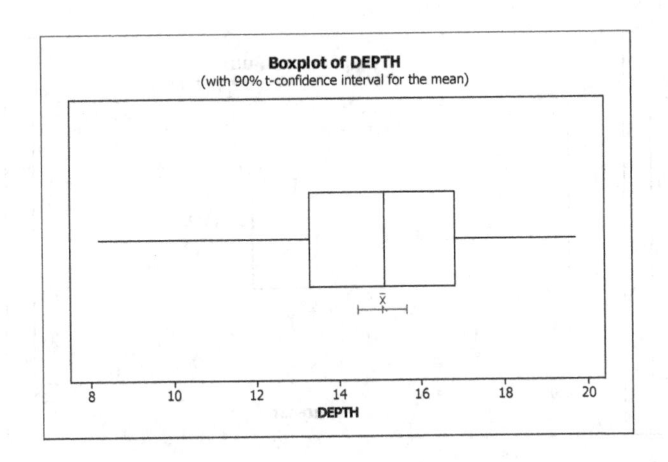

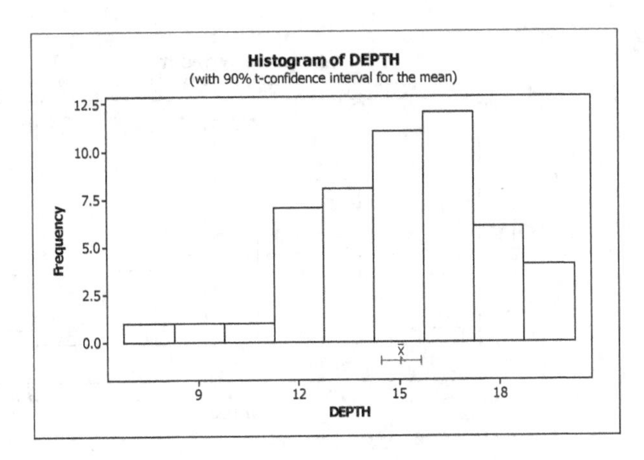


```
Stem-and-leaf of DEPTH  N  = 51
Leaf Unit = 0.10

  1   8   2
  2   9   7
  2  10
  4  11  08
 12  12  01123588
 16  13  3399
 23  14  0245799
(8)  15  00134689
 20  16  002567789
 11  17  24459
  6  18  2389
  2  19  37
```

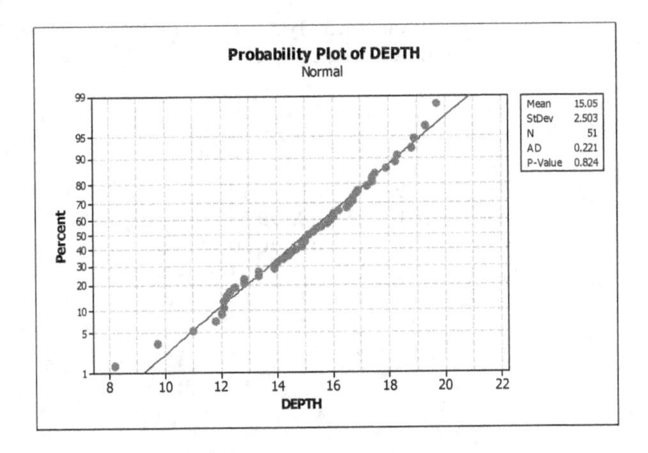

(b) Yes. The plots indicate no outliers, the sample size is quite large at 51, and the distribution of the data is only a little left skewed.

(c) The output for the confidence interval obtained in part (a) is

```
Variable   N     Mean   StDev  SE Mean        90% CI
DEPTH     51  15.0510  2.5028   0.3505  (14.4636, 15.6383)
```

We can be 90% confident that the mean, μ, for all burrow depths is somewhere between 14.46 cm and 15.64 cm.

8.141 (a) We will get the confidence interval for part (c) along with the histogram and boxplot. Using Minitab, retrieve the data from the WeissStats Resource Site. Then choose **Stat ▶ Basic statistics ▶ 1-Sample t.** Enter WITHOUT in the **Samples in Columns** text box. Click on the **Options** button and enter 95.0 in the **Confidence level** text box and click **OK**. Click on the **Graphs** button and check the boxes for **Histogram of data** and **Boxplot of data**, then click **OK** twice. To get the probability plot and stem-and leaf diagram, choose **Stat ▶ Basic statistics ▶ Normality test**, enter WITHOUT in the **Variable** text box and click **OK**. Then choose **Graph ▶ Stem-and-Leaf**, and enter WITHOUT in the **Graph Variables** text box and click **OK**. The results are

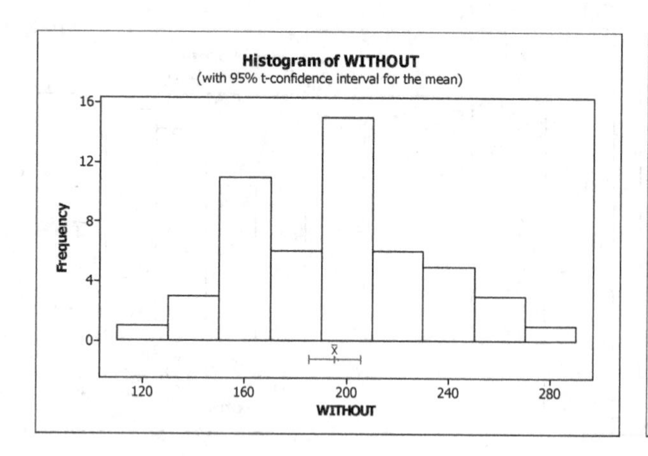

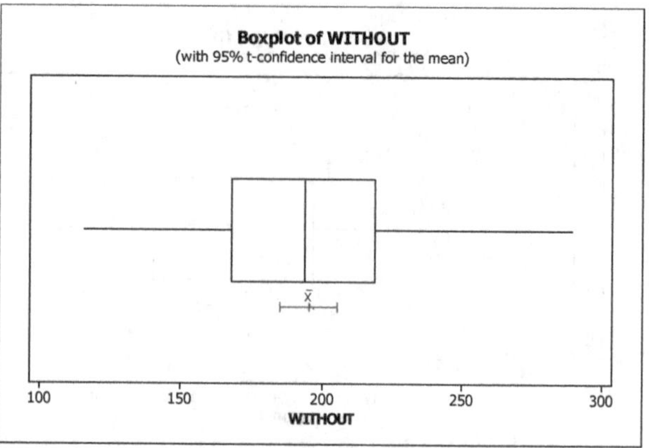


```
Stem-and-leaf of WITHOUT  N  = 51
Leaf Unit = 10

  1   1  1
  2   1  3
  9   1  4455555
 19   1  6666667777
 (8)  1  88999999
 24   2  000000000111
 12   2  2223333
  5   2  45
  3   2  66
  1   2  8
```

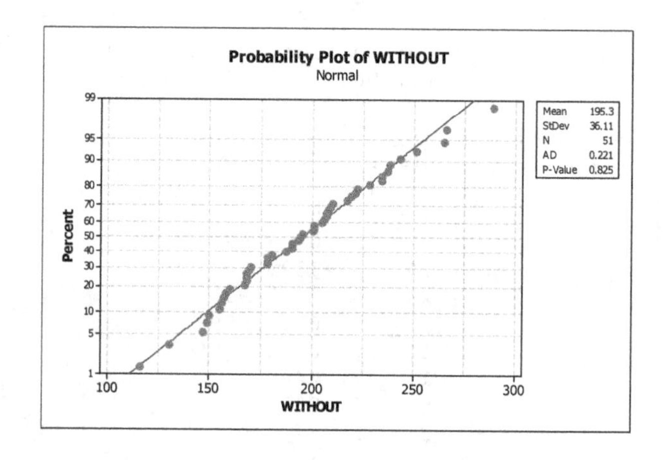

(b) Yes. The plots indicate no outliers, the sample size is quite large at 51, and the distribution of the data is quite symmetric.

(c) The output for the confidence interval obtained in part (a) is

```
Variable     N     Mean    StDev  SE Mean          95% CI
WITHOUT     51  195.275   36.110    5.056  (185.118, 205.431)
```

We can be 95% confident that the mean, μ, for all plasma cholesterol concentrations of patients without evidence of heart disease is somewhere between 185.118 mg/dl and 203.431 mg/dl.

(d) Repeating the process in part (a) using the WITH data, we obtain

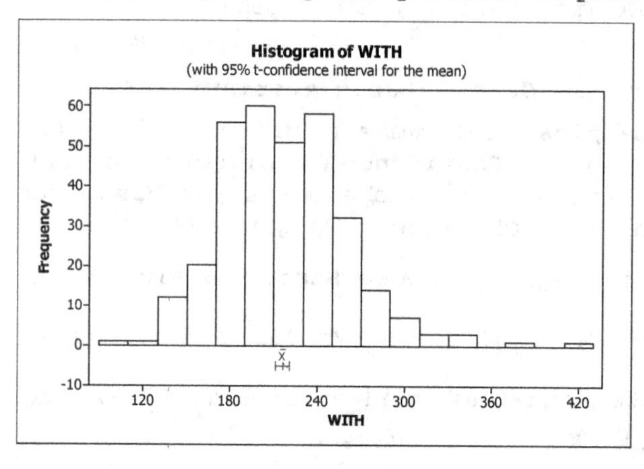

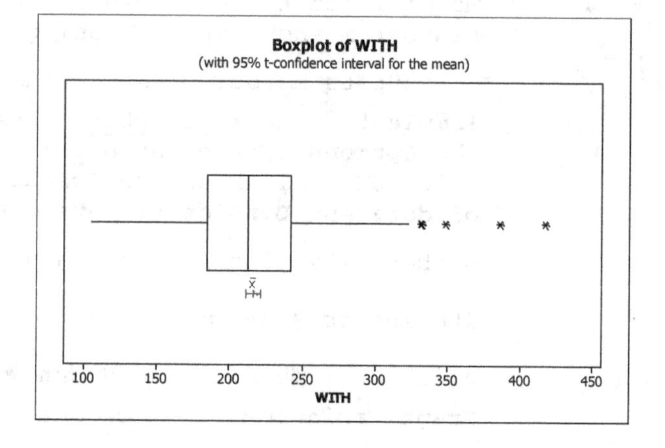

```
Stem-and-leaf of WITH  N  = 320
Leaf Unit = 10

  2    1  01
  8    1  333333
  19   1  44444455555
  63   1  66666666666666677777777777777777777777777777
  126  1  888888888888888888888888888999999999999999999999999999999999
 (50)  2  00000000000000000000000011111111111111111111111111
  144  2  222222222222222222222222223333333333333333333333333333333
  87   2  4444444444444444444444444455555555555555555
  44   2  66666666666666677777777
  21   2  8888889999
  11   3  00011
  6    3  2

 HI 33, 33, 34, 38, 41
```

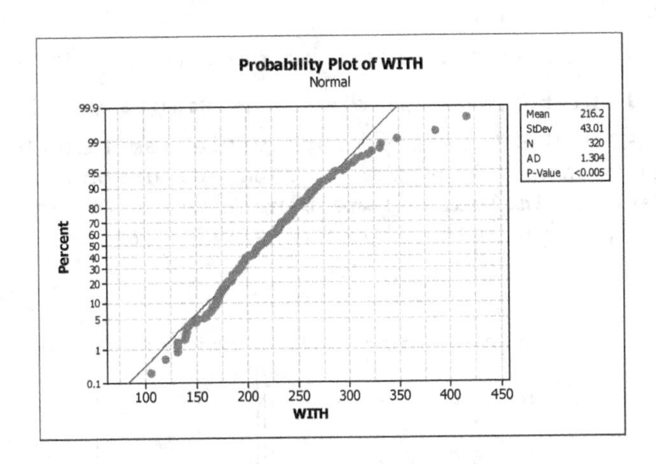

The data are quite symmetric except for five outliers on the high side. Nevertheless, the large sample size of 320 makes it reasonable to use the t-interval procedure with these data. The results of that procedure are

Variable	N	Mean	StDev	SE Mean	95% CI
WITH	320	216.191	43.015	2.405	(211.460, 220.921)

We can be 95% confident that the mean, μ, for all plasma cholesterol concentrations of patients with evidence of heart disease is somewhere between 211.460 mg/dl and 220.921 mg/dl.

8.143 (a) We could not provide entries for every possible degrees of freedom because the number of possibilities is infinite.

(b) As the degrees of freedom increase, the difference in consecutive entries becomes very small, making it unnecessary to list every possibility.

(c) Anytime the actual degrees of freedom lies between two consecutive entries in the table, we should use the table value associated with the lower number of degrees of freedom. Thus for $t_{0.05}$ with 87 df, we should use the value associated with 80 df, 1.664; for 125 df, use 1.660; for 650 df, use 1.660; and for 3000 df, use 1.645. This is a conservative approach, resulting in margins of error that are never smaller than we are entitled to have.

8.145 (a) Your results will vary. To obtain the 2000 samples using Minitab, Choose **Calc** ▶ **Random Data** ▶ **Normal...**, enter 2000 in the **Generate rows of data** text box, enter C1-C5 in the **Store in Column(s):** text box, enter .270 in the **Mean** text box, and enter .031 in the **Standard deviation:** text box. Click **OK.**

(b) To find the mean and sample standard deviation in each row (sample), choose **Calc** ▶ **Row statistics...** and click on **Mean**. Enter C1-C5 in the **Input variable(s):** text box and enter C6 in the **Store result in:** text box. Repeat this last process, selecting **Standard Deviation** instead of **Mean** and put the results in C7.

(c) To obtain the Standardized version of each mean in the sample, choose **Calc** ▶ **Calculator...**, enter STANDARD in the **Store results in variable:** text box, enter (C6-.270)/(.031/SQRT(5)) in the **Expression:** text box and click **OK**.

(d) To facilitate a comparison in part (k),., do part (h) now. Choose **Graph** ▶ **Histogram...**, select the **Simple** version, and click **OK**. Enter STANDARD STUDENT in the **Graph variables** text box. Click the **Multiple graphs** button, click the **On separate graphs** button and check the box for **Same X, including same bins**. Click **OK** and click **OK**. Our first graph is shown below, but yours will differ, yet look similar.

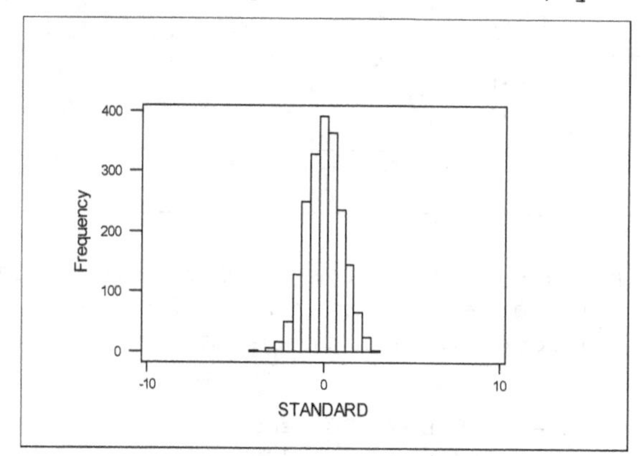

(e) In theory, the distribution of the standardized version of $\overline{x}$ is standard normal.

(f) The histogram in (d) appears to be very close to standard normal. Recall that the distribution is centered at zero and 99.7% of the data should be within 3 standard deviations of the mean. It does appear that this is so for this simulated data.

(g) To obtain the Studentized version of each $\overline{x}$ in the sample, choose **Calc** ▶ **Calculator...**, enter STUDENT in the **Store results in variable:** text box, enter (C6-.270)/(C7/SQRT(5) in the **Expression:** text box and click **OK**.

(h) This graph was produced in produced (d). Your graph should be similar to ours.

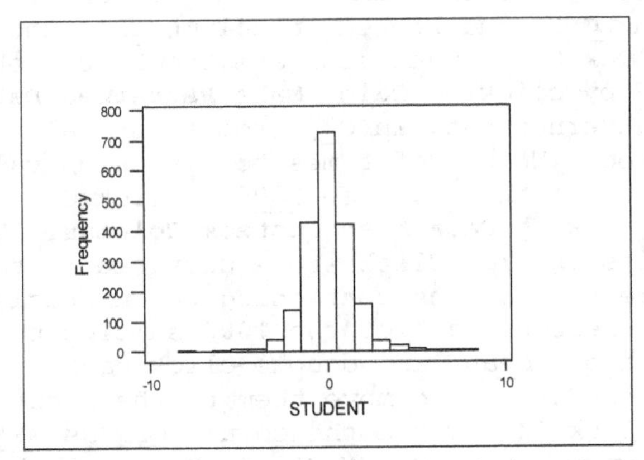

(i) In theory, the distribution of the studentized version of $\bar{x}$ is a t-distribution with 4 degrees of freedom.

(j) The distribution shown is symmetric about zero as is a t-distribution.

(k) The histogram of the studentized version in (h) is more spread out than that of the standardized version in (d). The reason is that there is more variability in the t-distribution due to the extra uncertainty arising from the use of s instead of σ.

8.147 (a) n=30; $\bar{x}$ = 27.97; s = 10.04; df = 29; $t_\alpha = t_{0.10} = 1.311$

$$\bar{x} + t_\alpha s / \sqrt{n}$$

$$27.97 + 1.311(10.04) / \sqrt{30}$$

$$30.37 \text{ minutes}$$

We can be 90% confident that the population mean commute time of all commuters in Washington, D.C. is less than 30.37 minutes.

(b) This upper confidence bound is less than the upper confidence limit of 31.08 found in Exercise 8.129. This is because the t-value used for a one-sided 90% confidence bound is 1.311, whereas, the t-value used for a two-sided 90% confidence interval is 1.699.

8.149 (a) n=30; $\bar{x}$ = 52.040; s = 2.807; df = 29; $t_\alpha = t_{0.05} = 1.699$

$$\bar{x} - t_\alpha s / \sqrt{n}$$

$$52.040 - 1.699(2.807) / \sqrt{30}$$

$$51.169 \text{ g}$$

(b) We can be 95% confident that the population mean weight of all small bags of peanut M&Ms is greater than 51.169 grams.

(c) According to our confidence interval in part (b), we are 95% confident that on average the small bags of peanut M&Ms will be greater than the advertised weight on the package of 49.3 grams. It seems like the bags are slightly heavier than advertised. This is good news for the consumer, rather than slightly lighter than advertised.

8.151 Using Minitab, we will generate 1000 samples of size 25 for the bootstrap confidence interval. First open the data file for Table 8.6 from Example 8.10 in Minitab from the WeissStats Resource Site. Generate all the data at once for the 1000 samples of size 25 (25,000 observations). Click **Calc, Random Data,** then **Sample from Columns.** The number of rows to sample is 25000, from column LOSS, and stored in column C2. Check Sample with replacement and click OK. Then separate the 25,000 observations into 1000 samples of size 25 by clicking **Calc, Make Patterned Data,** then **Simple Set of Numbers.** Store patterned data in C3, from first value 1 to last value 25, in steps of 1. Choose Number of times to list each value equal to 1 and Number of times to list the sequence 1000. Click OK. Separate the samples into 1000 rows by cicking **Data** then **Unstack Columns.** Unstack the data in C2, Using subscripts in C3. Click store unstacked data after last column in use, and check name the columns containing the unstacked data. Click OK. Now, calculate the mean of each of the 1000 samples by clicking **Calc** then **Row Statistics.** Choose Mean, and double click each of the 25 columns for the samples (columns C4-C28) to move them to the Input variables. Store the results in C29. Click OK. Title the column C29 as MEANS. Calculate the 2.5th and 97.5th percentile of the MEANS column by finding the 25th and 975th observation of the MEANS column after ordering them from smallest to largest. We calculated the 2.5th percentile as 418.92 and the 97.5th percentile as 613.72. Therefore, our 95% bootstrap confidence interval is 418.92 to 613.72. The 95% *t*-interval calculated in Example 8.10 was 405.07 to 621.57. The two intervals were close. However, the bootstrapping method resulted in a narrower and more accurate estimate of the population mean.

Review Problems for Chapter 8

1. A point estimate of a parameter consists of a single value with no indication of the accuracy of the estimate. A confidence interval consists of an interval of numbers obtained from a point estimate of the parameter together with a percentage that specifies how confident we are that the parameter lies in the interval.

2. False. We are 95% confident that the mean lies in the interval from 33.8 to 39.0, but about 5% of the time, the procedure will produce an interval that does not contain the population mean. Therefore, we cannot say that the mean must lie in the interval.

3. No. The z-interval procedure can be used almost anytime with large samples

because the sampling distribution of $\bar{x}$ is approximately normal for large n. The same is true for the t-interval procedure because when n is large, the t distribution is very similar to the normal distribution. However, when n is small, especially when n is 15 or less, the z-interval and t-interval procedures will not provide reliable estimates if the distribution of the underlying variable is not normal. For sample sizes in the range of 15 to 30, both procedures can be used if the data is roughly normal and has no outliers.

4. Approximately 950 of 1000 95% confidence intervals for a population mean would actually contain the true value of the mean.

5. Before applying a particular statistical inference procedure, we should look at graphical displays of the sample data to see if there appear to be any violations of the conditions required for the use of the procedure.

6. (a) Reducing the sample size from 100 to 50 will reduce the accuracy of the estimate (result in a longer confidence interval).

(b) Reducing the confidence level from .95 to .90 while maintaining the sample size will increase the accuracy of the estimate (result in a shorter confidence interval).

7. (a) The length of the confidence interval is twice the margin of error or 2 x 10.7 = 21.4.

(b) The confidence interval will be 75.2 $\pm$ 10.7 = (64.5, 85.9)

(c) The confidence interval will be 75.2 $\pm$ 10.7.

8. (a) $E = z_{\alpha/2}(\sigma/\sqrt{n}) = 1.645(12/\sqrt{9}) = 6.58$

(b) To obtain the confidence interval, you also need to know $\bar{x}$.

9. (a) The standardized value of $\bar{x}$ is $z = \dfrac{\bar{x}-\mu}{\sigma/\sqrt{n}} = \dfrac{262.1-266}{16/\sqrt{10}} = -0.77$

(b) The studentized value of $\bar{x}$ is $t = \dfrac{\bar{x}-\mu}{s/\sqrt{n}} = \dfrac{262.1-266}{20.4/\sqrt{10}} = -0.61$

10. (a) standard normal distribution

(b) t distribution with 14 degrees of freedom

11. The curve that looks more like the standard normal curve has the larger degrees of freedom because, as the number of degrees of freedom gets larger, *t*-curves look increasingly like the standard normal curve.

12. The t-interval procedure should be used.

13. The z-interval procedure should be used.

14. The z-interval procedure should be used.

15. Neither procedure should be used.

16. The z-interval procedure should be used.

17. Neither procedure should be used.

18. (a) $t_{0.025} = 2.101$ (b) $t_{0.05} = 1.734$

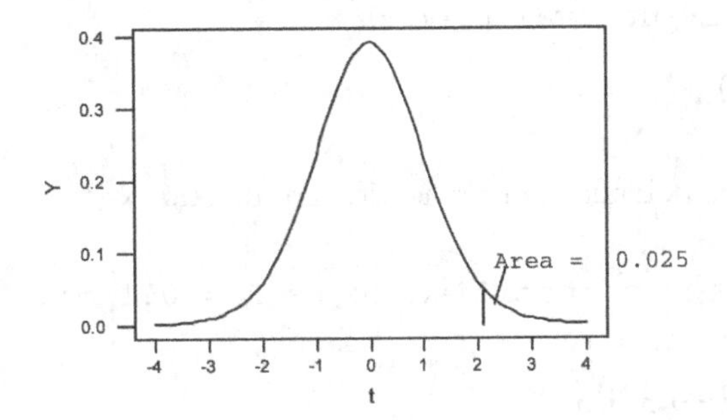

(c) $-t_{0.10} = -1.330$

(d) $\pm t_{0.005} = \pm 2.878$

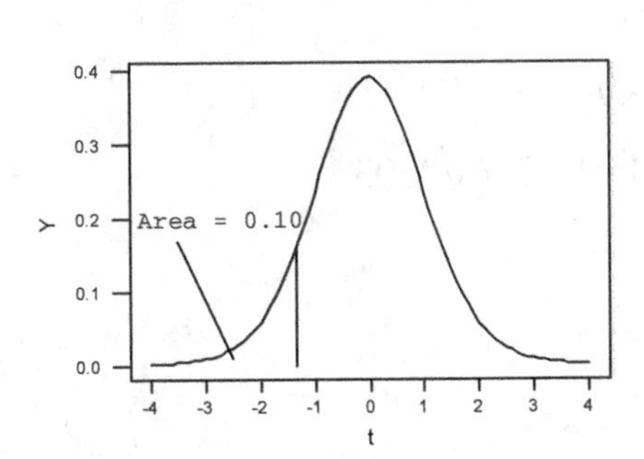

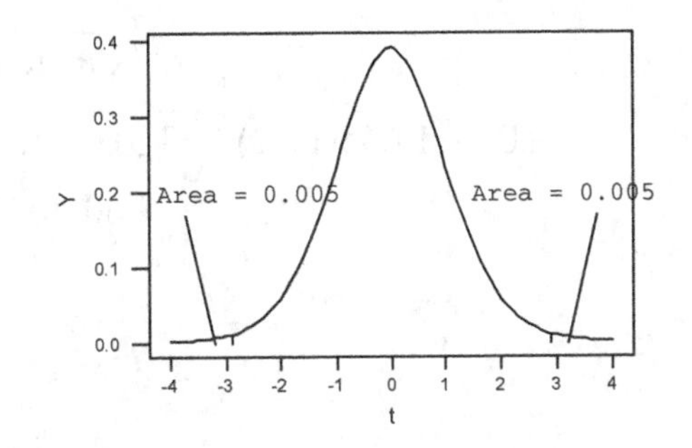

19. $n = 36$, $\bar{x} = 58.53$, $\sigma = 13.0$, $z_{\alpha/2} = z_{0.025} = 1.96$

$$\bar{x} - z_{\alpha/2}(\sigma/\sqrt{n}) \text{ to } \bar{x} + z_{\alpha/2}(\sigma/\sqrt{n})$$

$$58.53 - 1.96 \cdot (13.0/\sqrt{36}) \text{ to } 58.53 + 1.96 \cdot (13.0/\sqrt{36})$$

$$54.3 \text{ to } 62.8 \text{ years}$$

20. A confidence-interval estimate specifies how confident we are that a (unknown) parameter lies in the interval. This interpretation is presented correctly by (c). A *specific* confidence interval either will or will not contain the true value of the population mean μ; the *specific* interval is either sure to contain μ or sure not to contain μ. This interpretation is *not* presented correctly by (a), (b), and (d).

21. $n = 712$, $\bar{x} = 9.15$ years, $\sigma = 17.2$ years, $z_{\alpha/2} = z_{0.05} = 1.645$

(a)

$$\bar{x} - z_{\alpha/2}\sigma/\sqrt{n} \text{ to } \bar{x} + z_{\alpha/2}\sigma/\sqrt{n}$$

$$9.15 - 1.645(17.2)/\sqrt{712} \quad \text{to} \quad 9.15 + 1.645(17.2)/\sqrt{712}$$

$$8.090 \text{ to } 10.210$$

We can be 90% confident that the mean sentence length, μ, of all federally sentenced adult male prisoners is between 8.09 and 10.21 years.

(b) Since the sample size is very large, the distribution of sample means will be approximately normal regardless of the shape of the original distribution. It would be nice if the normal probability plot were roughly linear and did not indicate the presence of any extreme outliers, but some non-linearity and a few moderate outliers will not likely invalidate the use of the z-interval procedure.

22. (a) $E = z_{\alpha/2}\sigma/\sqrt{n} = 1.645(17.2)/\sqrt{712} = 1.060$

(b) We can be 90% confident that the maximum error made in using $\bar{x}$ to estimate μ is 1.06 years.

(c) The margin of error of the estimate is specified to be E = 0.1 mm.

$$n = \left[\frac{z_{\alpha/2}\ \sigma}{E}\right]^2 = \left[\frac{1.645\ (17.2)}{0.5}\right]^2 = 3202.2 \rightarrow 3203$$

(d)

$$\bar{x} - z_{\alpha/2}\sigma/\sqrt{n} \text{ to } \bar{x} + z_{\alpha/2}\sigma/\sqrt{n}$$

$$10.1 - 1.645(17.2)/\sqrt{3203} \quad \text{to} \quad 10.1 + 1.645(17.2)/\sqrt{3203}$$

$$9.60 \text{ to } 10.60$$

23. n = 16; df = 15; $t_{\alpha/2} = t_{0.025} = 2.131$; $\bar{x} = 85.99$ mm Hg; s = 8.08 mm Hg

(a)

$$\bar{x} - t_{\alpha/2} s/\sqrt{n} \text{ to } \bar{x} + t_{\alpha/2} s/\sqrt{n}$$

$$85.99 - 2.131(8.08)/\sqrt{16} \quad \text{to} \quad 85.99 + 2.131(8.08)/\sqrt{16}$$

$$81.69 \text{ to } 90.29$$

We can be 95% confident that the mean arterial blood pressure μ of all children of diabetic mothers is somewhere between 81.69 and 90.29 mm Hg.

(b) Using Minitab, choose **Stat ▶ Basic statistics ▶ 1-Sample t...**, enter PRESSURE in the **Samples in columns** text box. Click the **options...** button, type 95 in the **Confidence interval** text box, click the **Graphs** and check the boxes for **Histogram of data** and **Boxplot of data**, and click **OK** twice. Then choose **Stat ▶ Basic statistics ▶ Normality test**, enter PRESSURE in the **Variable** text box and click **OK**. Finally, choose **Graph ▶ Stem-and-Leaf**, enter PRESSURE in the **Graph Variables** text box and click **OK**. The results are

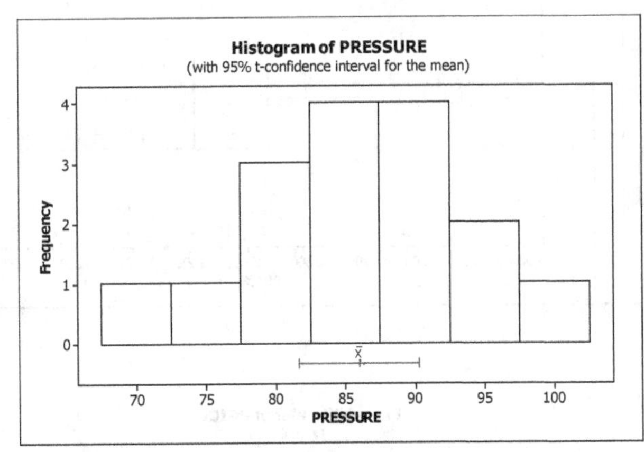

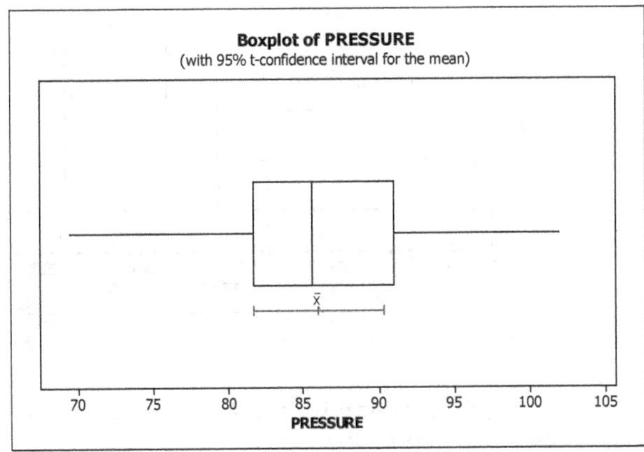


```
Stem-and-leaf of PRESSURE  N  = 16
        Leaf Unit = 1.0

         1  6   9
         1  7
         3  7   58
         8  8   12244
         8  8   678
         5  9   014
         2  9   6
         1  10  1
```

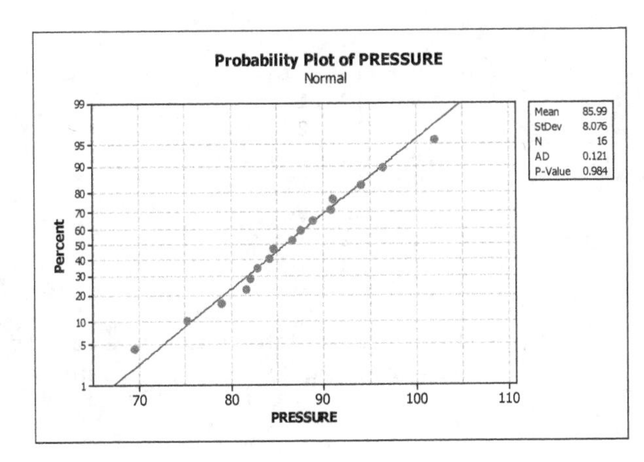

(c) Yes. There are no outliers and the distribution of the data is approximately normal.

24. (a) Using Minitab, choose **Stat ▶ Basic statistics ▶ 1-Sample t...**, enter PRICE in the **Samples in columns** text box. Click the **options...** button, enter 90 in the **Confidence interval** text box and click **OK**, click the **Graphs** button and check the boxes for **Histogram of data** and **Boxplot of data**, and click **OK** twice. Then choose **Stat ▶ Basic statistics ▶ Normality test**, enter PRICE in the **Variable** text box and click **OK**.

Finally, choose **Graph ▶ Stem-and-Leaf**, enter PRICE in the **Graph Variables** text box and click **OK**. The confidence interval is

```
Variable   N     Mean   StDev  SE Mean          90% CI
PRICE     18  1964.72  206.45    48.66  (1880.07, 2049.37)
```

We can be 90% confident that the interval ($1880.07, $2049.37) contains the mean diamond price for one-half carat diamonds.

(b) The graphs obtained by the procedures in part (a) are

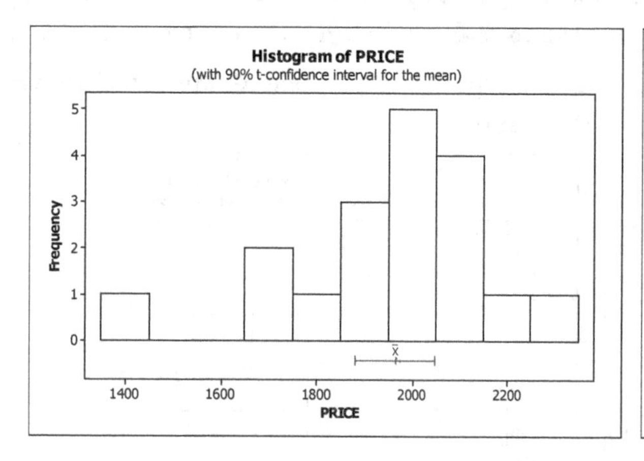

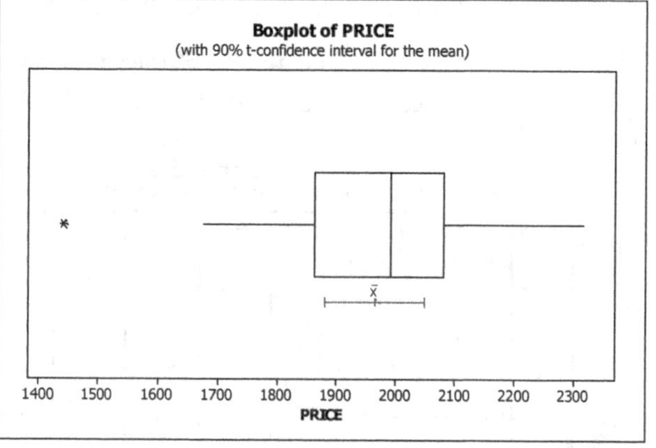


```
Stem-and-leaf of PRICE  N  = 18
Leaf Unit = 10

LO 144

 2   16  7
 3   17  1
 5   18  27
(6)  19  448899
 7   20  377
 4   21  04
 2   22  3
 1   23  1
```

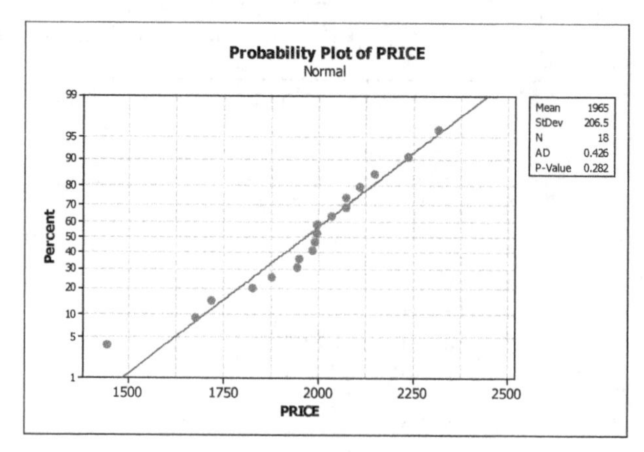

(c) No. Since the sample size is small (18), and there is an outlier (1442), it is not reasonable to use the t-interval procedure.

25. We used Minitab to obtain a boxplot and a normal probability plot of the data, which are shown as follows. The plots show that the value 907 is a potential outlier and the distribution is skewed right. Since the sample size is only 14, it is not reasonable to use the t-interval procedure to obtain a confidence interval for the population mean.

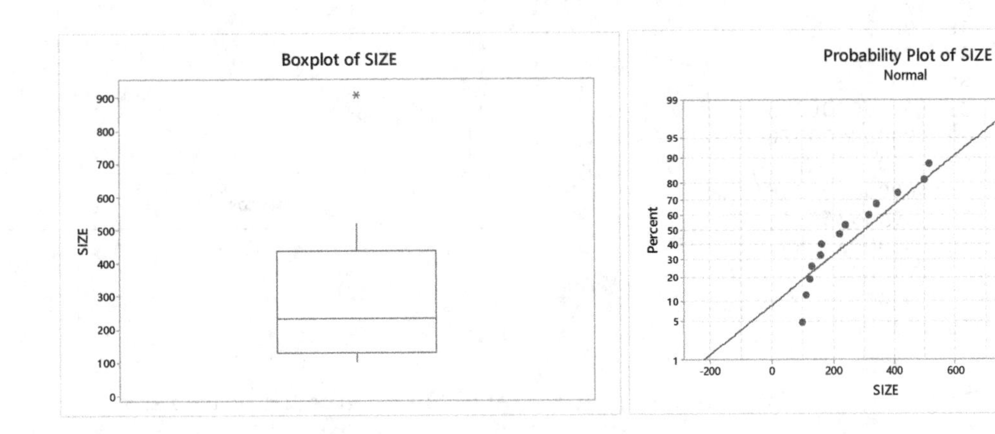

26. (a) We will import the data into Minitab.

(b) Choose **Stat ▶ Basic statistics ▶ 1-Sample t..**, enter DURATION in the **Samples in columns** text box. Click the **options...** button, enter 99 in the **Confidence interval** text box and click **OK**, click the **Graphs** button and check the boxes for **Histogram of data** and **Boxplot of data**, and click **OK** twice. Then choose **Stat ▶ Basic statistics ▶ Normality test**, enter DURATION in the **Variable** text box and click **OK**. Finally, choose **Graph ▶ Stem-and-Leaf**, enter DURATION in the **Graph Variables** text box and click **OK**. The graphs are

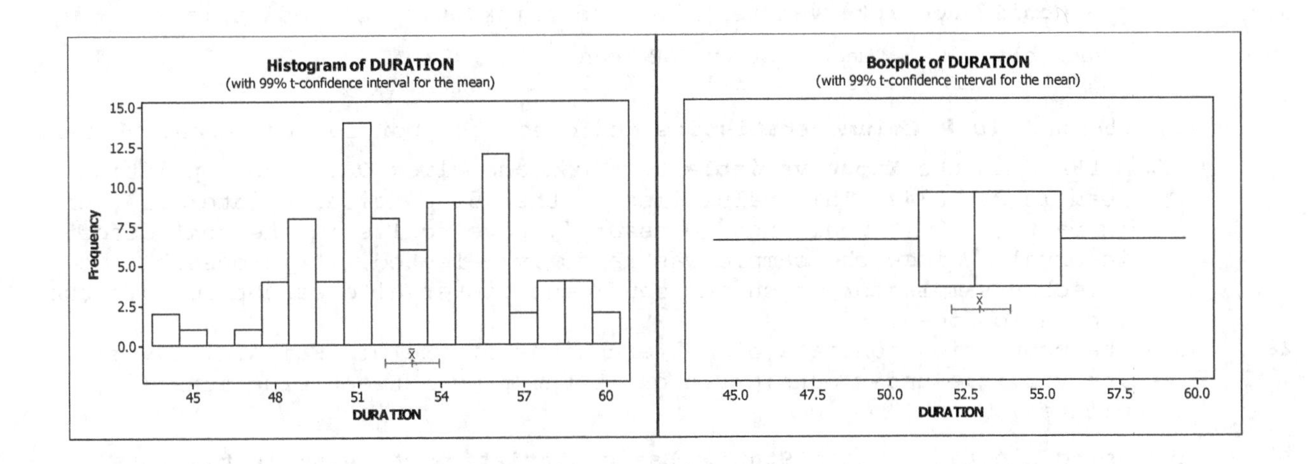

```
Stem-and-leaf of DURATION  N  = 90
Leaf Unit = 0.10

 2   44  33
 3   45  0
 3   46
 4   47  3
 9   48  00036
16   49  0000033
22   50  003366
34   51  000000003333
45   52  00000003666
45   53  000
42   54  000000033666
30   55  0000336666
20   56  000003336
11   57  0
10   58  00036
 5   59  00356
```

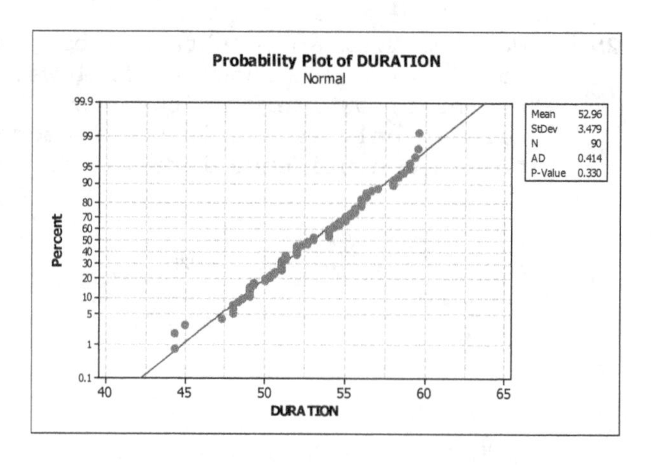

(c) Yes. The sample size is large (90) and there are no outliers.

(d) The procedure in part (a) resulted in the following confidence interval.

```
Variable   N     Mean    StDev  SE Mean        99% CI
DURATION  90  52.9633  3.4794   0.3668  (51.9979, 53.9287)
```

We can be 99% confident that the interval (51.9979, 53.9287) contains the population mean μ of the larval duration of convict surgeonfish in days.

27. (a) Using Minitab, choose **Calc ▶ Random data ▶ Sample from columns,** enter 35 in the **Sample ___ rows from column(s)** text box, enter MILEAGE in the first text box and enter SAMPLE in the **Store samples in** text box. Click **OK**.

(b) Choose **Stat ▶ Basic statistics ▶ 1-Sample t..**,enter SAMPLE in the **Samples in columns** text box. Click the **options...** button, enter 95 in the **Confidence interval** text box and click **OK.** The result is

```
Variable   N     Mean    StDev  SE Mean        95% CI
SAMPLE    35  24.4857  4.0174   0.6791  (23.1057, 25.8657)
```

(c) Choose **Calc ▶ Column statistics** and check the box for the mean. Enter MILEAGE in the **Input variable** text box and click **OK**. The population mean is 24.8234. This value lies in the 95% confidence interval found in part (b). It would not necessarily have to lie in the confidence interval. Since the sample was randomly selected, it is possible to select a sample for which the confidence interval does not contain the population mean.

28. (a) The population consists of all eruptions of the Old Faithful Geyser. The variable under consideration is the time between eruptions (in minutes).

(b) Using Minitab, choose **Stat ▶ Basic statistics ▶ 1-Sample t..**,enter TIME in the **Samples in columns** text box. Click the **options...** button, enter 99 in the **Confidence interval** text box and click **OK**. The resulting confidence interval is

```
Variable    N     Mean   StDev  SE Mean        99% CI
TIME      500  91.5780  9.2022   0.4115  (90.5139, 92.6421)
```

We can be 99% confident that the population mean time between eruptions of the Old Faithful Geyser lies in the interval from 90.5139 to 92.6421 minutes.

(c) Strictly speaking, the confidence interval applies to the population from which the sample was drawn. There were no observations in the sample from the future, so the interval is not relevant to the future. In geologic terms, however, five years is insignificant, so the confidence interval is probably still somewhat meaningful. One should consider that unforeseen events, such as earthquakes or volcanic eruptions in the region could have a substantial effect on the time between eruptions. In that case, the current sample would not be representative of the population in five years.

29. (a) Your results will vary. To obtain the 3000 samples using Minitab,

Choose **Calc ▶ Random Data ▶ Normal...**, enter 3000 in the **Generate rows of data** text box, type C1-C4 in the **Store in Column(s):** text box, enter 4.66 in the **Mean** text box, and enter 0.75 in the **Standard deviation:** text box. Click **OK.**

(b) To find the mean and sample standard deviation in each row (sample),

choose **Calc ▶ Row statistics...** and click on **Mean**. Enter C1-C4 in the **Input variable(s):** text box and enter XBAR in the **Store result in:** text box. Repeat this last process, selecting **Standard Deviation** instead of **Mean** and put the results in SD.

(c) To obtain the Standardized version of each $\bar{x}$ in the sample, choose **Calc**

▶ Calculator..., enter STANDARD in the **Store results in variable:** text box, enter ('XBAR'-4.66)/(0.75/SQRT(4)) in the **Expression:** text box and click **OK.**

(d) To facilitate a comparison in part (h), **do part (g) now**. Then choose

Graph ▶ Histogram..., select the **Simple** version, and click **OK**. Enter STANDARD and STUDENT in the **Graph variables** text box. Click on the **Multiple graphs** button, then click the **On separate graphs** button and check the boxes for **Same X, including same bins** and **Same Y**. Click **OK** twice. The first graph is shown below. Yours will differ, yet look similar.

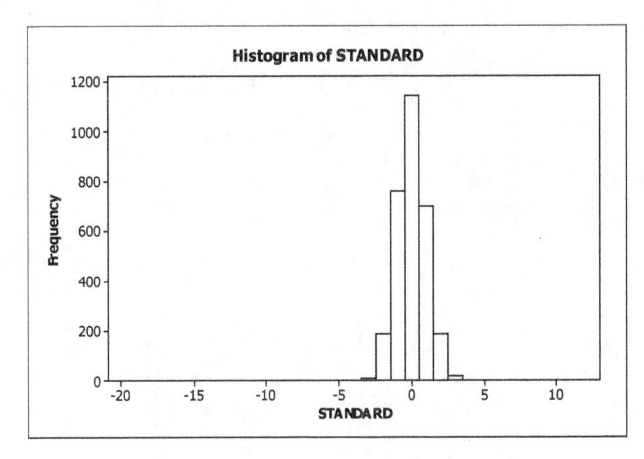

(e) In theory, the distribution of the standardized version of $\bar{x}$ is standard normal.

(f) The histogram in (e) appears to be very close to standard normal. Recall that the distribution is centered at zero and 99.7% of the data should be within 3 standard deviations of the mean. It does appear that this is so for this simulated data. (g) To obtain the Studentized version of each $\overline{x}$ in the sample, choose **Calc ▶ Calculator...**, enter STUDENT in the **Store results in variable** text box, enter ('SBAR'-4.66)/('SD'/SQRT(4) in the **Expression:** text box and click **OK**.

(h) This graph was produced by the procedure in part (d). Our second graph follows, but yours will differ, yet look similar.

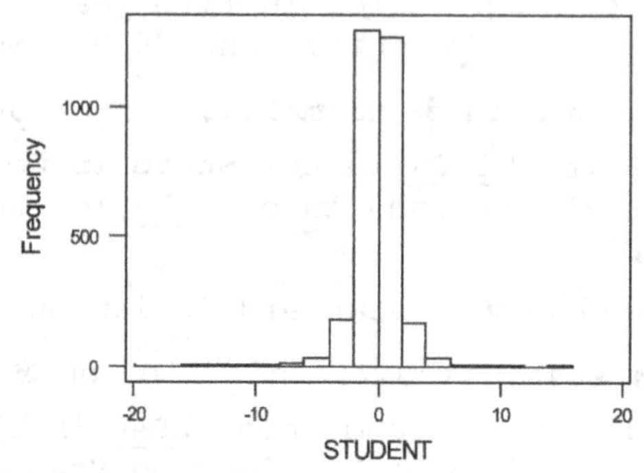

(i) In theory, the distribution of the studentized version of $\overline{x}$ is a t-distribution with 3 degrees of freedom.

(j) The distribution shown is symmetric about zero as is a t-distribution.

(k) The histogram of the studentized version in (h) is much more spread out than that of the standardized version in (d). The reason is that there is more variability in the t-distribution due to the extra uncertainty arising from the use of s instead of σ.

CHAPTER 9 SOLUTIONS

Exercises 9.1

9.1 A hypothesis is a statement that something is true.

9.3 The decision criterion specifies whether or not the null hypothesis should be rejected in favor of the alternative hypothesis.

9.5 (a) H_a: $\mu \neq \mu_0$; two-tailed

(b) H_a: $\mu < \mu_0$; left-tailed

(c) H_a: $\mu > \mu_0$; right-tailed

9.7 (a) If the null hypothesis is in fact false, you cannot possibly make a Type I error. You would either reject the null hypothesis when it is false, which is a correct decision, or you would not reject the null hypothesis when it is false, which is a Type II error.

(b) If the null hypothesis is in fact false, you could possibly make a Type II error. You would either reject the null hypothesis when it is false, which is a correct decision, or you would not reject the null hypothesis when it is false, which is a Type II error.

9.9 True. The significance level is equal to the probability of a Type I error, which is the probability of rejecting a true null hypothesis. Therefore, the smaller the significance level, the smaller the probability of rejecting a true null hypothesis.

9.11 A Type I error is made when a true null hypothesis is rejected. The probability of making this error is denoted by α. A Type II error is made when a false null hypothesis is not rejected. We denote the probability of a Type II error by β.

9.13 In this exercise, we are told that failing to reject the null hypothesis corresponds to approving the nuclear reactor for use. This action – approving the nuclear reactor suggests that the null hypothesis must be something like: "The nuclear reactor is safe." This further suggests that the alternative hypothesis is something like: "The nuclear reactor is unsafe." Putting things together, the Type II error in this situation is: "Approving the nuclear reactor for use when, in fact, it is unsafe." This type of error has consequences that are catastrophic. Thus, the property that we want the Type II error probability to exhibit is that it be small.

9.15 Let μ denote the mean cadmium level in *Boletus pinicola* mushrooms.

(a) H_0: $\mu = 0.5$ ppm (b) H_a: $\mu > 0.5$ ppm (c) right-tailed test

9.17 Let μ denote the mean daily intake of iron by adult females under age 51.

(a) H_0: $\mu = 18$ mg/day (b) H_a: $\mu < 18$ mg/day (c) left-tailed

9.19 Let μ denote the mean length of imprisonment for motor-vehicle theft offenders in Australia.

(a) H_0: $\mu = 16.7$ months (b) H_a: $\mu \neq 26.7$ months

(c) two-tailed

9.21 Let μ denote the mean body temperature of healthy humans.

(a) H_0: $\mu = 98.6$ºF (b) H_a: $\mu \neq 98.6$ºF (c) two-tailed

9.23 (a) A Type I error would occur if, in fact, $\mu = 0.5$ ppm, but the results of the sampling lead to the conclusion that $\mu > 0.5$ ppm.

(b) A Type II error would occur if, in fact, $\mu > 0.5$ ppm, but the results of the sampling fail to lead to that conclusion.

(c) A correct decision would occur if, in fact, $\mu = 0.5$ ppm and the results of the sampling do not lead to the rejection of that fact; or if, in fact, $\mu > 0.5$ ppm and the results of the sampling lead to that conclusion.

(d) If, in fact, the mean cadmium level in *Boletus pinicola* mushrooms is equal to 0.5 ppm, and we do not reject the null hypothesis that $\mu = 0.5$ ppm, we made a correct decision.

(e) If, in fact, the mean cadmium level in *Boletus pinicola* mushrooms is greater than to 0.5 ppm, and we do not reject the null hypothesis that $\mu = 0.5$ ppm, we made a Type II error.

9.25 (a) A Type I error would occur if, in fact, $\mu = 18$ mg, but the results of the sampling lead to the conclusion that $\mu < 18$ mg.

(b) A Type II error would occur if, in fact, $\mu < 18$ mg, but the results of the sampling fail to lead to that conclusion.

(c) A correct decision would occur if, in fact, $\mu = 18$ mg and the results of the sampling do not lead to the rejection of that fact; or if, in fact, $\mu < 18$ mg and the results of the sampling lead to that conclusion.

(d) If the mean iron intake equals the RDA of 18 mg, and we reject the null hypothesis that $\mu = 18$ mg, we made a Type I error.

(e) If, in fact, the mean iron intake is less than the RDA of 18 mg, and we reject the null hypothesis that $\mu = 18$ mg, we made a correct decision.

9.27 (a) A Type I error would occur if, in fact, $\mu = 16.7$ months, but the results of the sampling lead to the conclusion that $\mu \neq 16.7$ months.

(b) A Type II error would occur if, in fact, $\mu \neq 16.7$ months, but the results of the sampling fail to lead to that conclusion.

(c) A correct decision would occur if, in fact, $\mu = 16.7$ months and the results of the sampling do not lead to the rejection of that fact; or if, in fact, $\mu \neq 16.7$ months and the results of the sampling lead to that conclusion.

(d) If, in fact, the mean length of imprisonment equals 16.7 months, and we do not reject the null hypothesis that $\mu = 16.7$ months, we made a correct decision.

(e) If, in fact, the mean length of imprisonment does not equal 16.7 months, and we do not reject the null hypothesis that $\mu = 16.7$ months, we made a Type II error.

9.29 (a) A Type I error would occur if, in fact, $\mu = 98.6^{\circ}$ F, but the results of the sampling lead to the conclusion that $\mu \neq 98.6^{\circ}$ F.

(b) A Type II error would occur if, in fact, $\mu \neq 98.6^{\circ}$ F, but the results of the sampling fail to lead to that conclusion.

(c) A correct decision would occur if, in fact, $\mu = 98.6^{\circ}$ F and the results of the sampling do not lead to the rejection of that fact;

or if, in fact, $\mu \neq 98.6^{\circ}$ F and the results of the sampling lead to that conclusion.

(d) If the mean temperature of all healthy humans equals 98.6° F, and we reject the null hypothesis that $\mu = 98.6^{\circ}$ F, we made a Type I error.

(e) If, in fact, the temperature of all healthy humans is not equal to 98.6° F, and we reject the null hypothesis that $\mu = 98.6^{\circ}$ F, we made a correct decision.

Exercises 9.2

9.31 A test statistic is a statistic used to decide whether to reject the null hypothesis.

9.33 The nonrejection region is a set of values of the test statistic that do not lead to rejection of the null hypothesis.

9.35 (a) Rejection region: $z \geq 1.645$

(b) Nonrejection region: $z < 1.645$

(c) Critical value: $z = 1.645$

(d) Significance level: $\alpha = 0.05$

(e)

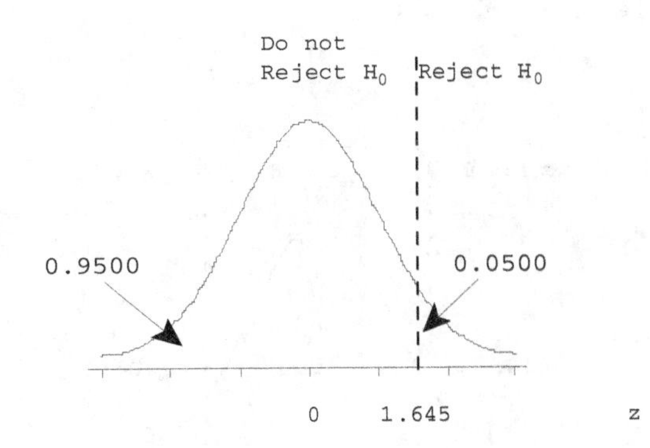

(f) Right-tailed test

9.37 (a) Rejection region: $z \leq -2.33$

(b) Nonrejection region: $z > -2.33$

(c) Critical value: $z = -2.33$

(d) Significance level: $\alpha = 0.01$

(e)

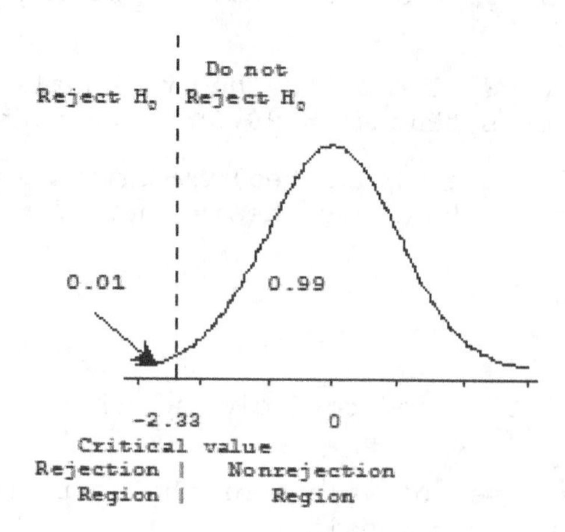

(f) Left-tailed test

9.39 (a) Rejection region: $z \leq -1.645$ or $z \geq -1.645$

(b) Nonrejection region: $-1.645 < z < -1.645$

(c) Critical values: $z = -1.645$ and $z = 1.645$

(d) Significance level: $\alpha = 0.10$

(e)

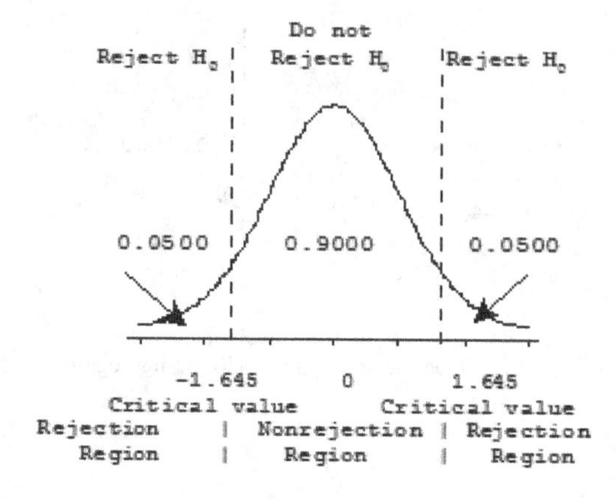

(f) Two-tailed test

9.41 Critical values: $\pm z_{0.05} = \pm 1.645$

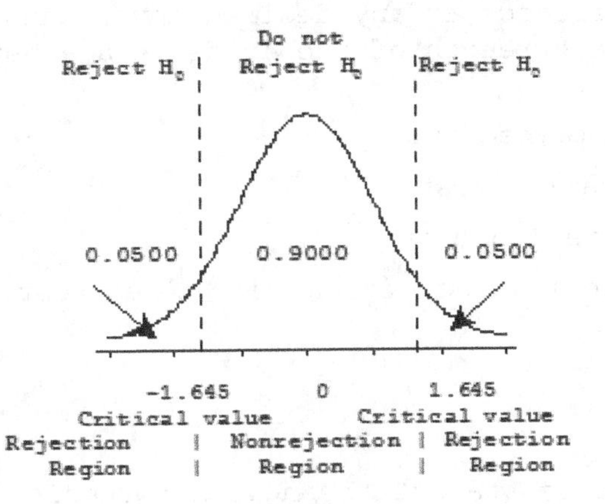

9.43 Critical value: $-z_{0.01} = -2.33$

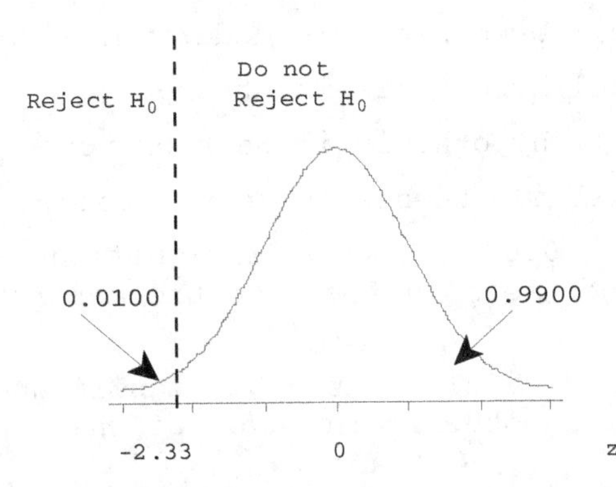

9.45 Critical value: $z_{0.01} = 2.33$

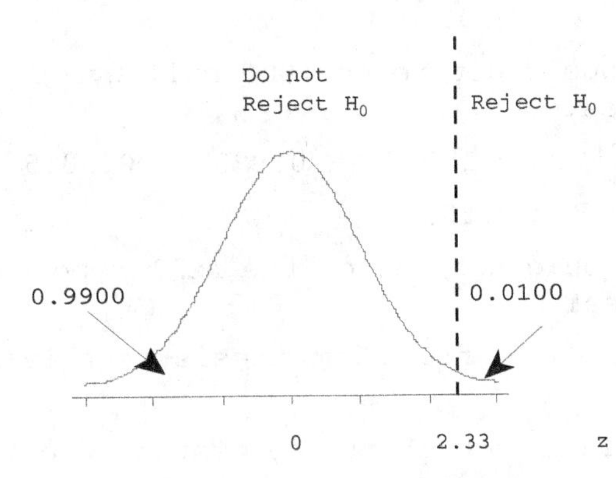

Exercises 9.3

9.47 (1) It allows the reader to assess significance at any desired level, and (2) it permits the reader to evaluate the strength of the evidence against the null hypothesis.

9.49 The P-value for a one-sample z-test is obtained as:

(a) $P(z \le \text{observed z value})$ for a left-tailed test

(b) $P(z \ge \text{observed z value})$ for a right-tailed test

(c) $2P(z \ge \text{absolute value of the observed z value})$ for a two-tailed test.

9.51 (a) Do not reject the null hypothesis.

(b) Reject the null hypothesis.

(c) Reject the null hypothesis.

9.53 A P-value of 0.02 provides stronger evidence against the null hypothesis than does a value of 0.03. It says that if the null hypothesis is true, the data are less likely than they are when the P-value is 0.03.

9.55 Strength of the evidence against the null hypothesis is moderate.

9.57 Strength of the evidence against the null hypothesis is strong.

9.59 Strength of the evidence against the null hypothesis is weak or none.

9.61 Strength of the evidence against the null hypothesis is very strong.

9.63 (a) $z = 2.03$, P-value $= 1.000 - 0.9788 = 0.0212$. At a 5% significance level, we would reject the null hypothesis in favor of the alternative hypothesis.

(b) $z = -0.31$, P-value $= 1.000 - 0.3783 = 0.6217$. At a 5% significance level, we would not reject the null hypothesis in favor of the alternative hypothesis.

9.65 (a) $z = -0.74$, P-value $= 0.2296$. At a 5% significance level, we would not reject the null hypothesis in favor of the alternative hypothesis.

(b) $z = 1.16$, P-value $= 0.8770$. At a 5% significance level, we would not reject the null hypothesis in favor of the alternative hypothesis.

9.67 (a) $z = -1.66$, Left-tail probability $= 0.0485$

P-value $= 0.0485 \times 2 = 0.0970$

At a 5% significance level, we would not reject the null hypothesis in favor of the alternative hypothesis.

(b) $z = 0.52$, Right-tail probability $= 1.0000 - 0.6985 = 0.3015$

P-value $= 0.3015 \times 2 = 0.6030$

At a 5% significance level, we would not reject the null hypothesis in favor of the alternative hypothesis.

9.69 (a) The P-value is expressed as $P(z \le z_0)$ if the hypothesis test is left-tailed.

(b) The P-value is expressed as $2 \cdot P(|z| \ge |z_0|)$ if the test is two-tailed.

9.71 Given that x can be transformed to z (and x_0 to z_0), we have:

1. $P(x \ge x_0) = P(z \ge z_0)$, for a right-tailed test
2. $P(x \le x_0) = P(z \le z_0)$, for a left-tailed test
3. $2 \cdot \min\{P(x \le x_0), P(x \ge x_0)\} = 2 \cdot \min\{P(z \le z_0), P(z \ge z_0)\}$

$$= \begin{cases} 2 \cdot P(z \le z_0) \text{ if } z_0 < 0 \\ 2 \cdot P(z \ge z_0) \text{ if } z_0 \ge 0 \end{cases}$$

$$\text{By symmetry} = \begin{cases} 2 \cdot P(z \ge -z_0) \text{ if } z_0 < 0 \\ 2 \cdot P(z \le z_0) \text{ if } z_0 \ge 0 \end{cases} = 2 \cdot P(z \ge |z_0|)$$

$$\text{By symmetry} = P(z \le -|z_0|) + P(z \ge |z_0|) = P(|z| \ge |z_0|)$$

Exercises 9.4

9.73 The z-test is not an appropriate method for highly skewed data when the sample size is less than 30.

9.75 The z-test is appropriate for large samples with no outliers even if the data are mildly skewed.

9.77 For the critical value approach, reject H_0 if $z < -1.645$;
$z = (20-22)/(4/\sqrt{32}) = -2.83$; therefore, reject H_0 and conclude that $\mu < 22$.
For the P-value approach, the P-value is 0.0023. Since the P-value is less than the significance level, reject H_0 and conclude that $\mu < 22$.

9.79 For the critical value approach, reject H_0 if $z > 1.645$;
$z = (24-22)/(4/\sqrt{15}) = 1.94$; therefore, reject H_0 and conclude that $\mu > 22$.
For the P-value approach, the P-value is 1.000 - 0.9738 = 0.0262. Since the P-value is less than the significance level, reject H_0 and conclude that $\mu > 22$.

9.81 For the critical value approach, reject H_0 if $z < -1.96$ or $z > 1.96$;
$z = (23-22)/(4/\sqrt{24}) = 1.22$; therefore, do not reject H_0. The data do not provide sufficient evidence to support H_a: $\mu \ne 22$.
For the P-value approach, the right tailed probability is 0.1112 and the P-value would be 0.2224. Since the P-value is greater than the significance level, do not reject H_0. The data do not provide sufficient evidence to support H_a: $\mu \ne 22$.

9.83 $n = 12$, $\sigma = 0.37$ ppm, $\bar{x} = 6.31/12 = 0.526$ ppm

Step 1: H_0: $\mu = 0.5$ ppm, H_a: $\mu > 0.5$ ppm

Step 2: $\alpha = 0.05$

Step 3: $z = (0.526 - 0.5)/(0.37/\sqrt{12}) = 0.24$

Step 4: Critical-Value Approach: Critical value = $z_\alpha = z_{0.05} = 1.645$.

P-Value Approach: P-value is $P(Z > 0.24) = 1 - 0.5948 = 0.4052$.

Step 5: Critical-Value Approach: Since 0.24 < 1.645, do not reject H_0.

P-value Approach: Since 0.4052 > 0.05, do not reject H_0.

Step 6: At the 5% significance level, the data do not provide sufficient evidence to conclude that the mean cadmium level μ of *Boletus pinicola* mushrooms is greater than the safety limit of 0.5 ppm.

9.85 $n = 45$, $\bar{x} = 14.68$, $\sigma = 4.2$

Step 1: H_0: $\mu = 18$ mg, H_a: $\mu < 18$ mg

Step 2: $\alpha = 0.01$

Step 3: $z = (14.68 - 18)/(4.2/\sqrt{45}) = -5.30$

Step 4: Critical-Value Approach: Critical value = $-z_{\alpha} = -z_{0.01} = -2.33$.

P-Value Approach: P-value is $P(Z < -5.30) \approx 0.0000$.

Step 5: Critical-Value Approach: Since -5.30 < -2.33, reject H_0.

P-value Approach: Since 0.0000 < 0.01, reject H_0.

Step 6: At the 1% significance level, the data provide sufficient evidence to conclude that adult females under the age of 51 are, on the average, getting less than the RDA of 18 mg of iron. Considering that iron deficiency causes anemia and that iron is required for transporting oxygen in the blood, this result could have practical significance as well.

9.87 n = 100, $\bar{x}$ = 17.8 months, σ = 6.0 months

Step 1: H_0: μ = 16.7 months, H_a: $\mu \neq$ 16.7 months

Step 2: α = 0.05

Step 3: $z = (17.8 - 16.7)/(6.0/\sqrt{100}) = 1.83$

Step 4: Critical-Value Approach: Critical value = $\pm z_{\alpha/2} = \pm z_{0.025} = \pm 1.96$.

P-Value Approach: P-value is $2 \cdot P(Z > 1.83) = 2 \cdot (1 - 0.9664) = 0.0672$.

Step 5: Critical-Value Approach: Since 1.83 < 1.96, do not reject H_0.

P-value Approach: Since 0.0672 > 0.05, do not reject H_0.

Step 6: At the 5% significance level, the data do not provide sufficient evidence to conclude that the mean length of imprisonment μ of motor-vehicle theft offenders in Sydney differs from the national mean in Australia.

9.89 (a) Using Minitab, with the data in a column named GAIN, we choose **Stat ▶ Basic Statistics ▶ 1-Sample z...**, click in the **Samples in columns** text box and specify GAIN, click in the **Standard deviation** text box and type 0.42, and click in the **Test mean** text box and type 0.2. Click the **Options...** button, enter 95 in the **Confidence level** text box, click the arrow button at the right of the **Alternative** drop-down list box and select **greater than** and click **OK.** Click on the **Graphs** button and check the boxes for **Histogram of Data** and **Boxplot of data.** Then click **OK** twice. The result of the test is

```
Test of mu = 0.2 vs > 0.2
The assumed standard deviation = 0.42

                                                    95%
                                                   Lower
Variable   N      Mean     StDev   SE Mean     Bound      Z      P
GAIN      20  0.295000  0.499974  0.093915  0.140524   1.01  0.156
```

(b) The histogram and boxplot were produced by the procedure in part (a). Now choose **Stat ▶ Basic Statistics ▶ Normality test** and enter GAIN in the **Variable** text box. Click **OK.** Then choose **Graph ▶ Stem-and-Leaf** , enter GAIN in the **Graph variables** text box and click **OK**. The results are

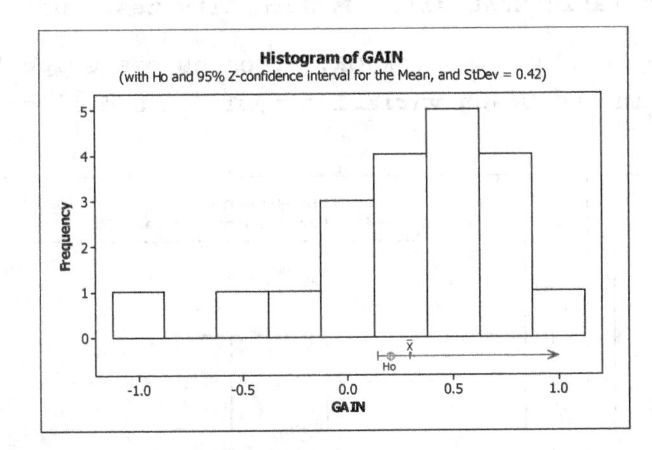

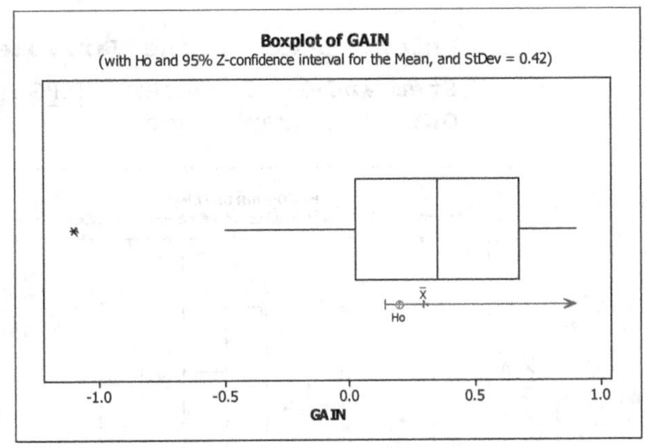


```
Stem-and-leaf of GAIN  N  = 20
Leaf Unit = 0.10

LO -11

 2   -0  5
 3   -0  2
 4   -0  1
 6    0  01
10    0  2233
10    0  45
 8    0  6667
 4    0  8889
```

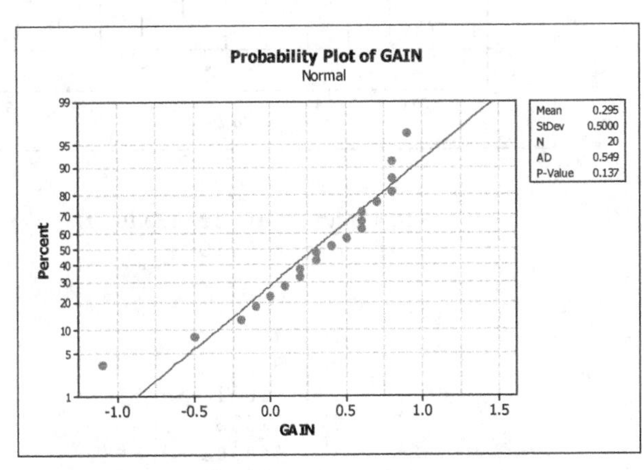

(c) Repeating the procedure of part (a), we obtain

```
Test of mu = 0.2 vs > 0.2
The assumed standard deviation = 0.42

                                                     95%
                                                    Lower
Variable   N      Mean     StDev   SE Mean     Bound     Z      P
GAIN      19  0.368421  0.387374  0.096355  0.209932  1.75  0.040
```

(d) The original sample size is only 20. The plots in part (b) indicate that the value -1.1 is a potential outlier. The z-test should not be used with the original data. This is further confirmed by the fact that using all of the data leads to z = 1.01, whereas, deleting the outlier leads to z = 1.75. This is enough of a change to alter our conclusion from not rejecting the null hypothesis to rejecting it. If there is no good reason for deleting the outlier, then the z-test is inappropriate for these data.

9.91 (a) Using Minitab, with the data in a column named TEMP, we choose **Stat ▶ Basic Statistics ▶ 1-Sample z...**, click in the **Samples in columns** text box and specify TEMP, click in the **Standard deviation** text box and enter 0.63, and click in the **Hypothesized** text box and enter 98.6. Click the **Options...** button, enter 95 in the **Confidence level** text box, click the arrow button at the right of the **Alternative** drop-down list box and select **not equal** and click **OK**. Click on the **Graphs** button and check the boxes for **Histogram of Data** and **Boxplot of data.** Then click **OK** twice. Now choose **Stat ▶ Basic Statistics ▶ Normality test** and enter CHARGE in the **Variable** text box. Click **OK**. Then choose **Graph ▶ Stem-and-Leaf**, enter CHARGE in the **Graph variables** text box and click **OK**. The graphs are

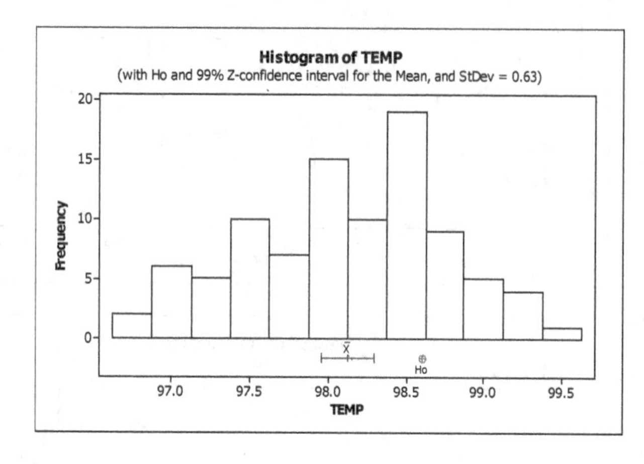

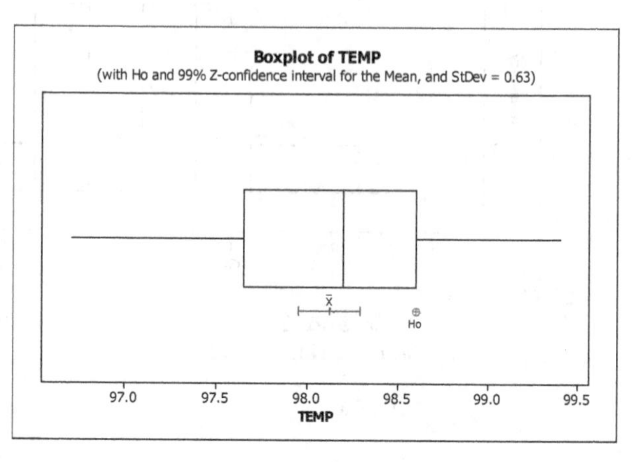


```
Stem-and-leaf of TEMP  N  = 93
Leaf Unit = 0.10

  1    96  7
  3    96  89
  8    97  00001
 13    97  22233
 19    97  444444
 26    97  6666777
 31    97  88889
 45    98  00000000000111
(10)   98  2222222233
 38    98  4444445555
 28    98  66666666677
 17    98  8888888
 10    99  00001
  5    99  2233
  1    99  4
```

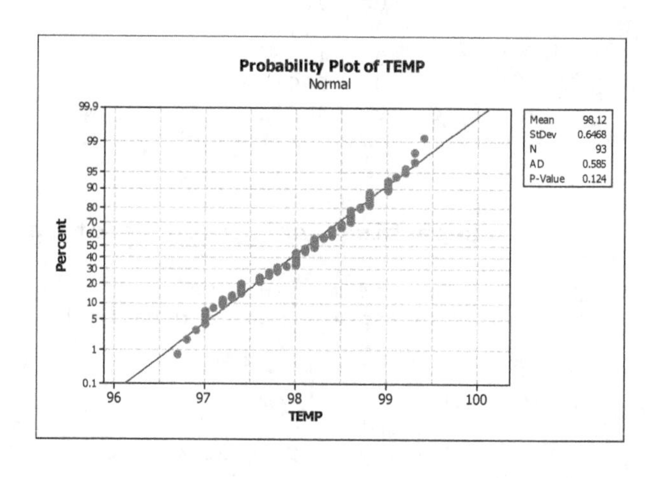

(b) Yes. The sample size 93 is large and the distribution of the data is quite symmetric.

(c) Yes. The procedure in part (a) also produced the results of the test which are

```
Test of mu = 98.6 vs not = 98.6
The assumed standard deviation = 0.63

Variable   N     Mean   StDev  SE Mean          99% CI              Z      P
TEMP      93  98.1237  0.6468   0.0653  (97.9554, 98.2919)  -7.29  0.000
```

The P-value is 0.000. Since the P-value is less than 0.01, we reject the null hypothesis and conclude that the mean body temperature of healthy humans is different from the generally accepted value of 98.6ºF.

9.93 (a) Using Minitab, with the data in a column named BILL, we choose **Stat ▶ Basic Statistics ▶ 1-Sample z...**, click in the **Samples in columns** text box and specify BILL, click in the **Standard deviation** text box and enter 25, and click in the **Hypothesized mean** text box and enter 48.16. Click the **Options…** button, enter 95 in the **Confidence level** text box, click the arrow button at the right of the **Alternative** drop-down list box and select **less than** and click **OK.** Click on the **Graphs** button and check the boxes for **Histogram of Data** and **Boxplot of data.** Then click **OK** twice. Now choose **Stat ▶ Basic Statistics ▶ Normality test** and enter BILL in the **Variable** text box. Click **OK.** Then choose **Graph ▶ Stem-and-Leaf**, enter BILL in the **Graph variables** text box and click **OK.** The graphs are

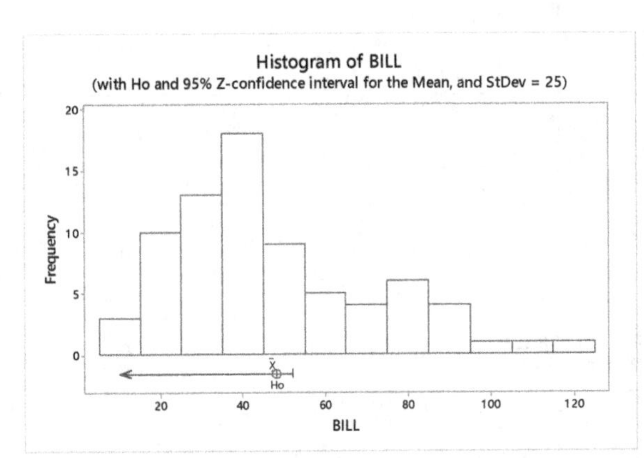

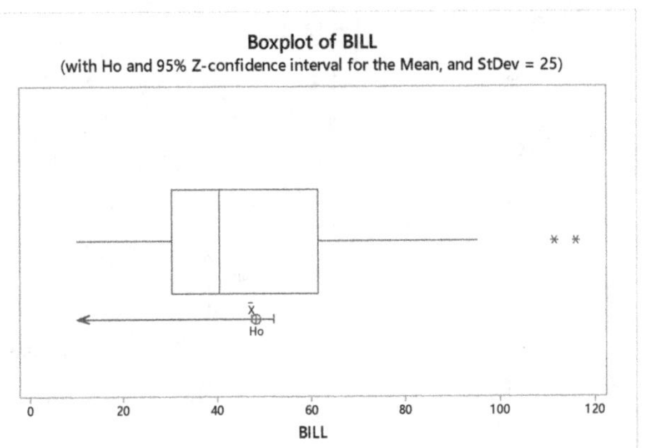

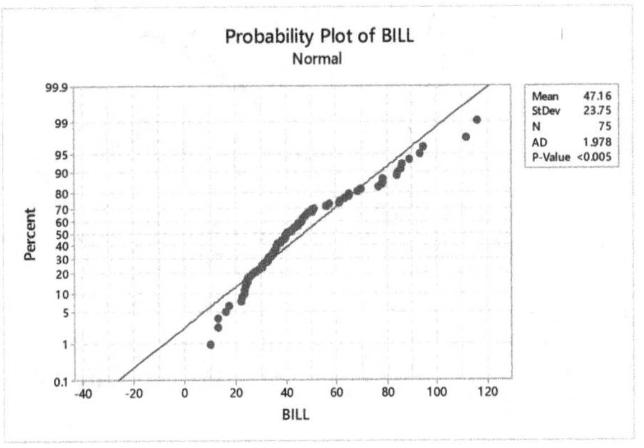


```
Stem-and-leaf of BILL  N  = 75
Leaf Unit = 1.0

  5    1   03367
 17    2   223334445679
 36    3   0012223445566678999
(14)   4   00223444666778
 25    5   00167
 20    6   0135589
 13    7   6888
  9    8   44569
  4    9   35
  2   10
  2   11   16
```

(b) The results of the test carried out by the procedure in part (a) are

```
Test of μ = 48.16 vs < 48.16
The assumed standard deviation = 25
```

Variable	N	Mean	StDev	SE Mean	95% Upper Bound	Z	P
BILL	75	47.16	23.75	2.89	51.91	-0.35	0.365

The P-value for the test is 0.365. Since the P-value is greater than 0.05, we do not reject the null hypothesis. The data do not provide sufficient evidence that the mean local monthly cell phone bill has decreased from the 2009 mean of $48.16.

(c) After deleting the two outliers 116.13 and 111.50, the graphs and test results are

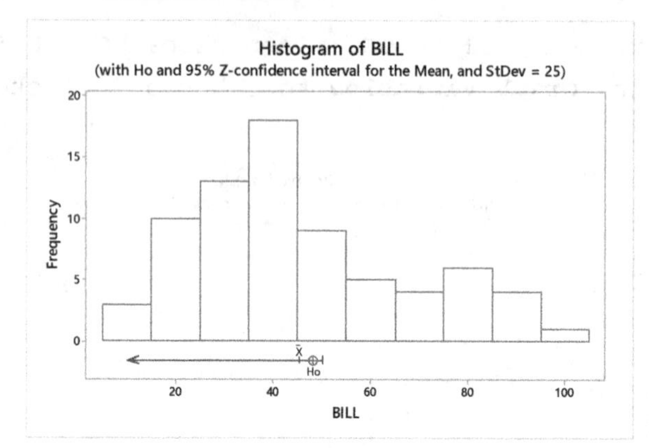

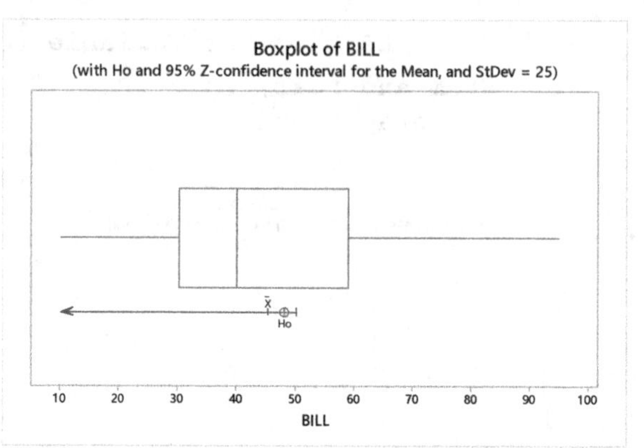

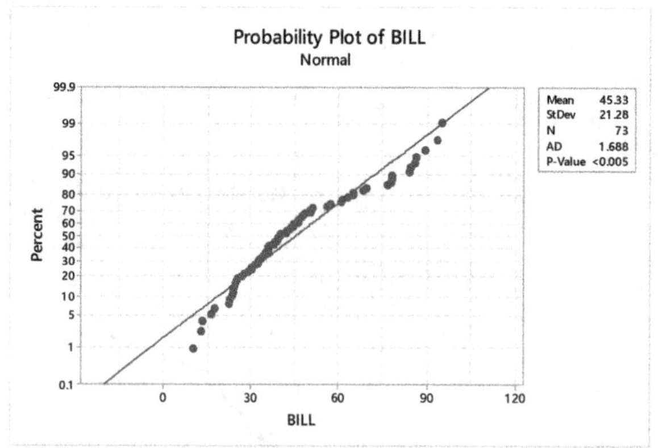


```
Stem-and-leaf of BILL  N  = 73
Leaf Unit = 1.0

  3   1  033
  5   1  67
 13   2  22333444
 17   2  5679
 26   3  001222344
 36   3  5566678999
(8)   4  00223444
 29   4  666778
 23   5  001
 20   5  67
 18   6  013
 15   6  5589
 11   7
 11   7  6888
  7   8  44
  5   8  569
  2   9  3
  1   9  5
```

Test of μ = 48.16 vs < 48.16
The assumed standard deviation = 25

Variable	N	Mean	StDev	SE Mean	95% Upper Bound	Z	P
BILL	73	45.33	21.28	2.93	50.15	-0.97	0.167

(d) Although the value of z has been changed from -0.35 to -0.97 by deleting the two outliers, the conclusion remains the same. We do not reject the null hypothesis.

9.95 (a) n = 28, σ = \$8.45, $\bar{x}$ = \$1788.62/28 = \$63.88

The 90% confidence interval is $63.88 \pm 1.645(8.45)/\sqrt{28} = (61.25, 66.51)$.

The hypothesized mean (\$66.52) lies outside of the confidence interval, so we should reject the null hypothesis. Since the test statistic is $z = (63.88 - 66.52)/(8.45/\sqrt{28}) = -1.653$, which is less than the lower critical value of -1.645, the hypothesis test also leads to the conclusion that we should reject the null hypothesis.

(b) n = 100, $\bar{x}$ = 17.8 months, σ = 6.0 months

The 90% confidence interval is $17.8 \pm 1.96(6.0)/\sqrt{100} = (16.6, 19.0)$.

The hypothesized mean (16.7) lies inside of the confidence interval, so we should not reject the null hypothesis. Since the test statistic is $z = (17.8 - 16.7)/(6.0/\sqrt{100}) = 1.83$, which is less than the upper critical value of 1.96, the hypothesis test also leads to the conclusion that we should not reject the null hypothesis.

9.97 (a) n = 12, σ = 0.37 ppm, $\bar{x}$ = 6.31/12 = 0.526 ppm

The 95% lower level confidence bound is $0.526 - 1.645(0.37)/\sqrt{12} = 0.350$.

The hypothesized mean (0.5) lies above the lower confidence bound, so we should not reject the null hypothesis. Since the test statistic is $z = (0.526 - 0.5)/(0.37/\sqrt{12}) = 0.24$, which is less than the critical value of 1.645, the hypothesis test also leads to the conclusion that we should not reject the null hypothesis.

(b) n = 30, $\bar{x}$ = 78.3, σ = 11.2

The 95% lower level confidence bound is $78.3 - 1.645(11.2)/\sqrt{30} = 74.94$.

The hypothesized mean (72) lies below the lower confidence bound, so we should reject the null hypothesis. Since the test statistic is $z = (78.3 - 72)/(11.2/\sqrt{30}) = 3.08$, which is greater than the critical value of 1.645, the hypothesis test also leads to the conclusion that we should reject the null hypothesis.

Exercises 9.5

9.99 (a) Neither the z-test nor the t-test is appropriate for small samples containing outliers or exhibiting extremely non-normal distributional shapes.

(b) A non-parametric test that is neither affected by a few outliers nor by non-normality might be appropriate.

9.101 (a) $0.01 < P < 0.025$

(b) Reject H_0 for $\alpha \geq 0.025$; Do not reject H_0 for $\alpha \leq 0.01$; Undecided for $0.01 < \alpha < 0.025$.

9.103 (a) $P < 0.005$

(b) Reject H_0 for $\alpha \geq 0.005$; Undecided for $\alpha < 0.005$.

9.105 (a) $0.01 < P < 0.02$

(b) Reject H_0 for $\alpha \geq 0.02$; Do not reject H_0 for $\alpha \leq 0.01$; Undecided for $0.01 < \alpha < 0.02$.

9.107 (a) df = 31; $t = -2.828$

(b) $P < 0.005$; Reject H_0; Evidence against H_0 is very strong.

9.109 (a) df = 14; $t = 1.936$

(b) $0.025 < P < 0.05$; Reject H_0; Evidence against H_0 is strong.

9.111 (a) df = 23; $t = 1.225$

(b) $P > 0.20$; Do not reject H_0; Evidence against H_0 is weak or none.

9.113 H_0: $\mu = 4.55$, H_a: $\mu \neq 4.55$, $\alpha = 0.10$; Critical values: ± 1.729

$$t = (\bar{x} - \mu)/(s/\sqrt{n}) = (4.760 - 4.55)/(2.297/\sqrt{20}) = 0.409 .$$

Since $-1.729 < 0.409 < 1.729$, we do not reject H_0. The P-value is $P > 0.20$. The data does not provide sufficient evidence to conclude that the amount of television watched per day last year by the average person differed from that in 2005.

9.115 $n = 20$, df = 19, $\bar{x} = 49.0$, $s = 10.0$

Step 1: H_0: $\mu = 45$, H_a: $\mu > 45$

Step 2: $\alpha = 0.05$

Step 3: $$t = \frac{49.0 - 45}{10.0/\sqrt{20}} = 1.789$$

Step 4: Critical Value Approach: Critical value = 1.729

P-Value Approach: $0.025 <$ P-value < 0.05

Step 5: Since $1.789 > 1.729$, reject H_0. Since the P-value $< \alpha$, reject H_0.

Step 6: At the 5% significance level, the data do provide sufficient evidence to conclude that the mean angle between the body and head of an alligator during a death roll is greater than 45 degrees.

9.117 $n = 187$, df = 186, $\bar{x} = 0.64$, $s = 0.15$

Step 1: H_0: $\mu = 0.9$, H_a: $\mu < 0.9$

Step 2: $\alpha = 0.01$

Step 3: $$t = \frac{0.64 - 0.90}{0.15/\sqrt{187}} = -23.703$$

Step 4: Critical Value Approach: Critical value = -2.345

P-Value Approach: $P < 0.005$

Step 5: Since $-23.703 < -2.345$, reject H_0. Since the p-value $< \alpha$, reject H_0.

Step 6: At the 1% significance level, the data provide sufficient evidence to conclude that the mean ABI for women with peripheral arterial disease is less than the healthy ABI of 0.9. Thus, we conclude that such women do have an unhealthy ABI. The practical significance of this result is that the ABI may be a good tool for determining the possibility of peripheral arterial disease in women. There could also be other causes of a low ABI, so this test by itself may not be able to determine the precise ailment.

9.119 We used Minitab to produce the following histogram.

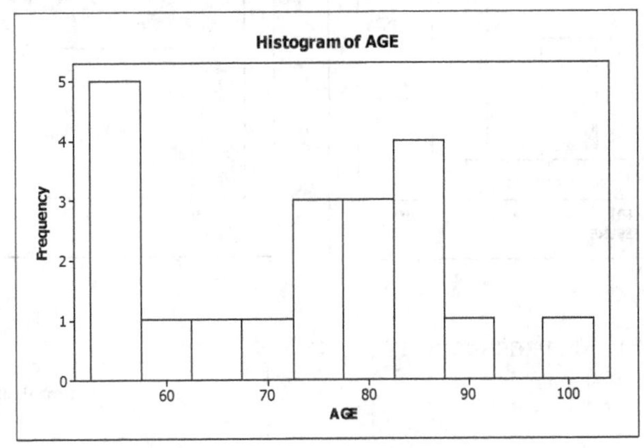

The sample size is only 20. Although there are no outliers, the distribution is not very close to being normally distributed. It does not appear to be reasonable to use a t-test with these data.

9.121 We used Minitab to produce the following normal probability plot.

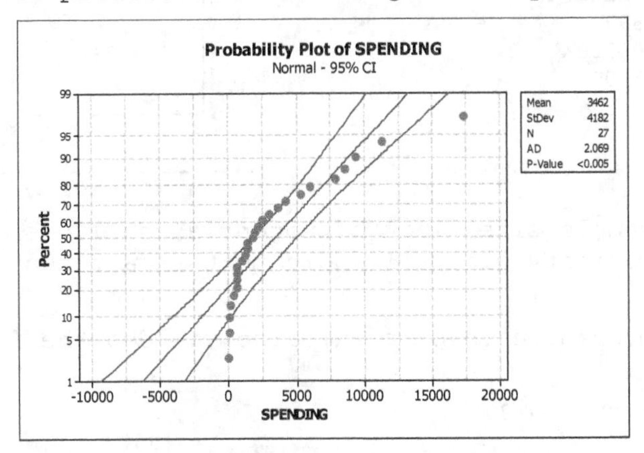

9.123 (a) Using Minitab, with the data in a column named PRESSURE, we choose **Stat ▶ Basic Statistics ▶ 1-Sample t...**, click in the **Samples in columns** text box and specify PRESSURE, click in the **Test mean** text box and enter 80. Click the **Options...** button, enter 90 in the **Confidence level** text box, click the arrow button at the right of the **Alternative** drop-down list box and select **greater than** and click **OK.** Click on the **Graphs** button and check the boxes for **Histogram of Data** and **Boxplot of data.** Then click **OK** twice. Now choose **Stat ▶ Basic Statistics ▶ Normality test** and enter PRESSURE in the **Variable** text box. Click **OK.** Then choose **Graph ▶ Stem-and-Leaf** , enter PRESSURE in the **Graph variables** text box and click **OK.** The results are

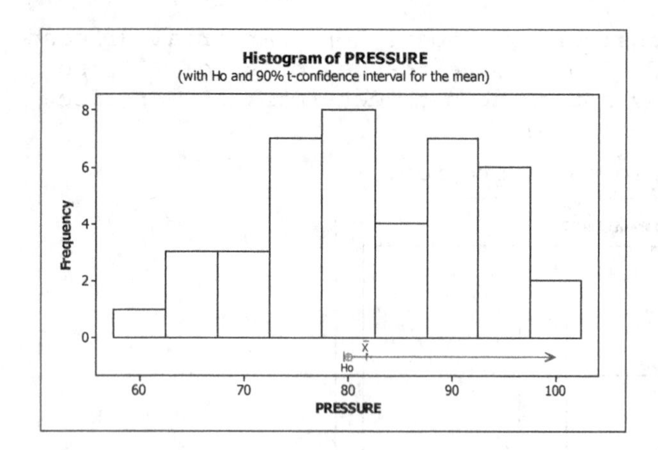

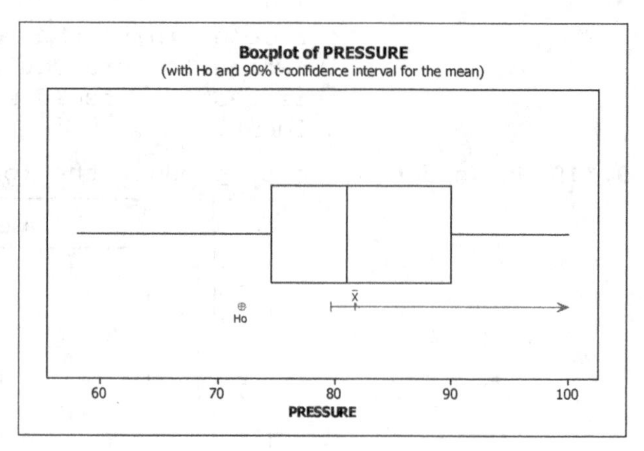


```
Stem-and-leaf of PRESSURE  N  = 41
Leaf Unit = 1.0


  1   5   8
  2   6   3
  5   6   569
 10   7   00334
 17   7   5677999
 (8)  8   00111334
 16   8   5899
 12   9   0001334
  5   9   5559
  1  10   0
```

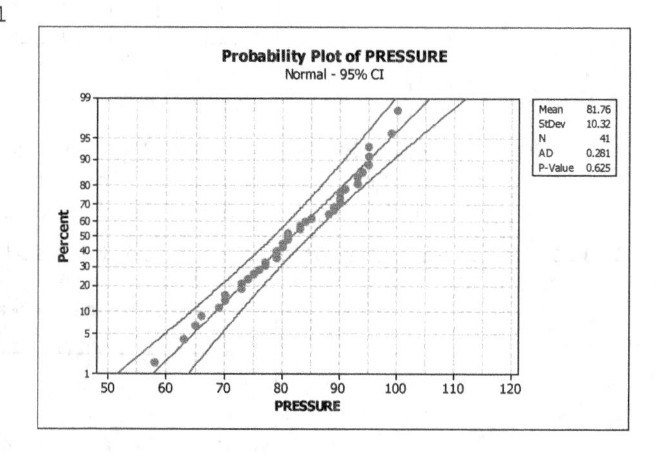

(b) The large sample size, lack of potential outliers, and the nearly linear probability plot all indicate that a t-test is reasonable for these data.

(c) The first procedure in part (a) also yielded the following test results:

```
Test of mu = 80 vs > 80

                                                   90%
                                                 Lower
Variable   N     Mean    StDev  SE Mean    Bound     T      P
PRESSURE  41  81.7561  10.3242   1.6124  79.6551  1.09  0.141
```

We see that t = 1.09 and the P-value is 0.141. Since the P-value is greater than the significance level 0.10, we do not reject the null hypothesis and conclude that the data do not provide evidence that the mean diastolic blood pressure of bus drivers in Stockholm exceeds the normal pressure of 88 mm Hg.

9.125 (a) Using Minitab, with the data in a column named RENT, we choose **Stat ▶ Basic Statistics ▶ 1-Sample t...**, click in the **Samples in columns** text box and specify RENT, click in the **Test mean** text box and enter 949. Click the **Options...** button, enter 95 in the **Confidence level** text box, click the arrow button at the right of the **Alternative** drop-down list box and select **greater than** and click **OK.**

```
Test of μ = 949 vs > 949

Variable     N    Mean  StDev  SE Mean  95% Lower Bound     T      P
 RENT      100  960.96  89.85     8.98           946.04  1.33  0.093
```

The P-value for the test is 0.093, larger than the significance lever 0.05, so we do not reject the null hypothesis. The data do not provide sufficient evidence to conclude that the mean rent for a two-bedroom unit is greater than the FMR of $949.

(b) After removing the outlier 662 and following the procedure in part (a), the results are

```
Test of μ = 949 vs > 949

Variable    N    Mean  StDev  SE Mean  95% Lower Bound     T      P
RENT       99  963.98  85.05     8.55           949.79  1.75  0.041
```

(c) Now the P-value is 0.041, leading to rejection of the null hypothesis. The sample mean increased by $3.02 and the standard deviation decreased by about $4.8.

(d) The sample size is large in both cases, yet the effect of the outlier is considerable. Caution should be used and perhaps the data should be analyzed using a method that is not influenced by outliers.

9.127 (a) n = 187, df = 186, $-t_{\alpha}$ = -2.345, s = 0.15, $\bar{x}$ = 0.64

The 95% upper confidence bound is $0.64 + 2.345(0.15)/\sqrt{187} = 0.666$.

The hypothesized mean (0.9) lies above the upper confidence bound, so we should reject the null hypothesis. Since the test statistic is $t = (0.64 - 0.9)/(0.15/\sqrt{187}) = -23.703$, which is less than the critical value of -2.345, the hypothesis test also leads to the conclusion that we should reject the null hypothesis.

(b) n = 30, df = 29, $-t_{\alpha}$ = -1.311, s = 0.74, $\bar{x}$ = 1.91

The 90% upper confidence bound is $1.91 + 1.311(0.74)/\sqrt{30} = 2.087$.

The hypothesized mean (2) lies below the upper confidence bound, so we should not reject the null hypothesis. Since the test statistic is $t = (1.91 - 2)/(0.74/\sqrt{30}) = -0.67$, which is greater than the critical value of -1.311, the hypothesis test also leads to the conclusion that we should not reject the null hypothesis.

Exercises 9.6

9.129 Technically, nonparametric methods are inferential methods that are not concerned with parameters. In practice, nonparametric methods are those that can be applied without assuming normality.

9.131 Because the D-value for such a data value equals 0, we cannot attach a sign to the rank of |D|.

9.133 t-test

9.135 Wilcoxon signed-rank test

9.137 Neither

9.139 (a) $W_{0.05} = 30$ (b) $W_{0.95} = 8(8 + 1)/2 - 30 = 6$

(c) $W_{0.025} = 32$ $W_{0.975} = 8(8 + 1)/2 - 32 = 4$

9.141 (a) $W_{0.10} = 128$ (b) $W_{0.90} = 19(19 + 1)/2 - 128 = 62$

(c) $W_{0.05} = 136$ $W_{0.995} = 19(19 + 1)/2 - 136 = 54$

9.143 H_0: $\mu = 5$, H_a: $\mu > 5$; $n = 8$; $\alpha = 0.10$,
$W_{0.10} = 28$

Observation	D	|D|	Rank |D|	R
12	7	7	8	8
7	2	2	2.5	2.5
11	6	6	7	7
9	4	4	6	6
3	-2	2	2.5	-2.5
2	-3	3	4.5	-4.5
8	3	3	4.5	4.5
6	1	1	1	1

The sum of the positive ranks is W = 29. This is greater than the critical value 28, so we reject the null hypothesis.

9.145 There is one observation that matches μ. We will remove the 6 and reduce the sample size to 7.
H_0: $\mu = 6$, H_a: $\mu \neq 6$; $n = 7$; $\alpha = 0.10$,
$W_{0.05} = 24$ and $W_{0.95} = 7(8)/2 - 24 = 4$

Observation	D	|D|	Rank |D|	R
4	-2	2	3.5	-3.5
8	2	2	3.5	3.5
4	-2	2	3.5	-3.5
1	-5	5	6.5	-6.5
1	-5	5	6.5	-6.5
4	-2	2	3.5	-3.5
7	1	1	1	1

The sum of the positive ranks is W = 4.5. This is within the critical values of 4 and 24, so we do not reject the null hypothesis.

9.147 H_0: $\mu = 12$, H_a: $\mu < 12$; $n = 9$; $\alpha = 0.10$,
$W_{0.90} = 9(10)/2 - 34 = 11$

Observation	D	|D|	Rank |D|	R
16	4	4	7.5	7.5
11	-1	1	2	-2
10	-2	2	4.5	-4.5
14	2	2	4.5	4.5
13	1	1	2	2
15	3	3	6	6
5	-7	7	9	-9
8	-4	4	7.5	-7.5
11	-1	1	2	-2

The sum of the positive ranks is W = 20. This is greater than the critical value of 11, so we do not reject the null hypothesis.

9.149 H_0: $\mu = 124.9$ days, H_a: $\mu < 124.9$ days; $n = 8$; $\alpha = 0.05$,
$W_{0.95} = 8(8 + 1)/2 - 30 = 6$

Days	D	|D|	Rank |D|	R
103	-21.9	21.9	4	-4
80	-44.9	44.9	5	-5
79	-45.9	45.9	7	-7
135	10.1	10.1	2	2
134	9.1	9.1	1	1
77	-47.9	47.9	8	-8
80	-44.9	44.9	5	-5
111	-13.9	13.9	3	-3

The sum of the positive ranks is W = 3. This is less than the critical value 6, so we reject the null hypothesis and conclude that the data provide sufficient evidence that the mean number of days of ice cover on Lake Wingra is less than it was in the late 1800s.

9.151 $\eta_0 = 37.2$, n = 10, $\alpha = 0.01$

Step 1: H_0: $\eta = 37.2$, H_a: $\eta > 37.2$
Step 2: $\alpha = 0.01$
Step 3:

x	$x-\eta_0=D$	|D|	Rank of |D|	Signed Rank R
44	6.8	6.8	3	3
47	9.8	9.8	5	5
64	26.8	26.8	10	10
51	13.8	13.8	6	6
16	-21.2	21.2	7	-7
41	3.8	3.8	2	2
59	21.8	21.8	8	8
13	-24.2	24.2	9	-9
38	0.8	0.8	1	1
28	-9.2	9.2	4	-4

W = sum of the + ranks = 35

Step 4: Critical value = 50. The P-value is 0.238.

Step 5: Since W < 50, do not reject H_0.

Step 6: At the 1% significance level, the data do not provide sufficient evidence to conclude that the median age has increased over the 2010 median age of 37.2 years.

9.153 μ_0 = \$18,000, n = 10, $\alpha = 0.01$

Step 1: H_0: μ = \$18,000, H_a: μ < \$18,000

Step 2: $\alpha = 0.01$

Step 3:

Price	D	|D|	Rank |D|	R
16594	-1406	1406	1	-1
15613	-2387	2387	8	-8
16106	-1894	1894	2	-2
14614	-3386	3386	9	-9
16102	-1898	1898	3	-3
13514	-4486	4486	10	-10
15914	-2086	2086	4	-4
15614	-2386	2386	7	-7
15713	-2287	2287	6	-6
15714	-2286	2286	5	-5

W = sum of the + ranks = 0

Step 4: Critical value = 10(11)/2 - 50 = 5. The P-value is 0.003.

Step 5: Since W < 5, reject H_0.

Step 6: At the 1% significance level, the data do provide sufficient evidence to conclude that the mean asking price for a 2-year-old Ford Mustang is less than the *Kelly Blue Book* value of $18,000.

9.155 $\mu_0 = 45$, n = 20, $\alpha = 0.05$

(a) Step 1: H_0: $\mu = 45$, H_a: $\mu > 45$

Step 2: $\alpha = 0.05$

Step 3:

x	$x-\mu_0=D$	$\lvert D\rvert$	Rank of $\lvert D\rvert$	Signed Rank R
58.6	13.6	13.6	15	15
59.5	14.5	14.5	17	17
42.7	-2.3	2.3	4	-4
51.5	6.5	6.5	7	7
58.7	13.7	13.7	16	16
29.4	-15.6	15.6	19	-19
34.8	-10.2	10.2	11	-11
42.8	-2.2	2.2	3	-3
57.3	12.3	12.3	12	12
43.4	-1.6	1.6	2	-2
39.2	-5.8	5.8	6	-6
57.5	12.5	12.5	13	13
54.5	9.5	9.5	10	10
31.8	-13.2	13.2	14	-14
61.3	16.3	16.3	20	20
43.6	-1.4	1.4	1	-1
52.9	7.9	7.9	9	9
52.3	7.3	7.3	8	8
60.4	15.4	15.4	18	18
47.6	2.6	2.6	5	5

W = sum of the + ranks = 150.0

Step 4: Critical value = 150

Step 5: Since $W \geq 150$, reject H_0.

Step 6: At the 5% significance level, the data provide sufficient evidence to conclude that the mean angle between the head and body of an alligator during a death roll is greater than 45 degrees. The P-value is 0.048.

(b) A Wilcoxon signed-rank test is permissible because a normally distributed population is symmetric.

9.157 n = 16, $\bar{x} = 306$, s = 8.671, $\alpha = 0.05$

(a) Step 1: H_0: $\mu = 310$, H_a: $\mu < 310$

Step 2: $\alpha = 0.05$

Step 3: $$t = \frac{306-310}{8.671/\sqrt{16}} = -1.845$$

Step 4: Critical Value Approach: Critical value = -1.753

P-Value Approach: $0.025 < \text{p-value} < 0.05$

Step 5: Since $-1.845 < -1.753$, reject H_0. Since P-value < α, reject H_0.

Step 6: At the 5% significance level, the data do provide sufficient evidence to conclude that the mean content, μ, is less than the advertised content of 310 ml.

(b) Step 1: H_0: $\mu = 310$, H_a: $\mu < 310$

Step 2: $\alpha = 0.05$

Step 3:

x	$x-\mu_0=D$	|D|	Rank of |D|	Signed Rank R
297	-13	13	14	-14
311	1	1	2	2
322	12	12	12.5	12.5
315	5	5	7	7
318	8	8	9	9
303	-7	7	8	-8
307	-3	3	5	-5
296	-14	14	15	-15
306	-4	4	6	-6
291	-19	19	16	-16
312	2	2	4	4
309	-1	1	2	-2
300	-10	10	10.5	-10.5
298	-12	12	12.5	-12.5
300	-10	10	10.5	-10.5
311	1	1	2	2

W = sum of the + ranks = 36.5

Step 4: Critical value = 36. The P-value is 0.054.

Step 5: Since W > 36, do not reject H_0.

Step 6: At the 5% significance level, the data do not provide sufficient evidence to conclude that the mean content, μ, is less than the advertised content of 310 ml.

(c) Since the population is normally distributed, the t-test is more powerful than the Wilcoxon signed-rank test; that is, the t-test is more likely to detect a false null hypothesis.

9.159 Using Minitab with the data in a column named HIC, we choose **Graph ▶ Histogram.** Select **Simple** and click **OK.** Select HIC in the **Variables** text box and click **OK.** Also choose **Graph ▶ Boxplot.** Select **Simple** and click **OK.** Select HIC in the **Variables** text box and click **OK.** The graphs are:

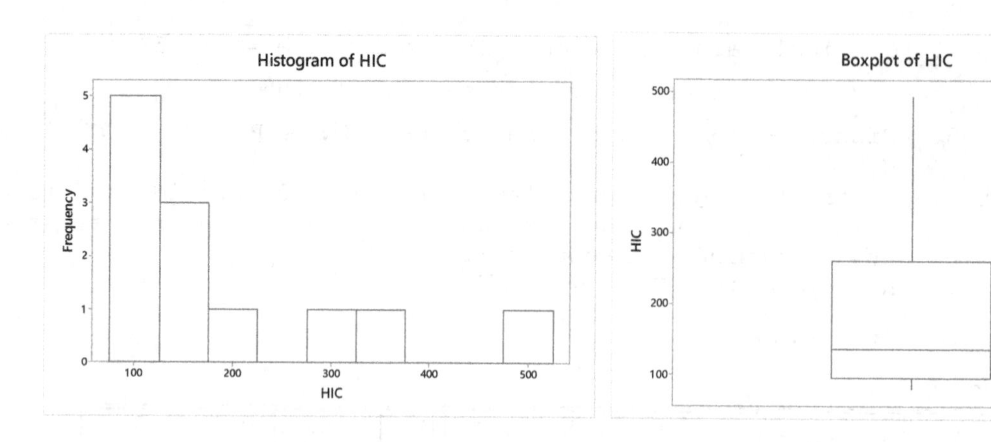

The graphs indicate that the distribution of the variable is not symmetric. Therefore, it is not reasonable to apply the Wilcoxon signed-rank test.

9.161 (a) Using Minitab, with the data in a column named DURATION, we choose **Stat ▶ Nonparametrics ▶ 1-Sample Wilcoxon...**, select DURATION in the **Variables** text box, click in the **Test median** box and type 52, click the arrow button at the right of the **Alternative** drop down list box, and select **greater than**, and click **OK**. The result is

```
Test of median = 52.00 versus median > 52.00

             N for  Wilcoxon            Estimated
          N   Test  Statistic      P       Median
DURATION  90    83     2309.5  0.005        53.00
```

Since the P-value of 0.005 is less than the significance level of 0.05, we reject the null hypothesis and conclude that there is strong evidence that the mean larval duration of convict surgeonfish exceeds 52 days.

(b) To perform the t-test with the original data, we choose **Stat ▶ Basic statistics ▶ 1-Sample t...**, select the **Samples in columns** box, enter DURATION in the **Samples in Columns** text box, enter 52 in the **Test Mean** text box, click on the **Options** button, enter 95.0 in the **Confidence level** text box, select **greater than** in the **Alternative** box, and click **OK** twice. The results in the Session Window are

```
Test of mu = 52 vs > 52

                                             95% Lower
Variable   N    Mean  StDev  SE Mean      Bound     T      P
DURATION  90  52.963  3.479    0.367     52.354  2.63  0.005
```

Again, since the P-value of 0.005 is less than the significance level of 0.05, we reject the null hypothesis and conclude that there is strong evidence that the mean larval duration of convict surgeonfish exceeds 52 days.

(c) The results are the same for the two tests. The *t*-test is more appropriate because the sample size is large.

9.163 (a) Using Minitab, with the data in a column named PRESSURE, we choose **Stat ▶ Nonparametrics ▶ 1-Sample Wilcoxon...**, select PRESSURE in the **Variables** text box, click in the **Test median** box and type 80, click the arrow button at the right of the **Alternative** drop down list box, and select **greater than**, and click **OK**. The result is

```
Test of median = 80.00 versus median > 80.00

                  N
                for   Wilcoxon              Estimated
           N   Test  Statistic       P        Median
PRESSURE  41     39      468.0   0.140         82.00
```

Since the P-value of 0.140 is greater than the significance level of 0.10, we do not reject the null hypothesis and conclude that there is weak or no evidence that the mean diastolic blood pressure of bus drivers in Stockholm exceeds the normal diastolic blood pressure of 80 mm Hg.

(b) In Exercise 9.123, we got a P-value of 0.141 for the *t*-test. This P-value is very close to the P-value that we got with the Wilcoxon signed-rank test. Both tests resulted in the same conclusion to not reject the null hypothesis and conclude that there is no evidence that the mean diastolic blood pressure in Stockholm exceeds the normal diastolic blood pressure of 80 mm Hg.

9.165 (a) Using Minitab, with the data in a column named RENT, we choose **Stat ▶ Nonparametrics ▶ 1-Sample Wilcoxon...**, select RENT in the **Variables** text box, click in the **Test median** box and type 949, click the arrow button at the right of the **Alternative** drop down list box, and select **greater than**, and click **OK**. The result is

```
Test of median = 949.0 versus median > 949.0

            N for   Wilcoxon              Estimated
       N    Test    Statistic      P        Median
RENT  100   99      2873.5       0.082      962.0
```

Since the P-value of 0.082 is greater than the significance level of 0.05, we do not reject the null hypothesis and conclude that the data do not provide sufficient evidence that the mean monthly rent for two bedroom units is greater than the FMR of $949.

(b) In Exercise 9.125, we got a P-value of 0.093 for the *t*-test. This P-value is very close to the P-value that we got with the Wilcoxon signed-rank test. Both tests resulted in the same conclusion to not reject the null hypothesis and conclude that the data do not provide sufficient evidence that the mean monthly rent for two bedroom units is greater than the FMR of $949. The dataset did have an outlier that the *t*-test would be more sensitive to than the Wilcoxon signed-rank test. This may explain why the *t*-test had a slightly larger P-value.

9.167 (a) Step 1: H_0: η = 7.5 lb, H_a: $\eta \neq$ 7.5 lb

Step 2: α = 0.05

Step 3: From Step 3 in the solution to Exercise 9.154, W = 57.5. Now:

$$z = \frac{W - n(n+1)/4}{\sqrt{n(n+1)(2n+1)/24}} = \frac{57.5 - 13(13+1)/4}{\sqrt{13(13+1)(2\cdot 13+1)/24}} = 0.84$$

Step 4: Critical values = ±1.96

Step 5: Since -1.96 < 0.84 < 1.96, do not reject H_0.

Step 6: At the 5% significance level, the data do not provide sufficient evidence to conclude that this year's median birth weight differs from that in 2005.

(b) Neither the Wilcoxon signed-rank test nor the normal approximation led to rejection of the null hypothesis.

9.169 (a) η_0 is the hypothesized median. The median is that value which has half of the population to its left and half to its right. Therefore, the probability that a value exceeds the median is 0.5.

(b) Each observation either exceeds η_0 or it doesn't. The observations are independent of each other and the probability p that an observation exceeds η_0 is 0.5 for each observation. The number of observations (trials) is fixed at n. Thus, the number of observations exceeding η_0 satisfies all of the criteria for a binomial distribution.

9.171 The sign test can be used with any distribution since the probability that an observation exceeds the median is always 0.5, regardless of the shape of the distribution. The Wilcoxon signed-rank test is based on an assumption that the underlying distribution of the data is symmetric, a slightly more restrictive assumption than for the sign test.

9.173 (a) For the sign test, x = 2 of the n = 8 observations are above 124.9. The P-value is $P(x \leq 2)$ for a binomial distribution with n = 8 and p = 0.5. From Table XII, this probability is 0.004 + 0.031 + 0.109 = 0.144. Since this is greater than the 0.05 significance level, the data do not provide sufficient evidence that the number of days the lake was frozen over is less now than in the late 1800s.

(b) The Wilcoxon signed rank test had W = 3.0 with a P-value of 0.021. Since this is less than the 0.05 significance level, the data do provide sufficient evidence that the number of days the lake was frozen over is less now than in the late 1800s.

9.175 (a) For the sign test, x = 7 of the n = 10 observations are above 37.2. The P-value is $P(x \geq 7)\}$ for a binomial distribution with n = 10 and p = 0.5. From Table XII, this probability is 0.117 + 0.044 + 0.010 + 0.001 = 0.172. Since this is greater than the 0.05 significance level, the data do not provide sufficient evidence that the median age of today's U.S. residents has increased from the 2010 median of 37.2 years.

(b) The Wilcoxon signed rank test had W = 35.0 with a P-value of 0.238. Since this is greater than the 0.05 significance level, the data do not provide sufficient evidence that the median age of today's U.S. residents has increased from the 2010 median of 37.2 years. Same conclusion, larger P-value.

9.177 (a) For the sign test, x = 0 of the n = 10 observations are above $18,000. The P-value is $P(x \leq 0)\}$ for a binomial distribution with n = 10 and p = 0.5. This probability is 0.001. Since this is less than the 0.01 significance level, the data do provide sufficient evidence that the mean asking price for a 2-year-old Ford Mustang coupes is less than the *Kelley Blue Book* retail value of $18000.

(b) The Wilcoxon signed rank test had W = 0.0 with a P-value of 0.003. Since this is less than the 0.01 significance level, the data do provide sufficient evidence that the mean asking price for 2-year-old Ford Mustang coupes is less than the *Kelley Blue Book* retail value of $18000. Same conclusion and similar P-value.

Exercises 9.7

9.179 Hypothesis tests have built-in margins of error. Errors will occur due to the uncontrollable randomness in the data observed.

9.181 (a) α = significance level = P(Type I error) = P(rejecting a true null hypothesis)

(b) β = P(Type II error) = P(not rejecting a false null hypothesis)

(c) 1 - β = Power of the test = P(rejecting a false null hypothesis)

9.183 Since μ is unknown, the power curve enables one to evaluate the effectiveness of a hypothesis test for a variety of values of μ.

9.185 If the significance level is decreased without changing the sample size, the rejection region is made smaller (in probability terms). This makes the nonrejection region larger, i.e., β gets larger. This, in turn, makes the power 1 - β smaller.

9.187 It can be very time consuming and heavy with computations to obtain Type II error probabilities by hand and can result in substantial rounding error.

9.189 (a)

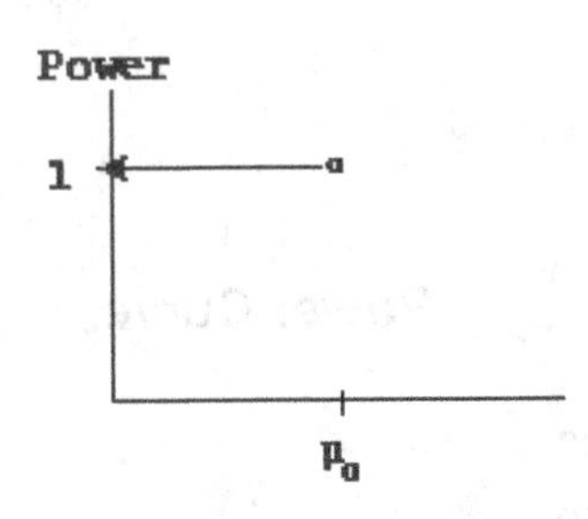

(b) The curve in part (a) portrays that, ideally, one desires the value for the power for any given μ_a to be as close to 1 as possible.

9.191 (a) P(Type I error) = α = 0.05

(b) Apply Procedure 9.4.

Step 1: The $\bar{x}$ critical value is $\mu_0 + z_\alpha \cdot \sigma/\sqrt{n} = 0.5 + 1.645(0.37)/\sqrt{12} = 0.676$.

Step 2: We have μ_a = 0.55, $\sigma/\sqrt{n} = (0.37)/\sqrt{12} = 0.107$, and from Step 1, the $\bar{x}$ critical value of 0.676. Therefore, we need to find the area under the normal curve with parameters 0.55 and 0.107 that lie to the left of 0.676. The z-score of the $\bar{x}$ critical value is $z = \dfrac{0.676 - 0.55}{0.107} = 1.18$.

The area to the left of 1.18 is 0.8810. Therefore, $\beta = 0.8810$.

(c) Apply Procedure 9.5

Step 1: The values of μ_a are given in the problem as 0.55, 0.60, 0.65, 0.70, 0.75, 0.80, and 0.85.

Steps 2-4:

True mean μ_a	z-score computation	P(Type II error) β	Power $1 - \beta$
0.55	$z = \frac{0.676 - 0.55}{0.37/\sqrt{12}} = 1.18$	0.8810	0.119 0
0.60	$z = \frac{0.676 - 0.60}{0.37/\sqrt{12}} = 0.71$	0.7611	0.2389
0.65	$z = \frac{0.676 - 0.65}{0.37/\sqrt{12}} = 0.24$	0.5948	0.4052
0.70	$z = \frac{0.676 - 0.70}{0.37/\sqrt{12}} = -0.22$	0.4129	0.5871
0.75	$z = \frac{0.676 - 0.75}{0.37/\sqrt{12}} = -0.69$	0.2451	0.7549
0.80	$z = \frac{0.676 - 0.80}{0.37/\sqrt{12}} = -1.16$	0.1230	0.8770
0.85	$z = \frac{0.676 - 0.85}{0.37/\sqrt{12}} = -1.63$	0.0516	0.9484

Steps 5-6:

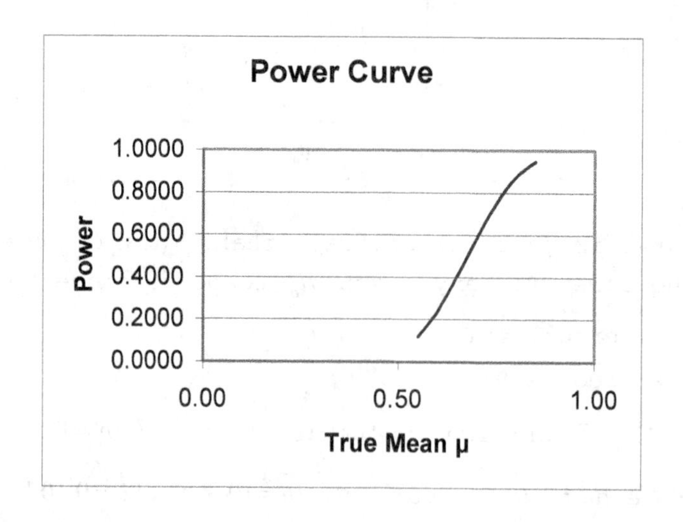

9.193 (a) P(Type I error) = α = 0.01

(b) Apply Procedure 9.4.

Step 1: The $\bar{x}$ critical value is $\mu_0 - z_\alpha \cdot \sigma/\sqrt{n} = 18 - 2.33(4.2)/\sqrt{45} = 16.54$.

Step 2: We have μ_a = 15.50, $\sigma/\sqrt{n} = (4.2)/\sqrt{45} = 0.626$, and from Step 1, the $\bar{x}$ critical value of 16.54. Therefore, we need to find the area under the normal curve with parameters 15.50 and 0.626 that lie to the right of 16.54. The z-score of the $\overline{x}$ critical value is $z=\frac{16.54-15.5}{0.626}=1.66$. The area to the right of 1.66 is 1 - 0.9515 = 0.0485. Therefore, $\beta=0.0485$. Answers may differ slightly from those in the text due to intermediate rounding.

(c) Apply Procedure 9.5

Step 1: The values of μ_a are given in the problem as 15.5, 15.75, 16.0, 16.25, 16.5, 16.75, 17.0, 17.25, 17.5, and 17.75.

Steps 2-4:

Answers may differ slightly from those in the text due to intermediate rounding.

True mean μ	z-score computation	P(Type II error) β	Power $1-\beta$
15.50	$z=\frac{16.54-15.50}{4.2/\sqrt{45}}=1.66$	1.000 - 0.9515 = 0.0485	0.9515
15.75	$z=\frac{16.54-15.75}{4.2/\sqrt{45}}=1.26$	1.000 - 0.8962 = 0.1038	0.8962
16.00	$z=\frac{16.54-16.00}{4.2/\sqrt{45}}=0.86$	1.000 - 0.8051 = 0.1949	0.8051
16.25	$z=\frac{16.54-16.25}{4.2/\sqrt{45}}=0.46$	1.000 - 0.6772 = 0.3228	0.6772
16.50	$z=\frac{16.54-16.50}{4.2/\sqrt{45}}=0.06$	1.000 - 0.5239 = 0.4761	0.5239
16.75	$z=\frac{16.54-16.75}{4.2/\sqrt{45}}=-0.34$	1.000 - 0.3669 = 0.6331	0.3669
17.00	$z=\frac{16.54-17.00}{4.2/\sqrt{45}}=-0.73$	1.000 - 0.2327 = 0.7673	0.2327
17.25	$z=\frac{16.54-17.25}{4.2/\sqrt{45}}=-1.13$	1.000 - 0.1292 = 0.8708	0.1292
17.50	$z=\frac{16.54-17.50}{4.2/\sqrt{45}}=-1.53$	1.000 - 0.0630 = 0.9370	0.0630
17.75	$z=\frac{16.54-17.75}{4.2/\sqrt{45}}=-1.93$	1.000 - 0.0268 = 0.9732	0.0268

Steps 5-6:

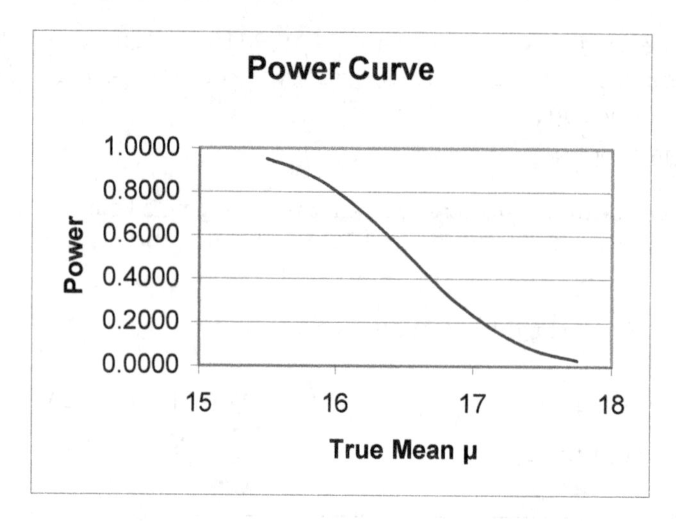

9.195 (a) P(Type I error) = α = 0.05

(b) Apply Procedure 9.4.

Step 1: The $\bar{x}$ critical values are
$\mu_0 \pm z_{\alpha/2} \cdot \sigma/\sqrt{n} = 16.7 \pm 1.96(6.0)/\sqrt{100} = 15.524$ *and* 17.876.

Step 2: We have μ_a = 14.0, $\sigma/\sqrt{n} = (6.0)/\sqrt{100} = 0.60$, and from Step 1, the $\bar{x}$ critical values of 15.524 and 17.876. Therefore, we need to find the area under the normal curve with parameters 14.0 and 0.60 that lie between 15.524 and 17.876. The z-score of the $\bar{x}$ critical values are $z = \dfrac{15.524 - 14.0}{0.60} = 2.54$ and $z = \dfrac{17.876 - 14.0}{0.60} = 6.46$. The area between z = 2.54 and z = 6.46 is 1.0000 - 0.9945 = 0.0055. Therefore, $\beta = 0.0055$.

(c) Apply Procedure 9.5

Step 1: The values of μ_a are given in the problem as 14.0, 14.5, 15.0, 15.5, 16.0, 16.5, 17.0, 17.5, 18.0, 18.5, 19.0

Steps 2-4:

Answers may differ slightly from those in the text due to intermediate rounding.

True mean μ	z-score computation	P(Type II error) β	Power $1 - \beta$
14.0	$z = \dfrac{15.524 - 14.0}{6.0/\sqrt{100}} = 2.54$ $z = \dfrac{17.876 - 14.0}{6.0/\sqrt{100}} = 6.46$	1.0000 - 0.9945 = 0.0055	0.9945

14.5	$z=\dfrac{15.524-14.5}{6.0/\sqrt{100}}=1.71$ $z=\dfrac{17.876-14.5}{6.0/\sqrt{100}}=5.63$	1.0000 - 0.9564 = 0.0436	0.9564
15.0	$z=\dfrac{15.524-15.0}{6.0/\sqrt{100}}=0.87$ $z=\dfrac{17.876-15.0}{6.0/\sqrt{100}}=4.79$	1.0000 - 0.8078 = 0.1922	0.8078
15.5	$z=\dfrac{15.524-15.5}{6.0/\sqrt{100}}=0.04$ $z=\dfrac{17.876-15.5}{6.0/\sqrt{100}}=3.96$	1.0000 - 0.5160 = 0.4540	0.5160
16.0	$z=\dfrac{15.524-16.0}{6.0/\sqrt{100}}=-0.79$ $z=\dfrac{17.876-16.0}{6.0/\sqrt{100}}=3.13$	0.9991 - 0.2148 = 0.7843	0.2157
16.5	$z=\dfrac{15.524-16.5}{6.0/\sqrt{100}}=-1.63$ $z=\dfrac{17.876-16.5}{6.0/\sqrt{100}}=2.29$	0.9890 - 0.0516 = 0.9374	0.0626
17.0	$z=\dfrac{15.524-17.0}{6.0/\sqrt{100}}=-2.46$ $z=\dfrac{17.876-17.0}{6.0/\sqrt{100}}=1.46$	0.9279 - 0.0069 = 0.9210	0.0790
17.5	$z=\dfrac{15.524-17.5}{6.0/\sqrt{100}}=-3.29$ $z=\dfrac{17.876-17.5}{6.0/\sqrt{100}}=0.63$	0.7357 - 0.0005 = 0.7352	0.2648
18.0	$z=\dfrac{15.524-18.0}{6.0/\sqrt{100}}=-4.13$ $z=\dfrac{17.876-18.0}{6.0/\sqrt{100}}=-0.21$	0.4168 - 0.0000 = 0.4168	0.5832
18.5	$z=\dfrac{15.524-18.5}{6.0/\sqrt{100}}=-4.96$ $z=\dfrac{17.876-18.5}{6.0/\sqrt{100}}=-1.04$	0.1492 - 0.0000 = 0.1492	0.8508

19.0	$z = \dfrac{15.524 - 19.0}{6.0/\sqrt{100}} = -5.79$ $z = \dfrac{17.876 - 19.0}{6.0/\sqrt{100}} = -1.87$	0.0307 - 0.0000 = 0.0307	0.9693

Steps 5-6:

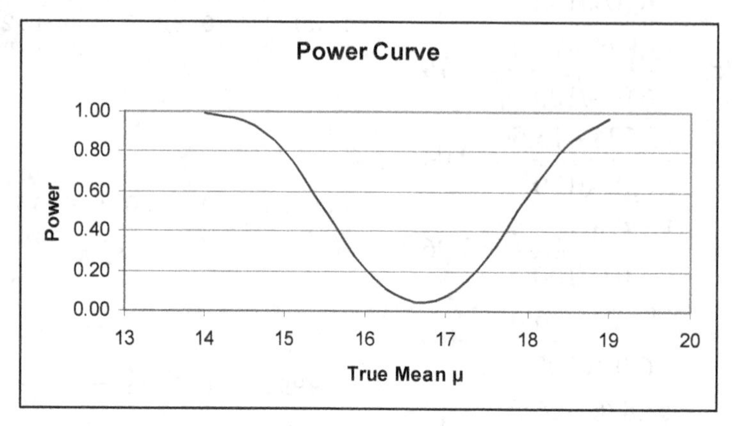

9.197 (a) P(Type I error) = α = 0.05

(b) Apply Procedure 9.4.

Step 1: The $\bar{x}$ critical value is $\mu_0 + z_\alpha \cdot \sigma/\sqrt{n} = 0.5 + 1.645(0.37)/\sqrt{20} = 0.636$.

Step 2: We have μ_a = 0.55, $\sigma/\sqrt{n} = (0.37)/\sqrt{20} = 0.083$, and from Step 1, the $\bar{x}$ critical value of 0.636. Therefore, we need to find the area under the normal curve with parameters 0.55 and 0.083 that lie to the left of 0.636. The z-score of the $\bar{x}$ critical value is $z = \dfrac{0.636 - 0.55}{0.083} = 1.04$. The area to the left of 1.04 is 0.8508. Therefore, $\beta = 0.8508$.

(c) Apply Procedure 9.5

Step 1: The values of μ_a are given in the problem as 0.55, 0.60, 0.65, 0.70, 0.75, 0.80, and 0.85.

Steps 2-4:

Answers may differ slightly from those in the text due to intermediate rounding.

True mean μ	z-score computation	P(Type II error) β	Power $1 - \beta$
0.55	$z = \dfrac{0.636 - 0.55}{0.37/\sqrt{20}} = 1.04$	0.8508	0.1492
0.60	$z = \dfrac{0.636 - 0.60}{0.37/\sqrt{20}} = 0.44$	0.6700	0.3300
0.65	$z = \dfrac{0.636 - 0.65}{0.37/\sqrt{20}} = -0.17$	0.4325	0.5675

0.70	$z=\dfrac{0.636-0.70}{0.37/\sqrt{20}}=-0.77$	0.2206	0.7794
0.75	$z=\dfrac{0.636-0.75}{0.37/\sqrt{20}}=-1.38$	0.0838	0.9162
0.80	$z=\dfrac{0.636-0.80}{0.37/\sqrt{20}}=-1.98$	0.0239	0.9761
0.85	$z=\dfrac{0.636-0.85}{0.37/\sqrt{20}}=-2.59$	0.0048	0.9952

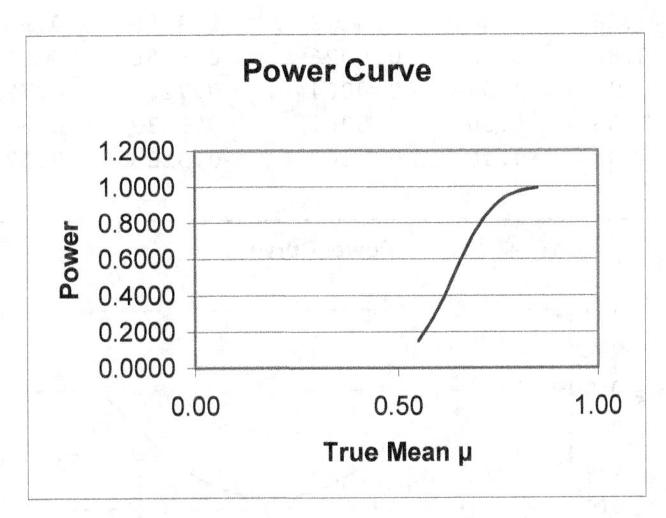

The power curve with n = 20 rises more quickly as the true mean μ increases, resulting in a higher power at any given value of μ than for n = 12. This illustrates the principle that a larger sample size has a higher power when the significance level remains the same. (a) P(Type I error) = α = 0.05

(b) Apply Procedure 9.4.

Step 1: The $\bar{x}$ critical values are

$\mu_0 \pm z_{\alpha/2}\cdot\sigma/\sqrt{n}=16.7\pm1.96(6.0)/\sqrt{40}=14.841$ *and* 18.559 .

Step 2: We have μ_a = 14.0, $\sigma/\sqrt{n}=(6.0)/\sqrt{40}=0.949$, and from Step 1, the $\bar{x}$ critical values of 14.841 and 18.559. Therefore, we need to find the area under the normal curve with parameters 14.0 and 0.949 that lie between 14.841 and 18.559. The z-score of the $\bar{x}$ critical values are $z=\dfrac{14.841-14.0}{0.949}=0.89$ and $z=\dfrac{18.559-14.0}{0.949}=4.80$. The area between z = 0.89 and z = 4.80 is 1.0000 - 0.8133 = 0.1867. Therefore, $\beta=0.1867$.

(c) Apply Procedure 9.5

Step 1: The values of μ_a are given in the problem as 14.0, 14.5, 15.0, 15.5, 16.0, 16.5, 17.0, 17.5, 18.0, 18.5, 19.0

Steps 2-4:

Answers may differ slightly from those in the text due to intermediate rounding.

The details of the z-score computation are the same as in Exercise 9.195, with 14.841 replacing 15.524, 18.559 replacing 17.876, and 40 replacing 100. The results of the computations are

μ	z-left	z-right	P(z<z-left)	P(z<z-right)	β	Power $1-\beta$
14.0	0.89	4.81	0.8133	1.0000	0.1867	0.8133
14.5	0.36	4.28	0.6406	1.0000	0.3594	0.6406
15.0	-0.17	3.75	0.4325	0.9999	0.5674	0.4326
15.5	-0.69	3.22	0.2451	0.9994	0.7543	0.2457
16.0	-1.22	2.70	0.1112	0.9965	0.8853	0.1147
16.5	-1.75	2.17	0.0401	0.9850	0.9449	0.0551
17.0	-2.28	1.64	0.0113	0.9495	0.9382	0.0618
17.5	-2.80	1.12	0.0026	0.8686	0.8661	0.1339
18.0	-3.33	0.59	0.0004	0.7224	0.7220	0.2780
18.5	-3.86	0.06	0.0001	0.5239	0.5239	0.4761
19.0	-4.38	-0.46	0.0000	0.3228	0.3228	0.6772

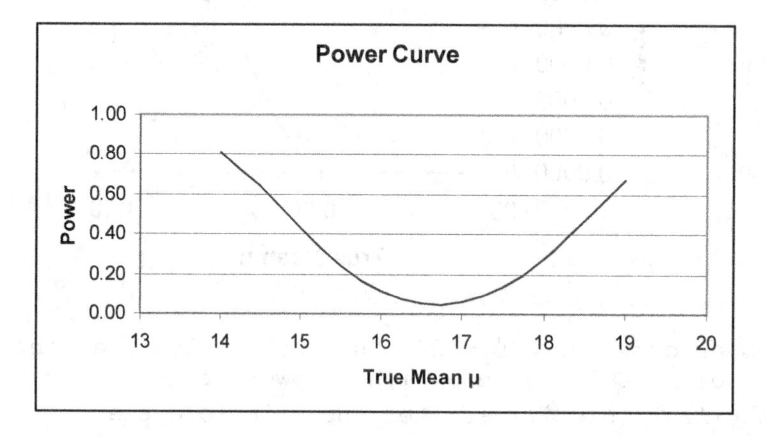

The power curve with n = 40 rises less quickly as the true mean μ increases or decreases from 16.7, resulting in a lower power at any given value of μ than for n = 100. This illustrates the principle that a larger sample size has a higher power when the significance level remains the same.

9.199 (a) P(Type I error) = α = 0.01

(b) Apply Procedure 9.4.

Step 1: The $\bar{x}$ critical value is $\mu_0 - z_\alpha \cdot \sigma/\sqrt{n} = 55 - 2.33(6.8)/\sqrt{15} = 50.91$.

Step 2: We have μ_a = 47, $\sigma/\sqrt{n} = (6.8)/\sqrt{15} = 1.756$, and from Step 1, the $\bar{x}$ critical value of 50.91. Therefore, we need to find the area under the normal curve with parameters 47 and 1.756 that lie to the right of 50.91. The z-score of the $\bar{x}$ critical value is $z = \dfrac{50.91-47}{1.756} = 2.23$. The area to the right of z = 2.23 is 1 - 0.9871 = 0.0129. Therefore, $\beta = 0.0129$. Answers may differ slightly from those in the text due to intermediate rounding.

(c) Apply Procedure 9.5

Step 1: The values of μ_a are given in the problem as 47, 48, 49, 50, 51, 52, 53, 54

Steps 2-4:

Answers may differ slightly from those in the text due to intermediate rounding.

The details of the z-score computation are the same as in Exercise 9.194, with 50.91 replacing 51.54 and 15 replacing 21. The results of the computations are

μ	z	P(Z<z)	P(Type II) β	Power $1-\beta$
47	2.23	0.9871	0.0129	0.9871
48	1.66	0.9515	0.0485	0.9515
49	1.09	0.8621	0.1378	0.8621
50	0.52	0.6985	0.3015	0.6985
51	-0.05	0.4801	0.5199	0.4801
52	-0.62	0.2676	0.7324	0.2676
53	-1.19	0.1170	0.8830	0.1170
54	-1.76	0.0392	0.9608	0.0392

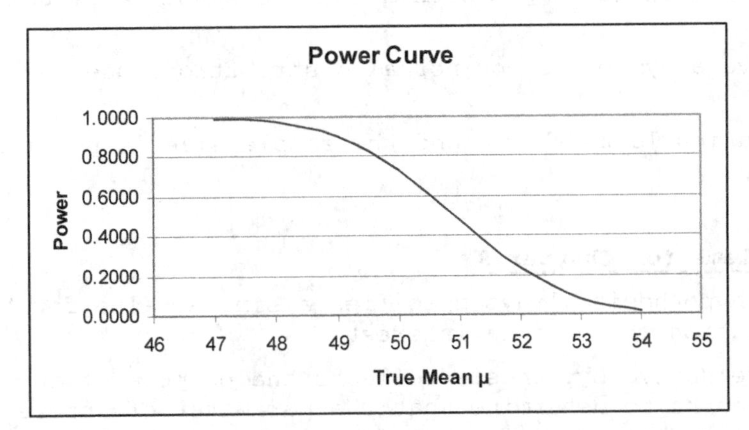

The power curve with n = 15 rises less quickly as the true mean μ or decreases from 55, resulting in a lower power at any given value of μ than for n = 21. This illustrates the principle that a larger sample size has a higher power when the significance level remains the same.

9.201 (a) Note: $z = \dfrac{\overline{x} - \mu_0}{\sigma / \sqrt{n}} \Rightarrow \overline{x} = \mu_0 + z \cdot \sigma / \sqrt{n}$

Since this is a left-tailed test, reject H_0 if $z \leq -1.645$; or equivalently if $\overline{x} \leq 26 - 1.645(1.4)/\sqrt{30} = 25.6$

So reject H_0 if $\overline{x} \leq 25.6$; otherwise do not reject H_0. If μ is really 25.4, then P(Type II error) = $P(\overline{x} \geq 25.6)$ =

$$P(z \geq \frac{25.6 - 25.4}{1.4/\sqrt{30}}) = P(z \geq 0.78) = 1.0000 - 0.7823 = 0.2177.$$

(b)-(c) Answers will vary. You can generate the samples using the Excel spreadsheet.

(d) Since the probability of a Type II error when μ = 25.4 (i.e., of failing to reject the null hypothesis when μ = 25.4) is 0.2177 from part (a), we would expect non-rejection of the null hypothesis about 21 or 22 times in the 100 samples.

(e)-(f) Answers will vary. Our simulation contained 20 samples in which the null hypothesis was not rejected when μ =25.4. This is very close to what we expected.

Exercises 9.8

9.203 (a) Yes. The t-test can be used when the sample size is large. It is almost equivalent to the z-test in this situation.

(b) Yes. The Wilcoxon signed-rank test can be used when the population distribution is symmetric.

(c) The Wilcoxon signed-rank test is preferable in this situation since it is more powerful (more likely to detect a false null hypothesis) when the population is symmetric, but nonnormal.

9.205 Since we have normality and σ is known, use the z-test.

9.207 Since we have a large sample with no outliers and σ is unknown, use the t-test.

9.209 Since we have a symmetric non-normal distribution, use the Wilcoxon signed-rank test.

9.211 The distribution looks skewed and the sample size is not large. Consult a statistician.

Review Problems for Chapter 9

1. (a) A null hypothesis always specifies a single value for the parameter of a population which is of interest.

(b) The alternative hypothesis reflects the purpose of the hypothesis test, which can be to determine that the parameter of interest is greater than, less than, or different from the single value specified in the null hypothesis.

(c) The test statistic is a quantity calculated from the sample, under the assumption that the null hypothesis is true, which is used as a basis for deciding whether or not to reject the null hypothesis.

(d) The significance level is the probability of making a Type I error. That is, it is the probability of rejecting a true null hypothesis.

2. (a) The statement is expressing the fact that there is variability in the net weights of the boxes' content and some boxes may actually contain less than the printed weight on the box. However, the net weights for each day's production will average a bit more than the printed weight.

(b) To test the truth of this statement, we would use a null hypothesis that stated that the population mean net weight of the boxes was equal to the printed weight and an alternative hypothesis that stated that the population mean net weight of the boxes was greater than the printed weight.

(c) Null hypothesis: Population mean net weight = 76 oz

Alternative hypothesis: Population mean net weight > 76 oz

or

H_0: $\mu = 76$ H_a: $\mu > 76$

3. (a) Roughly speaking, there is a range of values of the test statistic which one could reasonably expect to occur if the null hypothesis were true. If the value of the test statistic is one that would not be expected to occur when the null hypothesis is true, then we reject the null hypothesis.

   (b) To make this procedure objective and precise, we specify the probability with which we are willing to reject the null hypothesis when it is actually true. This is called the significance level of the test and is usually some small number like 0.05 or 0.01. Specifying the significance level allows us to determine the range of values of the test statistic that will lead to rejection of the null hypothesis. If the computed value of the test statistic falls in this "rejection region," then the null hypothesis is rejected. If it does not fall in the rejection region, then the null hypothesis is not rejected.

4. We would use the alternative hypothesis $\mu \neq \mu_0$ if we wanted to determine whether the population mean were <u>different from</u> the value μ_0 specified in the null hypothesis. We would use the alternative hypothesis $\mu > \mu_0$ if we wanted to determine whether the population mean were <u>greater than</u> from the value μ_0 specified in the null hypothesis. We would use the alternative hypothesis $\mu < \mu_0$ if we wanted to determine whether the population mean were <u>less than</u> the value μ_0 specified in the null hypothesis.

5. (a) A Type I error is made whenever the null hypothesis is true, but the value of the test statistic leads us to reject the null hypothesis. A Type II error is made whenever the null hypothesis is false, but the value of the test statistic leads us to not reject the null hypothesis.

   (b) The probability of a Type I error is represented by α and that of a Type II error by β.

   (c) If the null hypothesis is true, the test statistic can lead us to either reject or not reject the null hypothesis. The first is the correct decision, while the latter constitutes a Type I error. Thus a Type I error is the only type of error possible when the null hypothesis is true.

   (d) If the null hypothesis is not rejected, a correct decision has been made if the null hypothesis is, in fact, true. But if the null hypothesis is false, we have made a Type II error. Thus a Type II error is the only type of error possible when the null hypothesis is not rejected.

6. The probability of a Type II error is increased when the significance level is decreased for a fixed sample size.

7. (a) The rejection region is a set of values of the test statistic that lead to rejection of the null hypothesis.

   (b) The nonrejection region is a set of values of the test statistic that lead to not rejecting the null hypothesis.

   (c) The critical values are values of the test statistic that separate the rejection region from the nonrejection region.

8. True. You would reject the null hypothesis if your test statistic is within the rejection region or equal to the critical value.

9. Assuming that the null hypothesis is true, find the value of the test statistic for which the probability of obtaining a value less than this value (the critical value) is 0.05.

10. (a) For a right tailed one-mean z-test at 1% significance, the critical value is $z_\alpha = z_{0.01} = 2.33$.

(b) For a left tailed one-mean z-test at 1% significance, the critical value is $-z_\alpha = -z_{0.01} = -2.33$.

(c) For a two tailed one-mean z-test at 1% significance, the critical value is $\pm z_{\alpha/2} = \pm z_{0.005} = \pm 2.575$.

11. (a) Rejection region: $z \ge 1.28$

(b) Nonrejection region: $z < 1.28$

(c) Critical value: $z = 1.28$

(d) Significance level: $\alpha = 0.10$

(e)

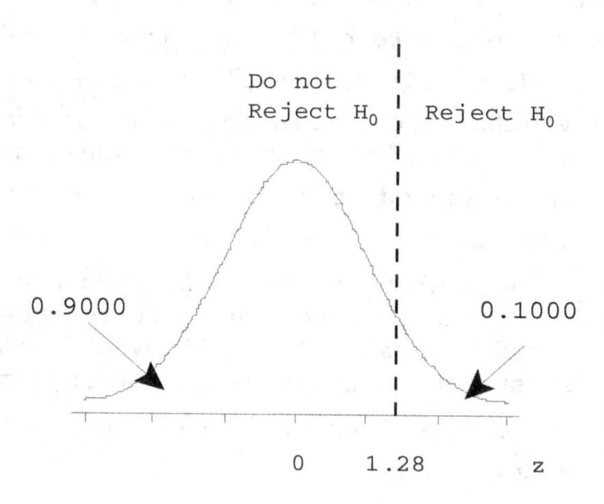

(f) Right-tailed test

12. Step 1: State the null and alternative hypothesis.

Step 2: Choose α.

Step 3: Calculate the test statistic.

Step 4: Calculate the critical value(s).

Step 5: If the test statistic falls within the rejection region, reject the null hypothesis. Otherwise, do not reject the null hypothesis.

Step 6: Interpret your conclusion in context of the problem.

13. The P-value of a hypothesis test is the probability, assuming that the null hypothesis is true, of getting a value of the test statistic that is as extreme or more extreme than the one actually obtained.

14. True. If the null hypothesis were true, a value of the test statistic with a P-value of 0.02 would be more extreme than one with a P-value of 0.03.

15. If the P-value is less than or equal to the significance level, reject the null hypothesis. Otherwise, do not reject the null hypothesis.

16. The P-value of a hypothesis test is also called the observed significance level since it represents the smallest possible significance level at which the null hypothesis could have been rejected.

17. The P-value is the probability, assuming that the null hypothesis is true, of seeing a test statistic at least as extreme than that observed.

18. (a) The P-value for a left-tailed test with a test statistic of z = -1.25 is $P(Z<-1.25)=0.1056$. This P-value of 0.1056 is greater than the significance level of 0.05. We would not reject the null hypothesis.

(b) The P-value for a right-tailed test with a test statistic of z = 2.36 is $P(Z>2.36)=0.0091$. This P-value of 0.0091 is less than the significance level of 0.05. We would reject the null hypothesis.

(c) The P-value for a two-tailed test with a test statistic of z = 1.83 is $2\cdot P(Z>1.83)=2\cdot(0.0336)=0.0672$. This P-value of 0.0672 is greater than the significance level of 0.05. We would not reject the null hypothesis.

19. Step 1: State the null and alternative hypothesis.

Step 2: Choose α.

Step 3: Calculate the test statistic.

Step 4: Calculate the P-value.

Step 5: If the P-value is less than or equal to the significance level, reject the null hypothesis. Otherwise, do not reject the null hypothesis.

Step 6: Interpret your conclusion in context of the problem.

20. The P-value of 0.062 is within the range of 0.05 < P-value < 0.10. Therefore, there is moderate evidence against the null hypothesis.

21. (a) A hypothesis test is exact if the actual significance level is the same as the one that is stated.

(b) A hypothesis test is approximately correct if the actual significance level only approximately equals α.

22. A statistically significant result occurs when the value of the test statistic falls in the rejection region. A result has practical significance when it is statistically significant and the result also is different enough from results expected under the null hypothesis to be important to the consumer of the results. By taking large enough sample sizes, almost any result can be made statistically significant due to the increased ability of the test to detect a false null hypothesis, but small differences from the conditions expressed by the null hypothesis may not be important, that is, they may not have practical significance.

23. (a) If the population standard deviation is unknown, and the population is normal or the sample size is large, we can use the one-mean t-statistic, $t=(\bar{x}-\mu_0)/(s/\sqrt{n})$.

(b) If the population standard deviation is known, and the population is normal or the sample size is large, we can use the one-mean z-statistic, $z=(\bar{x}-\mu_0)/(\sigma/\sqrt{n})$.

(c) If the population is symmetric, we can use the Wilcoxon signed-rank statistic W = the sum of the positive ranks.

24. Non-parametric methods have the advantages of involving fewer and simpler calculations than parametric methods and are more resistant to outliers and other extreme values. Parametric methods are preferred when the population is normal or the sample size is large since they are more powerful than non-parametric methods and thus tend to give more accurate results than non-parametric methods under those conditions.

25. (a) The power of a hypothesis test is the probability of rejecting the null hypothesis when the null hypothesis is false.

(b) The power of a test increases when the sample size is increased while keeping the significance level constant.

26. Let μ denote last year's mean cheese consumption by Americans.

(a) H_0: μ = 33.0 lb

(b) H_a: μ > 33.0 lb

(c) This is a right-tailed test.

27. (a) A Type I error would occur if, in fact, μ = 33.0 lb, but the results of the sampling lead to the conclusion that μ > 33.0 lb.

(b) A Type II error would occur if, in fact, μ > 33.0 lb, but the results of the sampling fail to lead to that conclusion.

(c) A correct decision would occur if, in fact, μ = 33.0 lb and the results of the sampling do not lead to the rejection of that fact; or if, in fact, μ > 33.0 lb and the results of the sampling lead to that conclusion.

(d) If, in fact, last year's mean consumption of cheese for all Americans has not increased over the 2010 mean of 33.0 lb, and we reject the null hypothesis that μ = 33.0 lb, we made a Type I error.

(e) If, in fact, last year's mean consumption of cheese for all Americans has increased over the 2010 mean of 33.0 lb, and we reject the null hypothesis that μ = 33.0 lb, we made a correct decision.

28. (a) n = 35, $\bar{x}$ = 1218/35 = 34.8, σ = 6.9

Step 1: H_0: μ = 33.0 lb, H_a: μ > 33.0 lb

Step 2: α = 0.10

Step 3: $z = (34.8 - 33.0)/(6.9/\sqrt{35}) = 1.54$

Step 4: Critical Value Approach: Critical value = 1.28

P-value Approach: P-value is $P(Z > 1.54) = 0.0618$

Step 5: Since 1.54 > 1.28, reject H_0. Since the P-value of 0.0618 is less than the significance level, reject H_0.

Step 6: At the 10% significance level, the data provide sufficient evidence to conclude that last year's mean cheese consumption μ for all Americans has increased over the 2010 mean of 33.0 lb.

(b) Given the conclusion in part (a), if an error has been made, it must be a Type I error. This is because, given that the null hypothesis was rejected, the only error that could be made is the error of rejecting a true null hypothesis.

29. n = 12, $\bar{x}$ = \$455.0, s = \$86.8

Step 1: H_0: μ = \$468, H_a: μ < \$468

Step 2: α = 0.05

Step 3: $t = (455.0 - 468)/(86.8/\sqrt{12}) = -0.52$

Step 4: Critical Value Approach: Critical value = -1.796

P-Value Approach: $P > 0.10$

Step 5: Since $-0.52 > -1.796$, do not reject H_0. Since the P-value is greater than the significance level, do not reject H_0.

Step 6: At the 5% significance level, the data do not provide sufficient evidence to conclude that the mean value lost because of purse snatching has decreased from the 2012 mean of $468.

30. (a) n=12

H_0: $\eta = 468$; H_a: $\eta < 468$

$\alpha = 0.05$

Critical value = 12(13)/2 - 61 = 17

x	$x-\eta_0=D$	$\lvert D \rvert$	Rank of $\lvert D \rvert$	Signed Rank R
415	-53	53	4	-4
572	104	104	11	11
539	71	71	6	6
487	19	19	3	3
365	-103	103	10	-10
550	82	82	7	7
479	11	11	1	1
481	13	13	2	2
375	-93	93	8	-8
371	-97	97	9	-9
303	-165	165	12	-12
523	55	55	5	5

W = sum of the positive signed ranks = 35

Since 35 is greater than the critical value of 17, we do not reject the null hypothesis and conclude that there is no evidence that last year's mean value lost to purse snatching has decreased from the 2012 mean.

(b) In performing the Wilcoxon signed-rank test, we are assuming that the distribution of last year's values lost to purse snatching is symmetric.

(c) If the distribution of values lost is, in fact, a normal distribution, it is permissible to use the Wilcoxon test since a normal distribution is also symmetric.

31. If the values lost last year do have a normal distribution, the t-test is the preferred procedure for performing the hypothesis test since it is the more powerful test when the distribution is normal, that is, it has a greater chance of rejecting a false null hypothesis.

32. (a) If the odds-makers are estimating correctly, the mean point-spread error is zero.

(b) It seems reasonable to assume that the distribution of point spread errors is approximately normal. In any case, the sample size of 2109 is very large, so the t-test of H_0: $\mu = 0$ vs. H_a: $\mu \neq 0$ is

appropriate. At the 5% significance level, the critical values are ±1.960. Since $t = (-0.2 - 0.0)/(10.9/\sqrt{2109}) = 0.843$, we do not reject H_0.

(c) There is not sufficient evidence to conclude that the mean point-spread is different from zero.

33. (a) P(Type I error) = significance level = α = 0.10

(b) The distribution of $\bar{x}$ will be approximately normal with a mean of 36.5 and a standard deviation of $6.9/\sqrt{35} = 1.166$.

(c) Apply Procedure 9.4.

Step 1: The $\bar{x}$ critical value is $33.0 + 1.28(6.9)/\sqrt{35} = 34.493$.

Step 2: We have μ_a = 36.5, $6.9/\sqrt{35} = 1.166$, and from Step 1, the $\bar{x}$ critical value of 34.493. Therefore, we need to find the area under the normal curve with parameters 36.5 and 1.166 that lie to the left of 34.493. The z-score of the $\bar{x}$ critical value is $z = \dfrac{34.493 - 36.5}{1.166} = -1.72$.

The area to the left of z = -1.72 is 0.0427. Therefore, $\beta = 0.0427$.

(d-e) Assuming that the true mean μ is one of the values listed, the distribution of $\bar{x}$ will be approximately normal with that mean and with a standard deviation of 1.166. The computations of β and the power $1 - \beta$ are shown in the table below.

True mean μ	z-score computation	P(Type II error) β	Power $1 - \beta$
33.5	$\dfrac{34.493 - 33.50}{6.9/\sqrt{35}} = 0.85$	0.8023	0.1977
34.0	$\dfrac{34.493 - 34.00}{6.9/\sqrt{35}} = 0.42$	0.6628	0.3372
34.5	$\dfrac{34.493 - 34.50}{6.9/\sqrt{35}} = -0.01$	0.4960	0.5040
35.0	$\dfrac{34.493 - 35.00}{6.9/\sqrt{35}} = -0.44$	0.3300	0.6700
35.5	$\dfrac{34.493 - 35.50}{6.9/\sqrt{35}} = -0.87$	0.1922	0.8078
36.0	$\dfrac{34.493 - 36.00}{6.9/\sqrt{35}} = -1.29$	0.0985	0.9015
37.0	$\dfrac{34.493 - 37.00}{6.9/\sqrt{35}} = -2.15$	0.0158	0.9842

(f)

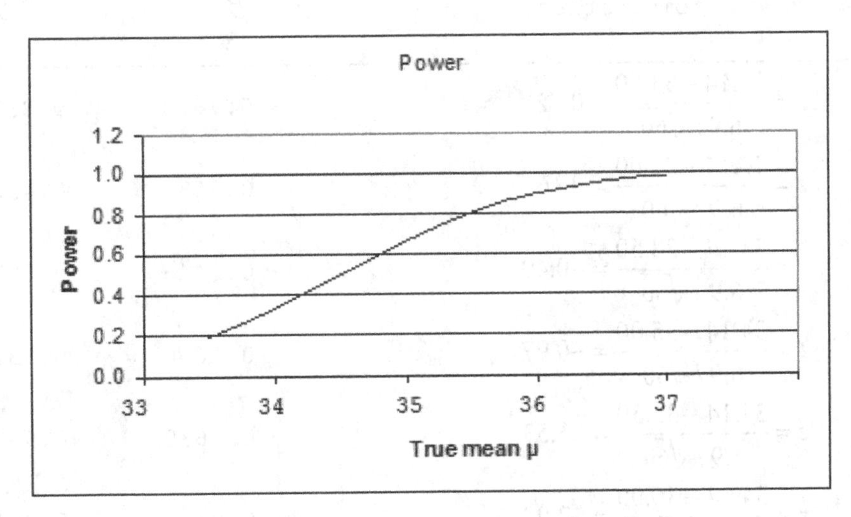

(g) The distribution of $\bar{x}$ will be approximately normal with a mean of 36.5 and a standard deviation of $6.9\ /\ \sqrt{60}\ =\ 0.891$.

(h) Apply Procedure 9.4.

Step 1: The $\bar{x}$ critical value is $33.0 + 1.28(6.9)/\sqrt{60} = 34.140$.

Step 2: We have μ_a = 36.5, $6.9/\sqrt{60} = 0.891$, and from Step 1, the $\bar{x}$ critical value of 34.140. Therefore, we need to find the area under the normal curve with parameters 36.5 and 0.891 that lie to the left of 34.140. The z-score of the $\bar{x}$ critical value is $z = \dfrac{34.140 - 36.5}{0.891} = -2.65$.

The area to the left of z = -2.65 is 0.0040. Therefore, $\beta = 0.0040$.

(i-j) Assuming that the true mean μ is one of the values listed, the distribution of $\bar{x}$ will be approximately normal with that mean and with a standard deviation of 0.891. The computations of β and the power 1 - β are shown in the following table.

True mean μ	z-score computation	P(Type II error) β	Power $1-\beta$
33.5	$z=\frac{34.14-33.50}{6.9/\sqrt{60}}=0.72$	0.7642	0.2358
34.0	$z=\frac{34.14-34.00}{6.9/\sqrt{60}}=0.16$	0.5636	0.4364
34.5	$z=\frac{34.14-34.50}{6.9/\sqrt{60}}=-0.40$	0.3446	0.6554
35.0	$z=\frac{34.14-35.00}{6.9/\sqrt{60}}=-0.97$	0.1660	0.8340
35.5	$z=\frac{34.14-35.50}{6.9/\sqrt{60}}=-1.53$	0.0630	0.9370
36.0	$z=\frac{34.14-36.00}{6.9/\sqrt{60}}=-2.09$	0.0183	0.9817
37.0	$z=\frac{34.14-37.00}{6.9/\sqrt{60}}=-3.21$	0.0007	0.9993

(k)

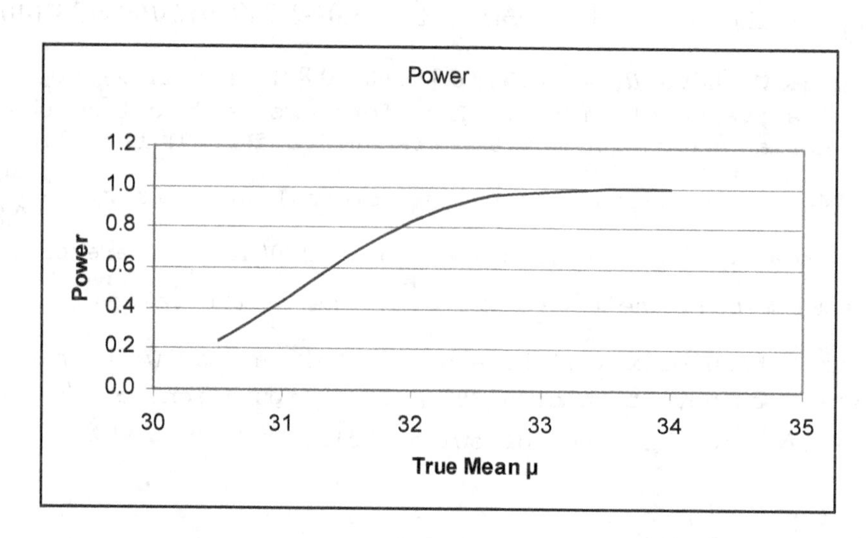

(l) The principle being illustrated is that increasing the sample size for a hypothesis test without changing the significance level α increases the power.

34. Since the distribution is symmetric, use the Wilcoxon signed-rank test. The sample size is 50 and σ is known, so the z-test may be appropriate. Use caution, however, since it appears that there may be outliers.

35. The distribution is far from normal, being left-skewed. However, since the sample size is 37 and σ is unknown, it is probably reasonable to use the t-test.

36. (a) The Wilcoxon signed-rank test is appropriate in Problem 34 since the distribution is symmetric. It is not appropriate in Problem 35 since the distribution is highly skewed to the left.

(b) The sample size is large (n=50) and the distribution is symmetric in Problem 34, so either the z-test or the Wilcoxon signed-rank test could be used. Since the distribution appears to be more peaked with longer tails than a normal distribution would have, the Wilcoxon test is preferable.

37. Using Minitab, we choose **Stat** ▶ **Basic Statistics** ▶ **1-Sample t...**, click in the **Samples in columns** text box, enter COST in the **Samples in columns** text box, click in the **Test mean** text box and enter 239 in the **Test mean** text box, click on the **Options** button, type 95.0 in the **Confidence level** text box, click on the arrow to the right of the **Alternative** box and select **greater than**, and click **OK**. Then choose **Stat** ▶ **Nonparametrics** ▶ **1-Sample Wilcoxon...**, enter COST in the **Variables** text box, click on the **Test median** button and type 239 in its text box, select **greater than** in the **Alternative** text box, and click **OK**. Click on the **Graphs** button and check the boxes for **Histogram of data** and **Boxplot of data**. Click **OK** twice. Then choose **Stat** ▶ **Basic Statistics** ▶ **Normality test**, enter COST in the **Variable** text box and click **OK**. Finally, choose **Graph** ▶ **Stem-and-Leaf** and enter COST in the **Graph Variables** text box and click **OK**. The results are

(a) Test of μ = 239 vs > 239

Variable	N	Mean	StDev	SE Mean	95% Lower Bound	T	P
COST	11	267.6	59.1	17.8	235.4	1.61	0.069

The p-value is greater than the significance level of 0.05, so we do not reject the null hypothesis. There is not sufficient evidence to claim that the average cost of a private room in a nursing home this year exceeded $239 per day.

(b) Test of median = 239.0 versus median > 239.0

	N	N for Test	Wilcoxon Statistic	P	Estimated Median
COST	11	11	53.5	0.038	270.8

The p-value is less than the significance level of 0.05, so we reject the null hypothesis. There is sufficient evidence to claim that the average cost of a private room in a nursing home this year exceeded $239 per day.

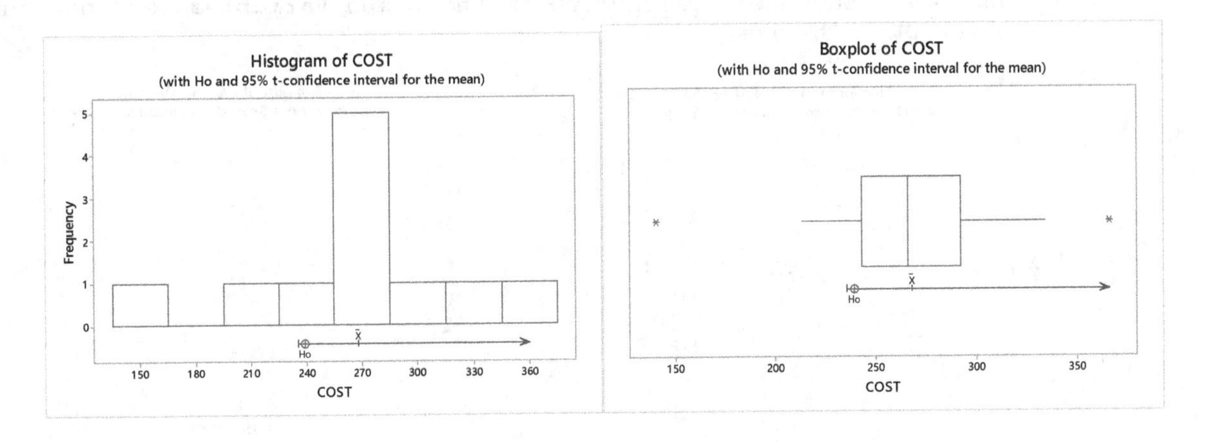

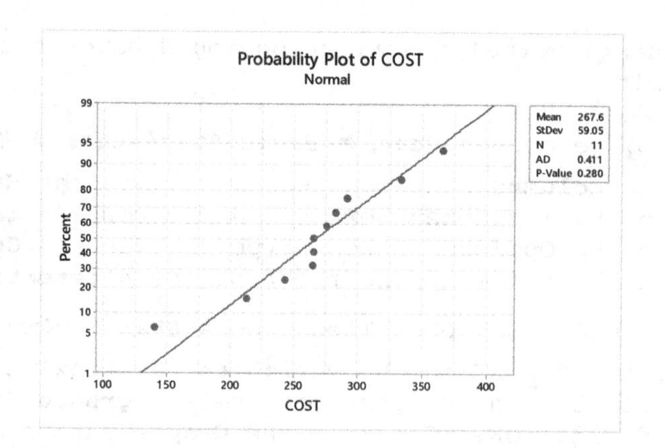


```
Stem-and-leaf of COST  N  = 11
Leaf Unit = 10

 1    1  4
 1    1
 3    2  14
(6)   2  666789
 2    3  3
 1    3  6
```

(d) Although the distribution may be symmetric, it is not bell-shaped and it contains two potential outliers (see boxplot). This explains the discrepancy between the two tests. Since the t-test should not be used when the sample size is small (11) and there are outliers, the Wilcoxon test is the more appropriate one to use.

38. (a,b) Using Minitab, choose **Stat ▶ Basic Statistics ▶ 1-Sample t,** enter CONSUMPTION in the **Samples in columns** text box and 57.5 in the **Test mean** text box. Click on the **Graphs** button and check the boxes for **Histogram of data** and **Boxplot of data**. Click **OK**. Click on the **Options** button, enter 95 in the **Confidence level** text box and select **Less than** from the **Alternative** drop down box. Click **OK** twice. Then choose **Stat ▶ Basic Statistics ▶ Normality test** and enter CONSUMPTION in the **Variable** text box and click **OK**. Finally, choose **Graph ▶ Stem-and-Leaf** and enter CONSUMPTION in the **Graph Variables** text box and click **OK.** The results are

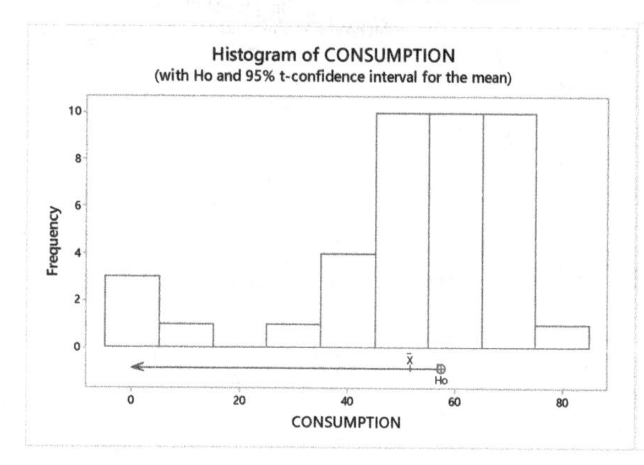

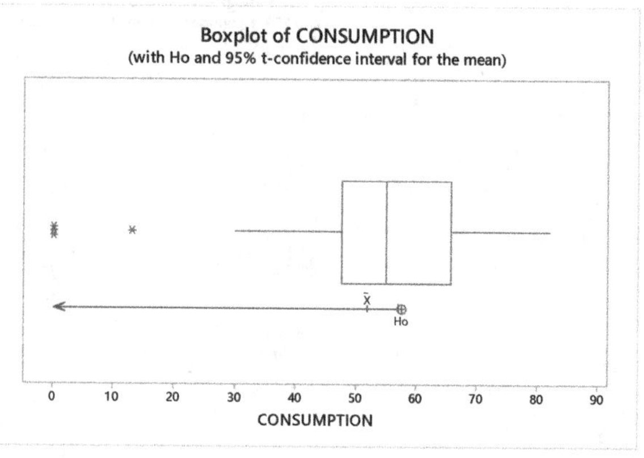

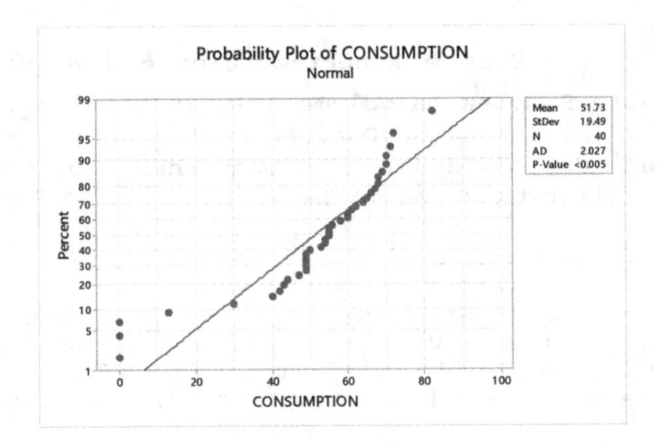


```
Stem-and-leaf of CONSUMPTION  N  = 40
Leaf Unit = 1.0

  3   0  000
  4   1  3
  4   2
  5   3  0
 15   4  0234799999
 (9)  5  034455568
 16   6  00124567889
  5   7  0012
  1   8  2
```

```
Test of μ = 57.5 vs < 57.5

Variable      N   Mean  StDev  SE Mean  95% Upper Bound      T      P
CONSUMPTION  40  51.73  19.49     3.08            56.92  -1.87  0.034
```

The P-value of the test is 0.034, which is less than the 0.05 significance level. We reject H_0. The data provide sufficient evidence to conclude that the mean beef consumption last year is less than the 2011 mean of 57.5 lbs.

(c) After removing the four outliers, the test results are

```
Test of μ = 57.5 vs < 57.5

Variable      N   Mean  StDev  SE Mean  95% Upper Bound      T      P
CONSUMPTION  36  57.11  11.02     1.84            60.21  -0.21  0.417
```

Now the P-value of the test is 0.417, which is greater than the 0.05 significance level. We do not reject H_0. The data do not provide sufficient evidence to conclude that the mean beef consumption last year is less than the 2011 mean of 57.5 lbs.

(d) The outliers had a very large effect on the test results. Although the sample size was 40, four outliers are too many for the t-test to be appropriate. If the outliers are not recording errors, they represent legitimate observations from the population and should not be deleted. If they were deleted, the test results would not yield valid conclusions about the population. With so many outliers, it is just possible that the population itself is quite left skewed as is the sample.

(e) This is more appropriately done with a non-parametric test that is insensitive to outliers.

39. (a) Using Minitab, choose **Stat ▶ Nonparametrics ▶ 1-Wilcoxon,** enter CONSUMPTION in the **Samples in columns** text box and 57.5 in the **Test mean** text box. Click on the **Options** button, enter 95 in the **Confidence level** text box and select **Less than** from the **Alternative** drop down box. Click **OK** twice. The results are

```
Test of median = 57.50 versus median < 57.50

                    N for   Wilcoxon              Estimated
                N   Test    Statistic       P       Median
 CONSUMPTION   40    40       324.5       0.127     55.00
```

After removing the outliers and repeating part (a), the results are

```
Test of median = 57.50 versus median < 57.50

                    N for   Wilcoxon              Estimated
                N   Test    Statistic       P       Median
CONSUMPTION    36    36       324.5       0.450     57.50
```

(b) The results are similar to those in Problem 38 in that the P-value increases substantially. The results are different in that the P-value increases from 0.034 to 0.417 for the t-test, and only from 0.127 to 0.450 for the Wilcoxon test when the four outliers are deleted. In both cases, we should expect the P-value to increase since the four smallest observations were deleted and the alternative hypothesis was $\mu < 57.5$. At the 5% significance level, the conclusion does not change in either case for the Wilcoxon test.

(c) While the Wilcoxon test is less sensitive to outliers, it is based on an assumption that the data are symmetrically distributed. That assumption does not appear to be reasonable for these data, so caution is advised in using the Wilcoxon test. The sign test would be a better choice.

40. (a) Using Minitab, choose **Stat ▶ Basic Statistics ▶ 1-Sample z,** enter BMI in the **Samples in columns** text box and 25 in the **Test mean** text box. Click on the **Graphs** button and check the boxes for **Histogram of data** and **Boxplot of data**. Click **OK**. Click on the **Options** button, enter 95 in the **Confidence level** text box and select **Greater than** from the **Alternative** drop down box. Click **OK** twice. Then choose **Stat ▶ Basic Statistics ▶ Normality test** and enter BMI in the **Variable** text box and click **OK**. Finally, choose **Graph ▶ Stem-and-Leaf** and enter BMI in the **Graph Variables** text box and click **OK.** The graphic results are

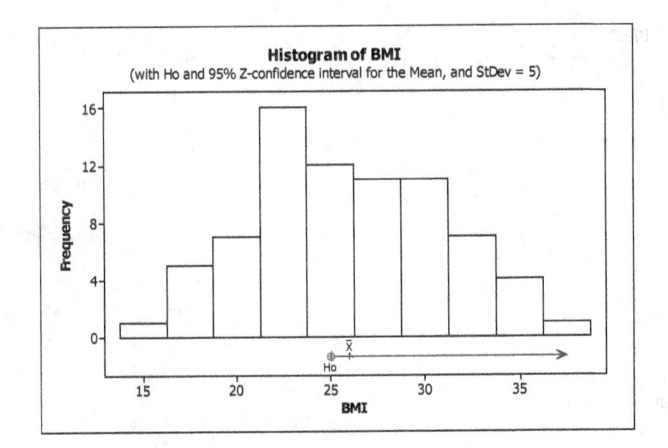

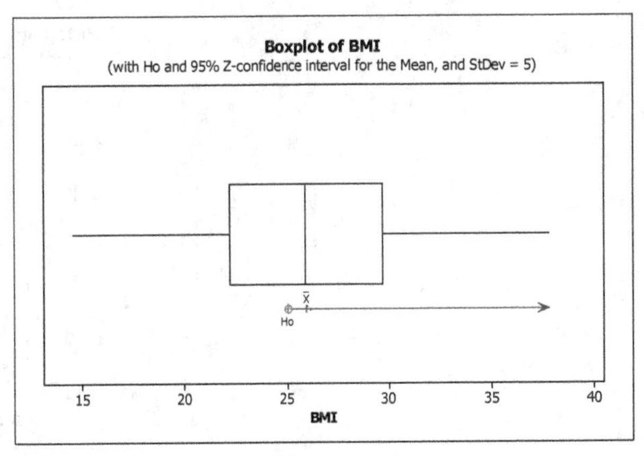


```
Stem-and-leaf of BMI  N  = 75
Leaf Unit = 1.0

  1   1  4
  4   1  677
  6   1  88
 16   2  0000111111
 29   2  2222222233333
 (9)  2  444444455
 37   2  666666777
 28   2  88888899999
 17   3  0000111
 10   3  22233
  5   3  4555
  1   3  7
```

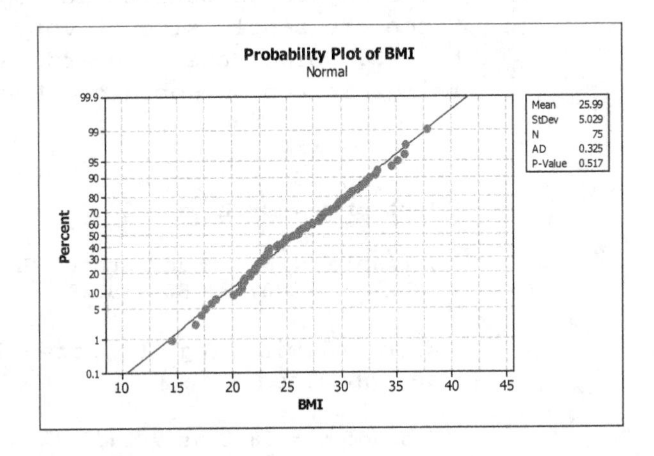

(b) Yes. There are no outliers, the sample size is large, and the data is reasonably normally distributed.

(c) The process in part (a) yielded the following test results:

```
Test of mu = 25 vs > 25
The assumed standard deviation = 5

                                                95%
                                               Lower
Variable   N     Mean    StDev  SE Mean      Bound     Z      P
BMI       75  25.9867   5.0293   0.5774    25.0370  1.71  0.044
```

Since the P-value of 0.044 is less than the significance level 0.05, we reject the null hypothesis. The data do provide sufficient evidence to conclude that the mean BMI of U.S. adults is greater than that for a healthy weight.

41. (a) Using Minitab, we choose **Graph ▶ Histogram,** choose the **Simple** version, enter BEER in the **Graph variables** text box, and click **OK**. The result is

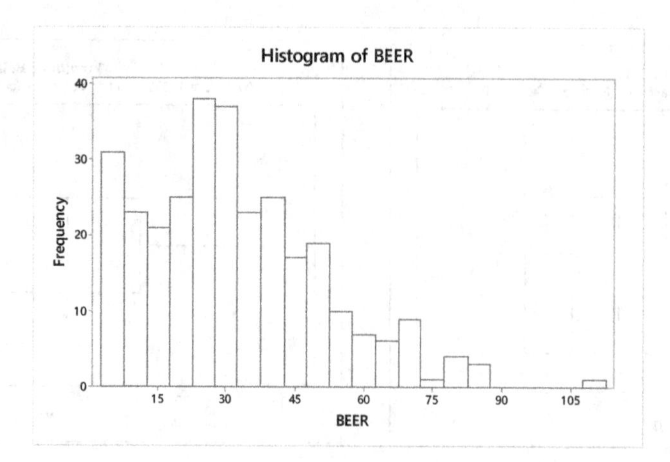

(b) There may be an outlier at about 110.

(c) Since the sample size is very large (300), we can use the z-test or the t-test. Since the standard deviation is unknown, we use the t-test. Using Minitab, we obtained the following result using all of the data.

```
Test of μ = 28.2 vs ≠ 28.2

Variable    N   Mean  StDev  SE Mean        99% CI            T      P
BEER      300  31.52  19.43     1.12  (28.61, 34.43)  2.96  0.003
```

After eliminating the potential outlier (112), we repeated the process and obtained

```
Test of μ = 28.2 vs ≠ 28.2

Variable    N   Mean  StDev  SE Mean        99% CI            T      P
BEER      299  31.26  18.92     1.09  (28.42, 34.09)  2.79  0.006
```

Clearly, elimination of the value 110 had little effect on the value of t or on the P-value. In either case, we reject the null hypothesis. The data do provide sufficient evidence to claim that the mean annual consumption of beer in Missouri is different from the national mean.

CHAPTER 10 SOLUTIONS

Exercises 10.1

10.1 Answers will vary.

10.3 (a) μ_1, σ_1, μ_2, and σ_2 are parameters; $\bar{x}_1$, s_1, $\bar{x}_2$, and s_2 are statistics.

(b) μ_1, σ_1, μ_2, and σ_2 are fixed numbers; $\bar{x}_1$, s_1, $\bar{x}_2$, and s_2 are random variables.

10.5 (a) $H_0: \mu_1 = \mu_2$ $H_a: \mu_1 \neq \mu_2$ (b) two tailed

10.7 (a) $H_0: \mu_1 = \mu_2$ $H_a: \mu_1 > \mu_2$ (b) right tailed

10.9 (a) $H_0: \mu_1 = \mu_2$ $H_a: \mu_1 < \mu_2$ (b) left tailed

10.11 It is the sampling distribution of the difference of the two sample means that allows us to determine whether the difference of the sample means can reasonably be attributed to sampling error or whether it is large enough for us to conclude that the population means are different.

10.13 We can be 95% confident that the mean for population one is between 15 to 20 units greater than the mean for population two.

10.15 We can be 90% confident that the mean for population one is between 5 to 10 units less than the mean for population two.

10.17 We can be 99% confident that the mean for population one is between 20 units less to 15 units greater than the mean for population two.

10.19 (a) The mean of $\bar{x}_1 - \bar{x}_2$ is $\mu_1 - \mu_2 = 40 - 40 = 0$

The standard deviation of $\bar{x}_1 - \bar{x}_2$ is $\sigma_{\bar{x}_1-\bar{x}_2} = \sqrt{\frac{\sigma_1^2}{n_1}+\frac{\sigma_2^2}{n_2}} = \sqrt{\frac{12^2}{9}+\frac{6^2}{4}} = 5$

(b) No. The determination of the mean of $\bar{x}_1 - \bar{x}_2$ is the same for all populations. The formula for the standard deviation of $\bar{x}_1 - \bar{x}_2$ is dependent on the samples being independent, but not on the populations from which they came.

(c) No. It is not known that the two populations are normally distributed (which would lead to $\bar{x}_1 - \bar{x}_2$ being normally distributed), and the sample sizes of 9 and 4 are too small to claim that $\bar{x}_1$ and $\bar{x}_2$ are normally distributed (which would also lead to $\bar{x}_1 - \bar{x}_2$ being normally distributed).

10.21 (a) The mean of $\bar{x}_1 - \bar{x}_2$ is $\mu_1 - \mu_2 = 40 - 40 = 0$

The standard deviation of $\bar{x}_1 - \bar{x}_2$ is $\sigma_{\bar{x}_1-\bar{x}_2} = \sqrt{\frac{\sigma_1^2}{n_1}+\frac{\sigma_2^2}{n_2}} = \sqrt{\frac{12^2}{9}+\frac{6^2}{4}} = 5$

(b) Yes. Since x_1 and x_2 are each normally distributed, then so are $\bar{x}_1$ and $\bar{x}_2$. Since the difference of two normally distributed random variables is also normally distributed, so is $\bar{x}_1 - \bar{x}_2$. See also Key Fact 10.1.

(c) Since $\bar{x}_1 - \bar{x}_2$ is normally distributed with mean 0 and standard deviation 5, $P(-10 < \bar{x}_1 - \bar{x}_2 < 10) = P([-10 - 0]/5 < z < [10 - 0]/5)$
$= P(-2 < z < 2) = 0.9772 - 0.0228 = 0.9544$, so 95.44% of the differences in sample means from samples of size 9 and 4 from the two populations will lie between -10 and 10.

10.23 H_0: $\mu_1 = \mu_2$, H_a: $\mu_1 > \mu_2$

10.25 (a) The variable is systolic blood pressure (SBP).

(b) The two populations are the adolescent offspring of diabetic mothers (ODM) and of nondiabetic mothers (ODN).

(c) H_0: $\mu_1 = \mu_2$, H_a: $\mu_1 > \mu_2$ where μ_1 is the mean SBP for ODM and μ_2 is the mean SBP for ODN.

(d) Right tailed

10.27 (a) The variable is household vehicle miles traveled (VMT).

(b) The two populations are households in the Midwest and in the South.

(c) H_0: $\mu_1 = \mu_2$, H_a: $\mu_1 \neq \mu_2$ where μ_1 is the mean VMT for households in the Midwest and μ_2 is the mean VMT for households in the South.

(d) two-tailed

10.29 (a) The variable is operative times in minutes.

(b) The two populations are operations performed using a dynamic system (Z-plate) and operations performed using a static system (ALPS plate).

(c) H_0: $\mu_1 = \mu_2$, H_a: $\mu_1 < \mu_2$ where μ_1 is the mean time of operations performed using the dynamic system and μ_2 is the mean time of operations performed using the static system.

(d) left-tailed

10.31 (a) We have Minitab take 2000 observations from a normally distributed population having mean 100 and standard deviation 16 by choosing **Calc ▶ Random Data ▶ Normal...**, typing 2000 in the **Generate rows of data** text box, typing X1 in the **Store in columns** text box, typing 100 in the **Mean** text box and 16 in the **Standard deviation** text box, and clicking **OK**.

(b) We repeat part (a) for a normally distributed population having mean 120 and standard deviation 12 by choosing **Calc ▶ Random Data ▶ Normal...**, typing 2000 in the **Generate rows of data** text box, typing X2 in the **Store in columns** text box, typing 120 in the **Mean** text box and 12 in the **Standard deviation** text box, and clicking **OK**.

(c) To determine the difference between each pair of observations in parts (a) and (b), we choose **Calc ▶ Calculator...**, type DIFF in the **Store results in variable** text box, type X1 - X2 in the **Expression** text box, and click **OK**. DIFF will be stored in C3.

(d) To obtain a histogram of the 2000 differences found in part (c), we choose **Graph ▶ Histogram...**, select the **Simple** version and click **OK**. Enter DIFF in the **Graph variables** text box and click **OK**.

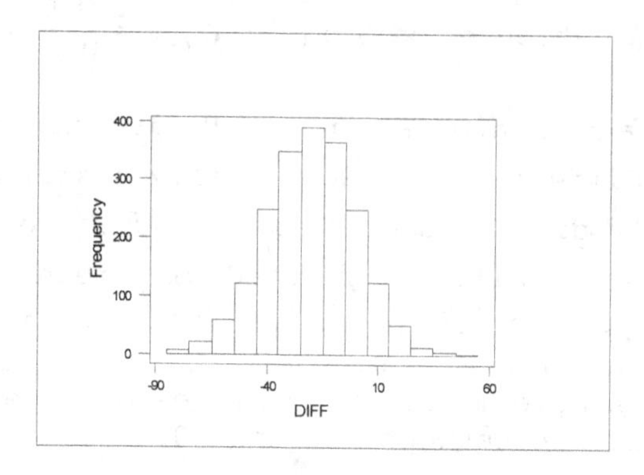

The histogram is bell-shaped because the random variable named DIFF is the difference of two random variables which themselves are normally distributed. A random variable which is the difference of two normally distributed random variables is itself normally distributed. Thus, its histogram is bell-shaped.

Section 10.2

10.33 (a) The four conditions required for using the pooled t-procedure are (i) both samples are simple random samples, (ii) independence of the two samples, (iii) the variable under consideration is normally distributed in each of the two populations or both samples are large, and (iv) the population standard deviations are equal.

(b) Independence and randomness are absolutely essential. Moderate violations of the normality assumption are not serious problems even for small or moderate size samples. Moderate violations of the equal standard deviation requirement are not serious if the two sample sizes are roughly equal.

10.35 Not Reasonable. One sample standard deviation is more than twice the other one. This is a good indication that the population standard deviations are not equal. Since the two sample sizes are not roughly equal (6 and 14), nor are both of them large, there appears to be a serious violation of the equal standard deviations requirement of the pooled t-test.

10.37 Reasonable. The standard deviations are roughly equal and both sample sizes are large.

10.40 (a) $$s_p = \sqrt{\frac{14(4)^2 + 14(5)^2}{15+15-2}} = \sqrt{20.5} = 4.5277$$

Step 1: H_0: $\mu_1 = \mu_2$, H_a: $\mu_1 \neq \mu_2$

Step 2: $\alpha = 0.05$

Step 3: $$t = \frac{10-12}{4.5277\sqrt{(1/15)+(1/15)}} = -1.210$$

Step 4: df = 28, Critical values = ± 2.048
P-value > 0.2.

Step 5: Since $-2.048 < -1.210 < 2.048$, do not reject H_0.

(b) 95% CI = $(10-12) \pm 2.048(4.527)\sqrt{(1/15+1/15)} = (-5.385, 1.385)$

10.42 (a) $$s_p = \sqrt{\frac{9(4)^2 + 14(5)^2}{10+15-2}} = \sqrt{21.47826} = 4.6345$$

Step 1: H_0: $\mu_1 = \mu_2$, H_a: $\mu_1 < \mu_2$

Step 2: $\alpha = 0.05$

Step 3: $$t = \frac{20-23}{4.6345\sqrt{(1/10)+(1/15)}} = -1.586$$

Step 4: df = 23, Critical value = -1.714
$0.05 <$ P-value < 0.10.

Step 5: Since $-1.586 > -1.714$, do not reject H_0.

(b) 90% CI = $(20-23) \pm 1.714(4.6345)\sqrt{(1/10+1/15)} = (-6.243, 0.243)$

10.44 (a) $$s_p = \sqrt{\frac{29(4)^2 + 39(5)^2}{30+40-2}} = \sqrt{21.16176} = 4.6002$$

Step 1: H_0: $\mu_1 = \mu_2$, H_a: $\mu_1 > \mu_2$

Step 2: $\alpha = 0.05$

Step 3: $t = \dfrac{20-18}{4.6002\sqrt{(1/30)+(1/40)}} = 1.800$

Step 4: df = 68, Critical value = 1.668
0.025 < P-value < 0.050.

Step 5: Since 1.800 > 1.671, reject H_0.

(b) 90% CI = $(20-18) \pm 1.671(4.6002)\sqrt{(1/30+1/40)} = (0.143, 3.857)$

10.46 Population 1: Males, $n_1 = 30$, $\bar{x}_1 = 37.6$, $s_1 = 38.5$

Population 2: Females, $n_2 = 30$, $\bar{x}_2 = 55.8$, $s_2 = 48.3$

$$s_p = \sqrt{\frac{29(38.5)^2 + 29(48.3)^2}{30+30-2}} = \sqrt{1907.57} = 43.6757$$

Step 1: H_0: $\mu_1 = \mu_2$, H_a: $\mu_1 < \mu_2$

Step 2: $\alpha = 0.01$

Step 3: $t = \dfrac{37.6-55.8}{43.6757\sqrt{(1/30)+(1/30)}} = -1.614$

Step 4: Critical value = -2.392
0.05 < P-value < 0.10

Step 5: Since -1.614 > -2.392, do not reject H_0.

Step 6: At the 1% significance level, the data do not provide sufficient evidence to conclude that males have a better sense of direction than females.

10.48 Population 1: Midwest, $n_1 = 15$, $\bar{x}_1 = 16.23$, $s_1 = 4.06$

Population 2: South, $n_2 = 14$, $\bar{x}_2 = 17.69$, $s_2 = 4.42$

$$s_p = \sqrt{\frac{14(4.06)^2 + 13(4.42)^2}{27}} = \sqrt{17.95347} = 4.2372$$

Step 1: H_0: $\mu_1 = \mu_2$, H_a: $\mu_1 \neq \mu_2$

Step 2: $\alpha = 0.05$

Step 3: $t = \dfrac{16.23-17.69}{4.2372\sqrt{1/15+1/14}} = -0.927$

Step 4: Critical values = ± 2.052
2{P(t < -0.927)} > 0.20.

Step 5: Since -2.052 < -0.927 < 2.052, do not reject H_0.

Step 6: At the 5% significance level, the data do not provide sufficient evidence that there is a difference in the average number of miles driven by Midwestern and Southern households.

10.50 Population 1: Lunch before recess, $n_1 = 889$, $\bar{x}_1 = 223.1$, $s_1 = 122.9$

Population 2: Lunch after recess, $n_2 = 1119$, $\bar{x}_2 = 156.6$, $s_2 = 108.1$

$$s_p = \sqrt{\frac{888(122.9)^2 + 1118(108.1)^2}{889+1119-2}} = \sqrt{13199.017} = 114.887$$

Step 1: H_0: $\mu_1 = \mu_2$, H_a: $\mu_1 > \mu_2$

Step 2: $\alpha = 0.01$

Step 3: $$t = \frac{223.1 - 156.6}{114.887\sqrt{1/889 + 1/1119}} = 12.88$$

Step 4: df = 2006, Critical value = 2.328 (using technology)
For the P-value approach, $P(t > 12.88) < 0.005$.

Step 5: Since 12.88 > 2.328, reject H_0. Because the P-value is less than the significance level, reject H_0.

Step 6: At the 1% significance level, the data provide sufficient evidence that the mean amount of food wasted for lunches before recess exceeds that for lunches after recess.

In each of Exercises 10.51-10.55, the value of s_p has been obtained in Exercise 10.45-10.49, respectively.

10.51 $t_{\alpha/2} = 1.734$, $90\%\text{CI} = (10.12\text{-}18.78) \pm 1.734(4.7718)\sqrt{1/10 + 1/10} = (-12.36, -4.96)$

10.53 $t_{\alpha/2} = 1.711$, $90\%\text{CI} = (9.0\text{-}1.6) \pm 1.711(36.1436)\sqrt{1/14 + 1/12} = (-16.93, 31.73)$

10.55 $t_{\alpha/2} = 2.601$, $99\%\text{CI} = (11.1\text{-}10.1) \pm 2.601(0.7972)\sqrt{1/91 + 1/109} = (0.71, 1.29)$

10.57 (a) Using Minitab, we choose **Graph ▶ Probability Plot**, select **Single** and click **OK**. Enter VEGETARIAN and OMNIVORE in the **Graph Variables** text box. Click on the **Multiple Graphs** button and select **On separate graphs**. Click **OK** twice. Note that the standard deviations are shown on the plots. Then choose **Graph ▶ Boxplot**, click on the **Simple** plot in the **Multiple Y's** row and click **OK**..Enter VEGETARIAN and OMNIVORE in the **Graph Variables** text box and click **OK**. The results are

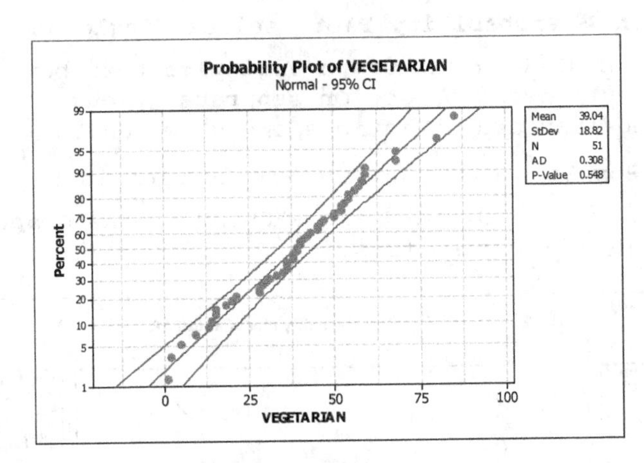

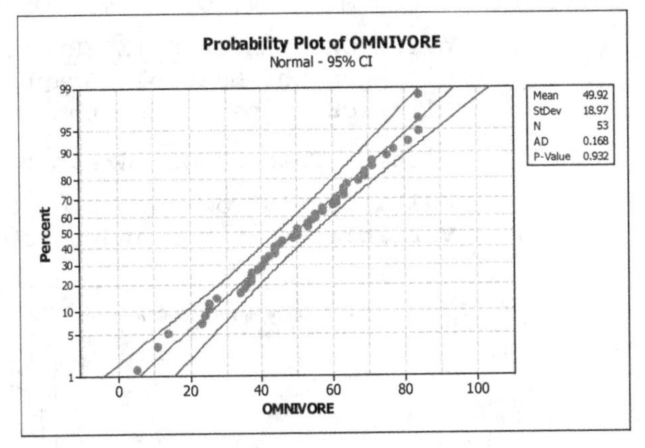

The standard deviations are 18.82 for Vegetarians and 18.97 for Omnivores.

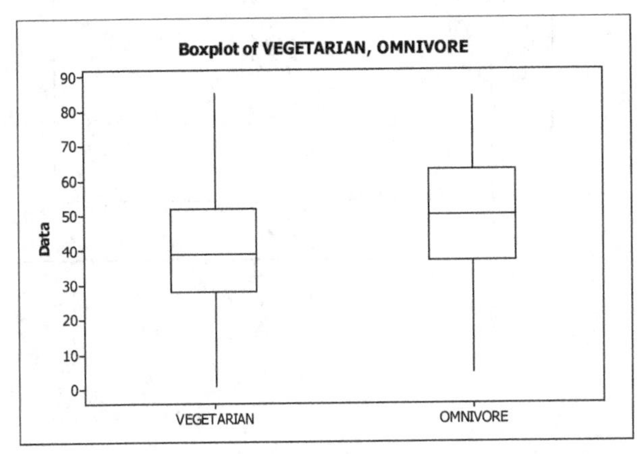

(b) Choose **Stat ▶ Basic Statistics ▶ 2-Sample t**, click on **Samples in different columns** and enter VEGETARIANS in the **First** textbox and OMNIVORES in the **Second** text box. Check the box for **Assume equal variances** and click on the **Options** button. Enter 99 in the **Confidence level** textbox, 0 in the **Test difference** text box and select **not equal** in the **Alternative drop down box**. Click **OK** twice. The results are

```
Two-sample T for VEGETARIAN vs OMNIVORE

             N  Mean  StDev  SE Mean
VEGETARIAN  51  39.0   18.8      2.6
OMNIVORE    53  49.9   19.0      2.6

Difference = mu (VEGETARIAN) - mu (OMNIVORE)
Estimate for difference:  -10.8853
99% CI for difference:  (-20.6147, -1.1559)
T-Test of difference = 0 (vs not =): T-Value = -2.94  P-Value = 0.004
DF = 102
Both use Pooled StDev = 18.8964
```

Since the P-value of 0.004 is less than the significance level of 0.01, the data do provide sufficient evidence that there is a difference between the mean daily protein intakes of female vegetarians and female omnivores.

(c) The 99% confidence interval is (-20.6147, -1.1559).

(d) Yes. The probability plots indicate that normality is reasonable, the sample sizes are large, and there are no outliers. Furthermore, the two sample standard deviations are almost equal, making the pooled t procedure a good choice.

10.59 (a) Using Minitab, we choose **Graph ▶ Probability Plot**, select **Single** and click **OK**. Enter STINGER and REGULAR in the **Graph Variables** text box. Click on the **Multiple Graphs** button and select **On separate graphs**. Click **OK** twice. Note that the standard deviations are shown on the plots. Then choose **Graph ▶ Boxplot**, click on the **Simple** plot in the **Multiple Y's** row and click **OK**...Enter STINGER and REGULAR in the **Graph Variables** text box and click **OK**. The results are

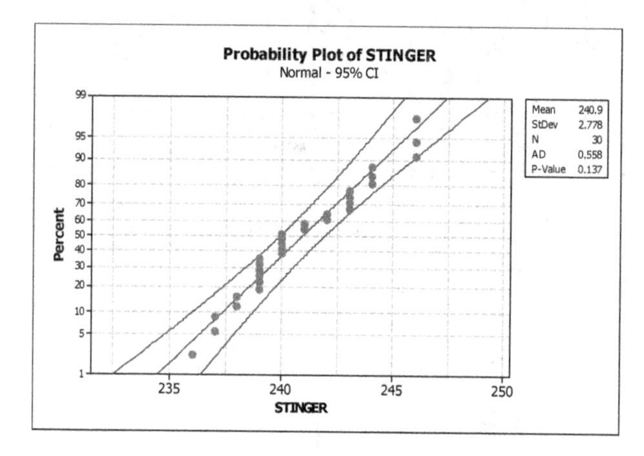

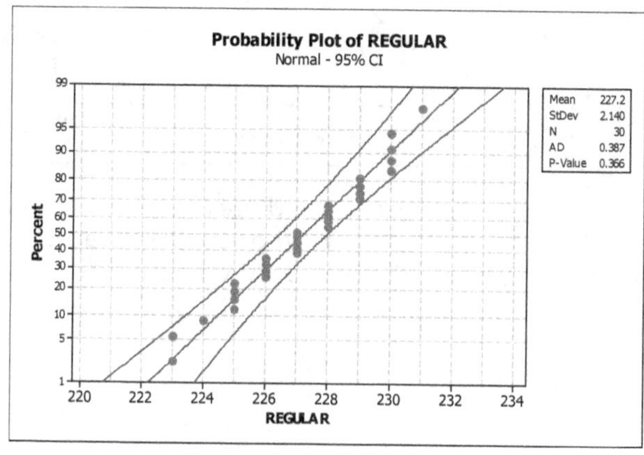

The standard deviations are
2.778 for STINGER and
2.140 for REGULAR.

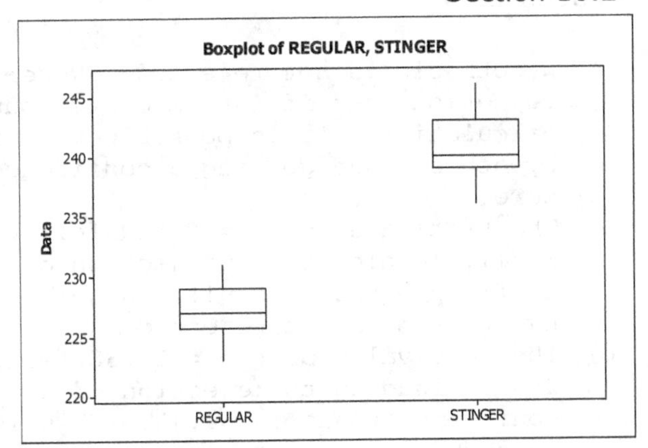

(b) Choose **Stat ▶ Basic Statistics ▶ 2-Sample t**, click on **Samples in different columns** and enter STINGER in the **First** text box and REGULAR in the **Second** text box. Check the box for **Assume equal variances** and click on the **Options** button. Enter 99 in the **Confidence level** textbox, 0 in the **Test difference** text box and select **greater than** in the **Alternative drop down** box. Click **OK** twice. The results are

```
Two-sample T for STINGER vs REGULAR

          N    Mean  StDev  SE Mean
STINGER  30  240.93   2.78     0.51
REGULAR  30  227.20   2.14     0.39

Difference = mu (STINGER) - mu (REGULAR)
Estimate for difference:  13.7333
99% lower bound for difference:  12.2015
T-Test of difference = 0 (vs >): T-Value = 21.45  P-Value = 0.000  DF = 58
Both use Pooled StDev = 2.4798
```

Since the P-value of 0.000 is less than the significance level of 0.01, the data do provide sufficient evidence that the Stinger tee improves total distance traveled.

(c) Repeat the process in part (b), but select **not equal** in the **Alternative drop down** box. The confidence interval is

```
99% CI for difference:  (12.0281, 15.4386)
```

(d) Yes. The probability plots indicate that normality is reasonable, the sample sizes are large, and there are no outliers. Furthermore, the two sample standard deviations are almost equal, making the pooled t procedure a good choice.

10.61 If $n_1 = n_2$, say both equal n, then

$$s_p^2 = \frac{(n_1-1)s_1^2+(n_2-1)s_2^2}{n_1+n_2-2} = \frac{(n-1)s_1^2+(n-1)s_2^2}{n+n-2}$$

$$= \frac{(n-1)(s_1^2+s_2^2)}{2n-2} = \frac{(n-1)(s_1^2+s_2^2)}{2(n-1)} = \frac{(s_1^2+s_2^2)}{2}$$

Thus, if $n_1 = n_2$, then s_p^2 is just the mean of s_1^2 and s_2^2.

10.63 (a) The test value of t = -0.927 fell between the critical values ±2.052, leading to nonrejection of the null hypothesis. The 95% confidence interval was (-4.69, 1.77). This includes zero, so the null hypothesis is not rejected.

(b) The test value of t = -1.977 fell between the critical values ±1.982, leading to nonrejection of the null hypothesis. The 95% confidence interval was (-2.603, 0.003). This includes zero, so the null hypothesis is not rejected. Note: These results are so close to rejection that if intermediate rounding is done during the calculations, it is possible to obtain results rejecting the null hypothesis and getting a confidence interval that does not include zero.

10.65 (a) The test value of t = 0.520 fell to the left of the critical value 1.711, leading to nonrejection of the null hypothesis. The 95% lower confidence bound is -16.928. This is not positive, so the null hypothesis is not rejected.

(b) The test value of t = = 12.88 fell to the right of the critical value 2.328, leading to rejection of the null hypothesis. The 99% lower confidence bound is 54.48. This is positive, so the null hypothesis is rejected.

Exercises 10.3

10.67 Pooled t-test. When the population standard deviations are equal, the pooled t-test is more powerful than the non-pooled t-test.

10.69 Nonpooled t-test. The sample standard deviations are enough different to raise doubt about the equality of the population standard deviations.

10.71 (a) Pooled t-test. Both populations are normally distributed and the population standard deviations are equal.

(b) Nonpooled t-test. Both populations are normally distributed, but the population standard deviations are unequal.

(c) Neither. Both populations are skewed and it is given that both sample sizes are small.

(d) Neither. Only one population is normally distributed; the other is skewed. Since the sample sizes are small, the nonnormality of one population rules out the use of either t-test.

10.73 (a) Step 1: H_0: $\mu_1 = \mu_2$, H_a: $\mu_1 \neq \mu_2$

Step 2: $\alpha = 0.05$

Step 3: $$t = \frac{10-12}{\sqrt{(2^2/15)+(5^2/15)}} = -1.438$$

Step 4: $$\Delta = \frac{\left[\left(\frac{s_1^2}{n_1}\right)+\left(\frac{s_2^2}{n_2}\right)\right]^2}{\frac{\left(\frac{s_1^2}{n_1}\right)^2}{n_1-1}+\frac{\left(\frac{s_2^2}{n_2}\right)^2}{n_2-1}} = \frac{\left[\left(\frac{2^2}{15}\right)+\left(\frac{5^2}{15}\right)\right]^2}{\frac{\left(\frac{2^2}{15}\right)^2}{15-1}+\frac{\left(\frac{5^2}{15}\right)^2}{15-1}} = 18.37;\ df = 18$$

Critical values = ± 2.101

0.100 < P-value < 0.200.

Step 5: Since -2.101 < -1.438 < 2.101, do not reject H_0.

(b) 95% CI = $(10-12) \pm 2.101\sqrt{(2^2/15+5^2/15)} = (-4.921, 0.921)$

10.75 (a) Step 1: H_0: $\mu_1 = \mu_2$, H_a: $\mu_1 > \mu_2$

Step 2: $\alpha = 0.05$

Step 3: $$t = \frac{20-18}{\sqrt{(4^2/10)+(5^2/15)}} = 1.107$$

Step 4: $$\Delta = \frac{\left[\left(\frac{s_1^2}{n_1}\right)+\left(\frac{s_2^2}{n_2}\right)\right]^2}{\frac{\left(\frac{s_1^2}{n_1}\right)^2}{n_1-1}+\frac{\left(\frac{s_2^2}{n_2}\right)^2}{n_2-1}} = \frac{\left[\left(\frac{4^2}{10}\right)+\left(\frac{5^2}{15}\right)\right]^2}{\frac{\left(\frac{4^2}{10}\right)^2}{10-1}+\frac{\left(\frac{5^2}{15}\right)^2}{15-1}} = 22.10;\ df = 22$$

Critical value = 1.717

P-value > 0.100.

Step 5: Since 1.107 < 1.717, do not reject H_0.

(b) 90% CI = $(20-18)\pm 1.717\sqrt{(4^2/10+5^2/15)} = (-1.103, 5.103)$

10.77 (a) Step 1: H_0: $\mu_1 = \mu_2$, H_a: $\mu_1 < \mu_2$

Step 2: $\alpha = 0.05$

Step 3: $$t = \frac{20-24}{\sqrt{(6^2/20)+(2^2/15)}} = -2.782$$

Step 4: $$\Delta = \frac{\left[\left(\frac{s_1^2}{n_1}\right)+\left(\frac{s_2^2}{n_2}\right)\right]^2}{\frac{\left(\frac{s_1^2}{n_1}\right)^2}{n_1-1}+\frac{\left(\frac{s_2^2}{n_2}\right)^2}{n_2-1}} = \frac{\left[\left(\frac{6^2}{20}\right)+\left(\frac{2^2}{15}\right)\right]^2}{\frac{\left(\frac{6^2}{20}\right)^2}{20-1}+\frac{\left(\frac{2^2}{15}\right)^2}{15-1}} = 24.322;\ df = 24$$

Critical value = -1.711

0.005 < P-value < 0.010.

Step 5: Since -2.782 < -1.711, reject H_0.

(b) 90% CI = $(20-24)\pm 1.711\sqrt{(6^2/20+2^2/15)} = (-6.460,\ -1.540)$

10.79 Population 1: Chronic, $n_1 = 32$, $\bar{x}_1 = 28.8$, $s_1 = 9.2$

Population 2: Remitted, $n_2 = 20$, $\bar{x}_2 = 22.1$, $s_2 = 5.7$

Step 1: H_0: $\mu_1 = \mu_2$, H_a: $\mu_1 \neq \mu_2$

Step 2: $\alpha = 0.10$

Step 3: $$t = \frac{25.8-22.1}{\sqrt{(9.2^2/32)+(5.7^2/20)}} = 1.791$$

Step 4: $$\Delta = \frac{\left[\left(\frac{s_1^2}{n_1}\right)+\left(\frac{s_2^2}{n_2}\right)\right]^2}{\frac{\left(\frac{s_1^2}{n_1}\right)^2}{n_1-1}+\frac{\left(\frac{s_2^2}{n_2}\right)^2}{n_2-1}} = \frac{\left[\left(\frac{9.2^2}{32}\right)+\left(\frac{5.7^2}{20}\right)\right]^2}{\frac{\left(\frac{9.2^2}{32}\right)^2}{32-1}+\frac{\left(\frac{5.7^2}{20}\right)^2}{20-1}} = 49.99995;\ df = 49$$

Critical values = ± 1.677

0.050 < P-value < 0.100.

Step 5: Since 1.791 > 1.677, reject H_0.

For the P-value approach, $0.05 < 2\{P(t > 1.791)\} < 0.10$.

Therefore, because the P-value is smaller than the significance level, reject H_0.

Step 6: At the 10% significance level, the data do provide sufficient evidence to conclude that there is a difference between the mean age at arrest.

10.81 Population 1: Dynamic, $n_1 = 14$, $\bar{x}_1 = 7.36$, $s_1 = 1.22$

Population 2: Static, $n_2 = 6$, $\bar{x}_2 = 10.50$, $s_2 = 4.59$

Step 1: H_0: $\mu_1 = \mu_2$, H_a: $\mu_1 < \mu_2$

Step 2: $\alpha = 0.05$

Step 3: $t = \dfrac{7.36 - 10.50}{\sqrt{(1.22^2/14) + (4.59^2/6)}} = -1.651$

Step 4: $$\Delta = \frac{\left[\left(\frac{s_1^2}{n_1}\right) + \left(\frac{s_2^2}{n_2}\right)\right]^2}{\frac{\left(\frac{s_1^2}{n_1}\right)^2}{n_1 - 1} + \frac{\left(\frac{s_2^2}{n_2}\right)^2}{n_2 - 1}} = \frac{\left[\left(\frac{1.22^2}{14}\right) + \left(\frac{4.59^2}{6}\right)\right]^2}{\frac{\left(\frac{1.22^2}{14}\right)^2}{14 - 1} + \frac{\left(\frac{4.59^2}{6}\right)^2}{6 - 1}} = 5.305;\ df = 5$$

Critical value = -2.015
0.050 < P-value < 0.100.

Step 5: Since -1.651 > -2.015, do not reject H_0.
For the P-value approach, 0.05 < P(t<-1.651) < 0.10. Therefore, since the P-value is larger than the significance level, do not reject H_0.

Step 6: At the 5% significance level, the data do not provide sufficient evidence to conclude that the mean number of acute postoperative days in the hospital is smaller with the dynamic system than with the static system.

10.83 Population 1: Psychotic patients, $n_1 = 10$, $\bar{x}_1 = 0.02426$, $s_1 = 0.00514$

Population 2: Non-psychotic patients, $n_2 = 15$, $\bar{x}_2 = 0.01643$, $s_2 = 0.00470$

(a) Step 1: H_0: $\mu_1 = \mu_2$, H_a: $\mu_1 > \mu_2$

Step 2: $\alpha = 0.01$

Step 3: $t = \dfrac{0.02426 - 0.01643}{\sqrt{(0.00514^2/10) + (0.00470^2/15)}} = 3.860$

Step 4: $$\Delta = \frac{\left[\left(\frac{s_1^2}{n_1}\right) + \left(\frac{s_2^2}{n_2}\right)\right]^2}{\frac{\left(\frac{s_1^2}{n_1}\right)^2}{n_1 - 1} + \frac{\left(\frac{s_2^2}{n_2}\right)^2}{n_2 - 1}} = \frac{\left[\left(\frac{0.00514^2}{10}\right) + \left(\frac{0.01643^2}{15}\right)\right]^2}{\frac{\left(\frac{1.00514^2}{10}\right)^2}{10 - 1} + \frac{\left(\frac{0.01643^2}{15}\right)^2}{15 - 1}} = 18.20;\ df = 18$$

Critical value = 2.552 (approximately)
P-value < 0.005.

Step 5: Since 3.860 > 2.552, reject H_0.
For the P-value approach, P(t > 3.860) < 0.005. Therefore, since the P-value is smaller than the significance level, reject H_0.

Step 6: At the 1% significance level, the data do provide sufficient evidence to conclude that dopamine activity is higher, on average, in psychotic patients than in non-psychotic patients.

10.85 $t_{\alpha/2} = 1.677$; 90% CI $= (25.8\text{-}22.1) \pm 1.677\sqrt{9.2^2/32 + 5.7^2/20} = (0.235, 7.165)$

10.87 $t_{\alpha/2} = 2.015$; 90% CI $= (7.36\text{-}10.50) \pm 2.015\sqrt{1.22^2/14 + 4.59^2/6} = (-6.973, 0.693)$

10.89 $t_{\alpha/2} = 2.552$; 98% CI $= (0.02426\text{-}0.01643) \pm 2.552\sqrt{0.00514^2/10 + 0.00470^2/15}$
$= (0.00265, 0.01301)$

10.91 (a) Choose nonpooled t-prodedures. There is clearly much more variability in the data for males than in the data for females.

(b) No. The two sample sizes are small and moderate. The data for the males is not normally distributed or has a large outlier at 39.2. Neither t-procedure is reasonable under these circumstances.

10.93 (a) Using Minitab, we choose **Graph ▶ Probability Plot**, select **Single** and click **OK**. Enter HEALTHCARE and TECHNOLOGY in the **Graph Variables** text box. Click on the **Multiple Graphs** button and select **On separate graphs**. Click **OK** twice. Note that the standard deviations are shown on the plots. Then choose **Graph ▶ Boxplot**, click on the **Simple** plot in the **Multiple Y's** row and click **OK**...Enter HEALTHCARE and TECHNOLOGY in the **Graph Variables** text box and click **OK**. The resulting standard deviations are 4.684% for HEALTHCARE and 1.504% for TECHNOLOGY.

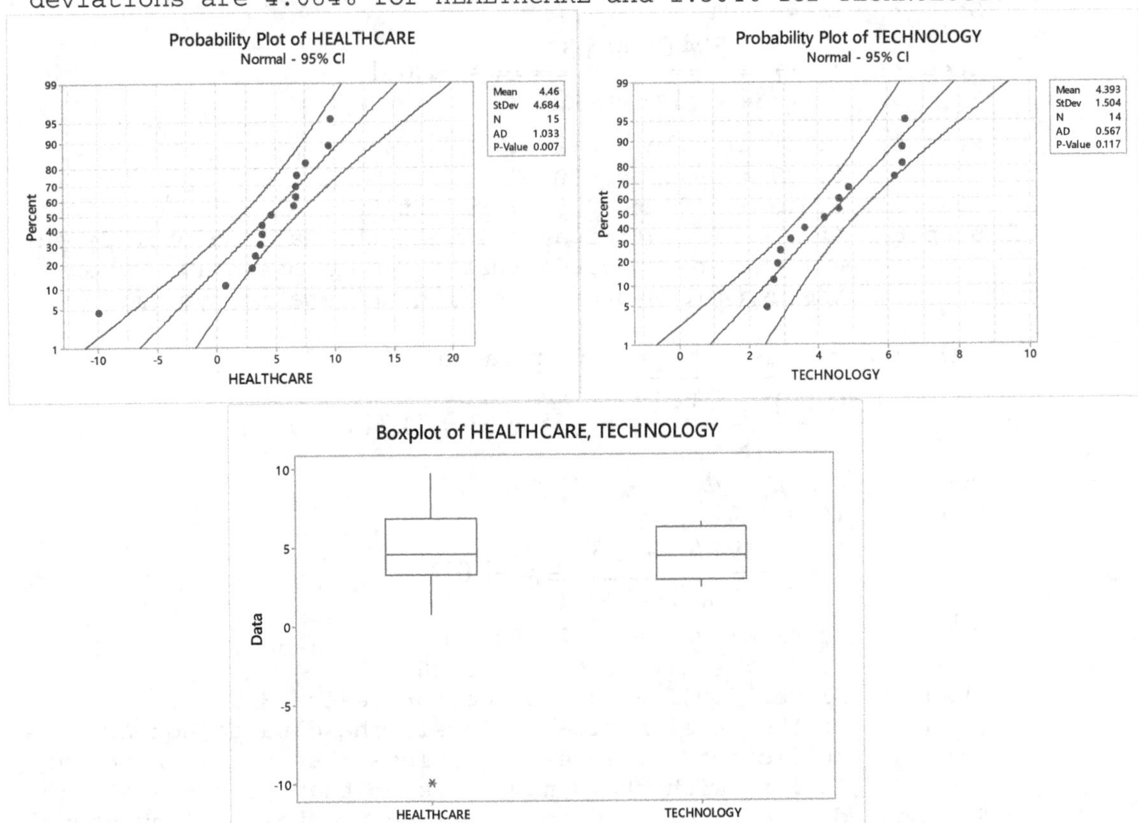

The sample standard deviations for Healthcare and Technology are very different from each other. If we had to choose between pooled and nonpooled, we would choose the nonpooled *t*-procedure.

(b) It is not reasonable to use the nonpooled *t*-procedure because the healthcare data is not normally distributed and contains an outlier.

10.95 Population 1: Dynamic, $n_1 = 14$, $\bar{x}_1 = 394.6$, $s_1 = 84.7$

Population 2: Static, $n_2 = 6$, $\bar{x}_2 = 468.3$, $s_2 = 38.2$

(a) $$s_p = \sqrt{\frac{13(84.7)^2 + 5(38.2)^2}{18}} = \sqrt{5586.63} = 74.7438$$

Step 1: H_0: $\mu_1 = \mu_2$, H_a: $\mu_1 < \mu_2$

Step 2: $\alpha = 0.05$

Step 3: $$t = \frac{394.6 - 468.3}{74.7438\sqrt{1/14 + 1/6}} = -2.021$$

Step 4: Critical value = -1.734

0.025 < P-value < 0.05

Step 5: Since -2.021 < -1.734, reject H_0.

Step 6: At the 5% significance level, the data do provide sufficient evidence to conclude that the mean operative time is less with the dynamic system than with the static system.

(b) Both tests ended up rejecting the null hypothesis test. However, the P-value for the non-pooled approach was smaller than the P-value for the pooled approach.

(c) The non-pooled test at 1% significance:

Step 1: H_0: $\mu_1 = \mu_2$, H_a: $\mu_1 < \mu_2$

Step 2: $\alpha = 0.01$

Step 3: $$t = \frac{394.6 - 468.3}{\sqrt{(84.7^2/14) + (38.2^2/6)}} = -2.681$$

Step 4: Critical value = -2.567

0.005 < P-value < 0.01

Step 5: Since -2.681 < -2.567, reject H_0.

Step 6: At the 1% significance level, the data do provide sufficient evidence to conclude that the mean operative time is less with the dynamic system than with the static system.

The pooled test at 1% significance:

$$s_p = \sqrt{\frac{13(84.7)^2 + 5(38.2)^2}{18}} = \sqrt{5586.63} = 74.7438$$

Step 1: H_0: $\mu_1 = \mu_2$, H_a: $\mu_1 < \mu_2$

Step 2: $\alpha = 0.01$

Step 3: $$t = \frac{394.6 - 468.3}{74.7438\sqrt{1/14 + 1/6}} = -2.021$$

Step 4: Critical value = -2.552

0.025 < P(t < -2.021) < 0.05

Step 5: Since -2.021 > -2.552, do not reject H_0.

Step 6: At the 1% significance level, the data do not provide sufficient evidence to conclude that the mean operative time is less with the dynamic system than with the static system.

The pooled t-test resulted in not rejecting the null hypothesis while the nonpooled t-test resulted in rejecting the null hypothesis.

(d) The nonpooled t-test is more appropriate. One sample standard deviation is more than two times as large as the other, making it unlikely that the two population standard deviations are equal. The fact that the two sample sizes are also quite different makes it essential that the pooled t-test not be used.

10.97 (a) Using Minitab, we choose **Graph ▶ Probability Plot**, select **Single** and click **OK**. Enter REGULAR and STINGER in the **Graph Variables** text box. Click on the **Multiple Graphs** button and select **On separate graphs**. Click **OK** twice. Note that the standard deviations are shown on the plots. Then choose **Graph ▶ Boxplot**, click on the **Simple** plot in the **Multiple Y's** row and click **OK**...Enter REGULAR and STINGER in the **Graph Variables** text box and click **OK**. The results are

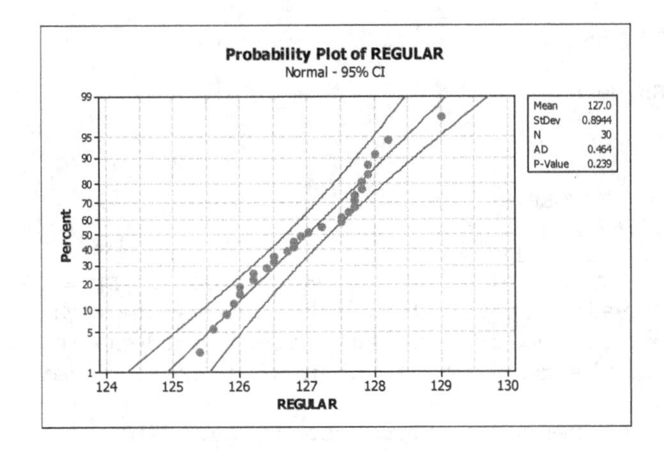

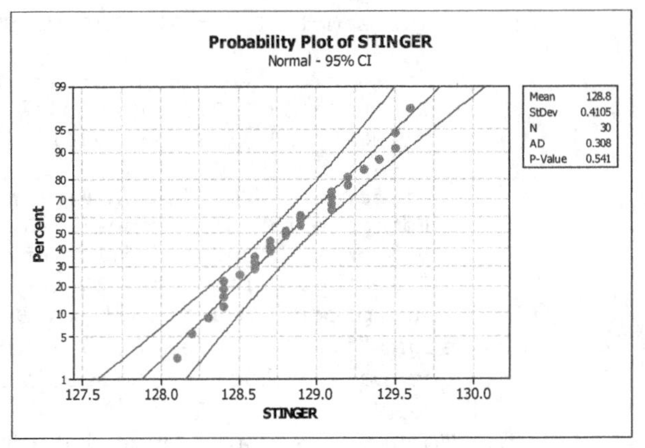

Standard deviations are 0.8944 for REGULAR and 0.4105 for STINGER.

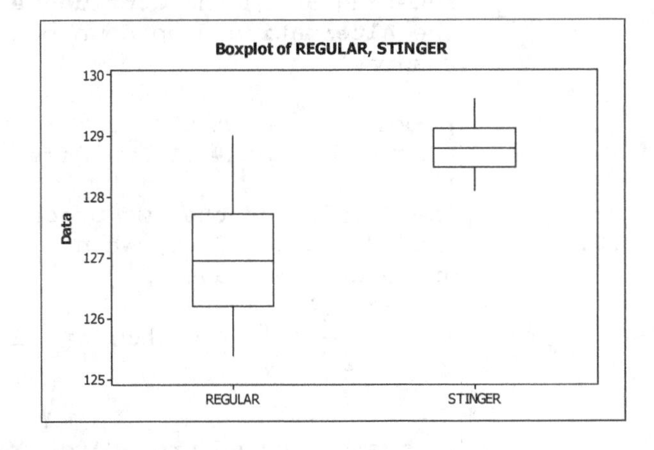

(b) Nonpooled t-procedure. The samples are both large enough (30), normality is reasonable, there are no potential outliers, but one standard deviation is more than twice the other, making it unlikely that the population standard deviations are equal.

(c) Choose **Stat ▶ Basic Statistics ▶ 2-Sample t**, click on **Samples in different columns** and enter REGULAR in the **First** text box and STINGER in the **Second** text box. Check the box for **Assume equal variances** and click on the **Options** button. Enter 95 in the **Confidence level** textbox, 0 in the **Test difference** text box and select **Less than** in the **Alternative** drop down box. Click **OK** twice. Then repeat this procedure with the **Assume equal variances** box unchecked. The results are

```
Two-sample T for REGULAR vs STINGER

          N    Mean   StDev  SE Mean
REGULAR  30  127.007  0.894     0.16
STINGER  30  128.833  0.410    0.075

Difference = mu (REGULAR) - mu (STINGER)
Estimate for difference:  -1.82667
95% upper bound for difference:  -1.52634
T-Test of difference = 0 (vs <): T-Value = -10.17  P-Value = 0.000  DF = 58
Both use Pooled StDev = 0.6959

Two-sample T for REGULAR vs STINGER

          N    Mean   StDev  SE Mean
REGULAR  30  127.007  0.894     0.16
STINGER  30  128.833  0.410    0.075

Difference = mu (REGULAR) - mu (STINGER)
Estimate for difference:  -1.82667
95% upper bound for difference:  -1.52413
T-Test of difference = 0 (vs <): T-Value = -10.17  P-Value = 0.000  DF = 40
```

Both procedures result in a P-value of 0.000, which is less than the significance level 0.05. The data do provide sufficient evidence that the ball velocity is less with the regular tee than with the Stinger tee.

(d) To get the confidence intervals, repeat the process in part (b), entering 90 in the **Confidence level** text box and selecting **not equal** in the **Alternative** drop down box. The confidence interval portions of the results are

```
Pooled       90% CI for difference:  (-2.12700, -1.52634)
Nonpooled    90% CI for difference:  (-2.12921, -1.52413)
```

The results of the two procedures are almost identical.

10.99 (a) From Exercise 10.61, when the sample sizes are equal, in the pooled procedure, we have

$$s_p^2 = \frac{s_1^2 + s_2^2}{2};\ \text{if } n_1 = n_2, \text{ then } t = \frac{\bar{x}_1 - \bar{x}_2}{\sqrt{\frac{s_1^2 + s_2^2}{2}\left(\frac{1}{n} + \frac{1}{n}\right)}} = \frac{\bar{x}_1 - \bar{x}_2}{\sqrt{\frac{s_1^2 + s_2^2}{2}\left(\frac{2}{n}\right)}} = \frac{\bar{x}_1 - \bar{x}_2}{\sqrt{\frac{s_1^2 + s_2^2}{n}}}.$$

In the nonpooled procedure, we have

$$t = \frac{\bar{x}_1 - \bar{x}_2}{\sqrt{\frac{s_1^2}{n} + \frac{s_2^2}{n}}} = \frac{\bar{x}_1 - \bar{x}_2}{\sqrt{\frac{s_1^2 + s_2^2}{n}}},$$ which is identical to the pooled t-statistic.

(b) If the standard deviations are not equal, the computed degrees of freedom in the nonpooled t-procedure could be quite different from the degrees of freedom in the pooled t-procedure. This will lead to different critical values and/or P-values and possibly to different conclusions.

10.101 (a) The confidence interval in Exercise 10.85 consists of all positive numbers and does not contain zero. Therefore, the hypothesis test for H_0: $\mu_1 = \mu_2$, H_a: $\mu_1 \neq \mu_2$ should reject the null hypothesis. The results of Exercise 10.79 confirm that by leading to a rejection of the null hypothesis.

(b) The confidence interval in Exercise 10.90 consists of all negative numbers and does not contain zero. Therefore, the hypothesis test for H_0: $\mu_1 = \mu_2$ vs. H_a: $\mu_1 \neq \mu_2$ should reject the null hypothesis. The results of Exercise 10.84 confirm that by leading to a rejection of the null hypothesis.

10.103 (a) Population 1: Malignant, $n_1 = 52$, $\bar{x}_1 = 48.5$, $s_1 = 38.5$
Population 2: Benign, $n_2 = 116$, $\bar{x}_2 = 33.7$, $s_2 = 26.6$
From Exercise 10.80, df = 73 and $\alpha = 0.01$

$$t_\alpha = 2.379;\ (48.5\text{-}33.7)\text{-}2.379\sqrt{38.5^2/52 + 26.6^2/116} = 0.805$$

The 99% lower confidence bound for $\mu_1 - \mu_2$ is greater than zero. Therefore, the right-tailed hypothesis test for H_0: $\mu_1 = \mu_2$ vs. H_a: $\mu_1 > \mu_2$ will be rejected at 1% significance. This compares to the result of the hypothesis test in Exercise 10.80.

(b) Population 1: Psychotic patients, $n_1 = 10$, $\bar{x}_1 = 0.02426$, $s_1 = 0.00514$
Population 2: Non-psychotic patients, $n_2 = 15$, $\bar{x}_2 = 0.01643$, $s_2 = 0.00470$
From Exercise 10.83, df = 18 and $\alpha = 0.01$

$$t_\alpha = 2.552;\ (0.02426\text{-}0.01643)\text{-}2.552\sqrt{0.00514^2/10 + 0.00470^2/15} = 0.0026$$

The 99% lower confidence bound for $\mu_1 - \mu_2$ is greater than zero. Therefore, the right-tailed hypothesis test for H_0: $\mu_1 = \mu_2$ vs. H_a: $\mu_1 > \mu_2$ will be rejected at 1% significance. This compares to the result of the hypothesis test in Exercise 10.83.

Exercises 10.4

10.105 Use the pooled t-test. It is slightly more powerful than the Mann-Whitney when the conditions for its use (normal distributions with equal stand deviations) are met.

10.107 (a) Since the populations do not have the same shape, use the nonpooled t-test. This is appropriate because the populations are normally distributed.
(b) Since the populations have the same shape, use the Mann-Whitney test.
(c) Since both samples are large and standard deviations are unequal, use the non-pooled t-test.

10.109 Since the populations have the same shape and there is some question about the normality of the distributions, use the Mann-Whitney test. The presence of outliers in one of the samples strengthens this decision since the outliers could have a significant effect on the standard deviation and hence, on either the pooled or nonpooled t-test, whereas the presence of outliers will not have a great effect on the rank sum.

10.111 All normal distributions are symmetric about the population mean, and the standard deviation determines the spread (and hence the shape). Thus two normal distributions with the same standard deviation have the same shape.

10.113 (a) 90 (b) 8(8 + 9 + 1) - 90 = 54
(c) Right tail = 93, Left tail = 8(8 + 9 + 1) - 93 = 51

10.115 (a) 95 (b) 9(9 + 8 + 1) - 95 = 67
(c) Right tail = 99, Left tail = 9(9 + 8 + 1) - 99 = 63

10.117 Step 1: H_0: $\mu_1 = \mu_2$, H_a: $\mu_1 > \mu_2$
Step 2: $\alpha = 0.10$

Step 3:

Sample 1	Overall Rank	Sample 2	Overall Rank
4	7	3	4
3	4	4	7
5	9	2	1.5
3	4	2	1.5
		4	7
	24		

M = 24

Step 4: The critical value is $M_{0.10}$=26.

Step 5: Since M = 24 < 26, we do not reject the null hypothesis.

10.119 Step 1: H₀: $\mu_1 = \mu_2$, Hₐ: $\mu_1 \neq \mu_2$

Step 2: α = 0.10

Step 3:

Sample 1	Overall Rank	Sample 2	Overall Rank
8	6	8	6
2	1	14	9.5
4	2	6	3
7	4	14	9.5
8	6	10	8
	19		

M = 19

Step 4: The critical values are $M_{0.05} = 36$ and $M_{0.95} = 5(11) - M_{0.05} = 55 - 36 = 19$.

Step 5: Since M = 19 = 19, we reject the null hypothesis.

10.121 Step 1: H₀: $\mu_1 = \mu_2$, Hₐ: $\mu_1 < \mu_2$

Step 2: α = 0.10

Step 3:

Sample 1	Overall Rank	Sample 2	Overall Rank
5	3	10	10
1	1	7	7.5
5	3	6	5.5
8	9	6	5.5
7	7.5	11	11
5	3		
	26.5		

M = 26.5

Step 4: The critical value is $M_{0.90} = 6(12) - M_{0.10} = 72 - 44 = 28$.

Step 5: Since M = 26.5 < 28, we reject the null hypothesis.

10.123 Step 1: H₀: $\mu_1 = \mu_2$, Hₐ: $\mu_1 \neq \mu_2$

Step 2: α = 0.05

Step 3:

Friese	Overall Rank	Cockerell	Overall Rank
188	7	169	1
190	8	178	2
225	9	180	3.5
235	10	180	3.5
		182	5
		185	6
	34		

M = 34

Step 4: The critical values are 32 and 4(4 + 6 + 1) - 32 = 12.

Step 5: Since M = 34 > 32, we reject the null hypothesis. The data do provide sufficient evidence that there is a difference between the mean wing stroke frequencies of the two species of Euglossine bees.

10.125 Population 1: Students with fewer than two years of high-school algebra;
Population 2: Students with two or more years of high-school algebra.

Step 1: H_0: $\mu_1 = \mu_2$, H_a: $\mu_1 < \mu_2$, $n_1 = 6$, $n_2 = 9$

Step 2: $\alpha = 0.05$

Step 3: Construct a worktable based upon the following: First, rank all the data from both samples combined. Adjacent to each column of data as it is presented in the Exercise, record the overall rank. Assign tied rankings the average of the ranks they would have had if there were no ties. For example, the two 81s in the table below are tied for eleventh smallest. Thus, each is assigned the rank (11 + 12)/2 = 11.5.

Fewer Than Two Years of High-School Algebra	Overall Rank	Two or More Years of High-School Algebra	Overall Rank
58	3	84	14
81	11.5	67	7
74	8.5	65	6
61	4	75	10
64	5	74	8.5
43	1	92	15
		83	13
		52	2
		81	11.5
	33		

The value of the test statistic is the sum of the ranks for the sample data from Population 1:

M = 3 + 11.5 + 8.5 + 4 + 5 + 1 = 33.

Step 4: We have $n_1 = 6$ and $n_2 = 9$. Since the hypothesis test is left-tailed with $\alpha = 0.05$, we use Table VI to obtain the critical value, which is $M_l = 6(6 + 9 + 1) - 63 = 33$. Thus, we reject H_0 if $M \le 33$.

Step 5: Since M = 33 equals the critical value, reject H_0.

Step 6: At the 5% significance level, the data provide sufficient evidence to conclude that students with fewer than two years of high-school algebra have a lower mean semester average in this teacher's chemistry courses than do students with two or more years of high-school algebra.

10.127 Population 1: Weekly earnings of males;
Population 2: Weekly earnings of females.

Step 1: H_0: $\mu_1 = \mu_2$, H_a: $\mu_1 > \mu_2$

Step 2: $\alpha = 0.05$

Step 3: To construct a worktable, first rank all the data from both samples combined. Adjacent to each column of data, record the overall rank. Assign tied rankings the average of the ranks they would have had if there were no ties.

Male	Overall Rank	Female	Overall Rank
382	4	353	1
401	5	369	2
411	7	370	3
571	9	407	6
812	11	505	8
920	12	589	10
1884	15	1176	13
2617	18	1440	14
		2073	16
		2188	17
	81		

The value of the test statistic is the sum of the ranks for the sample data from Population 1: M = 81

Step 4: We have $n_1 = 8$ and $n_2 = 10$. Since the hypothesis test is right-tailed with $\alpha = 0.05$, we use Table VI to obtain the critical value, which is $M_r = 95$. Thus, we reject H_0 if $M \geq 95$.

Step 5: Since M < 95, do not reject H_0.

Step 6: At the 5% significance level, the data do not provide sufficient evidence to conclude that the mean weekly earnings of male full-time wage and salary workers is greater than that for females.

10.129 (a) Population 1: Release times for prisoners with fraud offenses; Population 2: Release times for prisoners with firearms offenses.

Step 1: H_0: $\mu_1 = \mu_2$, H_a: $\mu_1 < \mu_2$

Step 2: $\alpha = 0.05$

Step 3: To construct a worktable, first rank all the data from both samples combined. Adjacent to each column of data, record the overall rank. Assign tied rankings the average of the ranks they would have had if there were no ties. For example, the two 17.9s in the following table are tied for thirteenth smallest. Thus, each is assigned the rank (13 + 14)/2 = 13.5.

Fraud	Overall Rank	Firearms	Overall Rank
3.6	1	10.4	6
5.3	2	13.3	9
5.9	3	16.1	11
7.0	4	17.9	13.5
8.5	5	18.4	15
10.7	7	19.6	16
11.8	8	20.9	17
13.9	10	21.9	18
16.6	12	23.8	19
17.9	13.5	25.5	20
	65.5		

The value of the test statistic is the sum of the ranks for the sample data from Population 1: $M = 65.5$

Step 4: We have $n_1 = 10$ and $n_2 = 10$. Since the hypothesis test is left-tailed with $\alpha = 0.05$, we use Table VI to obtain the critical value, which is $M_1 = 10(10 + 10 + 1) - 127 = 83$. Thus, we reject H_0 if $M \leq 83$.

Step 5: Since $M < 83$, reject H_0.

Step 6: At the 5% significance level, the data provide sufficient evidence to conclude that the mean time served for fraud offenses is less than that for firearms offenses.

(b) Normal distributions with the same standard deviations have the same shape, thus meeting the requirement for using the Mann-Whitney test. If the distributions are, in fact, normal with equal standard deviations, it is better to use the pooled t-test since it is slightly more powerful than the Mann-Whitney test in this situation.

10.131 (a) Using Minitab, we choose **Graph ▶ Probability Plot**, select **Single** and click **OK**. Enter MEN and WOMEN in the **Graph Variables** text box. Click on the **Multiple Graphs** button and select **On separate graphs**. Click **OK** twice. Then choose **Graph ▶ Boxplot**, click on the **Simple** plot in the **Multiple Y's** row and click **OK**...Enter MEN and WOMEN in the **Graph Variables** text box and click **OK**. The results are

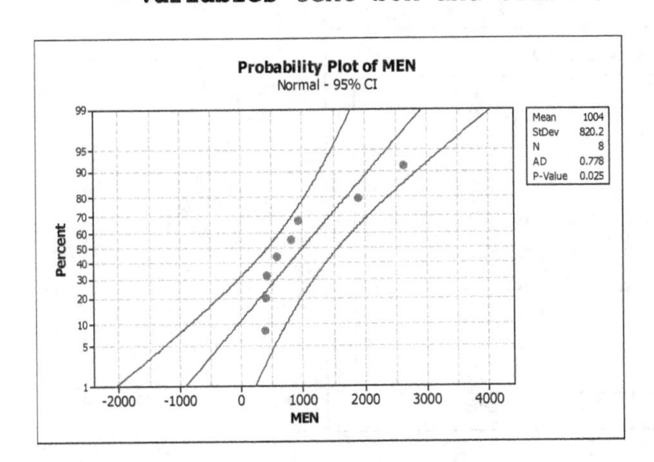

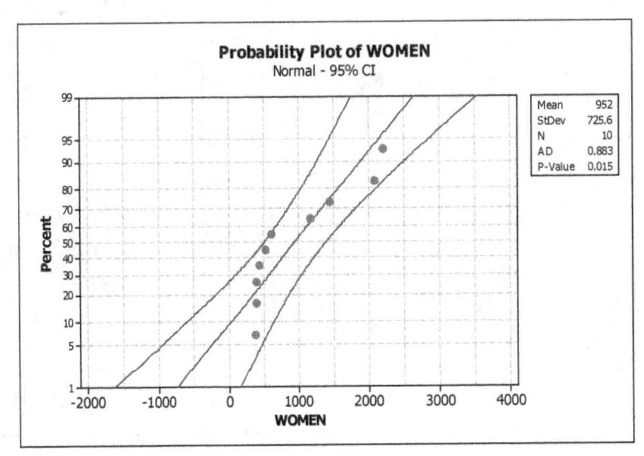

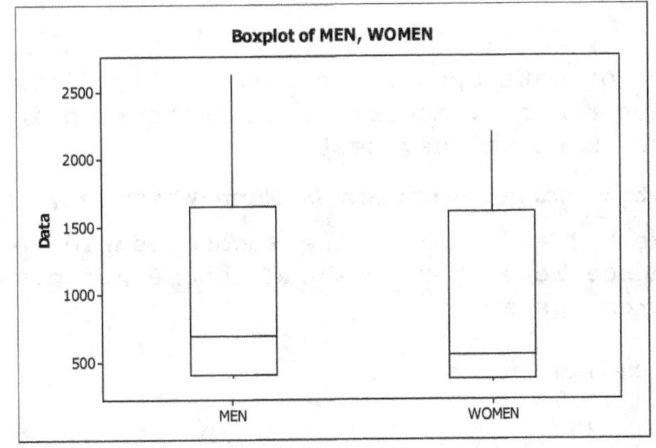

(b) No. The sample sizes are small - 10 and 8, and the samples are both right skewed.

(c) Yes. The sample distributions are both right skewed, about the same shape.

10.133 (a) Using Minitab with the data in two columns named LAB1 and LAB2, we choose **Graph ▶ Probability Plot...**, select the **Single** version and click **OK**. Enter LAB1 and LAB2 in the **Graph variables** text box and click **OK**. Then choose **Graph ▶ Boxplot**, click on the **Simple** plot in the **Multiple Y's** row and click **OK**...Enter LAB1 and LAB2 in the **Graph Variables** text box and click **OK**. The results are

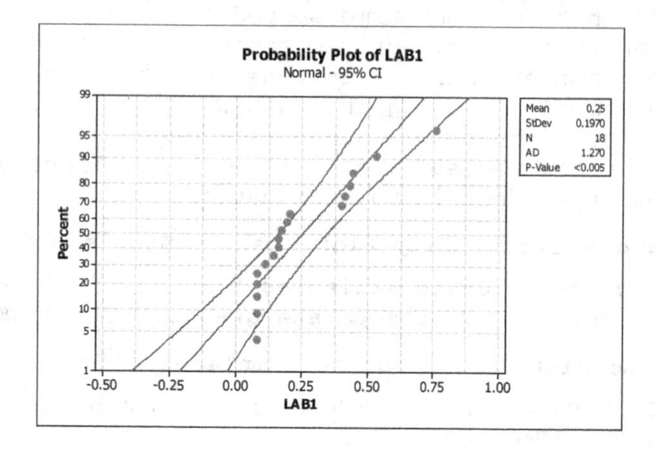

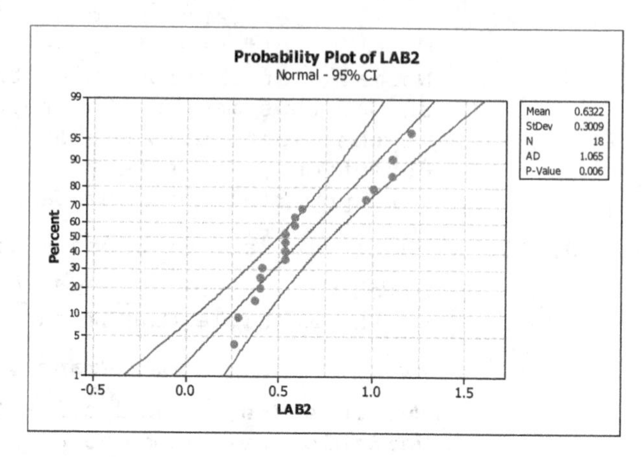

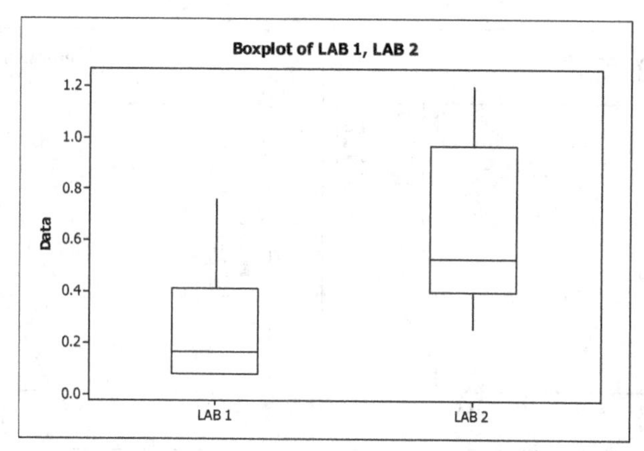

(b) Neither set of data appears to be normally distributed. Since the sample sizes are only moderate, the better choice for comparing the two means is the Mann Whitney test.

(c) Choose **Stat ▶ Nonparametrics ▶ Mann-Whitney...**, enter LAB1 in the **First sample** text box and LAB2 in the **Second sample** text box. Enter 95 in the **Confidence level** text box and choose **not equal** in the **Alternative** box. The results are

```
        N  Median
LAB1   18  0.1650
LAB2   18  0.5300

Point estimate for ETA1-ETA2 is -0.3500
95.2 Percent CI for ETA1-ETA2 is (-0.4799,-0.1999)

W = 213.5
Test of ETA1 = ETA2 vs ETA1 not = ETA2 is significant at 0.0002
The test is significant at 0.0002 (adjusted for ties)
```

Since the P-value is 0.0002, the data provide sufficient evidence that there is a difference in the median formaldehyde exposure in the two labs at the 5% significance level.

10.135 Step 1: State the null and alternative hypotheses.
Step 2: Decide on the significance level α.
Step 3: Construct a worktable of the form.

Sample from Population 1	Overall Rank	Sample from Population 2	Overall Rank
.	.	.	.
.	.	.	.
.	.	.	.

Compute the value of the test statistic

$$z = \frac{M - n_1(n_1 + n_2 + 1)/2}{\sqrt{n_1 n_2 (n_1 + n_2 + 1)/12}}$$

where M is the sum of the ranks for the sample data from Population 1.

Step 4: The critical value(s):

(a) for a two-tailed test are $\pm z_{\alpha/2}$.

(b) for a left-tailed test is $-z_{\alpha}$.

(c) for a right-tailed test is z_{α}.

Use Table II to find the critical value(s)

Step 5: If the value of the test statistic falls in the rejection region, reject H_0; otherwise, do not reject H_0.

Step 6: State the conclusion in words.

10.137 (a) (b) 5%

Rank						
1	2	3	4	5	6	M
A	A	A	B	B	B	6
A	A	B	A	B	B	7
A	A	B	B	A	B	8
A	A	B	B	B	A	9
A	B	A	A	B	B	8
A	B	A	B	A	B	9
A	B	A	B	B	A	10
A	B	B	A	A	B	10
A	B	B	A	B	A	11
A	B	B	B	A	A	12
B	A	A	A	B	B	9
B	A	A	B	A	B	10
B	A	A	B	B	A	11
B	A	B	A	A	B	11
B	A	B	A	B	A	12
B	A	B	B	A	A	13
B	B	A	A	A	B	12

M	P(M)
6	0.05
7	0.05
8	0.10
9	0.15
10	0.15
11	0.15
12	0.15
13	0.10
14	0.05
15	0.05

B	B	A	A	B	A	13		
B	B	A	B	A	A	14		
B	B	B	A	A	A	15		

(c) Each row in part (a) has a 1/20 = 0.05 chance of occurring. This results in the probability distribution above at the right for M for the case when $n_1 = 3$ and $n_2 = 3$.

(d) A histogram for the probability distribution of M is shown at the right for the case when $n_1 = 3$ and $n_2 = 3$.

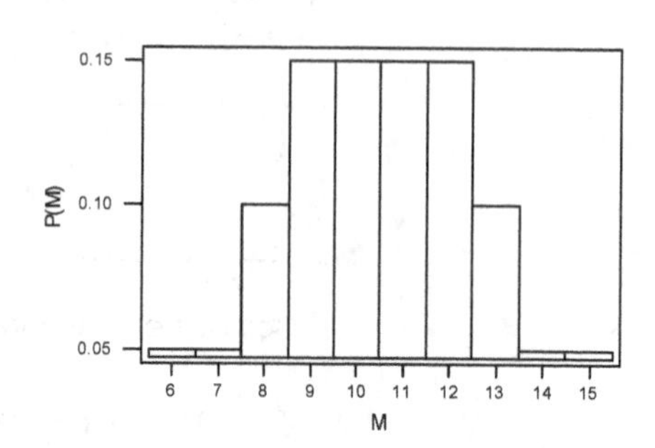

(e) From part (c), we see that $P(M \le 6) = 0.05$ and $P(M \ge 15) = 0.05$. These results correspond with the entries in Table VI for $n_1 = 3$ and $n_2 = 3$. That is, $M_l = 6$ at $\alpha = 0.05$, and $M_r = 15$ at $\alpha = 0.05$.

Exercises 10.5

10.139 Paired sampling helps to remove extraneous sources of variation, resulting in a smaller sampling error for the estimated difference between the sample means. This, in turn, makes it more likely that differences between the means will be detected when those differences actually exist.

10.141 The samples must be paired (this is essential), and the population of all paired differences must be normally distributed or the sample must be large (the procedure works reasonably well even for small or moderate size samples if the paired-difference variable is not normally distributed provided that the deviation from normality is small).

10.143 (a) TV viewing time (b) Married men and married women
(c) Married couples
(d) Paired difference variable = TV viewing time of Man – TV viewing time of Woman
(e) H_0: $\mu_M = \mu_W$; H_a: $\mu_M \neq \mu_W$ (f) Two-tailed

10.145 (a) VAS sensory ratings (b) People under and not under hypnosis
(c) pair is a person (1) under and (2) not under hypnosis (same person)
(d) Paired difference variable = VAS sensory rating under hypnosis – VAS sensory rating not under hypnosis.
(e) H_0: $\mu_1 = \mu_2$; H_a: $\mu_1 < \mu_2$ (f) Left-tailed

10.147 (a) total antioxidant capacity (b) fresh and stored breastmilk
(c) pair is a woman (1) stored breastmilk and (2) fresh breastmilk
(d) Paired difference variable = antioxidant capacity of stored breastmilk – antioxidant capacity of fresh breastmilk
(e) H_0: $\mu_1 = \mu_2$; H_a: $\mu_1 < \mu_2$ (f) Left-tailed

10.149 (a) Step 1: H_0: $\mu_1 = \mu_2$, H_a: $\mu_1 \neq \mu_2$
Step 2: $\alpha = 0.10$

Step 3: The paired differences, $d = x_1 - x_2$, are

2	1	3	6	4	-1	3

$$\bar{d} = \frac{\sum d_i}{n} = \frac{18}{7} = 2.571 \quad s_d = \sqrt{\frac{\sum d^2 - (\sum d)^2/n}{n-1}} = \sqrt{\frac{76 - 324/7}{6}} = 2.225$$

$$t = \frac{\bar{d}}{s_d/\sqrt{n}} = \frac{2.571}{2.225/\sqrt{7}} = 3.057$$

Step 4: df = 6; critical values = ± 1.943
$0.02 <$ P-value < 0.05

Step 5: Since $3.057 > 1.943$ and the P-value is less than 10%, reject H_0.

10.151 (a) Step 1: H_0: $\mu_1 = \mu_2$, H_a: $\mu_1 > \mu_2$

Step 2: $\alpha = 0.10$

Step 3: The paired differences, $d = x_1 - x_2$, are

4	-1	1	5	3	0	-5	-6

$$\bar{d} = \frac{\sum d_i}{n} = \frac{1}{8} = 0.125 \quad s_d = \sqrt{\frac{\sum d^2 - (\sum d)^2/n}{n-1}} = \sqrt{\frac{113 - 1/8}{7}} = 4.016$$

$$t = \frac{\bar{d}}{s_d/\sqrt{n}} = \frac{0.125}{4.016/\sqrt{8}} = 0.088$$

Step 4: df = 7; critical values = 1.415
P-value > 0.10

Step 5: Since $0.088 < 1.415$ and the P-value is greater than 10%, do not reject H_0.

10.153 (a) Step 1: H_0: $\mu_1 = \mu_2$, H_a: $\mu_1 < \mu_2$

Step 2: $\alpha = 0.10$

Step 3: The paired differences, $d = x_1 - x_2$, are

-3	-3	-2	3	-6	0	-2	-7	-1

$$\bar{d} = \frac{\sum d_i}{n} = \frac{-21}{9} = -2.333 \quad s_d = \sqrt{\frac{\sum d^2 - (\sum d)^2/n}{n-1}} = \sqrt{\frac{121 - 441/9}{8}} = 3.000$$

$$t = \frac{\bar{d}}{s_d/\sqrt{n}} = \frac{-2.333}{3.000/\sqrt{9}} = -2.333$$

Step 4: df = 8; critical values = -1.397
$0.01 <$ P-value < 0.025

Step 5: Since $-2.333 < -1.397$ and the P-value is less than 10%, reject H_0.

10.155 (a) The variable is the height of the plant.

(b) The two populations are the cross-fertilized plants and the self-fertilized plants.

(c) The paired-difference variable is the difference in heights of the cross-fertilized plants and the self-fertilized plants grown in the same pot.

(d) Yes. They represent the difference in heights of two plants grown under the same conditions (same pot), one from each category of plants.

(e) Step 1: H_0: $\mu_1 = \mu_2$, H_a: $\mu_1 \neq \mu_2$

Step 2: $\alpha = 0.05$

Step 3: The paired differences are given.

$$t = \frac{\bar{d}}{s_d / \sqrt{n}} = \frac{20.93}{37.74 / \sqrt{15}} = 2.148$$

Step 4: df = 14; critical values = ± 2.145
0.02 < P-Value < 0.05

Step 5: Since 2.148 > 2.145, reject H_0. The data do provide sufficient evidence at the 5% significance level that there is a difference between the mean heights of cross-fertilized and self-fertilized plants.

(f) For the 0.01 significance level, the critical values are ± 2.977 and the P-value is the same as in part (e). Since t = 2.148 is not in the rejection region, do not reject the null hypothesis at the 0.01 significance level.

10.157 Population 1: Weights after treatment for anorexia nervosa
Population 2: Weights before treatment for anorexia nervosa

Step 1: H_0: $\mu_1 = \mu_2$, H_a: $\mu_1 > \mu_2$

Step 2: $\alpha = 0.05$

Step 3: The paired differences 'Weight after - Weight before' are 11.0, 5.5, 9.4, 13.6, -2.9, 10.7, -0.1, 7.4, 21.5, -5.3, -3.8, 11.4, 13.4, 13.1, 9.0, 3.9, 5.7

For these differences, $\sum d = 123.5$ and $\sum d^2 = 1716.9$

$$\bar{d} = \sum d / n = 123.5 / 17 = 7.26$$

$$s_d = \sqrt{\frac{\sum d^2 - \left(\sum d\right)^2 / n}{n-1}} = \sqrt{\frac{1716.9 - 123.5^2 / 17}{17-1}} = 7.16$$

$$t = \frac{\bar{d}}{s_d / \sqrt{n}} = \frac{7.26}{7.16 / \sqrt{17}} = 4.181$$

Step 4: df = 16; critical values = 1.746
P-Value < 0.005

Step 5: Since 4.181 > 1.746, reject H_0. The data do provide sufficient evidence at the 5% significance level that family therapy is effective in helping anorexic young women gain weight.
For the P-value approach, since the P-value is smaller than the significance level, reject H_0.

10.159 Population 1: Corneal thickness in normal eyes
Population 2: Corneal thickness in glaucoma eyes

Step 1: H_0: $\mu_1 = \mu_2$, H_a: $\mu_1 > \mu_2$

Step 2: $\alpha = 0.10$

Step 3: The paired differences, d = $x_1 - x_2$, are
-4, 0 12, 18, -4, -12, 6, 16

For these differences, $\sum d = 32$ and $\sum d^2 = 936$

$$\bar{d} = \sum d / n = 32/8 = 4.0$$

$$s_d = \sqrt{\frac{\sum d^2 - \left(\sum d\right)^2 / n}{n-1}} = \sqrt{\frac{936 - 32^2/8}{8-1}} = 10.7438$$

$$t = \frac{\bar{d}}{s_d / \sqrt{n}} = \frac{4.0}{10.7438/\sqrt{8}} = 1.053$$

Step 4: df = 7, Critical value = 1.415
P-Value > 0.10

Step 5: Since 1.053 < 1415, do not reject H_0.
For the P-value approach, since the P-value is larger than the significance level, do not reject H_0.

Step 6: The data do not provide sufficient evidence at the 10% significance level that the mean corneal thickness is greater in normal eyes than in eyes with glaucoma.

10.161 From Exercise 10.155, df = 14.

(a)

$$20.93 \pm 2.145 \cdot \frac{37.74}{\sqrt{15}} = 20.93 \pm 20.90 = 0.03 \text{ to } 41.83 \text{ eighths of an inch}$$

We can be 95% confident that the difference, $\mu_1 - \mu_2$, between the mean heights of cross-fertilized and self-fertilized Zea mays is somewhere between 0.03 and 41.83 eighths of an inch.

(b)

$$20.93 \pm 2.977 \cdot \frac{37.74}{\sqrt{15}} = 20.93 \pm 29.01 = -8.08 \text{ to } 49.94 \text{ eighths of an inch}$$

We can be 99% confident that the difference, $\mu_1 - \mu_2$, between the mean heights of cross-fertilized and self-fertilized Zea mays is somewhere between -8.08 and 49.94 eighths of an inch.

10.163 From Exercise 10.157, df = 16.

$$7.26 \pm 1.746 \cdot \frac{7.16}{\sqrt{17}} = 7.26 \pm 3.03 = 4.23 \text{ to } 10.29 \text{ pounds}$$

We can be 90% confident that the mean weight gain, $\mu_1 - \mu_2$, resulting from family-therapy treatment by anorexic young women is somewhere between 4.23 and 10.29 pounds.

10.165 From Exercise 10.159, df = 7.

$$4.0 \pm 1.415 \cdot \frac{10.7434}{\sqrt{8}} = 4.0 \pm 5.37 = (-1.37, 9.37) \text{ microns}$$

We can be 80% confident that the difference, $\mu_1 - \mu_2$, in corneal thickness of patients who had glaucoma in one eye but not the other is between -1.37 and 9.37 microns.

10.167 Using Minitab, with the differences in a column named DIFF, choose **Graph ▶ Probability plot...**, select the **Single** version and click **OK**. Enter DIFF in the **Graph variables text box** and click **OK**. Then choose **Graph ▶ Boxplot**, select the **One Y Simple** version and click **OK**. Enter DIFF in **Graph variables** text box and click **OK**.

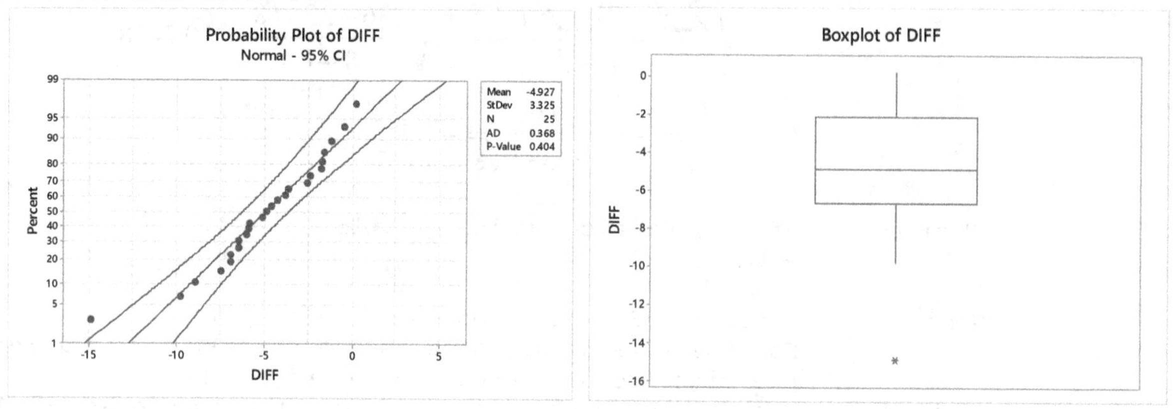

The sample size is moderate (25), and there is an outliers in the differences. Since the t-test is based on an assumption of normality, the t-test is not appropriate for this data. However, since the sample size is close to 30, perhaps an analysis of the differences with and without the outlier can be explored to see the effect of the outlier on the conclusion.

10.169 Using Minitab, enter DIFF at the top of column C3, then choose **Calc ▶ Calculator,** enter DIFF in the **Store result in variable** box, enter 'RESOLUTION'-'ONSET' in the Expression box, and click **OK**. Choose **Graph ▶ Probability plot...**, select the **Single** version and click **OK**. Enter ONSET RESOLUTION DIFF in the **Graph variables** text box and click **OK**. Then choose **Graph ▶ Boxplot**, select the **Multiple Y's Simple** version and click **OK**. Enter ONSET and RESOLUTION in the **Graph Variables** box and click **OK**. Finally, choose **Graph ▶ Boxplot**, select the **One Y Simple** version and click **OK**. Enter DIFF in **Graph variables** text box and click **OK**. The results are

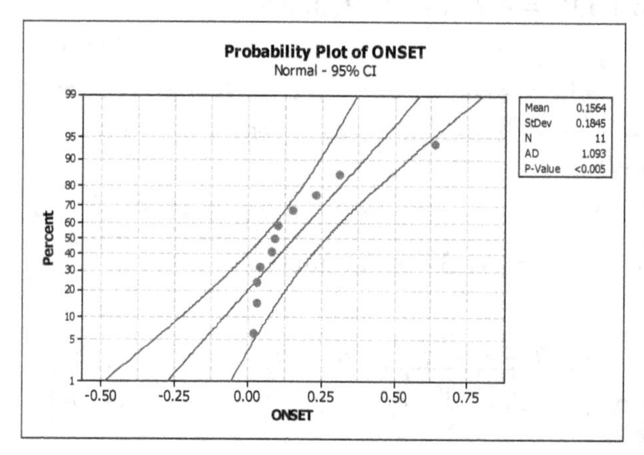

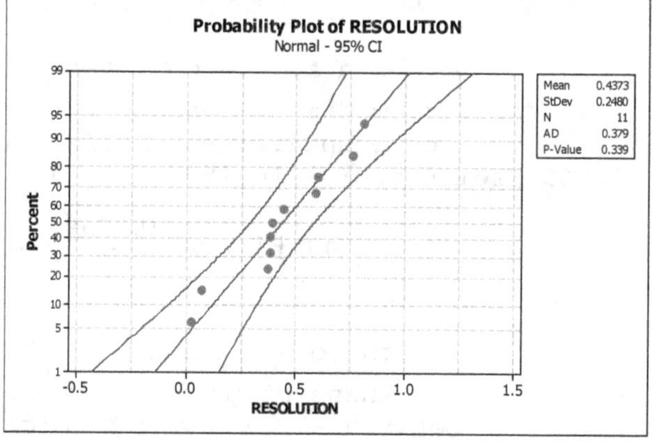

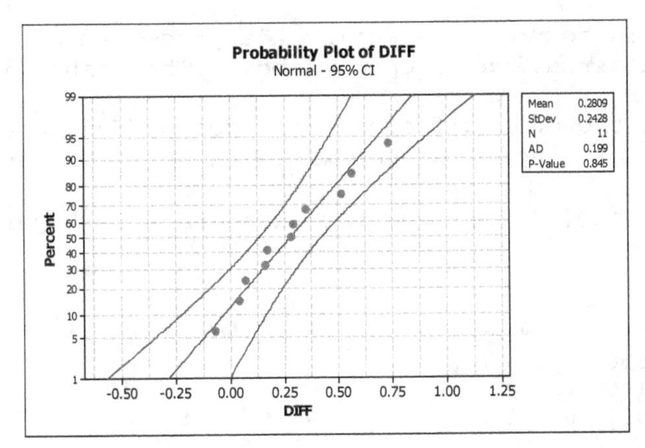

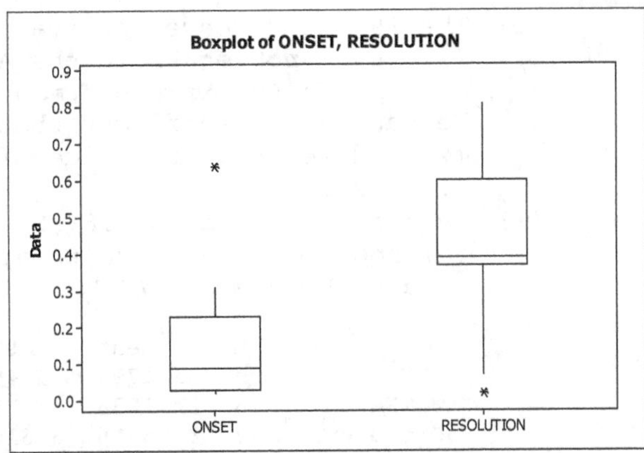

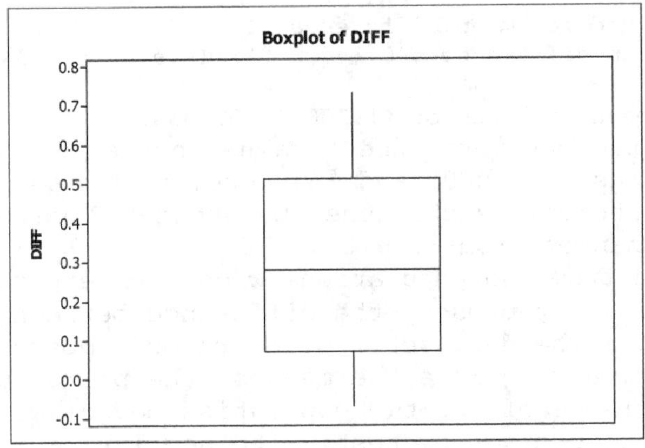

(b) No. The sample size is small, there is an outlier, and the normal probability plot is not linear.

(c) No. The sample size is small and there is an outlier. The probability plot is closer to linear than for the ONSET data, but overall, t-procedures are not advisable.

(d) Yes. There are no outliers and the probability plot is reasonably linear.

(e) They imply that the assumptions for using a paired t-procedure may be met even though the assumptions for using a t-procedure may not be met for the original variables.

10.171 (a) Using Minitab, choose **Stat ▶ Basic Statistics ▶ Paired t...**, enter MEN in the **First sample** box and WOMEN in the **Second Sample** box. Click on the **Options** button and enter 99 in the **Confidence** level box, 0.0 in the **Test mean** box, and **greater than** in the **Alternative** drop down box. Click **OK** twice. The results are

```
Paired T for MEN - WOMEN

             N     Mean    StDev  SE Mean
MEN         75  35.4800   9.7695   1.1281
WOMEN       75  32.0800   7.0265   0.8114
Difference  75  3.40000  7.23542  0.83547

99% lower bound for mean difference: 1.41341
T-Test of mean difference = 0 (vs > 0): T-Value = 4.07  P-Value = 0.000
```

The P-value is 0.000, which is less than the significance level 0.01. Therefore, the data do provide sufficient evidence that the mean age of Norwegian men at the time of marriage exceeds that of Norwegian women.

(b) The 99% confidence interval is found by repeating the procedure of part (a) using **not equal** in the **Alternative** droop down box. The result is

```
99% CI for mean difference: (1.19108, 5.60892)
```

We can be 99% confident that the mean difference in ages of Norwegian men and women at the time of marriage lies somewhere between 1.19 and 5.61 years.

(c) We remove the pairs (18,32) and (53,29) from the data and repeat parts (a) and (b). The results are

```
Paired T for MEN - WOMEN

              N     Mean    StDev  SE Mean
MEN          73  35.4795   9.4650   1.1078
WOMEN        73  32.1233   7.1140   0.8326
Difference   73  3.35616  6.61095  0.77375

99% lower bound for mean difference: 1.51520
T-Test of mean difference = 0 (vs > 0): T-Value = 4.34  P-Value = 0.000

99% CI for mean difference: (1.30893, 5.40340)
```

By removing the outliers, the t-value increases from 4.07 to 4.34, the P-value remains at 0.000, and the conclusion remains the same. The 99% confidence interval is shortened by about 0.3 years and the mean difference changes from 3.40 to 3.36.

10.173 Data is obtained from two populations whose members can be naturally paired. By letting d represent the difference between the values in each pair, we can reduce the data set from pairs of numbers to a single number representing each pair. Since the mean of the paired differences is the same as the difference of the two population means, we can test the equality of the two population means by testing to see if the mean of the paired differences is zero (or some other value if appropriate). If the paired differences are normally distributed, then the one-sample t-test can be used to carry out the test.

10.175 (a) Population 1: Additive used, $n_1 = 10$, $\bar{x}_1 = 18.910$, $s_1 = 7.4721$

Population 2: Additive not used, $n_2 = 10$, $\bar{x}_2 = 18.250$, $s_2 = 7.4188$

Step 1: H_0: $\mu_1 = \mu_2$, H_a: $\mu_1 > \mu_2$

Step 2: $\alpha = 0.05$

Step 3: The paired differences, $d = x_1 - x_2$, are

0.8	1.2	0.7	0.7	1.0	1.2	0.6	-0.1	0.0	0.5

$$\bar{d} = \frac{\sum d_i}{n} = \frac{6.600}{10} = 0.66 \qquad s_d = \sqrt{\frac{\sum d^2 - (\sum d)^2/n}{n-1}} = \sqrt{\frac{6.120 - 43.56/10}{9}} = 0.443$$

$$t = \frac{\bar{d}}{s_d/\sqrt{n}} = \frac{0.66}{0.443/\sqrt{10}} = 4.711$$

Step 4: df = 9; critical values = 1.833
P-Value < 0.005

Step 5: Since 4.711 > 1.833 and the P-value is less than 5%, reject H_0.

Step 6: At 5% significance level, the data provide evidence that the mean gas mileage with the gasoline additive is higher than the mean gas mileage without the additive.

(b) $$s_p = \sqrt{\frac{9(7.4721)^2 + 9(7.4188)^2}{10+10-2}} = \sqrt{55.4354} = 7.4454$$

Step 1: H_0: $\mu_1 = \mu_2$, H_a: $\mu_1 > \mu_2$

Step 2: $\alpha = 0.05$

Step 3: $t = \dfrac{18.91 - 18.25}{7.4454\sqrt{(1/10)+(1/10)}} = 0.198$

Step 4: df = 18, Critical values = 1.734
P-Value > 0.10

Step 5: Since 0.198 < 1.734 and since the P-value is greater than 5%, do not reject H_0.

Step 6: At the 5% significance level, the data do not provide sufficient evidence to conclude that the mean gas mileage when the additive is used is greater than the mean gas mileage when the additive is not used.

(c) It is inappropriate to perform the hypothesis test as presented in part (b) because the data are paired and so the samples are not independent.

(d) In part (a), where the appropriate procedure was used, we rejected H_0. On the other hand, in part (b), where an inappropriate procedure was used, we did not reject H_0.

Exercises 10.6

10.177 (a) No. Unless the sample size is large, the paired t-test should only be used when the distribution is normal.

(b) Yes. When the sample size is large, the distribution of the sample mean will be approximately normal regardless of the shape of the distribution and the t-test will be nearly the same as a z-test.

(c) Yes. The paired Wilcoxon signed-rank test requires only that the distribution of differences be symmetric.

(d) When the paired-difference variable is far from normally distributed, the paired Wilcoxon signed-rank test is preferred because it is usually more powerful in this situation.

10.179 (a) Paired t-test. With approximately normally distributed data, the t-test will be the more powerful test.

(b) Neither. The Wilcoxon signed-rank test is based on an assumption of symmetry in the distribution and the paired t-test is based on an assumption of normality or large samples. Neither assumption appears to be valid.

(c) Wilcoxon signed-rank test. The data are not normally distributed, but they are symmetric.

10.181 Step 1: H_0: $\mu_1 = \mu_2$, H_a: $\mu_1 \neq \mu_2$

Step 2: $\alpha = 0.10$

Step 3:

d	|d|	Rank	Signed-rank
2	2	3	3
1	1	1.5	1.5
3	3	4.5	4.5
6	6	7	7
4	4	6	6
-1	1	1.5	-1.5
3	3	4.5	4.5

W = 26.5

Step 4: The critical values are $W_{0.05}=24$ and $W_{0.95}=7(8)/2-24=4$

Step 5: Since 26.5 > 24, reject H_0.

10.183 Step 1: H_0: $\mu_1 = \mu_2$, H_a: $\mu_1 > \mu_2$

Step 2: $\alpha = 0.10$

Step 3:

d	|d|	Rank	Signed-rank
4	4	4	4
-1	1	1.5	-1.5
1	1	1.5	1.5
5	5	5.5	5.5
3	3	3	3
0	0		
-5	5	5.5	-5.5
-6	6	7	-7

There was one difference of zero which was discarded and the sample size was reduced to 7.

W = 14

Step 4: The critical value is $W_{.10}=22$

Step 5: Since 14 < 22, do not reject H_0.

10.185 Step 1: H_0: $\mu_1 = \mu_2$, H_a: $\mu_1 < \mu_2$

Step 2: $\alpha = 0.10$

Step 3:

d	|d|	Rank	Signed-rank
-3	3	5	-5
-3	3	5	-5
-2	2	2.5	-2.5
3	3	5	5
-6	6	7	-7
0	0		
-2	2	2.5	-2.5
-7	7	8	-8
-1	1	1	-1

There was one difference of zero which was discarded and the sample size was reduced to 8. W = 5

Step 4: The critical value is $W_{.90}= 8(9)/2 - 28 = 8$

Step 5: Since 5 < 8 , reject H_0.

10.187(a) Population 1: Cross-fertilized; Population 2: Self-fertilized

Step 1: H_0: $\mu_1 = \mu_2$, H_a: $\mu_1 \neq \mu_2$

Step 2: $\alpha = 0.05$

Step 3:

d	|d|	Rank	Signed-rank
49	49	11	11
-67	67	14	-14
8	8	2	2
16	16	4	4
6	6	1	1
23	23	5	5
28	28	7	7
41	41	9	9
14	14	3	3
29	29	8	8
56	56	12	12
24	24	6	6
75	75	15	15
60	60	13	13
-48	48	10	-10

The value of the test statistic is the sum of the positive ranks.

W = 11 + 2 + 4 + 1 + 5 + 7 + 9 + 3 + 8 + 12 + 6 + 15 + 13 = 96

Step 4: The critical values are W_l = 25, W_r = 95

Step 5: Since 96 > 95, reject H_0.

Step 6: At the 5% significance level, the data provide enough evidence to conclude that there is a difference in the mean heights of cross-fertilized Zea mays and self-fertilized Zea mays.

(b) All of the computations above are the same for a 1% significance level except for the critical values which are now W_l = 16 and W_r = 104. Since W = 96 is between the two critical values, we do not reject the null hypothesis at the 1% significance level. Thus at the 1% level, the data do not provide enough evidence to conclude that there is a difference in the mean heights of cross-fertilized Zea mays and self-fertilized Zea mays.

10.189 (a) Population 1: Weight before therapy

Population 2: Weight after therapy

Step 1: H_0: $\mu_1 = \mu_2$, H_a: $\mu_1 < \mu_2$

Step 2: α = 0.05

Step 3:

D	|d|	Rank	Signed-rank
-11.0	11.0	12	-12
-5.5	5.5	6	-6
-9.4	9.4	10	-10
-13.6	13.6	16	-16
2.9	2.9	2	2
-10.7	10.7	11	-11
0.1	0.1	1	1
-7.4	7.4	8	-8
-21.5	21.5	17	-17
5.3	5.3	5	5
3.8	3.8	3	3
-11.4	11.4	13	-13
-13.4	13.4	15	-15
-13.1	13.1	14	-14
-9.0	9.0	9	-9
3.9	3.9	4	-4
-5.7	5.7	7	-7

The value of the test statistic is the sum of the positive ranks.

W = 1 + 2 + 3 + 5 = 11

Step 4: The critical value: W_l = (17)(18)/2 - 112 = 41

Step 5: Since 11 < 41, reject H_0.

Step 6: At the 5% significance level, the data provide enough evidence to conclude that the family therapy is effective in helping anorexic young women gain weight.

10.191 Population 1: Corneal thickness in normal eyes

Population 2: Corneal thickness in eyes with Glaucoma

Step 1: H_0: $\mu_1 = \mu_2$, H_a: $\mu_1 > \mu_2$

Step 2: α = 0.10

Step 3:

d	|d|	Rank	Signed-rank
-4	4	1.5	-1.5
0	0	---	---
12	12	4.5	4.5
18	18	7	7
-4	4	1.5	-1.5
-12	12	4.5	-4.5
6	6	3	3
16	16	6	6

The value of the test statistic is the sum of the positive ranks. $W = 4.5 + 7 + 3 + 6 = 20.5$

Step 4: The critical value: $W_r = 22$

Step 5: Since 20.5 < 22, do not reject H_0.

Step 6: At the 10% significance level, the data do not provide enough evidence to conclude that the mean corneal thickness is greater in normal eyes than in eyes with glaucoma.

10.193 Using Minitab, with the differences in a column named DIFF, choose **Graph ▶ Probability plot...**, select the **Single** version and click **OK**. Enter DIFF in the **Graph variables text box** and click **OK**. Then choose **Graph ▶ Boxplot**, select the **One Y Simple** version and click **OK**. Enter DIFF in **Graph variables** text box and click **OK**.

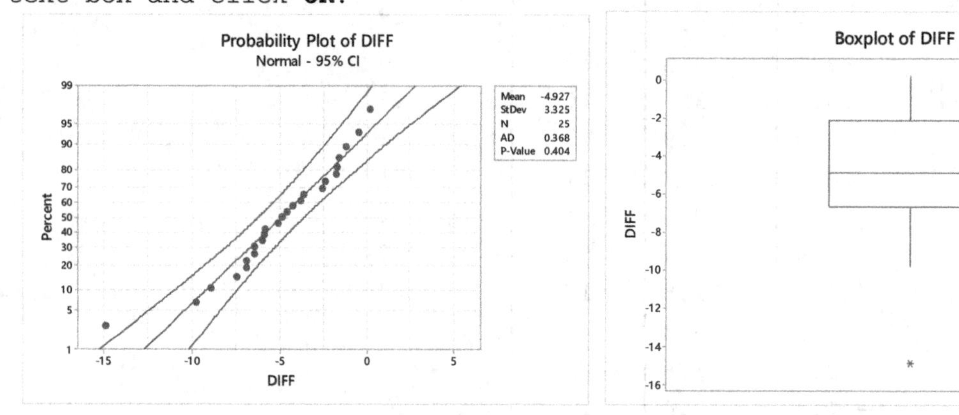

The sample size is moderate (25), and there is an outliers in the differences. Otherwise, the boxplot reveals that the distribution is symmetric. Since the sample size is close to 30, perhaps an analysis of the differences with and without the outlier can be explored to see the effect of the outlier on the conclusion. Most likely it is OK to use the paired Wilcoxon signed-rank test.

10.195(a) Using Minitab, we prepared a normal probability plot and a boxplot of the difference (Initial - Final) data.

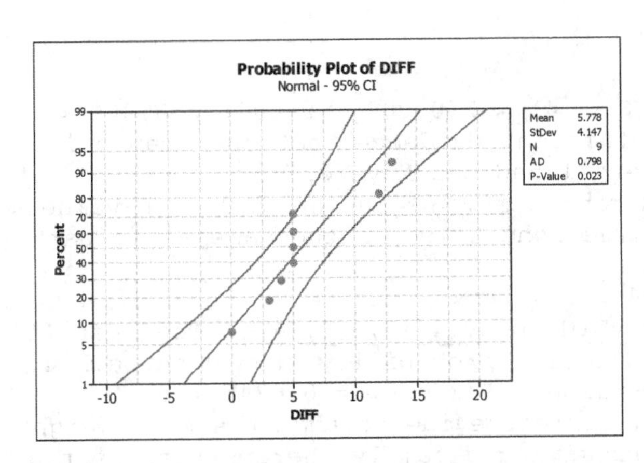

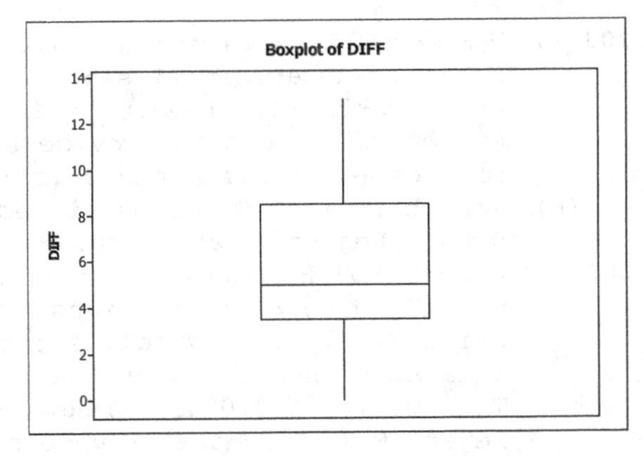

It is not reasonable to apply the paired-t test. The sample size is small and the probability plot is far from linear.

(b) The paired Wilcoxon signed-rank test is reasonable. We see from the box plot that the data is nearly symmetric.

10.197 (a) Using Minitab, choose **Calc ▶ Calculator,** enter 'DIFF' in the **Store result in variable** text box, and 'MEN' - 'WOMEN' in the Expression box and click **OK**. **Stat ▶ Nonparametrics ▶ 1-Sample Wilcoxon**, enter DIFF in the **Variables** text box, enter 0.0 in the **Test median** box and **greater than** in the **Alternative** drop down box, click **OK**. The result is

```
Test of median = 0.000000 versus median > 0.000000

              N
             for   Wilcoxon                Estimated
        N   Test   Statistic       P          Median
DIFF   75     68      1763.0   0.000           3.000
```

The P-value = 0.000, so reject the null hypothesis. The data do provide sufficient evidence that the mean age of Norwegian men at the time of marriage exceeds that of Norwegian women.

(b) In Exercise 10.171, the P-value was also zero, so the conclusions are the same.

(c) Paired Wilcoxon signed-rank test. The difference data have two outliers, one on the left and one on the right. These will not affect the mean difference much, but they will affect the standard deviation. They will have no effect on the Wilcoxon test. The sample size of 73 is quite large, so if the Wilcoxon test has any advantage over the paired t-test, that advantage is small.

10.199 Since the difference of the two population means, $\mu_1 - \mu_2$, is the same as the mean of the differences, μ_d, the null hypothesis that $\mu_1 = \mu_2$ or $\mu_1 - \mu_2 = 0$ is the same as $\mu_d = 0$.

10.201 (a) The paired Wilcoxon signed-rank test makes use of the magnitude of each difference as well as the sign of the difference. If the only criterion for using the Wilcoxon test, symmetry of the distribution, is satisfied, the paired Wilcoxon singed-rank test will be more powerful than the paired sign test, which does not require symmetry.

(b) The paired sign test can be used for any data, regardless of the shape of the distribution. It does not require symmetry for the distribution, as does the Wilcoxon signed-rank test.

10.203 (a) H_0: $\eta_d = 0$; H_0: $\eta_d \neq 0$; $\alpha = 0.05$
n = 15, Number of '+' signs = 14
Using Table XII, P-value = $2P(X \geq 14) = 2(0.000 + 0.000) = 0.000 < \alpha$
Reject H_0. The data provide sufficient evidence that the mean heights of cross-fertilized and self-fertilized Zea Mays differ.

(b) For the paired Wilcoxon signed-rank test, W = 96 and 0.02 < P-value < 0.05. Reject H_0 at $\alpha = 0.05$. The conclusion is the same as in part

10.205 (a) H_0: $\eta_d = 0$; H_0: $\eta_d < 0$; $\alpha = 0.05$
n = 17, Number of '+' signs = 4
Using Excel, in any cell enter =BINOMDIST(4,17,0.5,1) . The last '1' indicates that you want the cumulative probability from 0 through 4. The result is 0.0245. Thus, P-value = $P(X \leq 4) = 0.0245 < \alpha$.
Reject H_0. The data provide sufficient evidence that the mean weights of anorexic young women after receiving a family therapy treatment are greater than the mean weights before treatment.

(b) For the paired Wilcoxon signed-rank test, W = 11 and P-value < 0.005. Reject H_0 at $\alpha = 0.05$. The conclusion is the same as in part (a).

10.207 (a) H_0: $\eta_d = 0$; H_0: $\eta_d > 0$; $\alpha = 0.10$
n = 7, Number of '+' signs = 4 (There was one 0 which was dropped.)
Using Table XII, P-value = $P(X \geq 4) = (0.273 + 0.164 + 0.055 + 0.008) = 0.500 > \alpha$.
Do not reject H_0. The data do not provide sufficient evidence that the mean corneal thickness of normal eyes is greater than in eyes with glaucoma.

(b) For the paired Wilcoxon signed-rank test, W = 20.5 and P-value > 0.100. Do not reject H_0 at $\alpha = 0.10$. The conclusion is the same as in part (a).

Exercises 10.7

10.209 (a) Pooled t-test, nonpooled t-test, Mann-Whitney test

(b) **Pooled t-test**: Independent simple random samples, normal populations or large samples, and equal population standard deviations
Nonpooled t-test: Independent simple random samples, and normal populations or large samples
Mann-Whitney signed-rank test: Independent simple random samples, same shape populations, and $n_1 \leq n_2$

(c) **Pooled t-test**

$$t = \frac{\bar{x}_1 - \bar{x}_2}{s_p\sqrt{\frac{1}{n_1} + \frac{1}{n_2}}} \text{ where } s_p = \sqrt{\frac{(n_1 - 1)s_1^2 + (n_2 - 1)s_2^2}{n_1 + n_2 - 2}}$$

and the degrees of freedom are $n_1 + n_2 - 2$.

Nonpooled t-test

$$t = \frac{\bar{x}_1 - \bar{x}_2}{\sqrt{\frac{s_1^2}{n_1} + \frac{s_2^2}{n_2}}}$$

where the degrees of freedom are given by

$$\Delta = \frac{[s_1^2/n_1 + s_2^2/n_2]^2}{\frac{(s_1^2/n_1)^2}{n_1 - 1} + \frac{(s_2^2/n_2)^2}{n_2 - 1}}$$

. This number is truncated to an integer if it contains a fractional part.

Mann-Whitney test
M = the sum of the ranks for sample data from Population 1

10.211(a) One could use the pooled t-test, nonpooled t-test, or Mann-Whitney test.

(b) The pooled t-test is the most appropriate since its conditions are satisfied and it will be the most powerful under these circumstances.

10.213(a) Because the sample sizes are large, one could use the pooled t-test, nonpooled t-test, or Mann-Whitney test.

(b) Since the populations have the same shape but are not normally distributed, the Mann-Whitney test is the most appropriate since its conditions are satisfied and it will be the most powerful under these circumstances.

10.215(a) One could use the paired t-test or the paired Wilcoxon signed-rank test (paired W-test).

(b) The paired W-test is the more appropriate test when the distribution is symmetric, but nonnormal.

10.217 To determine which procedure should be used to perform the hypothesis test, ask the following questions in sequence according to the flowchart in Figure 10.19:

	Answer to question	
Question to ask	Yes	No
Are the samples paired?		x
Are the populations normal?	x	
Are the populations the same shape?		x
Is the sample large?		x

Since the two independent populations are normal, use the nonpooled t-test.

10.219 To determine which procedure should be used to find the required confidence interval, ask the following questions in sequence according to the flowchart in Figure 10.19:

	Answer to question	
Question to ask	Yes	No
Are the samples paired?		x
Are the populations normal?		x
Are the samples large?	x	

It appears that one of the populations is neither normal nor symmetric, but the sample size is large. Thus, use the nonpooled t-test.

10.221 To determine which procedure should be used to perform the hypothesis test, ask the following questions in sequence according to the flowchart in Figure 10.19:

	Answer to question	
Question to ask	Yes	No
Are the samples paired?	x	
Are the differences normal?		x
Are the differences symmetric?		x
Is the sample large?		x

It appears the paired differences are neither normal or symmetric. Thus, a statistician must be consulted.

Review Problems for Chapter 10

1. Randomly sample independently from both populations, compute the means of both samples and reject the null hypothesis if the sample means differ by too much. Otherwise, do not reject the null hypothesis.
2. Sample independent pairs of observations from the two populations, compute the difference of the two observations in each pair, compute the mean of the differences, and reject the null hypothesis if the sample mean of the differences differs from zero by too much. Otherwise, do not reject the null hypothesis.
3. (a) The pooled t-test requires that the population standard deviations be equal whereas the nonpooled t-test does not.
   (b) It is absolutely essential that the assumption of independence be satisfied.
   (c) The normality assumption is especially important for both t-tests for small samples. With large samples, the Central Limit Theorem applies and the normality assumption is less important.
4. Unless you are quite sure that the population standard deviations are equal, the nonpooled *t*-procedures should be used instead of the pooled *t*-procedures.
5. (a) No. If the two distributions are normal and have the same shape, then they have the same population standard deviations, and the pooled t-test should be used. If the two distributions are not normal, but have the same shape, the Mann-Whitney test is preferred.
   (b) The pooled t-test is preferred to the Mann-Whitney test if both populations are normally distributed with equal standard deviations.
6. A paired sample may reduce the estimate of the standard error of the mean of the differences, making it more likely that a difference of a given size will be judged significant.
7. The paired t-test is preferred to the paired Wilcoxon signed-rank test if the distribution of differences is normal, or if the sample size is large and the distribution of differences is not symmetric. If the distribution of differences is symmetric but not normal, then the paired Wilcoxon signed-rank test is preferred.
8. (a) Population 1: Male; $n_1 = 13$, $\bar{x}_1 = 2127$, $s_1 = 513$

   Population 2: Female; $n_2 = 14$, $\bar{x}_2 = 1843$, $s_2 = 446$

   Step 1: H_0: $\mu_1 = \mu_2$, H_a: $\mu_1 > \mu_2$

   Step 2: $\alpha = 0.05$,

   Step 3:

$$s_p = \sqrt{\frac{12(513)^2 + 13(446)^2}{25}} = 479.33$$

$$t = \frac{2127 - 1843}{479.33\sqrt{\frac{1}{13} + \frac{1}{14}}} = 1.538$$

   Step 4: df = 25, Critical value = 1.708
   $0.05 < P(t > 1.538) < 0.10$

   Step 5: Since $1.538 < 1.708$ and since the P-value is larger than the significance level, do not reject H_0.

Step 6: At the 5% significance level, the data do not provide enough evidence to conclude that the mean right-leg strength of males exceeds that of females.

9.

$$2127-1843 \pm 1.708(479.33)\sqrt{\frac{1}{13}+\frac{1}{14}}$$

$$284 \pm 315.33$$

$$-31.33 \text{ to } 599.33$$

We can be 90% confident that the difference, $\mu_1 - \mu_2$, between the mean right-leg strengths of males and females is between -31.3 and 599.3 newtons.

10. Population 1: Florida, $n_1 = 24$, $\bar{x}_1 = 5.46$, $s_1 = 1.59$

Population 2: Virginia, $n_2 = 44$, $\bar{x}_2 = 7.59$, $s_2 = 2.68$

$$\Delta = \frac{\left[\left(\frac{s_1^2}{n_1}\right)+\left(\frac{s_2^2}{n_2}\right)\right]^2}{\frac{\left(\frac{s_1^2}{n_1}\right)^2}{n_1-1}+\frac{\left(\frac{s_2^2}{n_2}\right)^2}{n_2-1}} = \frac{\left[\left(\frac{5.46^2}{24}\right)+\left(\frac{7.59^2}{44}\right)\right]^2}{\frac{\left(\frac{5.46^2}{24}\right)^2}{24-1}+\frac{\left(\frac{7.59^2}{44}\right)^2}{44-1}} = 65.45;\ df = 65$$

Step 1: $H_0: \mu_1 = \mu_2$, $H_a: \mu_1 < \mu_2$

Step 2: $\alpha = 0.01$

Step 3:

$$t = \frac{5.4583-7.5909}{\sqrt{\frac{1.5874^2}{24}+\frac{2.6791^2}{44}}} = -4.119$$

Step 4: Critical value = -2.385
$P(t < -4.119) < 0.005$

Step 5: Since -4.119 < -2.385 and also the P-value is smaller than the significance level, reject H_0.

Step 6: At the 1% significance level, the data provide sufficient evidence to conclude that the average litter size of cottonmouths in Florida is less than that in Virginia.

11.

$$(5.46-7.49) \pm 2.385\sqrt{\frac{1.59^2}{24}+\frac{2.68^2}{44}}$$

$$-2.13 \pm 1.24$$

$$-3.37 \text{ to } -0.89$$

We can be 98% confident that the average difference, $\mu_1 - \mu_2$, between cottonmouth litter sizes in Florida and Virginia is somewhere between -3.37 and -0.89.

12. Population 1: Home prices in Atlantic City;
Population 2: Home prices in Las Vegas.

Step 1: H_0: $\mu_1 = \mu_2$, H_a: $\mu_1 \neq \mu_2$

Step 2: $\alpha = 0.05$

Step 3: Construct a worktable based upon the following: First, rank all the data from both samples combined. Adjacent to each column of data as it is presented in the Exercise, record the overall rank.

Atlantic City	Overall Rank	Las Vegas	Overall Rank
207.2	10	177.65	7
186.2	9	165.95	5
596.3	20	418.15	18
265.2	14	121.25	1
209.3	11	300.65	16
166.0	6	182.75	8
229.6	13	162.15	4
223.4	12	125.85	2
370.7	17	488.95	19
275.1	15	129.75	3

The value of the test statistic is the sum of the ranks for the sample data from Population 1:
$M = 10 + 9 + 20 + 14 + 11 + 6 + 13 + 12 + 17 + 15 = 127$.

Step 4: We have $n_1 = 10$ and $n_2 = 10$. Since the hypothesis test is two-tailed with $\alpha = 0.05$, we use Table VI to obtain the critical values, which are $M_{0.975} = 10(21) - 131 = 79$ and $M_{0.025} = 131$. Thus, we reject H_0 if $M \geq 131$ or $M \leq 79$.

Step 5: Since $79 < M < 131$, do not reject H_0.

Step 6: At the 5% significance level, the data do not provide sufficient evidence to conclude that the mean costs for existing single-family homes differ in Atlantic City and Las Vegas.

13. (a) We used Minitab to produce the following normal probability plot and boxplot of the differences.

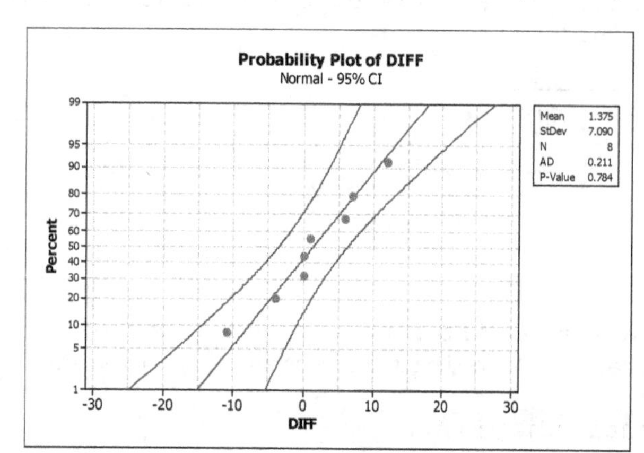

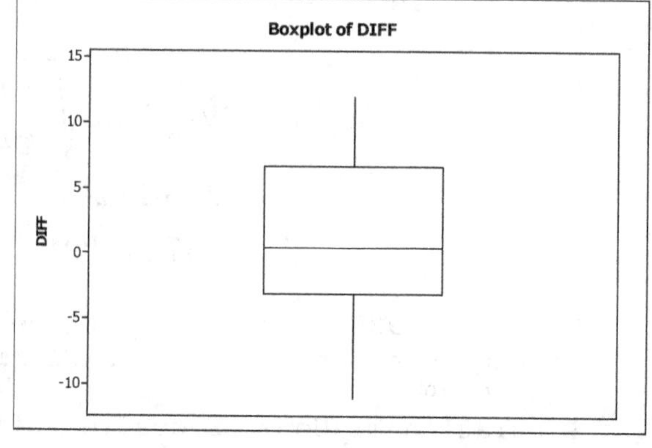

(b) Yes. The normal probability plot is reasonably linear and the boxplot shows no outliers.

(c) Step 1: H_0: $\mu_1 = \mu_2$, H_a: $\mu_1 \neq \mu_2$

Step 2: $\alpha = 0.10$

Step 3: Paired differences, $d = x_1 - x_2$:

12	7	1	0
-11	0	6	-4

From the probability plot, we see that the sample mean of the differences is 1.375 and the standard deviation is 7.090. Therefore,

$$t = \frac{1.375}{7.090/\sqrt{8}} = 0.549$$

Step 4: Critical values = ± 1.895

$2\{P(t > 0.549)\} > 0.20$

Step 5: Since $-1.895 < 0.549 < 1.895$ and also since the P-value is greater than the significance level, do not reject H_0.

Step 6: At the 10% significance level, the data do not provide sufficient evidence to conclude that there is a difference in the mean length of time that ice stays on the two lakes.

14.

$$1.375 \pm 1.860 \cdot 7.090/\sqrt{8} = 1.375 \pm 4.749 = -3.374 \text{ to } 6.124 \text{ days}$$

We can be 90% confident that the difference, $\mu_1 - \mu_2$, between the mean numbers of days ice stays on Lakes Mendota and Monona is somewhere between -3.374 and 6.124 days.

15. Population 1: Before; Population 2: After;

Step 1: H_0: $\mu_1 = \mu_2$, H_a: $\mu_1 > \mu_2$

Step 2: $\alpha = 0.05$

Step 3: Delete the one observation with a difference of zero

Before	After	Difference	Rank	Signed Rank
212	194	18	9	9
142	159	-17	8	-8
200	202	-2	1.5	-1.5
180	160	20	10	10
115	105	10	6	6
184	134	50	18	18
211	219	-8	4.5	-4.5
202	179	23	11	11
165	137	28	13	13
173	175	-2	1.5	-1.5
137	129	8	4.5	4.5
195	164	31	14	14
175	149	26	12	12
176	180	-4	3	-3
225	214	11	7	7
233	189	44	16	16
185	231	-46	17	-17
228	187	41	15	15

The value of the test statistic is the sum of the positive ranks.

$W = 9+10+6+18+11+13+4.5+14+12+7+16+15 = 135.5$

Step 4: The critical value, $W_r = 124$

Step 5: Since $135.5 > 124$, reject H_0.

Step 6: At the 5% significance level, the data do provide sufficient evidence to conclude that, for females who suffer from depression, the mean total cholesterol level is lower after a 4-week treatment of the antidepressant fluoxetine.

16. (a) Using Minitab, choose **Graph ▶ Probability plot...**, select the **Single** version and click **OK**. Enter MALE FEMALE in the **Graph Variables** text box, and click **OK**. Then choose **Graph ▶ Boxplot**, select the **Multiple Y's Simple** version and click **OK**. Enter MALE FEMALE **in the Graph variables** text box, and click **OK**. The standard deviations will be produced on the probability plot legend. The results are

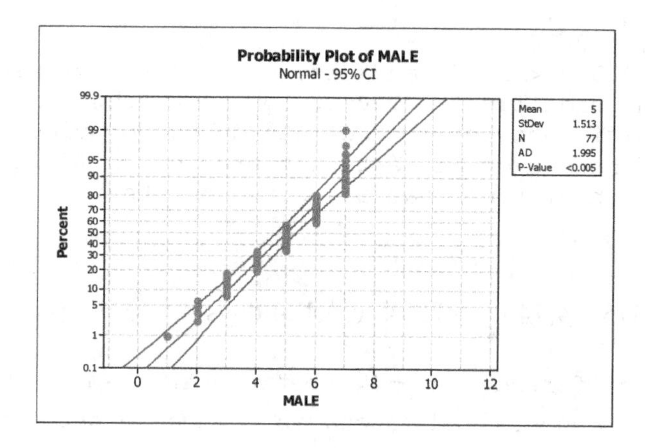

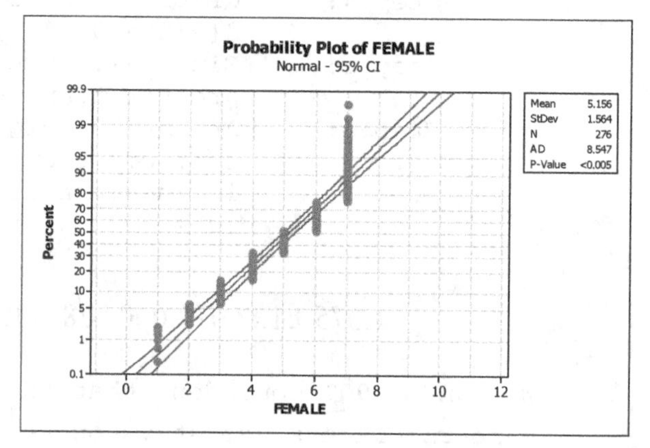

The standard deviations are 1.513 for MALE and 1.564 for FEMALE.

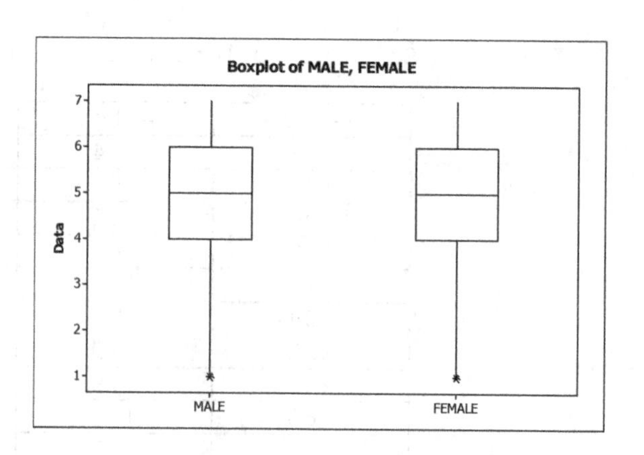

(b) Pooled t-procedures. The sample sizes are both large, there are no potential outliers in either sample, and the sample standard deviations are nearly equal.

(c) To carry out the pooled t-test, choose **Stat ▶ Basic Statistics ▶ 2-Sample t**, click on **Samples in different columns**, enter MALE in the **First** text box and FEMALE in the **Second** text box, select **not equal** in the **Alternative** box, enter 90 in the **Confidence level** text box, click on **Assume equal variances** to make certain that there is an **X** in the box, and click **OK**. The results are

```
Two-sample T for MALE vs FEMALE

          N  Mean  StDev  SE Mean
MALE     77  5.00   1.51     0.17
FEMALE  276  5.16   1.56    0.094


Difference = mu (MALE) - mu (FEMALE)
Estimate for difference:  -0.155797
95% CI for difference:  (-0.549387, 0.237793)
T-Test of difference = 0 (vs not =): T-Value = -0.78  P-Value = 0.437  DF =
351
Both use Pooled StDev = 1.5528
```

The P-value is 0.437, which is greater than the significance level 0.05. Do not reject the null hypothesis. The data do not provide sufficient evidence to conclude that there is a difference in mean dating satisfaction of male and female college students.

(d) A 95% confidence interval for the difference between mean dating satisfaction of male and female college students was produced as part of the process in part (b). The interval is (-0.55, 0.24).

(e) Yes. Although the data for both genders are left-skewed and discrete (possible values include only the whole numbers from 1 to 7), the large sample sizes justify the use of the pooled t-procedures.

17. (a) Using Minitab, choose **Graph ▶ Histogram...**, select the **Simple** version and click **OK**. Enter MALE FEMALE in the **Graph Variables** text box, clcik on the scale button and the Y-Scle type tab. Select Percent and click **OK** twice. The results are

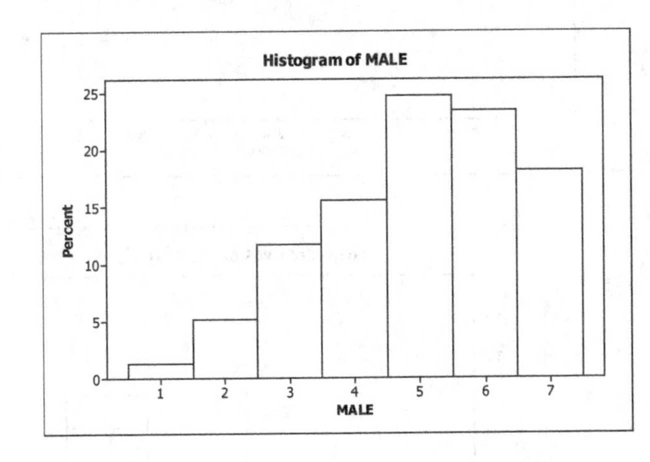

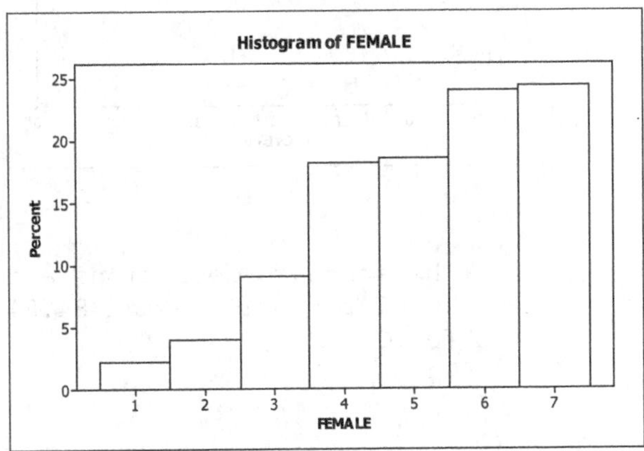

(b) Yes. All that is required for the Mann-Whitney test is that both populations be the same shape. We can see from the histograms that this is the case.

(c) Choose **Stat ▶ Nonparametrics ▶ Mann-Whitney,** enter MALE in the **First Sample** text box and FEMALE in the **Second Sample** text box. Enter 95 in the **Confidence level** text box and select **not equal** in the **Alternative** drop down box. The results are

Mann-Whitney Test and CI: MALE, FEMALE

```
          N  Median
MALE     77  5.0000
FEMALE  276  5.0000

Point estimate for ETA1-ETA2 is -0.0000
95.0 Percent CI for ETA1-ETA2 is (-1.0000,0.0001)
W = 12917.0
Test of ETA1 = ETA2 vs ETA1 not = ETA2 is significant at 0.3689
The test is significant at 0.3591 (adjusted for ties)
```

The P-value is 0.3591, which is greater than the significance level 0.05. Do not reject the null hypothesis. The data do not provide sufficient evidence that there is a difference in mean dating satisfaction for male and female college students. This is the same conclusion we reached in Problem 16.

18. (a) Using Minitab, choose **Graph ▶ Probability plot...**, select the **Single** version and click **OK**. Enter OVER 65 and CONTROL in the **Graph Variables** text box, and click **OK**. Then choose **Graph ▶ Boxplot**, select the **Multiple Y's Simple** version and click **OK**. Enter OVER 65 and CONTROL **in the Graph variables** text box, and click **OK**. The standard deviations will be produced on the probability plot legend. The results are

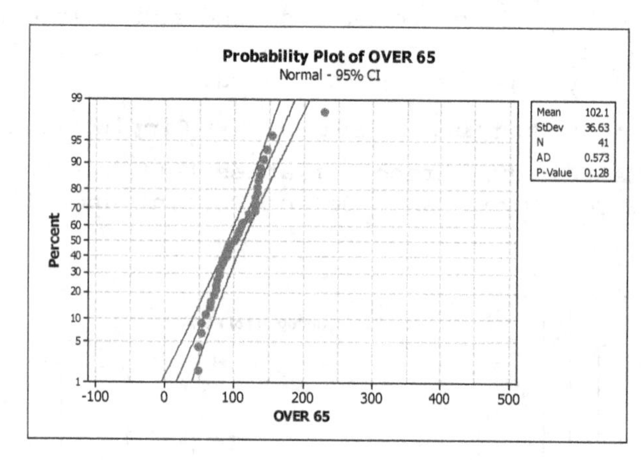

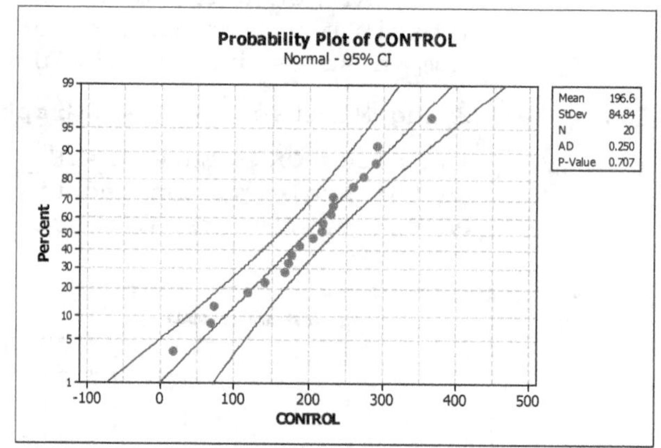

The standard deviations are 36.63 for OVER 65 and 84.84 for CONTROL,

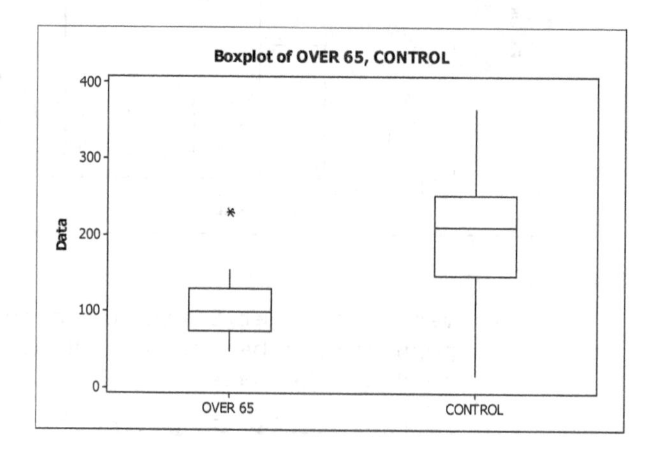

(b) The nonpooled t-procedure is preferable since the ratio of the standard deviations is greater than 2.

(c) To carry out the nonpooled t-test, choose **Stat ▶ Basic Statistics ▶ 2-Sample t**, click on **Samples in different columns**, enter 'OVER 65' in the

First text box and CONTROL in the **Second** text box, and leave the **Assume equal variances** box unchecked. Click on the **Options** button, enter 99 in the **Confidence level** text box, select **less than** in the **Alternative** box, and click **OK** twice. The results are

```
Two-sample T for OVER 65 vs CONTROL

          N   Mean  StDev  SE Mean
OVER 65  41  102.1   36.6      5.7
CONTROL  20  196.6   84.8       19

Difference = mu (OVER 65) - mu (CONTROL)
Estimate for difference:  -94.4768
99% upper bound for difference:  -44.7729
T-Test of difference = 0 (vs <): T-Value = -4.77  P-Value = 0.000  DF = 22
```

Since the P-value = 0.000, which is less than the significance level 0.01, reject the mull hypothesis. The data do provide sufficient evidence to conclude that, on average, men over 65 have a lower IGF-1 level than younger men.

(d) To obtain the 99% confidence interval, follow the procedure in part (c), but select not equal in the Alternative drop down box. The confidence interval is given by

```
99% CI for difference:  (-150.3322, -38.6215)
```

We can be 99% confident that the difference in IGF-1 levels, on average, between men over 65 and younger men is somewhere between -150.3 and -38.6.

(e) Yes. The CONTROL sample of size 20 is reasonably normally distributed with no outliers. The OVER 65 sample of size 41 is also close to normal with one outlier. The effect of the outlier is not great enough to alter the conclusion (Deleting it changes the value of t from -4.77 to -4.99).

19. (a) Using Minitab, choose **Stat ▶ Basic Statistics ▶ Paired t**, click on **Samples in columns**, enter MEN in the **First** text box and WOMEN in the **Second** text box. Click on the **Options** button, enter 95 in the **Confidence level** text box, select **greater than** in the **Alternative** box, and click **OK.** Click on the **Graphs** button, check the box for **Boxplot of differences**, and click **OK** twice. The results are

```
Paired T for MEN - WOMEN

             N   Mean  StDev  SE Mean
MEN         50   1030    820      116
WOMEN       50    899    818      116
Difference  50  131.1   93.8     13.3

95% lower bound for mean difference: 108.9
T-Test of mean difference = 0 (vs > 0): T-Value = 9.89  P-Value = 0.000
```

Since the P-value = 0.000, which is less than the significance level 0.05, reject the null hypothesis. The data do provide sufficient evidence to conclude that, on average, the weekly earnings of male full-time and salary workers exceed that of women.

(b) Follow the procedure in part (a), but enter 90 in the **Confidence level** text box and select **not equal** from the **Alternative** drop down box. The resulting interval is

```
90% CI for mean difference: (108.9, 153.3)
```

We can be 90% confident that the mean difference in weekly earnings of male and female full-time wage and salary workers is somewhere between $108.9 and $153.3.

(c) To create a column of differences, click in the **Sessions Window** and then click on the **Editor** tab and on **Enable Commands**. An MTB> prompt will appear at the bottom of the Sessions Window. After the prompt, enter Let c3=c1-c2 and press ENTER. Click at the top of Column 3 and name the column DIFF. Choose **Graph ▶ Probability plot...**, select the **Single** version and click **OK**. Enter DIFF in the **Graph Variables** text box, and click **OK**. The standard deviations will be produced on the probability plot legend. A boxplot was produced by the procedure in part (a). Now choose **Graph ▶ Stem-and-Leaf...**, enter DIFF in the **Graph Variables** text box, and click **OK**. The results are

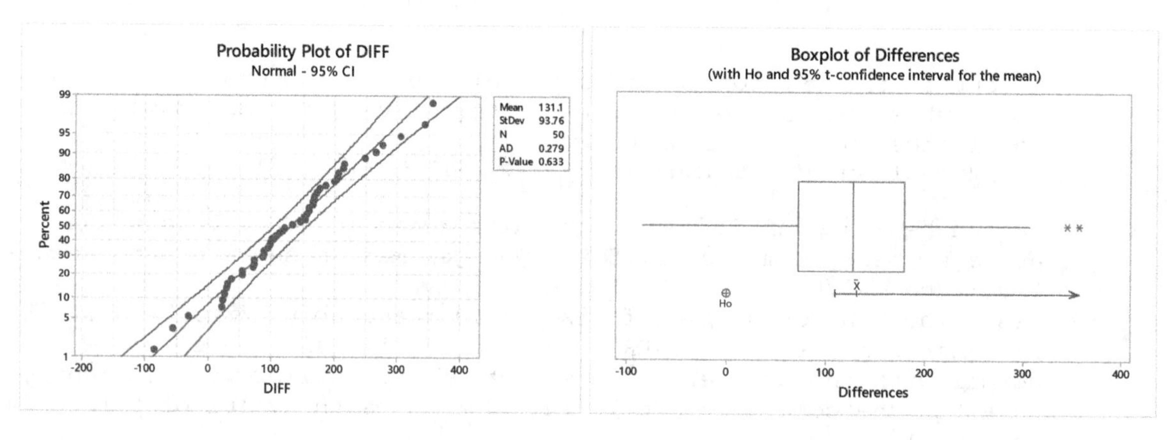


```
Stem-and-leaf of DIFF  N  = 50
Leaf Unit = 10

  2   -0   85
  3   -0   3
  9    0   222333
 19    0   5577788899
(8)    1   00011234
 23    1   555666666778
 11    2   000114
  5    2   67
  3    3   04
```

(d) Yes. The sample size is large (50), there are only two mild outliers on the right, and the normal probability plot is close to linear. The outliers may cause some concern since they are in direction that helps to reject the null hypothesis. If so, a nonparametric method may be used.

20. (a) Using Minitab, choose **Stat ▶ Nonparametrics ▶ 1-Sample Wilcoxon**, enter DIFF in the **Variables** text box, Click on the **Test median**, enter 0 in the **Test median** text box, select **greater than** in the **Alternative** box, and click **OK**. The results are

```
Test of median = 0.000000 versus median > 0.000000

            N for   Wilcoxon              Estimated
        N    Test  Statistic       P        Median
DIFF   50      50     1246.5   0.000         129.0
```

Since the P-value = 0.000, which is less than the significance level 0.05, reject the null hypothesis. The data do provide sufficient evidence to conclude that, on average, the weekly earnings of male full-time and salary workers exceed that of women.

(b) The P-value and conclusion are the same as those in Problem 19(a).

(c) See Problem 19(c) for the boxplot and stem-and-leaf diagram. For the histogram, choose **Graph ▶ Histogram...**, select the **Simple** version and click **OK**. Enter DIFF in the **Graph Variables** text box, and click **OK**. The result is

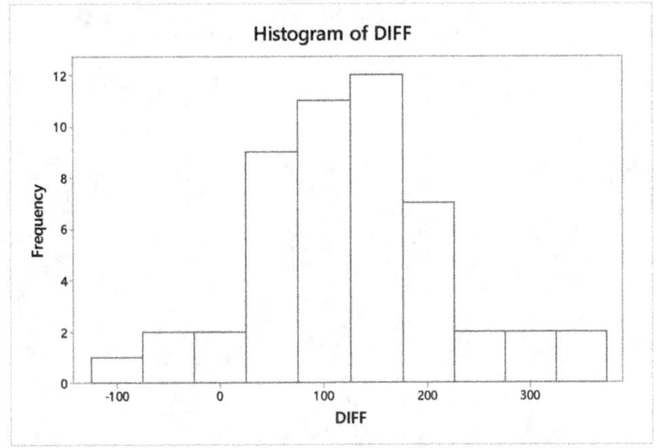

(d) Yes. The only assumption for the Wilcoxon signed-rank test is that the population be symmetric. All of the plots indicate that this is a reasonable assumption.

CHAPTER 11 SOLUTIONS

Exercises 11.1

11.1 A variable has a chi-square distribution if its distribution has the shape of a right-skewed curve called a chi-square curve.

11.3 The curve with 20 degrees of freedom more closely resembles a normal distribution. By Property 4 of Key Fact 11.1, as the degrees of freedom increase, the distributions look increasing like normal distribution curves.

11.5 (a) $\chi^2_{0.025} = 32.852$ (b) $\chi^2_{0.95} = 10.117$

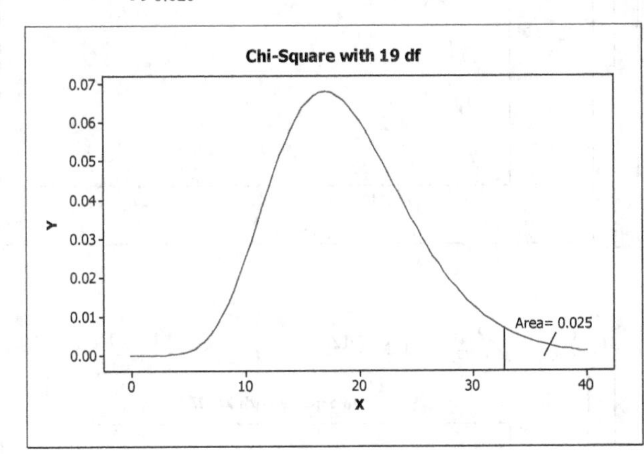

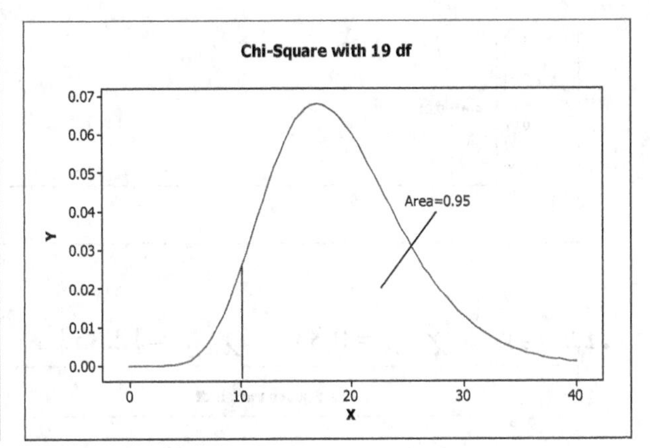

11.7 (a) $\chi^2_{0.05} = 18.307$ (b) $\chi^2_{0.975} = 3.247$

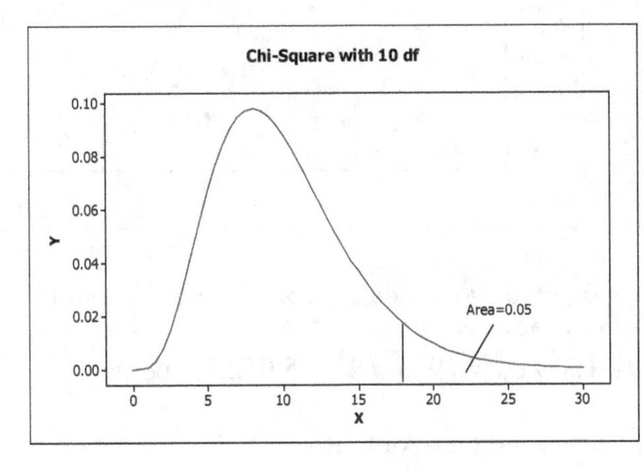

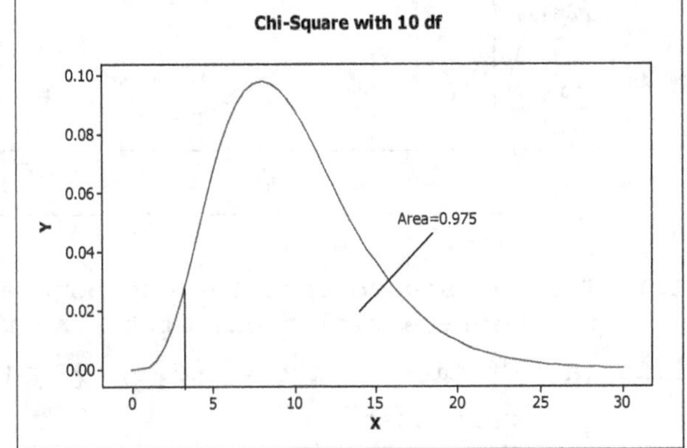

11.9

(a) A left area of 0.01 is equivalent to a right area of 0.99.

(b) A left area of 0.95 is equivalent to a right area of 0.05.

$\chi^2_{0.99} = 1.646$

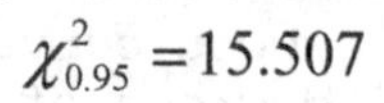

$\chi^2_{0.95} = 15.507$

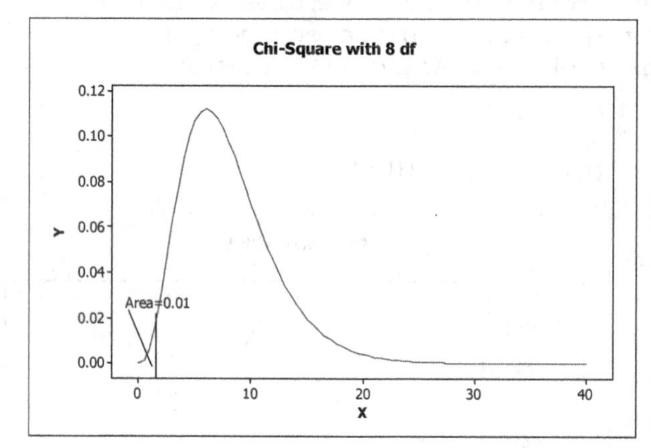

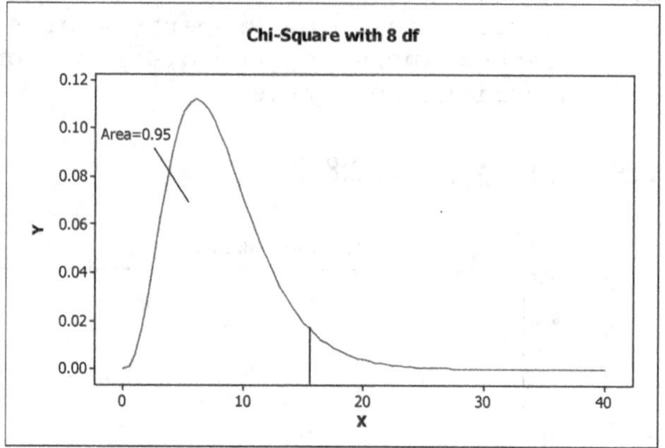

11.11 (a) $\chi^2_{0.975} = 0.831$ $\chi^2_{0.025} = 12.833$

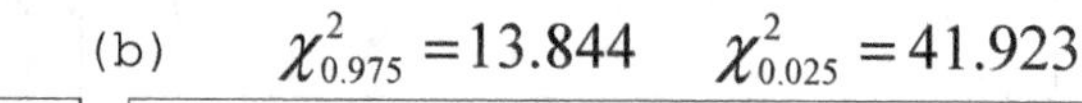

(b) $\chi^2_{0.975} = 13.844$ $\chi^2_{0.025} = 41.923$

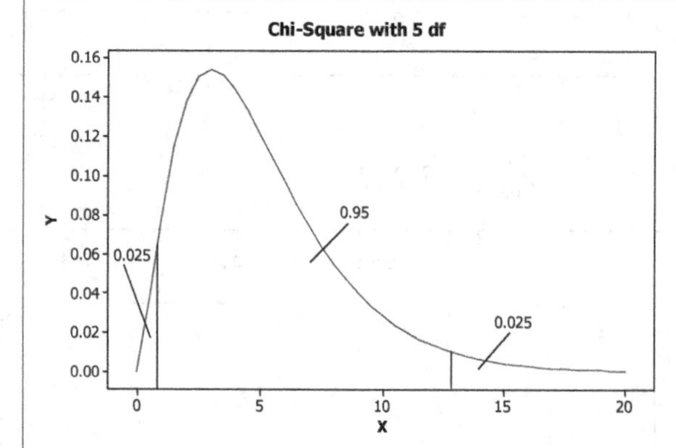

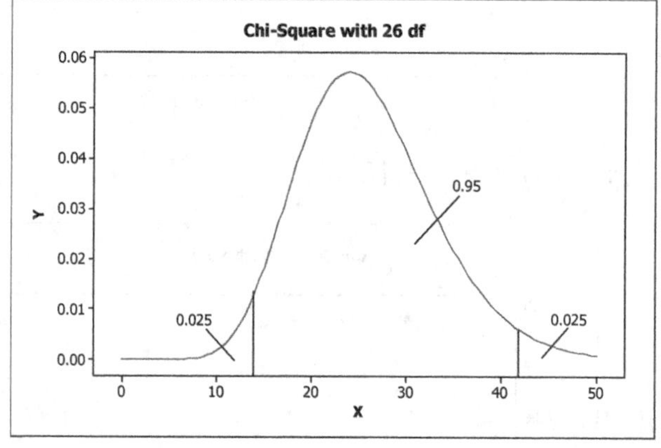

11.13 The chi-square test for one population standard deviation is not robust to moderate violations of the normality assumption.

11.15 (a) Critical value = 3.325; $\chi^2 = (n-1)s^2/\sigma_0^2 = (9)3^2/4^2 = 5.0625$; Left tail test

Since this is greater than 3.325, do not reject H_0.

(b) 90% Confidence Interval =

$$(s\sqrt{\frac{(n-1)}{\chi^2_{0.05}}}, s\sqrt{\frac{(n-1)}{\chi^2_{0.95}}}) = (3\sqrt{\frac{(10-1)}{16.919}}, 3\sqrt{\frac{(10-1)}{3.325}}) = (2.188, 4.936)$$

11.17 (a) Critical value = 44.314; $\chi^2 = (n-1)s^2/\sigma_0^2 = (25)7^2/5^2 = 49.00$; Right tail test

Since this is greater than 44.314, reject H_0.

(b) 98% Confidence Interval =

$$(s\sqrt{\frac{(n-1)}{\chi^2_{0.01}}}, s\sqrt{\frac{(n-1)}{\chi^2_{0.99}}}) = (7\sqrt{\frac{(26-1)}{44.314}}, 7\sqrt{\frac{(26-1)}{11.524}}) = (5.257, 10.310)$$

11.19 (a) Critical values = 8.907 and 32.852;

$\chi^2 = (n-1)s^2/\sigma_0^2 = (19)5^2/6^2 = 13.194$; Two tail test

Since 8.907 < 13.194 < 32.852, do not reject H_0.

(b) 95% Confidence Interval =

$$\left(s\sqrt{\frac{(n-1)}{\chi^2_{0.025}}}, s\sqrt{\frac{(n-1)}{\chi^2_{0.975}}}\right) = \left(5\sqrt{\frac{(20-1)}{32.852}}, 5\sqrt{\frac{(20-1)}{8.907}}\right) = (3.802, 7.303)$$

11.21 Step 1: H_0: σ = 3 H_a: $\sigma \neq 3$; df = 14 − 1 = 13

Step 2: α = 0.05

Step 3: $\chi^2 = (n-1)s^2/\sigma_0^2 = (13)2.501^2/3.0^2 = 9.035$

Step 4: df = 14 − 1 = 13. Critical values = 5.892 and 22.362; Two tail

Step 5: Since 5.892 < 9.035 < 22.362, do not reject H_0.

Step 6: At the 5% significance leve, there is no evidence to conclude that the population standard deviation is not equal to 3.

11.23 Step 1: H_0: σ = 0.27 H_a: σ > 0.27

Step 2: α = 0.01

Step 3: $\chi^2 = \frac{n-1}{\sigma_0^2}s^2 = \frac{9}{.27^2}0.756^2 = 70.56$

Note: Due to the rounding of *s*, answers may be slightly different.

Step 4: The critical value with n-1 = 9 df is 21.666.

Step 5: Since 70.560 > 21.666, reject H_0.

Using Excel, in any cell, enter =chidist(70.560,9) to obtain $P(\chi^2 > 70.560) = 1.18 \times 10^{-11}$. The P-value is 1.18×10^{-11}.

Step 6: There is sufficient evidence at the 0.01 level to claim that the product variability for this piece of equipment exceeds the analytical capability of 0.27

11.25 (a) Step 1: H_0: σ = 0.2 H_a: σ < 0.2

Step 2: α = 0.05

Step 3: $\chi^2 = \frac{n-1}{\sigma_0^2}s^2 = \frac{14}{0.2^2}0.154^2 = 8.301$

Note: Due to the rounding of *s*, answers may be slightly different.

Step 4: The critical value with n-1 = 14 df is 6.571.

Using Excel, in any cell, enter =1-chidist(8.301,14) to obtain $P(\chi^2 < 8.301) = 0.1269$. The P-value is 0.1269.

Step 5: Since 8.301 > 6.571, do not reject H_0.

Step 6: There is insufficient evidence at the 0.05 level to claim that the standard deviation of the amounts dispensed is less than 0.2 fluid ounces.

(b) So that the amount of coffee doesn't vary too much from cup to cup.

11.27 The 90% confidence interval for σ of lactation periods of grey seals is

$$\left(\sqrt{\frac{n-1}{\chi^2_{0.05}}}\cdot s, \sqrt{\frac{n-1}{\chi^2_{0.95}}}\cdot s\right) = \left(\sqrt{\frac{13}{22.362}}2.501, \sqrt{\frac{13}{5.892}}2.501\right) = (1.91, 3.71) \text{ days}$$

11.29 The 98% confidence interval for the product variability of the piece of equipment under consideration is

$$\left(\sqrt{\frac{n-1}{\chi^2_{0.01}}}\cdot s, \sqrt{\frac{n-1}{\chi^2_{0.99}}}\cdot s\right)=\left(\sqrt{\frac{9}{21.666}}\cdot 0.756, \sqrt{\frac{9}{2.089}}\cdot 0.756\right)=(0.487,1.570)$$

11.31 The 90% confidence interval for the standard deviation of coffee dispensed is

$$\left(\sqrt{\frac{n-1}{\chi^2_{0.05}}}\cdot s, \sqrt{\frac{n-1}{\chi^2_{0.95}}}\cdot s\right)=\left(\sqrt{\frac{14}{23.685}}\cdot 0.154, \sqrt{\frac{14}{6.571}}\cdot 0.154\right)=(0.1184,0.2248) \text{ fluid oz.}$$

In each of the next two exercises, we have used Minitab to produce the displayed normal probability plot. Choose **Graph ▶ Probability Plot...**, select the **Single** version and click **OK**. Enter the variable name in the **Graph variables** text box and click **OK**.

11.33

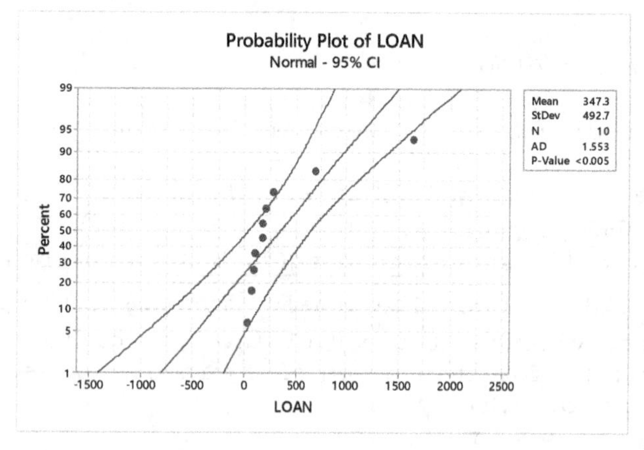

Applying one-standard deviation χ^2-procedures is not reasonable for these data in Exercise 11.33. The probability plot is not linear and there is at least one outlier.

11.35

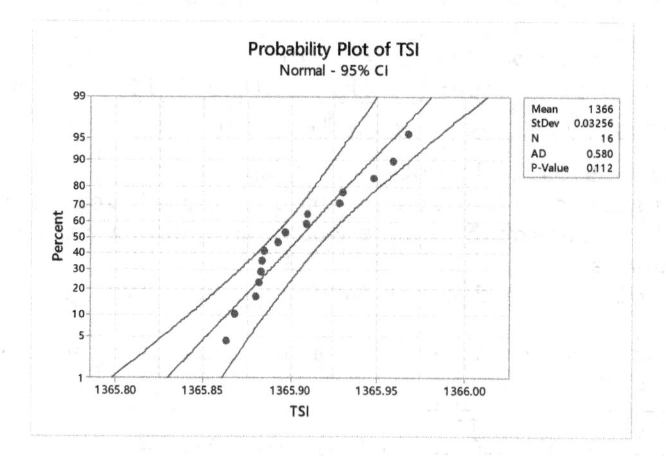

Applying one-standard deviation χ^2-procedures is reasonable (barely) for these data in Exercise 11.35. The probability plot is roughly linear and there are no outliers.

11.37 (a) Using Minitab, choose **Graph ▶ Probability Plot...**, select the **Single** version and click **OK**. Enter TEMP in the **Graph variables** text box and click **OK**. Then choose **Graph ▶ Boxplot...**, select the **One Y Simple** version and click **OK**. Enter TEMP in the **Graph variables** text box and click **OK**. Now choose **Graph ▶ Histogram...**, select the **Simple** version and click **OK**. Enter TEMP in the **Graph variables** text box and click **OK**. Then choose **Graph ▶ Stem-and-Leaf...** enter TEMP in the **Graph variables** text box and click **OK**. The results are

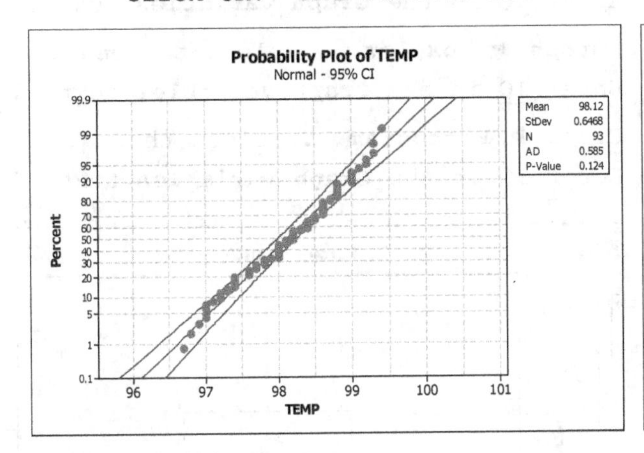

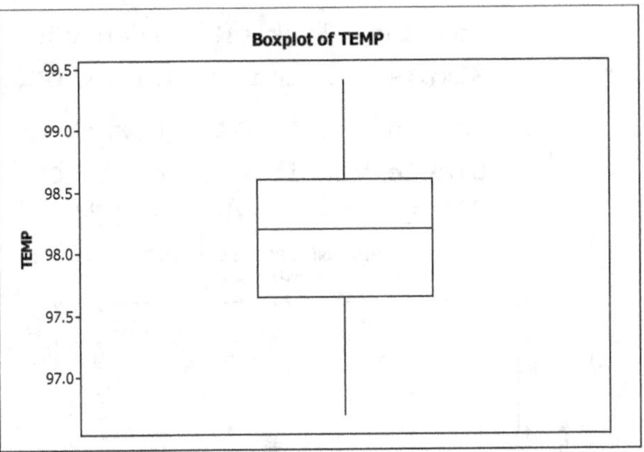


```
Stem-and-leaf of TEMP  N  = 93
Leaf Unit = 0.10

  1    96  7
  3    96  89
  8    97  00001
 13    97  22233
 19    97  444444
 26    97  6666777
 31    97  88889
 45    98  00000000000111
(10)   98  2222222233
 38    98  4444445555
 28    98  66666666677
 17    98  8888888
 10    99  00001
  5    99  2233
  1    99  4
```

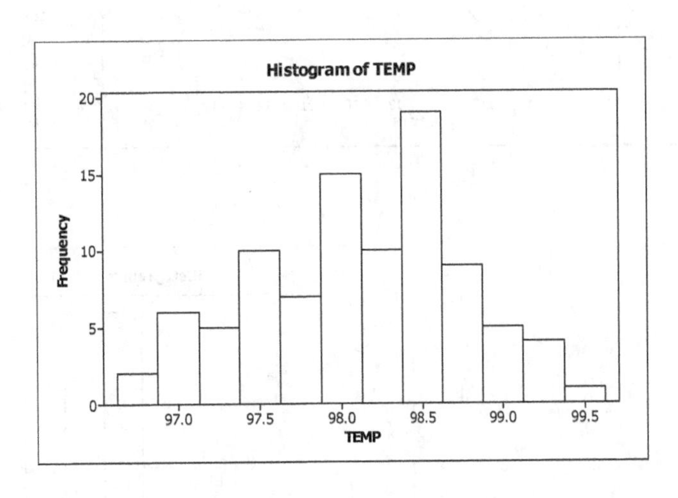

(b) Yes. The probability plot is very close to linear and there are no outliers.

(c) Choose **Stat ▸ Basic Statistics ▸ 1-Variance**, enter TEMP in the **Samples in Columns** text box, check the box for **Perform hypothesis test**, enter the variance 0.3969 (the square of 0.63) in the **Hypothesized Variance** text box, click on the **Options** button, enter 0.95 in the **Confidence level** text box and select **not equal** in the **Alternative** drop down box, and click **OK** twice. The result is

```
Test of sigma squared = 0.3969 vs not = 0.3969
Chi-Square Method (Normal Distribution)

Variable   N  Variance       95% CI        Chi-Square      P
TEMP      93     0.418  (0.320, 0.571)          96.97  0.683
```

Since the P-value = 0.683, which is larger than the significance level 0.05, do not reject the null hypothesis. The data do not provide sufficient evidence that the standard deviation differs from 0.63 degrees.

(d) The 95% confidence interval for the variance was produced in part (c) as (0.320,0.571). Taking the square root of each confidence limit, this is equivalent to a 95% confidence interval for the standard deviation of (0.57, 0.76). We can be 95% confident that the population standard deviation lies somewhere in the interval from 0.57 degrees to 0.76 degrees (F).

11.39 (a) Using Minitab, choose **Graph ▶ Probability Plot...**, select the **Single** version and click **OK**. Enter IQ in the **Graph variables** text box and click **OK**. Then choose **Graph ▶ Boxplot...**, select the **One Y Simple** version and click **OK**. Enter IQ in the **Graph variables** text box and click **OK**. Now choose **Graph ▶ Histogram...**, select the **Simple** version and click **OK**. Enter IQ in the **Graph variables** text box and click **OK**. The results are

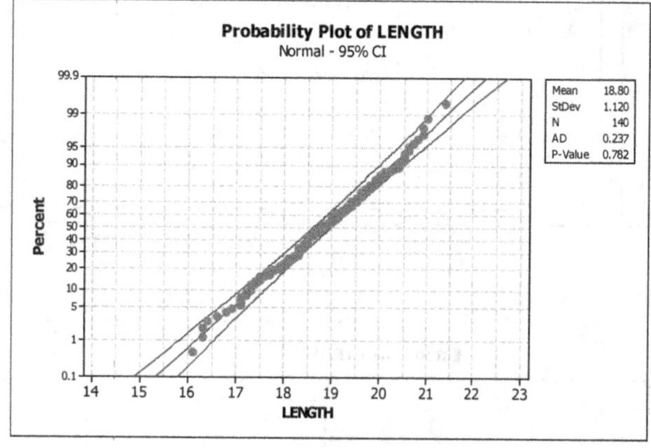

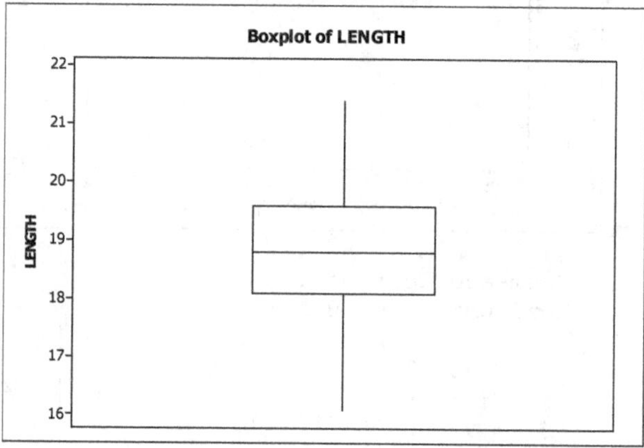

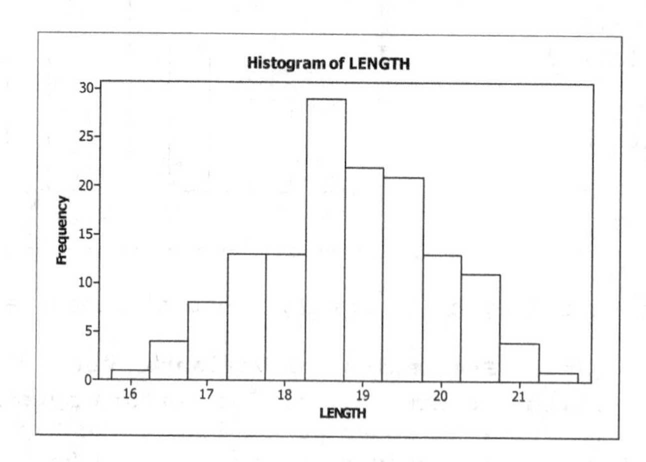

(b) Yes. None of the plots exhibit departures from normality and there are no potential outliers.

(c) Choose **Stat ▸ Basic Statistics ▸ 1-Variance**, enter IQ in the **Samples in Columns** text box, uncheck the box for **Perform hypothesis test**, click on the **Options** button, enter 0.95 in the **Confidence level** text box and select 0**less than** in the **Alternative** drop down box, and click **OK** twice.

The result is

```
Chi-Square Method (Normal Distribution)

Variable    N  Variance       99% CI
LENGTH    140      1.26   (1.01, 1.61)
```

The 95% confidence interval for the variance is (1.01, 1.61). Taking the square root of each confidence limit, this is equivalent to a 95% confidence interval for the standard deviation of (1.005, 1.269). We can be 95% confident that the population standard deviation σ lies somewhere in the interval from 1.005 to 1.269 inches.

11.41 (a) We will generate the 1000 samples as 1000 rows in Minitab. Choose **Calc ▶ Random data ▶ Normal...**, type 1000 in the **Generate rows of data** text box, enter IQ1 IQ2 IQ3 IQ4 in the **Store in columns** text box, enter 100 in the **Mean** text box, enter 16 in the **Standard deviation** text box, and click **OK**. Now compute the standard deviation of each of the 1000 rows.

(b) Choose **Calc ▶ Row statistics**, click on **Standard deviation**, enter IQ1, IQ2, IQ3, and IQ4 in the **Input variables** text box and SD in the **Store result in** text box. enter SD in the **Store result in variable** text box, and click **OK**.

(c) Choose **Calc ▶ Calculator**, enter CHI2 in the **Store result in variable** text box, click in the **Expression** text box, enter 3*SD**2/256, and click **OK**.

(d) Choose **Graph ▶ Histogram...**, select the **Simple** version and click **OK**. Enter CHI2 in the **Graph variables** text box and click **OK**. The result is

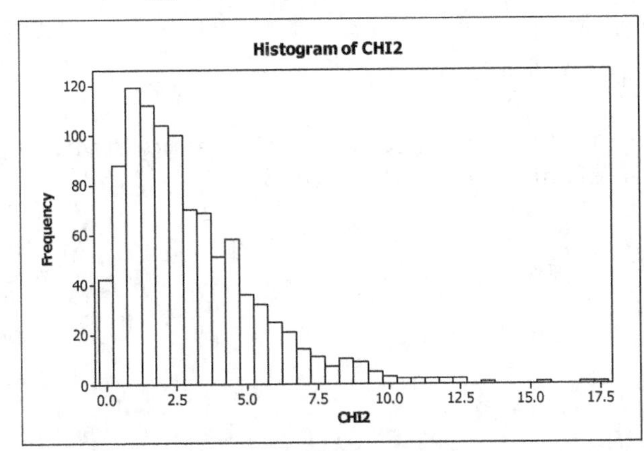

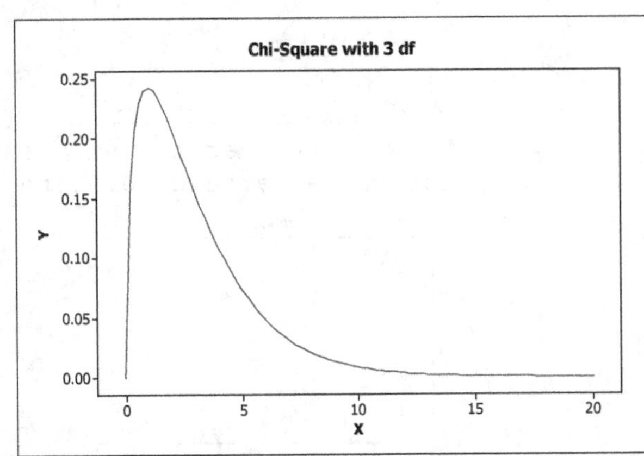

The distribution of the variable in part (c) is a χ^2 distribution with n - 1 = 4 - 1 = 3 degrees of freedom (as shown above right). A Chi-square distribution with 3 degrees of freedom is very right- skewed. The histogram in part (d) corresponds to such a distribution.

Exercises 11.2

11.43 $F_{0.05}$, $F_{0.025}$, F_{α}

11.45 (a) 12

(b) 7

11.47 (a) $F_{0.05} = 1.89$ (b) $F_{0.01} = 2.47$ (c) $F_{0.025} = 2.14$

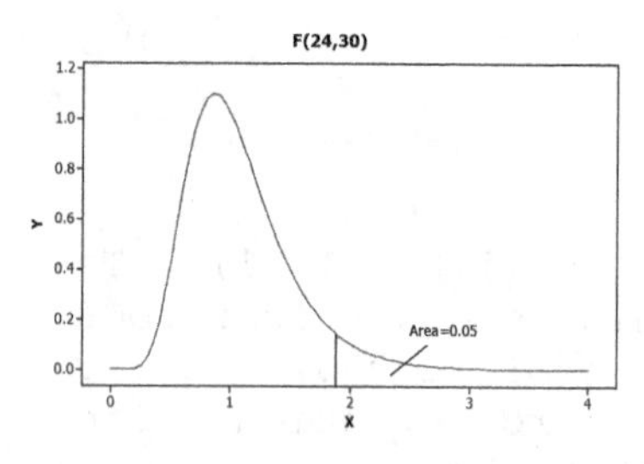

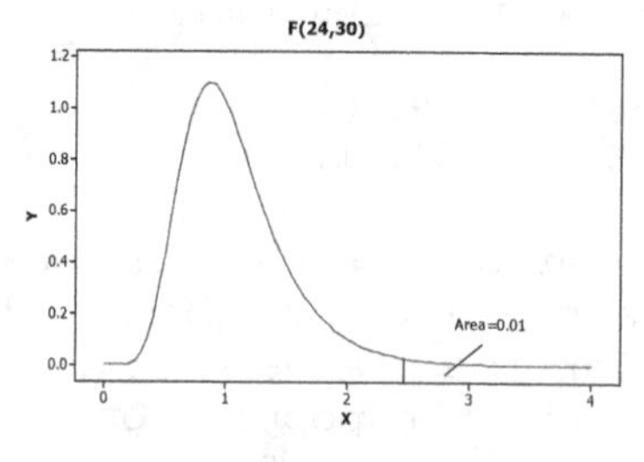

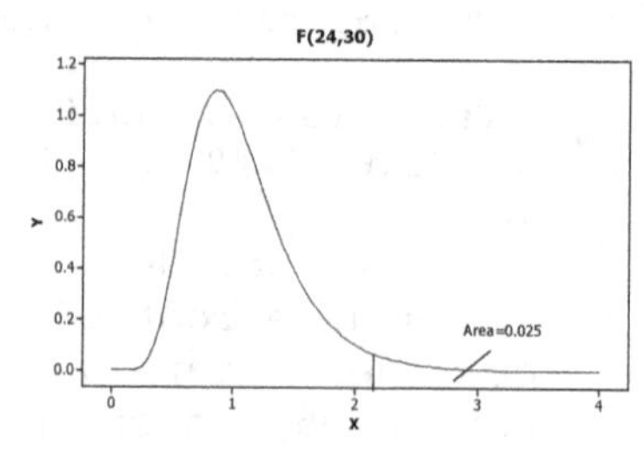

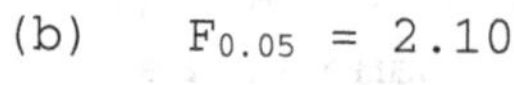

11.49 (a) $F_{0.01} = 2.88$ (b) $F_{0.05} = 2.10$ (c) $F_{0.10} = 1.78$

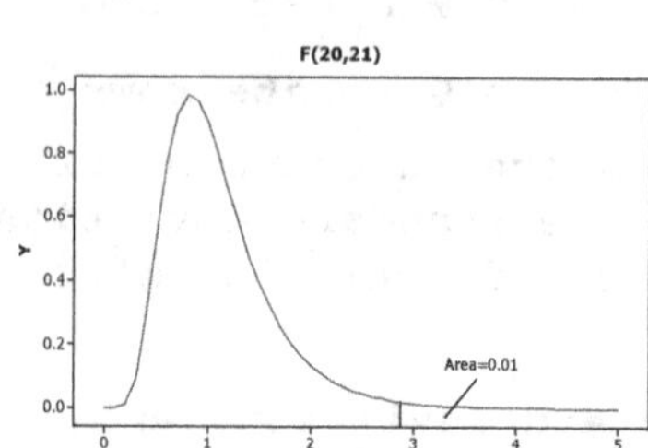

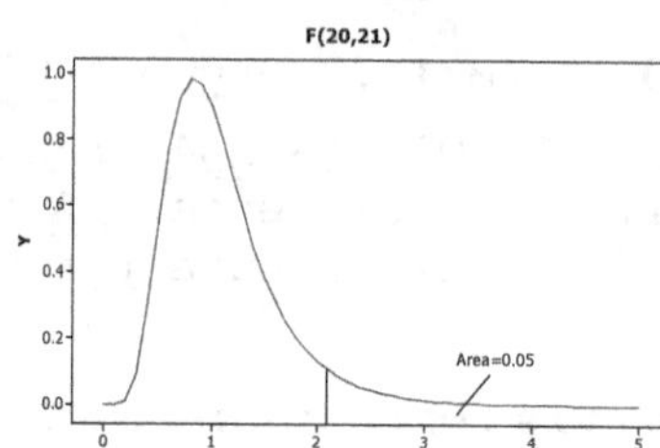

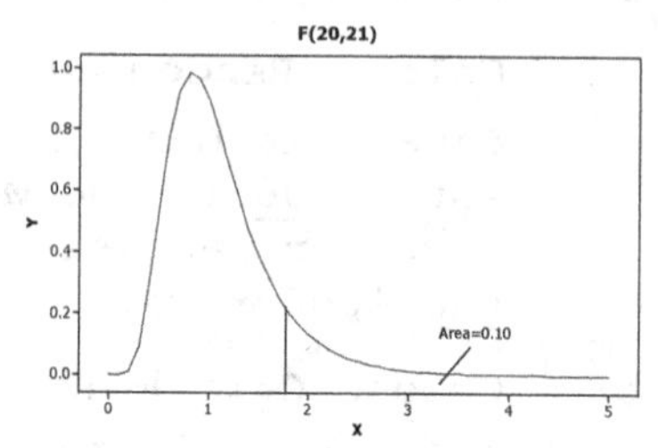

11.51 (a) Apply the Reciprocal Property of F-Curves stated in Key Fact 11.4. The F_1 value with an area of 0.01 to its left for df = (6,8) equals the reciprocal, $1/F_2$, of an F_2 value with area of 0.01 to its right for df = (8,6). The F_2 value with area of 0.01 to its right with df = (8,6) is 8.10. Thus the F_1 value with an area of 0.01 to its left for df = (6,8) is 1/8.10 = 0.12.

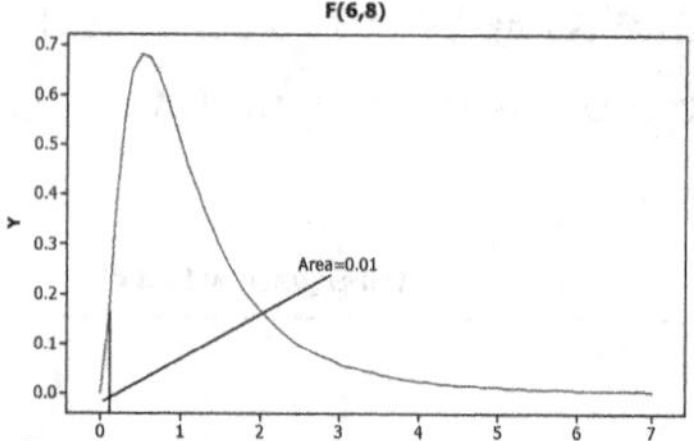

(b) The F value that has an area of 0.95 to its left is the same F value that has an area of 0.05 to its right. For df = (6,8), F = 3.58.

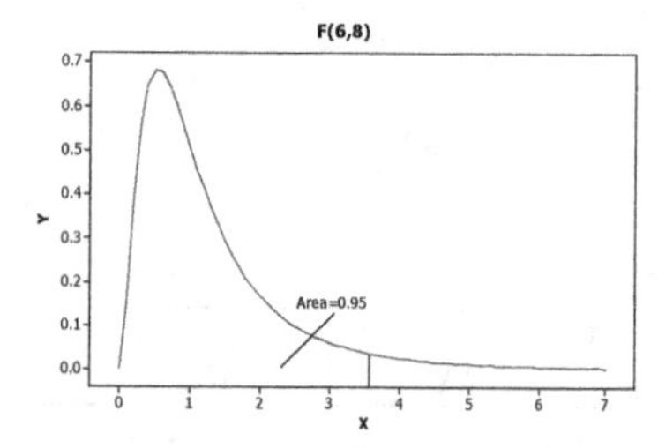

11.53 (a) The F value that has an area of 0.025 to its right with df = (7,4) is F = 9.07. To find the F value that has an area of 0.025 to its left with df = (7,4), apply the Reciprocal Property of F-Curves stated in Key Fact 11.4. The F_1 value with an area of 0.025 to its left for df = (7,4) equals the reciprocal, $1/F_2$, of an F_2 value with area of 0.025 to its right for df = (4,7). The F_2 value with area of 0.025 to its right with df = (4,7) is 5.52. Thus the F_1 value with an area of 0.025 to its left for df = (7,4) = 1/5.52 is 0.18. The two F-values that divide the area under the curve into a middle 0.95 area are 0.18 and 9.07.

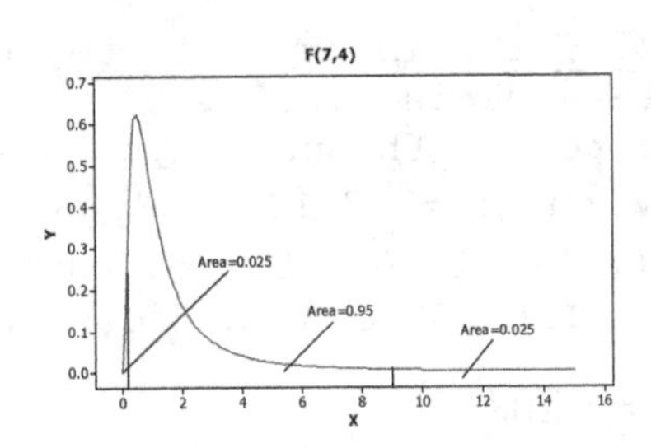

(b) The F value that has an area of 0.025 to its right with df = (12,20) is F = 2.68. To find the F value that has an area of 0.025 to its left with df = (12,20), apply the Reciprocal Property of F-Curves stated in Key Fact 11.4. The F_1 value with an area of 0.025 to its left for df = (12,20) equals the reciprocal, $1/F_2$, of an F_2 value with area of 0.025 to its right for df = (20,12). The F_2 value with area of 0.025 to its right with df = (20,12) is 3.07. Thus the F_1 value with an area of 0.025 to its left for df = (12,20) is 1/3.07 = 0.33. The two F-values that divide the area under the curve into a middle 0.95 area are 0.33 and 2.68.

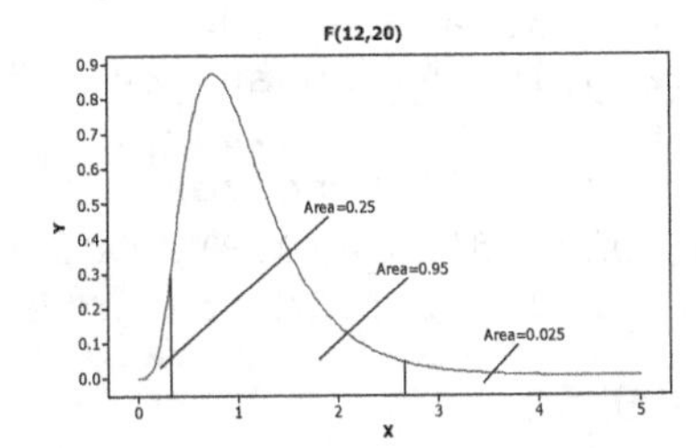

11.55 The F procedures are extremely nonrobust to even moderate violations of the normality assumption for both populations.

11.57 (a) Step 1: H_0: $\sigma_1 = \sigma_2$ H_a: $\sigma_1 > \sigma_2$

Step 2: α = 0.01

Step 3: $F = 19.4^2/10.5^2 = 3.414$

Step 4: The critical value is $F_{0.01}$ with df = (30,15) or 3.21

In Excel, enter =FDIST(3.414,30,15) in any cell to obtain P(F > 3.414) = 0.0074. The P-value is 0.0074.

Step 5: Since 3.414 > 3.21, reject the null hypothesis.

(b) Critical Values are $F_{0.01}$ and $F_{0.99}$ with df = (30,15). $F_{0.01}$ with df = (30,15) is 3.21. Apply the Reciprocal Property of F-Curves to get $F_{0.99}$ with df = (30,15). This property states that $F_{0.99}$ with df = (30,15) is equal to the reciprocal of $F_{0.01}$ with df = (15,30). $F_{0.01}$ with df = (15,30) is 2.70. Therefore, $F_{0.99}$ with df = (30,15) is 1/2.70 = 0.370. Critical values are 3.21 and 0.370. The 98% confidence interval is $\frac{1}{\sqrt{3.21}}\cdot\frac{19.4}{10.5}=1.031$ to $\frac{1}{\sqrt{0.370}}\cdot\frac{19.4}{10.5}=3.037$.

11.59 (a) Step 1: H_0: $\sigma_1 = \sigma_2$ H_a: $\sigma_1 < \sigma_2$

Step 2: α = 0.10

Step 3: $F = 28.82^2/38.97^2 = 0.547$

Step 4: The critical value is $F_{0.90}$ with df = (7,12). Apply the Reciprocal Property of F-Curves to get $F_{0.90}$ with df = (7,12). This property states that $F_{0.90}$ with df = (7,12) is equal to the reciprocal of $F_{0.10}$ with df = (12,7). $F_{0.10}$ with df = (12,7) is 2.67. Therefore, $F_{0.90}$ with df = (7,12) is 1/2.67 = 0.375

In Excel, enter =1-FDIST(0.547,7,12) in any cell to obtain $P(F < 0.547) = 0.2160$. The P-value is 0.2160.

Step 5: Since 0.547 > 0.375, do not reject the null hypothesis.

(b) Critical Values are $F_{0.10}$ and $F_{0.90}$ with df = (7,12). $F_{0.10}$ with df = (7,12) is 2.28 and $F_{0.90}$ with df = (7,12) is 0.375 from part (a). Critical values are 2.28 and 0.375. The 80% confidence interval is

$$\frac{1}{\sqrt{2.28}}\cdot\frac{28.82}{38.97}=0.490 \text{ to } \frac{1}{\sqrt{0.375}}\cdot\frac{28.82}{38.97}=1.208.$$

11.61 (a) Step 1: H_0: $\sigma_1 = \sigma_2$ H_a: $\sigma_1 \neq \sigma_2$

Step 2: $\alpha = 0.05$

Step 3: $F = 14.5^2/30.4^2 = 0.228$

Step 4: The critical values are $F_{0.025}$ and $F_{0.975}$ with df = (10,8). $F_{0.025}$ with df = (10,8) is 4.30. Apply the Reciprocal Property of F-Curves to get $F_{0.975}$ with df = (10,8). This property states that $F_{0.975}$ with df = (10,8) is equal to the reciprocal of $F_{0.025}$ with df = (8,10). $F_{0.025}$ with df = (8,10) is 3.85. Therefore, $F_{0.975}$ with df = (10,8) is 1/3.85 = 0.260. Critical values are 4.30 and 0.260.

In Excel, enter =2(1-FDIST(0.228,10,8)) in any cell to obtain $2(P(F < 0.228)) = 0.0327$. The P-value is 0.0327.

Step 5: Since 0.228 < 0.260, reject the null hypothesis.

(b) Critical Values are $F_{0.025}$ and $F_{0.975}$ with df = (10,8). Critical values are 4.30 and 0.260 from part (a). The 95% confidence interval is

$$\frac{1}{\sqrt{4.30}}\cdot\frac{14.5}{30.4}=0.230 \text{ to } \frac{1}{\sqrt{0.260}}\cdot\frac{14.5}{30.4}=0.935.$$

11.63 1=Control, 2=Experimental

Step 1: H_0: $\sigma_1 = \sigma_2$ H_a: $\sigma_1 > \sigma_2$

Step 2: $\alpha = 0.05$

Step 3: $F = 7.813^2/5.286^2 = 2.185$

Step 4: The critical value is $F_{0.05}$ with df = (40,19) or 2.03

In Excel, enter =FDIST(2.1853,40,19) in any cell to obtain $P(F > 2.185) = 0.0348$. The P-value is 0.0348.

Step 5: Since 2.185 > 2.03, reject the null hypothesis.

Step 6: There is sufficient evidence at the 0.05 significance level to claim that the variation in the control group is greater than that in the experimental group.

11.65 1=Relaxation tapes, 2=Neutral tapes

Step 1: H_0: $\sigma_1 = \sigma_2$ H_a: $\sigma_1 \neq \sigma_2$

Step 2: $\alpha = 0.10$

Step 3: $F = 10.154^2/9.197^2 = 1.2189$

Step 4: The critical values are $F_{0.05}$ and $F_{0.95}$ with df = (30,24). $F_{0.05}$ with df = (30,24) is 1.94. Apply the Reciprocal Property of F-Curves to get $F_{0.95}$ with df = (30,24). This property states that $F_{0.95}$ with df = (30,24) is equal to the reciprocal of $F_{0.05}$ with df = (24,30). $F_{0.05}$ with df = (24,30) is 1.89. Therefore, $F_{0.95}$ with df = (30,24) is 1/1.89 = 0.53. Critical values are 0.53 and 1.94.

In Excel, enter =FDIST(1.2189,30,24) in any cell to obtain $P(F > 1.2189) = 0.31217$. The P-value is 2(0.31217) = 0.62434.

Step 5: Since 0.53 < 1.2189 < 1.94, do not reject the null hypothesis.

Step 6: There is not sufficient evidence at the 0.10 significance level to claim that the variation in anxiety test scores for patients seeing videotapes showing progressive relaxation exercises is different from that in patients seeing neutral videotapes

11.67 1=Stinger, 2=Regular

Step 1: H_0: $\sigma_1 = \sigma_2$ H_a: $\sigma_1 < \sigma_2$

Step 2: α = 0.01

Step 3: F = $0.410^2/0.894^2$ = 0.210

Step 4: The critical value is $F_{0.99}$ with df = (29,29). Apply the Reciprocal Property of F-Curves to get $F_{0.99}$ with df = (29,29). This property states that $F_{0.99}$ with df = (29,29) is equal to the reciprocal of $F_{0.01}$ with df = (29,29). $F_{0.01}$ with df = (29,29) is 2.42. Therefore, $F_{0.99}$ with df = (29,29) is 1/2.42 = 0.41.

In Excel, enter =1-FDIST(0.210,29,29) in any cell to obtain P(F < 0.210) = 0.000035. The P-value is 0.000035.

Step 5: Since 0.210 < 0.41, reject the null hypothesis.

Step 6: There is sufficient evidence at the 0.01 significance level to claim that the standard deviation of ball velocity is less with the Stinger tee than with the regular tee.

11.69 Step 1: df = (40,19). $F_{0.05}$ = 2.03. $F_{0.95} = 1/F_{0.05}$ for df = (19,40) = 1/1.85 = 0.54.

Step 2: The 90% confidence interval for σ_1/σ_2 is

$$\frac{1}{\sqrt{F_{0.05}}} \cdot \frac{s_1}{s_2} \text{ to } \frac{1}{\sqrt{F_{0.95}}} \cdot \frac{s_1}{s_2} = \frac{1}{\sqrt{2.03}} \cdot \frac{7.813}{5.286} \text{ to } \frac{1}{\sqrt{0.54}} \cdot \frac{7.813}{5.286} = (1.037, 2.011)$$

We can be 90% confident that the ratio of population standard deviations for the final-exam scores for students taught by the conventional method and those taught be the new method is between 1.037 and 2.011. In other words, we can be 90% confident that the standard deviation for the conventional method is between 1.037 to 2.011 times larger than the standard deviation for the new method.

11.71 Step 1: The critical values are $F_{0.05}$ and $F_{0.95}$ with df = (30,24). $F_{0.05}$ with df = (30,24) is 1.94. Apply the Reciprocal Property of F-Curves to get $F_{0.95}$ with df = (30,24). This property states that $F_{0.95}$ with df = (30,24) is equal to the reciprocal of $F_{0.05}$ with df = (24,30). $F_{0.05}$ with df = (24,30) is 1.89. Therefore, $F_{0.95}$ with df = (30,24) is 1/1.89 = 0.529. Critical values are 0.529 and 1.94.

Step 2: The 90% confidence interval for σ_1/σ_2 is

$$\frac{1}{\sqrt{F_{0.05}}} \cdot \frac{s_1}{s_2} \text{ to } \frac{1}{\sqrt{F_{0.95}}} \cdot \frac{s_1}{s_2} = \frac{1}{\sqrt{1.94}} \cdot \frac{10.154}{9.197} \text{ to } \frac{1}{\sqrt{0.529}} \cdot \frac{10.154}{9.197} = (0.79, 1.52)$$

We can be 90% confident that the ratio of population standard deviations of scores for patients who are shown videotapes of progressive relaxation exercises and those who are shown neutral videotapes is between 0.79 and 1.52. In other words, we can be 90% confident that the standard deviation for those shown videotapes of progressive relaxation exercises is between 1.27 times smaller to 1.52 times larger than the standard deviation for those shown neutral videotapes.

11.73 Step 1: The critical values are $F_{0.01}$ and $F_{0.99}$ with df = (29,29). $F_{0.01}$ with df = (29,29) is 2.42. Apply the Reciprocal Property of F-Curves to get $F_{0.99}$ with df = (29,29). This property states that $F_{0.99}$ with df = (29,29) is equal to the reciprocal of $F_{0.01}$ with df = (29,29). Therefore, $F_{0.99}$ with df = (29,29) is 1/2.42 = 0.413. Critical values are 0.413 and 2.42.

Step 2: The 98% confidence interval for σ_1/σ_2 is

$$\frac{1}{\sqrt{F_{0.01}}}\cdot\frac{s_1}{s_2}\ to\ \frac{1}{\sqrt{F_{0.99}}}\cdot\frac{s_1}{s_2}=\frac{1}{\sqrt{2.42}}\cdot\frac{0.410}{0.894}\ to\ \frac{1}{\sqrt{0.413}}\cdot\frac{0.410}{0.894}=(0.295,0.714)$$

We can be 98% confident that the ratio of population standard deviations of ball velocity for the Stinger tee and the regular tee is between 0.295 and 0.714. In other words, we can be 98% confident that the standard deviation of ball velocity for the Stinger tee is between 1.401 to 3.390 times smaller than the standard deviation for ball velocity for the regular tee.

11.75 (a) **Using Minitab,** choose Stat ▶ Basic Statistics ▶ 2-Variances, click on **Samples in different columns**, enter ITALIANS in the **First** box and ETRUSCANS in the **Second** box, and click **OK**. The results are

Tests

Method	DF1	DF2	Test Statistic	P-Value
F Test (normal)	69	83	0.93	0.750
Levene's Test (any continuous)	1	152	0.05	0.826

Since the P-value for the F test = 0.750, which is larger than the significance level 0.05, do not reject the null hypothesis. The data do not provide sufficient evidence to conclude that there is a difference in the variation of skull measurements of the two populations.

(b) **Using Minitab,** choose Stat ▶ Basic Statistics ▶ 2-Variances, click on **Samples in different columns**, enter ITALIANS in the **First** box and ETRUSCANS in the **Second** box, and click **OK**. Click **Options**, enter 95 in the **Confidence level** text box. Select **StDev1/StDev2** from the **Hypothesized ratio** box. Enter 1 in the **value** box. Choose **not equal** in the **Alternative** test box. Click **OK** twice. The results are

95% Confidence Intervals

Distribution of Data	CI for StDev Ratio	CI for Variance Ratio
Normal	(0.769, 1.212)	(0.591, 1.469)
Continuous	(0.752, 1.265)	(0.566, 1.600)

The 95% confidence interval for σ_1/σ_2 is 0.769 to 1.212.

(c) Choose **Graph ▶ Probability Plot...**, select the **Single** version and click **OK**. Enter ITALIANS and ETRUSCANS in the **Graph** variables text box and click on the Multiple graphs button. Select **In separate panels of the same graph** and click **OK.** The result is

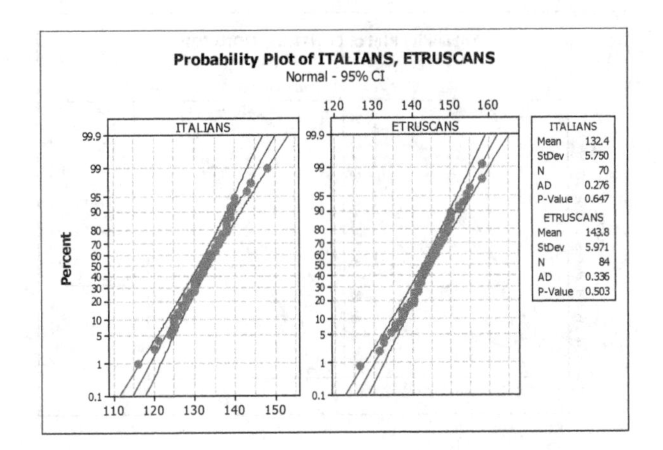

(d) Yes. Both plots are very close to linear with only one minor outlier at the low end. The sample sizes are also large (70 and 84).

11.77 (a) **Using Minitab,** choose Stat ▶ Basic Statistics ▶ 2-Variances, click on **Samples in different columns**, enter CONTROL in the **First** box and MONITOR in the **Second** box. Click the **Options** tab. Enter 95 in the **Confidence level** text box, choose **less than** in the **Alternative** textbox and click **OK** twice. The results are

```
Tests

                                                Test
Method                          DF1  DF2   Statistic  P-Value
F Test (normal)                  46   45        0.73    0.141
Levene's Test (any continuous)    1   91        0.25    0.309
```

The P-value = 0.141, which is larger than the significance level 0.05, so do not reject the null hypothesis. The data do not provide sufficient evidence that the variation in hemoglobin level is less without the in-line blood gas and chemistry monitor.

(b) **Using Minitab,** choose Stat ▶ Basic Statistics ▶ 2-Variances, click on **Samples in different columns**, enter CONTROL in the **First** box and MONITOR in the **Second** box, and click **OK**. Click **Options**, enter 90 in the **Confidence level** text box. Select **StDev1/StDev2** from the **Hypothesized ratio** box. Enter 1 in the **value** box. Choose **not equal** in the **Alternative** test box. Click **OK** twice. The results are

```
90% Confidence Intervals

                                      CI for
Distribution   CI for StDev         Variance
of Data            Ratio              Ratio
Normal         (0.666, 1.090)  (0.443, 1.187)
Continuous     (0.697, 1.229)  (0.486, 1.511)
```

The 90% confidence interval for σ_1/σ_2 is 0.666 to 1.090.

(c) Choose **Graph** ▶ **Probability Plot...**, select the **Single** version and click **OK**. Enter CONTROL and MONITOR in the **Graph** variables text box and click on the **Multiple graphs** button. Select **In separate panels of the same graph** and click **OK.** The result is

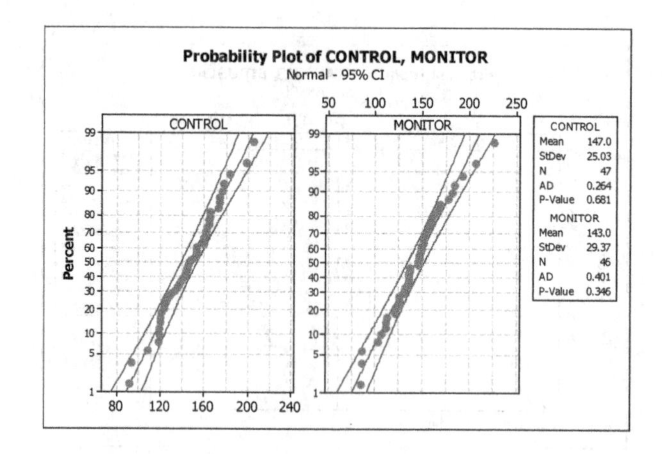

(d) Yes. Both probability plots are linear and there are no outliers.

11.79 In Example 11.14, the F test statistic of the two tailed test is 0.413 and the df = (9,9). For the P-value, we want 2(P(F < 0.413)). Looking at Table VIII, the F value of 0.413 is smaller than all of the given F values for df = (9,9). Therefore, the P(F < 0.413) is larger than all of the given values. So, P(F < 0.413) is greater than 0.10. Since it is two-tailed, the P-value is twice this area or 2(P(F < 0.413)). Therefore, the P-value exceeds 0.20.

Review problems for Chapter 11

1. Chi-square distribution (χ^2)

2. (a) A χ^2-curve is right skewed.

(b) A χ^2-curve looks increasingly like a normal curve as the number of degrees of freedom becomes larger.

3. The variable must be normally distributed. That assumption is very important because the χ^2 procedures are not robust to even moderate violations of the normality assumption.

4. 6.408

5. 33.409

6. 27.587

7. 8.672

8. 7.564 and 30.191

9. The F distribution is used when making inferences comparing two population standard deviations.

10. (a) An F-curve is right-skewed.

(b) reciprocal, (5,14)

(c) 0

11. Both variables must be normally distributed. This is very important since the F procedures are not robust to violations of the normality assumption.

12. 7.01

13. Apply the Reciprocal Property for F-curves stated in Key Fact 11.4. $F_{0.99} = 1/F_{0.01}$ where $F_{0.01}$ has df = (8,4). Thus $F_{0.99} = 1/14.80 = 0.068$

14. 3.84

15. Apply the Reciprocal Property for F-curves stated in Key Fact 11.4. The F value with 0.05 to its left = $F_{0.95} = 1/F_{0.05}$ where $F_{0.05}$ has df = (8,4). Thus $F_{0.95} = 1/6.04 = 0.166$

16. Apply the Reciprocal Property for F-curves stated in Key Fact 11.4. The F value with 0.025 to its left = $F_{0.975} = 1/F_{0.025}$ where $F_{0.025}$ has df = (8,4). Thus $F_{0.975} = 1/8.98 = 0.111$; $F_{0.025} = 5.05$.

17. (a) Step 1: H_0: $\sigma = 16$ H_a: $\sigma \neq 16$

Step 2: $\alpha = 0.10$

Step 3: $\chi^2 = \frac{n-1}{\sigma_0^2} \cdot s^2 = \frac{24}{16^2} \cdot 15.006^2 = 21.111$

Step 4: The critical values with n-1 = 24 degrees of freedom are 13.848 and 36.415. Using technology, the P-value is 0.7357.

Step 5: Since 13.848 < 21.111 < 36.415, we do not reject H_0.

Step 6: There is insufficient evidence at the 0.10 level to claim that σ for IQs measured on the Stanford revision of the Binet-Simon Intelligence Scale is different from 16.

(b) Normality is crucial for the hypothesis test in (a) since the procedure is not robust to moderate deviations from the normality assumption.

18. The 99% confidence interval for the standard deviation of the rebound distances of the bouncing rocks

$$\left(\sqrt{\frac{n-1}{\chi^2_{0.005}}} \cdot s, \sqrt{\frac{n-1}{\chi^2_{0.995}}} \cdot s\right) = \left(\sqrt{\frac{15}{32.801}} \cdot 3.054, \sqrt{\frac{15}{4.601}} \cdot 3.054\right) = (2.07, 5.51)$$

19. (a) F distribution with df = (14,19)

(b) 1=Runners, 2=Others

Step 1: H_0: $\sigma_1 = \sigma_2$ H_a: $\sigma_1 < \sigma_2$

Step 2: $\alpha = 0.01$

Step 3: $F = 1.798^2/6.606^2 = 0.074$

Step 4: Apply the Reciprocal Property of F-Curves to get $F_{0.99}$ with df = (14,19). This property states that $F_{0.99}$ with df = (14,19) is equal to the reciprocal of $F_{0.01}$ with df = (19,14). $F_{0.01}$ with df = (19,14) is 3.53. Therefore, $F_{0.99}$ with df = (14,19) is 1/3.43 = 0.28.

Using technology, the P-value is 0.0000.

Step 5: Since 0.074 < 0.28, reject the null hypothesis.

Step 6: There is sufficient evidence at the 0.01 significance level to claim that the variation in skinfold thickness among runners is less than that among others.

(c) We are assuming that skinfold thickness is a normally distributed variable. This assumption can be checked by looking at a normal probability plot for each set of data. If both plots are linear, the normality assumption is reasonable.

(d) We also assume that the samples are independent.

20. Step 1: The critical values are $F_{0.05}$ and $F_{0.95}$ with df = (7,19). $F_{0.05}$ with df = (7,19) is 2.54. Apply the Reciprocal Property of F-Curves to get $F_{0.95}$ with df = (7,19). This property states that $F_{0.95}$ with df = (7,19) is equal to the reciprocal of $F_{0.05}$ with df = (19,7). $F_{0.05}$ with df = (19,7) is 3.46. Therefore, $F_{0.95}$ with df = (7,19) is 1/3.46 = 0.289. Critical values are 0.289 and 2.54.

Step 2: The 90% confidence interval for σ_1/σ_2 is

$$\frac{1}{\sqrt{F_{0.05}}} \cdot \frac{s_1}{s_2} \text{ to } \frac{1}{\sqrt{F_{0.95}}} \cdot \frac{s_1}{s_2} = \frac{1}{\sqrt{2.54}} \cdot \frac{4.76}{3.72} \text{ to } \frac{1}{\sqrt{0.29}} \cdot \frac{4.76}{3.72} = (0.80, 2.38).$$

We can be 90% confident that the ratio of population standard deviations of heart-rate increase when laughing for male-male and female-female dyads is between 0.80 and 2.38. In other words, we can be 90% confident that the population standard deviation for male-male is between 1.25 times smaller to 2.38 times larger than the population standard deviation for female-female.

21. (a) Using Minitab, choose **Graph ▶ Probability Plot**, select the **Single** version and click **OK**. Enter BMI in the **Graph variables** text box, and click **OK.** Choose **Graph ▶ Boxplot**, select the **One-y Simple** version and click **OK**. Enter BMI in the **Graph variables** text box, and click **OK.** Choose **Graph ▶ Histogram**, select the **Simple** version and click **OK**. Enter BMI in the **Graph variables** text box, and click **OK.** The results are

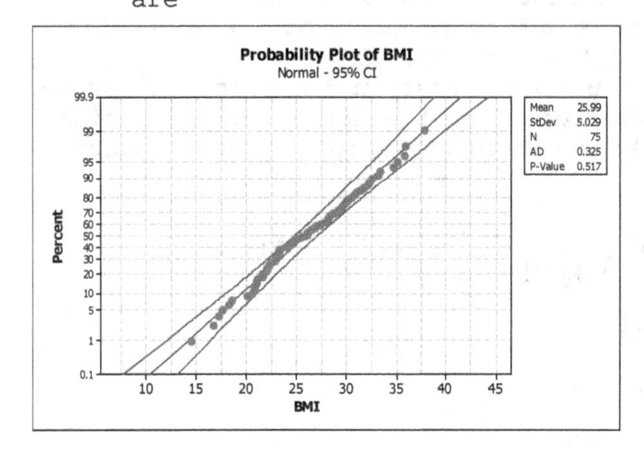

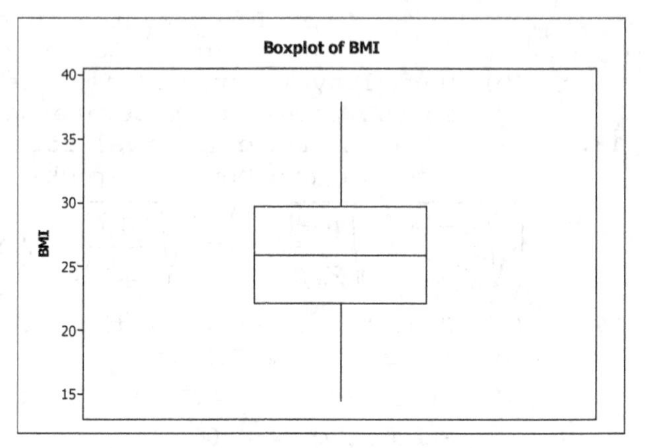

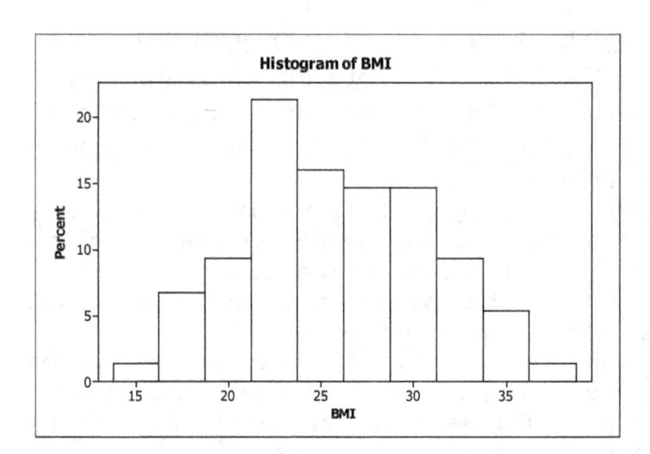

(b) Yes. The probability plot is nearly linear and there are no outliers.

(c) In Minitab, click on the Sessions Window. If there is no MTB> prompt showing, click on the Editor pull-down menu and on Enable commands. After the Mtb> prompt(and with the macro file 1stdev.mac saved from the WeissStats Resource Site), enter …\1stdev.mac 'BMI' and press ENTER. We have underlined your responses to the questions that appear in the following dialog.

```
MTB > %1stdev.mac 'bmi'
Executing from file: …\1stdev.mac

This macro performs a hypothesis test and/or obtains
a confidence interval for one population standard deviation.

Do you want to perform a hypothesis test (Y/N)?
Y

Enter the null hypothesis population standard deviation.

DATA> 5
```

```
Enter 0, 1, or -1, respectively, for a two-tailed, right-tailed,
or left-tailed test.

DATA> 0

Test of sigma =     5.00000 vs sigma not =     5.00000

Row  Variable  n   StDev  Chi-Sq  P
  1  BMI       75  5.029  74.871  0.900

Do you want a confidence interval (Y/N)?
Y

Enter the confidence level, as a percentage.

DATA> 95

Row  Variable  n   StDev  Level  CI for sigma
  1  BMI       75  5.029  95.0%  (4.333,5.994)
```

The P-value = 0.900, which is greater than the significance level 0.05, so do not reject the null hypothesis. The data do not provide sufficient evidence that the standard deviation is not 5.0.

22. The 95% confidence interval was obtained in Problem 21 as (4.333, 5.994).

23. We used Minitab to obtain a histogram of each set of data. With the data in columns named GSOD and PSOD, choose **Graph ▸ Histogram...**, select the **Simple** version, enter GSOD and PSOD in the **Graph variables** text box, click on the **Multiple graphs** button and select **On separate graphs**, **Same Y**, and **Same X, including bins**, and click **OK** twice. The results are

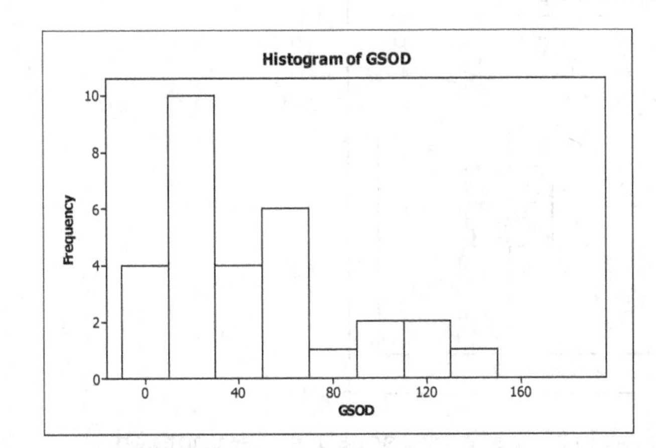

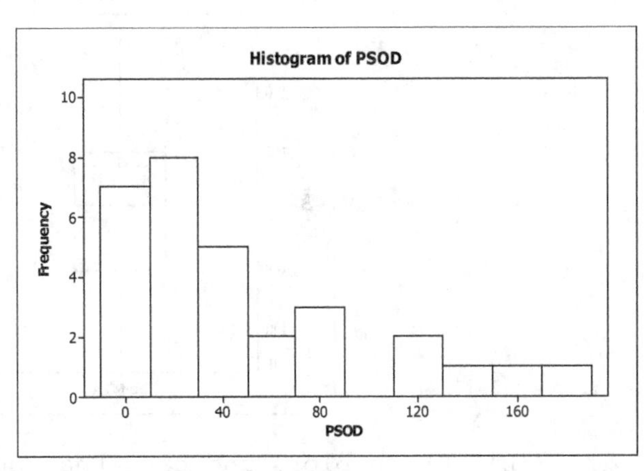

It is clear from both histograms that the data are very right-skewed. Since the F-test is very sensitive to non-normality, the F-test should not be used with these data.

24. First, check to ensure that the data distributions are reasonably normal. With the data in columns in Minitab named BRAND A and BRAND B, choose **Graph ▶ Probability Plot**, select the **Single** version and click **OK**. Enter BRAND A and BRAND B in the **Graph variables** text box, and click **OK**. Choose **Graph ▶ Histogram,** select the **Single** version and click **OK**. Enter BRAND A and BRAND B in the **Graph variables** text box, and click **OK**. Choose **Graph ▶ Boxplot,** select the **Single** version from the **Multiple Y's** and click **OK.** Enter BRAND A and BRAND B in the **Graph variables** text box, and click **OK.** The results are

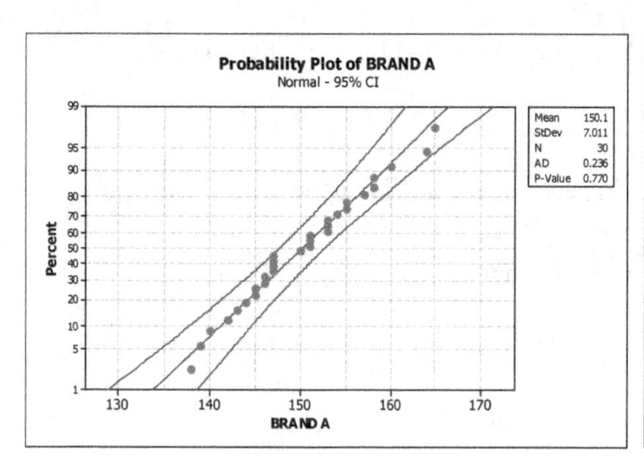

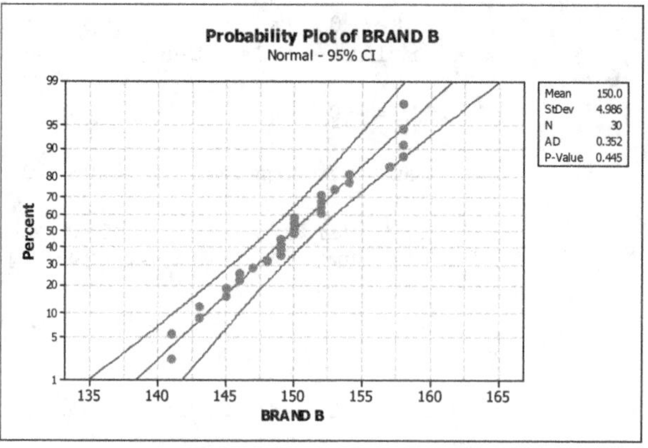

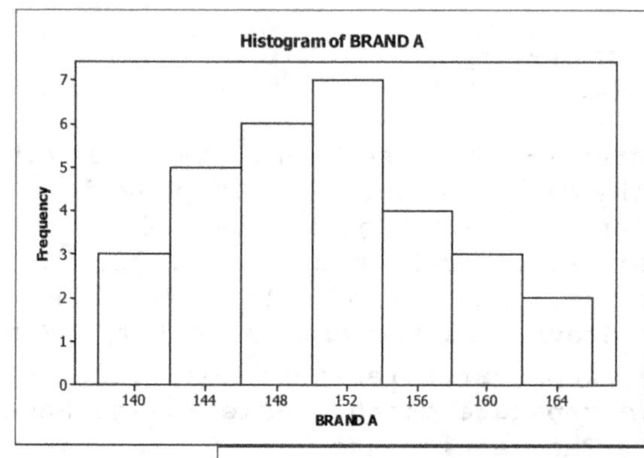

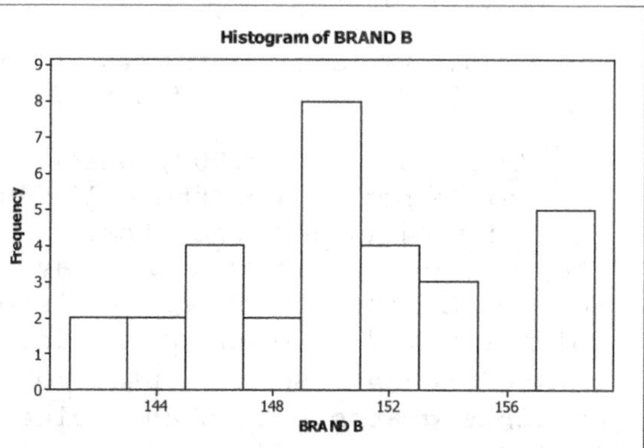

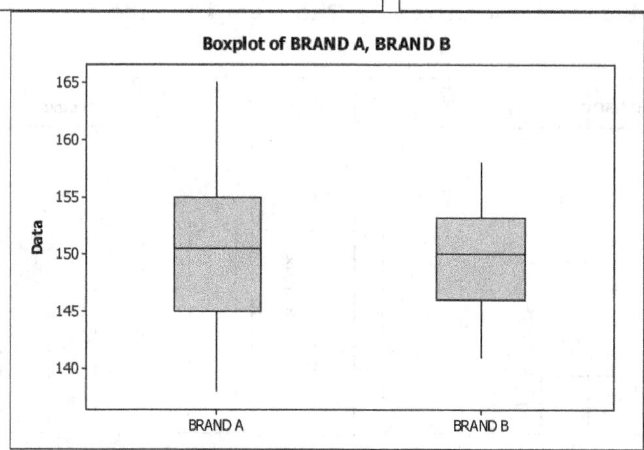

(b) Both plots indicate that normality is a reasonable assumption. Therefore we can proceed with an F test to determine whether BRAND A has a smaller standard deviation than BRAND B.

(c) **Using Minitab,** choose Stat ▶ Basic Statistics ▶ 2-Variances, click on **Samples in different columns**, enter BRAND A in the **First** box and BRAND B in the **Second** box. Click the **Options** tab. Enter 95 in the **Confidence level** text box, choose **greater than** in the **Alternative** textbox and click **OK** twice. The results are

Tests

Method	DF1	DF2	Test Statistic	P-Value
F Test (normal)	29	29	1.98	0.036
Levene's Test (any continuous)	1	58	4.13	0.023

The P-value for the test is 0.036, which is less than the significance level of 0.05. Therefore, we reject the null hypothesis of equal standard deviations. There is sufficient evidence at the 0.05 level that BRAND B has a smaller standard deviation than BRAND A and therefore has a more consistent popping time.

(d) **Using Minitab,** choose Stat ▶ Basic Statistics ▶ 2-Variances, click on **Samples in different columns**, enter BRAND A in the **First** box and BRAND B in the **Second** box, and click **OK**. Click **Options**, enter 90 in the **Confidence level** text box. Select **StDev1/StDev2** from the **Hypothesized ratio** box. Enter 1 in the **value** box. Choose **not equal** in the **Alternative** test box. Click **OK** twice. The results are

90% Confidence Intervals

Distribution of Data	CI for StDev Ratio	CI for Variance Ratio
Normal	(1.031, 1.918)	(1.063, 3.679)
Continuous	(1.073, 2.042)	(1.151, 4.168)

We can be 90% confident that the ratio of the standard deviations of popping times for Brand A and Brand B is between 1.031 to 1.918. In other words, we can be 90% confident that the standard deviation of popping times for Brand A is between 1.031 to 1.918 times larger than the standard deviation of popping times for Brand B.

CHAPTER 12 SOLUTIONS

Exercises 12.1

12.1 Answers will vary.

12.3 A population proportion p is a parameter since it is a descriptive measure for a population. A sample proportion $\hat{p}$ is a statistic since it is a descriptive measure for a sample.

12.5 (a) A sample proportion is the proportion (percentage) of a sample from the population that has the specified attribute.

(b) $\hat{p}$

12.7 The "number of failures" is an abbreviation for the number of members of the sample that do not have the specified attribute, and it is denoted by $n - x$, where n is the size of the sample.

12.8 $\hat{p} = x/n$. The sample proportion is equal to the number of successes divided by the sample size.

12.9 (a) $p = 2/5 = 0.4$

(b)

Sample	Number of females x	Sample proportion $\hat{p}$
J,G	1	0.5
J,P	0	0.0
J,C	0	0.0
J,F	1	0.5
G,P	1	0.5
G,C	1	0.5
G,F	2	1.0
P,C	0	0.0
P,F	1	0.5
C,F	1	0.5

(c) The population proportion is marked by the vertical line below.

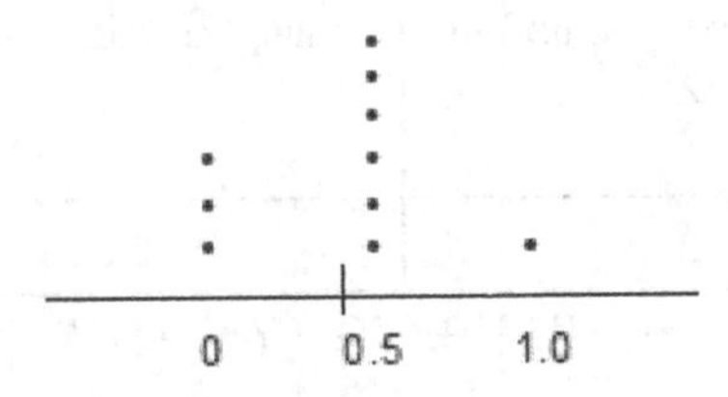

(d) $\mu_{\hat{p}} = \sum \hat{p}/10 = 4.0/10 = 0.4$

(e) The answers to (a) and (d) are the same. $\hat{p}$ is a sample mean. The mean of the sampling distribution of $\hat{p}$ is the same as the mean of the population, which is p.

12.11 (b)

Sample	Number of females x	Sample proportion $\hat{p}$
J,P,C	0	0
J,P,G	1	1/3
J,P,F	1	1/3
J,C,G	1	1/3
J,C,F	1	1/3
J,G,F	2	2/3
P,C,G	1	1/3
P,C,F	1	1/3
P,G,F	2	2/3
C,G,F	2	2/3

(c) The population proportion is marked by the vertical line below.

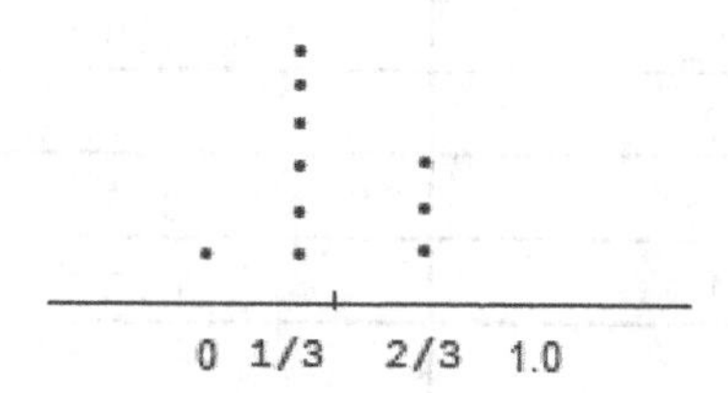

(d) $\mu_{\hat{p}} = \sum \hat{p}/10 = (12/3)/10 = 0.4$

(e) The answers to (a) and (d) are the same. $\hat{p}$ is a sample mean. The mean of the sampling distribution of $\hat{p}$ is the same as the mean of the population, which is p.

12.13 (b)

Sample	Number of females X	Sample proportion $\hat{p}$
J,P,C,G,F	2	0.4

(c) The population proportion is marked by the vertical line below.

0 0.25 0.5 1.0

(d) $\mu_{\hat{p}} = \sum \hat{p}/1 = 0.4/1 = 0.4$

(e) The answers to (a) and (d) are the same. $\hat{p}$ is a sample mean. The mean of the sampling distribution of $\hat{p}$ is the same as the mean of the population, which is p.

12.15 (a) The population consists of all #1 NBA draft picks since 1947.
(b) The specified attribute is being other than a U.S. national.
(c) The 10.4% is a population proportion since all of the #1 picks are known. There is no need to sample this group.

12.17 (a) $\hat{p}$ = 0.79; $z_{0.005}$ = 2.575; the margin of error is

$$E = z_{0.005} \cdot \sqrt{\hat{p}(1-\hat{p})/n} = 2.575\sqrt{0.79(0.21)/21355} = 0.00718$$

(b) The margin of error will be smaller for a 90% confidence interval. Specifically, 2.575 will be replaced by 1.645 in the formula for E and everything else stays the same. More generally speaking, in order to have a higher level of confidence in an interval, one needs to have a wider interval.

12.19 (a) 0.4 (b) $0.4 < \hat{p} < 0.6$

12.21 (a) 0.2 (b) $0.2 < \hat{p} < 0.8$

12.23 (a) 0.5 (b) none

12.25 (a) $\hat{p} = 8/40 = 0.2$
(b) x = 8 and n − x = 32. Both are at least 5, so the one-proportion z-interval procedure is appropriate.
(c) The 95% confidence interval is

$$\hat{p} \pm z_{\alpha/2}\sqrt{\hat{p}(1-\hat{p})/n} = 0.2 \pm 1.96\sqrt{0.2(1-0.2)/40} = (0.076, 0.324)$$

(d) The margin of error is $E = z_{\alpha/2}\sqrt{\hat{p}(1-\hat{p})/n} = 1.96\sqrt{0.2(1-0.2)/40} = 0.1240$. The confidence interval in terms of point estimate ± margin of error is 0.2 ± 0.1240.

12.27 (a) $\hat{p} = 35/50 = 0.70$
(b) x = 35 and n − x = 15. Both are at least 5, so the one-proportion z-interval procedure is appropriate.
(c) The 99% confidence interval is

$$\hat{p} \pm z_{\alpha/2}\sqrt{\hat{p}(1-\hat{p})/n} = 0.70 \pm 2.575\sqrt{0.70(1-0.70)/50} = (0.533, 0.867)$$

(d) The margin of error is $E = z_{\alpha/2}\sqrt{\hat{p}(1-\hat{p})/n} = 2.575\sqrt{0.70(1-0.70)/50} = 0.1669$. The confidence interval in terms of point estimate ± margin of error is 0.70 ± 0.1669.

12.29 (a) $\hat{p} = 16/20 = 0.8$
(b) x = 16 and n − x = 4. n − x is less than 5, so the one-proportion z-interval procedure is not appropriate.

12.31 $n = 0.25\left(\frac{z_{\alpha/2}}{E}\right)^2 = 0.25\left(\frac{1.96}{0.01}\right)^2 = 9604$

12.33 $n = 0.25\left(\frac{z_{\alpha/2}}{E}\right)^2 = 0.25\left(\frac{1.645}{0.02}\right)^2 = 1691.27$ Round this up to n = 1692.

12.35 $n = 0.25\left(\frac{z_{\alpha/2}}{E}\right)^2 = 0.25\left(\frac{2.575}{0.03}\right)^2 = 1841.84$ Round this up to n = 1842.

If you use $z_{\alpha/2} = 2.576$, you will get n = 1844.

12.37 (a) $n = \hat{p}_g(1-\hat{p}_g)\left(\frac{z_{\alpha/2}}{E}\right)^2 = 0.3(1-0.3)\left(\frac{1.96}{0.01}\right)^2 = 8067.36$ Round this up to n = 8068.

(b) This is smaller than the sample size in Exercise 12.31 because using an educated guess will give us a more realistic sample, while using the constant 0.25 will give us a sample that will be generally larger than necessary.

12.39 (a) $n = \hat{p}_g(1-\hat{p}_g)\left(\frac{z_{\alpha/2}}{E}\right)^2 = 0.1(1-0.1)\left(\frac{1.645}{0.02}\right)^2 = 608.86$ Round this up to n = 609.

(b) This is smaller than the sample size in Exercise 12.33 because using an educated guess will give us a more realistic sample, while using the constant 0.25 will give us a sample that will be generally larger than necessary.

12.41 (a) $n = \hat{p}_g(1-\hat{p}_g)\left(\frac{z_{\alpha/2}}{E}\right)^2 = 0.5(1-0.5)\left(\frac{2.575}{0.03}\right)^2 = 1841.84$ Round this up to n = 1842. If you use $z_{\alpha/2} = 2.576$, you will get n = 1844.

(b) This is the same as the sample size in Exercise 12.35 because an educated guess of 0.5 will give $\hat{p}_g(1-\hat{p}_g) = 0.5(1-0.5) = 0.25$ which is the same as using the constant 0.25.

12.43 Use $\hat{p}_g = 0.4$ since it is the closest in the likely range to 0.5.

$$n = \hat{p}_g(1-\hat{p}_g)\left(\frac{z_{\alpha/2}}{E}\right)^2 = 0.4(1-0.4)\left(\frac{1.96}{0.01}\right)^2 = 9219.84$$

Round this up to n = 9220.

12.45 Use $\hat{p}_g = 0.2$ since it is the closest in the likely range to 0.5.

$$n = \hat{p}_g(1-\hat{p}_g)\left(\frac{z_{\alpha/2}}{E}\right)^2 = 0.2(1-0.2)\left(\frac{1.645}{0.02}\right)^2 = 1082.41$$

Round this up to n = 1083.
Round this up to n = 5683.

12.47 Use $\hat{p}_g = 0.5$ since it is the closest in the likely range to 0.5.

$$n = \hat{p}_g(1-\hat{p}_g)\left(\frac{z_{\alpha/2}}{E}\right)^2 = 0.5(1-0.5)\left(\frac{2.575}{0.03}\right)^2 = 1841.84$$

Round this up to n = 1842. If you use $z_{\alpha/2} = 2.576$, you will get n = 1844.

12.49 $\hat{p} = 0.32$; The 95% confidence interval is

$$\hat{p} \pm z_{\alpha/2}\sqrt{\hat{p}(1-\hat{p})/n} = 0.32 \pm 1.96\sqrt{0.32(1-0.32)/1000} = (0.291, 0.349)$$

We are 95% confident that the proportion of all U.S. adults that never clothes-shop online is between 0.291 and 0.349.

12.51 n = 500, x = 38, and n - x = 462. Both x and n - x are at least 5.
$\hat{p} = x/n = 38/500 = 0.076$; $z_{\alpha/2} = z_{0.025} = 1.96$

$$0.076 - 1.96\sqrt{0.076(1-0.076)/500} \text{ to } 0.076 + 1.96\sqrt{0.076(1-0.076)/500}$$

$$0.0528 \text{ to } 0.0992$$

We can be 95% confident that the proportion of U.S. asthmatics who are allergic to sulfites is somewhere between 0.0528 and 0.0992.

12.53 n = 1000, x = 800, and n - x = 200. Both x and n - x are at least 5.

$\hat{p}$ = x/n = 800/1000 = 0.80; $z_{\alpha/2} = z_{0.005} = 2.575$

$$0.80 - 2.575\sqrt{0.80(1-0.80)/1000} \text{ to } 0.80 + 2.575\sqrt{0.80(1-0.80)/1000}$$

$$0.767 \text{ to } 0.833$$

We can be 99% confident that the percentage of all registered voters who favor the creation of standards on CAFO pollution is somewhere between 76.7% and 83.3%.

12.55 Here, n = 100, x = 96, and n - x = 4. To use Procedure 12.1, both x and n - x must be at least 5. Since n - x < 5, Procedure 12.1 should not have been used.

12.57 $\hat{p}$ = 0.50, E = 0.03

$\hat{p}$ - E to $\hat{p}$ + E

0.50 - 0.03 to 0.50 + 0.03
0.47 to 0.53
We can be 95% confident that the percentage of all American adults who favor a plan to break up the 12 megabanks is between 47% and 53%.

12.59 From Exercise 12.51, $\hat{p}$ = 0.076; $z_{\alpha/2}$ = 1.96; n = 500

(a) $E = z_{\alpha/2}\sqrt{\hat{p}(1-\hat{p})/n} = 1.96\sqrt{0.076(1-0.076)/500} = 0.0232$

(b) E = 0.01; $z_{\alpha/2} = z_{0.025} = 1.96$;

$$n = \hat{p}(1-\hat{p})\frac{z_{\alpha/2}^2}{E^2} = 0.5(1-0.5)\frac{1.96^2}{0.01^2} = 9604$$

(c) n = 9604; $\hat{p}$ = 0.071; $z_{\alpha/2} = z_{0.025} = 1.96$

$$0.071 - 1.96\sqrt{0.071(1-0.071)/9604} \text{ to } 0.3192 + 1.96\sqrt{0.071(1-0.071)/9604}$$

$$0.0659 \text{ to } 0.0761$$

(d) The margin of error for the estimate is 0.0051. As expected, this is less than the required margin of error of 0.01 specified in part (b).

(e) $\hat{p}_g$ = 0.10; $z_{\alpha/2} = z_{0.025} = 1.96$

part (b) E = 0.01, $z_{0.025}$ = 1.96; sample size is:

$$n = \hat{p}(1-\hat{p})\frac{z_{\alpha/2}^2}{E^2} = 0.10(1-0.10)\frac{1.96^2}{0.01^2} = 3457.44 \rightarrow 3458$$

Thus the required sample size is n = 3458.
part (c)

$$0.071 - 1.96\sqrt{0.071(1-0.071)/3458} \text{ to } 0.3192 + 1.96\sqrt{0.071(1-0.071)/3458}$$

$$0.0624 \text{ to } 0.0796$$

part (d) The margin of error is 0.0086. As expected, this is less than what was specified in part (b).

(f) By employing the guess for $\hat{p}$ in part (d) we can reduce the required sample size by more than 6000 (from 9604 to 3458), saving considerable time and money. Moreover, the margin of error only rises from 0.0051 to 0.0086. The risk of using the guess 0.10 for $\hat{p}$ is that if the actual value of $\hat{p}$ turns out to be between 0.10 and 0.90, then the achieved margin of error will exceed the specified 0.01.

12.61 (a) The confidence interval from Exercise 12.53 was (76.7%, 83.3%). To find the margin of error, take the width of the confidence interval and divide by 2. E = (83.3% - 76.7%)/2 = 3.3%.

(b) E = 0.015; $z_{\alpha/2} = z_{0.005} = 2.575$;

$$n = \hat{p}(1-\hat{p})\frac{z_{\alpha/2}^2}{E^2} = 0.5(1-0.5)\frac{2.575^2}{0.015^2} = 7367.36 \rightarrow 7368$$

(c) n = 7368; = 0.822; $z_{\alpha/2} = z_{0.005} = 2.575$

$$0.822 - 2.575\sqrt{0.822(1-0.822)/7368} \text{ to } 0.822 + 2.575\sqrt{0.822(1-0.822)/7368}$$

$$0.811 \text{ to } 0.833$$

(d) The margin of error for the estimate is 0.011, which is less than the 0.015 required in part (b).

(e) $\hat{p}_g$ is between 0.75 and .85; $z_{\alpha/2} = z_{0.005} = 2.575$

part (b) E = 0.015, $z_{0.005} = 2.575$; sample size is:

$$n = \hat{p}(1-\hat{p})\frac{z_{\alpha/2}^2}{E^2} = 0.75(1-0.75)\frac{2.575^2}{0.015^2} = 5525.52 \rightarrow 5526$$

Thus the required sample size is n = 5526.
part (c)

$$0.822 - 2.575\sqrt{0.822(1-0.822)/5526} \text{ to } 0.822 + 2.575\sqrt{0.822(1-0.822)/5526}$$

$$0.809 \text{ to } 0.835$$

part (d)
The margin of error is 0.013, which is less than the 0.015 specified in part (b).

(f) By employing the guess for $\hat{p}$ in part (d) we can reduce the required sample size by 1842, from 7368 to 5526, saving considerable time and money. Moreover, the margin of error only rises from 0.011 to 0.013. The risk of using the guess 0.75 for $\hat{p}$ is that if the actual value of $\hat{p}$ turns out to be between 0.25 and 0.75, then the achieved margin of error will exceed the specified 0.015.

12.63 (a) E = 0.01; $z_{\alpha/2} = z_{0.025} = 1.96$;

$$n = \hat{p}(1-\hat{p})\frac{z_{\alpha/2}^2}{E^2} = 0.5(1-0.5)\frac{1.96^2}{0.01^2} = 9604$$

(b) $\hat{p}_g$ is between 0.5% and 4.9%; $z_{\alpha/2} = z_{0.025} = 1.96$; E = 0.01

$$n = \hat{p}(1-\hat{p})\frac{z_{\alpha/2}^2}{E^2} = 0.049(1-0.049)\frac{1.96^2}{0.01^2} = 1790.15 \rightarrow 1791$$

Thus the required sample size is n = 1,791.

(c) The sample size has decreased from 9604 to 1791.

(d) If the actual value of $\hat{p}$ turns out to be between 0.049 and 0.951, then the achieved margin of error will exceed the specified error of 0.01.

12.65 We will assume that both polls used a 95% confidence level. Then at the 95% confidence level, both polls are giving a range of believable values for the true population (Americans) proportion that felt that Obama was doing his job. The Gallup poll gave its range to be 0.39 to 0.45. The Quinnipiac University poll gave its range to be 0.36 to 0.40. We note that both ranges of values overlap and have common believable values from 0.39 to 0.40. Thus, it is possible that both of these polls were correct in their conclusions.

12.67 The sample size is directly proportional to the quantity p(1 - p), which takes on its maximum value of 0.25 when p = .5. Sample sizes increase as p approaches 0.5 from either above or below. Thus to achieve a sample size adequate for any p in a given range, we should choose the largest sample that could result for any p in the range. This will always happen when we choose the value of p in the range that is closest to 0.5.

12.69 (a) $\tilde{p} = (8+2)/(40+4) = 0.227$

The 95% one-proportion plus-four *z*-interval is

$$\tilde{p} \pm z_{\alpha/2}\sqrt{\tilde{p}(1-\tilde{p})/(n+4)} = 0.227 \pm 1.96\sqrt{0.227(1-0.227)/44} = (0.103, 0.351)$$

(b) This interval has the same level of precision as the interval found in Exercise 12.25. However, it is centered at 0.227 rather than 0.20.

12.71 (a) $\tilde{p} = (35+2)/(50+4) = 0.685$

The 99% one-proportion plus-four *z*-interval is

$$\tilde{p} \pm z_{\alpha/2}\sqrt{\tilde{p}(1-\tilde{p})/(n+4)} = 0.685 \pm 2.575\sqrt{0.685(1-0.685)/54} = (0.522, 0.848)$$

(b) This interval is slightly more precise than the interval found in Exercise 12.27. Also, it is centered at 0.685 rather than 0.70.

12.73 (a) $\tilde{p} = (16+2)/(20+4) = 0.75$

The 90% one-proportion plus-four *z*-interval is

$$\tilde{p} \pm z_{\alpha/2}\sqrt{\tilde{p}(1-\tilde{p})/(n+4)} = 0.75 \pm 1.645\sqrt{0.75(1-0.75)/24} = (0.605, 0.895)$$

(b) It was not appropriate to find the confidence interval in Exercise 12.29.

12.75 $\tilde{p} = (707+2)/(1039+4) = 0.6798$

The 95% one-proportion plus-four *z*-interval is

$$\tilde{p} \pm z_{\alpha/2}\sqrt{\tilde{p}(1-\tilde{p})/(n+4)} = 0.6798 \pm 1.96\sqrt{0.6798(1-0.6798)/1043} = (0.651, 0.708)$$

We can be 95% confident that the proportion of all American adults who would continue working if they won 10 million dollars in the lottery is between 0.651 and 0.708.

12.77 77% of 434 is 334. Thus, x = 334 and n = 434.

$\tilde{p} = (334+2)/(434+4) = 0.767$

The 90% one-proportion plus-four *z*-interval is

$$\tilde{p} \pm z_{\alpha/2}\sqrt{\tilde{p}(1-\tilde{p})/(n+4)} = 0.767 \pm 1.645\sqrt{0.767(1-0.767)/438} = (0.734, 0.800)$$

We can be 90% confident that 73.4% to 80.0% of all new mothers breast-feed their infants at least briefly.

Exercises 12.2

12.79 (a) The sample proportion is $\hat{p}$ = x/n = 8/40 = 0.2.

(b) np_o = 40(0.3) = 12; $n(1 - p_0)$ = 40(1-0.30) = 28

Since both are at least 5, we can employ Procedure 12.2.

(c) $z = \dfrac{0.2-0.3}{\sqrt{.03(1-0.3)/40}} = -1.38$; For α = 0.10, the critical value is

$-z_\alpha$ = -1.28. Since -1.38 < -1.28, reject H_0.

12.81 (a) The sample proportion is $\hat{p}$ = x/n = 35/50 = 0.70.

(b) np_o = 50(0.6) = 30; $n(1 - p_0)$ = 50(1-0.60) = 20
Since both are at least 5, we can employ Procedure 12.2.

(c) $z = \dfrac{0.70 - 0.60}{\sqrt{0.6(1-0.6)/50}} = 1.44$; For α = 0.05, the critical value is

z_α = 1.645. Since 1.44 < 1.645, do not reject H_0.

12.83 (a) The sample proportion is $\hat{p}$ = x/n = 16/20 = 0.80.

(b) np_o = 20(0.7) = 14; $n(1 - p_0)$ = 20(1-0.70) = 6
Since both are at least 5, we can employ Procedure 12.2.

(c) $z = \dfrac{0.80 - 0.70}{\sqrt{0.7(1-0.7)/20}} = 0.98$; For α = 0.05, the critical values are

$\pm z_{\alpha/2}$ = $\pm$1.96. Since −1.96 < 0.98 < 1.96, do not reject H_0

12.85 (a) The sample proportion is $\hat{p}$ = x/n = 534/1008 = 0.5298. 52.98% of those sampled disapprove of the government program.

(b) α = 0.05, p_0 = 0.50
np_o = 1008(0.50) = 504; $n(1 - p_0)$ = 1008(1-0.50) = 504
Since both are at least 5, we can employ Procedure 12.2.

Step 1: H_0: p = 0.50, H_a: p > 0.50
Step 2: α = 0.05
Step 3: $z = \dfrac{0.5298 - 0.50}{\sqrt{0.50(1-0.50)/1008}} = 1.89$
Step 4: Since α = 0.05, the critical value is z_α = 1.645

For the p-value approach, P(z > 1.89) = 0.0294
Step 5: Since 1.89 > 1.645, reject H_0. P -value < α. Thus, we reject H_0.
Step 6: The test results are statistically significant at the 5% level; that is, at the 5% significance level, the data do provide sufficient evidence to conclude that a majority of American adults disapprove of this government surveillance program.

12.87 The sample proportion is $\hat{p}$ = 0.60.

α = 0.10, p_0 = 0.58
np_o = 1028(0.58) = 596.24; $n(1 - p_0)$ = 1028(1 - 0.58) = 431.76
Since both are at least 5, we can employ Procedure 12.2.
Step 1: H_0: p = 0.58, H_a: p ≠ 0.58
Step 2: α = 0.10
Step 3: $z = \dfrac{0.60 - 0.58}{\sqrt{0.58(1-0.58)/1028}} = 1.30$
Step 4: Since α = 0.10, the critical values are $\pm z_{\alpha/2}$ = $\pm$1.645

For the p-value approach, 2P(Z > 1.30) = 2(0.0968) = 0.1936
Step 5: Since 1.30 < 1.645, reject H_0. P-value > α. Thus, we do not reject H_0.
Step 6: The test results are not statistically significant at the 10% level; that is, at the 10% significance level, the data do not provide sufficient evidence to conclude that the percentage of all American adults who now think that a third major party is needed has changed from that in 2010.

12.89 (a) The sample proportion is $\hat{p} = 0.650$.

$\alpha = 0.05$, $p_0 = 0.72$
$np_o = 1003(0.72) = 722.2$; $n(1 - p_0) = 1003(1 - 0.72) = 280.8$
Since both are at least 5, we can employ Procedure 12.2.
Step 1: H_0: $p = 0.72$, H_a: $p < 0.72$
Step 2: $\alpha = 0.05$

Step 3: $$z = \frac{0.65 - 0.72}{\sqrt{0.72(1-0.72)/1003}} = -4.94$$

Step 4: Since $\alpha = 0.05$, the critical value is $-z_\alpha = -1.645$

For the p-value approach, $P(Z < -4.94) = 0.0000$

Step 5: Since $-4.94 < -1.645$, reject H_0. P-value < α. Thus, we reject H_0.

Step 6: The test results are statistically significant at the 5% level; that is, at the 5% significance level, the data do provide sufficient evidence to conclude that the percentage of Americans who approve of labor unions has decreased since 1936.

(b) The sample proportion is $\hat{p} = 0.650$.

$\alpha = 0.05$, $p_0 = 2/3 = 0.667$
$np_o = 1003(0.67) = 672$; $n(1 - p_0) = 1003(1 - 0.67) = 331$
Since both are at least 5, we can employ Procedure 12.2.
Step 1: H_0: $p = 0.67$, H_a: $p < 0.67$
Step 2: $\alpha = 0.05$

Step 3: $$z = \frac{0.65 - 0.67}{\sqrt{0.67(1-0.67)/1003}} = -1.35$$

Step 4: Since $\alpha = 0.05$, the critical value is $-z_\alpha = -1.645$

For the p-value approach, $P(Z < -1.35) = 0.0885$

Step 5: Since $-1.35 > -1.645$, do not reject H_0. P-value > α. Thus, we do not reject H_0.

Step 6: The test results are not statistically significant at the 5% level; that is, at the 5% significance level, the data do not provide sufficient evidence to conclude that the percentage of Americans who approve of labor unions has decreased since 1963.

12.91 (a) $n = 1053$, $p_0 = 0.5$, $x = 548$, $\hat{p} = 548/1053 = 0.5204$

$np_0 = 526.5$, $n(1 - p_0) = 526.5$
Since both are at least 5, we can employ Procedure 12.2.

Step 1: H_0: $p = 0.50$, H_a: $p > 0.50$
Step 2: $\alpha = 0.05$

Step 3: $$z = \frac{0.5204 - 0.50}{\sqrt{0.50(1-0.50)/1053}} = 1.324$$

Step 4: Since $\alpha = 0.05$, the critical value is $z_\alpha = 1.645$

For the p-value approach, $P(z > 1.32) = 0.0934$

Step 5: Since $1.32 < 1.645$, do not reject H_0. P-value > α. Thus, do not reject H_0.

Step 6: The test results are not statistically significant at the 5% level; that is, at the 5% significance level, the data do not provide sufficient evidence to conclude that a majority of U.S. adults favored passage of the legislation.

(b) The words "slim majority" still implies that there can be said a majority of U.S. adults favored passage. From part (a), we cannot conclude that a majority favor passage of the legislation.

(c) It could say that a slim majority of those polled favor passage of the legislation.

12.93 (a) Using Minitab, choose **Stat**, **Basic Statistics,** then **1 Proportion.** Choose **Summarized data.** Enter 921 in the **Number of Events** textbox. Enter 1001 in the **Number of trials** textbox. Check **Perform Hypothesis test.** Enter 0.90 in the **Hypothesized proportion** textbox. Click **Options.** Enter 95.0 in the **Confidence level** textbox. Choose **greater than** in the **Alternative** pull-down menu. Check **Use test and interval based on normal distribution.** Click **OK** twice. The results are

```
Test of p = 0.9 vs p > 0.9

                                 95% Lower
Sample     X     N  Sample p       Bound  Z-Value  P-Value
1        921  1001  0.920080    0.905982     2.12    0.017

Using the normal approximation.
```

The test-statistic is 2.12 and the P-value is 0.017. This P-value is less than a significance level of 5%. Therefore, we would reject the null hypothesis. At 5% significance, there is evidence that more than 9 out of 10 Americans always wash up after using the bathroom.

(b) Using the results in part (a) and changing the significance level to 1%, the test statistic and the P-value would still be 2.12 and 0.017 respectively. The P-value is greater than a significance level of 1%. Therefore, we would not reject the null hypothesis. At 1% significance, there is not significant evidence that more than 9 out of 10 Americance always wash up after using the bathroom.

Exercises 12.3

12.95 We need to decide whether the difference between two sample proportions can reasonably be attributed to sampling error or are the population proportions really different.

12.97 (a) Using sunscreen before going out in the sun

(b) Teen-age girls and teen-age boys

(c) The two proportions are sample proportions. The reference is specifically to those teen-age girls and boys who were surveyed

12.99 (a) The parameters are p_1 and p_2. The rest are statistics.

(b) The fixed numbers are p_1 and p_2. The rest are variables.

12.101 (a) $\hat{p}_1 = x_1/n_1 = 18/40 = 0.45$; $\hat{p}_2 = x_2/n_2 = 30/40 = 0.75$;

$\hat{p}_p = (x_1 + x_2)/(n_1 + n_2) = 48/80 = 0.6$

(b) $x_1 = 18$, $n_1 - x_1 = 22$, $x_2 = 30$, $n_2 - x_2 = 10$.

All are at least 5, so z-procedures are appropriate.

(c) $$z = \frac{0.45 - 0.75}{\sqrt{0.6(1-0.6)[(1/40)+(1/40)]}} = -2.74$$

$\alpha = 0.10$, critical value is -1.28; since $-2.74 < -1.28$, reject H_0.

(d) The 80% confidence interval is

$$\hat{p}_1 - \hat{p}_2 \pm z_{\alpha/2}\sqrt{\hat{p}_1(1-\hat{p}_1)/n_1 + \hat{p}_2(1-\hat{p}_2)/n_2}$$

$$(0.45-0.75) \pm 1.28\sqrt{0.45(1-0.4)/40 + 0.75(1-0.75)/40} = (-0.43,\ -0.17)$$

12.103 (a) $\hat{p}_1 = x_1/n_1 = 15/20 = 0.75;\ \hat{p}_2 = x_2/n_2 = 18/30 = 0.60;$

$\hat{p}_p = (x_1 + x_2)/(n_1 + n_2) = 33/50 = 0.66$

(b) $x_1 = 15,\ n_1 - x_1 = 5,\ x_2 = 18,\ n_2 - x_2 = 12.$

All are at least 5, so z-procedures are appropriate.

(c) $$z = \frac{0.75-0.60}{\sqrt{0.66(1-0.66)[(1/20)+(1/30)]}} = 1.10$$

$\alpha = 0.05$, critical value is 1.645; since $1.10 < 1.645$, do not reject H_0.

(d) The 90% confidence interval is

$$\hat{p}_1 - \hat{p}_2 \pm z_{\alpha/2}\sqrt{\hat{p}_1(1-\hat{p}_1)/n_1 + \hat{p}_2(1-\hat{p}_2)/n_2}$$

$$(0.75-0.60) \pm 1.645\sqrt{0.75(1-0.75)/20 + 0.60(1-0.60)/30} = (-0.067,\ 0.367)$$

12.105 (a) $\hat{p}_1 = x_1/n_1 = 30/80 = 0.375;\ \hat{p}_2 = x_2/n_2 = 15/20 = 0.750;$

$\hat{p}_p = (x_1 + x_2)/(n_1 + n_2) = 45/100 = 0.45$

(b) $x_1 = 30,\ n_1 - x_1 = 50,\ x_2 = 15,\ n_2 - x_2 = 5.$

All are at least 5, so z-procedures are appropriate.

(c) $$z = \frac{0.375-0.750}{\sqrt{0.45(1-0.45)[(1/80)+(1/20)]}} = -3.02$$

$\alpha = 0.05$, critical values are ± 1.96; since $-3.02 < -1.96$, reject H_0.

(d) The 95% confidence interval is

$$\hat{p}_1 - \hat{p}_2 \pm z_{\alpha/2}\sqrt{\hat{p}_1(1-\hat{p}_1)/n_1 + \hat{p}_2(1-\hat{p}_2)/n_2}$$

$$(0.375-0.750) \pm 1.96\sqrt{0.375(1-0.375)/80 + 0.750(1-0.750)/20} =$$

$$(-0.592,\ -0.158)$$

12.107 (a) Population 1: Women who took multivitamins containing folic acid

$\hat{p}_1 = 35/2701 = 0.01296$

Population 2: Women who received only trace elements

$\hat{p}_2 = 47/2052 = 0.02290$

$\hat{p}_p = (35 + 47)/(2,701 + 2,052) = 0.01725$

Step 1: H_0: $p_1 = p_2$, H_a: $p_1 < p_2$

Step 2: $\alpha = 0.01$

Step 3: $$z = \frac{0.01296-0.0220}{\sqrt{0.01725(1-0.01725)[(1/2701)+(1/2052)]}} = -2.61$$

Step 4: Since $\alpha = 0.01$, the critical value is $-z_\alpha = -2.33$

For the P-value approach, $P(z < -2.61) = 0.0045$

Step 5: Since $-2.61 < -2.33$, reject H_0. P-value $< \alpha$. Thus, we reject H_0.

Step 6: The test results are significant at the 1% level; that is, at the 1% significance level, the data do provide sufficient evidence to conclude that the women who take folic acid are at lesser risk of having children with major birth defects than those women who do not.

(b) This is a designed experiment. The researchers decided which women would take daily multivitamins.

(c) Yes. By using the basic principles of design (control, randomization, and replication) the doctors can conclude that the reduction in the rates of major birth defects in the folic acid group is likely caused by the folic acid.

12.109 Population 1: Drivers of age 16-24, $\hat{p}_1 = 0.790 = 790/1000$

Population 2: Drivers of age 25-69, $\hat{p}_2 = 924/1100 = 0.840$

$\hat{p}_p = (790 + 924)/(1000 + 1100) = 0.8162$

Step 1: H_0: $p_1 = p_2$, H_a: $p_1 \neq p_2$

Step 2: $\alpha = 0.01$

Step 3: $$z = \frac{0.790 - 0.840}{\sqrt{0.8162(1-0.8162)[(1/1000)+(1/1100)]}} = -2.95$$

Step 4: Since $\alpha = 0.01$, the critical values is $\pm z_{\alpha} = \pm 2.575$

For the P-value approach, $2P(z < -2.95) = 0.0032$

Step 5: Since $-2.95 < -2.575$, reject H_0. P-value $< \alpha$. Thus, we reject H_0.

Step 6: The test results are significant at the 1% level; that is, at the 1% significance level, the data provide sufficient evidence to conclude that there is a difference in seat-belt usage between drivers 16-24 years old and those 25-69 years old.

12.111 Population 1: Bachelors degree, $\hat{p}_1 = 386/750 = 0.5147$

Population 2: Graduate degree, $\hat{p}_2 = 237/500 = 0.4740$

$\hat{p}_p = (386 + 237)/(750 + 500) = 0.4984$

(a) The assumptions for using the two-sample z-test are simple random samples, independent samples, and x_1, $n_1 - x_1$, x_2, and $n_2 - x_2$ must all be greater than or equal to 5.

(b) Step 1: H_0: $p_1 = p_2$, H_a: $p_1 > p_2$

Step 2: $\alpha = 0.05$

Step 3: $$z = \frac{0.5147 - 0.4740}{\sqrt{0.4984(1-0.4984)[(1/750)+(1/500)]}} = 1.41$$

Step 4: Since $\alpha = 0.05$, the critical value is $z_{\alpha} = 1.645$

For the P-value approach, $P(z > 1.41) = 0.0793$

Step 5: Since $1.41 < 1.645$, do not reject H_0. P-value $> \alpha$. Thus, we do not reject H_0.

Step 6: The test results are not significant at the 5% level; that is, at the 5% significance level, the data do not provide sufficient evidence to conclude that a higher percentage of adults with Bachelors degrees are overweight than of adults with graduate degrees.

12.113 From Exercise 12.107, the 99% confidence interval is

$$(0.012958-0.022904)\pm 2.33\sqrt{\frac{0.012958(1-0.012958)}{2701}+\frac{0.022904(1-0.022904)}{2052}}$$

$$-0.009946\pm 0.009215 \ or \ -0.019161 \ to \ -0.000731$$

We can be 98% confident that the difference $p_1 - p_2$ between the rates of major birth defects for babies born to women who have taken folic acid and those born to women who have not taken folic acid is somewhere between -0.019161 and -0.000731.

12.115 From Exercise 12.109, the 99% confidence interval is

$$\hat{p}_1-\hat{p}_2\pm z_{\alpha/2}\sqrt{\hat{p}_1(1-\hat{p}_1)/n_1+\hat{p}_2(1-\hat{p}_2)/n_2}$$

$$(0.790-0.840)\pm 2.575\sqrt{0.790(1-0.790)/1000+0.840(1-0.840)/1100}$$

$$=(-0.0937,\ \ -0.0063)$$

We can be 90% confident that the difference between proportions of seat belt users for drivers in the age groups 16-24 years and 25-69 years is somewhere between -0.0937 and -0.0063.

12.117 From Exercise 12.111, the 90% confidence interval is

$$\hat{p}_1-\hat{p}_2\pm z_{\alpha/2}\sqrt{\hat{p}_1(1-\hat{p}_1)/n_1+\hat{p}_2(1-\hat{p}_2)/n_2}$$

$$(0.515-0.474)\pm 1.645\sqrt{0.515(1-0.515)/750+0.474(1-0.474)/500}$$

$$=(-0.007,\ \ 0.088)$$

We can be 90% confident that the difference between percentages of adults in the two degree categories who have an above healthy weight is somewhere between -0.007 and 0.088.

12.119 (a) Using Minitab, choose **Stat ▶ Basic statistics ▶ 2 Proportions...**, select the **Summarized data** option button, click in the **Trials** text box for **First sample** and enter 300, click in the **Events** text box for **First sample** and type 215, click in the **Trials** text box for **Second sample** and type 250, click in the **Events** text box for **Second sample** and type 186, click the **Options...** button, click in the **Confidence level** text box and type 95, click in the **Test difference** text box and type 0, click the arrow button at the right of the **Alternative** drop-down list box and select **not equal**, select the **Use pooled estimate of p for test** check box, click **OK**, and click **OK**. The resulting output is

```
Sample     X    N  Sample p
1        215  300  0.716667
2        186  250  0.744000

Difference = p (1) - p (2)
Estimate for difference:  -0.0273333
95% CI for difference:  (-0.101675, 0.0470087)
Test for difference = 0 (vs not = 0):  Z = -0.72  P-Value = 0.473
```

Since the P-value = 0.473, which is greater than the significance level 0.05, do not reject the null hypothesis. The data do not provide sufficient evidence to conclude that there is a difference between the labor-force participation rates of U.S. and Canadian women.

(b) The confidence interval is part of the output for part (a). We can be 95% confident that the difference in labor-force participation rates between of U.S. and Canadian women is somewhere between -0.102 and 0.047.

12.121 (a) Using Minitab, choose **Stat ▶ Basic statistics ▶ 2 Proportions...**, select the **Summarized data** option button, click in the **Trials** text box for **First sample** and enter 1020, click in the **Events** text box for **First sample** and enter 35, click in the **Trials** text box for **Second sample** and enter 1008, click in the **Events** text box for **Second sample** and enter 69, click the **Options...** button, click in the **Confidence level** text box and type 99, click in the **Test difference** text box and type 0, click the arrow button at the right of the **Alternative** drop-down list box and select **less than**, select the **Use pooled estimate of p for test** check box, click **OK**, and click **OK**. The resulting output is

```
Test and CI for Two Proportions

Sample    X     N  Sample p
1        35  1020  0.03431
2        69  1008  0.06845

Difference = p (1) - p (2)
Estimate for difference:  -0.03414
Test for difference = 0 (vs < 0):  Z = -3.48  P-Value = 0.0002
```

Since the P-value = 0.0002, which is less than the significance level 0.01, reject the null hypothesis. The data provide sufficient evidence to conclude that college graduates have a lower unemployment rate than high school graduates.

(b) Repeat part (a), selecting not equal in the Alternative drop down box and click in the **Confidence level** text book and type 98. The confidence interval is

```
98% CI for difference:  (-0.0569, -0.0114)
```

We can be 98% confident that the difference in the unemployment rates between college and high school graduates is somewhere between -0.0569 and -0.0114.

12.123 (a) $\tilde{p}_1 = (18+1)/(40+2) = 0.452$ and $\tilde{p}_2 = (30+1)/(40+2) = 0.738$

The 80% two-proportions plus-four z-interval is

$$(\tilde{p}_1 - \tilde{p}_2) \pm z_{\alpha/2}\sqrt{\tilde{p}_1(1-\tilde{p}_1)/(n+2) + \tilde{p}_2(1-\tilde{p}_2)/(n+2)}$$

$$= (0.452 - 0.738) \pm 1.28\sqrt{0.452(1-0.452)/42 + 0.738(1-0.738)/42} = (-0.417, -0.155)$$

(b) This interval is close to the same level of precision as the interval found in Exercise 12.101. However, it is centered at -0.286 rather than -0.300.

12.125 (a) $\tilde{p}_1 = (15+1)/(20+2) = 0.727$ and $\tilde{p}_2 = (18+1)/(30+2) = 0.594$

The 90% two-proportions plus-four z-interval is

$$(\tilde{p}_1 - \tilde{p}_2) \pm z_{\alpha/2}\sqrt{\tilde{p}_1(1-\tilde{p}_1)/(n+2) + \tilde{p}_2(1-\tilde{p}_2)/(n+2)}$$

$$= (0.727 - 0.594) \pm 1.645\sqrt{0.727(1-0.727)/22 + 0.594(1-0.594)/32} = (-0.079, 0.345)$$

(b) This interval is slightly more precise than the interval found in Exercise 12.103. In addition, it is centered at 0.133 rather than 0.150.

12.127 (a) $\tilde{p}_1 = (30+1)/(80+2) = 0.378$ and $\tilde{p}_2 = (15+1)/(20+2) = 0.727$

The 95% two-proportions plus-four z-interval is

$$(\tilde{p}_1 - \tilde{p}_2) \pm z_{\alpha/2}\sqrt{\tilde{p}_1(1-\tilde{p}_1)/(n+2) + \tilde{p}_2(1-\tilde{p}_2)/(n+2)}$$

$$= (0.378 - 0.727) \pm 1.96\sqrt{0.378(1-0.378)/82 + 0.727(1-0.727)/22} = (-0.563, -0.135)$$

(b) This interval is more precise than the interval found in Exercise 12.105. In addition, it is centered at -0.349 rather than -0.375.

12.129 Let population 1 be the civilian labor forces of Finland and population 2 be the civilian labor forces of Denmark.

$\tilde{p}_1 = (7+1)/(100+2) = 0.078$ and $\tilde{p}_2 = (3+1)/(75+2) = 0.052$

The 95% two-proportions plus-four *z*-interval is

$$(\tilde{p}_1 - \tilde{p}_2) \pm z_{\alpha/2}\sqrt{\tilde{p}_1(1-\tilde{p}_1)/(n+2) + \tilde{p}_2(1-\tilde{p}_2)/(n+2)}$$

$$= (0.078 - 0.052) \pm 1.96\sqrt{0.078(1-0.078)/102 + 0.052(1-0.052)/77} = (-0.046, 0.098)$$

We can be 95% confident that the difference between the unemployment rates in Finland and Denmark is between -4.6% and 9.8%. In other words, we can be 95% confident that the unemployment rate in Finland is between 4.6% smaller to 9.8% larger than the unemployment rate in Denmark.

12.131 Let population 1 be elderly people taking calcium channel blockers and population 2 be elderly people taking beta blockers.

$\tilde{p}_1 = (27+1)/(202+2) = 0.137$ and $\tilde{p}_2 = (28+1)/(424+2) = 0.068$

The 95% two-proportions plus-four *z*-interval is

$$(\tilde{p}_1 - \tilde{p}_2) \pm z_{\alpha/2}\sqrt{\tilde{p}_1(1-\tilde{p}_1)/(n+2) + \tilde{p}_2(1-\tilde{p}_2)/(n+2)}$$

$$= (0.137 - 0.068) \pm 1.645\sqrt{0.137(1-0.137)/204 + 0.068(1-0.068)/426}$$

$$= (0.025, 0.113)$$

We can be 90% confident that the difference between the cancer rates of elderly people taking calcium channel blockers and those taking beta blockers is between 2.5% and 11.3%. In other words, we can be 90% confident that the cancer rate of elderly people taking calcium channel blockers is between 2.5% to 11.3% larger than the cancer rate of elderly people taking beta blockers.

12.133 In the previous equation, replace the n_1 and n_2 with n.

$E = z_{\alpha/2}\sqrt{\hat{p}_1(1-\hat{p}_1)/n + \hat{p}_2(1-\hat{p}_2)/n}$. Divide both sides by $z_{\alpha/2}$, square both sides, and replace each of the $\hat{p}_1(1-\hat{p}_1)$ and $\hat{p}_2(1-\hat{p}_2)$ with 0.25 since that is the maximum value for those expressions. $\left(\frac{E}{z_{\alpha/2}}\right)^2 = \frac{0.25}{n} + \frac{0.25}{n} = \frac{0.50}{n}$.

Reciprocate this equation and you have $\frac{n}{0.50} = \left(\frac{z_{\alpha/2}}{E}\right)^2$. Multiplying by 0.50 results in $n = n_1 = n_2 = 0.50\left(\frac{z_{\alpha/2}}{E}\right)^2$.

12.135 (a) The confidence interval found in Example 12.10 is (-0.129, -0.031). The length of this confidence interval is 0.098. Divide this in half to get the margin of error or 0.098/2 = 0.049. This is the amount of error that you add and subtract from the point estimate of the differences between the two sample proportions in order to form the confidence interval.

(b)

$$E = z_{\alpha/2}\sqrt{\hat{p}_1(1-\hat{p}_1)/n_1 + \hat{p}_2(1-\hat{p}_2)/n_2}$$

$$= 1.645\sqrt{0.369(1-0.369)/747 + 0.449(1-0.449)/434} = 0.049$$

12.137 (a)

$$n = n_1 = n_2 = \left(\hat{p}_{1g}\left(1-\hat{p}_{1g}\right)+\hat{p}_{2g}\left(1-\hat{p}_{2g}\right)\right)\left(\frac{z_{\alpha/2}}{E}\right)^2$$

$$= (0.41(1-0.41)+0.49(1-0.49))\left(\frac{1.645}{0.01}\right)^2 = 13308.23 \rightarrow 13{,}309$$

(b)

$$(\hat{p}_1-\hat{p}_2) \pm z_{\alpha/2}\sqrt{\hat{p}_1\left(1-\hat{p}_1\right)/n_1+\hat{p}_2\left(1-\hat{p}_2\right)/n_2}$$

$$(0.383-0.437) \pm 1.645\sqrt{0.383(1-0.383)/13309+0.437(1-0.437)/13309}$$

$$(-0.0639, -0.0441)$$

(c) The margin of error in part (b) is (0.0639-0.0441)/2 = 0.00975. This is less than the required error specified in part (a). It is also only slightly less than the margin of error we obtained in part (c) of Exercise 12.136. This is due to the fact that the educated guess values of 0.41 and 0.49 are both pretty close to 0.50.

Chapter 12 Review Problems

1. (a) Feeling that the overall medicinal benefits of marijuana outweigh the risks and potential harms.
(b) All international physicians
(c) Proportion of all international physicians who feel that the overall medicinal benefits of marijuana outweigh the risks and potential harms.
(d) Proportion of international physicians in the sample who feel that the overall medicinal benefits of marijuana outweigh the risks and potential harms. Clearly the population of all international physicians is much larger than 1,446.

2. It is often impossible to take a census of an entire population. It is also expensive and time-consuming.

3. (a) "Number of successes" stands for the number of members of the sample that exhibit the specified attribute.
(b) "Number of failures" stands for the number of members of the sample that do not exhibit the specified attribute.

4. (a) population proportion
(b) normal
(c) number of successes, number of failures, 5

5. The margin of error for the estimate of a population proportion tells us what the maximum difference between the sample proportion and the population proportion is likely to be. It is not an absolute maximum, and how likely it is depends on the confidence level used.

6. (a) Getting the "holiday blues"
(b) Men and women
(c) The proportion of men in the population who get the "holiday blues" and the proportion of women in the population who get the "holiday blues"
(d) The proportion of men in the sample who get the "holiday blues" and the proportion of women in the sample who get the "holiday blues"
(e) They are sample proportions since the information came from a poll, not a census. Also, it could not be a population proportion because I was not asked.

7. (a) The mean of all possible differences between the two sample proportions equals the difference of the population proportions.
(b) For large samples, the possible differences between the two sample proportions have approximately a normal distribution.

8. The 95% confidence interval for the percentage of U.S. adults who would get a smallpox shot if it were available is 40% $\pm$ 3.0% or from 37% to 43%.

9. (a) n = 1218, $\hat{p}$ = 733/1218 = 0.6018, $z_{\alpha/2} = z_{0.025} = 1.96$

$$0.6018-1.96\sqrt{0.6018(1-0.6018)/1218} \text{ to } 0.6018+1.96\sqrt{0.6018(1-0.6018)/1218}$$

$$0.5743 \text{ to } 0.6293$$

We can be 95% confident that the proportion, p, of students who expect difficulty finding a job is somewhere between 0.5743 and 0.6293.

(b) The error is found by taking the width of the confidence interval and dividing by 2. So E = (0.6293 - 0.5743)/2 = 0.0275.

(c) In terms of point estimate $\pm$ margin of error, the confidence interval is 0.6018 $\pm$ 0.0275.

10. (a) E = 0.02; $z_{\alpha/2} = z_{0.025} = 1.96$; $\hat{p}_g = 0.50$

$$n = 0.5^2\frac{z^2_{\alpha/2}}{E^2} = 0.25\cdot\frac{1.96^2}{0.02^2} = 2401$$

(b) n = 2401; $\hat{p}$ = 0.587; $z_{\alpha/2} = z_{0.025} = 1.96$

$$0.587-1.96\sqrt{0.587(1-0.587)/2401} \text{ to } 0.587+1.96\sqrt{0.587(1-0.587)/2401}$$

$$0.5673 \text{ to } 0.6067$$

(c) The margin of error for the estimate is 0.0197, slightly less than what is required in part (a).

(d) $\hat{p}_g = 0.56$; $z_{\alpha/2} = z_{0.025} = 1.96$

part (a) E = 0.02, $z_{0.025} = 1.96$; sample size is:

$$n = 0.56(1-0.56)\frac{z^2_{\alpha/2}}{E^2} = 0.2464\cdot\frac{1.96^2}{0.02^2} = 2366.42 \rightarrow 2367$$

Thus the required sample size is n = 2367.

part (b)

$$0.587-1.96\sqrt{0.587(1-0.587)/2367} \text{ to } 0.587+1.96\sqrt{0.587(1-0.587)/2367}$$

$$0.5672 \text{ to } 0.6068$$

part (c) The margin of error is 0.0198, which is the same as what is specified in part (a).

(e) By employing the guess for $\hat{p}$ in part (d) we can reduce the required sample size by 34, from 2401 to 2367, saving a little time and money. Moreover, the margin of error stays the same. The risk of using the guess 0.56 for $\hat{p}$ is that if the actual value of $\hat{p}$ turns out to be between .44 and .56, then the achieved margin of error will exceed the specified 0.02.

11. n = 2512, x = 578, α = 0.05, $\hat{p}$ =578/2512 = 0.2301

np_o = 2512(0.25) = 628; $n(1 - p_0)$ = 2512(1-0.25) = 1884
Since both are at least 5, we can employ Procedure 12.2.

(a) Step 1: H_0: p = 0.25, H_1: p < 0.25
Step 2: α = 0.05

Step 3: $$z = \frac{0.23-0.25}{\sqrt{0.25(1-0.25)/2512}} = -2.31$$

Step 4: Since α = 0.05, the critical value is $-z_\alpha$ = -1.645

For the P-value approach: $P(z < -2.31) = 0.0104$.

Step 5: Since $-2.31 < -1.645$, reject H_0. Since P-value $< \alpha$, reject H_0.

Step 6: The test results are statistically significant at the 5% level; that is, at the 5% significance level, the data do provide sufficient evidence to conclude that less than one in four Americans believe that juries "almost always" convict the guilty and free the innocent

12. (a) Observational study. The researchers had no control over any of the factors of the study.

(b) Height may not be the only factor to be considered. Although there does appear to be an association, we don't know if it is a direct association.

13. Population 1: men, $\hat{p}_1 = 607/1029 = 0.5899$,

Population 2: women, $\hat{p}_2 = 648/1223 = 0.5298$

$$\hat{p}_p = (607 + 648)/(1029 + 1223) = 0.5573$$

Step 1: H_0: $p_1 = p_2$, H_a: $p_1 \neq p_2$

Step 2: $\alpha = 0.01$

Step 3: $$z = \frac{0.5899 - 0.5298}{\sqrt{0.5573(1-0.5573)((1/1029)+(1/1223))}} = 2.86$$

Step 4: Since $\alpha = 0.01$, the critical values are $\pm z_{\alpha} = \pm 2.575$

$2*P(z > 2.86) = 0.0042$

Step 5: Since $2.86 > 2.575$, reject H_0. P-value $< \alpha$. Thus, we reject H_0.

Step 6: The test is significant at the 1% level; that is the data do provide evidence to conclude that a difference exists in the percentages of smartphone owners between men and women.

14. (a) From Problem 13

$$(0.5899 - 0.5298) \pm 2.576\sqrt{\frac{0.5899(1-0.5899)}{1029} + \frac{0.5298(1-0.5298)}{1223}}$$

$$0.0601 \pm 0.0540 \text{ or } 0.0061 \text{ to } 0.1141$$

(b) We can be 99% confident that the difference $p_1 - p_2$ between the proportions of smartphone owners between men and women is somewhere between 0.0061 to 0.1141. In other words, we can be 99% confident that the percentage of men who own a smartphone is somewhere between 0.61% to 11.41% greater than the percentage of women who own a smartphone.

15. (a) $\hat{p} = 4/103 = 0.039$

The 95% one-proportion z-interval is

$$\hat{p} \pm z_{\alpha/2}\sqrt{\hat{p}(1-\hat{p})/n} = 0.039 \pm 1.96\sqrt{0.039(1-0.039)/103} = (0.002, 0.076)$$

(b) $\tilde{p} = (4+2)/(103+4) = 0.056$

The 95% one-proportion plus-four z-interval is

$$\tilde{p} \pm z_{\alpha/2}\sqrt{\tilde{p}(1-\tilde{p})/(n+4)} = 0.056 \pm 1.96\sqrt{0.056(1-0.056)/107} = (0.012, 0.100)$$

(c) The discrepancy between the two methods is primarily due to using a z-interval when the conditions for using such an interval are not met. To get a good approximation to an exact interval, both x and n − x should be at least 5. In this case, x is only 4.

(d) I would use the one-proportion plus-four z-interval because the conditions to use the one-proportion z-interval are not met.

CHAPTER 13 SOLUTIONS

Exercises 13.1

13.1 A variable has a chi-square distribution if its distribution has the shape of a special type of right-skewed curve, called a chi-square curve.

13.3 The χ^2-curve with 20 degrees of freedom more closely resembles a normal curve. This follows from Property 4 of Key Fact 13.1, "As the number of degrees of freedom becomes larger, χ^2-curves look increasingly like normal curves."

13.5 (a) $\chi^2_{0.025} = 32.852$

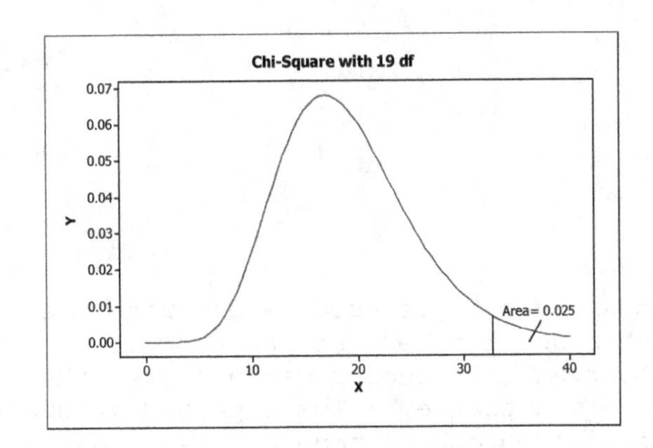

(b) $\chi^2_{0.01} = 36.191$

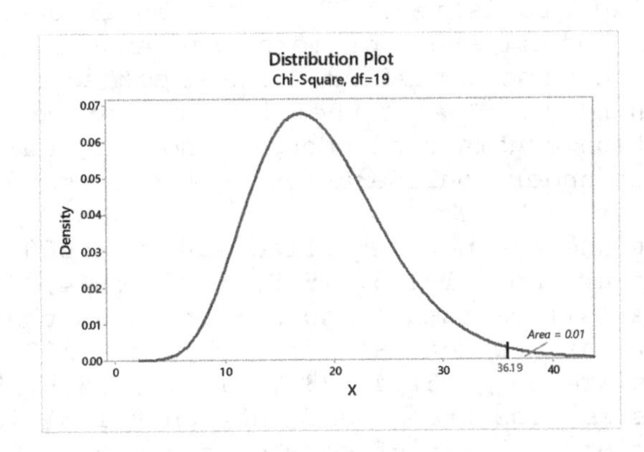

13.7 (a) $\chi^2_{0.05} = 18.307$

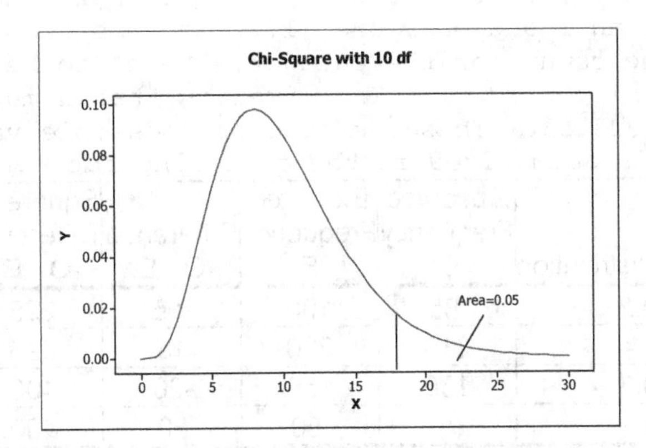

(b) $\chi^2_{0.025} = 20.483$

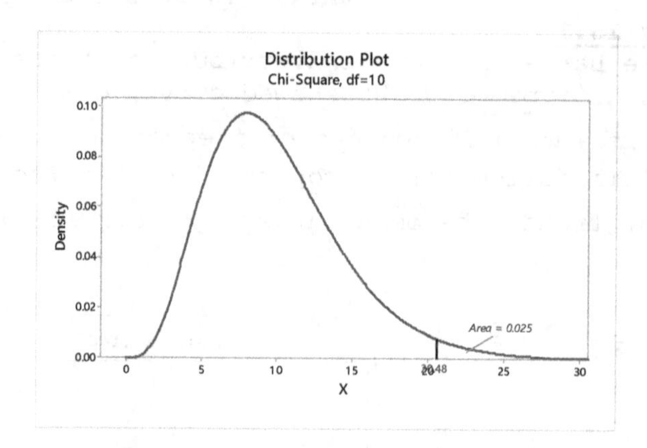

Exercises 13.2

13.9 The term "goodness-of-fit" is used to describe the type of hypothesis test considered in this section because the test is carried out by determining how well the observed frequencies match or fit the expected frequencies.

13.11 The assumptions are satisfied. The expected frequencies are np = 65, 30, and 5. All are 1 or more, and none are less than 5.

13.13 The assumptions are satisfied. The expected frequencies are np = 10, 10, 12.5, 15, and 2.5. All are 1 or more, and exactly 20% are less than 5.

13.15 The assumptions are not satisfied. The expected frequencies are np = 11, 11, 12.5, 15, and 0.5. One of them is not 1 or more.

13.17 (a) The population consists of occupied housing units built after 2010. The variable under consideration is the primary heating fuel of each unit.

(b) The assumptions are not satisfied when n = 200. The expected frequencies are np = 101.6, 15.8, 67.8, 10.4, 3.8, 0.6. One of the frequencies is less than 1 and 33% are less than 5.
The assumptions are not satisfied when n = 300. The expected frequencies are np = 152.4, 23.7, 101.7, 15.6, 5.7, 0.9. One of the frequencies is less than 1 and only one (17%) is less than 5.
The assumptions are satisfied when n = 400. The expected frequencies are np = 203.2, 31.6, 135.6, 20.8, 7.6, 1.2. None of the frequencies are less than 1 and only one (17%) is less than 5.

(c) We need the count for the category "None" to be at least 1. Thus we must have n(0.003) > 1. This implies that n must be greater than 1/0.003 = 333.33. Thus the smallest possible value of n is 334.

13.19 The procedure is summarized in the following table.

Distribution	Observed Frequency O	Expected Frequency E	Difference O - E	Square of Difference $(O - E)^2$	Chi-square subtotal $(O - E)^2/E$
0.2	85	100	-15	225	2.250
0.4	215	200	15	225	1.125
0.3	130	150	-20	400	2.667
0.1	70	50	20	400	8.000
	500				14.042

df = 3; From Table VII, critical Value = 7.815. χ^2 = 14.0422, which is greater than the critical value, so reject the null hypothesis. The data provide sufficient evidence that the variable differs from the given distribution.

13.21 The procedure is summarized in the following table.

Distribution	Observed Frequency O	Expected Frequency E	Difference O - E	Square of Difference $(O - E)^2$	Chi-square subtotal $(O - E)^2/E$
0.2	9	10	-1	1	0.100
0.1	7	5	2	4	0.800
0.1	1	5	-4	16	3.200
0.3	12	15	-3	9	0.600
0.3	21	15	6	36	2.400
	50				7.100

df = 4; From Table VII, critical Value = 7.779. χ^2 = 7.100, which is less than the critical value, so do not reject the null hypothesis. The data do not provide sufficient evidence that the variable differs from the given distribution.

13.23 The procedure is summarized in the following table.

Distribution	Observed Frequency O	Expected Frequency E	Difference O - E	Square of Difference $(O - E)^2$	Chi-square subtotal $(O - E)^2/E$
0.5	147	175	-28	784	4.480
0.3	115	105	10	100	0.952
0.2	88	70	18	324	4.629
	350				10.061

df = 2; From Table VII, critical Value = 9.210. χ^2 = 10.061, which is greater than the critical value, so reject the null hypothesis. The data provide sufficient evidence that the variable differs from the given distribution.

13.25 (a) The population consists of this year's incoming college freshmen in the U.S. The variable under consideration is their political view.

(b)

Political View	Distribution	Observed Frequency O	Expected Frequency E	Difference O - E	Square of Difference $(O - E)^2$	Chi-square subtotal $(O - E)^2/E$
Liberal	0.277	147	138.500	8.500	72.250	0.522
Moderate	0.519	237	259.500	-22.500	506.250	1.951
Conservative	0.204	116	102.000	14.000	196.000	1.922
		500				4.395

Step 1: H_0: The distribution of political views for this year's incoming freshmen is the same as the 2000 distribution.

H_a: The distribution of political views for this year's incoming freshmen is different from the 2000 distribution.

Step 2: $\alpha = 0.05$

Step 3: $\chi^2 = 4.395$ (See column 7 of the table.)

Step 4: df = 2; From Table VII, critical value = 5.991

For the P-value approach, $P(\chi^2 > 4.395) > 0.10$

Step 5: Since 4.395 < 5.991, do not reject H_0. Since the P-value is greater than the significance level, do not reject H_0.

Step 6: The data do not provide sufficient evidence at the 5% level to conclude that the distribution of political views for this year's incoming freshmen is different from the 2000 distribution.

(c) Since the P-value is greater than 0.10, do not reject H_0 at the 10% significance level. The data do not provide sufficient evidence at the 10% level to conclude that the distribution of political views for this year's incoming freshmen is different from the 2000 distribution.

13.27

Color	Distribution	Observed Frequency O	Expected Frequency E	Difference O - E	Square of Difference $(O-E)^2$	Chi-square subtotal $(O-E)^2/E$
Brown	0.30	152	152.7	-0.3	0.49	0.003
Yellow	0.20	114	101.8	12.2	148.84	1.462
Red	0.20	106	101.8	4.2	17.64	0.173
Orange	0.10	51	50.9	0.1	0.01	0.000
Green	0.10	43	50.9	-7.9	62.41	1.226
Blue	0.10	43	50.9	-7.9	62.41	1.226
		509				4.091

Step 1: H_0: The color distribution of M&Ms is the same as that reported by M&M/Mars consumer affairs.

H_a: The color distribution of M&Ms is different from that reported by M&M/Mars consumer affairs.

Step 2: $\alpha = 0.05$

Step 3: $\chi^2 = 4.091$ (See column 7 of the table.)

Step 4: df = 5; From Table VII, the critical value = 11.070

For the P-value approach, $P(\chi^2 > 4.091) > 0.10$

Step 5: Since 4.091 < 11.070, do not reject H_0. Since the P-value is larger than the significance level, do not reject H_0.

Step 6: There is not sufficient evidence to conclude that the color distribution of M&Ms is different from that reported by M&M/Mars consumer affairs.

13.29

Number	Distribution	Observed Frequency O	Expected Frequency E	Difference O - E	Square of Difference $(O-E)^2$	Chi-square subtotal $(O-E)^2/E$
1	0.167	23	25	-2	4	0.160
2	0.167	26	25	1	1	0.040
3	0.167	23	25	-2	4	0.160
4	0.167	21	25	-4	16	0.640
5	0.167	31	25	6	36	1.440
6	0.167	26	25	1	1	0.040
		150				2.480

Step 1: H_0: The die is not loaded.

H_a: The die is loaded.

Step 2: $\alpha = 0.05$

Step 3: $\chi^2 = 2.480$ (See column 7 of the table.)

Step 4: df = 5; From Table VII, the critical value = 11.071

For the P-value approach, $P(\chi^2 > 2.480) > 0.10$.

Step 5: Since 2.480 < 11.071, do not reject H_0. Since the P-value is larger than the significance level, do not reject H_0.

Step 6: The data do not provide sufficient evidence to conclude that the die is loaded.

13.31 (a) Using Minitab, choose **Stat/Tables/Chi-Square Goodness-of-Fit Test**, select the **Observed Counts** option button, enter O in the **Observed counts** text box, enter GAMES in the **Category names** text box, select the **Specific proportions** option button from the **Test** list, enter P in the **Specific proportions** text box, and click **OK**. The results are

```
Using category names in GAMES

                         Test               Contribution
Category  Observed  Proportion  Expected     to Chi-Sq
4               21      0.1250   13.1250       4.72500
5               24      0.2500   26.2500       0.19286
6               24      0.3125   32.8125       2.36679
7               36      0.3125   32.8125       0.30964

  N  DF   Chi-Sq  P-Value
105   3  7.59429    0.055
```

The P-value = 0.055, which is larger than the significance level 0.05, so do not reject H_0. At the 5% significance, the data do not provide sufficient evidence to conclude that World Series teams are not evenly matched.

(b) If $\alpha = 0.10$, the P-value is less than the significance level. Now we reject H_0. At th 10% significance level, the data provide sufficient evidence to conclude that World Series teams are not evenly matched.

(c) The usual assumption is that the data comprise a random sample from some population about which we want to make some inference. In this case, the data include the entire population of World Series records. We can think of these records as a sample from the population of all World Series that have been played and have yet to be played, but they are not a random sample. Nevertheless, each World Series is independent of the others, and the procedure does allow us to gain some insight into whether the overall records of the number of games played in each series follow a theoretical model based on the teams being evenly matched.

13.33 (a)

Gender	Distribution	Observed Frequency O	Expected Frequency E	Difference O – E	Square of Difference $(O-E)^2$	Chi-square subtotal $(O-E)^2/E$
Two girls	0.25	9523	10722	-1199	1437601	134.080
One girl and one boy	0.50	22031	21444	587	344569	16.068
Two boys	0.25	11334	10722	612	374544	34.932
		42888				185.080

Step 1: H_0: The distribution of genders follows the model
H_a: The distribution of genders does not follow the model

Step 2: $\alpha = 0.01$

Step 3: $\chi^2 = 185.080$ (See column 7 of the table.)

Step 4: df = 2; From Table VII, the critical value = 9.210
For the P-value approach, $P(\chi^2 > 185.080) < 0.005$.

Step 5: Since 185.080 > 9.210, reject H_0. Since the P-value is smaller than the significance level, reject H_0.

Step 6: At the 1% the data do provide sufficient evidence to conclude that the distribution of genders in two-children families differs from the distribution predicted by the model described.

(b) There seems to be strong evidence from part (a) that for child gender it is not true that a boy or a girl is equally likely to be born.

13.35 (a) When c =2, the chi-square goodness-of-fit test is equivalent to the one-sample z-test for one proportion (Procedure 12.2) since when there are only two categories in the chi-square test, one of them can be thought of as a success and the other as a failure.

(b) The values of p in the chi-square test correspond to p and (1 - p) in the z-test. The two observed frequencies in the chi-square test correspond to x and n - x in the z-test. In other words, knowing one of the probabilities in the chi-square test when c = 2 allows us to know the other one. The same is true for the observed frequencies and for the expected frequencies. Finally, we compare the z and chi-square statistics when c = 2 and p = p_0.

$$\chi^2 = \Sigma\frac{(O-E)^2}{E} = \frac{(x-np)^2}{np} + \frac{[(n-x)-n(1-p)]^2}{n(1-p)}$$

$$= \frac{(x-np)^2}{np} + \frac{(np-x)^2}{n(1-p)} = (x-np)^2[\frac{1}{np} + \frac{1}{n(1-p)}]$$

$$= (x-np)^2[\frac{1}{np(1-p)}] = \frac{(x-np)^2}{np(1-p)}$$

Note that since $z = \dfrac{\frac{x}{n} - p}{\sqrt{\frac{p(1-p)}{n}}} = \dfrac{x-np}{\sqrt{np(1-p)}}$, $z^2 = \dfrac{(x-np)^2}{np(1-p)} = \chi^2$. Thus the computed value of the χ^2 statistic is the same as the square of the z-test statistic. Since the χ^2 statistic is always positive, either a positive or a negative z-value will lead to the same χ^2 value, making the two tests equivalent when the alternative hypothesis for the z-test is two tailed.

Exercises 13.3

13.37 Cells

13.39 To obtain the total number of observations of bivariate data in a contingency table, one can sum the individual cell frequencies, sum the row subtotals, or sum the column subtotals.

13.41 Yes. If there were no association between the gender of the physician and specialty of the physician, then the same percentage of male and female physicians would choose internal medicine. Since different percentages of male and female physicians chose internal medicine, there is an association between the variables "gender" and "specialty." In other words, knowing the gender of a physician imparts information about the likelihood that the physician specialized in internal medicine.

13.43 (a)

	M	F	Total
BUS	2	7	9
ENG	10	2	12
LIB	3	1	4
Total	15	10	25

(b)

	BUS	ENG	LIB	Total
M	0.222	0.833	0.750	0.600
F	0.778	0.167	0.250	0.400
Total	1.000	1.000	1.000	1.000

(c)

	M	F	Total
BUS	0.133	0.700	0.360
ENG	0.667	0.200	0.480
LIB	0.200	0.100	0.160
Total	1.000	1.000	1.000

(d) Yes. The conditional distributions of gender are different within each college.

13.45 (a)

	Fresh	Soph	Jun	Sen	Total
Dem	2	6	8	4	20
Rep	3	9	12	6	30
Other	1	3	4	2	10
Total	6	18	24	12	60

(b)

	Fresh	Soph	Jun	Sen
Dem	0.333	0.333	0.333	0.333
Rep	0.500	0.500	0.500	0.500
Other	0.167	0.167	0.167	0.167
Total	1.000	1.000	1.000	1.000

(c) There is no association between party affiliation and class level. All of the conditional distributions of political party within class level are identical.

(d) The marginal distribution of party affiliation will be identical to each of the conditional distributions, that is

Party	Frequency
Dem	0.333
Rep	0.500
Other	0.167
Total	1.000

(e) True. Since party affiliation and class level are not associated, all of the conditional distributions of class level within political party will be identical and will be the same as the marginal distribution of class level.

13.47 (a) 12

(b)

Region\Race	White	Black	Other	Total
Northeast	1,100	2,493	1,524	5,117
Midwest	1,137	1,580	504	3,221
South	2,761	7,848	2,258	12,867
West	1,700	764	1,766	4,230
Total	6,698	12,685	6,052	25,435

(c) 25,435 (d) 6,698 (e) 12,867 (f) 764

13.49 (a) 1,056,500 (b) 48,600 (c) 10,600

(d) 88,800 + 226,500 - 10,600 = 304,700

(e) 51,900 (f) 51,900 (g) 1,648,900 - 88,800 = 1,560,100

13.51 (a) Complete the first column, then the second column, the third row, and then the remaining totals.

	Full Owner	Part Owner	Tenant	Total
Under 50	**739**	70	44	**853**
50 - 179	492	130	38	660
180 - 499	198	**143**	**27**	368
500 - 999	51	84	14	149
1000 & Over	41	114	17	172
Total	1521	541	**140**	**2202**

(b) 15 (c) 853,000 (d) 140,000 (e) 84,000 (f) 681,000

(g) 58,000

13.53

	Northeast	**Midwest**	**South**	**West**	**Total**
White	0.164	0.170	0.412	0.254	1.000
Black	0.197	0.125	0.619	0.060	1.000
Other	0.252	0.083	0.373	0.292	1.000
Total	0.201	0.127	0.506	0.166	1.000

(a) The conditional distributions of region within each race are given in rows 2, 3, and 4 of the above table. The conditional distributions of region by race indicate that in all populations, the largest percentage of AIDS cases is in the South, but there are differences in the percentages.

(b) The marginal distribution of region is given in the last row of the above table. The largest percentage of AIDS cases, regardless of race, is from the South.

(c) Yes. The conditional distributions for region are different for the three race categories.

(d) 50.6% of the AIDS cases were from the South.

(e) 41.2% of the AIDS cases among whites were from the South.

(f) True. Since there is an association between region and race category, the conditional distributions of race by region cannot be identical (If they were identical, there would be no association between the two variables).

(g)

	Northeast	Midwest	South	West	Total
White	0.215	0.353	0.215	0.402	0.263
Black	0.487	0.491	0.610	0.181	0.499
Other	0.298	0.156	0.175	0.417	0.238
Total	1.000	1.000	1.000	1.000	1.000

The conditional distributions of race by region are given in columns 2, 3, 4, and 5 of the table. The marginal distribution of race is given in the last column of the table. The conditional distributions of race by region indicate that among all the regions, the largest percentage of cases is Black except for in the West where the largest race category is Other.

13.55 (a)

	State	Federal	Local	Total
8th grade or less	0.142	0.119	0.131	0.137
Some high school	0.255	0.145	0.334	0.273
GED	0.285	0.226	0.141	0.238
High school diploma	0.205	0.270	0.259	0.225
Post secondary	0.090	0.158	0.103	0.098
College grad or more	0.024	0.081	0.032	0.029
Total	1.001	0.999	1.000	1.000

(b) Yes. The conditional distributions of educational attainment within the facility categories (columns 2, 3, and 4 of the table in part a) are not identical.

(c) The marginal distribution of educational attainment is given in the last column of the table in part (a).

(d)

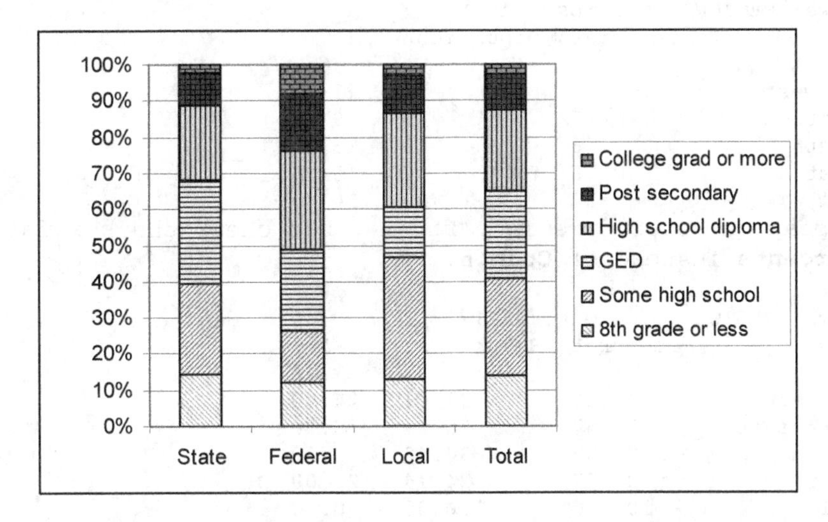

If the conditional distributions of educational attainment within the facility categories (first three bars) had been identical, each of the bars would have been segmented identically and would be identical to the fourth bar, the marginal distribution of educational attainment. The fact that they are not segmented identically means that there is an association between educational attainment and type of facility.

(e) False. Since educational attainment and type of facility are associated, the conditional distributions of facility type within educational attainment will be different.

(f) We interchanged the rows and columns of the table in Exercise 13.49 and then divided each cell entry by its associated column total to obtain the following table.

	8th grade or less	Some high school	GED	High school diploma	Post secondary	College grad or more	Total
State	0.662	0.598	0.768	0.584	0.590	0.521	0.641
Federal	0.047	0.029	0.051	0.065	0.087	0.148	0.054
Local	0.291	0.374	0.181	0.352	0.323	0.331	0.305
Total	1.000	1.001	1.000	1.001	1.000	1.000	1.000

The conditional distributions of facility type within educational attainment are given in columns 2, 3, 4, 5, 6, and 7. The marginal distribution of facility type is given by the last column.

(g) From the table in part (f), 5.4% of the prisoners are in federal facilities.

(h) From the table in part (f), 4.7% of the prisoners with at most an eighth grade education are in federal facilities.

(i) From the table in part (a), 11.9% of the prisoners in federal facilities have at most an 8th grade education.

13.57 Using Minitab, choose **Stat ▶ Tables ▶ Cross Tabulation and Chi-Square**, enter REGION in the **Rows** text box, enter PARTY in the **Columns** text box, check the **Display** box for **Counts** and click **OK**. The following table is produced.

```
Rows: REGION   Columns: PARTY
              Democrat  Republican  All

Midwest              3           9   12
Northeast            6           3    9
South                6          10   16
West                 6           7   13
All                 21          29   50
```

(b) Repeat the procedure in part (a), but check the box for **Column percents** instead of **Counts**.

```
Rows: REGION   Columns: PARTY
              Democrat  Republican     All

Midwest          14.29       31.03   24.00
Northeast        28.57       10.34   18.00
South            28.57       34.48   32.00
West             28.57       24.14   26.00
All             100.00      100.00  100.00
```

(c) Repeat the procedure in part (a), but check the box for **Row percents** instead of **Counts**.

```
Rows: REGION   Columns: PARTY
              Democrat  Republican     All

Midwest          25.00       75.00  100.00
Northeast        66.67       33.33  100.00
South            37.50       62.50  100.00
West             46.15       53.85  100.00
All              42.00       58.00  100.00
```

(d) Yes. The conditional distributions of party within each region are not identical.

13.59 Using Minitab, choose **Stat ▶ Tables ▶ Cross Tabulation and Chi-Square**, enter PARTY in the **Rows** text box, enter CLASS in the **Columns** text box, check the **Display** box for **Counts** and click **OK**. The following table is produced.

Rows: PARTY Columns: CLASS

	I	II	III	All
Democrat	23	20	10	53
Independent	2	0	0	2
Republican	8	13	24	45
All	33	33	34	100

(b) Repeat the procedure in part (a), but check the box for **Column percents** instead of **Counts**.

Rows: PARTY Columns: CLASS

	I	II	III	All
Democrat	69.70	60.61	29.41	53.00
Independent	6.06	0.00	0.00	2.00
Republican	24.24	39.39	70.59	45.00
All	100.00	100.00	100.00	100.00

(c) Repeat the procedure in part (a), but check the box for **Row percents** instead of **Counts**.

Rows: PARTY Columns: CLASS

	I	II	III	All
Democrat	43.40	37.74	18.87	100.00
Independent	100.00	0.00	0.00	100.00
Republican	17.78	28.89	53.33	100.00
All	33.00	33.00	34.00	100.00

(d) Yes. The conditional distributions of party within each class are not identical.

13.61 (a) If there were no association between age group and gender, the conditional distribution of age group within gender would be the same as the marginal distribution of age group. Therefore 7.3% of the resident population would be in the age group 20-24 years.

(b) If there were no association between age group and gender, the conditional distribution of age group within gender would be the same for each gender. Therefore 7.3% of the female residents would be in the age group 20-24 years.

(c) 7.3% of 158.0 million, i.e., 11,534,000, females would be in the age group 20-24 years.

(d) The number of female residents in the age group 20-24 years is not what we would expect IF there were NO association between age group and gender. Therefore there must be an association between age group and gender.

Exercises 13.4

13.63 H_0: The two variables under consideration are statistically independent.
H_a: The two variables under consideration are statistically dependent.

13.65 The degrees of freedom equal (r - 1)(c - 1) = (6 - 1)(4 - 1) = 15.

13.67 (a) No. The existence of an association between two variables when both variables are observed does not imply a causal relationship. For example, there might exist another variable that causes the frequencies of both variables to rise and fall at the same time.

(b) Most people believe that the positive association between educational level has a causative effect on salary. Obviously, such a relationship is not perfect. There are those who never finished high school who are millionaires, and there are college graduates without jobs, but, in general, additional education enables people to qualify for better-paying jobs and to more easily adjust to changing job conditions.

13.69 If there are four values for the first variable and five values for the second variable, there will be 20 expected frequencies. No more than 20% of these can be less than 5. 20% of 20 is 4. Therefore, a maximum number of 4 expected frequencies can be less than 5.

13.71 If there are two values for the first variable and three values for the second variable, there will be 6 expected frequencies. No more than 20% of these can be less than 5. 20% of 6 is 1.2. Therefore, a maximum number of 1 expected frequency can be less than 5.

13.73 If there are two values for the first variable and two values for the second variable, there will be 4 expected frequencies. No more than 20% of these can be less than 5. 20% of 4 is 0.8. Therefore, no frequencies can be less than 5.

13.75 (a) Calculate the expected frequencies using the formula E=RC/n where R = row total, C = column total, and n = sample size. The following table on the left computes the row totals, column totals, and grand total. The following table on the right reports the expected frequencies.

y \ *x*	*A*	*B*	*Total*
a	10	20	30
b	30	40	70
Total	40	60	100

y \ *x*	*A*	*B*	*Total*
a	12	18	30
b	28	42	70
Total	40	60	100

(b) Compute the value of the test statistic , $\chi^2 = \sum \frac{(O-E)^2}{E}$ where O and E represent the observed and expected frequencies respectively. We show the contributions to this sum from each of the cells of the contingency table in the following table.

	A	B
a	0.333	0.222
b	0.143	0.095

The total of the 4 table entries above is the value of the chi-square statistic, that is, $\chi^2 = 0.793$ Depending on rounding, your answer could differ slightly.

(c) The degrees of freedom are (2 - 1)(2 - 1) = 1, so the critical value from Table VII is $\chi^2_{0.05} = 3.841$.
For the P-value approach, P-value > 0.10
Since 0.793 < 3.841, we do not reject the null hypothesis. Also, since the P-value is greater than the significance level, we do not reject the null hypothesis. At the 5% significance level, the data do not provide sufficient evidence to conclude that the two variables are associated.

13.77 (a) Calculate the expected frequencies using the formula E=RC/n where R = row total, C = column total, and n = sample size. The following table on the left computes the row totals, column totals, and grand total. The following table on the right reports the expected frequencies.

y \ *x*	*A*	*B*	*C*	*Total*
a	10	15	75	100
b	0	25	75	100
Total	10	40	150	200

y \ *x*	*A*	*B*	*C*	*Total*
a	5	20	75	100
b	5	20	75	100
Total	10	40	150	200

(b) Compute the value of the test statistic , $\chi^2 = \sum \frac{(O-E)^2}{E}$ where O and E represent the observed and expected frequencies respectively. We show the contributions to this sum from each of the cells of the contingency table in the following table.

	A	B	C
a	5.000	1.250	0.000
b	5.000	1.250	0.000

The total of the 4 table entries above is the value of the chi-square statistic, that is, $\chi^2 = 12.500$ Depending on rounding, your answer could differ slightly.

(c) The degrees of freedom are (2 - 1)(3 - 1) = 2, so the critical value from Table VII is $\chi^2_{0.05} = 5.991$

For the P-value approach, P-value < 0.005

Since 12.500 > 5.991, we reject the null hypothesis. Also, since the P-value is less than the significance level, we reject the null hypothesis. At the 5% significance level, the data does provide sufficient evidence to conclude that the two variables are associated.

13.79 Step 1: H_0: Siskel's ratings and Ebert's ratings of movies are not associated.

H_a: Siskel's ratings and Ebert's ratings of movies are associated.

Step 2: α = 0.01

Step 3: Calculate the expected frequencies using the formula E=RC/n where R = row total, C = column total, and n = sample size. The results are shown in the following table.

		Ebert's rating			
		Thumbs down	**Mixed**	**Thumbs up**	**Total**
Siskel's Rating	Thumbs down	11.8	8.4	24.8	**45.0**
	Mixed	8.4	6.0	17.6	**32.0**
	Thumbs up	21.8	15.6	45.6	**83.0**
	Total	**42.0**	**30.0**	**88.0**	**160.0**

Compute the value of the test statistic , $\chi^2 = \sum \frac{(O-E)^2}{E}$

where O and E represent the observed and expected frequencies respectively. We show the contributions to this sum from each of the cells of the contingency table in the following table.

	Thumbs down	**Mixed**	**Thumbs up**
Thumbs down	12.574	0.023	5.578
Mixed	0.019	8.167	2.475
Thumbs up	6.377	2.767	7.376

The total of the 9 table entries above is the value of the chi-square statistic, that is, $\chi^2 = 45.357$. Depending on rounding, your answer could differ slightly.

Step 4: The degrees of freedom are (3 - 1)(3 - 1) = 4, so the critical value from Table VII is $\chi^2_{0.01} = 13.277$.

For the P-value approach, P-value < 0.005

Step 5: Since 45.357 > 13.277, we reject the null hypothesis. Since the P-value is less than the significance level, we reject the null hypothesis

Step 6: We conclude that there is an association between Siskel's ratings and Ebert's ratings of movies.

13.81 (a) The expected frequencies are shown below the observed frequencies in the following contingency table.

Social Class	A few	Some	Lots	Total
Middle	4 4.36	13 11.64	15 16.00	32
Working	5 4.64	11 12.36	18 17.00	34
Total	9	24	33	66

None of the expected values is less than one, but two of the six (33%) are less than 5. Thus Assumptions 1 and 2 are not both satisfied. The chi-square independence test should not be used.

(b)

Social Class	Never	Sometimes	Often	Total
Middle	2 6.303	8 8.727	22 16.970	32
Working	11 6.697	10 9.273	13 18.030	34
Total	13	18	35	66

None of the expected values is less than one, and none of the six are less than 5. Thus Assumptions 1 and 2 are both satisfied. The chi-square independence test may be used.

The following table gives the contributions from each cell to the chi-square statistic.

Row, Column	O	E	$(O - E)^2/E$
1,1	2	6.303	2.9376
1,2	8	8.727	0.0606
1,3	22	16.970	1.4909
2,1	11	6.697	2.7648
2,2	10	9.273	0.0570
2,3	13	18.030	1.4033
	66		8.7142

Step 1: H_0: The frequency with which parents play "I Spy" games with their children is independent of their economic class.

H_a: The frequency with which parents play "I Spy" games with their children is dependent of their economic class.

Step 2: $\alpha = 0.05$

Step 3: Observed and expected frequencies are presented in the frequency contingency table. Each expected frequency is placed below its corresponding observed frequency. Expected frequencies are calculated using the formula E = (R x C)/n.
The same information about the Os and Es is presented in the table below the contingency table.

$\chi^2 = 8.7142$ (See column 4 of the table below the contingency table.)

Step 4: Critical value = 5.991

For the P-value approach, $0.01 < P(\chi^2 > 8.7142) < 0.025$.

Step 5: Since 8.7142 > 5.991, reject H_0. Since the P-value is less than the significance level, reject H_0.

Step 6: The data do provide sufficient evidence to conclude that the frequency with which parents play "I Spy" with their children is dependent on economic class.

13.83 The expected frequencies are shown below the observed frequencies in the following contingency table.

Worry

Victim	Yes	No	**Total**
Yes	18 7.323	21 31.676	**39**
No	22 32.676	152 141.324	**174**
Total	**40**	**173**	**213**

The following table gives the contributions from each cell to the chi-square statistic.

Victim	Worry: Yes	Worry: NO
Yes	15.562	3.598
No	3.488	0.807

Step 1: H_0: Worry about a gang attack and actually being a victim of a gang attack are not associated.
H_a: Worry about a gang attack and actually being a victim of a gang attack are associated.

Step 2: $\alpha = 0.01$

Step 3: Observed and expected frequencies are presented in the frequency contingency table. Each expected frequency is placed below its corresponding observed frequency. Expected frequencies are calculated using the formula E = (R · C)/n.

$\chi^2 = 23.455$ (Find the sum of all the entries in the table above)
Note: Answers may vary slightly due to rounding.

Step 4: df = (2 - 1)(2 - 1) = 1; Critical value $\chi^2_{0.01} = 6.635$

For the P-value approach, $P(\chi^2 > 23.455) < 0.005$.

Step 5: Since 23.455 > 6.635, reject H_0. Since the P-value is smaller than the significance level, reject H_0.

Step 6: We conclude that an association exists between worry about a gang attack and actually being a victim of a gang attack are associated.

13.85 (a) The expected frequencies are shown below the observed frequencies in the following contingency table.

Education

Religiosity	Basic	Secondary	Advanced	**Total**
Religious	77 68.086	149 145.729	78 90.185	**304**
Not Religious	23 25.756	56 55.128	36 34.116	**115**
Atheist	8 13.662	24 29.242	29 18.096	**61**
Don't Know	6 6.495	15 13.902	8 8.603	**29**
Total	**114**	**244**	**151**	**509**

The following table gives the contributions from each cell to the chi-square statistic.

Education

Religiosity	Basic	Secondary	Advanced
Religious	1.167	0.073	1.646
Not Religious	0.295	0.014	0.104
Atheist	2.347	0.940	6.570
Don't Know	0.038	0.087	0.042

Step 1: H_0: Religiosity and education are not associated.
H_a: Religiosity and education are associated.

Step 2: $\alpha = 0.05$

Step 3: Observed and expected frequencies are presented in the frequency contingency table. Each expected frequency is placed below its corresponding observed frequency. Expected frequencies are calculated using the formula $E = (R \cdot C)/n$.
$\chi^2 = 13.322$ (Find the sum of all the entries in the table above)
Note: Answers may vary slightly due to rounding.

Step 4: df = (4 - 1)(3 - 1) = 6; Critical value $\chi^2_{0.05} = 12.592$

For the P-value approach, $0.025 < P(\chi^2 > 13.322) < 0.05$.

Step 5: Since 13.322 > 12.592, reject H_0. Since the P-value is smaller than the significance level, reject H_0.

Step 6: At the 5% significance level, we conclude that an association exists between religiosity and education.

(b) The test statistic remains the same as part (a); $\chi^2 = 13.322$.

For the P-value approach, $0.025 < P(\chi^2 > 13.322) < 0.05$

Critical Value $\chi^2_{0.01} = 16.812$

Since 13.322 < 16.812 and since the P-value is greater than the significance level, do not reject H_0. At the 1% significance level, we do not conclude that an association exists between religiosity and education.

13.87 Using Minitab, choose **Stat ▶ Tables ▶ Chi-Square Test (Table in Worksheet).** Choose **Summarized data in a two-way table,** and enter Male and Female in the **Columns containing the table** window. Click **OK.** The results are shown below using the P-value approach. Expected counts are printed below observed counts.

```
Chi-Square contributions are printed below expected counts
          Male  Female  Total
    1       77      77    154
         82.49   71.51
         0.366   0.422

    2      263     215    478
        256.05  221.95
         0.189   0.218

    3       28      27     55
         29.46   25.54
         0.072   0.084

Total      368     319    687
Chi-Sq = 1.350, DF = 2, P-Value = 0.509
```

Since the P-value = 0.509, which is greater than the significance level 0.05, do not reject H_0, the data do not provide evidence that gender and response are associated.

13.89 Using Minitab, choose **Stat ▶ Tables ▶ Chi-Square Test (Table in Worksheet).** Choose **Summarized data in a two-way table,** and enter Urban, Suburban, and Rural in the **Columns containing the table** window. Click **OK.** Expected counts are printed below observed counts.

```
Chi-Square contributions are printed below expected counts

         Urban  Suburban   Rural  Total
    1       62        47      42    151
         60.44     53.80   36.75
         0.040     0.860   0.749

    2      197       171     109    477
        190.94    169.96  116.10
         0.192     0.006   0.435

    3       14        25      15     54
         21.62     19.24   13.14
         2.683     1.724   0.262
Total      273       243     166    682
Chi-Sq = 6.952, DF = 4, P-Value = 0.138
```

Since the P-value = 0.138, which is greater than the significance level 0.05, do not reject H_0, the data do not provide evidence that type of community and response are associated.

13.91 Using Minitab, choose **Stat ▶ Tables ▶ Chi-Square Test (Table in Worksheet).** Choose **Summarized data in a two-way table,** and enter Under \$20,000, \$20,000-\$29,000, \$30,000-\$49,999 and \$50,000 and over in the **Columns containing the table** window. Click **OK.** Expected counts are printed below observed counts.

Chi-Square contributions are printed below expected counts

	Under $20,000	$20,000-$29,999	$30,000-$49,999	$50,000 and over	Total
1	40	32	41	34	147
	29.18	34.52	40.09	43.21	
	4.014	0.184	0.021	1.963	
2	80	116	131	138	465
	92.30	109.20	126.82	136.68	
	1.638	0.423	0.138	0.013	
3	11	7	8	22	48
	9.53	11.27	13.09	14.11	
	0.228	1.620	1.980	4.413	
Total	131	155	180	194	660

Chi-Sq = 16.634, DF = 6, P-Value = 0.011

Since the P-value = 0.011, which is less than the significance level 0.05, reject H_0, the data provide evidence that income and response are associated.

Exercises 13.5

13.93 For the chi-square homogeneity test, the null hypothesis is that the distributions of the variable are the same for all the populations. This means there is one variable for two or more populations. For the chi-square independence test, the null hypothesis is that the two variables are independent. This means that there are two variables for one population.

13.95 If populations are *homogeneous* with respect to a variable, it means that the distribution of the variable is similar among the different populations. If the populations are *nonhomogeneous,* it means the distribution of the variable is different among the different populations.

13.97 If a variable has only two possible values, the chi-square homogeneity test provides a procedure for comparing several population proportions.

13.99 If the variable has five possible values among six populations, the degrees of freedom for the chi-square statistic will be $(5-1)\cdot(6-1)=20$.

13.101 (a) The expected frequencies are shown below the observed frequencies in the following contingency table.

	Self Concept			
	High	Moderate	Low	Total
Sighted	13	73	14	100
	10	70.625	19.375	
Blind	3	40	17	60
	6	42.375	11.625	
Total	16	113	31	160

The table below gives the contributions from each cell to the chi-square statistic.

	High	Moderate	Low
Sighted	0.900	0.080	1.491
Blind	1.500	0.133	2.485

Step 1: H_0: The distribution of self-concept is the same among the sightedness.
H_a: There is a difference in the distributions of self-concept among the sightedness.

Step 2: $\alpha = 0.05$

Step 3: Observed and expected frequencies are presented in the frequency contingency table. Each expected frequency is placed below its corresponding observed frequency. Expected frequencies are calculated using the formula $E = (R \cdot C)/n$.

$\chi^2 = 6.589$ (The sum of the cells in the table.)

Note: Answers may vary slightly due to rounding.

Step 4: df = (2 − 1)(3 − 1) = 2; Critical value = $\chi^2_{0.05} = 5.991$

For the P-value approach, $0.025 < P(\chi^2 > 6.589) < 0.05$.

Step 5: Since 6.589 > 5.991, reject H_0. Since the P-value is smaller than the significance level, reject H_0.

Step 6: At the 5% significance level, we conclude that there is a difference in self-concept among those that are sighted and those that are blind.

(b) The test statistic remains the same as part (a); $\chi^2 = 6.589$.

For the P-value approach, $0.025 < P(\chi^2 > 6.589) < 0.05$

Critical Value $\chi^2_{0.01} = 9.210$

Since 6.589 < 9.210 and since the P-value is greater than the significance level, do not reject H_0. At the 1% significance level, we do not conclude that there is not a difference in self-concept among those that are sighted and those that are blind.

13.103 The expected frequencies are shown below the observed frequencies in the following contingency table.

	Race			
Region	White	Black	Other	Total
Northeast	93 92.47	14 14.50	6 6.03	113
Midwest	118 111.29	14 17.45	4 7.25	136
South	167 176.76	42 27.72	7 11.52	216
West	113 110.48	7 17.33	15 7.20	135
Total	491	77	32	600

The following table gives the contributions from each cell to the chi-square statistic.

Row, Column	O	E	$(O - E)^2/E$
1,1	93	92.47	0.003
1,2	14	14.50	0.017
1,3	6	6.03	0.000
2,1	118	111.29	0.405
2,2	14	17.45	0.682
2,3	4	7.25	1.457
3,1	167	176.76	0.539
3,2	42	27.72	7.356
3,3	7	11.52	1.773
4,1	113	110.48	0.057
4,2	7	17.33	6.157
4,3	15	7.20	8.450
	600	600.00	26.896

Step 1: H_0: The distribution of race is the same among the four U.S. regions.
H_a: There is a difference in the distributions of race among the four U.S. regions.

Step 2: $\alpha = 0.01$

Step 3: Observed and expected frequencies are presented in the frequency contingency table. Each expected frequency is placed below its corresponding observed frequency. Expected frequencies are calculated using the formula $E = (R \cdot C)/n$.
The same information about the Os and Es is presented in the table below the contingency table.

$\chi^2 = 26.896$ (See column 4 of the last table above.)

Note: Answers may vary slightly due to rounding.

Step 4: Critical value = 16.812

For the P-value approach, $P(\chi^2 > 26.896) < 0.005$.

Step 5: Since 26.896 > 16.812, reject H_0. Since the P-value is smaller than the significance level, reject H_0.

Step 6: At the 1% significance level, we conclude that there is a difference in race distributions among the four U.S. regions.

13.105 The expected frequencies are shown below the observed frequencies in the following contingency table.

	Region			
Response	Urban	Suburban	Rural	Total
Support	335 333.968	348 334.870	318 332.162	1001
Oppose	35 36.032	23 36.130	50 35.838	108
Total	370	371	369	1109

The table below gives the contributions from each cell to the chi-square statistic.

Response	Urban	Suburban	Rural
Support	0.003	0.515	0.604
Oppose	0.030	4.771	5.597

Step 1: H_a: There is no difference between the proportion of supporters among the three regions.
H_a: There is a difference between the proportion of supporters among the three regions.

Step 2: $\alpha = 0.01$

Step 3: Observed and expected frequencies are presented in the frequency contingency table. Each expected frequency is placed below its corresponding observed frequency. Expected frequencies are calculated using the formula E = (R · C)/n.

χ^2 = 11.520 (The sum of the cells in the table)

Note: Answers may vary slightly due to rounding.

Step 4: df = (2 - 1)(3 - 1) = 2; Critical value = $\chi^2_{0.01}$ = 9.210

For the P-value approach, P(χ^2 > 11.520) < 0.005.

Step 5: Since 11.520 > 9.210, reject H_0. Since the P-value is less than the significance level, reject H_0.

Step 6: At the 1% significance level, we conclude that there is a difference between the proportion of supporters among the three regions.

13.107 (a) Population 1: Men, $\hat{p}_1$ = 20/500 = 0.0400

Population 2: Women, $\hat{p}_2$ = 36/512 = 0.0703

$\hat{p}_P$ = (20 + 36)/(500 + 512) = 0.0553

Step 1: H_0: $p_1 = p_2$, H_a: $p_1 \neq p_2$

Step 2: α = 0.05

Step 3: $$z = \frac{0.04 - 0.0703}{\sqrt{0.0553(1-0.0553)[(1/500)+(1/512)]}} = -2.108$$

Step 4: Since α = 0.05, the critical values are $\pm z_{\alpha/2} = \pm 1.96$

For the P-value approach, 2*P(z < -2.11) = 0.0348

Step 5: Since -2.108 < -1.96, reject H_0. P-value < α. Thus, reject H_0.

Step 6: The test results are significant at the 5% level; that is, at the 5% significance level, the data provide sufficient evidence to conclude that a difference exists in the proportion of male and female vegetarians.

(b) The expected frequencies are shown below the observed frequencies in the following contingency table.

	Vegetarian		
	Yes	No	Total
Men	20 27.668	480 472.332	500
Women	36 28.332	476 483.668	512
Total	56	956	1012

The table below gives the contributions from each cell to the chi-square statistic.

	Yes	No
Men	2.125	0.124
Women	2.075	0.122

Step 1: H_0: The distribution of vegetarian preference is the same for men and women.

H_a: The distribution of vegetarian preference is not the same for men and women.

Step 2: α = 0.05

Step 3: Observed and expected frequencies are presented in the frequency contingency table. Each expected frequency is placed below its corresponding observed frequency. Expected frequencies are calculated using the formula $E = (R \cdot C)/n$.

$\chi^2 = 4.446$ (sum of the cells in the table)

Note: Answers may vary slightly due to rounding.

Step 4: df = (2-1)(2-1) = 1; Critical value $\chi^2_{0.05} = 3.841$

For the P-value approach, $0.025 < P(\chi^2 > 4.446) < 0.05$.

From technology, the exact P-value is 0.03497

Step 5: Since 4.446 > 3.841, reject H_0. Since the P-value is smaller than the significance level, reject H_0.

Step 6: At the 5% significance level, we conclude that there is a difference in distribution of vegetarian preferences between the two genders.

(c) Both the homogeneity test and the two proportions z - test resulted in the same conclusion and had the exact same P-values.

(d) The homogeneity test is the same as a two-tailed two proportion z-test if there are only two populations.

Review Problems for Chapter 13

1. The distributions and curves are distinguished by their numbers of degrees of freedom.

2. (a) zero (b) Right skewed (c) Normal curve

3. (a) No. The degrees of freedom for the χ-square goodness-of-fit test depends on the number of categories, not the number of observations.

(b) No. The degrees of freedom for the χ-square independence test is (r -1)(c-1) where r and c are the number of rows and columns, respectively, in the contingency table.

(c) No. The degrees of freedom for the χ-square homogeneity test is (r -1)(c-1) where r and c are the number of rows and columns, respectively, in the contingency table.

4. Values of the test statistic near zero arise when the observed and expected frequencies are in close agreement. It is only when these frequencies differ enough to produce large values of the test statistic that the null hypothesis is rejected. These values are in the right tail of the χ-square distribution.

5. The value would be zero since O - E = 0 for pair of expected and observed frequencies.

6. (a) All expected frequencies are 1 or greater, and at most 20% of the expected frequencies are less than 5.

(b) Very important. If either of these assumptions are not met, the test should not be carried out by these procedures.

7. (a) 18%

(b) 18% of 37 million, or 6.66 million

(c) Since the observed number of Americans of Irish ancestry who reside in the Northeast is not equal to the number expected if there were no association between ancestry and region, we conclude that there *is* an association between the ancestry and region of residence.

8. (a) Compare the conditional distributions of one of the variables within categories of the other variable. If all of the conditional distributions are identical, there is no association between the variables; if not, there is an association.

(b) No. Since the data are for an entire population, we are not making an inference from a sample to the population. The association (or non-association) is a fact.

9. (a) Perform a chi-square test of independence. If the null hypothesis (of non-association) is rejected, we conclude that there is an association between the variables.

(b) Yes. It is possible (with probability α) that we could reject the null hypothesis when it is, in fact, true. It is also possible, that, even though there is actually an association between the variables, the evidence is not strong enough to draw that conclusion. Either one of these types of errors is due to randomness in selecting a sample which does not exactly reflect the characteristics of the population.

10. For df = 17:

(a) $\chi^2_{0.10} = 24.769$ (b) $\chi^2_{0.01} = 33.409$ (c) $\chi^2_{0.05} = 27.587$

11.

Highest level	O	p	E = np	(O-E)²/E
Not HS graduate	45	0.129	64.5	5.895
HS graduate	163	0.312	156.0	0.314
Some college	76	0.168	84.0	0.762
Associate's degree	53	0.091	45.5	1.236
Bachelor's degree	102	0.195	97.5	0.208
Advanced degree	61	0.105	52.5	1.376
Total	500		500.0	9.792

Step 1: H_0: The educational attainment distribution for adults 25 years old and over this year is the same as in 2010.

H_a: The educational attainment distribution for adults 25 years old and over this year differs from that of 2010.

Step 2: $\alpha = 0.05$

Step 3: $\chi^2 = 9.792$ (See column 5 of the table.)

Step 4: df = 6 - 1 = 5; Critical value = 11.070

For the P-value approach, $0.05 < P(\chi^2 > 9.792) < 0.10$

Step 5: Since 9.792 < 11.070, do not reject H_0.
Since the P-value is larger than the significance level, do not reject H_0.

Step 6: At the 5% significance level, data does not provide evidence that the educational attainment distribution for adults 25 years old and over this year differs from that of 2010.

12. (a) The population consists of U.S. presidents.

(b) The two variables are the presidents' region of birth and their political party.

(c)

	D	DR	F	R	U	W	TOTAL
NE	7	1	1	5	0	1	15
MW	1	0	0	10	0	0	11
SO	6	3	1	2	1	3	16
WE	1	0	0	1	0	0	2
TOTAL	15	4	2	18	1	4	44

13. (a) Dividing each cell entry by the row total in the table in part (a), we obtain the conditional distributions of party for each birth region and in the bottom row, the marginal distribution of party.

	D	DR	F	R	U	W	TOTAL
NE	0.467	0.067	0.067	0.333	0.000	0.067	1.000
MW	0.091	0.000	0.000	0.909	0.000	0.000	1.000
SO	0.375	0.188	0.063	0.125	0.063	0.188	1.000
WE	0.500	0.000	0.000	0.500	0.000	0.000	1.000
TOTAL	0.341	0.093	0.046	0.409	0.023	0.091	1.000

(b) Yes. The conditional distributions of party for each birth region are not all the same.
(c) 40.9% of Presidents are Republicans.
(d) 40.9%
(e) 12.5% of Presidents born in the South are Republicans.

14. (a) Dividing each cell entry by the column total in the table in part (a), we obtain the conditional distributions of birth region for each party and in the right hand column, the marginal distribution of birth region.

	D	DR	F	R	U	W	TOTAL
NE	0.467	0.250	0.500	0.277	0.000	0.250	0.341
MW	0.067	0.000	0.000	0.556	0.000	0.000	0.250
SO	0.400	0.750	0.500	0.111	1.000	0.750	0.364
WE	0.000	0.000	0.000	0.056	0.000	0.000	0.045
TOTAL	1.000	1.000	1.000	1.000	1.000	1.000	1.000

(b) Yes. The conditional distributions of birth region for each party are not all the same.
(c) 36.4% of presidents were born in the South.
(d) 36.4%
(e) 11.1% of Republican presidents were born in the South.

15. (a) 2046 (b) 737 (c) 266 (d) 3046 (e) 6580 - 1167 = 5413
(f) 5403 + 1167 - 660 = 5910

16. (a)

	General	**Psychiatric**	**Chronic**	**Tuberculosis**	**Other**	**Total**
GOV	0.314	0.361	0.808	0.750	0.144	0.311
PROP	0.122	0.486	0.038	0.000	0.361	0.177
NP	0.564	0.153	0.154	0.250	0.495	0.512
Total	1.000	1.000	1.000	1.000	1.000	1.000

The conditional distributions of control type with facility type are given in columns 2, 3, 4, 5, and 6 of the table.
(b) Yes. The conditional distributions of control type within facility type are not all identical.
(c) The marginal distribution of control type is given by the last column of the table above.
(d)

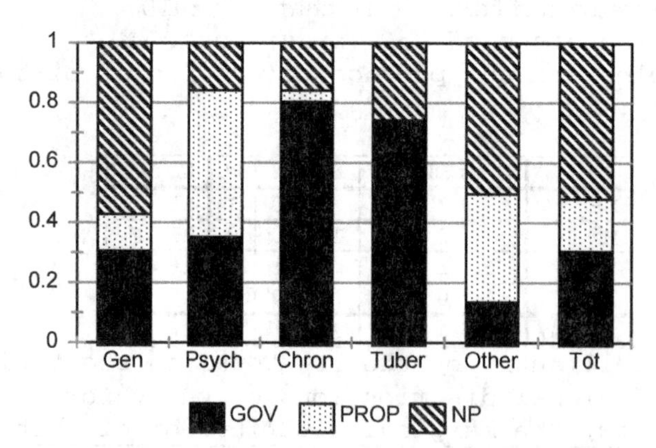

The conditional distributions of control type within facility type are shown be the first five bars of the graph. The marginal distribution of control type is given by the last bar. Since the bars are not identically shaded, control type and facility type are associated variables.

17. (a) False. Since we have established that facility type and control type are associated, the conditional distributions of facility type within control types will not be identical.

(b) After interchanging the rows and columns of the table given in Problem 15, we divided each cell entry by its associated column total to obtain the table below.

	GOV	PROP	NP	Total
General	0.829	0.566	0.905	0.821
Psychiatric	0.130	0.307	0.034	0.112
Chronic	0.010	0.001	0.001	0.004
Tuberculosis	0.001	0.000	0.000	0.001
Other	0.029	0.127	0.060	0.062
Total	1.000	1.000	1.000	1.000

The conditional distributions of facility type within control type are given in columns 2, 3, and 4 of the table.

(c) The marginal distribution of facility type is given by the last column of the table.

18. (a) 17.7% of the hospitals are under proprietary control.
(b) 48.6% of the psychiatric hospitals are under proprietary control.
(c) 11.2% of hospitals are psychiatric hospitals.
(d) 30.7% of the hospitals under proprietary control are psychiatric hospitals.

19. The expected frequencies are shown below the observed frequencies in the following contingency table.

	Age			
Usage	18-24	25-64	65+	Total
Never	6 22.926	38 38.974	31 13.100	75
Sometimes	14 14.978	31 25.463	4 8.559	49
Every day	50 32.096	50 54.563	5 18.341	105
Total	70	119	40	229

The table below gives the contributions from each cell to the chi-square statistic.

Usage	18-24	25-64	65+
Never	12.496	0.024	24.457
Sometimes	0.064	1.204	2.428
Every day	9.987	0.382	9.704

Step 1: H_a: There is no association between age and internet usage.
H_a: There is an association between age and internet usage.

Step 2: $\alpha = 0.01$

Step 3: Observed and expected frequencies are presented in the frequency contingency table. Each expected frequency is placed below its corresponding observed frequency. Expected frequencies are calculated using the formula E = (R · C)/n.

$\chi^2 = 60.746$ (The sum of the cells in the table above

Note: Answers may vary slightly due to rounding.

Step 4: df = (3 − 1)(2 − 1) = 4; Critical value = $\chi^2_{0.01} = 13.277$

For the P-value approach, $P(\chi^2 > 60.746) < 0.005$.

Step 5: Since 60.746 > 13.277, reject H_0. Since the P-value is less than the significance level, reject H_0.

Step 6: At the 1% significance level, we conclude that there is an association between age and internet usage.

20. (a) The populations are they types of residence of people. The can reside inside principal cities, outside principal cities but within metropolitan areas, or outside metropolitan areas.

(b) The variable is income level.

(c) The expected frequencies are shown below the observed frequencies in the following contingency table.

	IPC	OPC	OMA	Total
Under $15,000	75 70.560	106 119.110	46 37.330	227
$15,000 - $34,999	106 101.954	161 172.107	61 53.939	328
$35,000 - $74,999	98 103.509	183 174.730	52 54.761	333
$75,000 & over	48 50.977	102 86.053	14 26.970	164
Total	327	552	173	1052

The table below gives the contributions from each cell to the chi-square statistic.

	IPC	OPC	OMA
Under $15,000	0.279	1.443	2.014
$15,000 - $34,999	0.161	0.717	0.924
$35,000 - $74,999	0.293	0.391	0.139
$75,000 & over	0.174	2.955	6.237

Step 1: H_0: People residing in the three types of residence are homogeneous with respect to income level.
H_a: People residing in the three types of residence are nonhomogeneous with respect to income level.

Step 2: $\alpha = 0.05$

Step 3: Observed frequencies are presented in the frequency contingency table. Expected frequencies are calculated using the formula $E = (R \cdot C)/n$ and are included in the third column of the second table.

$\chi^2 = 15.728$ (sum up the cells in the table above)

Note: Answers may vary slightly due to rounding.

Step 4: df = (4 - 1)(3 - 1) = 6; Critical value = $\chi^2_{0.05} = 12.592$

For the P-value approach, $0.01 < P(\chi^2 > 15.728) < 0.025$.

Step 5: Since 15.728 > 12.592, reject H_0. Since the P-value is smaller than the significance level, reject H_0.

Step 6: At the 5% significance level, the data provide sufficient evidence to conclude that people residing in the three types of residence are nonhomogeneous with respect to income level.

21.

	Democrat	Republican	Independent	Total
Yes	220	349	342	911
No	408	122	304	834
Total	628	471	646	1745

Row, Column	O	E	$(O - E)^2/E$
1,1	220	327.86	35.482
1,2	349	245.89	43.236
1,3	342	337.25	0.067
2,1	408	300.14	38.761
2,2	122	225.11	47.228
2,3	304	308.75	0.073
	1745	1745.00	164.847

Step 1: H_0: The percentages of Democrats, Republicans, and Independents who thought the U.S. economy was in a recession are not different

H_a: There is a difference in the percentages of Democrats, Republicans, and Independents who thought the U.S. economy was in a recession.

Step 2: $\alpha = 0.01$

Step 3: Observed frequencies are presented in the frequency contingency table. Expected frequencies are calculated using the formula $E = (R \cdot C)/n$ and are included in the third column of the second table.

$\chi^2 = 164.847$ (See column 4 of the table below the contingency table.)

Note: Answers may vary slightly due to rounding.

Step 4: Critical value = 9.210

For the P-value approach, $P(\chi^2 > 164.847) < 0.005$.

Step 5: Since 164.847 > 9.210, reject H_0. Since the P-value is smaller than the significance level, reject H_0.

Step 6: At the 1% significance level, the data provide sufficient evidence to conclude that there is a difference in the percentages of Democrats, Republicans, and Independents who thought the U.S. economy was in a recession.

22. (a) Using Minitab, **Stat ▶ Tables ▶ Cross Tabulation and Chi-Square,** enter LOT SIZE in the **For rows:** text box and HOUSE SIZE in the **For columns:** text box, click on the **Counts** box under **Display** to check it. Click **OK**. The results are

```
Rows: LOT SIZE   Columns: HOUSE SIZE

       H1  H2  H3  All

L1     18   3   6   27
L2      6   4   1   11
L3      2   2   1    5
L4      2   2   1    5
All    28  11   9   48
```

(b) Repeat the steps above, except click on the **Column Percents** box under **Display.** Click **OK.** The results are

Rows: LOT SIZE   Columns: HOUSE SIZE

	H1	H2	H3	All
L1	64.29	27.27	66.67	56.25
L2	21.43	36.36	11.11	22.92
L3	7.14	18.18	11.11	10.42
L4	7.14	18.18	11.11	10.42
All	100.00	100.00	100.00	100.00

The marginal distribution for lot size is given in the last column above.

(c) Repeat the steps, except click on the **Row Percents** box under **Display.** Click **OK.** The results are

Rows: LOT SIZE   Columns: HOUSE SIZE

	H1	H2	H3	All
L1	66.67	11.11	22.22	100.00
L2	54.55	36.36	9.09	100.00
L3	40.00	40.00	20.00	100.00
L4	40.00	40.00	20.00	100.00
All	58.33	22.92	18.75	100.00

The marginal distribution for House Size is given in the last row above.

(d) By looking at the distributions, you could say that an association exists between the lot size and the house size, because the conditional distributions of house size by lot size are different. The assumptions are not met to conduct a chi-square independence test because too many of the expected counts are less than 5.

23. Using Minitab, **Stat ▶ Tables ▶ Cross Tabulation and Chi-Square,** enter REGION in the **For rows:** text box and AGE in the **For columns:** text box, click on the **Counts** box under **Display** to check it. Click on the Chi-square button, check the box for Chi-Square analysis and click **OK** twice. The results are

Rows: OPINION   Columns: EDUCATION

	College grad	Some college	HS grad	Not HS grad	All
Favor	264	205	461	290	1220
Oppose	17	26	81	81	205
No opinion	6	7	34	56	103
All	287	238	576	427	1528

Cell Contents: Count

Pearson Chi-Square = 77.837, DF = 6, P-Value = 0.000
Likelihood Ratio Chi-Square = 79.327, DF = 6, P-Value = 0.000

Since the Pearson Chi-Square P-value = 0.000, which is less than the significance level 0.01, reject the null hypothesis. The data provide sufficient evidence to conclude that opinion on this issue and educational level are associated.

CHAPTER 14 SOLUTIONS

Exercises 14.1

14.1 (a) $y = b_0 + b_1x$ (b) Constants are b_0, b_1; variables are x, y.
(c) The independent variable is x and the dependent variable is y.

14.3 (a) b_0 is the y-intercept; it is the value of y where the line crosses the y-axis.
(b) b_1 is the slope; it indicates the change in the value of y for every 1 unit increase in the value of x.

14.5 (a) $b_0 = 3$, $b_1 = 4$
(b) slopes upward
(c)

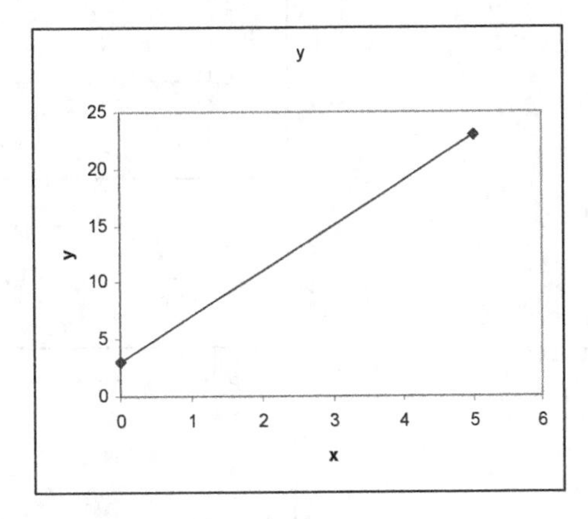

14.7 (a) $b_0 = 6$, $b_1 = -7$
(b) slopes downward
(c)

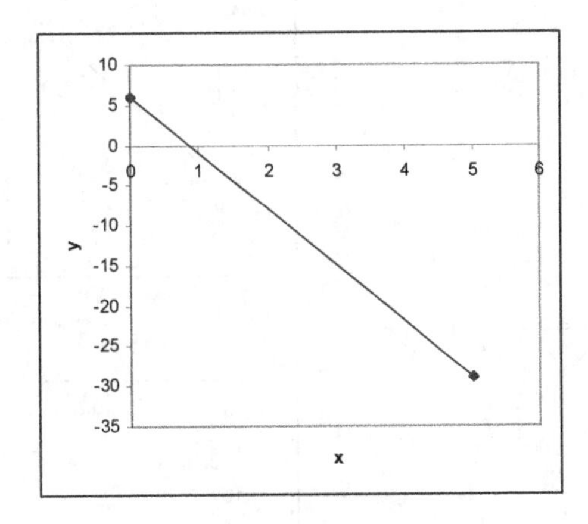

14.9 (a) $b_0 = -2$, $b_1 = 0.5$
(b) slopes upward
(c)

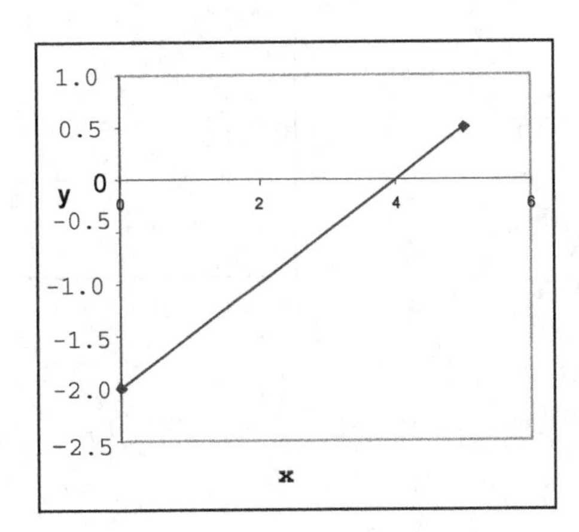

14.11 (a) $b_0 = 2$, $b_1 = 0$
(b) horizontal
(c)

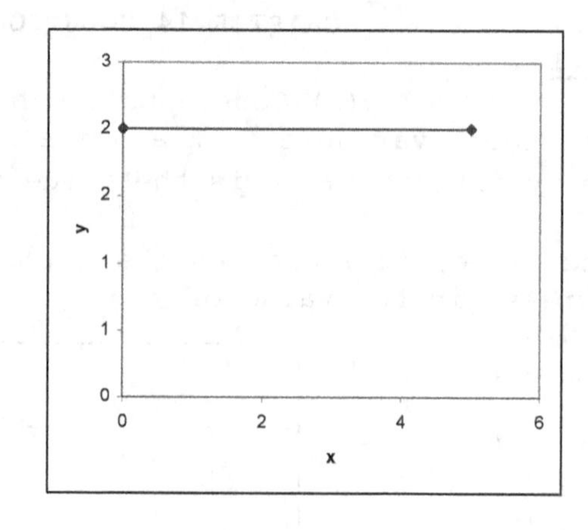

14.13 (a) $b_0 = 0$, $b_1 = 1.5$
(b) slopes upward
(c)

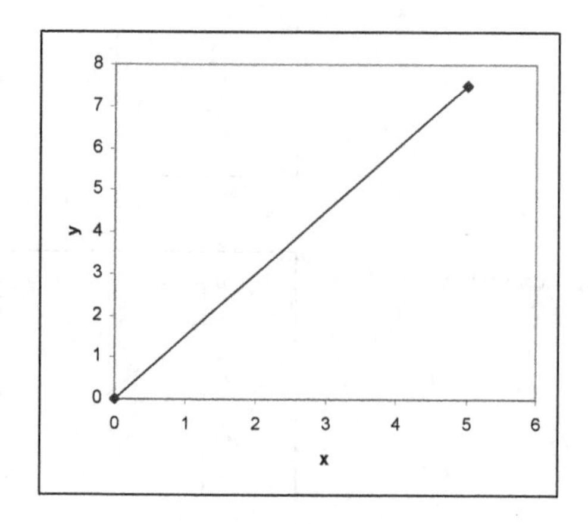

14.15 (a) slopes upward
(b) y = 5 + 2x
(c)

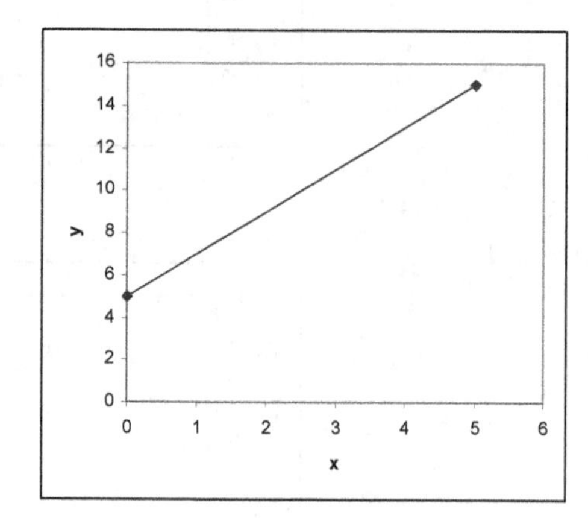

14.17 (a) slopes downward
(b) y = -2 - 3x
(c)

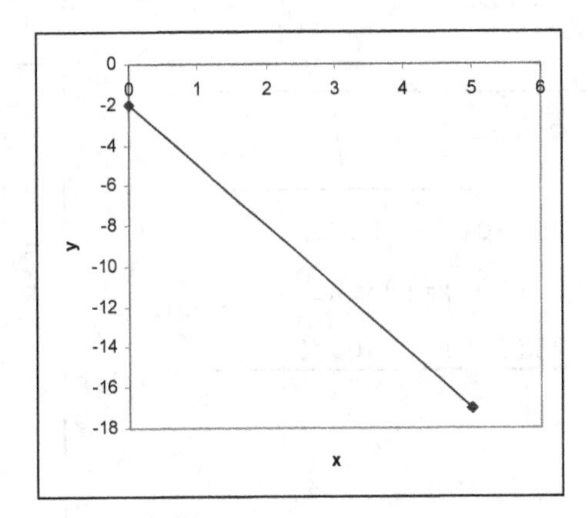

14.19 (a) slopes downward
(b) *y* = -0.5x
(c)

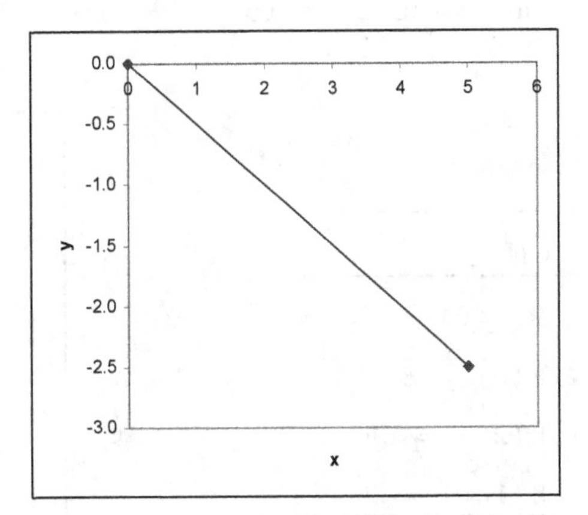

14.21 (a) horizontal
(b) y = 3
(c)

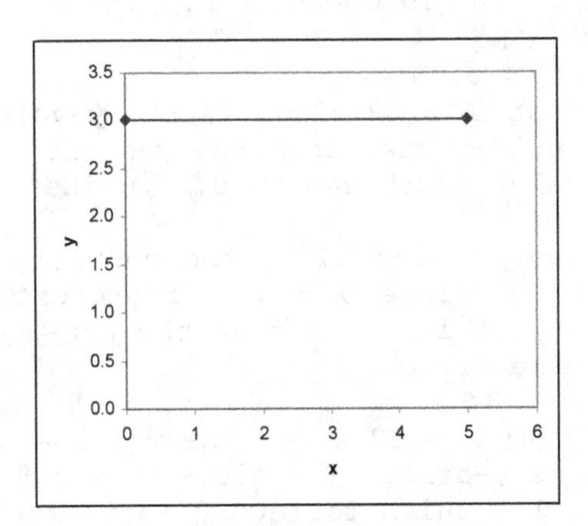

14.23 (a) $y = 68.22 + 0.25x$ (b) $b_0 = 68.22$, $b_1 = 0.25$

(c)

Miles x	Cost ($) Y
50	68.22 + 0.25(50) = 80.72
100	68.22 + 0.25(100) = 93.22
250	68.22 + 0.25(250) = 130.72

(d)

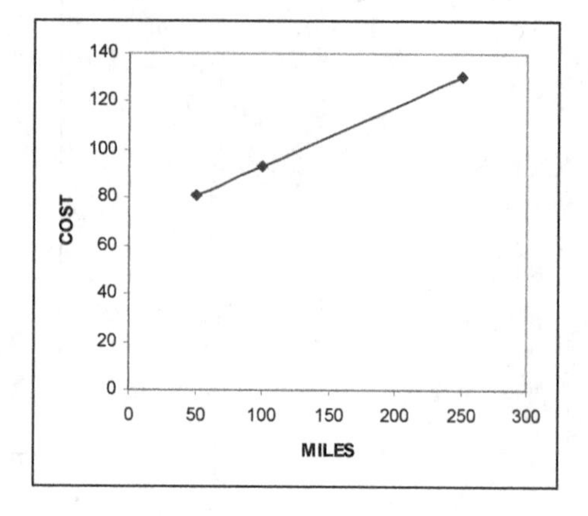

(e) The visual cost estimate of driving the car 150 miles is a little more than $100. The exact cost is
$y = 68.22 + 0.25(150) = 105.72$

14.25 (a) $b_0 = 32$, $b_1 = 1.8$

(b)

X(°C)	y(°F)
-40	32 + 1.8(-40) = -40
0	32 + 1.8(0) = 32
20	32 + 1.8(20) = 68
100	32 + 1.8(100) = 212

(c)

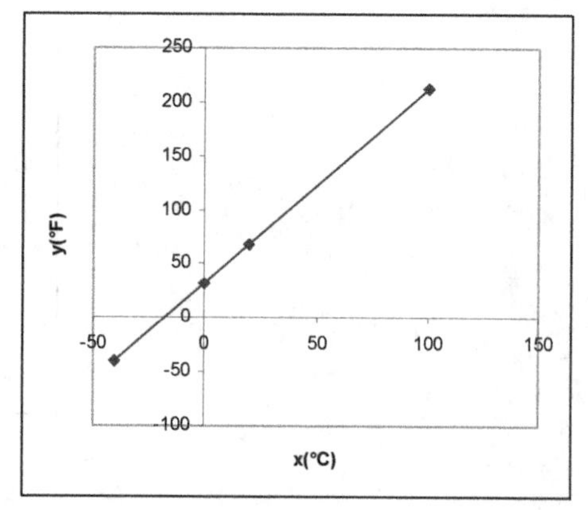

(d) The visual Fahrenheit temperature estimate corresponding to a Celsius temperature of 28° is about 80°F. The exact temperature is $y = 32 + 1.8(28) = 82.4$°F.

14.27 (a) $b_0 = 68.22$, $b_1 = 0.25$

(b) The y-intercept $b_0 = 68.22$ gives the y-value at which the straight line $y = 68.22 + 0.25x$ intersects the y-axis. The slope $b_1 = 0.25$ indicates that the y-value increases by 0.25 units for every increase in x of one unit.

(c) The y-intercept $b_0 = 68.22$ is the cost (in dollars) for driving the car zero miles. The slope $b_1 = 0.25$ represents the fact that the cost per mile is $0.25; it is the amount the total cost increases for each additional mile driven.

14.29 (a) $b_0 = 32$, $b_1 = 1.8$

(b) The y-intercept $b_0 = 32$ gives the y-value where the line $y = 32 + 1.8x$ intersects the y-axis. The slope $b_1 = 1.8$ indicates that the y-value increases by 1.8 units for every increase in x of one unit.

(c) The y-intercept $b_0 = 32$ is the Fahrenheit temperature corresponding to 0°C. The slope $b_1 = 1.8$ represents the fact that Fahrenheit temperature increases by 1.8° for every increase of the Celsius temperature of 1°.

14.31 The line defined by the equation must pass through the two points (x,F) - (2,32 and (3,16). We need to solve for k and x_0. We have

$$\begin{Bmatrix} 32 = -k(2 - x_0) \\ 16 = -k(3 - x_0) \end{Bmatrix} \quad or \quad \begin{Bmatrix} 32 = -2k + kx_0 \\ 16 = -3k + kx_0 \end{Bmatrix}$$

Subtracting the second equation from the first, we obtain 16 = k. Substituting 16 for k in the first equation, we find

$32 = -2(16) + 16x_0$ *or* $64 = 16\,x_0$., from which we find $x_0 = 4$.

(a) Thus, the linear equation that relates the force exerted to the length compressed is F = -16(x - 4).

(b) The spring constant k = 16.

(c) The natural length of the spring is $x_0 = 4$ feet.

14.33 (a) If we can express a line in the form $y = b_0 + b_1x$, then there will be one and only one y-value corresponding to each *x*-value. However, that is not the case for a vertical line; one value of *x* results in an infinite number of y-values.

(b) The form of the equation of a vertical line is $x = x_0$, where x_0 is the *x*-coordinate of the vertical line.

(c) For a linear equation, the slope indicates how much the y-value on the line increases (or decreases) when the *x*-value increases by one unit. We cannot apply this concept to a vertical line, since the *x*-value is *not* permitted to change. Thus, a vertical line has no slope.

Exercises 14.2

14.35 (a) The criterion used to decide on the line that best fits a set of data points is called the least squares criterion.

(b) The criterion is that the line that best fits a set of data points is the one that has the smallest possible sum of the squares of the errors (errors are the differences between and actual and predicted y values).

14.37 (a) The dependent variable is called the response variable.

(b) The independent variable is called the predictor variable or the explanatory variable.

14.39 (a) In the context of regression, an outlier is a data point that lies far from the regression line, relative to the other data points.

(b) In regression analysis, an influential observation is a data point whose removal causes the regression equation to change considerably.

14.41 The idea behind finding a regression line is based on the assumption that the data points are actually scattered about a line. Only the second data set appears to be scattered about a line. Thus, it is reasonable to determine a regression line only for the second set of data.

14.43 (a) Line A: $y = -1 + 3x$ Line B: $y = 1 + 2x$

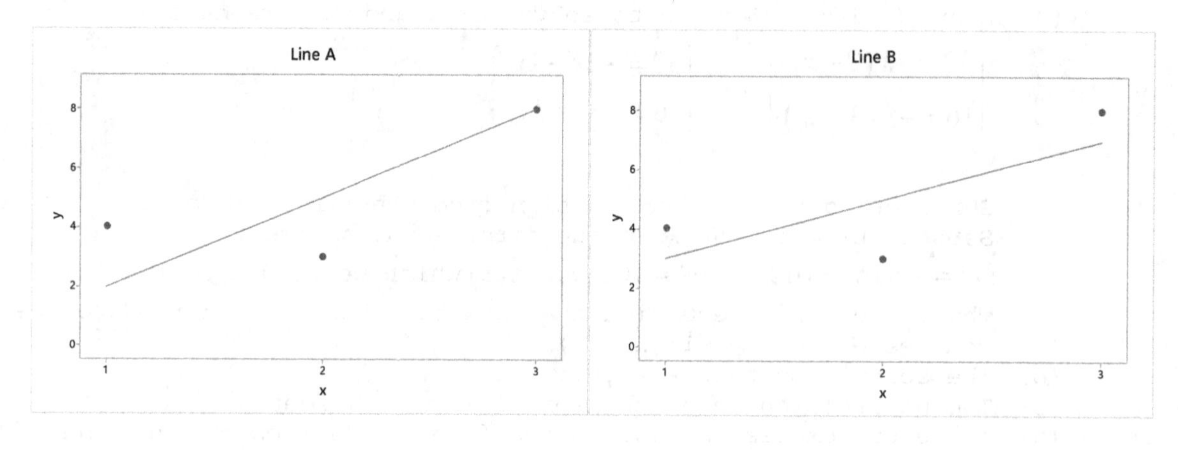

(b)

Line A: $y = -1 + 3x$

x	y	$\hat{y}$	e	e^2
1	4	2	2	4
2	3	5	-2	4
3	8	8	0	0
				8

Line B: $y = 1 + 2x$

x	y	$\hat{y}$	e	e^2
1	4	3	1	1
2	3	5	-2	4
3	8	7	1	1
				6

(c) According to the least-squares criterion, Line B fits the set of data points better than Line A. This is because the sum of squared errors, i.e., $\sum e^2$, is smaller for Line B than for Line A.

14.45 (a) Line A: $y = 3 - 0.6x$ Line B: $y = 4 - x$

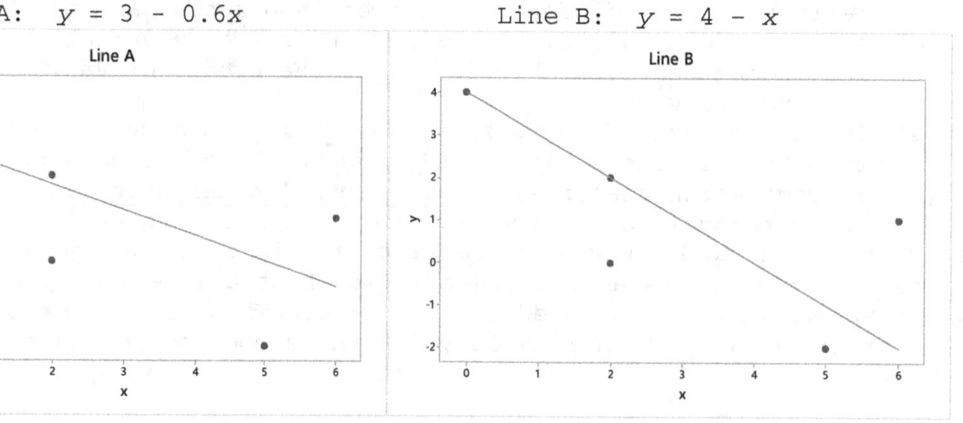

(b)

Line A: $y = 3 - 0.6x$

x	y	$\hat{y}$	e	e^2
0	4	3.0	1.0	1.00
2	2	1.8	0.2	0.04
2	0	1.8	-1.8	3.24
5	-2	0.0	-2.0	4.00
6	1	-0.6	1.6	2.56
				10.84

Line B: $y = 4 - x$

x	y	$\hat{y}$	e	e^2
0	4	4	0	0
2	2	2	0	0
2	0	2	-2	4
5	-2	-1	-1	1
6	1	-2	3	9
				14

(c) According to the least-squares criterion, Line A fits the set of data points better than Line B. This is because the sum of squared errors, i.e., $\sum e^2$, is smaller for Line A than for Line B.

14.47 (a)

x	y	xy	x^2
2	1	2	4
4	3	12	16
6	4	14	20

Using Formula 14.1,

$$S_{xx} = 20 - 6^2/2 = 2,\ S_{xy} = 14 - (6)(4)/2 = 2,\ b_1 = S_{xy}/S_{xx} = 2/2 = 1$$

$$b_0 = \bar{y} - b_1\,\bar{x} = 2 - 1(3) = -1$$

Thus, the regression line is $\hat{y} = -1 + x$.

Without using Formula 14.1, both points must be on the line $y = b_0 + b_1 x$,

so $\left\{\begin{matrix} 1 = b_0 + b_1 2 \\ 3 = b_0 + b_1 4 \end{matrix}\right\}$. Subtracting the first equation from the second, we get $2 = b_1 2$, or $b_1 = 1$. Substituting 1 back in the first equation for b_1, $1 = b_0 + 1(2)$, so $b_0 = -1$, and the equation of the line is y = -1 + x.

(b)

x	y	xy	x^2
1	3	3	1
5	-3	-15	25
6	0	-12	26

Using Formula 14.1,

$$S_{xx} = 20 - 6^2/2 = 2,\ S_{xy} = -12 - (6)(0)/2 = -12,\ b_1 = S_{xy}/S_{xx} = -12/8 = -1.5$$

$$b_0 = \bar{y} - b_1\,\bar{x} = 0 - (-1.5)(3) = 4.5$$

Thus, the regression line is $\hat{y} = 4.5 - 1.5\,x$.

Without using Formula 14.1, both points must be on the line $y = b_0 + b_1 x$,

so $\left\{\begin{matrix} 3 = b_0 + b_1 1 \\ -3 = b_0 + b_1 5 \end{matrix}\right\}$. Subtracting the first equation from the second, we get $-6 = 4b_1$, or $b_1 = -1.5$. Substituting -1.5 back in the first equation for b_1, $3 = b_0 + (-1.5)(1)$, so $b_0 = 4.5$, and the equation of the line is y = 4.5 - 1.5x.

14.49

x	y	xy	x^2
3	-4	-12	9
1	0	0	1
2	-5	-10	4
6	-9	-22	14

(a)

$$S_{xx} = \sum x^2 - \left(\sum x\right)^2 / n = 14 - 6^2 / 3 = 2;$$

$$S_{xy} = \sum xy - \left(\sum x\right)\left(\sum y\right) / n = -22 - (6)(-9) / 3 = -4$$

$$b_1 = S_{xy} / S_{xx} = -4 / 2 = -2; \quad b_0 = \bar{y} - b_1 \bar{x} = -3 - (-2)(2) = 1$$

$$\hat{y} = 1 - 2x$$

(b) Applying the regression equation for x = 1 and x = 3,

x	$\hat{y}$
1	–1
3	–5

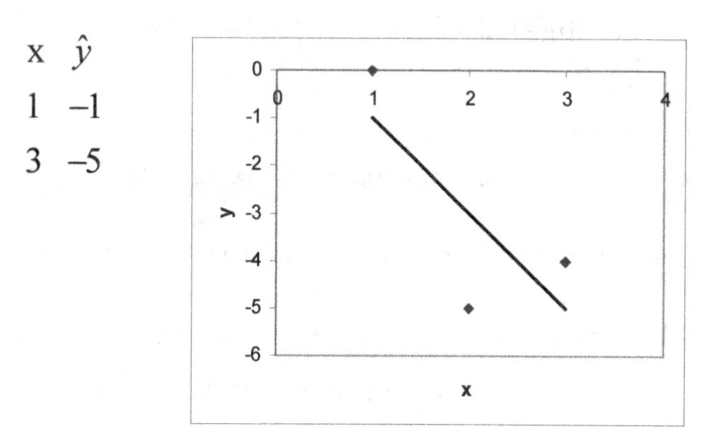

14.51

x	y	xy	x^2
3	4	12	9
4	5	20	16
1	0	0	1
2	-1	-2	4
10	8	30	30

(a)

$$S_{xx} = \sum x^2 - \left(\sum x\right)^2 / n = 30 - 10^2 / 4 = 5;$$

$$S_{xy} = \sum xy - \left(\sum x\right)\left(\sum y\right) / n = 30 - (10)(8) / 4 = 10$$

$$b_1 = S_{xy} / S_{xx} = 10 / 5 = 2; \quad b_0 = \bar{y} - b_1 \bar{x} = 2 - (2)(2.5) = -3$$

$$\hat{y} = -3 + 2x$$

(b) Applying the regression equation for x = 1 and x = 4

x	$\hat{y}$
1	–1
4	5

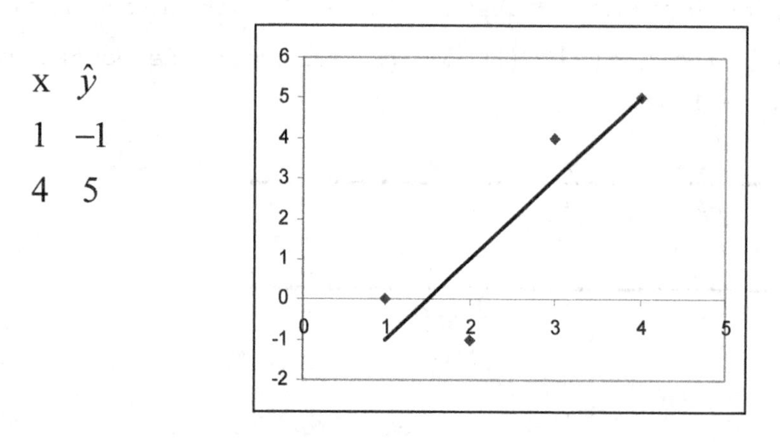

14.53

x	y	xy	x^2
2	3	6	4
2	4	8	4
3	0	0	9
4	2	8	16
4	1	4	16
15	10	26	49

(a)

$$S_{xx} = \sum x^2 - \left(\sum x\right)^2 / n = 49 - 15^2 / 5 = 4;$$
$$S_{xy} = \sum xy - \left(\sum x\right)\left(\sum y\right) / n = 26 - (15)(10) / 5 = -4$$
$$b_1 = S_{xy} / S_{xx} = -4 / 4 = -1; \quad b_0 = \bar{y} - b_1 \bar{x} = 2 - (-1)(3) = 5$$
$$\hat{y} = 5 - x$$

(b) Applying the regression equation for x = 2 and x = 4

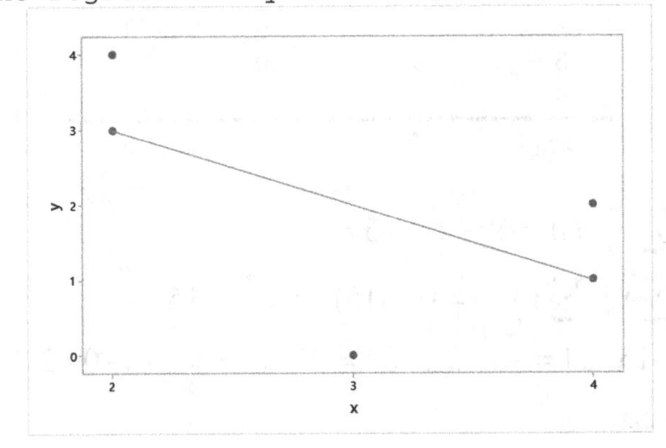

x	$\hat{y}$
2	3
4	1

14.55

x	y	xy	x^2
1	4	4	1
2	3	6	4
3	8	24	9
6	15	34	14

(a)

$$S_{xx} = \sum x^2 - \left(\sum x\right)^2 / n = 14 - 6^2 / 3 = 2;$$
$$S_{xy} = \sum xy - \left(\sum x\right)\left(\sum y\right) / n = 34 - (6)(15) / 3 = 4$$
$$b_1 = S_{xy} / S_{xx} = 4 / 2 = 2; \quad b_0 = \bar{y} - b_1 \bar{x} = 5 - (2)(2) = 1$$
$$\hat{y} = 1 + 2x$$

(b) Applying the regression equation for x = 1 and x = 3

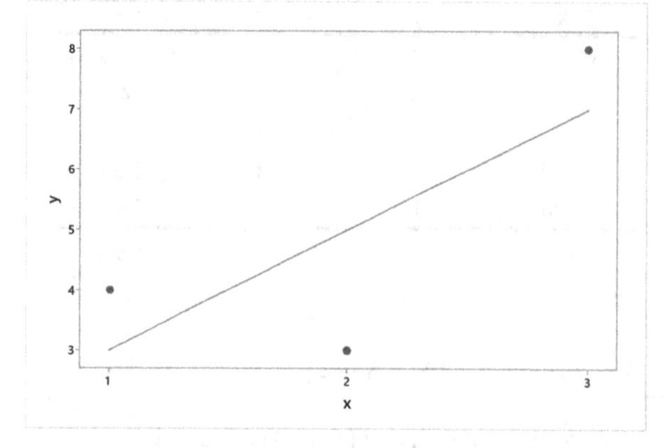

x	$\hat{y}$
1	3
3	7

14.57

x	y	xy	x2
0	4	0	0
2	2	4	4
2	0	0	4
5	-2	-10	25
6	1	6	36
15	5	0	69

(a)

$$S_{xx} = \sum x^2 - \left(\sum x\right)^2 / n = 69 - 15^2/5 = 24;$$

$$S_{xy} = \sum xy - \left(\sum x\right)\left(\sum y\right)/n = 0 - (15)(5)/5 = -15$$

$$b_1 = S_{xy}/S_{xx} = -15/24 = -0.625;\ b_0 = \bar{y} - b_1\bar{x} = 1 - (-0.625)(3) = 2.875$$

$$\hat{y} = 2.875 - 0.625x$$

(b) Applying the regression equation for x = 0 and x = 6

x	$\hat{y}$
0	2.875
6	–0.875

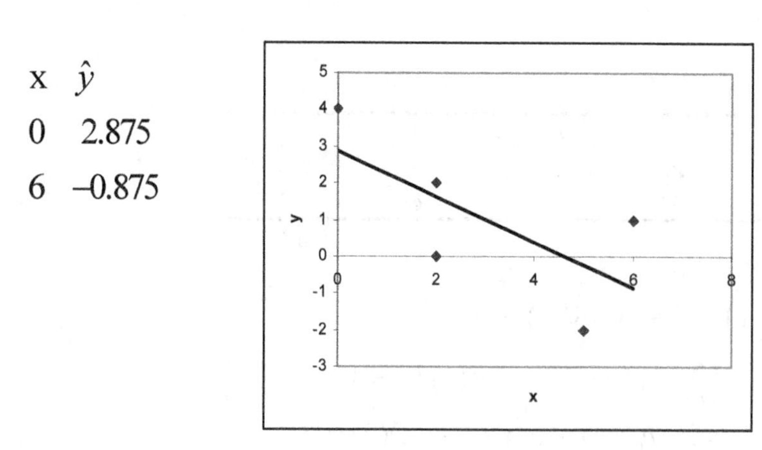

14.59 (a) Formulas for the slope (b_1) and intercept (b_0) of the regression equation are, respectively:

$$b_1 = \frac{\sum xy - \sum x \sum y / n}{\sum x^2 - (\sum x)^2 / n} \qquad b_0 = \frac{1}{n}\left(\sum y - b_1 \sum x\right)$$

To compute b_0 and b_1, construct a table of values for x, y, xy, x^2, and their sums:

x	y	xy	x^2
6	290	1740	36
6	280	1680	36
6	295	1770	36
2	425	850	4
2	384	768	4
5	315	1575	25
4	355	1420	16
5	328	1640	25
1	425	425	1
4	325	1300	16
41	3422	13168	199

Thus: $b_1 = \dfrac{13168-41(3422)/10}{199-41^2/10} = -27.9029 \quad b_0 = \dfrac{1}{10}(3422-(-27.9029)41) = 456.6019$

and the regression equation is $\hat{y} = 456.6019 - 27.9029x$

(b) Begin by selecting two x-values within the range of the x-data. For the x-values 1 and 6, the calculated values for y (in $hundreds) are, respectively:

$$\hat{y} = 456.6019 - 27.9029(1) = 428.6990$$

$$\hat{y} = 456.6019 - 27.9029(6) = 289.1845$$

The regression equation can be graphed by plotting the pairs (1,428.699) and (6, 289.185) and connecting these points with a line. This equation, the original set of 10 data points, and all of the predicted values are shown in the following graph:

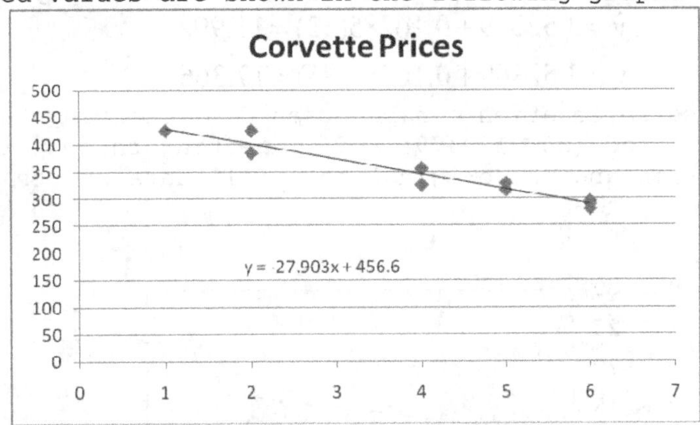

(c) Price tends to decrease as age increases.

(d) Corvettes depreciate an estimated $2,790.29 per year.

(e) The predictor variable is age. The response variable is price.

(f) There are no outliers or potential influential observations.

(g) Price of two-year old Corvette: $\hat{y} = 456.6019 - 27.9029(2) = 400.7961$ or $40,080.

Price of three-year old Corvette: $\hat{y} = 456.6019 - 27.9029(3) = 372.8932$ or $37,289.

14.61 (a) Formulas for the slope (b_1) and intercept (b_0) of the regression equation are, respectively:

$$b_1 = \frac{\sum xy - \sum x \sum y / n}{\sum x^2 - (\sum x)^2 / n} \qquad b_0 = \frac{1}{n}(\sum y - b_1 \sum x)$$

To compute b_0 and b_1, construct a table of values for x, y, xy, x^2, and their sums:

x	y	xy	x^2
57	8.0	456.0	3249
85	22.0	1870.0	7225
57	10.5	598.5	3249
65	22.5	1462.5	4225
52	12.0	624.0	2704
67	11.5	770.5	4489
62	7.5	465.0	3844
80	13.0	1040.0	6400
77	16.5	1270.5	5929
53	21.0	1113.0	2809
68	12.0	816.0	4624
723	156.5	10486.0	48747

$$b_1 = \frac{10486 - 723(1565)/11}{48747 - 723^2/11} = 0.16285 \quad b_0 = \frac{1}{11}(1565 - 0.16285(723) = 3.52369$$

and the regression equation is $\hat{y} = 3.52369 + 0.16285x$.

(b) Begin by selecting two *x*-values within the range of the *x*-data. For the *x*-values 52 and 85, the calculated values for *y* are, respectively:

$$\hat{y} = 3.52369 + 0.16285(52) = 11.992$$

$$\hat{y} = 3.52369 + 0.16285(85) = 17.366$$

The regression equation can be graphed by plotting the pairs (52,12.000) and (85, 17.379) and connecting these points with a line. This equation and the original set of 11 data points are presented as follows:

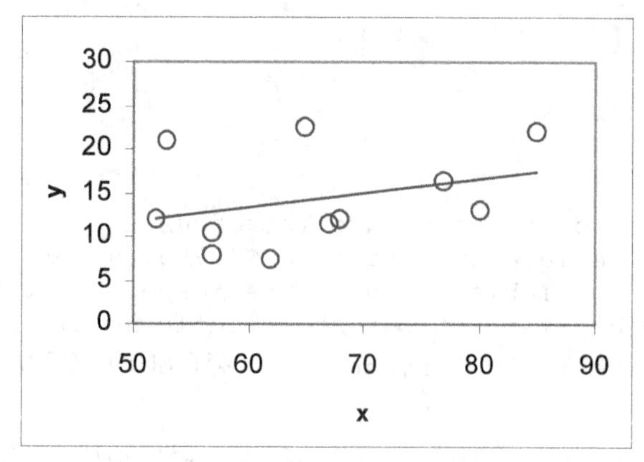

(c) The quantity of volatile compounds emitted tends to decrease as plant weight increases.

(d) The quantity of volatile compounds emitted increases by an estimated 0.163 hundred nanograms for each increase in plant weight of one gram.

(e) The predictor variable is plant weight. The response variable is hundreds of nanograms of volatile compounds emitted by the plant.

(f) There are no outliers or potential influential observations.

(g) The quantity of volatile compounds emitted by a 75 gram plant:

$\hat{y} = 3.52369 + 0.16285(75) = 15.737$ hundred nanograms or 1574 nanograms.

14.63 (a) Formulas for the slope (b_1) and intercept (b_0) of the regression equation are, respectively:

$$b_1 = \frac{\sum xy - \sum x \sum y / n}{\sum x^2 - (\sum x)^2 / n} \qquad b_0 = \frac{1}{n}(\sum y - b_1 \sum x)$$

To compute b_0 and b_1, construct a table of values for x, y, xy, x^2, and their sums:

x	y	xy	x^2
10	92	920	100
15	81	1215	225
12	84	1008	144
20	74	1480	400
8	85	680	64
16	80	1280	256
14	84	1176	196
22	80	1760	484
117	660	9519	1869

$$b_1 = \frac{9519 - 117(660)/8}{1869 - 117^2/8} = -0.84561 \quad b_0 = \frac{1}{8}(660 - (-0.84561)(117) = 94.86698$$

and the regression equation is $\hat{y} = 94.86698 - 0.84561x$.

(b) Begin by selecting two x-values within the range of the x-data. For the x-values 10 and 28, the calculated values for y are, respectively:

$$\hat{y} = 94.86698 - 0.84561(10) = 86.41$$

$$\hat{y} = 94.86698 - 0.84561(22) = 76.26$$

The regression equation can be graphed by plotting the pairs (10,68.40) and (28, 278.16) and connecting these points with a line. This equation and the original set of 10 data points are presented as follows:

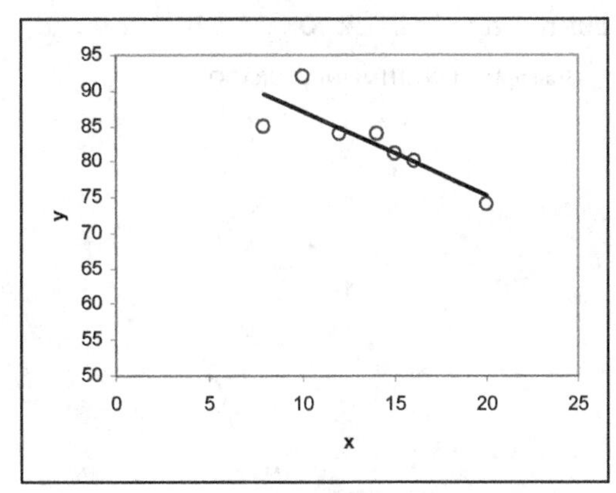

(c) The exam scores tend to decrease as the number of study hours increases.

(d) The exam score decreases, on average, about 0.85 points for each hour of study time.

(e) The predictor variable is study time, in hours. The response variable is the exam score.

(f) There may be two outliers, (8,85) and (10,92). There are no potential influential observations.

(g) For a student who studies 15 hours the predicted exam score is $\hat{y} = 94.86698 - 0.84561(15) = 82.18$

14.65 (a) It is acceptable to use the regression equation to predict the price of a four-year-old Corvette since that age lies within the range of the ages in the sample data. It is not acceptable (and would be extrapolation) to use the regression equation to predict the price of a 10-year-old Corvette since that age falls outside the range of the ages in the sample data.

(b) It is reasonable to use the regression equation to predict price for ages between one and six years old, inclusive.

14.67 One possibility is that there are students who understand the material well after only a short period of study. These students do not need to study as long as students who have a more difficult time comprehending the material and possibly never reach the same level of understanding. There are other possibilities. Perhaps some students are more organized and do their studying when they are wide awake while others put it off until they are tired and less efficient.

14.69 (a)

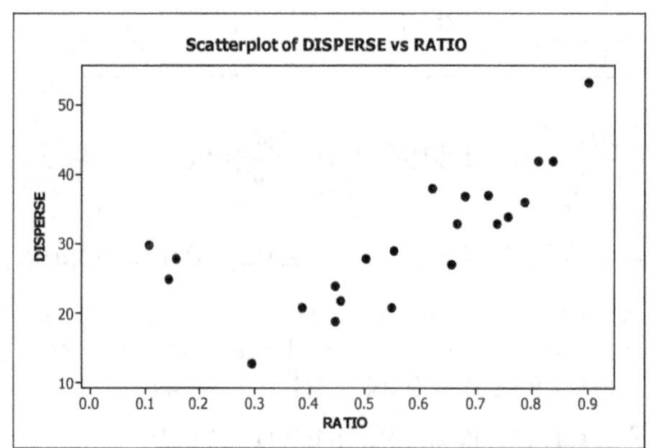

(b) No. The data obviously do not fall in a line.

14.71 (a) Using Minitab, with the data in columns named INAUGURATION and DEATH, choose **Graph ▶ Scatterplot...**, select the **Simple** version and click **OK**. Enter <u>DEATH</u> in Row 1 of the **Y variables** column and <u>INAUGURATION</u> in the **X variables** column, and click **OK**. The result is

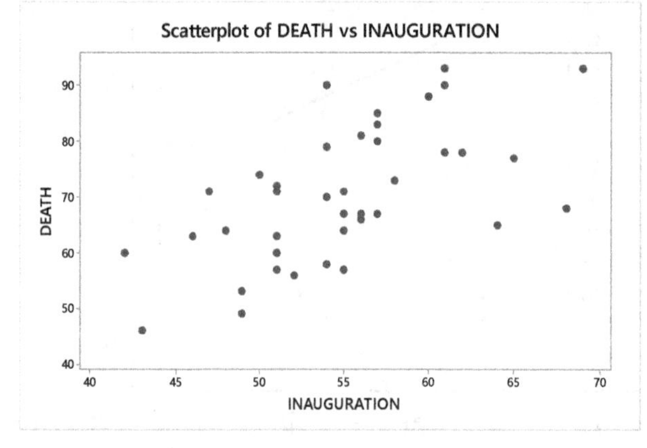

(b) A line appears to be reasonable, although there may be an outlier or two and an influential observation. There does not appear to be any curvature in the data.

(c) Choose **Stat ▶ Regression ▶ Regression,** enter DEATH in the **Response** text box, and INAUGURATION in the **Predictors** text box. Click **OK**. The results are

```
Analysis of Variance

Source           DF  Adj SS   Adj MS  F-Value  P-Value
Regression        1    2021  2020.57    21.45    0.000
  INAUGURATION    1    2021  2020.57    21.45    0.000
Error            37    3485    94.19
  Lack-of-Fit    19    2179   114.68     1.58    0.168
  Pure Error     18    1306    72.56
Total            38    5506

Model Summary

      S    R-sq  R-sq(adj)  R-sq(pred)
9.70514  36.70%     34.99%      28.71%


Coefficients

Term            Coef  SE Coef  T-Value  P-Value   VIF
Constant         5.7     14.1     0.40    0.690
INAUGURATION   1.179    0.255     4.63    0.000  1.00

Regression Equation

DEATH = 5.7 + 1.179 INAUGURATION

Fits and Diagnostics for Unusual Observations
                                  Std
Obs  DEATH    Fit  Resid  Resid
 31  90.00  69.35  20.65   2.16  R
 39  93.00  87.04   5.96   0.67     X

R  Large residual
X  Unusual X
```

The regression equation (bold-faced in the output) indicates that the age at death is 5.7 years plus an increase of 1.179 years for every year later that inauguration occurs.

(d) There is one outlier, the point (54,90) and a potential influential observation (69,93)

(e) After removing the outlier, the regression equation becomes
DEATH = 4.4 + 1.193 INAUGURATION
The effect of removing the outlier is to reduce the constant and increase the slope, both slightly.

(f) After replacing the outlier and removing the potential influential observation (69,93), the regression equation becomes
DEATH = 9.3 + 1.110 INAUGURATION
Removing the influential observation increases the constant and reduces the slope, both significantly. In addition, part of the output is

Unusual Observations

Obs	DEATH	Fit	Resid	Std Resid	
9	68.00	84.77	-16.77	-1.88	X
31	90.00	69.23	20.77	2.15	R

This indicates that the point (54,90) is still an outlier, and there is now a new potential influential observation at (68,68).

14.73 (a) Using Minitab, with the data in columns named LOT SIZE and VALUE EXPECTANCY, choose **Graph ▶ Scatterplot...**, select the **Simple** version and click **OK**. Enter VALUE in Row 1 of the **Y variables** column and 'LOT SIZE' in the **X variables** column, and click **OK**. The result is

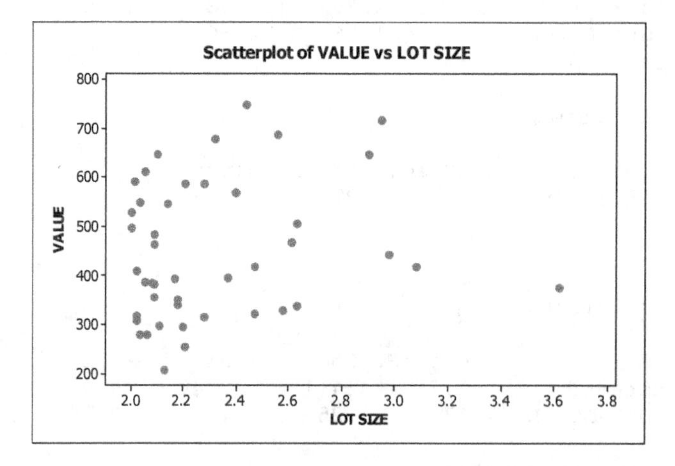

(b) The data points do not have a linear pattern, there is at least one influential observation, and there are likely some outliers. Therefore, find a regression line for these data is not reasonable.

14.75 (a) Using Minitab, with the data in columns named HIGH and LOW, choose **Graph ▶ Scatterplot...**, select the **Simple** version and click **OK**. Enter LOW in Row 1 of the **Y variables** column and HIGH in the **X variables** column, and click **OK**. The result is

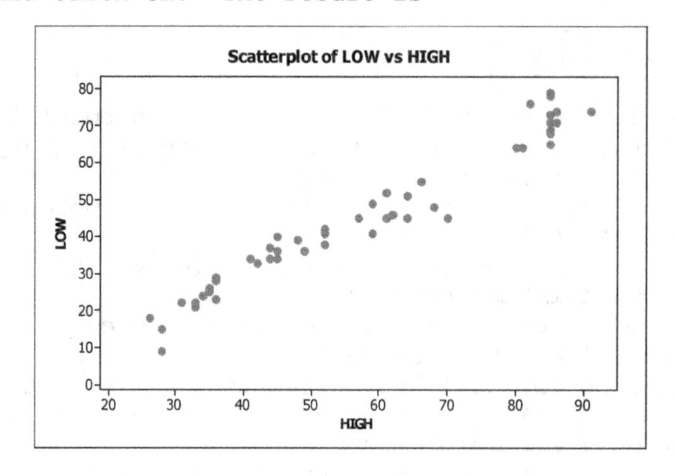

(b) It is reasonable to find a regression line for these data. There is a linear pattern without influential observations.

(c) Choose **Stat ▶ Regression ▶ Regression,** enter LOW in the **Response** text box, and HIGH in the **Predictors** text box. Click **OK**. The results are

```
The regression equation is
LOW = - 7.57 + 0.917 HIGH

Predictor      Coef   SE Coef      T      P
Constant     -7.572     1.786  -4.24  0.000
HIGH        0.91685   0.02967  30.91  0.000

S = 4.15111   R-Sq = 95.2%   R-Sq(adj) = 95.1%
Analysis of Variance

Source          DF     SS     MS       F      P
Regression       1  16459  16459  955.17  0.000
Residual Error  48    827     17
Total           49  17286

Unusual Observations

Obs  HIGH     LOW     Fit  SE Fit  Residual  St Resid
 15  70.0  45.000  56.607   0.705   -11.607    -2.84R
 17  82.0  76.000  67.610   0.949     8.390     2.08R
 30  85.0  79.000  70.360   1.021     8.640     2.15R
 45  28.0   9.000  18.100   1.038    -9.100    -2.26R

R denotes an observation with a large standardized residual.
```

As one would expect, there is positive linear relationship between the high and low average temperatures.

(d) Observations 15, 17, 30 and 45 are all potential outliers. There are no influential observations.

(e) Delete the four data points mentioned in part (d), starting with number 45 and ending with number 15. Repeat the regression procedure. The result is

```
The regression equation is
LOW = - 5.66 + 0.884 HIGH

Predictor      Coef   SE Coef      T      P
Constant     -5.655     1.442  -3.92  0.000
HIGH        0.88407   0.02433  36.34  0.000

S = 3.18323   R-Sq = 96.8%   R-Sq(adj) = 96.7%

Analysis of Variance

Source          DF     SS     MS        F      P
Regression       1  13381  13381  1320.50  0.000
Residual Error  44    446     10
Total           45  13826

Unusual Observations

Obs  HIGH     LOW     Fit  SE Fit  Residual  St Resid
 32  68.0  48.000  54.462   0.552    -6.462    -2.06R
 37  85.0  78.000  69.491   0.847     8.509     2.77R

R denotes an observation with a large standardized residual.
```

Deleting the outliers resulted in an increase in the intercept and a slight decrease in the slope of the regression line. It also resulted in two new potential outliers.

(f) Not applicable

14.77 (a) Using Minitab, with the data in columns named INCOME and BEER, choose **Graph ▶ Scatterplot...**, select the **Simple** version and click **OK**. Enter BEER in Row 1 of the **Y variables** column and INCOME in the **X variables** column, and click **OK**. The result is

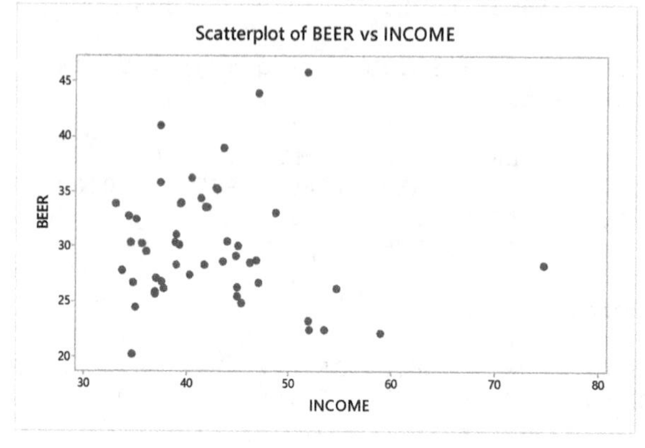

(b) There is no linear pattern in these data, so it is inappropriate to carry out a linear regression. Parts (c)-(f) are omitted.

14.79 (a) Using Minitab, with the data in columns named MANAGERS and ASSETS, choose **Graph ▶ Scatterplot...**, select the **Simple** version and click **OK**. Enter MANAGERS in Row 1 of the **Y variables** column and ASSETS in the **X variables** column, and click **OK**. When you see the scatterplot, you might wonder just where the regression line will go. To see, put the mouse cursor on the graph and right click. Select Add ▶ Regression fit…, select the button for Linear and check the box for Fit intercept. Click OK twice. The result is (You could reverse the roles of ASSETS and MANAGERS, but the basic conclusion will remain the same.)

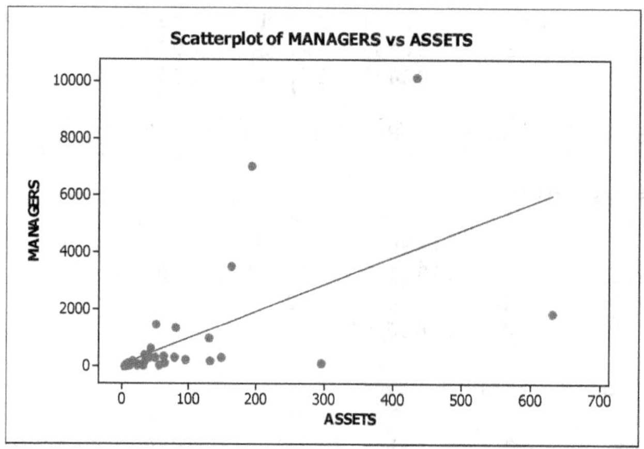

(b) Clearly, there will be outliers and influential data points if we fit a line to these data. A linear regression fit will be of little use, so we omit parts (c-f). [You can do them if you wish just to confirm this opinion.]

14.81 (a) From Exercise 14.45 and 14.57,

$n = 5;\ \sum x_i = 15; \sum y_i = 5;\ \sum x_i y_i = 0;\ \sum x_i^2 = 69$

Therefore, $s_{xy} = \dfrac{\sum x_i y_i - (\sum x_i)(\sum y_i)/n}{n-1} = \dfrac{0-(15)(5)/5}{4} = -15/4 =$ Covariance

(b) $s_x^2 = \dfrac{\sum x_i^2 - (\sum x_i)^2/n}{n-1} = \dfrac{69-(15)^2/5}{4} = 6$

$b_1 = s_{xy}/s_x^2 = (-15/4)/(6) = -0.625;\ b_0 = \bar{y} - b_1\bar{x} = 1-(-0.625)(3) = 2.875$

$\hat{y} = 2.875 - 0.625x$

This is the same result as was obtained in Exercise 14.57.

Exercises 14.3

14.83 (a) The coefficient of determination is a descriptive measure of the utility of the regression equation for making predictions. The symbol for the coefficient of determination is r^2.

(b) Two interpretations of the coefficient of determination are
- (i) It represents the percentage reduction obtained in the total squared error by using the regression equation to predict the observed values, instead of simply using the mean of the y-values.
- (ii) It represents the percentage of variation in the observed *y*-values that is explained by the regression.

14.85 A measure of the amount of variation in the observed values of the response variable that is explained by the regression is the regression sum of squares. The mathematical abbreviation for it is SSR.

14.87 (a) The coefficient of determination is r^2 = SSR/SST = 7626.6/8291.0 = 0.9199. 91.99% of the variation in the observed values of the response variable is explained by the regression. The fact that r^2 is near 1 indicates that the regression equation is extremely useful for making predictions.

(b) SSE = SST - SSR = 8291.0 - 7626.6 = 664.4

14.89 To use the defining formulas, we begin with the following table.

x	y	$(y-\bar{y})$	$(y-\bar{y})^2$	$\hat{y}$	$(\hat{y}-\bar{y})$	$(\hat{y}-\bar{y})^2$	$(y-\hat{y})$	$(y-\hat{y})^2$
3	-4	-1.0	1	-5	-2.0	4.00	1	1
1	0	3.0	9	-1	2.0	4.00	1	1
2	-5	-2.0	4	-3	0.0	0.00	-2	4
6	-9	0.0	14		0.0	8.00	0	6

Each $\hat{y}$ value is obtained by substituting the respective value of *x* into the regression equation $\hat{y} = 1 - 2x$.

(a) $SST = \sum(y-\bar{y})^2 = 14 \qquad SSR = \sum(\hat{y}-\bar{y})^2 = 8 \qquad SSE = \sum(y-\hat{y})^2 = 6$

(b) 14 = 8 + 6

(c) $r^2 = 1 - \dfrac{SSE}{SST} = 1 - \dfrac{6}{14} = \dfrac{4}{7} = 0.571 \quad or \quad r^2 = \dfrac{SSR}{SST} = \dfrac{8}{14} = 0.571$

(d) The percentage of variation in the observed *y*-values that is explained by the regression is 57.1%.

(e) Based on the answer in part (d), the regression equation is somewhat useful for making predictions.

14.91 To use the defining formulas, we begin with the following table.

x	y	$(y-\bar{y})$	$(y-\bar{y})^2$	$\hat{y}$	$(\hat{y}-\bar{y})$	$(\hat{y}-\bar{y})^2$	$(y-\hat{y})$	$(y-\hat{y})^2$
3	4	2.0	4	3	1.0	1.00	1	1
4	5	3.0	9	5	3.0	9.00	0	0
1	0	-2.0	4	-1	-3.0	9.00	1	1
2	-1	-3.0	9	1	-1.0	1.00	-2	4
10	8	0.0	26		0.0	20.00	0	6

Each $\hat{y}$ value is obtained by substituting the respective value of x into the regression equation $\hat{y}=-3+2x$.

(a) $SST=\sum(y-\bar{y})^2=26 \qquad SSR=\sum(\hat{y}-\bar{y})^2=20 \qquad SSE=\sum(y-\hat{y})^2=6$

(b) 46 = 40 + 6

(c) $r^2=1-\dfrac{SSE}{SST}=1-\dfrac{6}{26}=\dfrac{10}{13}=0.769 \quad or \quad r^2=\dfrac{SSR}{SST}=\dfrac{20}{26}=0.769$

(d) The percentage of variation in the observed y-values that is explained by the regression is 76.9%.

(e) Based on the answer in part (d), the regression equation is useful for making predictions.

14.93 To use the defining formulas, we begin with the following table.

x	y	$(y-\bar{y})$	$(y-\bar{y})^2$	$\hat{y}$	$(\hat{y}-\bar{y})$	$(\hat{y}-\bar{y})^2$	$(y-\hat{y})$	$(y-\hat{y})^2$
2	3	1	1	3	1	1	0	0
2	4	2	4	3	1	1	1	1
3	0	-2	4	2	0	0	-2	4
4	2	0	0	1	-1	1	1	1
4	1	-1	1	1	-1	1	0	0
15	10	0	10		0	4	0	6

Each $\hat{y}$ value is obtained by substituting the respective value of x into the regression equation $\hat{y}=5-x$.

(a) $SST=\sum(y-\bar{y})^2=10 \qquad SSR=\sum(\hat{y}-\bar{y})^2=4 \qquad SSE=\sum(y-\hat{y})^2=6$

(b) 10 = 4 + 6

(c) $r^2=1-\dfrac{SSE}{SST}=1-\dfrac{6}{10}=0.4 \quad or \quad r^2=\dfrac{SSR}{SST}=\dfrac{4}{10}=0.4$

(d) The percentage of variation in the observed y-values that is explained by the regression is 40%.

(e) Based on the answer in part (c), the regression equation is moderately useful for making predictions.

14.95 To use the defining formulas, we begin with the following table.

x	y	$(y-\bar{y})$	$(y-\bar{y})^2$	$\hat{y}$	$(\hat{y}-\bar{y})$	$(\hat{y}-\bar{y})^2$	$(y-\hat{y})$	$(y-\hat{y})^2$
1	4	-1	1	3	-2	4	1	1
2	3	-2	4	5	0	0	-2	4
3	8	3	9	7	2	4	1	1
6	15	0	14		0	8	0	6

Each $\hat{y}$ value is obtained by substituting the respective value of x into the regression equation $\hat{y}=1+2x$.

(a) $SST=\sum(y-\bar{y})^2=14 \quad SSR=\sum(\hat{y}-\bar{y})^2=8 \quad SSE=\sum(y-\hat{y})^2=6$

(b) 14 = 8 + 6

(c) $r^2=1-\frac{SSE}{SST}=1-\frac{6}{14}=0.571$ *or* $r^2=\frac{SSR}{SST}=\frac{8}{14}=0.571$

(d) The percentage of variation in the observed y-values that is explained by the regression is 57.1%.

(e) Based on the answer in part (c), the regression equation is moderately useful for making predictions.

14.97 To use the defining formulas, we begin with the following table.

x	Y	$(y-\bar{y})$	$(y-\bar{y})^2$	$\hat{y}$	$(\hat{y}-\bar{y})$	$(\hat{y}-\bar{y})^2$	$(y-\hat{y})$	$(y-\hat{y})^2$
0	4	3	9	2.875	1.875	3.51563	1.125	1.26563
2	2	1	1	1.625	0.625	0.39063	0.375	0.14063
2	0	-1	1	1.625	0.625	0.39063	-1.625	2.64062
5	-2	-3	9	-0.250	-1.250	1.56250	-1.750	3.06250
6	1	0	0	-0.875	-1.875	3.51563	1.875	3.51563
15	5	0	20		0	9.375	0	10.625

Each y value is obtained by substituting the respective value of x into the regression equation $\hat{y}=2.875-0.625x$.

(a) $SST=\sum(y-\bar{y})^2=20 \quad SSR=\sum(\hat{y}-\bar{y})^2=9.375 \quad SSE=\sum(y-\hat{y})^2=10.625$

(b) 20 = 9.365 + 10.625

(c) $r^2=1-\frac{SSE}{SST}=1-\frac{10.625}{20}=0.46875$ *or* $r^2=\frac{SSR}{SST}=\frac{6.375}{20}=0.46875$

(d) The percentage of variation in the observed y-values that is explained by the regression is 46.875%.

(e) Based on the answer in part (c), the regression equation appears to be moderately useful for making predictions.

14.99 To use the computing formulas, we begin with the following table. Notice that columns 1-4 are presented in the solution to part (a) of Exercise 14.59, so that only the column for y^2 needs to be developed.

x	y	xy	x^2	y^2
6	290	1740	36	84100
6	280	1680	36	78400
6	295	1770	36	87025
2	425	850	4	180625
2	384	768	4	147456
5	315	1575	25	99225
4	355	1420	16	126025
5	328	1640	25	107584
1	425	425	1	180625
4	325	1300	16	105625
41	3422	13168	199	1196690

(a) Using the last row of the table and Formula 14.2 of the text, we obtain the three sums of squares as follows.

$$SST = S_{yy} = \sum y^2 - (\sum y)^2 / n = 1196690 - 3422^2 / 10 = 25681.6$$

$$SSR = \frac{S_{xy}^2}{S_{xx}} = \frac{[\sum xy - (\sum x)(\sum y)/n]^2}{\sum x^2 - (\sum x)^2 / n} = \frac{[13168 - (41)(3422)/10]^2}{199 - 41^2/10} = 24057.9$$

$$SSE = S_{yy} - \frac{S_{xy}^2}{S_{xx}} = 25681.6 - 24057.9 = 1623.7$$

(b) $r^2 = \dfrac{SSR}{SST} = \dfrac{24057.9}{25681.6} = 0.9368$

(c) The percentage of variation in the observed y-values that is explained by the regression is 93.68% In words, 93.68% of the variation in the tax efficiency data is explained by the percentage of the mutual funds investments that are in energy securities.

(d) Based on the answers to parts (b) and (c), the regression equation appears to be very useful for making predictions.

14.101 To use the computing formulas, we begin with the following table. Notice that columns 1-4 are presented in the solution to part (a) of Exercise 14.61, so that only the column for y^2 needs to be developed.

x	y	xy	x^2	y^2
57	8.0	456.0	3249	64.00
85	22.0	1870.0	7225	484.00
57	10.5	598.5	3249	110.25
65	22.5	1462.5	4225	506.25
52	12.0	624.0	2704	144.00
67	11.5	770.5	4489	132.25
62	7.5	465.0	3844	56.25
80	13.0	1040.0	6400	169.00
77	16.5	1270.5	5929	272.25
53	21.0	1113.0	2809	441.00
68	12.0	816.0	4624	144.00
723	156.5	10486.0	48747	2523.25

(a) Using the last row of the table and Formula 14.2 of the text, we obtain the three sums of squares as follows.

$$SST = S_{yy} = \sum y^2 - (\sum y)^2 / n = 2523.25 - 156.5^2 / 11 = 296.68$$

$$SSR = \frac{S_{xy}^2}{S_{xx}} = \frac{[\sum xy - (\sum x)(\sum y)/n]^2}{\sum x^2 - (\sum x)^2/n} = \frac{[10486-(723)(1565)/11]^2}{48747-723^2/11} = 32.52$$

$$SSE = S_{yy} - \frac{S_{xy}^2}{S_{xx}} = 296.68 - 32.52 = 264.16$$

(b) $r^2 = \frac{SSR}{SST} = \frac{32.52}{7296.68} = 0.1096$

(c) The percentage of variation in the observed y-values that is explained by the regression is 10.96%. In words, only 10.96% of the variation in the price data is explained by size.

(d) Based on the answers to parts (b) and (c), the regression equation appears to be useless for making predictions.

14.103 To use the computing formulas, we begin with the following table. Notice that columns 1-4 are presented in the solution to part (a) of Exercise 14.63, so that only the column for y^2 needs to be developed.

x	y	xy	x^2	y^2
10	92	920	100	8464
15	81	1215	225	6561
12	84	1008	144	7056
20	74	1480	400	5476
8	85	680	64	7225
16	80	1280	256	6400
14	84	1176	196	7056
22	80	1760	484	6400
117	660	9519	1869	54638

(a) Using the last row of the table and Formula 14.2 of the text, we obtain the three sums of squares as follows.

$$SST = S_{yy} = \sum y^2 - (\sum y)^2/n = 54638 - 660^2/8 = 188$$

$$SSR = \frac{S_{xy}^2}{S_{xx}} = \frac{[\sum xy - (\sum x)(\sum y)/n]^2}{\sum x^2 - (\sum x)^2/n} = \frac{[9519-(117)(660)/8]^2}{1869-117^2/8} = 112.89$$

$$SSE = S_{yy} - \frac{S_{xy}^2}{S_{xx}} = 188 - 112.89 = 75.11$$

(b) $r^2 = \frac{SSR}{SST} = \frac{112.89}{188} = 0.6005$

(c) The percentage of variation in the observed y-values that is explained by the regression is 60.05%. In words, 60.05% of the variation in the score data is explained by study-time.

(d) Based on the answers to parts (b) and (c), the regression equation appears to be somewhat useful for making predictions.

14.105 (a) In Exercise 14.71, we saw from the scatterplot that finding a regression line for these data was reasonable.

(b) From the regression output in that exercise, the coefficient of determination (shown as R-Sq) was found to be 0.367.

(c) The percentage of variation in the observed scores explained by the regression line is 36.7%; i.e., 36.7% of the variation in the death ages was explained by inauguration ages.

(d) Although the pattern is linear, the variation around the fitted line is so great that the regression equation is only marginally useful for making predictions.

14.107(a) In Exercise 14.73, we saw from the scatterplot that finding a regression line for these data was not reasonable. Parts (b-d) are omitted.

14.109(a) In Exercise 14.75, we saw from the scatterplot that finding a regression line for these data was very reasonable.

(b) From the regression output in that exercise, the coefficient of determination (shown as R-Sq) was found to be 0.952.

(c) The percentage of variation in the observed values explained by the regression line is 95.2%; i.e., 95.2% of the variation in the average low temperatures was explained by the average high temperatures.

(d) The regression equation is very useful for making predictions.

14.111(a) Using Minitab, with the data in columns named INCOME and BEER, we choose **Graph ▶ Scatterplot...**, select the **Simple** version and click **OK**. Enter BEER in the first row of the **Y variables** column and INCOME in the first row of the **X variables** column. Click **OK**. The result is

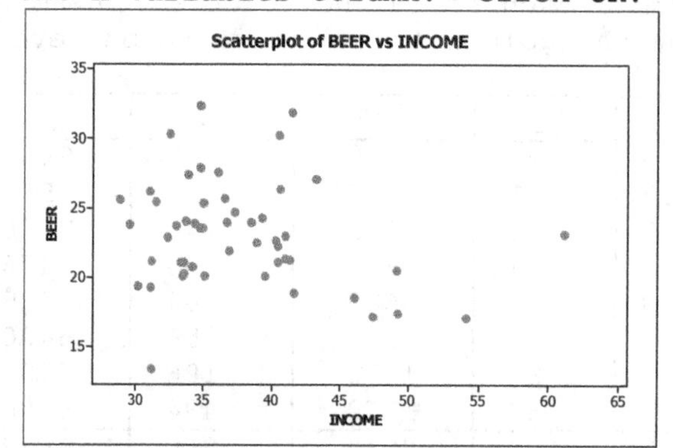

There does not appear to be a linear relationship between beer consumption and income. It is not reasonable to find a regression line. Parts (b-d) are omitted.

14.113(a) In Exercise 14.80, we saw from the scatterplot that finding a regression line for these data was not reasonable because the data exhibited an increasing concave upward curved patternn. Parts (b-d) are omitted.

14.115(a) Using Minitab, with the data in columns named ESTRIOL and WEIGHT, we choose **Graph ▶ Scatterplot...**, select the **Simple** version and click **OK**. Enter WEIGHT in the first row of the **Y variables** column and ESTRIOL in the first row of the **X variables** column. Click **OK**. The result is

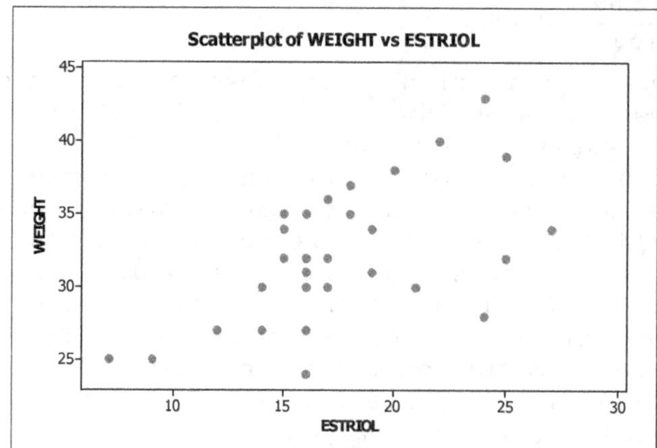

It appears that there is a linear relationship between Estriol and Weight. It is reasonable to find a linear regression equation.

(b) Using Minitab, with the data in columns named ESTRIOL and WEIGHT, we choose **Stat ▶ Regression ▶ Regression...**, select WEIGHT in the **Response** text box, select ESTRIOL in the **Predictors** text box, and click **OK**. The results are

```
The regression equation is
WEIGHT = 21.5 + 0.608 ESTRIOL

Predictor        Coef       StDev        T        P
Constant       21.523       2.620     8.21    0.000
ESTRIOL        0.6082      0.1468     4.14    0.000

S = 3.821        R-Sq = 37.2%      R-Sq(adj) = 35.0%

Analysis of Variance

Source           DF          SS          MS         F        P
Regression        1      250.57      250.57     17.16    0.000
Residual Error   29      423.43       14.60
Total            30      674.00
```

Thus the coefficient of determination is $r^2 = 0.372$.

(c) Only 37.2% of the variation in the observed values of birth weight is explained by a linear relationship with the predictor variable, estriol level.

(d) Based solely on the coefficient of determination, the regression equation is not very useful for making predictions.

14.117 (a) $r^2 = 1 - SSE/SST = (SST - SSE)/SST$. If the mean were used to predict the observed values of the response variable, the total squared error would be SST. By using the regression line to predict the observed values of the response variable, the sum of squares of the differences between the predicted values and the mean is SSR = SST - SSE. This is a reduction in the error sum of squares. Dividing this quantity by SST (and converting to a percentage) gives us the percentage reduction in the squared error when we use the regression equation instead of the mean to predict the observed values of the response variable.

(b) From Exercise 14.99, $r^2 = 0.9368$, so the percentage reduction obtained in the total squared error by using the regression equation instead of the mean of observed prices to predict the observed prices is 93.68%.

Exercises 14.4

14.119 Pearson product moment correlation coefficient

14.121 strong

14.123 ± 1

14.125 positively

14.127 uncorrelated

14.129 negative

14.131 False. It is possible that both variables are effects from a third variable. Correlation does not imply causation.

14.133 The sign of the slope and of r are always the same. This can be seen from $r = \frac{S_{xy}}{\sqrt{S_{xx}S_{yy}}} = \frac{S_{xy}}{S_{xx}}\frac{\sqrt{S_{xx}}}{\sqrt{S_{yy}}} = b_1\frac{\sqrt{S_{xx}}}{\sqrt{S_{yy}}}$.

Since the ratio of square roots in the last term is always positive, r will have the same sign as b_1. Since the slope of the regression line is -3.58, r will also be negative. Therefore, $r = -\sqrt{r^2} = -\sqrt{0.709} = -0.842$

14.135 (a) To compute the linear correlation coefficient using the defining formula, begin with the following table.

x	y	$x-\bar{x}$	$y-\bar{y}$	$(x-\bar{x})^2$	$(y-\bar{y})^2$	$(x-\bar{x})(y-\bar{y})$
3	-4	1	-1	1	1	-1
1	0	-1	3	1	9	-3
2	-5	0	-2	0	4	0
6	-9			2	14	-4

The columns after the first two have used $\bar{x} = 2$ and $\bar{y} = -3$.
We will need s_x and s_y. These are given by

$$s_x = \sqrt{\frac{\sum(x_i - \bar{x})^2}{n-1}} = \sqrt{\frac{2}{2}} = 1 \text{ and } s_y = \sqrt{\frac{\sum(y_i - \bar{y})^2}{n-1}} = \sqrt{\frac{14}{2}} = 2.646$$

$$r = \frac{\frac{1}{n-1}\sum(x_i - \bar{x})(y_i - \bar{y})}{s_x s_y} = \frac{\frac{1}{2}(-4)}{(1)(2)} = \frac{-2}{2.646} = -0.76$$

(b) To compute the linear correlation coefficient using the computing formula, begin with the following table.

x	y	xy	x^2	y^2
3	-4	-12	9	16
1	0	0	1	0
2	-5	-10	4	25
6	-9	-22	14	41

$$r = \frac{\sum x_i y_i - \left(\sum x_i\right)\left(\sum y_i\right)/n}{\sqrt{\left[\sum x_i^2 - \left(\sum x_i\right)^2/n\right]\left[\sum y_i^2 - \left(\sum y_i\right)^2/n\right]}}$$

$$= \frac{-22-(6)(-9)/3}{\sqrt{\left[14-6^2/3\right]\left[41-(-9)^2/3\right]}} = \frac{-4}{\sqrt{28}} = -0.76$$

The answers to both the defining and computing formula are equivalent.

14.137 (a) To compute the linear correlation coefficient using the defining formula, begin with the following table.

x	y	$x-\bar{x}$	$y-\bar{y}$	$(x-\bar{x})^2$	$(y-\bar{y})^2$	$(x-\bar{x})(y-\bar{y})$
3	4	0.5	2	0.25	4	1.0
4	5	1.5	3	2.25	9	4.5
1	0	-1.5	-2	2.25	4	3.0
2	-1	-0.5	-3	0.25	9	1.5
10	8			5.00	26	10.0

The columns after the first two have used $\bar{x} = 2.5$ and $\bar{y} = 2$.
We will need s_x and s_y. These are given by

$$s_x = \sqrt{\frac{\sum(x_i - \bar{x})^2}{n-1}} = \sqrt{\frac{5}{3}} = 1.291 \text{ and } s_y = \sqrt{\frac{\sum(y_i - \bar{y})^2}{n-1}} = \sqrt{\frac{26}{3}} = 2.944$$

$$r = \frac{\frac{1}{n-1}\sum(x_i - \bar{x})(y_i - \bar{y})}{s_x s_y} = \frac{\frac{1}{3}(10)}{(1.291)(2.944)} = \frac{10/3}{3.801} = 0.877$$

(b) To compute the linear correlation coefficient using the computing formula, begin with the following table.

x	y	xy	x^2	y^2
4	5	20	16	25
3	4	12	9	16
1	0	0	1	0
2	-1	-2	4	1
10	8	30	30	42

$$r = \frac{\sum x_i y_i - \left(\sum x_i\right)\left(\sum y_i\right)/n}{\sqrt{\left[\sum x_i^2 - \left(\sum x_i\right)^2/n\right]\left[\sum y_i^2 - \left(\sum y_i\right)^2/n\right]}}$$

$$= \frac{30-(10)(8)/4}{\sqrt{\left[30-10^2/4\right]\left[42-(8)^2/4\right]}} = \frac{10}{\sqrt{130}} = 0.877$$

The answers to both the defining and computing formula are equivalent.

14.139 (a) To compute the linear correlation coefficient using the defining formula, begin with the following table.

x	y	$x-\bar{x}$	$y-\bar{y}$	$(x-\bar{x})^2$	$(y-\bar{y})^2$	$(x-\bar{x})(y-\bar{y})$
2	3	-1	1	1	1	-1
2	4	-1	2	1	4	-2
3	0	0	-2	0	4	0
4	2	1	0	1	0	0
4	1	1	-1	1	1	-1
15	10			4	10	-4

The columns after the first two have used $\bar{x}=3$ and $\bar{y}=2$.

We will need s_x and s_y. These are given by

$$s_x = \sqrt{\frac{\sum (x_i-\bar{x})^2}{n-1}} = \sqrt{\frac{4}{4}} = 1.0 \text{ and } s_y = \sqrt{\frac{\sum (y_i-\bar{y})^2}{n-1}} = \sqrt{\frac{10}{4}} = 1.581$$

$$r = \frac{\frac{1}{n-1}\sum (x_i-\bar{x})(y_i-\bar{y})}{s_x s_y} = \frac{\frac{1}{4}(-4)}{(1.0)(1.581)} = -0.633$$

(b) To compute the linear correlation coefficient using the computing formula, begin with the following table.

x	y	xy	x^2	y^2
2	3	6	4	9
2	4	8	4	16
3	0	0	9	0
4	2	8	16	4
4	1	4	16	1
15	10	26	49	30

$$r = \frac{\sum x_i y_i - \left(\sum x_i\right)\left(\sum y_i\right)/n}{\sqrt{\left[\sum x_i^2 - \left(\sum x_i\right)^2/n\right]\left[\sum y_i^2 - \left(\sum y_i\right)^2/n\right]}}$$

$$= \frac{26-(15)(10)/5}{\sqrt{\left[49-15^2/5\right]\left[30-(10)^2/5\right]}} = -0.632$$

The answers to the defining and computing formula are only slightly different due to rounding.

14.141 (a) To compute the linear correlation coefficient using the defining formula, begin with the following table.

x	y	$x-\bar{x}$	$y-\bar{y}$	$(x-\bar{x})^2$	$(y-\bar{y})^2$	$(x-\bar{x})(y-\bar{y})$
1	4	-1	-1	1	1	1
2	3	0	-2	0	4	0
3	8	1	3	1	9	3
6	15			2	14	4

The columns after the first two have used $\bar{x}=2$ and $\bar{y}=5$. We will need s_x and s_y. These are given by

$$s_x=\sqrt{\frac{\sum(x_i-\bar{x})^2}{n-1}}=\sqrt{\frac{2}{2}}=1.0 \text{ and } s_y=\sqrt{\frac{\sum(y_i-\bar{y})^2}{n-1}}=\sqrt{\frac{14}{2}}=2.646$$

$$r=\frac{\frac{1}{n-1}\sum(x_i-\bar{x})(y_i-\bar{y})}{s_x s_y}=\frac{\frac{1}{2}(4)}{(1.0)(2.646)}=0.756$$

(b) To compute the linear correlation coefficient using the computing formula, begin with the following table.

x	y	xy	x^2	y^2
1	4	4	1	16
2	3	6	4	9
3	8	24	9	64
6	15	34	14	89

$$r=\frac{\sum x_i y-(\sum x_i)(\sum y_i)/n}{\sqrt{[\sum x_i^2-(\sum x_i)^2/n][\sum y_i^2-(\sum y_i)^2/n]}}$$

$$=\frac{34-(6)(15)/3}{\sqrt{[14-6^2/3][89-(15)^2/3]}}=0.756.$$

The answers to the defining and computing formula are equivalent.

14.143 (a) To compute the linear correlation coefficient using the defining formula, begin with the following table.

x	y	$x-\bar{x}$	$y-\bar{y}$	$(x-\bar{x})^2$	$(y-\bar{y})^2$	$(x-\bar{x})(y-\bar{y})$
0	4	-3	3	9	9	-9
2	2	-1	1	1	1	-1
2	0	-1	-1	1	1	1
5	-2	2	-3	4	9	-6
6	1	3	0	9	0	0
15	5			24	20	-15

The columns after the first two have used $\bar{x}=3$ and $\bar{y}=1$. We will need s_x and s_y. These are given by

$$s_x=\sqrt{\frac{\sum(x_i-\bar{x})^2}{n-1}}=\sqrt{\frac{24}{4}}=2.449 \text{ and } s_y=\sqrt{\frac{\sum(y_i-\bar{y})^2}{n-1}}=\sqrt{\frac{20}{4}}=2.236$$

$$r=\frac{\frac{1}{n-1}\sum(x_i-\bar{x})(y_i-\bar{y})}{s_x s_y}=\frac{\frac{1}{4}(-15)}{(2.449)(2.236)}=\frac{-3.75}{5.476}=-0.685$$

(b) To compute the linear correlation coefficient using the computing formula, begin with the following table.

x	y	xy	x^2	y^2
0	4	0	0	16
2	2	4	4	4
2	0	0	4	0
5	-2	-10	25	4
6	1	6	36	1
15	5	0	69	25

$$r = \frac{\sum x_i y_i - \left(\sum x_i\right)\left(\sum y_i\right)/n}{\sqrt{\left[\sum x_i^2 - \left(\sum x_i\right)^2/n\right]\left[\sum y_i^2 - \left(\sum y_i\right)^2/n\right]}}$$

$$= \frac{0-(15)(5)/5}{\sqrt{\left[69-15^2/5\right]\left[25-(5)^2/5\right]}} = \frac{-15}{\sqrt{480}} = -0.685$$

The answers to both the defining and computing formula are equivalent.

14.145 To compute the linear correlation coefficient, return to the table presented at the beginning of the solution to Exercise 14.99. Use the last row of this table to perform the calculations in part (a).

(a) The linear correlation coefficient r is computed using the formula in Formula 14.3 of the text:

$$r = \frac{S_{xy}}{\sqrt{S_{xx}S_{xy}}} = \frac{\sum xy - (\sum x)(\sum y)/n}{\sqrt{[\sum x^2 - (\sum x)^2/n][\sum y^2 - (\sum y)^2/n]}}$$

$$= \frac{13168-(41)(3422)/10}{\sqrt{[199-41^2/10][1196690-3422^2/10]}} = -0.967872$$

(b) The value of r in part (a) suggests a strong negative linear relationship between age and price of Corvettes.

(c) Data points are clustered closely about the regression line.

(d) $r^2 = (-0.967872)^2 = 0.9368$. This matches the coefficient of determination that was calculated in part (b) of Exercise 14.99.

14.147 To compute the linear correlation coefficient, return to the table presented at the beginning of the solution to Exercise 14.101. Use the last row of this table to perform the calculations in part (a).

(a) The linear correlation coefficient r is computed using the formula in Formula 14.3 of the text:

$$r = \frac{S_{xy}}{\sqrt{S_{xx}S_{xy}}} = \frac{\sum xy - (\sum x)(\sum y)/n}{\sqrt{[\sum x^2 - (\sum x)^2/n][\sum y^2 - (\sum y)^2/n]}}$$

$$= \frac{10486-(723)(156.5)/11}{\sqrt{[48747-723^2/11][2523.252-156.5^2/11]}} = 0.331067$$

(b) The value of r in part (a) suggests a weak positive linear relationship between potato plant weight and quantity of volatile emissions.

(c) Data points are clustered very loosely about the regression line.

(d) $r^2 = (0.331037)^2 = 0.1096$. This matches the coefficient of determination that was calculated in part (b) of Exercise 14.101.

14.149 To compute the linear correlation coefficient, return to the table presented at the beginning of the solution to Exercise 14.103. Use the last row of this table to perform the calculations in part (a).

(a) The linear correlation coefficient r is computed using the formula in Formula 14.3 of the text:

$$r = \frac{S_{xy}}{\sqrt{S_{xx}S_{xy}}} = \frac{\sum xy - (\sum x)(\sum y)/n}{\sqrt{[\sum x^2 - (\sum x)^2/n][\sum y^2 - (\sum y)^2/n]}}$$

$$= \frac{9519 - (117)(660)/8}{\sqrt{[1869 - 117^2/8][54638 - 660^2/8]}} = -0.7749$$

(b) The value of r in part (a) suggests a fairly strong negative linear relationship between study-time and score for calculus students.

(c) Data points are clustered fairly closely about the regression line.

(d) $r^2 = (-0.7749)^2 = 0.6005$. This matches the coefficient of determination that was calculated in part (b) of Exercise 14.103.

14.151 To compute the linear correlation coefficient, begin with the following table.

x	y	xy	x^2	y^2
-3	9	-27	9	81
-2	4	-8	4	16
-1	1	-1	1	1
0	0	0	0	0
1	1	1	1	1
2	4	8	4	16
3	9	27	9	81
0	28	0	28	196

(a) The linear correlation coefficient r is computed using the formula in Formula 14.3 of the text:

$$r = \frac{S_{xy}}{\sqrt{S_{xx}S_{xy}}} = \frac{\sum xy - (\sum x)(\sum y)/n}{\sqrt{[\sum x^2 - (\sum x)^2/n][\sum y^2 - (\sum y)^2/n]}}$$

$$= \frac{0 - (0)(282)/7}{\sqrt{[28 - 0^2/7][196 - 28^2/7]}} = 0.0$$

(b) We cannot conclude from the result in part (a) that x and y are unrelated. We can conclude only that there is no *linear* relationship between x and y.

(c) Graph for part (e)

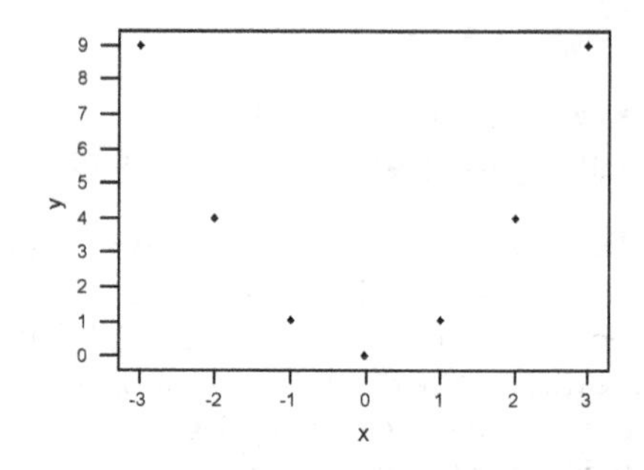

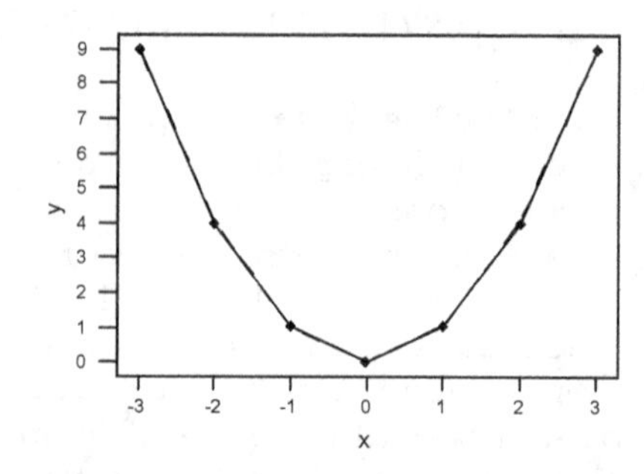

(d) It is not appropriate to use the linear correlation coefficient as a descriptive measure for the data because the data points are not scattered about a line.

(e) For each data point (x, y), we have $y = x^2$. (See columns 1 and 2 of the previous table.) See the previous graph at the right.

14.153 (a) From the data, we obtain the following sums:

$$\sum x_i = 21; \ \sum y_i = 91; \ \sum x_i^2 = 91; \ \sum y_i^2 = 2275; \ \sum x_i y_i = 441$$

Then

$$r = \frac{S_{xy}}{\sqrt{S_{xx}S_{xy}}} = \frac{\sum xy - (\sum x)(\sum y)/n}{\sqrt{[\sum x^2 - (\sum x)^2/n][\sum y^2 - (\sum y)^2/n]}}$$

$$= \frac{441 - (21)(91)/7}{\sqrt{[91 - (21)^2/7][2275 - (91)^2/7]}} = 0.961$$

(b) No. A graphical analysis is also needed before you can conclude that they are linearly related.

(c) 

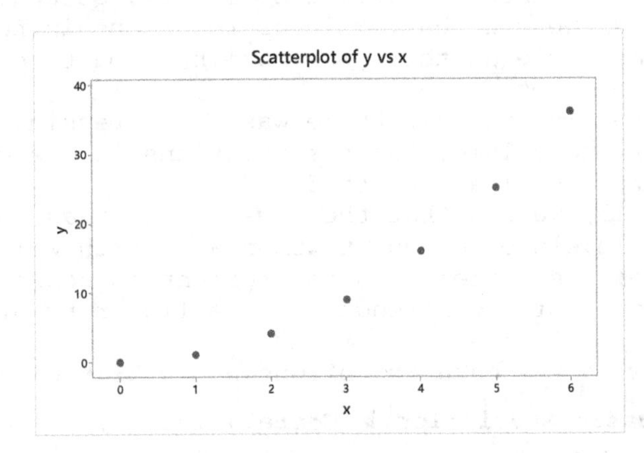

(d) No. The scatterplot reveals that the graph is not linear.

(e) Each of the y-values is the solution of the respective x-value plugged into the equation. For example, plugging in x = 3 in the equation yields (3)^2 = 9.

(f) The following is the graph of the equation.

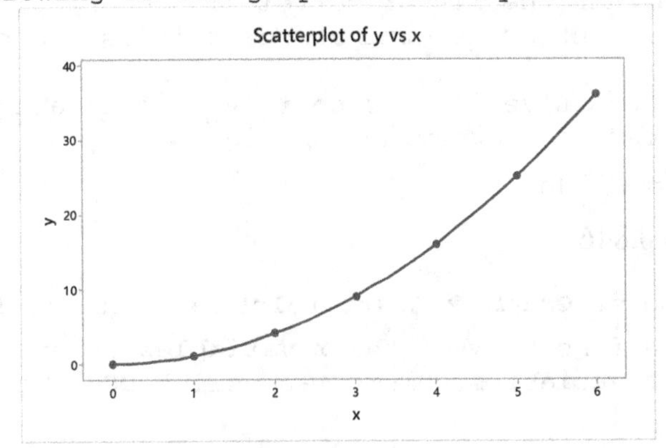

14.155 (a) In Exercise 14.71, we saw that a linear relationship between the age at inauguration and the age at death was reasonable. The scatterplot showed that the line relating the two variables had a positive slope.

(b) From the output in Exercise 14.71, the coefficient of determination was 0.367. Since this is r^2 and r is positive because of the positive slope of the line, $r = \sqrt{0.367} = 0.606$.

You can also obtain the correlation coefficient in Minitab by choosing **Stat ▶ Basic Statistics ▶ Correlation...**, select INAUGURATION and DEATH in the **Variables** text box, and click **OK**. The result is

Pearson correlation of INAUGURATION and DEATH = 0.606

(c) The correlation coefficient is positive, indicating that the linear relationship has a positive slope. The coefficient is not very close to 1, indicating that the linear relationship is not very strong.

14.157 (a) In Exercise 14.73, we saw that a linear relationship between LOT SIZE and VALUE was not reasonable. Parts (b-c) are omitted.

14.159 (a) In Exercise 14.75, we saw that a linear relationship between HIGH and LOW was reasonable.

(b) In Minitab, choose **Stat ▶ Basic Statistics ▶ Correlation...**, select HIGH and LOW in the **Variables** text box, and click **OK**. The result is

Pearson correlation of HIGH and LOW = 0.976

(c) The correlation coefficient is positive, indicating that the linear relationship has a positive slope. The coefficient is very close to 1, indicating that the linear relationship is very strong.

14.161 (a) In Exercise 14.77, we saw that a linear relationship between INCOME and BEER was not reasonable as there was no linear pattern in the data. Parts (b-c) are omitted.

14.163 (a) In Exercise 14.80, we saw that there was an increasing relationship between diameter and volume, but the relationship was concave upward, not linear. Parts (b-c) are omitted.

14.165 (a) In Exercise 14.115, we saw that there was an increasing relationship between estriol levels of pregnant women and birth weights of their children. Use of the correlation coefficient is appropriate as a descriptive measure of the strength of the linearity of that relationship.

(b) Using Minitab, with the data in columns named ESTRIOL and WEIGHT, we choose **Stat ▶ Basic Statistics ▶ Correlation...**, select ESTRIOL and WEIGHT in the **Variables** text box, and click **OK**. The result is

Pearson correlation of ESTRIOL and WEIGHT = 0.610, P-Value = 0.000

(c) There is a moderately positive linear relationship between estriol concentration and birth weight. The data points will be moderately clustered about the regression line.

14.167 (a) No. We only know that the linear correlation coefficient is the square root of the coefficient of determination or it is the negative of that square root.

(b) No. The slope is positive if r is positive and negative if r is negative, but we can not determine the sign of r.

(c) Yes. $r = -\sqrt{0.716} = -0.846$

(d) Yes. $r = \sqrt{0.716} = 0.846$

14.169 (a) Using Minitab, choose **Graph ▶ Scatterplot**, select the **Simple** version, enter SCORE in the first row of the **Y variables** column and TIME in the first row of the **X variables** column, and click **OK**. The result is

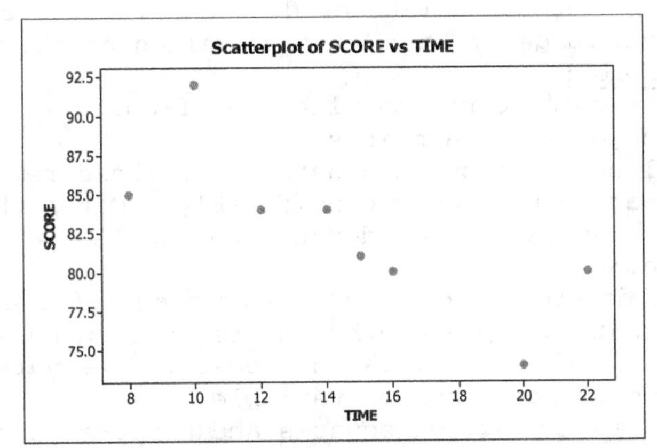

(b) Score decreases as time increases, so the using the rank correlation coefficient is reasonable.
(c) Score decreases linearly as time increases, so the using the linear correlation coefficient is reasonable.
(d) To find the rank correlation coefficient using Minitab, we must first find the ranks for each of the data values in the SCORE and TIME columns. This is done for TIME by choosing **Data ▶ Rank**, entering TIME in the **Rank data in** text box and RANKTIME in the **Store ranks in** text box and clicking **OK**. Then choose **Data ▶ Rank**, enter SCORE in the **Rank data in** text box and RANKSCORE in the **Store ranks in** text box and click **OK**. Now choose **Stat ▶ Basic Statistics ▶ Correlation...**, select RANKTIME and RANKSCORE in the **Variables** text box, and click **OK**. The result is

```
Pearson correlation of RANKTIME and RANKSCORE = -0.928
P-Value = 0.001
```

The rank correlation coefficient is -0.928, indicating that the ranks are strongly negatively correlated and that SCORE decreases as TIME increases.

Review Problems for Chapter 14

1. (a) x (b) y (c) b_1 (d) b_0
2. (a) It intersects the y-axis when x = 0. Therefore y = 4.
   (b) It intersects the x-axis when y = 0. Therefore x = 4/3.
   (c) slope = -3
   (d) The y-value decreases by 3 units when x increases by 1 unit.
   (e) The y-value increases by 6 units when x decreases by 2 units.
3. True. The y-intercept is determined by b_0 and that value is independent of b_1, the slope.
4. False. A horizontal line has a slope of zero.
5. True. If the slope is positive, the x-values and y-values increase and decrease together.
6. Scatterplot or scatter diagram
7. A regression equation can be used to make predictions of the response variable for specific values of the predictor variable within the range of the observed values of the predictor variable.
8. (a) predictor variable or explanatory variable
   (b) response variable
9. Based on the least-squares criterion, the line that best fits a set of data points is the one have the smallest possible sum of squared errors.
10. The line that best fits a set of data points according to the least-squares criterion is called the regression line.

11. Using a regression equation to make predictions for values of the predictor variable outside the range of the observed values of the predictor variable is called extrapolation.

12. (a) An outlier is a data point that lies far from the regression line relative to the other data points.

(b) An influential observation is a data point whose removal causes the regression equation to change considerably. Often this is a data point for which the x-value is considerably to the left or right of the rest of the data points.

13. The coefficient of determination is the percentage of the total variation in the y-values that is explained by the regression equation.

14. (a) SST is the total sum of squares and measures the variation in the observed values of the response variable.

(b) SSR is the regression sum of squares and measures the variation in the observed values of the response variable that is explained by the regression. It can also be thought of as the variation in the predicted values of the response variable corresponding to the x-values in the data points.

(c) SSE is the error sum of squares and measures the variation in the observed values of the response variable that is not explained by the regression.

15. One use of the linear correlation coefficient is as a descriptive measure of the strength of the linear relationship between two variables.

16. A positive linear relationship between two variables means that one variable tends to increase linearly as the other increases.

17. A value of r close to -1 suggests a strong negative linear relationship between the variables.

18. A value of r close to zero suggests at most a weak linear relationship between the variables.

19. True. It is quite possible that both variables are strongly affected by one (or more) other variables (called lurking variables).

20. (a) $y = 72 - 12x$ (b) $b_0 = 72$, $b_1 = -12$

(c) The line slopes downward since $b_1 < 0$.

(d) After two years: $y = 72 - 12(2) = \$48$ hundred $= \$4800$

After five years: $y = 72 - 12(5) = \$12$ hundred $= \$1200$.

(e)

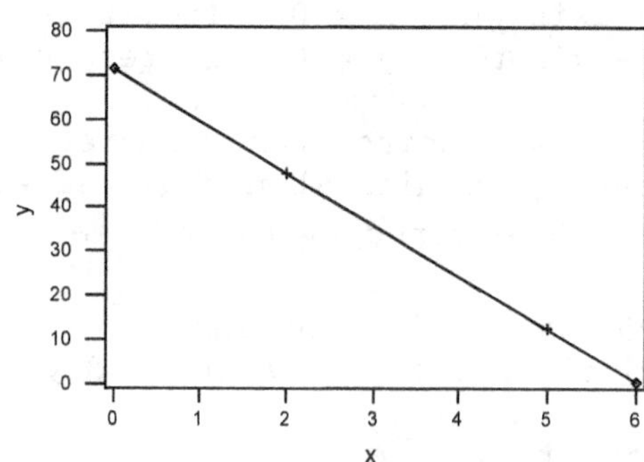

(f) From the graph, we estimate the value to be about \$2500 after 4 years. The actual value is $y = 7200 - 1200(4) = \$2400$.

21. (a)

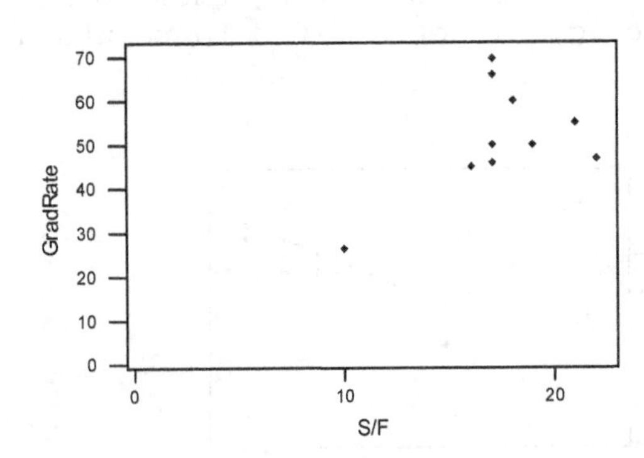

(b) It is moderately reasonable to find a regression line for the data because the data points appear to be scattered about a line. This perception is, however, heavily influenced by the single data point at x = 10.

(c) The regression equation can be determined by calculating its slope and intercept. Formulas for the slope (b_1) and intercept (b_0) of the regression equation are, respectively:

$$b_1 = \frac{\sum xy - \sum x \sum y / n}{\sum x^2 - (\sum x)^2 / n} \qquad b_0 = \frac{1}{n}(\sum y - b_1 \sum x)$$

To compute b_0 and b_1, construct a table of values for x, y, xy, x^2, and their sums. (Note: A column for y^2 is also presented. This is used in Problems 22 and 23.)

x	y	xy	x^2	y^2
16	45	720	256	2,025
20	55	1,100	400	3,025
17	70	1,190	289	4,900
19	50	950	361	2,500
22	47	1,034	484	2,209
17	46	782	289	2,116
17	50	850	289	2,500
17	66	1,122	289	4,356
10	26	260	100	676
18	60	1,080	324	3,600
173	515	9,088	3,081	27,907

Thus,

$$b_1 = \frac{9088 - 173(515)/10}{3081 - 173^2/10} = 2.02611 \quad b_0 = \frac{1}{11}(515 - 2.02611(173) = 16.448$$

and the regression equation is: $\hat{y} = 16.448 + 2.02611x$.

To graph the regression equation, begin by selecting two *x*-values within the range of the *x*-data. For the *x*-values 10 and 22, the calculated values for *y* are, respectively:

$$\hat{y} = 16.448 + 2.02611(10) = 36.71$$

$$\hat{y} = 16.448 + 2.02611(22) = 61.02$$

The regression equation can be graphed by plotting the pairs (10, 36.71) and (22, 61.02) and connecting these points with a line. This equation and the original set of 10 data points are presented as follows:

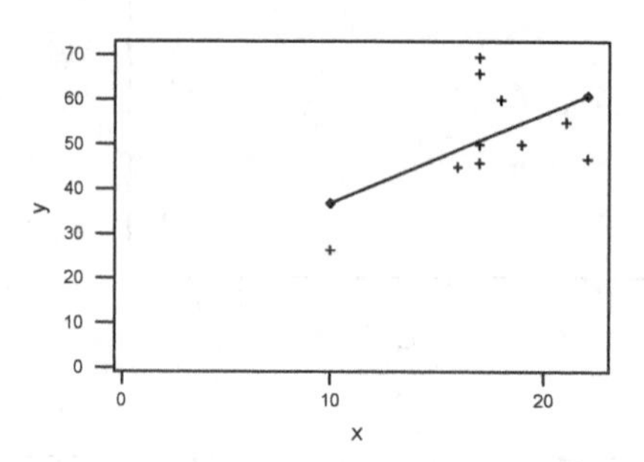

(d) Graduation rate tends to increase as the student-to-faculty ratio increases.

(e) Graduation rate increases an estimated 2.026% for each additional 1 unit increase in the student-to-faculty ratio.

(f) The predicted graduation rate for a university with a student-to-faculty ratio of 17 is 16.448 + 2.0261(17) = 50.89%.

(g) There is one potential influential observation, (10, 26) and no outliers. Looking at the scatterplot in part (c), we suspect that there may be little relationship between the student-to-faculty ratio and the graduation rate if the influential observation is removed from the data.

22. (a) To compute SST, SSR, and SSE using the computing formulas, begin with the table presented in Problem 21(c). Using the last row of this table and the formula in Definition 14.7 of the text, we obtain the three sums of squares and the coefficient of determination as follows.

$$SST = S_{yy} = \sum y^2 - (\sum y)^2 / n = 27907 - 515/10 = 1384.5$$

$$SSR = \frac{S_{xy}^2}{S_{xx}} = \frac{[\sum xy - (\sum x)(\sum y)/n]^2}{\sum x^2 - (\sum x)^2 / n} = \frac{[9088 - (173)(515)/10]^2}{3081 - 173^2/10} = 361.66$$

$$SSE = S_{yy} - \frac{S_{xy}^2}{S_{xx}} = 1384.50 - 361.66 = 1022.846$$

$$r^2 = \frac{SSR}{SST} = \frac{361.66}{1384.50} = 0.261$$

(b) The percentage reduction obtained in the total squared error by using the regression equation, instead of the sample mean y, to predict the observed costs is 26.1%.

(c) The percentage of the variation in the graduation rate that is explained by the student-to-faculty ratio is 26.1%.

(d) The regression equation appears to be not very useful for making predictions.

23. (a) To compute the linear correlation coefficient, begin with the table presented in Problem 21(c). Using the last row of this table and the formula in Formula 14.3 of the text, we get

$$r = \frac{S_{xy}}{\sqrt{S_{xx}S_{xy}}} = \frac{\sum xy - (\sum x)(\sum y)/n}{\sqrt{[\sum x^2 - (\sum x)^2/n][\sum y^2 - (\sum y)^2/n]}}$$

$$= \frac{9088 - (173)(515)/10}{\sqrt{[3081 - 173^2/10][27907 - 515^2/10]}} = 0.511$$

(b) The value of r suggests a weak to moderate positive linear correlation.

(c) Data points are clustered about the regression line, but not very closely.

(d) $r^2 = (0.511)^2 = 0.261$

24. Using Minitab, with the data in three columns named POPULATION, AREA, and PLANTS, choose **Stat ▶ Basic statistics ▶ Correlation, enter** POPULATION, AREA, and PLANTS in the **Variables** text **box.** Make sure that the **Display P-values** box is checked. **Click OK.** The output is

```
             POPULATION      AREA
AREA              0.109
                  0.453

PLANTS            0.721    -0.309
                  0.000     0.029
Cell Contents: Pearson correlation
               P-Value
```

(a) The correlation coefficient between population and area is 0.109.

(b) The correlation coefficient between population and number of exotic plants is 0.721.

(c) The correlation coefficient between area and number of exotic plants is -0.309.

(d) Any linear association between population and area is very weak since the correlation coefficient is near zero. There is a moderately strong positive linear relationship between population and number of exotic plants. There is a fairly weak negative linear association between area and number of exotic plants. One possible explanation of the stronger positive relationship between population and number of exotic plants might be that the more people there are, the greater the possibility that exotic plants are introduced into the state by those people, intentionally or unintentionally.

25. (a) Using Minitab, with the data in columns named IMR and LE, choose **Graph ▶ Scatterplot…**, select the **Simple** version, enter 'LE' in row 1 of the **Y variables** column and IMR in the **X variables** column and click **OK.**

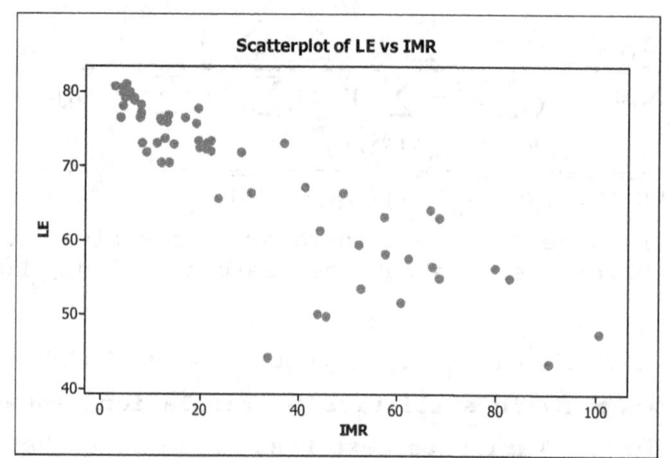

Life expectancy appears to decrease linearly as IMR increases.

(b) It appears to be reasonable to find a regression line for the data, although we expect that there will be both outliers and influential observations.

(c) Choose **Stat ▶ Regression ▶ Regression,** enter 'LE' in the **Response box** and IMR in the **Predictor box.** Click on the Options button and enter 30 in the Prediction intervals for new observations text box. Click **OK** twice. The results are

```
The regression equation is
LE = 79.4 - 0.354 IMR

Predictor      Coef  SE Coef       T      P
Constant     79.396    1.009   78.72  0.000
IMR        -0.35374  0.02601  -13.60  0.000

S = 5.17598   R-Sq = 76.1%   R-Sq(adj) = 75.7%

Analysis of Variance

Source          DF      SS      MS       F      P
Regression       1  4956.2  4956.2  185.00  0.000
Residual Error  58  1553.9    26.8
Total           59  6510.1

Unusual Observations

Obs  IMR      LE     Fit  SE Fit  Residual  St Resid
 11   46  49.890  63.251   0.795   -13.361    -2.61R
 25  100  47.430  43.894   1.971     3.536     0.74 X
 34   34  44.280  67.418   0.680   -23.138    -4.51R
 48   44  50.160  63.828   0.773   -13.668    -2.67R
 52   91  43.450  47.365   1.734    -3.915    -0.80 X

R denotes an observation with a large standardized residual.
X denotes an observation whose X value gives it large leverage.

Predicted Values for New Observations

New Obs     Fit  SE Fit        95% CI              95% PI
      1  68.784   0.669  (67.445, 70.122)  (58.337, 79.231)

Values of Predictors for New Observations

New Obs   IMR
      1  30.0
```

The regression equation, **LE = 79.4 - 0.354 IMR,** indicates that life expectancy decreases as the infant mortality rate increases.

(d) For a country with an IMR = 30, the predicted life expectancy is 79.4 - 0.354(30) = 68.78 years. The output also has the predicted value of 68.784 years

(e) Choose **Stat ▶ Basic statistics ▶ Correlation, enter** 'LE' and IMR in the **Variables** text **box.** Click **OK.** The results are

```
Pearson correlation of IMR and LE = -0.873
P-Value = 0.000
```

Life expectancy and IMR are highly negatively correlated. Thus, there is strong negative linear association between the two variables.

(f) The output shown in part (c) indicates that observations 11, 34, and 48 are potential outliers and observations 25 and 52 are influential observations.

26. (a) Using Minitab, with the data in columns named HIGH and PRECIPITAION, choose **Graph ▶ Scatterplot…**, select the **Simple** version, enter 'PRECIPITATIONN' in row 1 of the **Y variables** column and HIGH in the **X variables** column and click **OK.**

The graph at the right does not show any relationship between average July precipitation and average July high temperature.

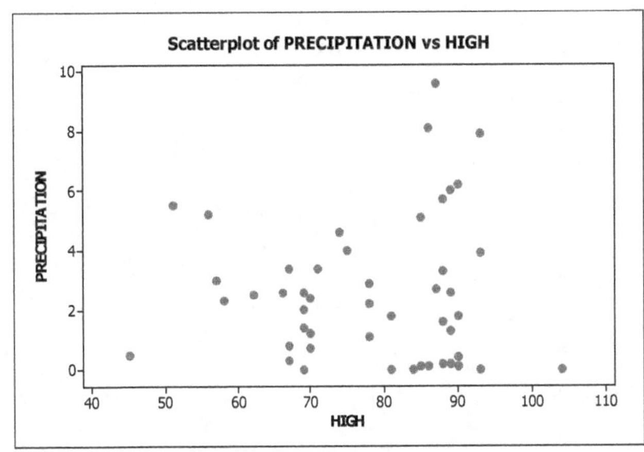

(b) The graph in part (a) does not indicate that there is any linear relationship between average precipitation and average high temperature in July for 48 cities. Parts (c-f) are omitted.

27. (a) Using Minitab, with the data in columns named FIRST and SECOND, choose **Graph ▶ Scatterplot…**, select the **Simple** version, enter SECOND in row 1 of the **Y variables** column and FIRST in the **X variables** column and click **OK.**

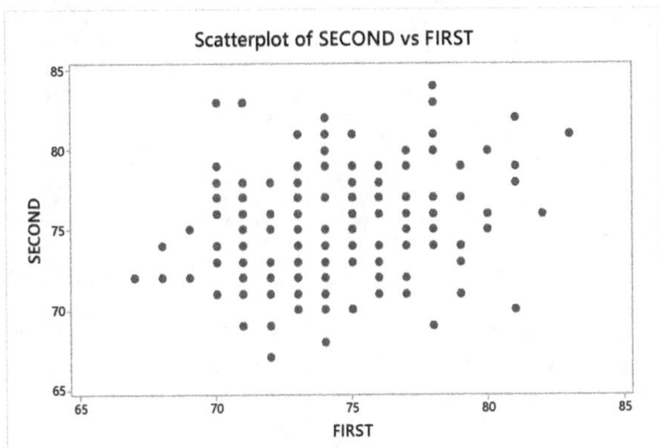

(b) Obtaining a linear regression does not seem reasonable. There is no noticeable linear relationship in the data. Therefore, we will skip steps (c) - (f).

CHAPTER 15 SOLUTIONS

Exercises 15.1

15.1 Conditional distribution, conditional mean, and conditional standard deviation

15.3 (a) population regression line (b) σ

(c) normal, β_0 + 6 β_1, σ

15.5 The sample regression line is the best estimate of the population regression line.

15.7 Residual

15.9 The plot of the residuals against the values of the predictor variable provides the same information as a scatter diagram of the data points. However, it has the advantage of making it easier to spot patterns such as curvature and non-constant standard deviation.

15.11 (a) The assumption of linearity (Assumption 1) may be violated since the band is not horizontal, as may be the assumption of equal standard deviations (Assumption 2) since there is more variation in the residuals for small x than for large x.

(b) It appears that the standard deviation does not remain constant; thus, Assumption 2 is violated.

(c) The graph does not suggest violation of one or more of the assumptions for regression inferences.

(d) The normal probability plot appears to be more curved than linear; thus, the assumption of normality is violated.

15.13 (a) We first create the table below.

x	y	$\hat{y}$	$e = y - \hat{y}$	e^2
3	-4	-5	1	1
1	0	-1	1	1
2	-5	-3	-2	4
				6

$$s_e = \sqrt{\frac{SSE}{n-2}} = \sqrt{\frac{6}{1}} = 2.449$$

(b-c) For the residual plot, we plot e against x in the graph below at the left. We can use Excel for the normal probability plot. With the values of e in a column named 'e', highlight the data with the mouse, choose **Charts and Plots**, click on the **Function Type** drop down box, and click on **Normal Probability Plot**. Drag the 'e' from the **Names and Columns** box into the **Quantitative Variable** box and click **OK**. The graph is shown below at the right.

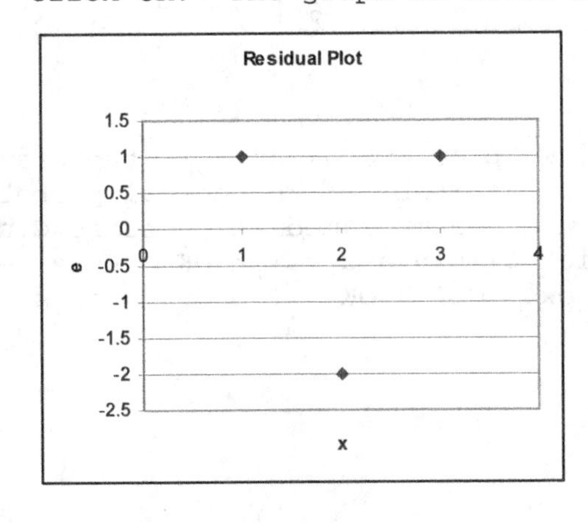

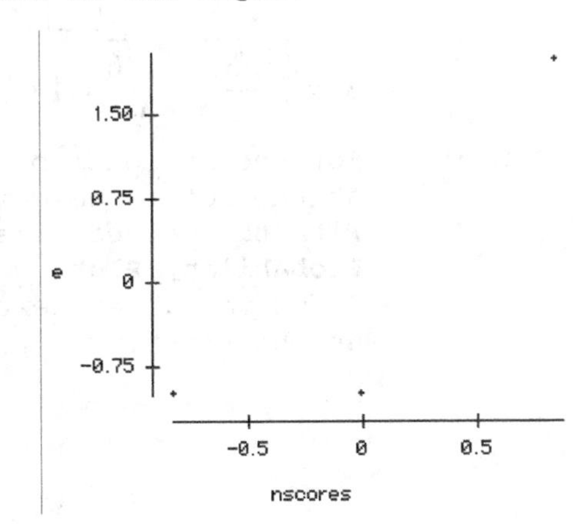

15.15 (a) We first create the table below.

x	y	$\hat{y}$	$e = y - \hat{y}$	e^2
3	4	3	1	1
4	5	5	0	0
1	0	-1	1	1
2	-1	1	-2	4
				6

$$s_e = \sqrt{\frac{SSE}{n-2}} = \sqrt{\frac{6}{2}} = 1.732$$

(b-c) For the residual plot, we plot e against x in the graph following at the left. We can use Excel for the normal probability plot. With the values of e in a column named 'e', highlight the data with the mouse, choose **Charts and Plots**, click on the **Function Type** drop down box, and click on **Normal Probability Plot**. Drag the 'e' from the **Names and Columns** box into the **Quantitative Variable** box and click **OK**. The graph is shown following at the right.

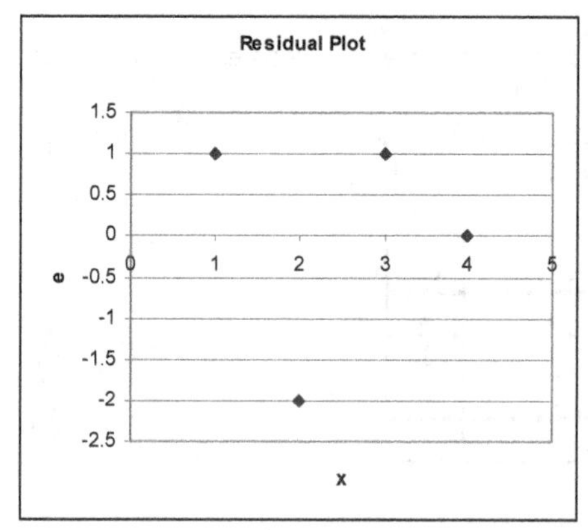

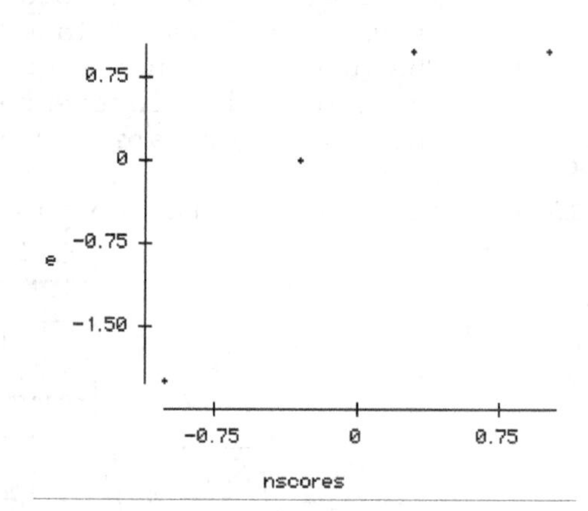

15.17 (a) We first create the table below.

x	y	$\hat{y}$	$e = y - \hat{y}$	e^2
2	3	3	0	0
2	4	3	1	1
3	0	2	-2	4
4	2	1	1	1
4	1	1	0	0
				6

$$s_e = \sqrt{\frac{SSE}{n-2}} = \sqrt{\frac{6}{3}} = 1.414$$

(b-c) For the residual plot, we plot e against x in the graph following at the left. We can use Minitab for the normal probability plot. With the values of e in a column named 'e', choose **Graph, Probability Plot.** Click **Simple** and click **OK**. Enter e into the Graph variables text box. Click **OK.** The graph is shown following at the right.

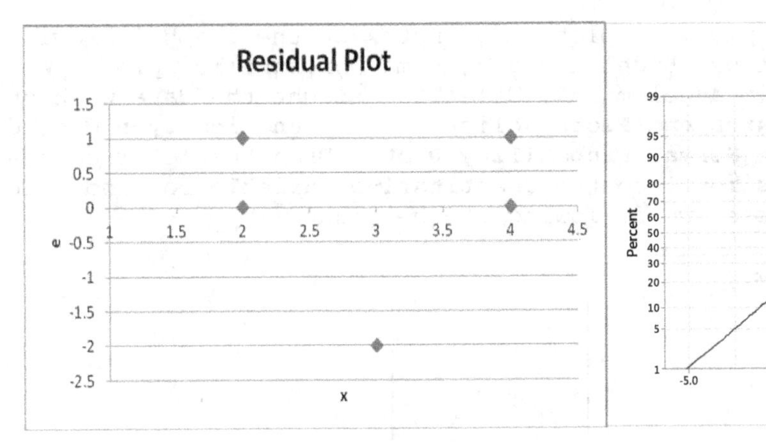

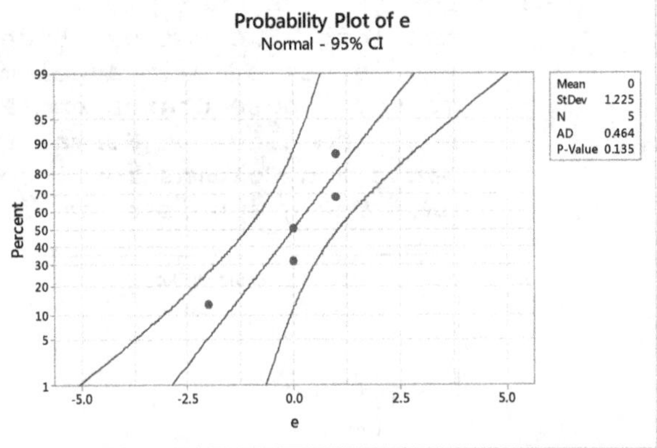

15.19 (a) We first create the table below.

x	y	$\hat{y}$	$e = y - \hat{y}$	e^2
1	4	3	1	1
2	3	5	-2	4
3	8	7	1	1
				6

$$s_e = \sqrt{\frac{SSE}{n-2}} = \sqrt{\frac{6}{1}} = 2.449$$

(b-c) For the residual plot, we plot e against x in the graph following at the left. We can use Minitab for the normal probability plot. With the values of e in a column named 'e', choose **Graph, Probability Plot.** Click **Simple** and click **OK**. Enter e into the Graph variables text box. Click **OK.** The graph is shown following at the right.

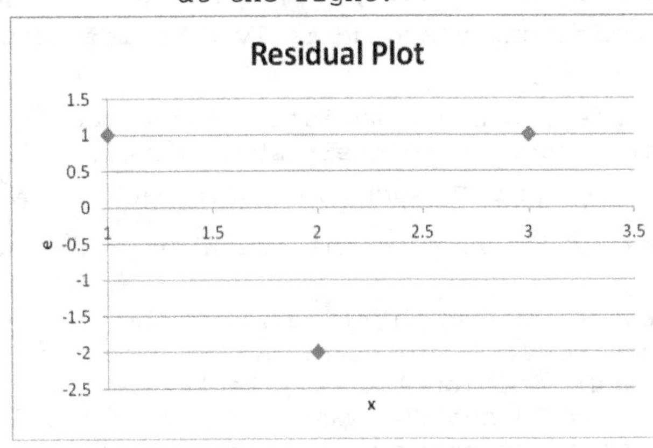

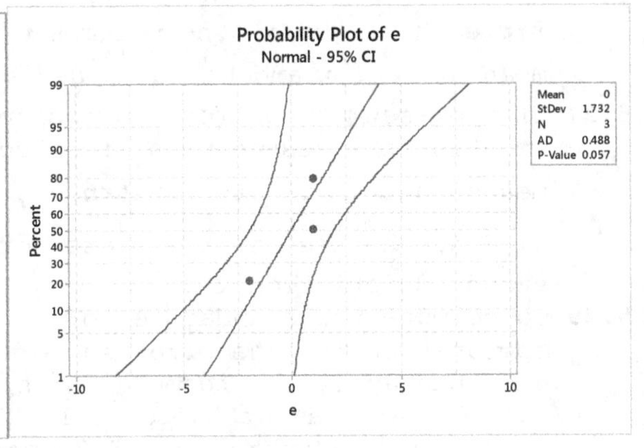

15.21 (a) We first create the table below.

x	y	$\hat{y}$	$e = y - \hat{y}$	e^2
0	4	2.875	1.125	1.265625
2	2	1.625	0.375	0.140625
2	0	1.625	-1.625	2.640625
5	-2	-0.250	-1.750	3.062500
6	1	-0.875	1.875	3.515625
				10.625000

$$s_e = \sqrt{\frac{SSE}{n-2}} = \sqrt{\frac{10.625}{3}} = 1.882$$

(b-c) For the residual plot, we plot e against x in the graph following at the left. We can use Excel for the normal probability plot. With the values of e in a column named 'e', highlight the data with the mouse, choose **Charts and Plots**, click on the **Function Type** drop down box, and click on **Normal Probability Plot**. Drag the 'e' from the **Names and Columns** box into the **Quantitative Variable** box and click **OK**. The graph is shown following at the right.

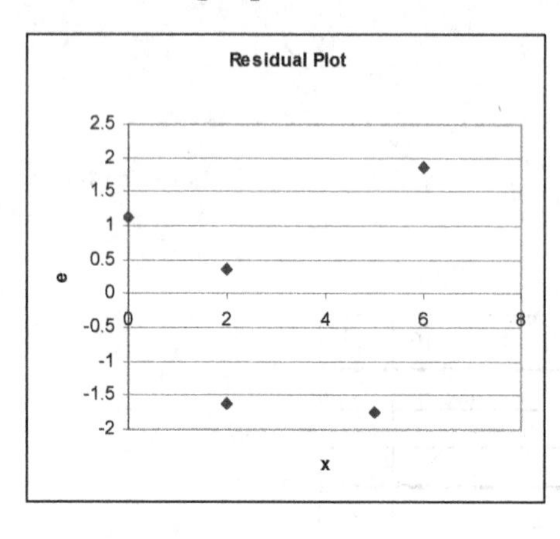

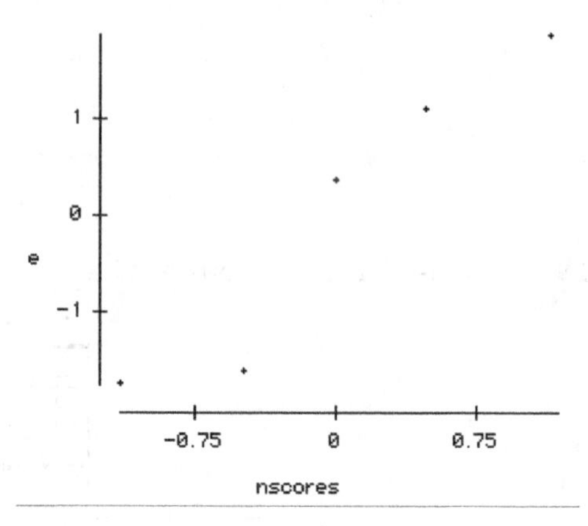

15.23 If the assumptions for regression inferences are satisfied for a model relating a Corvette's age to its price, this means that there are constants β_0, β_1, and σ such that, for each age x, the prices for Corvettes of that age are normally distributed with mean $\beta_0 + \beta_1 x$ and standard deviation σ.

15.25 If the assumptions for regression inferences are satisfied for a model relating the volume of plant emissions of volatile compounds to the weight of the plant, this means that there are constants β_0, β_1, and σ such that, for each weight x, the volumes of emissions y are normally distributed with mean $\beta_0 + \beta_1 x$ and standard deviation σ.

15.27 If the assumptions for regression inferences are satisfied for a model relating the test scores of calculus students to their study-times, this means that there are constants β_0, β_1, and σ such that, for each study-time x, the test scores y are normally distributed with mean $\beta_0 + \beta_1 x$ and standard deviation σ.

15.29 To compute the standard error of the estimate, first retrieve the computation for the error sum of squares (SSE) in part (a) of Exercise 14.99 and then apply the formula for s_e in Definition 15.2 of the text.

(a) In part (a) of Exercise 14.99, n = 10 and SSE was computed as 1,623.7. Applying Definition 15.2, the standard error of the estimate is

$$s_e = \sqrt{\frac{SSE}{n-2}} = \sqrt{\frac{1623.7}{10-2}} = 14.2465$$

Roughly speaking, the predicted prices for the Corvettes differ, on average, from the observed prices by $1424.65 .

(b) Presuming that the variables age (x) and price (y) for Corvettes satisfy Assumptions (1) - (3) for regression inferences, the standard error of the estimate s_e = 14.2465 provides an estimate for the common population standard deviation σ of prices for all Corvettes of any given age.

(c) Obtain the necessary normal scores from Table III in the text.

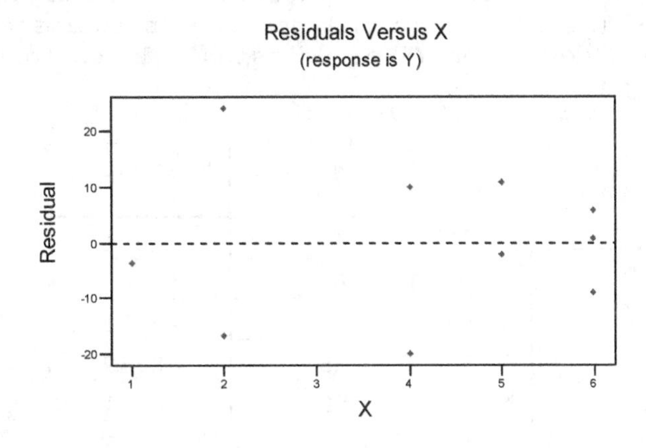

Age x	Residual e
6	0.82
6	-9.18
6	5.82
2	24.20
2	-16.80
5	-2.09
4	10.01
5	10.91
1	-3.70
4	-19.99

Normal Probability Plot of the Residuals
(response is Y)

Residual e	Normal score n
-19.99	-1.55
-16.80	-1.00
-9.18	-0.65
-3.70	-0.37
-2.09	-0.12
0.82	0.12
5.82	0.37
10.01	0.65
10.91	1.00
24.20	1.55

(d) Taking into account the small sample size, we can say that the residuals fall roughly in a horizontal band centered and symmetric about the x-axis. We can also say that the normal probability plot for residuals is very roughly linear. Therefore, it appears reasonable to consider the assumptions for regression inferences for the variables age and price of Corvettes to be met.

15.31 To compute the standard error of the estimate, first retrieve the computation for the error sum of squares (SSE) in part (a) of Exercise 14.101 and then apply the formula for s_e in Definition 15.2 of the text.

(a) In part (a) of Exercise 14.101, n = 10 and SSE was computed as 88.5. Applying Definition 15.2, the standard error of the estimate is

$$s_e = \sqrt{\frac{SSE}{n-2}} = \sqrt{\frac{264.16}{11-2}} = 5.4177 \text{ hundred nanograms}$$

Roughly speaking, the predicted volumes of volatile emissions from the plants differ, on average, from the observed volumes by 541.77 nanograms.

(b) Presuming that the weight (x) of the potato plants *Solanum tubersom* and volume of volatile emissions (y) from the plants satisfy Assumptions (1) - (3) for regression inferences, the standard error of the estimate s_e = 5.4177 hundred nanograms (541.77 nanograms) provides an estimate for the common population standard deviation σ of emission volumes for all plants of any particular weight.

(c)

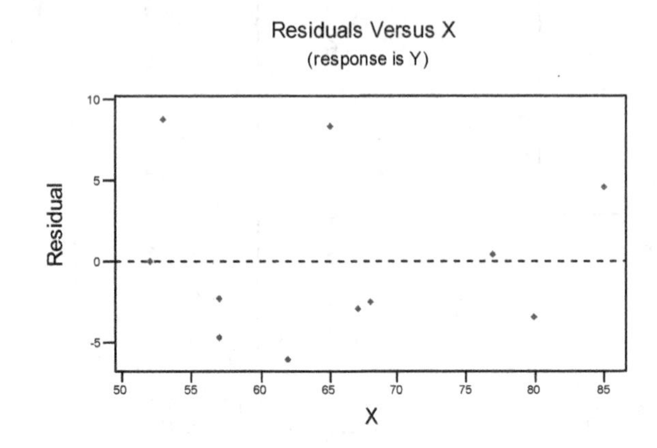

Weight x	Residual e
57	-4.81
85	4.63
57	-2.31
65	8.39
52	0.01
67	-2.93
62	-6.12
80	-3.55
77	0.44
53	8.85
68	-2.60

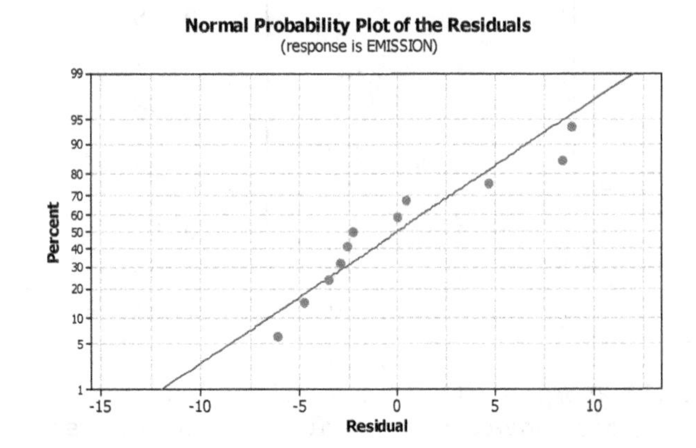

Residual e	Normal score n
-6.12	-1.59
-4.81	-1.06
-3.55	-0.73
-2.93	-0.46
-2.60	-0.22
-2.31	0.00
0.01	0.22
0.44	0.46
4.63	0.73
8.39	1.06
8.85	1.59

(d) Taking into account the small sample size, we can say that the residuals fall roughly in a horizontal band centered and symmetric about the x-axis. We can also say that the normal probability plot for residuals is approximately linear.
Therefore, based on the sample data, there are no obvious violations of the assumptions for regression inferences for the variables plant weight and volume of volatile emissions.

15.33 To compute the standard error of the estimate, first retrieve the computation for the error sum of squares (SSE) in part (a) of Exercise 14.103 and then apply the formula for s_e in Definition 15.2 of the text.

(a) In part (a) of Exercise 14.103, n = 8 and SSE was computed as 75.11. Applying Definition 15.2, the standard error of the estimate is

$$s_e = \sqrt{\frac{SSE}{n-2}} = \sqrt{\frac{75.11}{8-2}} = 3.5382$$

Roughly speaking, the predicted test scores differ, on average, from the observed scores by 3.5382 points.

(b) Presuming that the study-time (x) of the calculus students and their test scores (y) satisfy Assumptions (1) - (3) for regression inferences, the standard error of the estimate s_e = 3.5382 provides an estimate for the common population standard deviation σ of test scores for all calculus students with any particular study-time.

(c) Obtain the necessary normal scores from Table III in the text.

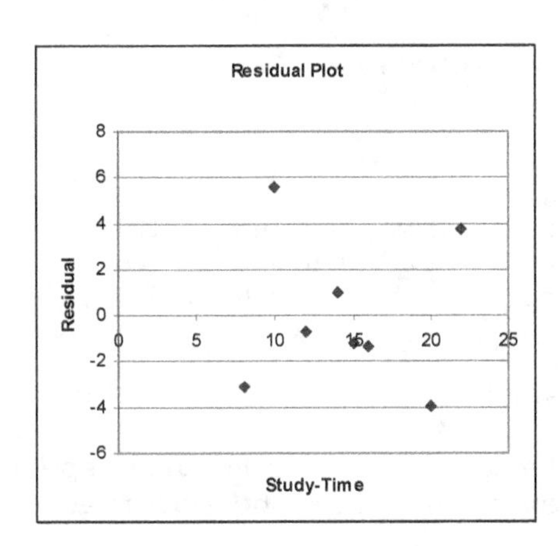

Study Time x	Residual e
10	5.59
15	-1.18
12	-0.72
20	-3.95
8	-3.10
16	-1.34
14	0.97
22	3.74

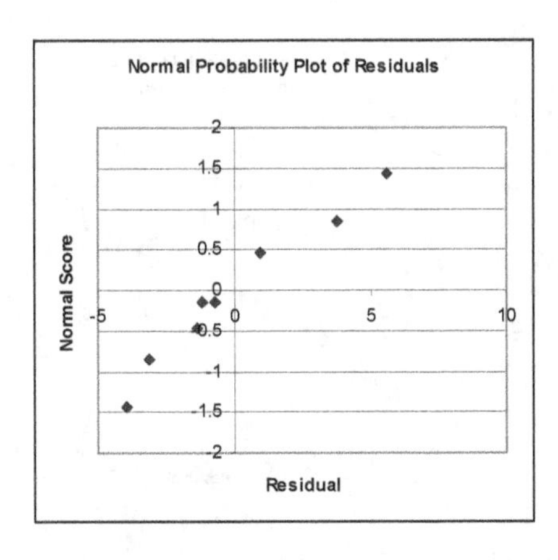

Residual e	Normal score n
-3.95	-1.43
-3.10	-0.85
-1.34	-0.47
-1.18	-0.15
-0.72	0.15
0.97	0.47
3.74	0.85
5.59	1.43

(d) Taking into account the small sample size, we can say that the residuals fall roughly in a horizontal band centered and symmetric about the x-axis. We can also say that the normal probability plot for residuals is approximately linear. Therefore, based on the sample data, there are no obvious violations of the assumptions for regression inferences for the variables study-time and test scores of calculus students.

15.35 Using Minitab, with the data in columns named INAUGURATION and DEATH, we choose **Stat ▶ Regression ▶ Regression...**, select DEATH in the **Response** text box, and select INAUGURATION in the **Predictors** text box. Click the **Graphs** button, select the **Regular** option button from the **Residuals for Plots** list, select the **Normal plot of residuals** check box from the **Residual Plots** list, click in the **Residuals versus the variables** text box and specify INAUGURATION, click **OK**, and click **OK**. The results are

Analysis of Variance

Source	DF	Adj SS	Adj MS	F-Value	P-Value
Regression	1	2021	2020.57	21.45	0.000
INAUGURATION	1	2021	2020.57	21.45	0.000
Error	37	3485	94.19		
Lack-of-Fit	19	2179	114.68	1.58	0.168
Pure Error	18	1306	72.56		
Total	38	5506			

Model Summary

S	R-sq	R-sq(adj)	R-sq(pred)
9.70514	36.70%	34.99%	28.71%

Coefficients

Term	Coef	SE Coef	T-Value	P-Value	VIF
Constant	5.7	14.1	0.40	0.690	
INAUGURATION	1.179	0.255	4.63	0.000	1.00

Regression Equation

DEATH = 5.7 + 1.179 INAUGURATION

(a) The standard error of the estimate is 9.70514. Roughly speaking, the predicted death ages differ, on average, from the observed death ages by 9.70514 years.

(b)

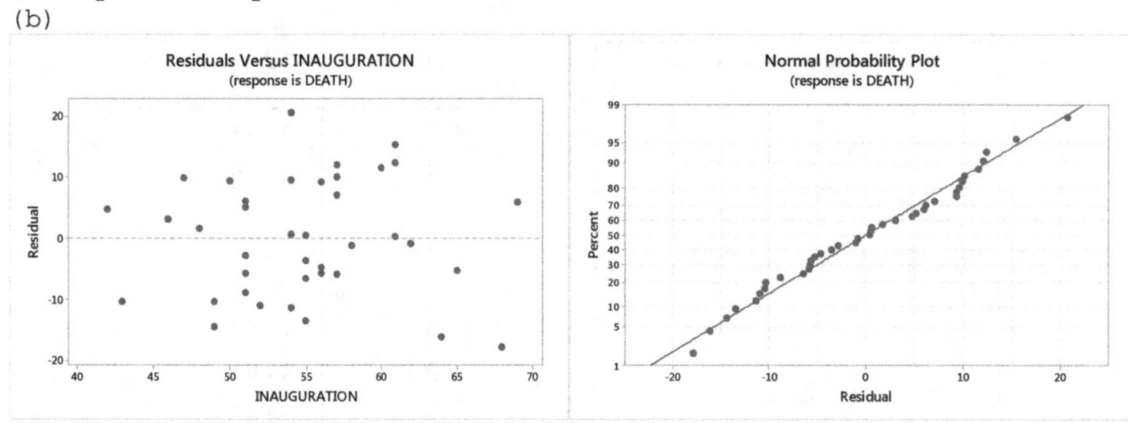

(c) The residuals fall roughly in a horizontal band centered and symmetric about the x-axis. We can also say that the normal probability plot for residuals is approximately linear. Therefore, based on the sample data, there are no obvious violations of the assumptions for regression inferences for the variables inauguration age and death age of U.S. Presidents.

15.37 Using Minitab, with the data in columns named LOT SIZE and VALUE, we choose **Stat ▶ Regression ▶ Regression...**, select VALUE in the **Response** text box, and select 'LOT SIZE' in the **Predictors** text box. Click the **Graphs** button, select the **Regular** option button from the **Residuals for Plots** list, select the **Normal plot of residuals** check box from the **Residual Plots** list, click in the **Residuals versus the variables** text box and specify 'LOT SIZE', click **OK**, and click **OK**. The results are

```
The regression equation is
VALUE = 292 + 67.1 LOT SIZE

Predictor    Coef  SE Coef     T      P
Constant    292.0    140.3  2.08  0.043
LOT SIZE    67.11    59.87  1.12  0.269

S = 139.012   R-Sq = 2.9%   R-Sq(adj) = 0.6%

Analysis of Variance
Source          DF      SS     MS     F      P
Regression       1   24286  24286  1.26  0.269
Residual Error  42  811619  19324
Total           43  835905
```

(a) The standard error of the estimate is 139.012 thousand dollars. Roughly speaking, the predicted values differ, on average, from the observed values by 139.012 thousand dollars.

(b)

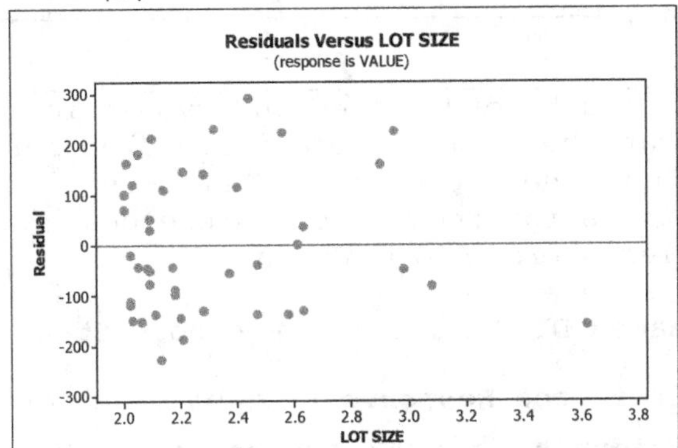

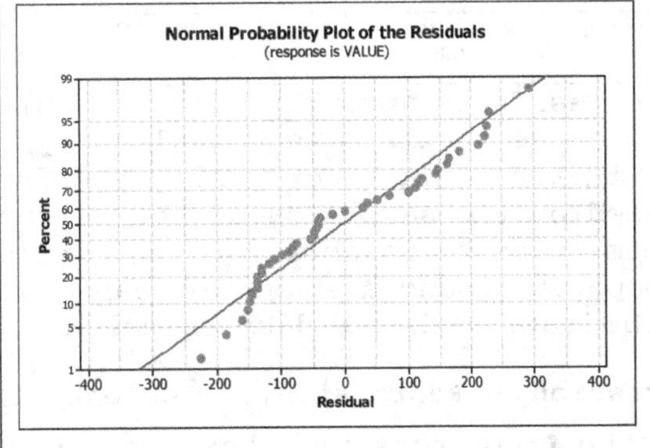

(c) The residuals appear to fall roughly in a band centered and symmetric about the x-axis. We can also say that the normal probability plot for residuals deviates from a linear pattern, but it is difficult to say whether the deviation is enough to rule out normality. Therefore, for the time being, based on the sample data, the assumptions for regression inferences seem to be reasonable for the variables lot size and value. More on these data later.

15.39 Using Minitab, with the data in columns named HIGH and LOW, we choose **Stat ▶ Regression ▶ Regression...**, select LOW in the **Response** text box, and select HIGH in the **Predictors** text box. Click the **Graphs** button, select the **Regular** option button from the **Residuals for Plots** list, select the **Normal plot of residuals** check box from the **Residual Plots** list, click in the **Residuals versus the variables** text box and specify HIGH, click **OK**, and click **OK**. The results are

```
The regression equation is
LOW = - 7.57 + 0.917 HIGH

Predictor      Coef  SE Coef      T      P
Constant     -7.572    1.786  -4.24  0.000
HIGH        0.91685  0.02967  30.91  0.000

S = 4.15111   R-Sq = 95.2%   R-Sq(adj) = 95.1%

Analysis of Variance

Source          DF     SS     MS       F      P
Regression       1  16459  16459  955.17  0.000
Residual Error  48    827     17
Total           49  17286
```

(a) The standard error of the estimate is 4.15111 degrees. Roughly speaking, the predicted low temperatures differ, on average, from the observed low temperatures by 4.15111 degrees.

(b)

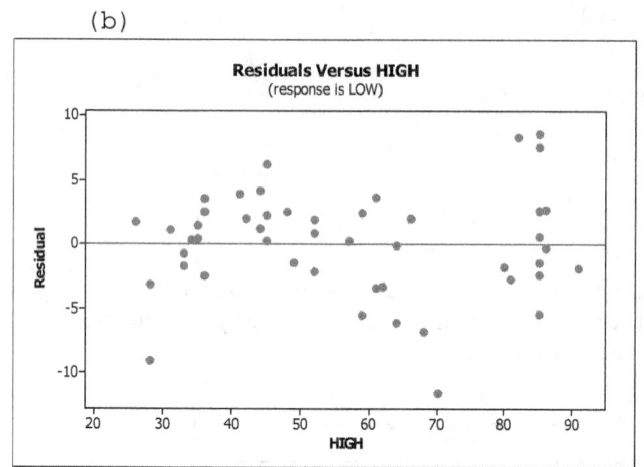

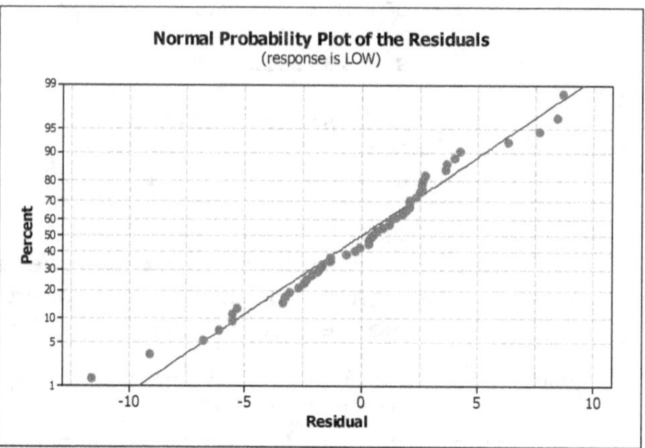

(c) The residuals appear to fall roughly in a band centered and symmetric about the x-axis. We can also say that the normal probability plot for residuals deviates only slightly from a linear pattern. Therefore, based on the sample data, the assumptions for regression inferences seem to be reasonable for the variables average high January temperature and average low January temperature.

15.41 Using Minitab, with the data in columns named DISP and MPG, we choose **Stat ▶ Regression ▶ Regression...**, select DISP in the **Response** text box, select MPG in the **Predictors** text box, select **Residuals** from the **Storage** check-box list. Click the **Graphs...** button, select the **Regular** option button from the **Residuals for Plots** list, select the **Normal plot of residuals** check box from the **Residual Plots** list, click in the **Residuals versus the variables** text box and specify DISP, click **OK**, and click **OK**. The results are

(a) The regression equation is
MPG = 32.3 - 3.58 DISP

Predictor	Coef	StDev	T	P
Constant	32.2732	0.6704	48.14	0.000
DISP	-3.5762	0.2099	-17.03	0.000

S = 2.248   R-Sq = 70.9%   R-Sq(adj) = 70.7%

Analysis of Variance

Source	DF	SS	MS	F	P
Regression	1	1465.8	1465.8	290.18	0.000
Residual Error	119	601.1	5.1		
Total	120	2067.0			

The standard error of the estimate is the first entry in the sixth line of the computer output. It is reported as s = 2.248. Presuming that the variables DISP and MPG satisfy the assumptions for regression inferences, the standard error of the estimate, 2.248, provides an estimate for the common population standard deviation, σ, of miles per gallon for all cars with a particular engine displacement.

(b) The two graphs that result follow.

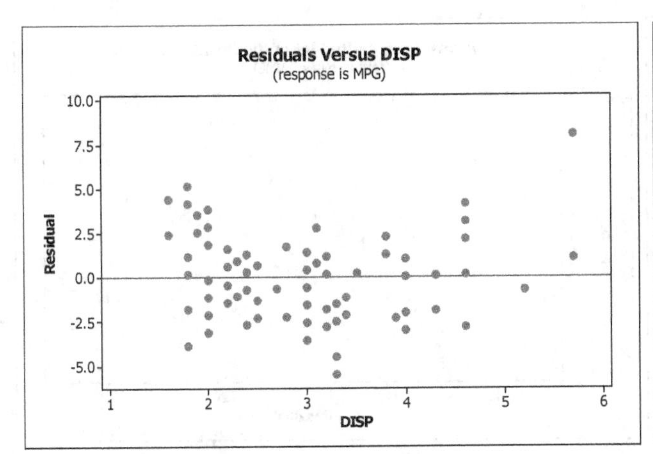

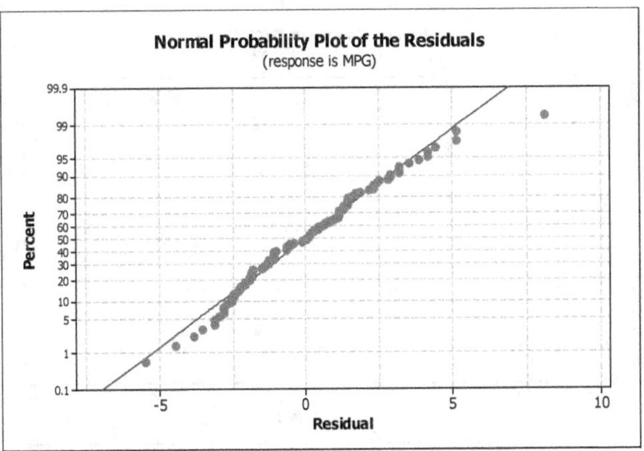

(c) The residuals plotted against engine displacement display a concave upward curve, indicating that the linear model is not appropriate for the data. Thus, Assumption 1 is violated and the statement made in part (a) interpreting the meaning of the standard error does not hold.

15.43 Using Minitab, with the data in columns named DIAMETER and VOLUME, we choose **Stat ▶ Regression ▶ Regression...**, select VOLUME in the **Response** text box, select DIAMETER in the **Predictors** text box, select **Residuals** from the **Storage** check-box list. Click the **Graphs...** button, select the **Regular** option button from the **Residuals for Plots** list, select the **Normal plot of residuals** check box from the **Residual Plots** list, click in the **Residuals versus the variables** text box and specify DIAMETER, click **OK**, and click **OK**. The results are

(a)

```
The regression equation is
VOLUME = - 41.6 + 6.84 DIAMETER

Predictor        Coef       StDev          T        P
Constant      -41.568       3.427     -12.13    0.000
DIAMETER       6.8367      0.2877      23.77    0.000

S = 9.875        R-Sq = 89.3%      R-Sq(adj) = 89.1%

Analysis of Variance

Source            DF          SS          MS         F        P
Regression         1       55083       55083    564.88    0.000
Residual Error    68        6631          98
Total             69       61714
```

The standard error of the estimate is the first entry in the sixth line of the computer output. It is reported as s = 9.875. Presuming that the variables DIAMETER and VOLUME satisfy the assumptions for regression inferences, the standard error of the estimate, 9.875, provides an estimate for the common population standard deviation, σ, of volumes for all trees having a particular diameter at breast height.

(b) The two graphs that result follow.

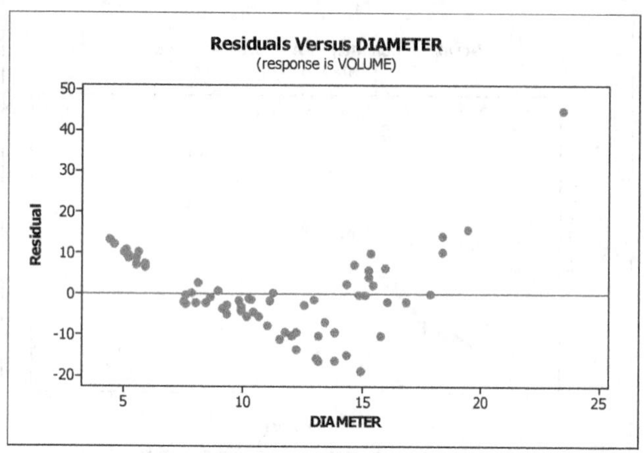

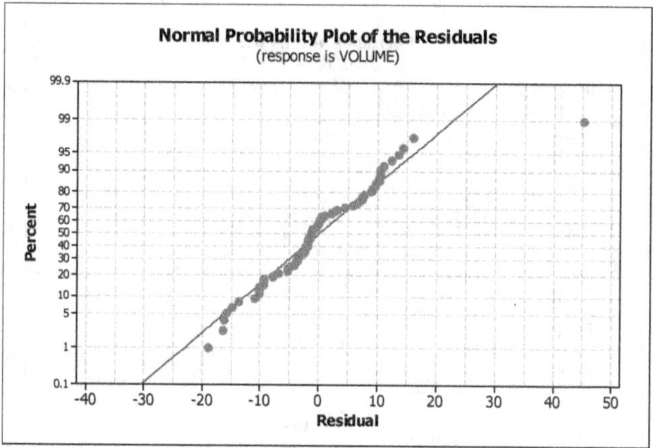

(c) The residuals plotted against diameter are displayed in a curved concave upward pattern, indicating that the linear model is not appropriate. This is a violation of Assumption 1. In addition, both plots indicate the presence of an outlier. Finally, it appears that there is more variability in the center of the first graph than at the left side, so Assumption 2 is also violated. Removal of the outlier will not change the curved pattern of the other residuals or the unequal variation in the residuals, so this data should not be analyzed using a linear model.

Exercises 15.2

15.45 normal distribution with mean β_1 = -3.5

15.47 We can also use the coefficient of determination, r^2, and the linear correlation coefficient, r, as a basis for a test to decide whether a regression equation is useful for prediction.

15.49 From Exercise 15.13, we have $s_e = 2.44949$.

Also, $\sum x_i = 6;\ \sum x_i^2 = 14;\ n=3;\ S_{xx} = \sum x_i^2 - \left(\sum x_i\right)^2/n = 14 - 6^2/3 = 2$

(a) $t = \dfrac{b_1}{s_e/\sqrt{S_{xx}}} = \dfrac{-2}{2.44949/\sqrt{2}} = -1.155$; the critical values with 1 df are ± 6.314. Since $-6.314 < t < 6.314$, we cannot reject $\beta_1 = 0$ and conclude that x is not useful for predicting y.

(b) The 90% confidence interval for the slope of the regression line is

$$b_1 \pm t_{\alpha/2}\frac{s_e}{\sqrt{S_{xx}}} = -2 \pm 6.314\frac{2.44949}{\sqrt{2}} = (-12.936, 8.936)$$

15.51 From Exercise 15.15, we have $s_e = 1.732$.

Also, $\sum x_i = 10;\ \sum x_i^2 = 30;\ n=4;\ S_{xx} = \sum x_i^2 - \left(\sum x_i\right)^2/n = 30 - 10^2/4 = 5$

(a) $t = \dfrac{b_1}{s_e/\sqrt{S_{xx}}} = \dfrac{2}{1.732/\sqrt{5}} = 2.582$; the critical values with 2 df are ± 2.920.

Since $-2.920 < t < 2.920$, we cannot reject $\beta_1 = 0$ and conclude that x is not useful for predicting y.

(b) The 90% confidence interval for the slope of the regression line is

$$b_1 \pm t_{\alpha/2}\frac{s_e}{\sqrt{S_{xx}}} = 2 \pm 2.920\frac{1.732}{\sqrt{5}} = (-0.260, 4.262)$$

15.53 From Exercise 15.17, we have $s_e = 1.414$.

Also, $\sum x_i = 15$; $\sum x_i^2 = 49$; n=5; $S_{xx} = \sum x_i^2 - \left(\sum x_i\right)^2/n = 49 - 15^2/5 = 4$

(a) $t = \dfrac{b_1}{s_e/\sqrt{S_{xx}}} = \dfrac{-1}{1.414/\sqrt{4}} = -1.414$; the critical values with 3 df are ± 2.353.

Since $-2.353 < t < 2.353$, we cannot reject $\beta_1 = 0$ and conclude that x is not useful for predicting y.

(b) The 90% confidence interval for the slope of the regression line is

$$b_1 \pm t_{\alpha/2}\frac{s_e}{\sqrt{S_{xx}}} = -1 \pm 2.353\frac{1.414}{\sqrt{4}} = (-2.664, 0.664)$$

15.55 From Exercise 15.19, we have $s_e = 2.449$.

Also, $\sum x_i = 6$; $\sum x_i^2 = 14$; n=3; $S_{xx} = \sum x_i^2 - \left(\sum x_i\right)^2/n = 14 - 6^2/3 = 2$

(a) $t = \dfrac{b_1}{s_e/\sqrt{S_{xx}}} = \dfrac{2}{2.449/\sqrt{2}} = 1.154$; the critical values with 1 df are ± 6.314.

Since $-6.314 < t < 6.314$, we cannot reject $\beta_1 = 0$ and conclude that x is not useful for predicting y.

(b) The 90% confidence interval for the slope of the regression line is

$$b_1 \pm t_{\alpha/2}\frac{s_e}{\sqrt{S_{xx}}} = 2 \pm 6.314\frac{2.449}{\sqrt{2}} = (-8.938, 12.938)$$

15.57 From Exercise 15.21, we have $s_e = 1.88193$.

Also, $\sum x_i = 15$; $\sum x_i^2 = 69$; n=5; $S_{xx} = \sum x_i^2 - \left(\sum x_i\right)^2/n = 69 - 15^2/5 = 24$

(a) $t = \dfrac{b_1}{s_e/\sqrt{S_{xx}}} = \dfrac{-0.625}{1.88193/\sqrt{24}} = -1.627$; the critical values with 3 df are ± 2.353. Since $-2.353 < t < 2.353$, we cannot reject $\beta_1 = 0$ and conclude that x is not useful for predicting y.

(b) The 90% confidence interval for the slope of the regression line is

$$b_1 \pm t_{\alpha/2}\frac{s_e}{\sqrt{S_{xx}}} = -0.625 \pm 2.353\frac{1.88193}{\sqrt{24}} = (-1.512, 0.282)$$

15.59 From Exercise 14.59, $\sum x = 41$, $\sum x^2 = 199$, and $b_1 = -27.9029$. From Exercise 15.29, $s_e = 14.2464$.

Step 1: H_0: $\beta_1 = 0$, H_a: $\beta_1 \neq 0$

Step 2: $\alpha = 0.10$

Step 3: $t = \dfrac{b_1}{s_e/\sqrt{\sum x^2 - (\sum x)^2/n}} = \dfrac{-27.9029}{14.2465/\sqrt{199 - 41^2/10}} = -10.887$

Step 4: df = n - 2 = 8; critical values = ± 1.860
For the p-value approach, $p < 0.01$.

Step 5: Since -10.887 < -1.860, reject H_0.
Because the p-value is less than the significance level of 0.10, we can reject H_0.

Step 6: At 10% significance level, the data provide sufficient evidence to conclude that the slope of the population regression line is not zero and, hence, that age is useful as a predictor of price for Corvettes.

15.61 From Exercise 14.61, $\sum x = 723$, $\sum x^2 = 48747$, and $b_1 = 0.16285$. From Exercise 15.31, $s_e = 5.418$.

Step 1: H_0: $\beta_1 = 0$, H_a: $\beta_1 \neq 0$

Step 2: $\alpha = 0.05$

Step 3: $$t = \frac{b_1}{s_e / \sqrt{\sum x^2 - (\sum x)^2 / n}} = \frac{0.16285}{5.418 / \sqrt{48747 - 723^2 / 11}} = 1.053$$

Step 4: df = n - 2 = 9; critical values = ± 2.262
For the p-value approach, p > 0.20.

Step 5: Since 1.053 < 2.262, do not reject H_0.
Because the p-value is greater than the significance level of 0.05, we cannot reject H_0.

Step 6: At the 5% significance level, we conclude that plant weight is not useful as a predictor of volume of volatile emissions.

15.63 From Exercise 14.63, $\sum x = 117$, $\sum x^2 = 1869$, and $b_1 = -0.846$. From Exercise 15.33, $s_e = 3.538$.

Step 1: H_0: $\beta_1 = 0$, H_a: $\beta_1 \neq 0$

Step 2: $\alpha = 0.01$

Step 3: $$t = \frac{b_1}{s_e / \sqrt{\sum x^2 - (\sum x)^2 / n}} = \frac{-0.846}{3.538 / \sqrt{1869 - 117^2 / 8}} = -3.004$$

Step 4: df = n - 2 = 6; critical values = ± 3.707
For the p-value approach, 0.02 <p < 0.05.

Step 5: Since -3.707 < -3.004 < 3.707, do not reject H_0.
Because the p-value is less than the significance level of 0.01, we cannot reject H_0.

Step 6: At the 1% significance level, we conclude that study-time is not useful as a predictor of test scores of calculus students.

15.65 From Exercise 14.59, $\sum x = 41$, $\sum x^2 = 199$, and $b_1 = -27.9029$. From Exercise 15.29, $s_e = 14.2465$.

Step 1: For a 90% confidence interval, $\alpha = 0.10$. With df = n - 2 = 8, $t_{\alpha/2} = t_{0.05} = 1.860$.

Step 2: The endpoints of the confidence interval for β_1 are

$$b_1 \pm t_{\alpha/2} \cdot s_e / \sqrt{\sum x^2 - (\sum x)^2 / n}$$

$$-27.9029 \pm 1.860 \cdot 14.2465 / \sqrt{199 - 41^2 / 10}$$

$$-27.9029 \pm 4.7670$$

$$-32.6699 \text{ to } -23.1359$$

We can be 90% confident that the yearly decrease in mean price for Corvettes is somewhere between \$2314 and \$3267.

15.67 From Exercise 14.61, $\sum x = 723$, $\sum x^2 = 48747$, and $b_1 = 0.16285$. From Exercise 15.31, $s_e = 5.4177$.

Step 1: For a 95% confidence interval, $\alpha = 0.05$. With df = n - 2 = 9, $t_{\alpha/2} = t_{0.025} = 2.262$.

Step 2: The endpoints of the confidence interval for β_1 are

$$b_1 \pm t_{\alpha/2} \cdot s_e / \sqrt{\sum x^2 - (\sum x)^2 / n}$$

$$0.16285 \pm 2.262 \cdot 5.4177 / \sqrt{48747 - 723^2 / 11}$$

$$0.16285 \pm 0.34997$$

$$-0.187 \ to \ 0.513$$

We can be 95% confident that the increase in mean volatile plant emissions per one gram increase in weight is somewhere between -0.187 and 0.513 hundred nanograms.

15.69 From Exercise 14.63, $\sum x = 117$, $\sum x^2 = 1869$, and $b_1 = -0.846$. From Exercise 15.33, $s_e = 3.53816$.

Step 1: For a 99% confidence interval, $\alpha = 0.01$. With df = n - 2 = 6, $t_{\alpha/2} = t_{0.005} = 3.707$.

Step 2: The endpoints of the confidence interval for β_1 are

$$b_1 \pm t_{\alpha/2} \cdot s_e / \sqrt{\sum x^2 - (\sum x)^2 / n}$$

$$-0.846 \pm 3.707 \cdot 3.53816 / \sqrt{1869 - 117^2 / 8}$$

$$-0.846 \pm 1.044$$

$$-1.890 \ to \ 0.198$$

We can be 99% confident that the change in test score per one hour increase in study-time for calculus students is somewhere between -1.890 and 0.198.

15.71 (a) In Exercise 15.35, part (c), we determined that it was reasonable to consider Assumptions 1-3 for regression inferences met by these variables.

(b) Part of the Minitab output from Exercise 15.35 is reproduced below.

Term	Coef	SE Coef	T-Value	P-Value	VIF
Constant	5.7	14.1	0.40	0.690	
INAUGURATION	1.179	0.255	4.63	0.000	1.00

Regression Equation

The t value and the P-value for the regression t-test are 4.63 and 0.000, respectively. Since 0.000 < 0.05, we conclude that the inauguration ages are useful for predicting the presidents' death ages.

15.73 (a) In Exercise 15.37, part (c), we determined that it was temporarily reasonable to consider Assumptions 1-3 for regression inferences met by these variables, although Assumption 3 was questionable. Note: In Exercise 14.73, we concluded that the observations did not follow a linear pattern. This exercise will help to resolve the conflict.

(b) Part of the Minitab output from Exercise 15.37 is reproduced below.

Predictor	Coef	SE Coef	T	P
Constant	292.0	140.3	2.08	0.043
LOT SIZE	67.11	59.87	1.12	0.269

The t value and the P-value for the regression t-test are 1.12 and 0.269, respectively. Since 0.269 > 0.05, we conclude that the lot size is not useful for predicting the home values.

15.75 (a) In Exercise 15.39, part (c), we determined that it was reasonable to consider Assumptions 1-3 for regression inferences met by these variables.

(b) Part of the Minitab output from Exercise 15.39 is reproduced below.

Predictor	Coef	SE Coef	T	P
Constant	-7.572	1.786	-4.24	0.000
HIGH	0.91685	0.02967	30.91	0.000

The t value and the P-value for the regression t-test are 30.91 and 0.000, respectively. Since 0.000 < 0.05, we conclude that average high temperatures are useful for predicting average low temperatures in January.

15.77 (a) In Exercise 15.41, part (c), we determined that it was not reasonable to consider Assumption 1 for regression inferences met by these variables. The residual plot was convex upward. Therefore, parts (b) and (c) are omitted.

15.79 (a) In Exercise 15.43, part (c), we determined that it was not reasonable to consider Assumptions 1 and 2 for regression inferences met by these variables. Therefore, parts (b) and (c) are omitted.

Exercises 15.3

15.81 \$11,443

15.83 From Exercise 15.13, $\hat{y}=1-2x$. From Exercise 15.49, $s_e=2.44949$ *and* $S_{xx}=2$.

(a) The point estimate for the conditional mean of y at x = 2 is $\hat{y}_p=1-2x=1-2(2)=-3$.

(b) The 95% confidence interval for the conditional mean at x = 2 is

$$\hat{y}_p \pm t_{\alpha/2}s_e\sqrt{\frac{1}{n}+\frac{(x_p-\sum x_i/n)^2}{S_{xx}}}=-3\pm 12.706(2.44949)\sqrt{\frac{1}{3}+\frac{(2-6/3)^2}{2}}=-3\pm 17.97$$

$$=(-20.97,14.97)$$

(c) The predicted value of y at x = 2 is $\hat{y}_p=1-2x=1-2(2)=-3$.

(d) The 95% prediction interval for y at x = 2 is

$$\hat{y}_p \pm t_{\alpha/2}s_e\sqrt{1+\frac{1}{n}+\frac{(x_p-\sum x_i/n)^2}{S_{xx}}}=-3\pm 12.706(2.44949)\sqrt{1+\frac{1}{3}+\frac{(2-6/3)^2}{2}}$$

$$=-3\pm 35.94=(-38.94,32.94)$$

15.85 From Exercise 15.15, $\hat{y}=-3+2x$. From Exercise 15.51, $s_e=1.73205$ *and* $S_{xx}=24$.

(a) The point estimate for the conditional mean of y at x = 4 is $\hat{y}_p=-3+2x=-3+2(4)=5$.

(b) The 95% confidence interval for the conditional mean at x = 4 is

$$\hat{y}_p \pm t_{\alpha/2}s_e\sqrt{\frac{1}{n}+\frac{(x_p-\sum x_i/n)^2}{S_{xx}}}=5\pm 4.303(1.73205)\sqrt{\frac{1}{4}+\frac{(4-10/4)^2}{5}}=5\pm 6.24$$

$$=(-1.24,11.24)$$

(c) The predicted value of y at x = 4 is $\hat{y}_p=-3+2x=-3+2(4)=5$.

(d) The 95% prediction interval for y at x = 4 is

$$\hat{y}_p \pm t_{\alpha/2} s_e \sqrt{1+\frac{1}{n}+\frac{(x_p-\sum x_i/n)^2}{S_{xx}}} = 5 \pm 4.303(1.73205)\sqrt{1+\frac{1}{4}+\frac{(4-10/4)^2}{5}}$$
$$= 5 \pm 9.72 = (-4.72, 14.72)$$

15.87 From Exercise 15.17, $\hat{y} = 5 - x$. From Exercise 15.53, $s_e = 1.414$ *and* $S_{xx} = 4$.

(a) The point estimate for the conditional mean of y at x = 3 is $\hat{y}_p = 5 - x = 5 - 3 = 2$.

(b) The 95% confidence interval for the conditional mean at x = 3 is

$$\hat{y}_p \pm t_{\alpha/2} s_e \sqrt{\frac{1}{n}+\frac{(x_p-\sum x_i/n)^2}{S_{xx}}} = 2 \pm 3.182(1.414)\sqrt{\frac{1}{5}+\frac{(3-15/5)^2}{4}} = 2 \pm 2.01$$
$$= (-0.01, 4.01)$$

(c) The predicted value of y at x = 3 is $\hat{y}_p = 5 - x = 5 - 3 = 2$.

(d) The 95% prediction interval for y at x = 3 is

$$\hat{y}_p \pm t_{\alpha/2} s_e \sqrt{1+\frac{1}{n}+\frac{(x_p-\sum x_i/n)^2}{S_{xx}}} = 2 \pm 3.182(1.414)\sqrt{1+\frac{1}{5}+\frac{(3-15/5)^2}{4}}$$
$$= 2 \pm 4.93 = (-2.93, 6.93)$$

15.89 From Exercise 15.19, $\hat{y} = 1 + 2x$. From Exercise 15.55, $s_e = 2.449$ *and* $S_{xx} = 2$.

(a) The point estimate for the conditional mean of y at x = 2 is $\hat{y}_p = 1 + 2x = 1 + 2(2) = 5$.

(b) The 95% confidence interval for the conditional mean at x = 2 is

$$\hat{y}_p \pm t_{\alpha/2} s_e \sqrt{\frac{1}{n}+\frac{(x_p-\sum x_i/n)^2}{S_{xx}}} = 5 \pm 12.706(2.449)\sqrt{\frac{1}{3}+\frac{(2-6/3)^2}{2}} = 5 \pm 17.97$$
$$= (-12.97, 22.97)$$

(c) The predicted value of y at x = 2 is $\hat{y}_p = 1 + 2x = 1 + 2(2) = 5$.

(d) The 95% prediction interval for y at x = 2 is

$$\hat{y}_p \pm t_{\alpha/2} s_e \sqrt{1+\frac{1}{n}+\frac{(x_p-\sum x_i/n)^2}{S_{xx}}} = 5 \pm 12.706(2.449)\sqrt{1+\frac{1}{3}+\frac{(2-6/3)^2}{2}}$$
$$= 5 \pm 35.95 = (-30.95, 40.95)$$

15.91 From Exercise 15.21, $\hat{y} = 2.875 + 0.625x$. From Exercise 15.57, $s_e = 1.88193$ *and* $S_{xx} = 24$.

(a) The point estimate for the conditional mean of y at x = 3 is $\hat{y}_p = 2.875 + 0.625x = 2.875 + 0.625(3) = 4.75$.

(b) The 95% confidence interval for the conditional mean at x = 3 is

$$\hat{y}_p \pm t_{\alpha/2} s_e \sqrt{\frac{1}{n}+\frac{(x_p-\sum x_i/n)^2}{S_{xx}}} = 4.75 \pm 3.182(1.88193)\sqrt{\frac{1}{5}+\frac{(3-15/5)^2}{24}} = 4.75 \pm 2.68$$
$$= (2.07, 7.43)$$

(c) The predicted value of y at x = 3 is
$\hat{y}_p = 2.875 + 0.625x = 2.875 + 0.625(3) = 4.75$.

(d) The 95% prediction interval for y at x = 3 is

$$\hat{y}_p \pm t_{\alpha/2} s_e \sqrt{1 + \frac{1}{n} + \frac{(x_p - \sum x_i / n)^2}{S_{xx}}} = 4.75 \pm 3.182(1.88193)\sqrt{1 + \frac{1}{5} + \frac{(3 - 15/5)^2}{24}}$$

$$= 4.75 \pm 6.56 = (-1.81, 11.31)$$

15.93 From Exercise 14.59, $\hat{y}$ = 456.602 - 27.9029x, $\sum x = 41$, and $\sum x^2 = 199$. From Exercise 15.29, s_e = 14.2465.

(a) $\hat{y}_p$ = 456.602 - 27.9029(4) = 344.9904 = $34,499.

(b) Step 1: For a 90% confidence interval, α = 0.10.
With df = n - 2 = 8, $t_{\alpha/2} = t_{0.05}$ = 1.860.

Step 2: $\hat{y}_p$ = 344.9904

Step 3: The endpoints of the confidence interval are

$$\hat{y}_p \pm t_{\alpha/2} \cdot s_e \cdot \sqrt{\frac{1}{n} + \frac{(x_p - \sum x / n)^2}{\sum x^2 - (\sum x)^2 / n}}$$

$$= 344.9904 \pm 1.860 \cdot 14.2465 \sqrt{\frac{1}{10} + \frac{(4 - 41/10)^2}{199 - 41^2/10}}$$

$$= 344.9904 \pm 8.3931 = (336.60, 353.38)$$

The interpretation of this interval is as follows: We can be 90% confident that the mean price of all four-year-old Corvettes is somewhere between $33,660 and $35,338.

(c) This is the same as the answer in part (a): $\hat{y}_p$ = 344.9904.

(d) For a 90% prediction interval, Steps 1 and 2 are the same as Steps 1 and 2, respectively, in part (b). Thus, they are not repeated here. Only Step 3 is presented.

Step 3: The endpoints of the prediction interval are

$$\hat{y}_p \pm t_{\alpha/2} \cdot s_e \cdot \sqrt{1 + \frac{1}{n} + \frac{(x_p - \sum x / n)^2}{\sum x^2 - (\sum x)^2 / n}}$$

$$= 344.9904 \pm 1.860 \cdot 14.265 \sqrt{1 + \frac{1}{10} + \frac{(4 - 41/10)^2}{199 - 41^2/10}}$$

$$= 344.9904 \pm 27.7959 = (317.20, 372.78)$$

The interpretation of this interval is as follows: We can be 90% certain that the price of a randomly selected four-year-old Corvette will be somewhere between $31,720 and $37,278.

(e)

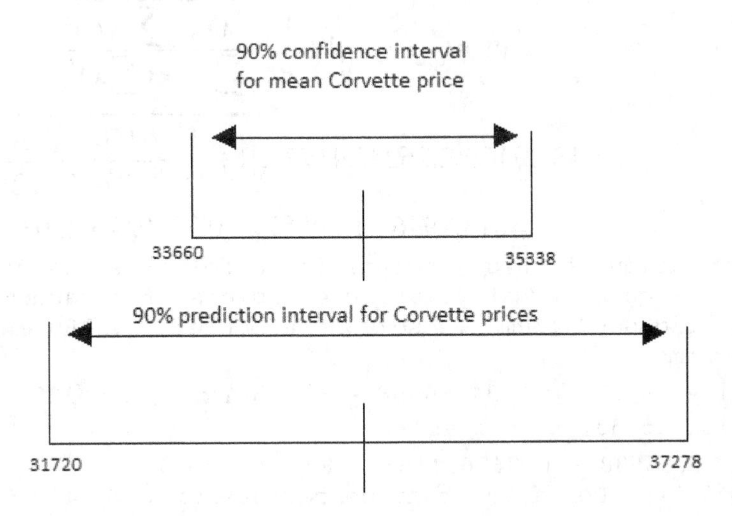

(f) The error in the estimate of the mean price of four-year-old Corvettes is due only to the fact that the population regression line is being estimated by a sample regression line; whereas, the error in the prediction of the price of a randomly selected four-year-old Corvette is due to that fact plus the variation in prices for four-year-old Corvettes.

15.95 From Exercise 14.61, $\hat{y}$ = 3.523688 + 0.162848x, $\sum x$ = 726, and $\sum x^2$ = 48747. From Exercise 15.31, s_e = 5.4177.

(a) $\hat{y}_p$ = 3.523688 + 0.162848(60) = 13.2946

(b) Step 1: For a 95% confidence interval, α = 0.05. With df = n - 2 = 9, $t_{\alpha/2}$ = $t_{0.025}$= 2.262.

Step 2: $\hat{y}_p$ = 13.2946

Step 3: The endpoints of the confidence interval are

$$\hat{y}_p \pm t_{\alpha/2} \cdot s_e \cdot \sqrt{\frac{1}{n} + \frac{(x_p - \sum x/n)^2}{\sum x^2 - (\sum x)^2/n}}$$

$$= 13.2946 \pm 2.262 \cdot 5.4177 \cdot \sqrt{\frac{1}{11} + \frac{(60 - 723/11)^2}{48747 - 723^2/11}}$$

$$= 813.2946 \pm 4.2036 = (9.0910, 17.4962)$$

The interpretation of this interval is as follows: We can be 95% confident that the mean quantity of volatile emissions of all plants that weigh 60 grams is somewhere between 9.0910 and 17.4982 hundred nanograms.

(c) This is the same as the answer in part (a): $\hat{y}_p$ = 13.2946.

(d) For a 95% prediction interval, Steps 1 and 2 are the same as Steps 1 and 2, respectively, in part (b). Thus, they are not repeated here. Only Step 3 is presented.

Step 3: The endpoints of the prediction interval are

$$\hat{y}_p \pm t_{\alpha/2} \cdot s_e \cdot \sqrt{1+\frac{1}{n}+\frac{(x_p-\sum x/n)^2}{\sum x^2-(\sum x)^2/n}}$$

$$=13.2946 \pm 2.262 \cdot 5.4177 \cdot \sqrt{1+\frac{1}{11}+\frac{(60-723/11)^2}{48747-723^2/11}}$$

$$=813.2946 \pm 12.9557=(0.3389,\ 26.2503)$$

The interpretation of this interval is as follows: We can be 95% certain that the quantity of volatile emissions of a randomly selected plant that weighs 60 grams is somewhere between 0.3389 and 26.2503 hundred nanograms.

15.97 From Exercise 14.63, $\hat{y}$ = 94.86698 - 0.84561x, $\sum x$ = 117, and $\sum x^2$ = 1869. From Exercise 15.33, s_e = 3.5382.

(a) $\hat{y}_p$ = 94.86698 - 0.84561(15) = 82.1828

(b) Step 1: For a 99% confidence interval, α = 0.01.

With df = n - 2 = 6, $t_{\alpha/2}$ = $t_{0.005}$= 3.707.

Step 2: $\hat{y}_p$ = 82.1828

Step 3: The endpoints of the confidence interval are

$$\hat{y}_p \pm t_{\alpha/2} \cdot s_e \cdot \sqrt{\frac{1}{n}+\frac{(x_p-\sum x/n)^2}{\sum x^2-(\sum x)^2/n}}$$

$$=82.1828 \pm 3.707 \cdot (3.5382) \cdot \sqrt{\frac{1}{8}+\frac{(15-117/8)^2}{1869-117^2/8}}$$

$$=82.1828 \pm 4.6537=(77.5291, 86.8365)$$

The interpretation of this interval is as follows: We can be 99% confident that the mean calculus test score of students who study 15 hours is somewhere between 77.53 and 86.84 mm.

(c) This is the same as the answer in part (a): $\hat{y}_p$ = 82.1828.

(d) For a 99% prediction interval, Steps 1 and 2 are the same as Steps 1 and 2, respectively, in part (b). Thus, they are not repeated here. Only Step 3 is presented.

Step 3: The endpoints of the prediction interval are

$$\hat{y}_p \pm t_{\alpha/2} \cdot s_e \cdot \sqrt{1+\frac{1}{n}+\frac{(x_p-\sum x/n)^2}{\sum x^2-(\sum x)^2/n}}$$

$$=82.1828 \pm 3.707 \cdot (3.5382) \cdot \sqrt{1+\frac{1}{8}+\frac{(15-117/8)^2}{1869-117^2/8}}$$

$$=82.1828 \pm 13.9172=(68.2656, 96.0100)$$

The interpretation of this interval is as follows: We can be 99% certain that a randomly selected calculus student who studies 125 hours will have a test score somewhere between 68.27 and 96.01 mm.

15.99 (a) In Exercise 15.35, part (c), we determined that it was reasonable to consider Assumptions 1-3 for regression inferences met by these variables.

(b) Using Minitab, with the data in columns named INAUGURATION and DEATH, we choose **Stat ▶ Regression ▶ Regression...**, select DEATH in the **Response** text box, and select INAUGURATION in the **Predictors** text box. Click the **Options** button, select the **Regular** option button from the **Residuals for Plots** list, enter 53 in the **Prediction intervals for new observations** box and 95 in the **Confidence level** box, click **OK**, and click **OK**. Since much of the output was shown in Exercise 15.35, we show below only that part of the output pertinent to this exercise.

```
Variable      Setting
INAUGURATION       53

     Fit   SE Fit        95% CI              95% PI
68.1679  1.62938  (64.8665, 71.4694)  (48.2282, 88.1076)
```

The point estimate for the conditional mean of DEATH for inauguration age 53 is found under the word 'Fit' in the output as 68.17. This is the same as the predicted score for inauguration at age 53.

(c) The 95% confidence interval for the conditional mean death age for presidents inaugurated at age 53 is (64.87, 71.47). We can be 95% confident that the mean death age of U.S. Presidents inaugurated at age 53 lies somewhere between 64.87 and 71.47 years.

(d) The predicted death age for Presidents inaugurated at age 53 is found under the word 'Fit' in the output as 68.17.

(e) The 95% prediction interval for the death age of a President inaugurated at age 53 is (48.23, 88.11). We can be 95% certain that the death age of a president inaugurated at age 53 will fall somewhere between 48.23 and 88.11. Note: The lower end of this prediction interval can be safely increased to 53 since a president cannot die prior to being inaugurated.

(f) The confidence interval and the prediction interval are both centered at the predicted value for an inauguration age of 53. The prediction interval is longer because there is variation associated with the additional observation that is not present in the confidence interval for the mean.

15.101(a) In Exercise 15.37, part (c), we determined that it was reasonable to consider Assumptions 1-3 for regression inferences met by these variables. Note: In Exercise 14.73, we concluded that the pattern of the observations was NOT linear, meaning that Assumption 1 for regressions inferences was not met by these data. In Exercise 15.73, we found that the evidence for a linear pattern was weak or none. Thus we will not continue with parts (b)-(f).

15.103(a) In Exercise 15.39, part (c), we determined that it was reasonable to consider Assumptions 1-3 for regression inferences met by these variables.

(b) Using Minitab, with the data in columns named HIGH and LOW, we choose **Stat ▶ Regression ▶ Regression...**, select LOW in the **Response** text box, and select HIGH in the **Predictors** text box. Click the **Options** button, select the **Regular** option button from the **Residuals for Plots** list, enter 55 in the **Prediction intervals for new observations** box and 95 in the **Confidence level** box, click **OK**, and click **OK**. Since much of the output was shown in Exercise 15.39, we show below only that part of the output pertinent to this exercise.

```
Predicted Values for New Observations

New
Obs     Fit  SE Fit        95% CI              95% PI
  1  42.855   0.590  (41.669, 44.040)  (34.425, 51.285)
```

The point estimate for the conditional mean January low temperature for an average high temperature of 55 degrees is found under the word 'Fit' in the output as 42.855 degrees. This is the same as the predicted value for an average high temperature of 55 degrees.

(c) The 95% confidence interval for the conditional mean low January temperature for cities with mean high temperature of 55 degrees is (41.669, 44.040). We can be 95% confident that the mean low January temperature for cities with a mean January high temperature of 55 degrees lies somewhere between 41.669 and 44.040 degrees.

(d) The predicted mean low January temperature for a city with a mean high January temperature of 55 degrees is found under the word 'Fit' in the output as 42.855 degrees.

(e) The 95% prediction interval for the mean January low temperature for a city with a mean January high temperature is (34.425, 51.285). We can be 95% certain that the mean low January temperature for a city with a mean January high temperature of 55 degrees will lie somewhere between 34.425 and 51.285 degrees.

(f) The confidence interval and the prediction interval are both centered at the predicted value for a city with a mean January high temperature of 55 degrees. The prediction interval is longer because there is variation associated with the additional observation that is not present in the confidence interval for the mean.

15.105 (a) In Exercise 15.41, part (c), we determined that it was not reasonable to consider Assumption 1 for regression inferences met by these variables. The residual plot was convex upward. Therefore, parts (b)-(f) are omitted.

15.107 (a) In Exercise 15.43, part (c), we determined that it was not reasonable to consider Assumptions 1 and 2 for regression inferences met by these variables. Therefore, parts (b)-(f) are omitted.

15.109 (a)-(b) In Example 15.9 of the text, the 95% confidence interval for the mean price of all Orions of a given age is given by

$$134.69 \pm 2.262 \cdot 12.58 \cdot \sqrt{\frac{1}{11} + \frac{(x_p - 58/11)^2}{326 - 58^2/11}}$$

Substituting the values 2 through 7 into this formula for x_p, we obtain the following confidence intervals and a plot showing the lower and upper confidence limits for the six different values of x_p. The margin of error (half of the difference between the upper and lower confidence limits) is shown in the last column of the table and is plotted against x_p in the second graph.

x_p	Lower Confidence Limit	Upper Confidence Limit	Margin of Error
2	112.25	157.13	22.44
3	117.93	151.45	16.76
4	122.92	146.46	11.77
5	125.94	143.44	8.75
6	124.95	144.43	9.74
7	120.79	148.59	13.90

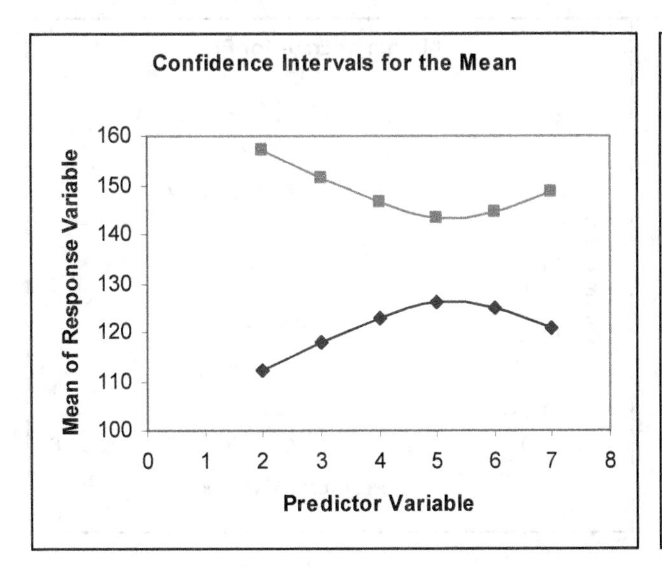

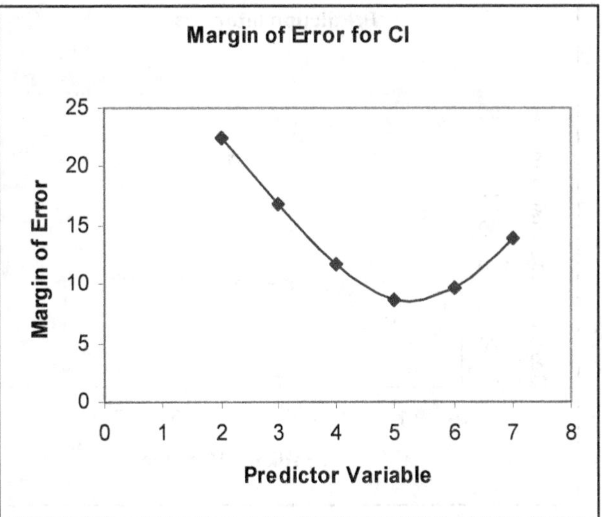

We see from the table and the first plot that the confidence intervals get wider as the predictor variable gets farther from the mean of the predictor variable. This is also seen in the second plot, which shows that the margin of error first decreases as x_p increases to the mean 5.27 of the predictor variable and then increases again as x_p moves away from 5.27.

(c) In Example 15.10 of the text, the 95% prediction interval for an Orion of a given age is given by

$$=134.69 \pm 2.262 \cdot 12.58 \cdot \sqrt{1+\frac{1}{11}+\frac{(x_p-58/11)^2}{326-58^2/11}}$$

Substituting the values 2 through 7 into this formula for x_p, we obtain the following confidence intervals and a plot showing the lower and upper confidence limits for the six different values of x_p. The margin of error (half of the difference between the upper and lower confidence limits) is shown in the last column of the table and is plotted against x_p in the second graph.

x_p	Lower Prediction Limit	Upper Prediction Limit	Margin of Error
2	98.45	170.93	36.24
3	101.67	167.71	33.02
4	103.89	165.49	30.80
5	104.92	164.46	29.77
6	104.61	164.77	30.08
7	103.02	166.36	31.67

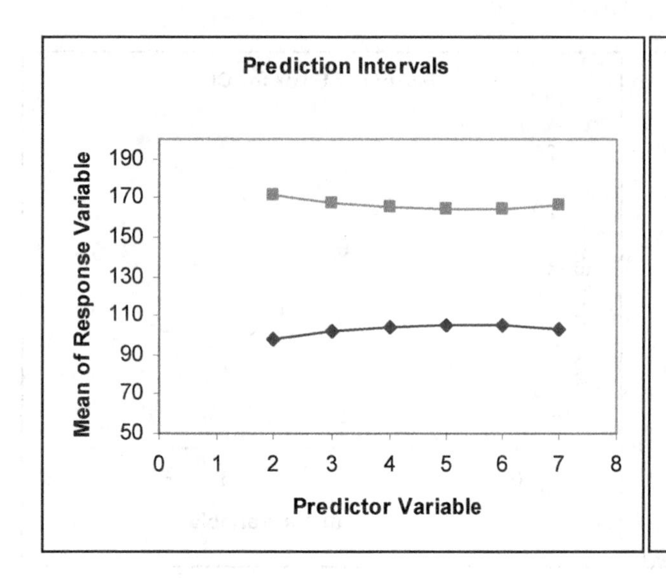

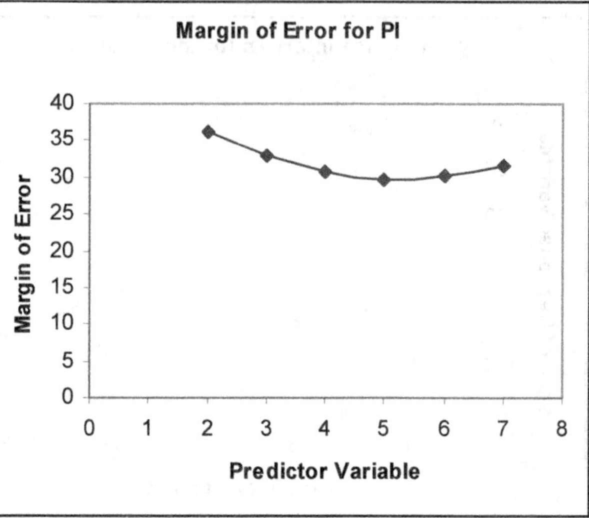

We see from the table and the first plot that the prediction intervals get wider as the predictor variable gets farther from the mean of the predictor variable. This is also seen in the second plot which shows that the margin of error first decreases as x_p increases to the mean 5.27 of the predictor variable and then increases again as x_p moves away from 5.27.

Exercises 15.4

15.111 r, the sample linear correlation coefficient

15.113 No. The variables could be negatively correlated, but not necessarily in a linear pattern.

15.115 uncorrelated

15.117 negatively

15.119 Step 1 $H_0: \rho = 0$, $H_a: \rho < 0$

Step 2 $\alpha = 0.10$

Step 3 From Exercise 14.135, $r = -0.76$, $n = 3$

$$t = \frac{r}{\sqrt{\frac{1-r^2}{n-2}}} = \frac{-0.76}{\sqrt{\frac{1-(-0.76)^2}{1}}} = -1.169$$

Step 4 The critical value is -3.0784
For the P-value approach, $P > 0.10$.

Step 5 Since $-1.169 > -3.078$, do not reject H_0.
Since the P-value is greater than 0.10, do not reject H_0.

15.121 Step 1 $H_0: \rho = 0$, $H_a: \rho > 0$

Step 2 $\alpha = 0.10$

Step 3 From Exercise 14.137, $r = 0.877$, $n = 4$

$$t = \frac{r}{\sqrt{\frac{1-r^2}{n-2}}} = \frac{0.877}{\sqrt{\frac{1-0.877^2}{2}}} = 2.581$$

Step 4 The critical value is 1.886
For the P-value approach, $0.05 < P < 0.10$

Step 5 Since $2.581 > 1.886$, reject H_0.
Since the P-value is less than 0.10, reject H_0.

15.123 Step 1 $H_0: \rho = 0$, $H_a: \rho \neq 0$
Step 2 $\alpha = 0.10$
Step 3 From Exercise 14.139, $r = -0.633$, $n = 5$

$$t = \frac{r}{\sqrt{\frac{1-r^2}{n-2}}} = \frac{-0.633}{\sqrt{\frac{1-(-0.633)^2}{3}}} = -1.41$$

Step 4 The critical values are ± 2.353
For the P-value approach, $P > 0.20$
Step 5 Since $-1.41 > -2.353$, do not reject H_0.
Since the P-value is greater than 0.10, do not reject H_0.

15.125 Step 1 $H_0: \rho = 0$, $H_a: \rho > 0$
Step 2 $\alpha = 0.10$
Step 3 From Exercise 14.141, $r = 0.756$, $n = 3$

$$t = \frac{r}{\sqrt{\frac{1-r^2}{n-2}}} = \frac{0.756}{\sqrt{\frac{1-(0.756)^2}{1}}} = 1.15$$

Step 4 The critical value is 3.078
For the P-value approach, $P > 0.10$
Step 5 Since $1.15 < 3.078$, do not reject H_0.
Since the P-value is greater than 0.10, do not reject H_0.

15.127 Step 1 $H_0: \rho = 0$, $H_a: \rho \neq 0$
Step 2 $\alpha = 0.10$
Step 3 From Exercise 14.143, $r = -0.685$, $n = 5$

$$t = \frac{r}{\sqrt{\frac{1-r^2}{n-2}}} = \frac{-0.685}{\sqrt{\frac{1-(-0.685)^2}{3}}} = -1.629$$

Step 4 The critical values are ± 2.353
For the P-value approach, $P > 0.20$.
Step 5 Since $-2.353 < -1.629 < 2.353$, do not reject H_0.
Since the P-value is greater than 0.10, do not reject H_0.

15.129 From Exercise 14.145, $r = -0.96787$ and $n = 10$.
Step 1: $H_0: \rho = 0$ $H_a: \rho < 0$
Step 2: $\alpha = 0.05$
Step 3:
$$t = \frac{r}{\sqrt{\frac{1-r^2}{n-2}}} = \frac{-0.96787}{\sqrt{\frac{1-(-0.96787)^2}{10-2}}} = -10.8870$$

Step 4: The critical value for $n - 2 = 8$ df is -1.860.
For the p-value approach, $p < 0.005$.
Step 5: Since $-10.8870 < -1.860$, reject H_0 and conclude that $\rho < 0$.
Since $p < 0.05$, reject H_0 and conclude that $\rho < 0$.

15.131 From Exercise 14.147, $r = 0.33107$ and $n = 11$.
Step 1: $H_0: \rho = 0$ $H_a: \rho \neq 0$
Step 2: $\alpha = 0.05$
Step 3:
$$t = \frac{r}{\sqrt{\frac{1-r^2}{n-2}}} = \frac{0.33107}{\sqrt{\frac{1-(0.33107)^2}{11-2}}} = 1.0526$$

Step 4: The critical values for $n - 2 = 9$ df are ± 2.262.
For the p-value approach, $p > 0.200$.

Step 5: Since $-2.262 < 1.0526 < 2.262$, do not reject H_0 and conclude that $\rho = 0$ is reasonable.

Since $p > 0.05$, do not reject H_0 and conclude that $\rho = 0$ is reasonable.

15.133 From Exercise 14.149, $r = -0.7749$ and $n = 8$.

Step 1: H_0: $\rho = 0$ H_a: $\rho < 0$

Step 2: $\alpha = 0.01$

Step 3: $$t = \frac{r}{\sqrt{\frac{1-r^2}{n-2}}} = \frac{-0.7749}{\sqrt{\frac{1-(-0.7749)^2}{8-2}}} = -3.003$$

Step 4: The critical value for $n - 2 = 6$ df is -3.143.
For the p-value approach, $0.010 < p < 0.025$.

Step 5: Since $-3.003 > 3.143$, do not reject H_0. There is not sufficient evidence to conclude that $\rho \neq 0$.

Since $p > 0.01$, do not reject H_0.

15.135 In Exercise 15.35, part (c), we determined that it was reasonable to consider Assumptions 1-3 for regression inferences met by these variables.

Using Minitab, choose **Stat ▶ Basic Statistics ▶ Correlation...**, select INAUGURATION and DEATH in the **Variables** text box, and click **OK**. The result is

```
Pearson correlation of INAUGURATION and DEATH = 0.606
P-Value = 0.000
```

Note: The p-value shown is for a two-tailed test. The p-value for the right-tail test would be even smaller. Since the p-value < 0.05, reject H_0 and conclude that the inauguration ages and death ages of U.S. Presidents are positively linearly correlated.

15.137 In Exercise 15.73, part (a), we determined that it was not reasonable to consider Assumptions 1 for regression inferences met by these variables. The correlation t-test should not be used. Over several exercises, there was some controversy regarding this decision. If you choose to continue with the correlation t-test, the results from Minitab are

```
Pearson correlation of LOT SIZE and VALUE = 0.170
P-Value = 0.269
```

The high P-value confirms our decision not to reject $\rho = 0$.

15.139 In Exercise 15.35, part (c), we determined that it was reasonable to consider Assumptions 1-3 for regression inferences met by these variables.

Using Minitab, choose **Stat ▶ Basic Statistics ▶ Correlation...**, select HIGH and LOW in the **Variables** text box, and click **OK**. The result is

```
Pearson correlation of HIGH and LOW = 0.976
P-Value = 0.000
```

Since the P-value < 0.05, reject $\rho = o$.

15.141 In Exercise 15.37, part (c), we determined that it was not reasonable to consider Assumption 1 for regression inferences met by these variables. The correlation t-test should not be used.

15.143 In Exercise 15.39, part (c), we determined that it was not reasonable to consider Assumptions 1 and 2 for regression inferences met by these variables. The correlation t-test should not be used.

Review Problems for Chapter 15

1. (a) conditional
   (b) The four assumptions for regression inferences are:
      (1) *Population regression line:* There is a line $y = \beta_0 + \beta_1 x$ such that, for each x-value, the mean of the corresponding population of y-values lies on that line.
      (2) *Equal standard deviations:* The standard deviation σ of the population of y-values corresponding to a particular x-value is the same, regardless of the x-value.
      (3) *Normality:* For each x-value, the corresponding population of y-values is normally distributed.
      (4) *Independence:* The observations of the response variable are independent of one another.
2. (a) The slope of the sample regression line, b_1
   (b) The y-intercept of the sample regression line, b_0
   (c) s_e
3. We used a plot of the residuals against the values of the predictor variable x and a normal probability plot of the residuals. In the first plot, the residuals should lie in a horizontal band centered on and symmetric about the x-axis. In the second plot, the points should lie roughly in a line.
4. (a) A residual plot showing curvature indicates that the first assumption, that of linearity, is probably not valid.
   (b) This type of plot indicates that the second assumption, that of constant standard deviation, is probably not valid.
   (c) A normal probability plot with extreme curvature indicates that the third assumption, that the conditional distribution of the response variable is normally distributed, is not valid.
   (d) A normal probability plot that is roughly linear, but shows outliers may be indicating that: the linear model is appropriate for most values of x, but not for all; or, that the standard deviation is not constant for all values of x, allowing for a few y values to lie far from the regression line; or, that a few data values are 'faulty', that is, the experimental conditions represented by the value of the x variable were not as they should have been or that the y value was in error.
5. If we reject the null hypothesis, we are claiming that β_1 is not zero. This means that different values of x will lead to different values of y. Hence, the regression equation is useful for making predictions.
6. t (for b_1), r, r^2
7. No. The best estimate of conditional mean of the response variable and the best prediction of a single future value of y are the same.
8. A confidence interval estimates the value of a parameter; a prediction interval is used to predict a future value of a random variable.
9. ρ
10. (a) If $\rho > 0$, the variables are positively correlated, that is, there is a tendency for one variable to increase linearly as the other one increases.
    (b) If we know only that $\rho \neq 0$, the two variables are linearly correlated, that is, there is a linear relationship between them.
    (c) If $\rho < 0$, the variables are negatively correlated, that is, there is a tendency for one variable to decrease linearly as the other one increases.
11. If the regression assumptions are satisfied, then there is a linear relationship between the student/faculty ratio and the graduation rate, the conditional standard deviation of the graduation rate is the same for all values of the student/faculty ratio, the conditional distribution of the

graduation rate is a normal distribution for all values of the student/faculty ratio, and the values of the graduation rate are independent of each other.

12. (a) To determine b_0 and b_1 for the regression line, construct a table of values for x, y, x^2, xy and their sums. A column for y^2 is also presented for later use.

x	y	x^2	Xy	y^2
16	45	256	720	2025
20	55	400	1100	3025
17	70	289	1190	4900
19	50	361	950	2500
22	47	484	1034	2209
17	46	289	782	2116
17	50	289	850	2500
17	66	289	1122	4356
10	26	100	260	676
18	60	324	1080	3600
173	515	3081	9088	27907

$S_{xx} = 3081 - 173^2/10 = 88.10$

$S_{xy} = 9088 - (173)(515)/10 = 178.50$

$b_1 = 178.50/88.10 = 2.0261$

$b_0 = 515/10 - 2.0261(173/10) = 16.4484$

$\hat{y}_p = 16.448 + 2.0261x$ is the equation of the regression line.

(b) $S_{yy} = 27907 - 515^2/10 = 1384.50$

$SSE = S_{yy} - S^2_{xy}/S_{xx} = 1384.50 - 178.50^2/88.10 = 1022.8400$

$$s_e = \sqrt{\frac{SSE}{n-2}} = \sqrt{\frac{1922.8400}{10-2}} = 11.3073$$

This value of s_e indicates that, roughly speaking, the predicted values of the graduation rate differ from the observed values of the graduation rate by about 11.31.

(c) Presuming that the variables Student/Faculty ratio and Graduation rate satisfy the assumptions for regression inferences, the standard error of the estimate, $s_e = 11.31$, provides an estimate for the common population standard deviation, σ, of graduation rates of entering freshmen at universities with any particular student/faculty ratio.

13. For each value of x, we compute $e = y - \hat{y}$ in the following table and plot e against x.

x	e
16	-3.87
20	-1.97
17	19.11
19	-4.94
22	-14.02
17	-4.89
17	-0.89
17	15.11
10	-10.71
18	7.08

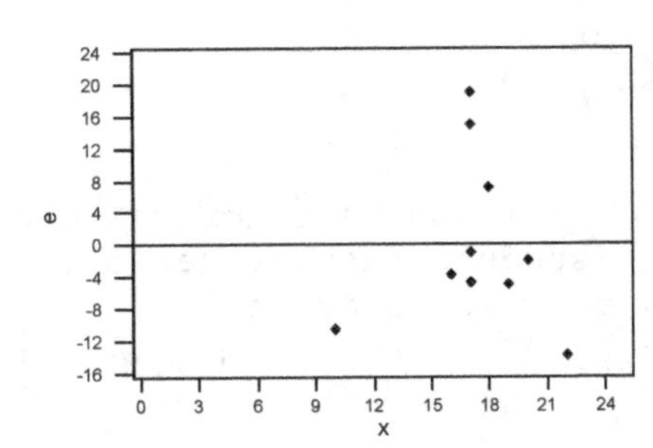

Residual	Normal score
-14.02	-1.55
-10.71	-1.00
-4.94	-0.65
-4.89	-0.37
-3.87	-0.12
-1.97	0.12
-0.89	0.37
7.08	0.65
15.11	1.00
19.11	1.55

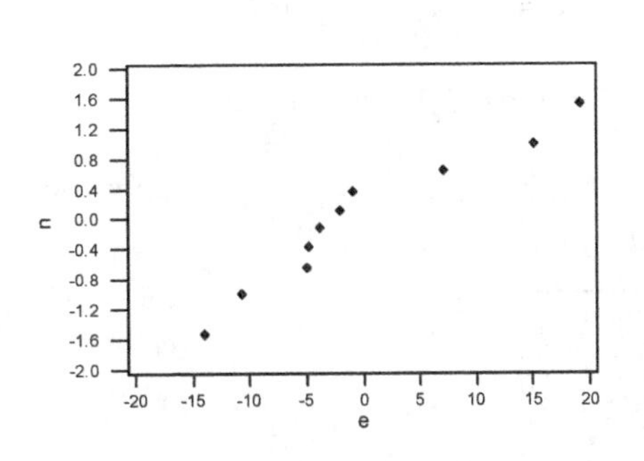

The small sample size makes it difficult to evaluate the plot of e against x. While there are no obvious patterns in the first plot, there are two points (in the lower left and right corners of the plot) that are cause for concern. The normal probability plot is not far from linear.

14. (a) From Review Problem 11, $\sum x = 173$, $\sum x^2 = 3{,}081$, and $b_1 = 2.0261$. From Review Problem 12, $s_e = 11.3073$.

Step 1: H_0: $\beta_1 = 0$

H_a: $\beta_1 \neq 0$

Step 2: $\alpha = 0.05$

Step 3: $$t = \frac{b_1}{s_e / \sqrt{\sum x^2 - (\sum x)^2 / n}} = \frac{2.0261}{11.3073 / \sqrt{3081 - 173^2 / 10}} = 1.68$$

Step 4: df = n - 2 = 8; critical values = ±2.306
For the P-value approach, note that $0.10 < p < 0.20$.

Step 5: Since $-2.306 < 1.68 < 2.306$, do not reject the null hypothesis at the 5% significance level.
Since $p > \alpha$, do not reject the null hypothesis.

Step 6: The data do not provide sufficient evidence to conclude that the student-to-faculty ratio is useful as a predictor of graduation rate.

(b) Step 1: For a 95% confidence interval, $\alpha = 0.05$. With df = n - 2 = 8, $t_{\alpha/2} = t_{0.025} = 2.306$.

Step 2: The endpoints of the confidence interval for β_1 are

$$b_1 \pm t_{\alpha/2} \cdot s_e / \sqrt{\sum x^2 - (\sum x)^2 / n}$$

$$2.0261 \pm 2.306 \cdot 11.3073 / \sqrt{3081 - 173^2 / 10}$$

$$2.0261 \pm 2.7780$$

$$-0.7519 \ to \ 4.8041$$

We can be 95% confident that the change in graduation rate for universities per 1 unit increase in the student/faculty ratio is somewhere between -0.7519 and 4.8041 percent.

15. From Review Problem 12, $\hat{y} = 16.4484 + 2.0261x$, $\sum x = 173$, and $\sum x^2 = 3,081$. From Review Problem 12, $s_e = 11.3073$.

(a) $\hat{y}_p = 16.4484 + 2.0261(17) = 50.89$

(b) Step 1: For a 95% confidence interval, $\alpha = 0.05$. With df = n - 2 = 8, $t_{\alpha/2} = t_{0.025} = 2.306$.

Step 2: $\hat{y}_p = 50.89$

Step 3: The endpoints of the confidence interval are:

$$\hat{y}_p \pm t_{\alpha/2} \cdot s_e \cdot \sqrt{\frac{1}{n} + \frac{(x_p - \sum x / n)^2}{\sum x^2 - (\sum x)^2 / n}}$$

$$= 50.89 \pm 2.306 \cdot 11.3073 \cdot \sqrt{\frac{1}{10} + \frac{(17 - 173/10)^2}{3081 - 173^2/10}}$$

$$= 50.89 \pm 8.29 = (42.60, 59.18)$$

The interpretation of this interval is as follows: We can be 95% confident that the graduation rate of universities with a student/faculty ratio of 17 is somewhere between 42.60% and 59.18%.

16. (a) This is the same as the answer in Problem 15 part (a): $\hat{y}_p = 50.89$.

(b) For a 95% prediction interval, Steps 1 and 2 are the same as Steps 1 and 2, respectively, in Problem 15 part (b). Thus, they are not repeated here. Only Step 3 is presented.

Step 3: The endpoints of the prediction interval are:

$$\hat{y}_p \pm t_{\alpha/2} \cdot s_e \cdot \sqrt{1 + \frac{1}{n} + \frac{(x_p - \sum x / n)^2}{\sum x^2 - (\sum x)^2 n}}$$

$$= 50.89 \pm 2.306 \cdot 11.3073 \cdot \sqrt{1 + \frac{1}{10} + \frac{(17 - 173/10)^2}{3081 - 173^2/10}}$$

$$= 50.89 \pm 27.36 = (23.53, 78.25)$$

The interpretation of this interval is as follows: We can be 95% certain that the graduation rate of a randomly selected university with a student/faculty ratio of 17 will be somewhere between 23.53% and 78.25%.

(c) The error in the estimate of the mean graduation rate for universities with a student/faculty ratio of 17 is due only to the fact that the population regression line is being estimated by a sample regression line. The error in the prediction of the graduation rate of a randomly chosen university with a student/faculty ratio is due to the estimation error mentioned above plus the variation in graduation rates of universities with a student/faculty ratio of 17.

17. From Review Problem 12,

S_{xx} = 88.10

S_{xy} = 178.50

S_{yy} = 1384.50

$$r = \frac{S_{xy}}{\sqrt{S_{xx}S_{yy}}} = \frac{178.50}{\sqrt{(88.10)(1384.50)}} = 0.5111$$

Step 1: H_0: ρ = 0, H_a: ρ > 0

Step 2: α = 0.025

Step 3:

$$t = \frac{r}{\sqrt{\frac{1-r^2}{n-2}}} = \frac{0.5111}{\sqrt{\frac{1-(0.5111)^2}{10-2}}} = 1.682$$

Step 4: df = n - 2 = 8; critical value = 2.306
For the P-value approach, 0.05 < p < 0.10.

Step 5: Since 1.682 < 2.306, do not reject H_0.
Because the p-value is larger than the significance level of 0.025, we cannot reject H_0.

Step 6: The data do not provide sufficient evidence to conclude that graduation rate and student /faculty ratio are positively linearly correlated.

18. Using Minitab, with the data in columns named IMR and LE, we choose **Stat ▶ Regression ▶ Regression...**, select 'LE' in the **Response:** text box, select IMR in the **Predictors:** text box, click on the **Graphs** button, click in the **Residuals versus the variables** text box and select IMR, and click **OK**. Looking ahead to parts (b-h) and (j), click on the **Options...** button, enter 30 in the **Prediction intervals for new observations:** text box, enter 95 in the **Confidence level:** text box, and click **OK**. Click on the **Storage...** button, check the **Residuals** box, click **OK**, and click **OK**. Then choose **Stat ▶ Basic statistics ▶ Normality test...**, select RESI1 in the **Variable** text box, select the **Ryan-Joiner** option button from the **Tests for Normality** field, and click **OK**. We will first show all of the written output and then answer the questions. The graphical output will be shown in part (k).

```
The regression equation is
LE = 79.4 - 0.354 IMR

Predictor       Coef  SE Coef       T      P
Constant      79.396    1.009   78.72  0.000
IMR         -0.35374  0.02601  -13.60  0.000

S = 5.17598   R-Sq = 76.1%   R-Sq(adj) = 75.7%

Analysis of Variance

Source          DF      SS      MS       F      P
Regression       1  4956.2  4956.2  185.00  0.000
Residual Error  58  1553.9    26.8
Total           59  6510.1

Unusual Observations

Obs  IMR      LE     Fit  SE Fit  Residual  St Resid
 11   46  49.890  63.251   0.795   -13.361    -2.61R
 25  100  47.430  43.894   1.971     3.536     0.74 X
```

```
 34   34  44.280  67.418   0.680   -23.138      -4.51R
 48   44  50.160  63.828   0.773   -13.668      -2.67R
 52   91  43.450  47.365   1.734    -3.915      -0.80 X

R denotes an observation with a large standardized residual.
X denotes an observation whose X value gives it large leverage.

Predicted Values for New Observations

New Obs     Fit  SE Fit         95% CI              95% PI
      1  68.784   0.669  (67.445, 70.122)  (58.337, 79.231)

Values of Predictors for New Observations

New Obs   IMR
      1  30.0
```

(a) LE = 79.4 - 0.354 IMR

(b) The standard error of the estimate is 5.17598. Roughly speaking, this indicates how much, on average, the predicted life expectancies differ from the observed values of life expectancies.

(c) The coefficient of determination is 0.761, indicating that 76.1% of the variation in life expectancy is explained by a linear relationship with the infant mortality rate. The t-test for the slope results in $t = -13.60$ with a P-value of 0.000. Thus, we would reject the null hypothesis that $\beta_1 = 0$. This is a fairly strong relationship and indicates that the IMR may be useful for predicting life expectancy if the assumptions for making such inferences are valid.

(d) The point estimate for the conditional mean of life expectancy for IMR = 30 is 68.784 years. This is the point on the sample regression line that best estimates the corresponding point on the population regression line.

(e) The 95% confidence interval for the conditional mean of life expectancy for IMR = 30 is 67.445 to 70.122 years. We are 95% confident that the mean life expectancy is somewhere between these limits when the IMR is 30.

(f) The predicted value of life expectancy for IMR = 30 is 68.784 years. This is our best estimate of life expectancy for a country with an IMR equal to 30.

(g) The 95% prediction interval for the life expectancy of a country with IMR = 30 is 58.337 to 79.231 years. We are 95% certain that the life expectancy is somewhere between these limits when the IMR is 30.

(h) The error in the estimate of the mean life expectancy of countries with IMR = 30 is due only to the fact that the population regression line is being estimated by a sample regression line; whereas, the error in the prediction of the life expectancy of a single country with an IMR = 30 is due to that fact plus the variation in life expectancies for countries with IMR = 30.

(i) Choose **Stat ▶ Basic Statistics ▶ Correlation,** enter IMR and 'LE' in the **Variables** text box, check the box to **Display P-values**, and click **OK**. The result is

```
Pearson correlation of IMR and LE = -0.873
P-Value = 0.000
```

The P-value for a left-tail test is half of that shown (still 0.000). Since the P-value is less than 0.05, we reject the null hypothesis that $\rho = 0$ and since r is negative, the data provide evidence that there is a negative correlation between IMR and life expectancy assuming that the assumptions for making regression inferences are reasonable.

(j) For the residual analysis, we look at the graph produced by the procedure at the beginning of this solution and also the results of a Ryan-Joiner normality test on the residuals, which were stored earlier.

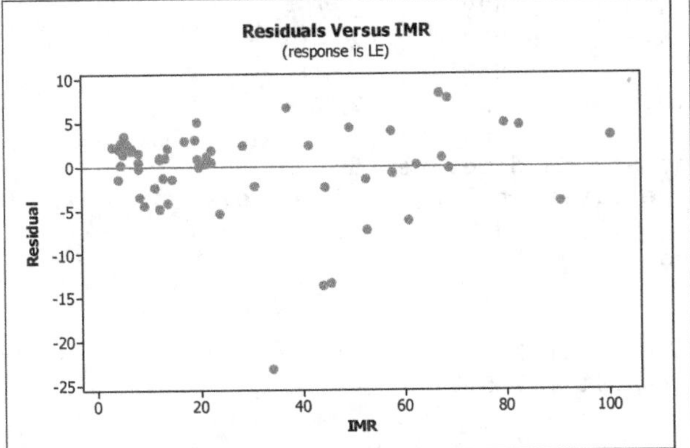

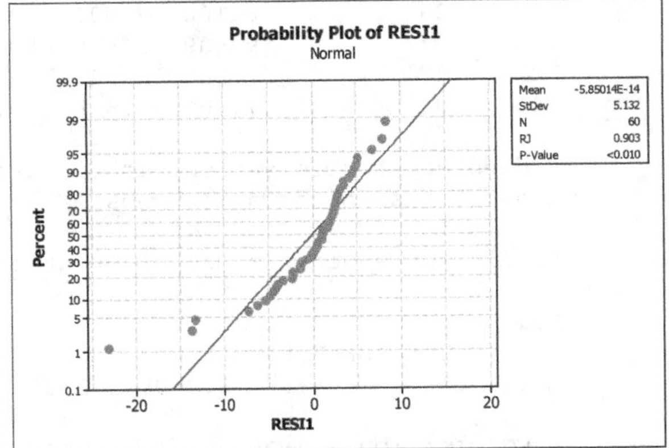

The first graph shows several residuals somewhat farther from the horizontal axis than the rest of the points and there are two residuals at the far right that indicate potential influential observations. This graph does not show a band of residuals centered on and symmetric about the horizontal axis. The normal probability plot shows several points well off the line and the Ryan-Joiner test has a P-value that is less than 0.01. Thus, Assumption 3 for making regression inferences appears to be violated. In addition, the concern about several outliers and potential influential observations indicates that the tests, confidence interval and prediction interval produced earlier may not be valid.

19. Using Minitab, with the data in columns named HIGH and PRECIPITATION, we choose **Stat ▶ Regression ▶ Regression...**, select PRECIPITATION in the **Response:** text box, select HIGH in the **Predictors:** text box, click on the **Graphs** button, click in the **Residuals versus the variables** text box and select HIGH, and click **OK**. Looking ahead to parts (b-h) and (j), click on the **Options...** button, enter 83 in the **Prediction intervals for new observations:** text box, enter 95 in the **Confidence level:** text box, and click **OK**. Click on the **Storage...** button, check the **Residuals** box, click **OK**, and click **OK**. Then choose **Stat ▶ Basic statistics ▶ Normality test...**, select RESI1 in the **Variable** text box, select the **Ryan-Joiner** option button from the **Tests for Normality** field, and click **OK**. We will first show all of the written output and then answer the questions. The graphical output will be shown in part (k).

The regression equation is
PRECIPITATION = 2.05 + 0.0067 HIGH

Predictor	Coef	SE Coef	T	P
Constant	2.048	2.170	0.94	0.350
HIGH	0.00667	0.02742	0.24	0.809

S = 2.41336 R-Sq = 0.1% R-Sq(adj) = 0.0%

Analysis of Variance

Source	DF	SS	MS	F	P
Regression	1	0.344	0.344	0.06	0.809
Residual Error	46	267.919	5.824		
Total	47	268.263			

```
Unusual Observations

Obs  HIGH  PRECIPITATION    Fit  SE Fit  Residual  St Resid
  6    86          8.100  2.621   0.410     5.479     2.30R
 10    87          9.600  2.628   0.425     6.972     2.93R
 22    45          0.500  2.348   0.972    -1.848    -0.84 X
 45    93          7.900  2.668   0.537     5.232     2.22R

R denotes an observation with a large standardized residual.
X denotes an observation whose X value gives it large influence.

Predicted Values for New Observations

New
Obs    Fit  SE Fit       95% CI              95% PI
  1  2.601   0.373  (1.850, 3.353)  (-2.314, 7.517)
```

(a) PRECIPITATION = 2.05 + 0.0067 HIGH

(b) The standard error of the estimate is 2.41336. Roughly speaking, this indicates how much, on average, the predicted average July precipitations differ from the observed values of precipitations.

(c) The coefficient of determination is 0.001, indicating that 0.1% of the variation in average July precipitation is explained by a linear relationship with the average July high temperature. The t-test for the slope results in t = 0.24 with a P-value of 0.809. Thus, we would not reject the null hypothesis that $\beta_1 = 0$. This is a very weak relationship and indicates that the average July high temperature is useless for predicting average July precipitation if the assumptions for making such inferences are valid.

(d) The point estimate for the conditional mean of average July precipitation for an average July high temperature of 83 degrees is 0.373 inches. This is the point on the sample regression line that best estimates the corresponding point on the population regression line.

(e) The 95% confidence interval for the conditional mean of average July precipitation for an average July high temperature of i3 degrees is 1.850 to 3.353 inches. We are 95% confident that the mean average July precipitation is somewhere between these limits when the average July high temperature is 83 degrees.

(f) The predicted value of average July precipitation for a city with an average July high temperature of 83 degrees is 0.373 inches. This is our best estimate of average July precipitation for a city with an average July high temperature of 83 degrees.

(g) The 95% prediction interval for the average July precipitation of a city with an average high July temperature of 83 degrees is -2.314 to 7.517 inches. The lower prediction limit can be changed to 0.00 since negative precipitation amounts are not possible. We are 95% certain that the average July precipitation is somewhere between these limits for a city with an average July high temperature of 83 degrees.

(h) The error in the estimate of the mean average July precipitation of cities with average July high temperature of 83 degrees is due only to the fact that the population regression line is being estimated by a sample regression line; whereas, the error in the prediction of the average July precipitation of a single city with an average July high temperature of 83 degrees is due to that fact plus the variation in average July precipitation for cities with average high July temperatures of 83 degrees.

(i) Choose **Stat ▶ Basic Statistics ▶ Correlation,** enter HIGH and PRECIPITATION in the **Variables** text box, check the box to **Display P-values**, and click **OK**. The result is

```
Pearson correlation of HIGH and PRECIPITATION = 0.036
P-Value = 0.809
```

Since the P-value is greater than 0.05, we do not reject the null hypothesis that $\rho = 0$.

(j) For the residual analysis, we look at the graph produced by the procedure at the beginning of this solution and also the results of a Ryan-Joiner normality test on the residuals, which were stored earlier.

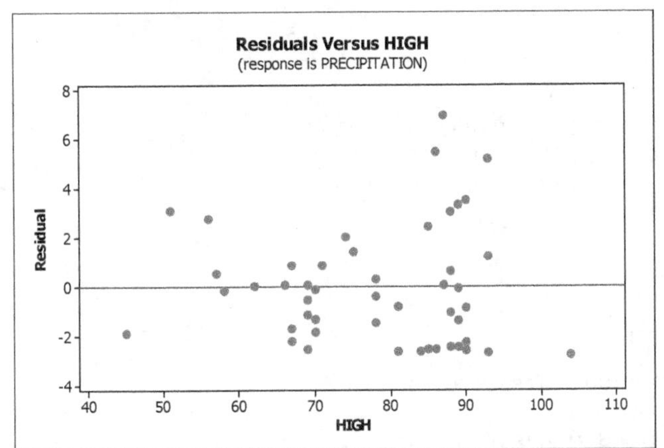

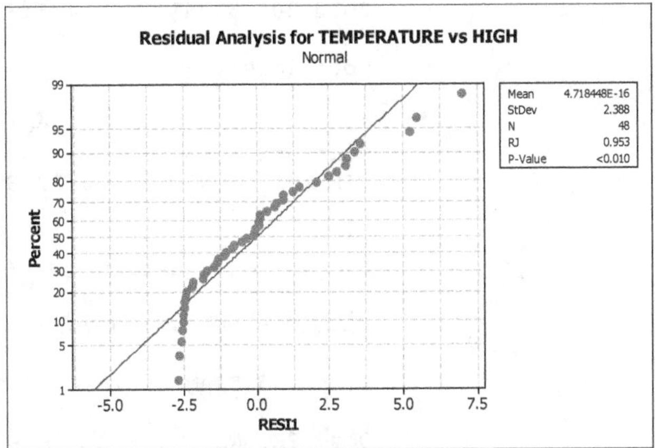

The first graph does not show a band of residuals centered on and symmetric about the horizontal axis. They may even be somewhat U-shaped. The normal probability plot shows several points well off the line and the Ryan-Joiner test has a P-value that is less than 0.01. Thus, Assumption 3 for making regression inferences appears to be violated. In addition, since a linear model accounts for only 0.1% of the variation in average July precipitation, it appears that Assumption 1 (of a linear model) is also violated. The regression inferences above are very unlikely to be valid and are pretty useless.

20. Using Minitab, with the data in columns named FIRST and SECOND, we choose **Stat ▶ Regression ▶ Regression...**, select SECOND in the **Response:** text box, select FIRST in the **Predictors:** text box, click on the **Graphs** button, click in the **Residuals versus the variables** text box and select FIRST, and click **OK**. Looking ahead to parts (b-h) and (j), click on the **Options...** button, enter 72 in the **Prediction intervals for new observations:** text box, enter 95 in the **Confidence level:** text box, and click **OK**. Click on the **Storage...** button, check the **Residuals** box, click **OK**, and click **OK**. Then choose **Stat ▶ Basic statistics ▶ Normality test...**, select RESI1 in the **Variable** text box, select the **Ryan-Joiner** option button from the **Tests for Normality** field, and click **OK**. We will first show all of the written output and then answer the questions. The graphical output will be shown in part (k).

```
Analysis of Variance

Source          DF   Adj SS   Adj MS  F-Value  P-Value
Regression       1    95.13   95.132     8.29    0.005
  FIRST          1    95.13   95.132     8.29    0.005
Error          150  1721.07   11.474
  Lack-of-Fit   15   113.54    7.569     0.64    0.841
  Pure Error   135  1607.53   11.908
Total          151  1816.20

Model Summary

      S    R-sq  R-sq(adj)  R-sq(pred)
3.38730   5.24%      4.61%       2.66%

Coefficients

Term         Coef  SE Coef  T-Value  P-Value   VIF
Constant    56.09     6.60     8.50    0.000
FIRST      0.2557   0.0888     2.88    0.005  1.00


Regression Equation

SECOND = 56.09 + 0.2557 FIRST

Fits and Diagnostics for Unusual Observations

Obs  SECOND     Fit   Resid  Std Resid
  2  72.000  73.215  -1.215      -0.37     X
  5  67.000  74.494  -7.494      -2.22  R
 21  68.000  75.005  -7.005      -2.07  R
 26  69.000  76.028  -7.028      -2.09  R
103  70.000  76.795  -6.795      -2.04  R
121  83.000  73.982   9.018       2.69  R
127  83.000  74.238   8.762       2.60  R
136  82.000  75.005   6.995       2.07  R
142  76.000  77.050  -1.050      -0.32     X
149  83.000  76.028   6.972       2.08  R
150  84.000  76.028   7.972       2.37  R
152  81.000  77.306   3.694       1.12     X
R  Large residual
X  Unusual X

Variable  Setting
FIRST          72

    Fit    SE Fit          95% CI              95% PI
74.4938  0.340386  (73.8212, 75.1663)  (67.7671, 81.2205)
```

(a) The regression equation is SECOND = 56.09 + 0.2557 FIRST

(b) The standard error of the estimate is 3.38730. Roughly speaking, this indicates how much, on average, the predicted values of the second round differ from the observed values of the second round.

(c) The coefficient of determination is 0.0524, indicating that 5.24% of the variation in second round scores is explained by a linear relationship with the first round scores. The t-test for the slope results in $t = 2.88$ with a P-value of 0.005. Thus, we would reject the null hypothesis that $\beta_1 = 0$. Even though the P-value is small, the low coefficient of determination indicates that this is a fairly weak relationship and that first round score is not very useful in making predictions of the second round score if the assumptions for making such inferences are valid.

(d) The point estimate for the conditional mean second round score for a player with a first round score of 72 is 74.4938. This is the point on the sample regression line that best estimates the corresponding point on the population regression line.

(e) The 95% confidence interval for the conditional mean second round score for players with a first round score of 72 is 73.8212 to 75.1663. We are 95% confident that the mean second round score is somewhere between these limits for players with a first round score of 72.

(f) The predicted second round score for a player with a first round score of 72 is 74.4938. This is our best estimate of the second round score for a player with a first round score of 72.

(g) The 95% prediction interval for the second round score for a player with a first round score of 72 is 67.7671 to 81.2205. We are 95% certain that the second round score is somewhere between these limits for a player with a first round score of 72.

(h) The error in the estimate of the mean second round score for players with a first round score of 72 is due only to the fact that the population regression line is being estimated by a sample regression line; whereas, the error in the prediction of the second round score of a player with a first round score of 72 is due to that fact plus the variation in second round scores for players with a first round score of 72.

(i) Choose **Stat ▶ Basic Statistics ▶ Correlation,** enter FIRST and SECOND in the **Variables** text box, check the box to **Display P-values**, and click **OK.** The result is

```
Pearson correlation of FIRST ROUND and SECOND ROUND = 0.229
P-Value = 0.005
```

The P-value for a right-tailed test is half of that shown (0.005/2 = 0.0025) if r is positive. Since the P-value is less than 0.05, we reject the null hypothesis that $\rho = 0$.

(j) For the residual analysis, we look at the graph produced by the procedure at the beginning of this solution and also the results of a Ryan-Joiner normality test on the residuals, which were stored earlier.

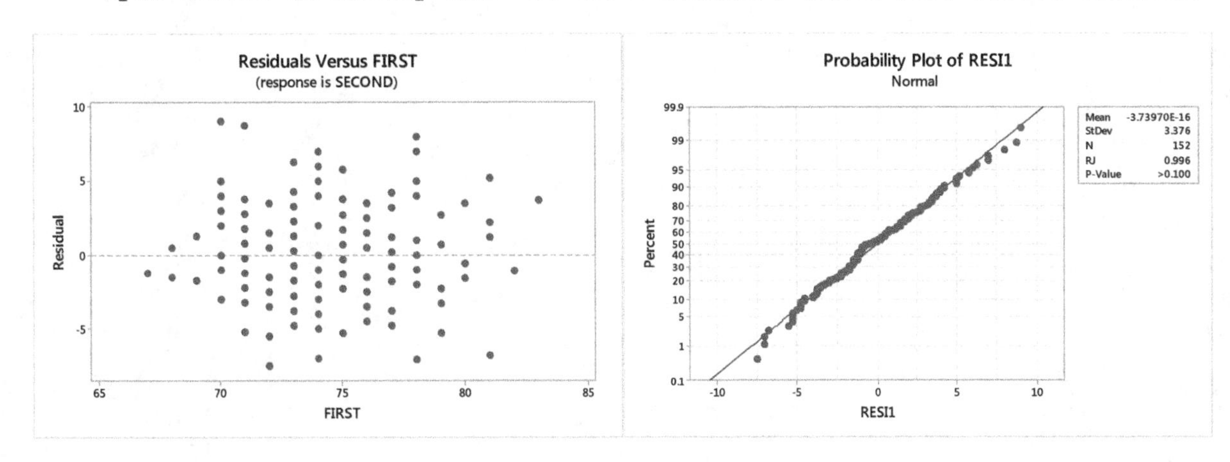

In the left-hand graph, the residuals are located in a horizontal band centered on and somewhat symmetrical about the horizontal axis. The P-value for the Ryan-Joiner test is > 0.100 indicating that normality of the residuals is a reasonable assumption. Since the linear model also appears to be appropriate, although not very useful, the regression inferences made earlier in this problem are likely to be reasonable.

CHAPTER 16 SOLUTIONS

Exercises 16.1

16.1 We state the two numbers of degrees of freedom.

16.3 $F_{0.05}$, $F_{0.025}$, F_{α}

16.5 (a) The first number in parentheses—12—is the number of degrees of freedom for the numerator.
(b) The second number in parentheses—7—is the number of degrees of freedom for the denominator.

16.7 (a) $F_{0.05} = 1.89$ (b) $F_{0.01} = 2.47$ (c) $F_{0.025} = 2.14$

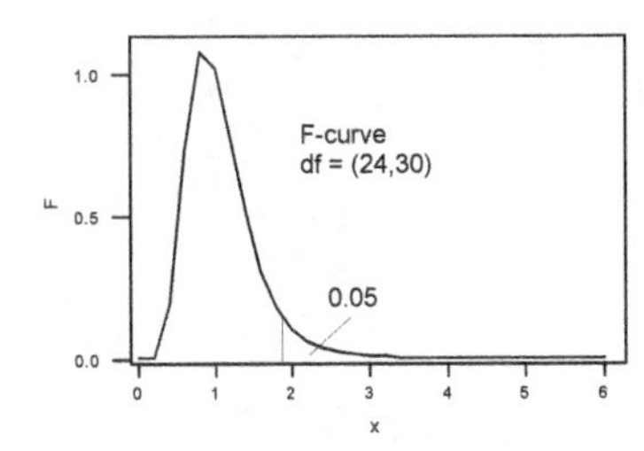

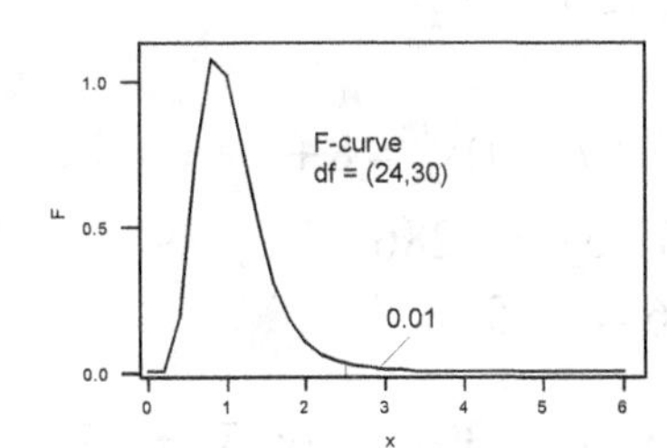

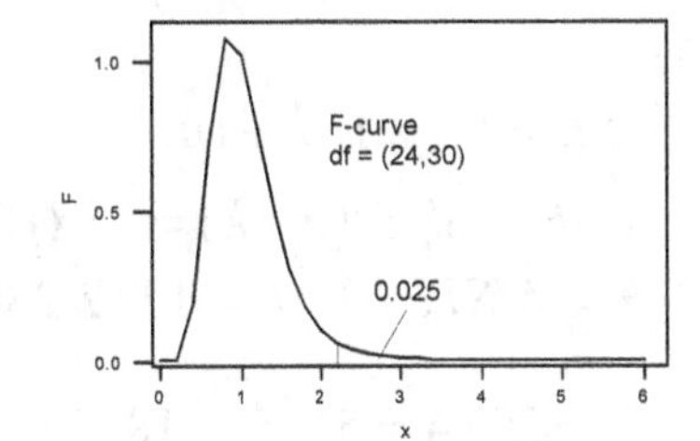

16.9 (a) $F_{0.01} = 2.88$ (b) $F_{0.05} = 2.10$ (c) $F_{0.10} = 1.78$

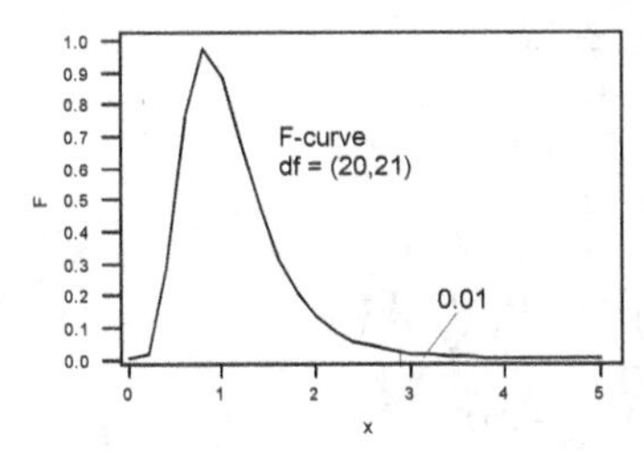

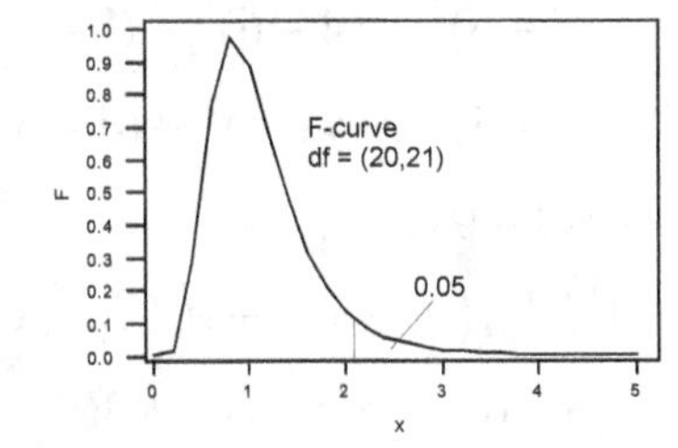

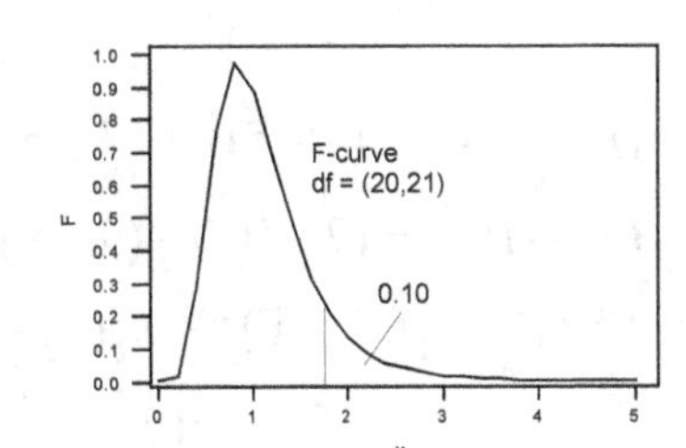

Exercises 16.2

16.11 The pooled-t procedure (Procedure 10.1) is a method for comparing the means of two populations. One-way ANOVA is a procedure for comparing the means of several populations. Thus, one-way ANOVA is a generalization of the pooled-t procedure.

16.13 The reason for the word "variance" in the phrase "analysis of variance" is because the analysis-of-variance procedure for comparing means involves analyzing the *variation* in the sample data.

16.15 A measure of variation among the sample means is called the treatment mean square. The mathematical abbreviation for it is *MSTR*.

16.17 To compare the variation among the sample means to the variation within the samples, we use the ratio of *MSTR* to *MSE*. This ratio is called the F-statistic.

16.19 The term *one-way* is used because the type of ANOVA compares the means of a variable for populations that result from a classification by *one* other variable, call the factor.

16.21 No, because the variation among the sample means is not large relative to the variation within the samples.

16.23 In a one-way ANOVA, the residual of an observation is the difference between an observation and the mean of the sample containing it.

16.25

$$\bar{x}_1 = 18/3 = 6;\ \bar{x}_2 = 6/3 = 2;\ \bar{x}_3 = 16/4 = 4$$

$$\bar{x} = \sum x_i / n = (18+6+16)/(3+3+4) = 40/10 = 4$$

$$(n_1 - 1)s_1^2 = (8-6)^2 + (4-6)^2 + (6-6)^2 = 8$$

$$(n_2 - 1)s_2^2 = (2-2)^2 + (1-2)^2 + (3-2)^2 = 2$$

$$(n_3 - 1)s_3^2 = (4-4)^2 + (3-4)^2 + (6-4)^2 + (3-4)^2 = 6$$

(a) $SSTR = n_1(\bar{x}_1 - \bar{x})^2 + n_2(\bar{x}_2 - \bar{x})^2 + n_3(\bar{x}_3 - \bar{x})^2 = 3(6-4)^2 + 3(2-4)^2 + 4(4-4)^2 = 24$

(b) $MSTR = SSTR/(k-1) = 24/(3-1) = 12$

(c-d)

$$SSE = (n_1 - 1)s_1^2 + (n_2 - 1)s_2^2 + (n_3 - 1)s_3^2 = 8 + 2 + 6 = 16$$

$$MSE = SSE/(n-k) = 16/(10-3) = 2.286$$

(e) $F = MSTR/MSE = 12/2.286 = 5.25$

16.27

$$\bar{x}_1 = 20/4 = 5;\ \bar{x}_2 = 18/3 = 6;\ \bar{x}_3 = 30/5 = 6;\ \bar{x}_4 = 21/5 = 5;\ \bar{x}_5 = 27/3 = 9$$

$$\bar{x} = \sum x_i / n = (20+18+30+21+27)/(4+3+5+5+3) = 120/20 = 6$$

$$(n_1 - 1)s_1^2 = (7-5)^2 + (4-5)^2 + (5-5)^2 + (4-5)^2 = 6$$

$$(n_2 - 1)s_2^2 = (5-6)^2 + (9-6)^2 + (4-6)^2 = 14$$

$$(n_3 - 1)s_3^2 = (6-6)^2 + (7-6)^2 + (5-6)^2 + (4-6)^2 + (8-6)^2 = 10$$

$$(n_4 - 1)s_4^2 = (3-5)^2 + (7-5)^2 + (7-5)^2 + (4-5)^2 + (4-5)^2 = 14$$

$$(n_5 - 1)s_5^2 = (7-9)^2 + (9-9)^2 + (11-9)^2 = 8$$

(a)

$$SSTR = n_1(\bar{x}_1 - \bar{x})^2 + n_2(\bar{x}_2 - \bar{x})^2 + n_3(\bar{x}_3 - \bar{x})^2 + n_4(\bar{x}_4 - \bar{x})^2 + n_5(\bar{x}_5 - \bar{x})^2$$
$$= 4(5-6)^2 + 3(6-6)^2 + 5(6-6)^2 + 5(5-6)^2 + 3(9-6)^2 = 36$$

(b) $MSTR = SSTR/(k-1) = 36/(5-1) = 9$

(c-d)

$$SSE = (n_1 - 1)s_1^2 + (n_2 - 1)s_2^2 + (n_3 - 1)s_3^2 + (n_4 - 1)s_4^2 + (n_5 - 1)s_5^2$$
$$= 6 + 14 + 10 + 14 + 8 = 52$$

$$MSE = SSE/(n-k) = 52/(20-5) = 3.467$$

(e) $F = MSTR/MSE = 9/3.467 = 2.600$

16.29

$$\bar{x}_1 = 24/3 = 8;\ \bar{x}_2 = 15/3 = 5;\ \bar{x}_3 = 36/3 = 12;\ \bar{x}_4 = 9/3 = 3$$

$$\bar{x} = \sum x_i / n = (24+15+36+9)/(3+3+3+3) = 84/12 = 7$$

$$(n_1 - 1)s_1^2 = (11-8)^2 + (6-8)^2 + (7-8)^2 = 14$$

$$(n_2 - 1)s_2^2 = (9-5)^2 + (2-5)^2 + (4-5)^2 = 26$$

$$(n_3 - 1)s_3^2 = (16-12)^2 + (10-12)^2 + (10-12)^2 = 24$$

$$(n_4 - 1)s_4^2 = (5-3)^2 + (1-3)^2 + (3-3)^2 = 8$$

(a)

$$SSTR = n_1(\bar{x}_1 - \bar{x})^2 + n_2(\bar{x}_2 - \bar{x})^2 + n_3(\bar{x}_3 - \bar{x})^2 + n_4(\bar{x}_4 - \bar{x})^2 + n_5(\bar{x}_5 - \bar{x})^2$$
$$= 3(8-7)^2 + 3(5-7)^2 + 3(12-7)^2 + 3(3-7)^2 = 138$$

(b) $MSTR = SSTR/(k-1) = 138/(4-1) = 46$

(c-d)

$$SSE = (n_1-1)s_1^2 + (n_2-1)s_2^2 + (n_3-1)s_3^2 + (n_4-1)s_4^2 + (n_5-1)s_5^2$$
$$= 14 + 26 + 24 + 8 = 72$$

$$MSE = SSE/(n-k) = 72/(12-4) = 9$$

(e) $F = MSTR/MSE = 46/9 = 5.11$

16.31 If the hypothesis test situation abides by the characteristics outlined in this exercise, we can use the pooled-t test or we can use the one-way ANOVA test discussed in this section using k = 2.

Exercises 16.3

16.33 A small value of F results when MSTR is small compared to MSE, i.e., when the variation between sample means is small compared to the variation within samples. This describes what should happen when the null hypothesis is true; thus it does not comprise evidence that the null hypothesis is false. Only when the variation between sample means is large compared to the variation within samples, i.e., when F is large, do we have evidence that the null hypothesis is not true.

16.35 SST = SSTR + SSE. This means that the total variation can be partitioned into a component representing variation among the sample means and another component representing variation within the samples.

16.37 (a) One-way ANOVA
(b) Two-way ANOVA

16.39 Since 0.708 = 2.124/df(Treatment), we have df(Treatment) = 3
It follows that df (Total) = 3 + 20 = 23.
Since 0.75 = 0.708/MSE, we have MSE = 0.708/0.75 = 0.944.
Then since 0.944 = SSE/20, we have SSE = 0.944(20) = 18.88.
Finally SST = SSTR + SSE = 2.124 + 18.880 = 21.004
Thus, the completed table is

Source	df	SS	MS=SS/df	F-statistic
Treatment	3	2.124	0.708	0.75
Error	20	18.880	0.944	
Total	23	21.004		

16.41 Since the Total df is 16, the df for Treatment is 14 - 12 = 2.
Since MSTR = SSTR/2, SSTR = 2(MSTR) = 2(1.4) = 2.8
Similarly, SSE = 12(MSE) = 12(0.9) = 10.8
Now SST = SSTR + SSE = 2.8 + 10.8 = 13.6
Finally, F = MSTR/MSE = 1.4/0.9 = 1.556
Thus, the completed table is

Source	df	SS	MS=SS/df	F-statistic
Treatment	2	2.8	1.4	1.556
Error	12	10.8	0.9	
Total	14	13.6		

16.43 We have the following:

$k = 3$

$n_1 = 3$ $n_2 = 3,$ $n_3 = 4$

$T_1 = 18$ $T_2 = 6$ $T_3 = 16$

$n = \sum n_j = 3 + 3 + 4 = 10$

$\sum x_i = \sum T_j = 18 + 6 + 16 = 40$

$\sum x_i^2 = (8)^2 + (4)^2 + (6)^2 + ... + (6)^2 + (3)^2 = 200$

(a) Consequently,

$$SST = \sum x_i^2 - (\sum x_i)^2 / n = 200 - 40^2 / 10 = 200 - 160 = 40$$

$$SSTR = \sum (T_j^2 / n_j) - (\sum x_i)^2 / n = (18)^2 / 3 + (6)^2 / 3 + (16)^2 / 4 - (40)^2 / 10 = 184 - 160 = 24$$

$$SSE = SST - SSTR = 40 - 24 = 16$$

(b) The results are the same as in Exercise 16.25.

(c)

Source	df	SS	MS=SS/df	F-statistic
Treatment	2	24	12.00	5.25
Error	7	16	2.29	
Total	9	40		

(d) The df are (2,7). The critical value is 4.74. The P-value is $0.025 < P < 0.50$. Since $5.25 > 4.74$ and since the p-value is less than the significance level, we reject the null hypothesis that the population means are equal.

16.45 We have the following:

$k = 5$

$n_1 = 4$ $n_2 = 3$ $n_3 = 5$ $n_4 = 5$ $n_5 = 3$

$T_1 = 20$ $T_2 = 18$ $T_3 = 30$ $T_4 = 25$ $T_5 = 27$

$n = \sum n_j = 4 + 3 + 5 + 5 + 3 = 20$

$\sum x_i = \sum T_j = 20 + 18 + 30 + 25 + 27 = 120$

$\sum x_i^2 = (7)^2 + (4)^2 + (5)^2 + ... + (9)^2 + (11)^2 = 808$

(a) Consequently,

$$SST = \sum x_i^2 - (\sum x_i)^2 / n = 808 - 120^2 / 20 = 808 - 720 = 88$$

$$SSTR = \sum (T_j^2 / n_j) - (\sum x_i)^2 / n$$

$$= (20)^2 / 4 + (18)^2 / 3 + (30)^2 / 5 + (25)^2 / 5 + (27)^2 / 3 - (120)^2 / 20 = 756 - 720 = 36$$

$$SSE = SST - SSTR = 88 - 36 = 52$$

(b) The results are the same as in Exercise 16.27.

(c)

Source	df	SS	MS=SS/df	F-statistic
Treatment	4	36	9.00	2.60
Error	15	52	3.47	
Total	19	88		

(d) The df are (4,15). The critical value is 3.06. The P-value is $0.05 < P < 0.10$. Since $2.60 > 3.06$ and since the P-value is greater than the significance level, we do not reject the null hypothesis that the population means are equal.

16.47 We have the following:

$k = 4$

$n_1 = 3$	$n_2 = 3$	$n_3 = 3$	$n_4 = 3$
$T_1 = 24$	$T_2 = 15$	$T_3 = 36$	$T_4 = 9$

$n = \sum n_j = 3 + 3 + 3 + 3 = 12$

$\sum x_i = \sum T_j = 24 + 15 + 36 + 9 = 84$

$\sum x_i^2 = (11)^2 + (6)^2 + (7)^2 + ... + (1)^2 + (3)^2 = 798$

(a) Consequently,

$$SST = \sum x_i^2 - (\sum x_i)^2 / n = 798 - 84^2 / 12 = 798 - 588 = 210$$

$$SSTR = \sum (T_j^2 / n_j) - (\sum x_i)^2 / n$$

$$= (24)^2 / 3 + (15)^2 / 3 + (36)^2 / 3 + (9)^2 / 3 - (84)^2 / 12 = 726 - 588 = 138$$

$$SSE = SST - SSTR = 210 - 138 = 72$$

(b) The results are the same as in Exercise 16.29.

(c)

Source	df	SS	MS=SS/df	F-statistic
Treatment	3	138	46.0	5.11
Error	8	72	9.0	
Total	11	210		

(d) The df are (3,8). The critical value is 4.07. The P-value is $0.025 < P < 0.05$. Since $5.11 > 4.07$ and since the P-value is less than the significance level, we reject the null hypothesis that the population means are equal.

16.49 The total number of populations being sampled is k = 3. Let the subscripts 1, 2, and 3 refer to diatoms, bacteria, and macroalgae, respectively. The total number of observations is n = 12. Also, $n_1 = 4$, $n_2 = 4$, and $n_3 = 4$.

Step 1: H_0: $\mu_1 = \mu_2 = \mu_3$ (population means are equal)

H_a: population means are different

Step 2: $\alpha = 0.05$

Step 3: Let T_1, T_2 and T_3 refer to the sum of the data values in each of the three samples, respectively. Thus:

$T_1 = 1828$ $\quad T_2 = 1225$ $\quad T_3 = 1175$.

Also, the sum of all the data values is $\sum x_i = 4228$, and their sum of squares is $\sum x_i^2 = 1,561,154$. Then,

$$SST = \sum x_i^2 - (\sum x_i)^2 / n = 1561.154 - 4228^2 / 12 = 71488.67$$

$$SSTR = \frac{T_1^2}{n_1} + \frac{T_2^2}{n_2} + \frac{T_3^2}{n_3} - \frac{(\sum x_i)^2}{n} = \frac{1828^2}{4} + \frac{12255^2}{4} + \frac{1175^2}{4} - \frac{4228^2}{12} = 66043.17$$

and SSE = SST - SSTR = 71488.67 - 66043.17 = 5445.50

MSTR = SSTR/(k - 1) = 66043.17/(3 - 1) = 33021.585

MSE = SSE/(n - k) = 5445.50/(12 - 3) = 605.056

F = MSTR/MSE = 33021.585/605.056 = 54.98

The one-way ANOVA table is

Source	Df	SS	MS = SS/df	F-statistic
Treatment	2	66043.17	33021.585	54.58
Error	9	5445.50	605.056	
Total	11	71488.67		

Step 4: df = (k - 1, n - k) = (2, 9); critical value = 4.26
For the p-value approach, P(F > 54.58) < 0.005

Step 5: Since 54.58 > 4.26, we reject the null hypothesis. Since P-value < 0.05, we reject the null hypothesis.

Step 6: At the 5% significance level, we reject the null hypothesis and conclude that there is a difference in the mean number of copepods among the three different diets.

16.51 The number of populations being sampled is k = 5. Let the subscripts 1-5 refer to strains A-E, respectively, The total number observations is n = 25. Also, $n_1 = n_2 = n_3 = n_4 = n_5 = 5$.

Step 1: H_0: $\mu_1 = \mu_2 = \mu_3 = \mu_4 = \mu_5$ (population means are equal)
H_a: Not all of the population means are equal

Step 2: $\alpha = 0.05$

Step 3: Let $T_1 - T_5$ refer to the sum of the data values in each of the five samples, respectively. Thus,
$T_1 = 104$ $T_2 = 129$ $T_3 = 185$ $T_4 = 98$ $T_5 = 194$.

Also, the sum of all the data values is $\sum x_i = 710$, and their sum of squares is $\sum x_i^2 = 25424$.

Consequently, $SST = \sum x_i^2 - (\sum x_i)^2/n = 25424 - 710^2/25 = 5260$

$$SSTR = \frac{T_1^2}{n_1} + \frac{T_2^2}{n_2} + \frac{T_3^2}{n_3} + \frac{T_4^2}{n_4} + \frac{T_5^2}{n_5} - \frac{(\sum x_i)^2}{n}$$

$$= \frac{104^2}{5} + \frac{129^2}{5} + \frac{185^2}{5} + \frac{98^2}{5} + \frac{194^2}{5} - \frac{710^2}{25} = 1620$$

and SSE = SST - SSTR = 5260 - 1620 = 3640.
MSTR = SSTR/(k - 1) = 1620/(5 - 1) = 405
MSE = SSE/(n - k) = 3640/(25 - 5) = 182
F = MSTR/MSE = 405/182 = 2.23

The one-way ANOVA table is:

Source	df	SS	MS = SS/df	F-statistic
Treatment	4	1620	405	2.23
Error	20	3640	182	
Total	24	5260		

Step 4: df = (k - 1, n - k) = (4, 20); critical value = 2.87
For the P-value approach, P(F > 2.23) > 0.10.

Step 5: From Step 3, F = 2.23. From Step 4, $F_{0.05} = 2.87$. Since 2.23 < 2.87, do not reject H_0.
Since the P-value is larger than the significance level, do not reject H_0.

Step 6: At the 5% significance level, the data do not provide sufficient evidence to conclude that the population means are different.

16.53 The total number of populations being sampled is $k = 4$. The total number of observations is $n = 34$. Also, $n_1 = 8$, $n_2 = 8$, $n_3 = 10$, $n_4 = 8$.

Step 1: H_0: $\mu_1 = \mu_2 = \mu_3 = \mu_4$ (mean battery life among the four brands are equal)
H_a: Not all the means are equal.

Step 2: $\alpha = 0.05$

Step 3: Let T_1, T_2, T_3 and T_4, refer to the sum of the data values in each of the four samples, respectively. Thus,
$T_1 = 61.50$ $T_2 = 83.50$ $T_3 = 85.50$ $T_4 = 47.25$
Also, the sum of all the data values is $\sum x_i = 277.75$, and their sum of squares is $\sum x_i^2 = 2483.8125$.
Consequently, $SST = \sum x_i^2 - (\sum x_i)^2 / n = 2483.8125 - 277.75^2 / 34 = 214.840$

$$SSTR = \frac{T_1^2}{n_1} + \frac{T_2^2}{n_2} + \frac{T_3^2}{n_3} + \frac{T_4^2}{n_4} - \frac{(\sum x_i)^2}{n}$$

$$= \frac{61.5^2}{8} + \frac{83.5^2}{8} + \frac{85.5^2}{10} + \frac{47.25^2}{8} - \frac{277.75^2}{34} = 85.435$$

and $SSE = SST - SSTR = 214.840 - 85.435 = 129.405$
$MSTR = SSTR/(k - 1) = 85.435/(4 - 1) = 28.478$
$MSE = SSE/(n - k) = 129.405/(34 - 4) = 4.314$
$F = MSTR/MSE = 28.478/4.314 = 6.602$

The one-way ANOVA table is:

Source	df	SS	MS = SS/df	F-statistic
Treatment	3	85.435	28.478	6.602
Error	30	129.405	4.314	
Total	33	214.840		

Step 4: df = $(k - 1, n - k) = (3, 30)$; critical value = 2.92
For the p-value approach, $P(F > 6.602) < 0.005$.

Step 5: From Step 3, $F = 6.602$. From Step 4, $F_{0.05} = 2.92$. Since $6.602 > 2.92$, reject H_0.
Since the p-value is smaller than the significance level, reject H_0.

Step 6: At the 5% significance level, the data provide sufficient evidence to conclude that a difference exists in mean battery life among the four brands of laptops.

16.55 (a) To carry out the Analysis of variance and residual analysis, we will use the data in a single column, which is named RENT. In the next column, which is named REGION, are the names of each of the regions corresponding to each data value in RENT. The residual analysis consists of plotting the residuals against the means (or fits) and constructing a normal probability plot. This is done at the same time as the analysis of variance in Minitab. To do this, we choose **Stat ▶ Anova ▶One-way...**, select RENT for the **Response:** text box and REGION in the **Factor:** text box. Then click on the **Graphs** button and click to place check marks in the boxes for **Normal plot of residuals** and **Residuals versus fits**, and click **OK**. The printed results are

One-way ANOVA: RENT versus REGION

```
Source  DF      SS      MS     F      P
REGION   3  400513  133504  7.54  0.002
Error   16  283265   17704
Total   19  683778
S = 133.1   R-Sq = 58.57%   R-Sq(adj) = 50.81%

                                Individual 99% CIs For Mean Based on
                                Pooled StDev
Level      N    Mean  StDev  -+---------+---------+---------+--------
Midwest    6   743.0   92.1  (-------*-------)
Northeast  5  1055.2  150.2                (--------*-------)
South      4   854.5  168.0      (---------*--------)
West       5  1064.4  128.4                 (-------*--------)
                             -+---------+---------+---------+--------
                             600       800       1000      1200

Pooled StDev = 133.1
```

(b) The F-value of 7.54 is significant at the 0.05 level (P-Value = 0.002). If the assumptions of normal populations and equal standard deviations are met, we reject the null hypothesis of equal population means and conclude that there is a difference in the mean monthly rents among newly completed apartments in the four U.S. regions.

(c) The graphs produced by the procedure in part (a) are shown following.

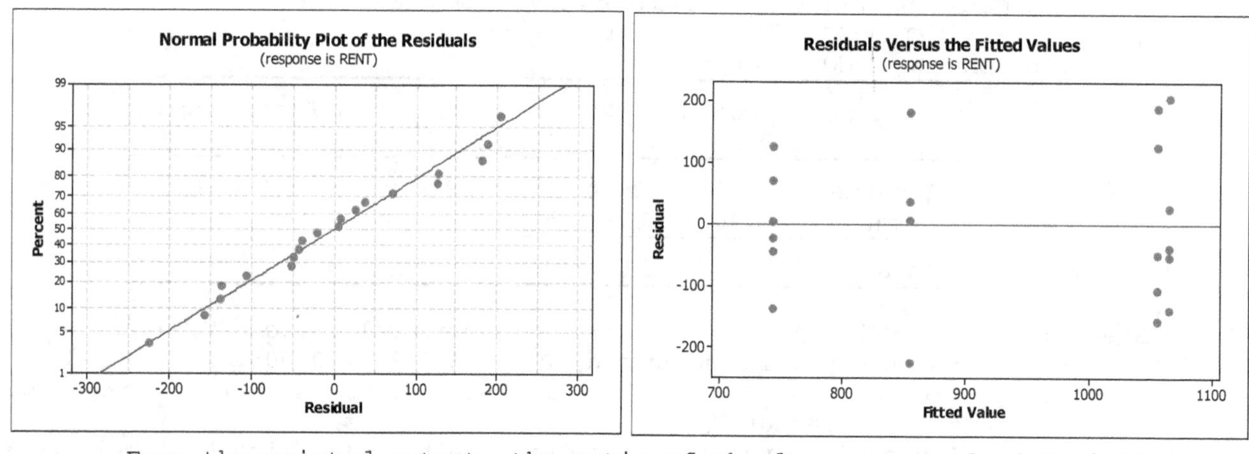

From the printed output, the ratio of the largest standard deviation to the smallest is 168.0/92.1 = 1.82. This is less than 2, indicating that the rule of 2 is not violated, leading us to conclude that the assumption of equal standard deviations is reasonable. This can also be seen in the right-hand graph above. There are no outliers in the normal probability plot and the points are linear, indicating that there are no problems with the normality assumption as well. [A Ryan-Joiner normality test of the residuals has a P-value of >0.10, leading us to conclude that the normality assumption is reasonable.]

16.57 (a) To carry out the Analysis of variance and residual analysis, we will use the data in the single column named RATE. In the next column, which is named BREAST, are names of each of the three feather treatments corresponding to each data value in RATE. The residual analysis consists of plotting the residuals against the means (or fits) and constructing a normal probability plot. This is done at the same time as the analysis of variance in Minitab. To do this, we choose

Stat ▶ Anova ▶One-way..., select RATE for the **Response:** text box and BREAST in the **Factor:** text box. Enter 99 in the **Confidence level** box.

Then click on the **Graphs** button and click to place check marks in the boxes for **Normal plot of residuals** and **Residuals versus fits**, and click **OK**. The printed results are

One-way ANOVA: RATE versus BREAST

```
Source  DF      SS     MS     F      P
BREAST   2   960.3  480.1  6.09  0.008
Error   21  1655.3   78.8
Total   23  2615.6

S = 8.878   R-Sq = 36.71%   R-Sq(adj) = 30.69%

                                   Individual 99% CIs For Mean Based on
                                   Pooled StDev
Level     N    Mean   StDev  ----+---------+---------+---------+-----
Control   8  14.838   6.556            (--------*--------)
Enlarged  8  19.775  13.048                 (--------*--------)
Reduced   8   4.588   4.821  (--------*-------)
                             ----+---------+---------+---------+-----
                                 0        10        20        30

Pooled StDev = 8.878
```

(b) The F-value of 6.09 is significant at the 0.01 level (P-Value = 0.008). If the assumptions of normal populations and equal standard deviations are met, we reject the null hypothesis of equal population means. There is sufficient evidence to conclude that there is a difference in the mean singing rates among male rock sparrows exposed to the three types of breast treatments.

(c) The graphical results are

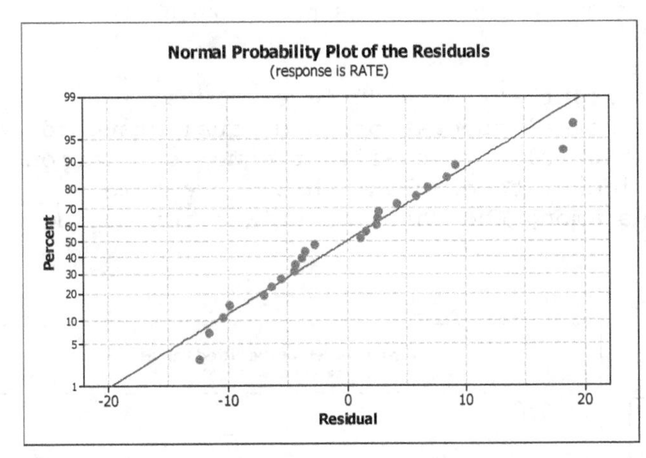

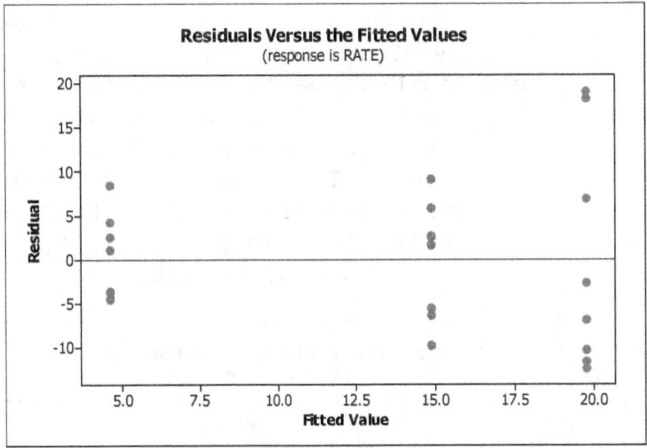

From the printed output, the ratio of the largest standard deviation to the smallest is 13.048/4.821 = 2.71. This is greater than 2, indicating that the rule of 2 is violated and leading us to conclude that the assumption of equal standard deviations may not be reasonable. The small sample sizes may be partly responsible for this occurrence, which can also be seen the right-hand graph above. The points in the probability plot are linear, indicating no problems with the normality assumption. [A Ryan-Joiner normality test of the residuals has a P-value of >0.10, leading us to conclude that the normality assumption is reasonable.]

16.59 (a) To carry out the analysis of variance and residual analysis, we will use the data in the single column named VOLUME. In the next column, which is named MATERIAL, are names of each of the three feather treatments corresponding to each data value in VOLUME. The residual analysis consists of plotting the residuals against the means (or fits) and constructing a normal probability plot. This is done at the same time as the analysis of variance in Minitab. To do this, we choose

Stat ▶ Anova ▶One-way..., select VOLUME for the **Response:** text box and MATERIAL in the **Factor:** text box. Enter 95 in the **Confidence level** box. Then click on the **Graphs** button and click to place check marks in the boxes for **Normal plot of residuals** and **Residuals versus fits**, and click **OK**. The printed results are

One-way ANOVA: HARDNESS versus MATERIAL

```
Source    DF      SS      MS       F      P
MATERIAL   2  805.31  402.66  114.71  0.000
Error     15   52.65    3.51
Total     17  857.97

S = 1.874   R-Sq = 93.86%   R-Sq(adj) = 93.04%


                                  Individual 95% CIs For Mean Based on
                                  Pooled StDev
Level      N    Mean  StDev  -----+---------+---------+---------+----
Duracross  6  39.633  3.120                                (--*---)
Duradent   6  24.017  0.615  (--*--)
Endura     6  27.533  0.647         (--*--)
                             -----+---------+---------+---------+----
                                25.0      30.0      35.0      40.0

Pooled StDev = 1.874
```

(b) The F-value of 114.71 is significant at the 0.05 level (P-Value = 0.000). If the assumptions of normal populations and equal standard deviations are met, we reject the null hypothesis of equal population means. There is sufficient evidence to conclude that there is a difference in the mean hardness among the three materials for making artificial teeth.

(c) The graphical results are

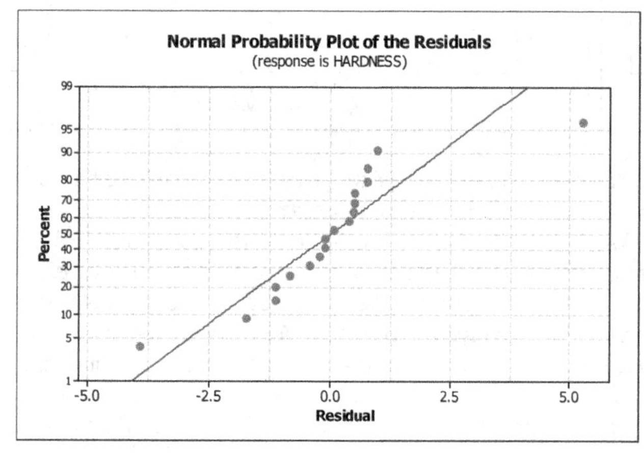

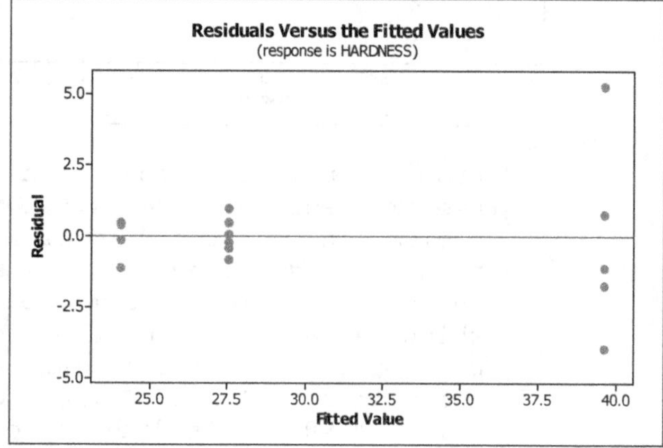

From the printed output, the ratio of the largest standard deviation to the smallest is 3.120/0.615 = 5.07. This is greater than 2. The rule of 2 is violated, leading us to conclude that the assumption of equal standard deviations may not be reasonable. The small sample sizes may be partly responsible for this occurrence, which can also be seen the previous right-hand graph. The points in the probability plot are not linear, indicating that there may also be problems with the normality assumption. [A Ryan-Joiner normality test of the residuals has a P-value of < 0.010, so we conclude that the normality assumption is not reasonable.] Although the F value is so large that it is unlikely that either violation would have any serious effect on our conclusions, these data should probably be analyzed with a more robust method.

16.61 The total number of populations being sampled is k = 5. The total number of observations is n = 236. Also, $n_1 = 46$, $n_2 = 54$, $n_3 = 47$, $n_4 = 42$, $n_5 = 47$.

Step 1: H_0: $\mu_1 = \mu_2 = \mu_3 = \mu_4 = \mu_5$ (mean BMIs are equal among the five treatment groups)
H_a: Not all the means are equal.

Step 2: $\alpha = 0.10$

Step 3: We can determine the mean of all the observations by the following

$$\bar{x} = \frac{n_1\bar{x}_1 + n_2\bar{x}_2 + n_3\bar{x}_3 + n_4\bar{x}_4 + n_5\bar{x}_5}{n_1 + n_2 + n_3 + n_4 + n_5}$$

$$= \frac{46(25.9) + 54(25.8) + 47(27.5) + 42(26.0) + 47(25.9)}{46 + 54 + 47 + 42 + 47} = 26.2136$$

Then, use the defining formulas to get the SSTR and SSE.

$$SSTR = \sum n_i(\bar{x}_i - \bar{x})^2 = 46(25.9 - 26.2136)^2 + 54(25.8 - 26.2136)^2 + 47(27.5 - 26.2136)^2$$
$$+ 42(26.0 - 26.2136)^2 + 47(25.9 - 26.2136)^2$$
$$= 98.0766$$

$$SSE = \sum (n_i - 1)s_i^2 = 45(4.3)^2 + 53(5.3)^2 + 46(5.8)^2 + 41(4.6)^2 + 46(4.3)^2$$
$$= 5586.36$$

and SST = SSE + SSTR = 5586.36 + 98.0766 = 5684.4366
MSTR = SSTR/(k - 1) = 98.0766/(5 - 1) = 24.5192
MSE = SSE/(n - k) = 5586.36/(236 - 5) = 24.1834
F = MSTR/MSE = 24.5192/24.1834 = 1.0139

The one-way ANOVA table is:

Source	df	SS	MS = SS/df	F-statistic
Treatment	4	98.0766	24.5192	1.0139
Error	231	5586.36	24.1834	
Total	235	5684.4366		

Step 4: df = (k - 1, n - k) = (4, 231); From the table in the problem, critical value = 1.97
For the p-value approach, P(F > 1.0139) > 0.10.

Step 5: From Step 3, F = 1.0139. From Step 4, $F_{0.10}$ = 1.97. Since 1.0139 < 1.97, do not reject H_0.
Since the p-value is larger than the significance level, do not reject H_0.

Step 6: At the 10% significance level, the data do not provide sufficient evidence to conclude that the mean BMI is different among the women in the five different treatment groups.

16.63 The total number of populations being sampled is k = 6. The total number of observations is n = 160.

Step 1: H_0: $\mu_1 = \mu_2 = \mu_3 = \mu_4 = \mu_5 = \mu_6$ (mean starting salaries are the same)

H_a: Not all the means are equal.

Step 2: $\alpha = 0.01$

Step 3: We can determine the mean of all the observations by the following

$$\bar{x} = \frac{n_1\bar{x}_1 + n_2\bar{x}_2 + n_3\bar{x}_3 + n_4\bar{x}_4 + n_5\bar{x}_5 + n_5\bar{x}_5 + n_6\bar{x}_6}{n_1 + n_2 + n_3 + n_4 + n_5 + n_6}$$

$$= \frac{46(55.1) + 11(44.6) + 30(59.1) + 11(40.6) + 44(62.6) + 18(43.0)}{46 + 11 + 30 + 11 + 44 + 18} = 54.833$$

Then, use the defining formulas to get the SSTR and SSE.

$$SSTR = \sum n_i(\bar{x}_i - \bar{x})^2$$

$$= 46(55.1 - 54.833)^2 + 11(44.6 - 54.833)^2 + 30(59.1 - 54.833)^2$$

$$+ 11(40.6 - 54.833)^2 + 44(62.6 - 54.833)^2 + 18(43.0 - 54.833)^2$$

$$= 9104.431$$

$$SSE = \sum (n_i - 1)s_i^2 = 45(5.6)^2 + 10(4.7)^2 + 29(4.0)^2 + 10(5.0)^2 + 43(5.7)^2 + 17(4.8)^2$$

$$= 4134.850$$

and SST = SSE + SSTR = 4134.850 + 9104.431 = 13239.281

MSTR = SSTR/(k - 1) = 9104.431/(6 - 1) = 1820.886

MSE = SSE/(n - k) = 4134.850/(160 - 6) = 26.850

F = MSTR/MSE = 1820.886/26.850 = 67.817

The one-way ANOVA table is:

Source	df	SS	MS = SS/df	F-statistic
Treatment	5	9104.431	1820.886	67.817
Error	154	4134.850	26.850	
Total	159	13239.281		

Step 4: df = (k - 1, n - k) = (5, 154); From the table in the problem, critical value = 3.14

For the p-value approach, $P(F > 67.817) < 0.005$.

Step 5: From Step 3, F = 67.817. From Step 4, $F_{0.01} = 3.14$. Since 67.817 > 3.14, reject H_0.

Since the p-value is smaller than the significance level, reject H_0.

Step 6: At the 1% significance level, the data provide sufficient evidence to conclude that a difference exists in mean starting salaries among bachelor's degree candidates in the six fields.

16.65 (a) We will obtain the plots and standard deviations using the normal probability plot in Minitab. Choose **Graph ▶ Probability plot...**, select the **Simple** version, enter WEIGHT_Lahna, WEIGHT_Siika, WEIGHT_Saerki, and WEIGHT_Parkki in the **Variables** text box, Click on the **Multiple graphs** button and select **In separate panels of the same graph**, and click **OK**. Repeat this process with LENGTH_Lahna, LENGTH_Siika, LENGTH_Saerki, and LENGTH Parkki.

The two sets of graphs are

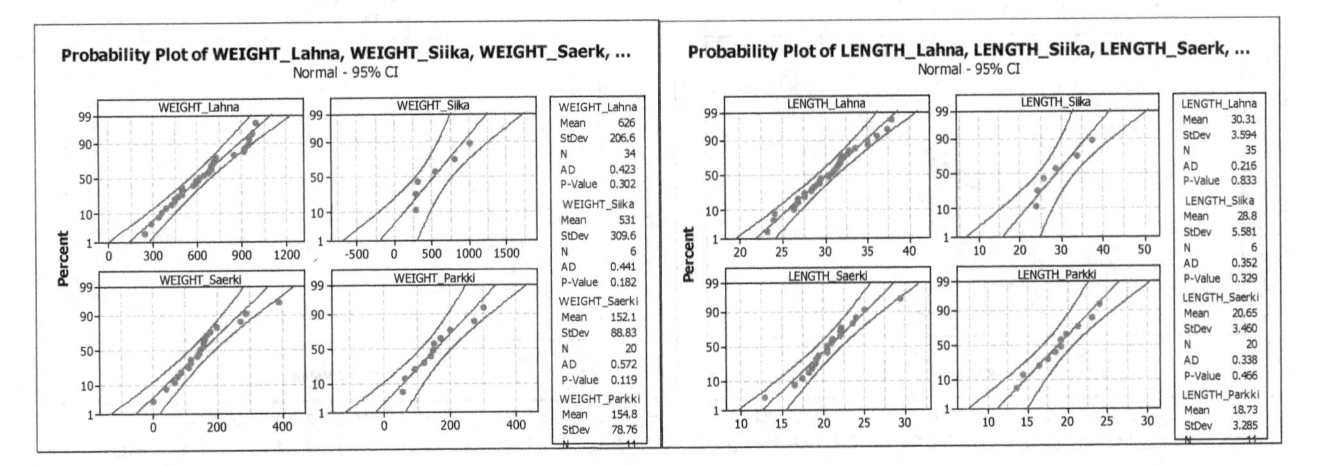

The standard deviations shown at the right of each set of graphs are for the weights: 206.6, 309.6, 88.83, and 78.76. For the lengths they are 3.594, 5.581, 3.460, and 3.285

(b) To carry out the residual analysis (and possibly the ANOVA), we will use the data in the single columns named WEIGHT and LENGTH. In the next column, which is named SPECIES, are names of each of the four species corresponding to each data value in WEIGHT and LENGTH. The residual analysis consists of plotting the residuals against the means (or fits) and constructing a normal probability plot. This is done at the same time as the analysis of variance in Minitab. To do this, we choose **Stat ▶ Anova ▶One-way...**, select WEIGHT for the **Response:** text box and SPECIES in the **Factor:** text box. Enter 95 in the **Confidence level** box. Then click on the **Graphs** button and click to place check marks in the boxes for **Normal plot of residuals** and **Residuals versus fits**, and click **OK** twice. Repeat the process for LENGTH. The residual analysis results for WEIGHT and LENGTH are

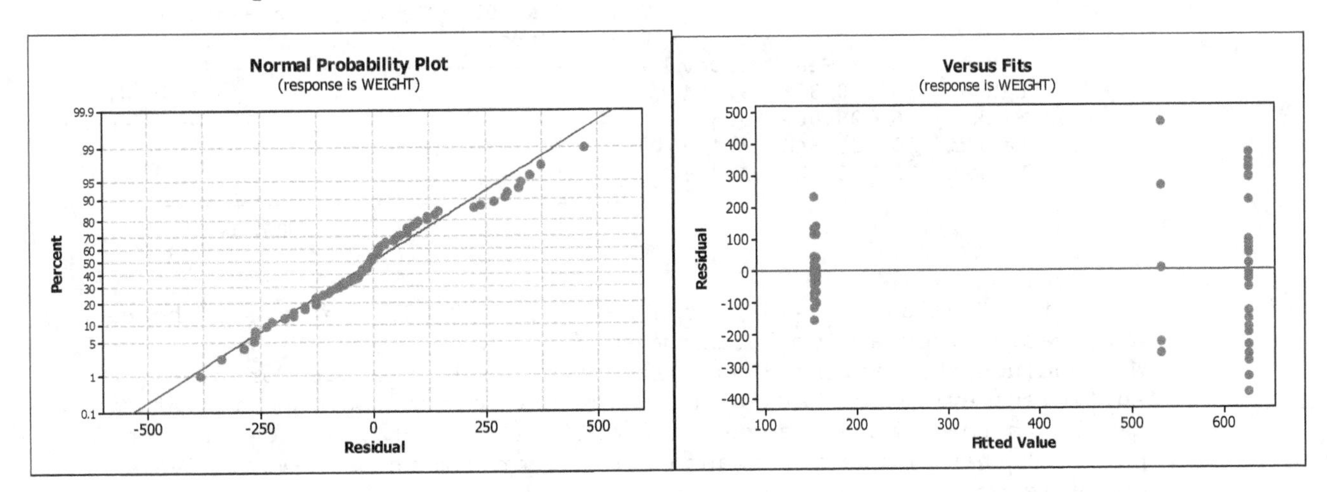

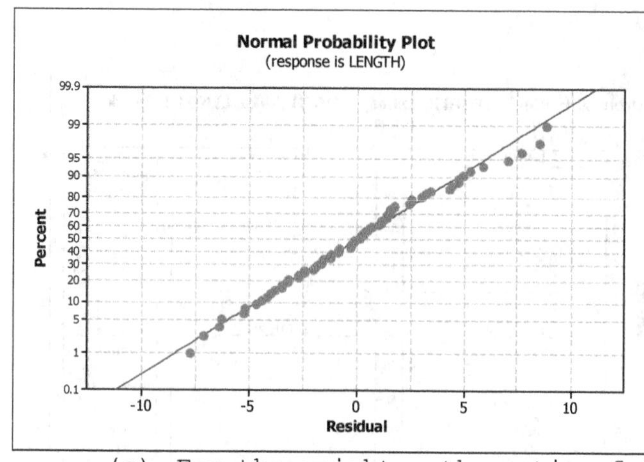

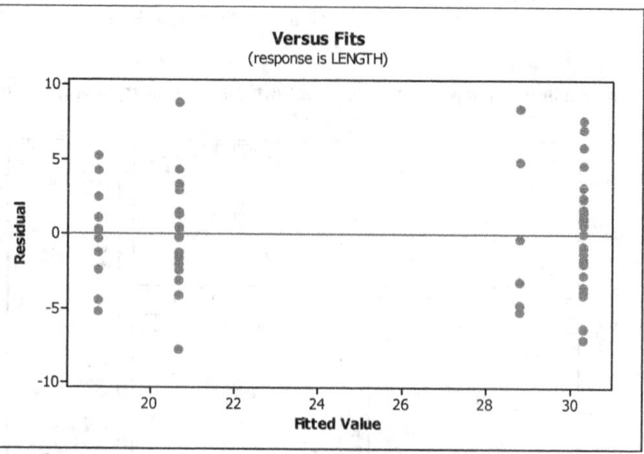

(c) For the weights, the ratio of the largest to smallest standard deviation is 309.6/78.76 = 3.93. The rule of 2 is violated for the weights. Conducting a one-way ANOVA test on the weights data does not appear to be reasonable. Parts (d)-(e) are omitted for the weights. For the lengths, the ratio of the largest to smallest standard deviation is 5.581/3.285 = 1.70. The rule of 2 is not violated for the lengths. The normal probability plots of the individual samples and of the residuals are approximately linear in every plot. Conducting a one-way ANOVA test on these data does appear to be reasonable. Parts (d)-(e) will be performed for the lenths data.

(d) The printed portion of the output from the procedure in part (b) is

One-way ANOVA: LENGTH versus SPECIES

```
Source   DF      SS     MS      F      P
SPECIES   3  1845.9  615.3  44.98  0.000
Error    68   930.2   13.7
Total    71  2776.1

S = 3.699   R-Sq = 66.49%   R-Sq(adj) = 65.02%

                                Individual 95% CIs For Mean Based on
                                Pooled StDev
Level    N    Mean  StDev  ---------+---------+---------+---------+
Lahna   35  30.306  3.594                                  (--*--)
Siika    6  28.800  5.581                         (-------*-------)
Saerki  20  20.645  3.460       (----*---)
Parkki  11  18.727  3.285  (-----*----)
                           ---------+---------+---------+---------+
                                 20.0      24.0      28.0      32.0

Pooled StDev = 3.699
```

Since the F-value of 44.98 has a P-value of 0.000, we reject the null hypothesis of equal population means.

(e) We conclude that at the 5% significance level, the data provide sufficient evidence that the mean lengths of the four species of fish caught in Lake Laengelmaevesi are different. Note: we determined in part (c) that it was not reasonable to perform the one-way ANOVA analysis on the weight data.

16.67 (a) We will obtain the plots and standard deviations using the normal probability plot in Minitab. Choose **Graph ▶ Probability plot...**, select the **Simple** version, enter Meadow Pipit, Tree Pipit, and Hedge Sparrow in the **Variables** text box, Click on the **Multiple graphs** button and select **In separate panels of the same graph**, and click **OK**. Then repeat with Robin, Pied Wagtail, and Wren. The graphs are

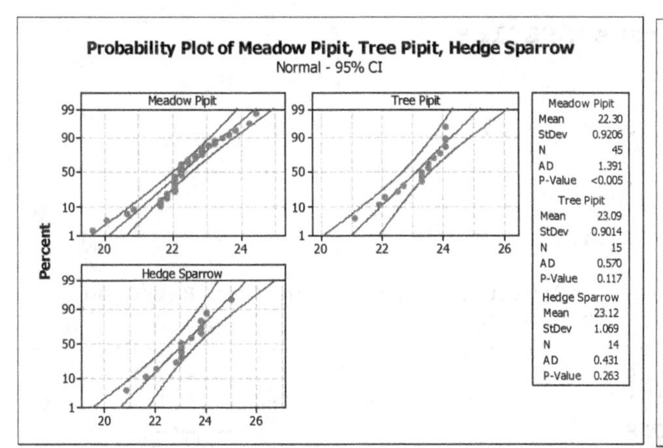

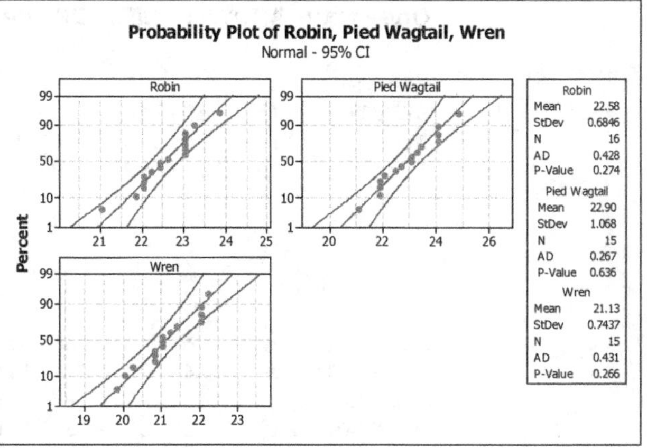

The standard deviations for the six birds are, respectively, 0.9206, 0.9014, 1.069, 0.6846, 1.068, and 0.7437.

(b) To carry out the residual analysis (and possibly the ANOVA), we will use the data in the single column named LENGTH. In the next column, which is named SPECIES, are the names of each of the six species corresponding to each data value in LENGTH. The residual analysis consists of plotting the residuals against the means (or fits) and constructing a normal probability plot. This is done at the same time as the analysis of variance in Minitab. To do this, we choose **Stat ▶ Anova ▶One-way...**, select LENGTH for the **Response:** text box and SPECIES in the **Factor:** text box. Enter 95 in the **Confidence level** box. Then click on the **Graphs** button and click to place check marks in the boxes for **Normal plot of residuals** and **Residuals versus fits**, and click **OK**. The residual analysis results are

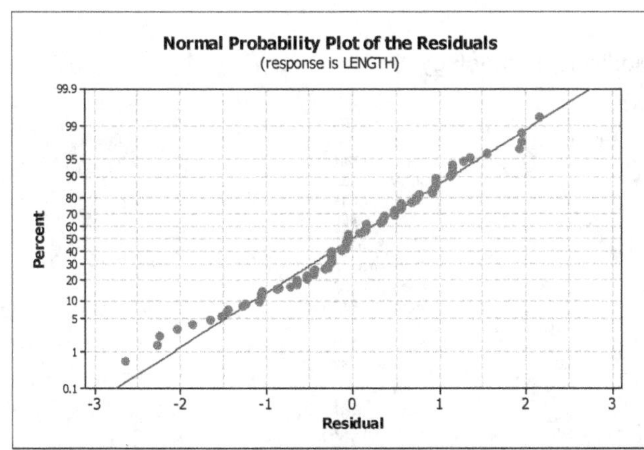

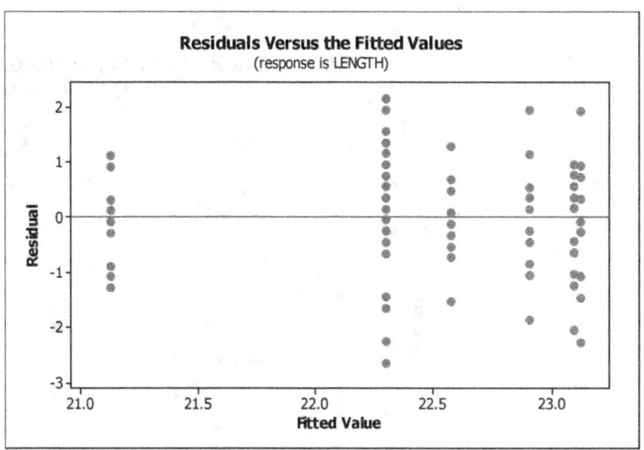

(c) The ratio of the largest to smallest standard deviation is 1.069/0.7437 = 1.44. The rule of 2 is not violated. This is also seen in the graph at the right above. The normal probability plots of the individual samples and of the residuals are all linear except for that of the Meadow Pipit, which appears to have two possible outliers with low values. Conducting a one-way ANOVA test on these data appears to be reasonable, although a more robust alternative procedure may be preferable.

(d) The printed portion of the output from the procedure in part (b) is

One-way ANOVA: LENGTH versus SPECIES

```
Source     DF       SS     MS      F      P
SPECIES     5   42.940  8.588  10.39  0.000
Error     114   94.248  0.827
Total     119  137.188

S = 0.9093   R-Sq = 31.30%   R-Sq(adj) = 28.29%

                                        Individual 95% CIs For Mean Based on
                                        Pooled StDev
Level            N    Mean   StDev   --+---------+---------+---------+-------
Hedge Sparrow   14  23.121   1.069                             (-----*-----)
Meadow Pipit    45  22.299   0.921                    (---*--)
Pied Wagtail    15  22.903   1.068                          (-----*-----)
Robin           16  22.575   0.685                      (----*-----)
Tree Pipit      15  23.090   0.901                             (-----*----)
Wren            15  21.130   0.744   (-----*-----)
                                     --+---------+---------+---------+-------
                                     20.80     21.60     22.40     23.20

Pooled StDev = 0.909
```

Since the F-value of 10.39 has a P-value of 0.000, we reject the null hypothesis of equal population means.

(e) At the 5% significance level, we conclude that the data provide sufficient evidence that the mean egg lengths for the eggs laid by cuckoos in the nests of the six species are different.

16.69 (a) We will obtain the plots and standard deviations using the normal probability plot in Minitab. Choose **Graph ▶ Probability plot...**, select the **Simple** version, enter <u>Law</u>, <u>Science</u>, <u>Medicine</u>, and <u>Technology</u> in the **Variables** text box, Click on the **Multiple graphs** button and select **In separate panels of the same graph**, and click **OK**. The graphs are

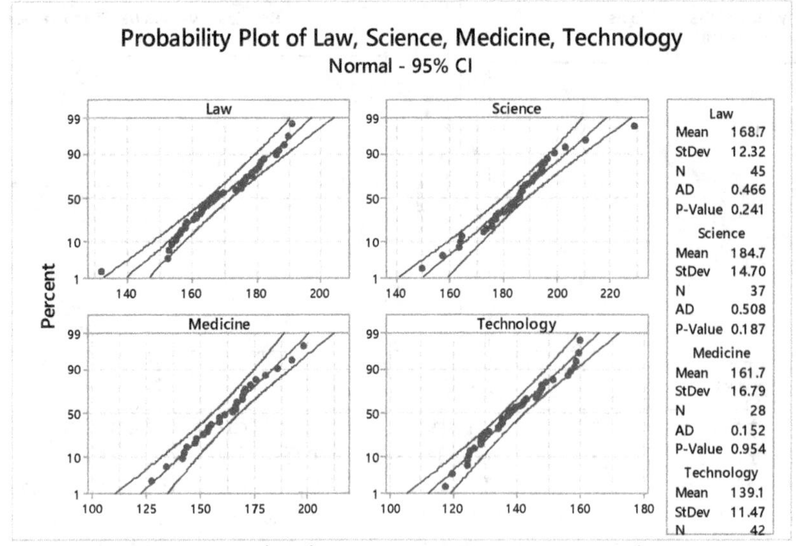

The standard deviations for the four subject areas are, respectively, 12.32, 14.70, 16.79, and 11.47.

(b) To carry out the residual analysis (and possibly the ANOVA), we will use the data in the single column named PRICE. In the next column, which is named SUBJECT, are the names of each of the four subject areas corresponding to each data value in PRICE. The residual analysis consists of plotting the residuals against the means (or fits) and constructing a normal probability plot. This is done at the same time as the analysis of variance in Minitab. To do this, we choose **Stat ▶ Anova ▶One-way...**, select PRICE for the **Response:** text box and SUBJECT in the **Factor:** text box. Enter 95 in the **Confidence level** box. Then click on the **Graphs** button and click to place check marks in the boxes for **Normal plot of residuals** and **Residuals versus fits**, and click **OK**. The residual analysis results are

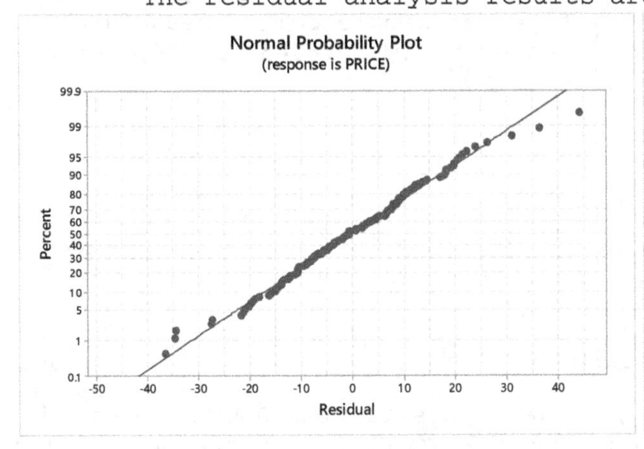

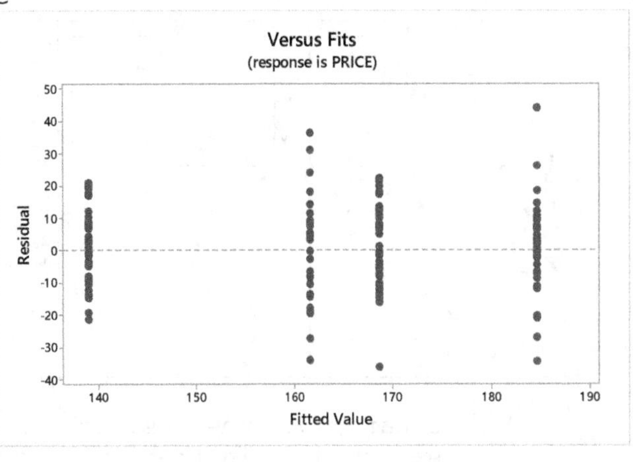

(c) The ratio of the largest to smallest standard deviation is 16.79/11.47 = 1.46. The rule of 2 is not violated. This is also seen in the graph at the right above. The normal probability plots of the individual samples and of the residuals are all linear. Conducting a one-way ANOVA test on these data appears to be reasonable.

(d) The printed portion of the output from the procedure in part (b) is

```
Method
        Null hypothesis          All means are equal
        Alternative hypothesis   At least one mean is different
        Significance level       α = 0.05

        Equal variances were assumed for the analysis.


        Factor Information

        Factor   Levels  Values
        SUBJECT       4  Law, Medicine, Science, Technology


        Analysis of Variance

        Source     DF  Adj SS   Adj MS  F-Value  P-Value
        SUBJECT     3   42930  14310.0    77.13    0.000
        Error     148   27458    185.5
        Total     151   70388

        Model Summary

              S    R-sq  R-sq(adj)  R-sq(pred)
        13.6208  60.99%     60.20%      58.74%
```

Since the F-value of 77.13 has a P-value of 0.000, we reject the null hypothesis of equal population means.

(e) At 5% significance, we conclude that the data provide sufficient evidence that the mean prices of hardcover books in the four subject areas are different.

16.71 (a) We will obtain the plots and standard deviations using the normal probability plot in Minitab. Choose **Graph ▶ Probability plot...**, select the **Simple** version, enter HB SC, HB SS, and HB ST in the **Variables** text box, click on the **Multiple graphs** button and select **In separate panels of the same graph**, and click **OK**. The graphs are

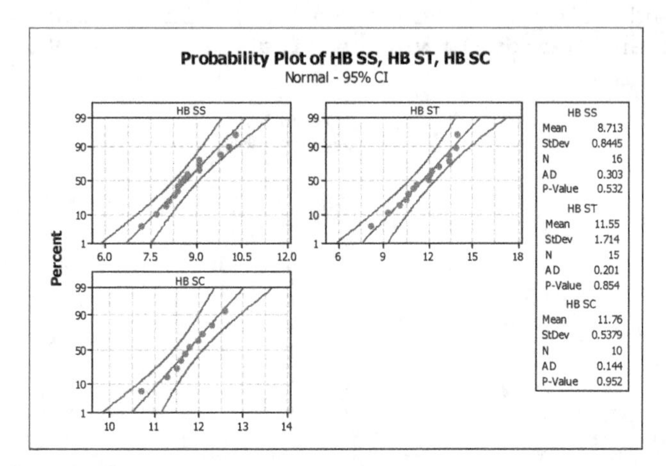

The standard deviations for the three types of sickle cell disease are, respectively, 0.8445, 1.714, and 0.5379.

(b) To carry out the residual analysis (and possibly the ANOVA), we will use the data in the single column named HEMOLEVEL. In the next column, which is named TYPE, are the names of each of the three sickle cell types corresponding to each data value in HEMOLEVEL. The residual analysis consists of plotting the residuals against the means (or fits) and constructing a normal probability plot. This is done at the same time as the analysis of variance in Minitab. To do this, we choose **Stat ▶ Anova ▶One-way...**, select HEMOLEVEL for the **Response:** text box and TYPE in the **Factor:** text box. Enter 95 in the **Confidence level** box. Then click on the **Graphs** button and click to place check marks in the boxes for **Normal plot of residuals** and **Residuals versus fits**, and click **OK**. The residual analysis results are

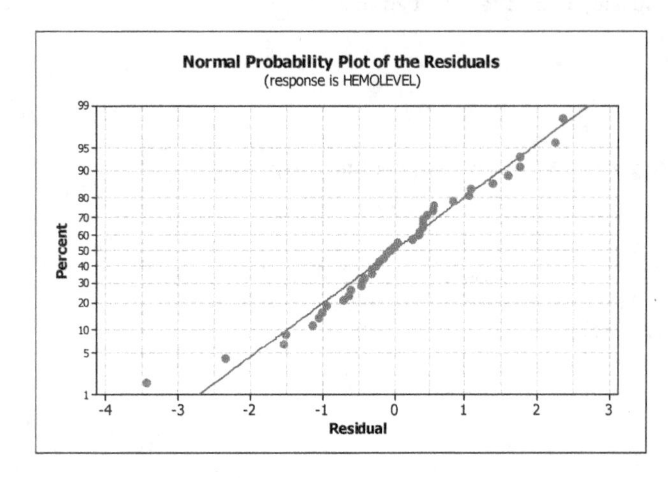

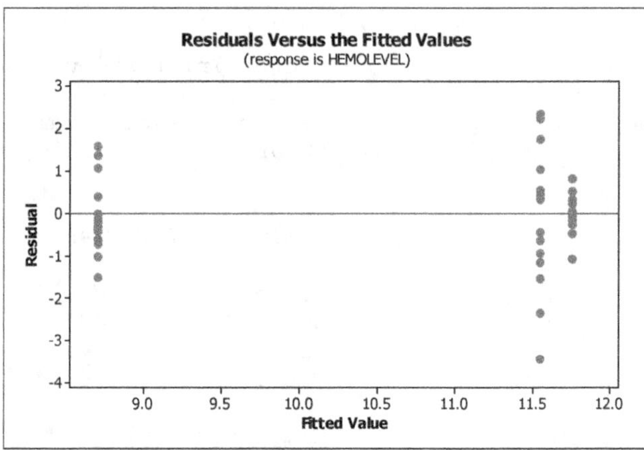

(c) The ratio of the largest to smallest standard deviation is 1.714/0.5379 = 3.19. The rule of 2 is violated. This is also seen in the previous graph at the right. The normal probability plots of the individual samples and of the residuals are all linear. Conducting a one-way ANOVA test on these data may be questionable. We will proceed with the ANOVA and later we will compare the results with those of a more robust procedure.

(d) The printed portion of the output from the procedure in part (b) is

One-way ANOVA: HEMOLEVEL versus TYPE

```
Source  DF      SS     MS      F      P
TYPE     2   83.43  41.71  29.13  0.000
Error   38   54.42   1.43
Total   40  137.85
S = 1.197   R-Sq = 60.52%   R-Sq(adj) = 58.44%

                                 Individual 95% CIs For Mean Based on
                                 Pooled StDev
Level   N    Mean  StDev  --+---------+---------+---------+-------
HB SC  10  11.760  0.538                         (-----*-----)
HB SS  16   8.713  0.844  (----*----)
HB ST  15  11.547  1.714                        (----*----)
                          --+---------+---------+---------+-------
                          8.4       9.6      10.8      12.0

Pooled StDev = 1.197
```

Since the F-value of 29.13 has a P-value of 0.000, we reject the null hypothesis of equal population means, assuming that the assumptions for an ANOVA test are met.

(e) We conclude that the data provide sufficient evidence that the mean hemoglobin levels in the three types of sickle cell disease are different.

16.73 (a) If you are given summary statistics and wish to conduct a one-way ANOVA, you can use the following formula to obtain the mean of all the observations.

$$\bar{x} = \frac{n_1\bar{x}_1 + n_2\bar{x}_2 + \cdots + n_k\bar{x}_k}{n_1 + n_2 + \cdots + n_k}$$

To verify this formula, recall that for the individual means for the samples from the separate populations, $\bar{x}_j = \frac{\sum x_{ji}}{n_j}$. Let the notation

$\sum x_{1i}$ mean to sum up all of the individual observations in the sample from the first population. We can substitute this formula for the individual sample means into the formula for the overall sample mean.

$$\bar{x} = \frac{n_1\frac{\sum x_{1i}}{n_1} + n_2\frac{\sum x_{2i}}{n_2} + \cdots + n_k\frac{\sum x_{ki}}{n_k}}{n_1 + n_2 + \cdots + n_k}.$$

The sample sizes cancel out of the numerator and we are left with

$$\bar{x} = \frac{\sum x_{1i} + \sum x_{2i} + \cdots + \sum x_{ki}}{n_1 + n_2 + \cdots + n_k} = \frac{\sum x_i}{n}.$$

The numerator of this expression means that we will add up all of the observations from each of the individual samples. In other words, add up all of the observations. The denominator of the expression means that we will add up all of the individual sample sizes, which is the overall sample size. Therefore, we have verified the formula for obtaining the mean of all the observations from the summary statistics.

(b) If all of the sample sizes are the same then we can simplify the formula by the following:

$$\bar{x} = \frac{n_1\bar{x}_1 + n_2\bar{x}_2 + \cdots + n_k\bar{x}_k}{n_1 + n_2 + \cdots + n_k} = \frac{n\bar{x}_1 + n\bar{x}_2 + \cdots + n\bar{x}_k}{kn}.$$

We can factor an n out of the numerator which will then cancel out with the n in the denominator. We are left with the following:

$$\bar{x} = \frac{\bar{x}_1 + \bar{x}_2 + \cdots + \bar{x}_k}{k}.$$

This means that, if all the sample sizes are equal, the mean of all the observations is found by adding up all of the sample means and dividing by the number of populations. In other words, it is just the mean of the sample means.

(c) To get the value of the F-statistic from the summary statistics, we use the defining formulas for the SSTR and SSE calculations. First, calculate SSTR. $SSTR = \sum n_i(\bar{x}_i - \bar{x})^2$. For each of the sample means, find its deviation from the overall sample mean found using the formula in part (a), square this deviation, multiply it by the respective sample size, and find the sum of all these terms for each of the samples. For SSE, $SSE = \sum (n_i - 1)s_i^2$, square each sample standard deviation, multiply it by the sample size minus one, and find the sum of all these terms for each of the samples. Then, the MSTR value is found by dividing the SSTR term by the degrees of freedom for the treatment: (k - 1). The MSE term is found by dividing the SSE term by the degrees of freedom for the error: (n - k). Then, finally, the F- statistic is found by dividing MSTR by MSE.

16.75 Suppose that we define the sample events A and B as follows:

A: the interval constructed around the difference $\mu_1 - \mu_2$

B: the interval constructed around the difference $\mu_1 - \mu_3$.

The 95% confidence interval for each event above is defined as:

P(A) = 0.95 and P(B) = 0.95.

The probability of both A and B occurring simultaneously is written P(A and B). Since A and B are realistically not independent, the general multiplication rule applies; i.e., P(A and B) = P(A)•P(B|A).

A difficulty arises in calculating a precise number for P(A and B) because we need additional information about P(B|A), which we do not have. However, P(B|A) is by no means equal to 1.00, which results in the product of P(A) and P(B|A) being less than 0.95 (because 0.95 times a probability less than 1.00 is clearly less than 0.95). Thus, the probability of both A and B occurring simultaneously is not 0.95.

Exercises 16.4

16.77 zero

16.79 (a) The family confidence level is smaller because the family confidence level is the confidence we have that all the confidence intervals contain the true differences between the population means. On the other hand, the individual confidence level pertains to the confidence we have that any particular confidence interval contains the true difference between the two corresponding population means.

(b) They are the same since there is only one difference in the family.

16.81 ν = n - k, the degrees of freedom for the denominator in the F-distribution.

16.83 (a) 4.69 (b) 5.98

16.85 (a) 5.65 (b) 4.72

16.87 Since the ANOVA test was not significant at the 5% level, we conclude that that there are no differences between the population means. Thus, the Tukey multiple comparison will show that each confidence interval for the difference between two means contains zero.

16.89 From Exercise 16.25, we have MSE = 2.286, κ = 3, and n = 10.

Step 1: Family confidence level = 0.95.

Step 2: κ = 3 and ν = n - k = 10 - 3 = 7. Consulting Table X, we find that $q_{0.05}$ = 4.16.

Step 3: Before obtaining all possible confidence intervals for $\mu_i - \mu_j$, we construct a table giving the sample means and sample sizes.

j	1	2	3
$\bar{x}_j$	6	2	4
n_j	3	3	4

Now we are ready to obtain the required confidence intervals.

The endpoints for $\mu_1 - \mu_2$ are

$$(6-2) \pm \frac{4.16}{\sqrt{2}} \cdot \sqrt{2.286} \cdot \sqrt{\frac{1}{3}+\frac{1}{3}} = 4 \pm 3.63 = 0.37 \quad to \quad 7.63$$

The endpoints for $\mu_1 - \mu_3$ are:

$$(6-4) \pm \frac{4.16}{\sqrt{2}} \cdot \sqrt{2.286} \cdot \sqrt{\frac{1}{3}+\frac{1}{4}} = 2 \pm 3.40 = -1.40 \quad to \quad 5.40$$

The endpoints for $\mu_2 - \mu_3$ are:

$$(2-4) \pm \frac{4.16}{\sqrt{2}} \cdot \sqrt{2.286} \cdot \sqrt{\frac{1}{3}+\frac{1}{4}} = -2 \pm 3.40 = -5.40 \quad to \quad 1.40$$

Step 4: Based on the confidence intervals in Step 3, μ_1 is different from μ_2. Since the null hypothesis of equal means was rejected in exercise 16.43, there must be at least one of the three confidence intervals that does not contain zero. The other two intervals do contain zero, indicating that μ_1 is not different from μ_3, and μ_2 is not different from μ_3.

16.91 From Exercise 16.27, we have MSE = 3.467, κ = 5, and n = 20.

Step 1: Family confidence level = 0.95.

Step 2: κ = 5 and ν = n - k = 20 - 5 = 15. Consulting Table X, we find that $q_{0.05}$ = 4.41.

Step 3: Before obtaining all possible confidence intervals for $\mu_i - \mu_j$, we construct a table giving the sample means and sample sizes.

j	1	2	3	4	5
$\bar{x}_j$	5	6	6	5	9
n_j	4	3	5	5	3

Now we are ready to obtain the required confidence intervals.

The endpoints for $\mu_1 - \mu_2$ are

$$(5-6)\pm\frac{4.41}{\sqrt{2}}\cdot\sqrt{3.467}\cdot\sqrt{\frac{1}{4}+\frac{1}{3}}=-1\pm 4.43=-5.43 \quad to \quad 3.43$$

The endpoints for $\mu_1 - \mu_3$ are

$$(5-6)\pm\frac{4.41}{\sqrt{2}}\cdot\sqrt{3.467}\cdot\sqrt{\frac{1}{4}+\frac{1}{5}}=-1\pm 3.89=-4.89 \quad to \quad 2.89$$

The endpoints for $\mu_1 - \mu_4$ are

$$(5-5)\pm\frac{4.41}{\sqrt{2}}\cdot\sqrt{3.467}\cdot\sqrt{\frac{1}{4}+\frac{1}{5}}=0\pm 3.89=-3.89 \quad to \quad 3.89$$

The endpoints for $\mu_1 - \mu_5$ are

$$(5-9)\pm\frac{4.41}{\sqrt{2}}\cdot\sqrt{3.467}\cdot\sqrt{\frac{1}{4}+\frac{1}{3}}=-4\pm 4.43=-8.43 \quad to \quad 0.43$$

The endpoints for $\mu_2 - \mu_3$ are

$$(6-6)\pm\frac{4.41}{\sqrt{2}}\cdot\sqrt{3.467}\cdot\sqrt{\frac{1}{3}+\frac{1}{5}}=0\pm 4.24=-4.24 \quad to \quad 4.24$$

The endpoints for $\mu_2 - \mu_4$ are

$$(6-5)\pm\frac{4.41}{\sqrt{2}}\cdot\sqrt{3.467}\cdot\sqrt{\frac{1}{3}+\frac{1}{5}}=1\pm 4.24=-3.24 \quad to \quad 5.24$$

The endpoints for $\mu_2 - \mu_5$ are

$$(6-9)\pm\frac{4.41}{\sqrt{2}}\cdot\sqrt{3.467}\cdot\sqrt{\frac{1}{3}+\frac{1}{3}}=-3\pm 4.74=-7.74 \quad to \quad 1.74$$

The endpoints for $\mu_3 - \mu_4$ are

$$(6-5)\pm\frac{4.41}{\sqrt{2}}\cdot\sqrt{3.467}\cdot\sqrt{\frac{1}{5}+\frac{1}{5}}=1\pm 3.6=-2.67 \quad to \quad 4.67$$

The endpoints for $\mu_3 - \mu_5$ are

$$(6-9)\pm\frac{4.41}{\sqrt{2}}\cdot\sqrt{3.467}\cdot\sqrt{\frac{1}{5}+\frac{1}{3}}=-3\pm 4.24=-7.24 \quad to \quad 1.24$$

The endpoints for $\mu_4 - \mu_5$ are

$$(5-9)\pm\frac{4.41}{\sqrt{2}}\cdot\sqrt{3.467}\cdot\sqrt{\frac{1}{5}+\frac{1}{3}}=-4\pm 4.24=-8.24 \quad to \quad 0.24$$

Step 4: Based on the confidence intervals in Step 3, all of which contain zero, none of the means is different from the others. This is consistent with the result of Exercise 16.45 in which the null hypothesis of equal means was not rejected.

16.93 From Exercise 16.29, we have MSE = 9, κ = 4, and n = 12.

Step 1: Family confidence level = 0.95.

Step 2: $\kappa = 4$ and $\nu = n - k = 12 - 4 = 8$. Consulting Table X, we find that $q_{0.05} = 4.53$.

Step 3: Before obtaining all possible confidence intervals for $\mu_i - \mu_j$, we construct a table giving the sample means and sample sizes.

j	1	2	3	4
$\bar{x}_j$	8	5	12	3
n_j	3	3	3	3

Now we are ready to obtain the required confidence intervals.

The endpoints for $\mu_1 - \mu_2$ are

$$(8-5) \pm \frac{4.53}{\sqrt{2}} \cdot \sqrt{9} \cdot \sqrt{\frac{1}{3}+\frac{1}{3}} = 3 \pm 7.846 = -4.846 \quad to \quad 10.846$$

The endpoints for $\mu_1 - \mu_3$ are

$$(8-12) \pm \frac{4.53}{\sqrt{2}} \cdot \sqrt{9} \cdot \sqrt{\frac{1}{3}+\frac{1}{3}} = -4 \pm 7.846 = -11.846 \quad to \quad 3.846$$

The endpoints for $\mu_1 - \mu_4$ are

$$(8-3) \pm \frac{4.53}{\sqrt{2}} \cdot \sqrt{9} \cdot \sqrt{\frac{1}{3}+\frac{1}{3}} = 5 \pm 7.846 = -2.846 \quad to \quad 12.846$$

The endpoints for $\mu_2 - \mu_3$ are

$$(5-12) \pm \frac{4.53}{\sqrt{2}} \cdot \sqrt{9} \cdot \sqrt{\frac{1}{3}+\frac{1}{3}} = -7 \pm 7.846 = -14.846 \quad to \quad 0.846$$

The endpoints for $\mu_2 - \mu_4$ are

$$(5-3) \pm \frac{4.53}{\sqrt{2}} \cdot \sqrt{9} \cdot \sqrt{\frac{1}{3}+\frac{1}{3}} = 2 \pm 7.846 = -5.846 \quad to \quad 9.846$$

The endpoints for $\mu_3 - \mu_4$ are

$$(12-3) \pm \frac{4.53}{\sqrt{2}} \cdot \sqrt{9} \cdot \sqrt{\frac{1}{3}+\frac{1}{3}} = 9 \pm 7.846 = 1.154 \quad to \quad 16.846$$

Step 4: Based on the confidence intervals in Step 3, μ_3 is different from μ_4. Since the null hypothesis of equal means was rejected in exercise 16.47, there must be at least one of the confidence intervals that does not contain zero. The other intervals do contain zero, indicating that μ_1 is not different from μ_2 or μ_3 or μ_4, and μ_2 is not different from μ_3 or μ_4,

16.95 Step 1: Family confidence level = 0.95.

Step 2: $\kappa = 3$ and $\nu = n - k = 12 - 3 = 9$. Consulting Table X, we find that $q_{0.05} = 3.95$.

Step 3: Before obtaining all possible confidence intervals for $\mu_i - \mu_j$, we construct a table giving the sample means and sample sizes.

j	1	2	3
$\bar{x}_j$	457.00	306.25	293.75
n_j	4	4	4

In Exercise 16.49, we found that MSE = 605.056. Now we are ready to obtain the required confidence intervals.

The endpoints for $\mu_1 - \mu_2$ are

$$(457.00-306.25) \pm \frac{3.95}{\sqrt{2}} \cdot \sqrt{605.056} \cdot \sqrt{\frac{1}{4}+\frac{1}{4}} = 150.75 \pm 48.58 = 102.17 \quad to \quad 199.33$$

The endpoints for $\mu_1 - \mu_3$ are:

$$(457.00-293.75) \pm \frac{3.95}{\sqrt{2}} \cdot \sqrt{605.056} \cdot \sqrt{\frac{1}{4}+\frac{1}{4}} = 163.25 \pm 48.58 = 114.67 \quad to \quad 211.83$$

The endpoints for $\mu_2 - \mu_3$ are:

$$(306.5-293.75) \pm \frac{3.95}{\sqrt{2}} \cdot \sqrt{605.056} \cdot \sqrt{\frac{1}{4}+\frac{1}{4}} = 12.50 \pm 48.58 = -36.08 \quad to \quad 61.08$$

Step 4: Based on the confidence intervals in Step 3, μ_1 is different from μ_2, and μ_1 is different from μ_3; μ_2 and μ_3 cannot be declared different.

Step 5: We summarize the results with the following diagram.

Microalgae	Bacteria	Diatoms
(3)	(2)	(1)
293.75	306.25	457.00

(A line is drawn under 293.75 and 306.25.)

Interpreting this diagram, we conclude with 95% confidence that μ_1 is different from μ_2, and μ_1 is different from μ_3.

16.97 (a) From Exercise 16.51, we have MSE = 182, κ = 5, and n = 25.

Step 1: Family confidence level = 0.95.

Step 2: $\kappa = 5$ and $\nu = n - k = 25 - 5 = 20$. Consulting Table X, we find that $q_{0.05} = 4.23$.

Step 3: Before obtaining all possible confidence intervals for $\mu_i - \mu_j$, we construct a table giving the sample means and sample sizes.

j	1	2	3	4	5
$\bar{x}_j$	20.8	25.8	37.0	19.6	38.8
n_j	5	5	5	5	5

Now we are ready to obtain the required confidence intervals.

The endpoints for $\mu_1 - \mu_2$ are

$$(20.8-25.8) \pm \frac{4.23}{\sqrt{2}} \cdot \sqrt{182} \cdot \sqrt{\frac{1}{5}+\frac{1}{5}} = -5.0 \pm 25.5 = -30.5 \quad to \quad 20.5$$

The endpoints for $\mu_1 - \mu_3$ are

$$(20.8-37.0) \pm \frac{4.23}{\sqrt{2}} \cdot \sqrt{182} \cdot \sqrt{\frac{1}{5}+\frac{1}{5}} = -16.2 \pm 25.5 = -41.7 \quad to \quad 9.3$$

The endpoints for $\mu_1 - \mu_4$ are

$$(20.8-19.6) \pm \frac{4.23}{\sqrt{2}} \cdot \sqrt{182} \cdot \sqrt{\frac{1}{5}+\frac{1}{5}} = 1.2 \pm 25.5 = -24.3 \quad to \quad 26.7$$

The endpoints for $\mu_1 - \mu_5$ are

$$(20.8-38.8)\pm\frac{4.23}{\sqrt{2}}\cdot\sqrt{182}\cdot\sqrt{\frac{1}{5}+\frac{1}{5}}=-18.0\pm25.5=-43.8 \ \textit{to} \ 7.5$$

The endpoints for $\mu_2 - \mu_3$ are

$$(25.8-37.0)\pm\frac{4.23}{\sqrt{2}}\cdot\sqrt{182}\cdot\sqrt{\frac{1}{5}+\frac{1}{5}}=-11.2\pm25.5=-36.7 \ \textit{to} \ 14.3$$

The endpoints for $\mu_2 - \mu_4$ are

$$(25.8-19.6)\pm\frac{4.23}{\sqrt{2}}\cdot\sqrt{182}\cdot\sqrt{\frac{1}{5}+\frac{1}{5}}=6.2\pm25.5=-19.3 \ \textit{to} \ 31.7$$

The endpoints for $\mu_2 - \mu_5$ are

$$(25.8-38.8)\pm\frac{4.23}{\sqrt{2}}\cdot\sqrt{182}\cdot\sqrt{\frac{1}{5}+\frac{1}{5}}=-13.0\pm25.5=-38.5 \ \textit{to} \ 12.5$$

The endpoints for $\mu_3 - \mu_4$ are

$$(37.0-19.6)\pm\frac{4.23}{\sqrt{2}}\cdot\sqrt{182}\cdot\sqrt{\frac{1}{5}+\frac{1}{5}}=17.4\pm25.5=-8.1 \ \textit{to} \ 42.9$$

The endpoints for $\mu_3 - \mu_5$ are

$$(37.0-38.8)\pm\frac{4.23}{\sqrt{2}}\cdot\sqrt{182}\cdot\sqrt{\frac{1}{5}+\frac{1}{5}}=-1.8\pm25.5=-27.3 \ \textit{to} \ 23.7$$

The endpoints for $\mu_4 - \mu_5$ are

$$(19.6-38.8)\pm\frac{4.23}{\sqrt{2}}\cdot\sqrt{182}\cdot\sqrt{\frac{1}{5}+\frac{1}{5}}=-19.2\pm25.5=-44.7 \ \textit{to} \ 6.3$$

Step 4: Based on the confidence intervals in Step 3, all of which contain zero, none of the means is different from the others.

(b) The data do not provide sufficient evidence at the 5% significance level to conclude that a difference exists in mean bacteria counts among the five strains of *Staphylococcus aureus*. If there were such evidence, then at least one of the intervals found in part (a) would not have contained zero.

16.99 Step 1: Family confidence level = 0.95.

Step 2: κ = 4 and ν = n - k = 34 - 4 = 30. Consulting Table X, we find that $q_{0.05}$ = 3.85 for the entry with df = (4,30).

Step 3: Before obtaining all possible confidence intervals for $\mu_i - \mu_j$, we construct a table giving the sample means and sample sizes.

j	1	2	3	4
$\bar{x}_j$	7.69	10.44	8.55	5.91
n_j	8	8	10	8

In Exercise 16.53, we found that MSE = 4.314. Now we are ready to obtain the required confidence intervals.

The endpoints for $\mu_1 - \mu_2$ are

$$(7.69-10.44)\pm\frac{3.85}{\sqrt{2}}\cdot\sqrt{4.314}\sqrt{\frac{1}{8}+\frac{1}{8}}=-2.75\pm2.83=-5.58 \text{ to } 0.08$$

The endpoints for $\mu_1 - \mu_3$ are:

$$(7.69-8.55)\pm\frac{3.85}{\sqrt{2}}\cdot\sqrt{4.314}\sqrt{\frac{1}{8}+\frac{1}{10}}=-0.86\pm2.68=-3.54 \text{ to } 1.82$$

The endpoints for $\mu_1 - \mu_4$ are:

$$(7.69-5.91)\pm\frac{3.85}{\sqrt{2}}\cdot\sqrt{4.314}\sqrt{\frac{1}{8}+\frac{1}{8}}=1.78\pm2.83=-1.05 \text{ to } 4.61$$

The endpoints for $\mu_2 - \mu_3$ are:

$$(10.44-8.55)\pm\frac{3.85}{\sqrt{2}}\cdot\sqrt{4.314}\sqrt{\frac{1}{8}+\frac{1}{10}}=1.89\pm2.68=-0.79 \text{ to } 4.57$$

The endpoints for $\mu_2 - \mu_4$ are:

$$(10.44-5.91)\pm\frac{3.85}{\sqrt{2}}\cdot\sqrt{4.314}\sqrt{\frac{1}{8}+\frac{1}{8}}=4.53\pm2.83=1.70 \text{ to } 7.36$$

The endpoints for $\mu_3 - \mu_4$ are:

$$(8.55-5.91)\pm\frac{3.85}{\sqrt{2}}\cdot\sqrt{4.314}\sqrt{\frac{1}{10}+\frac{1}{8}}=2.64\pm2.68=-0.04 \text{ to } 5.32$$

Step 4: Based on the confidence intervals in Step 3, only μ_2 can be determined different from μ_4.

Step 5: We summarize the results with the following diagram.

Brand D	Brand A	Brand C	Brand B
(4)	(1)	(3)	(2)
5.91	7.69	8.55	10.44

(A line underlines Brand D, Brand A and Brand C; a second line underlines Brand A, Brand C and Brand B.)

Interpreting this diagram, we conclude with 95% confidence that μ_2 is different from μ_4 and that the mean battery life of laptops from Brand B is different than that from Brand D.

16.101 Using Minitab, with all of the data in a column named RENT and the regions of the U.S. in a column named REGION, we choose **Stat ▶ ANOVA ▶ Oneway...**, select RENT in the **Response** text box and REGION in the **Factor** text box, click on the **Comparisons** button, click on **Tukey's, family error rate** so that an X shows in its check-box, type 5 in its text box, and click **OK**. The result is [The first part of the output will be identical to the output given in Exercise 16.55.]

```
Tukey 95% Simultaneous Confidence Intervals
All Pairwise Comparisons among Levels of REGION

Individual confidence level = 98.87%

REGION = Midwest subtracted from:

REGION       Lower  Center  Upper  ------+---------+---------+---------+---
Northeast     81.5   312.2  542.9                    (------*-------)
South       -134.5   111.5  357.5              (-------*-------)
West          90.7   321.4  552.1                    (-------*------)
                                   ------+---------+---------+---------+---
                                      -300         0       300       600
```

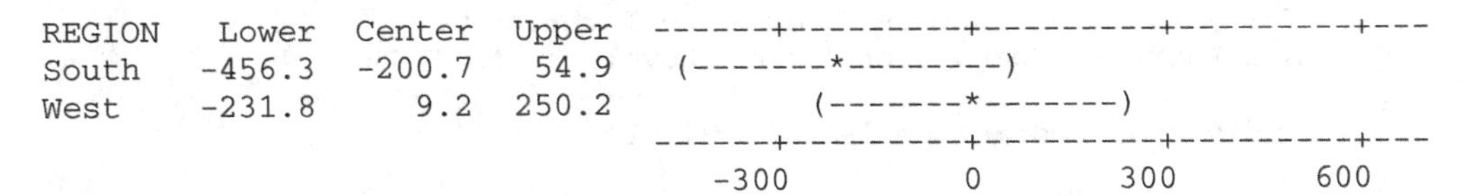

```
REGION = Northeast subtracted from:

REGION   Lower  Center  Upper  ------+---------+---------+---------+---
South   -456.3  -200.7   54.9   (-------*--------)
West    -231.8     9.2  250.2         (-------*-------)
                               ------+---------+---------+---------+---
                                  -300         0       300       600
```

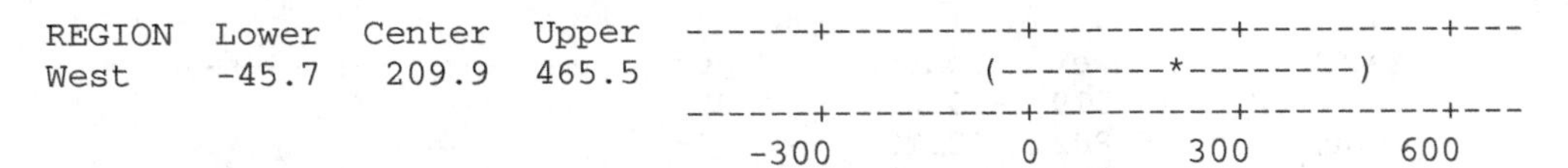

```
REGION = South subtracted from:

REGION  Lower  Center  Upper  ------+---------+---------+---------+---
West    -45.7   209.9  465.5            (--------*--------)
                              ------+---------+---------+---------+---
                                 -300         0       300       600
```

Looking at the intervals that do not include zero, we conclude that the mean monthly rent in the Midwest is different from that in the Northeast and from that in the West. All other intervals contain zero; therefore, we cannot declare the means for those pairs to be different.

16.103 Using Minitab, with all of the data in a column named RATE and the patch size manipulations in a column named BREAST, we choose **Stat ▶ ANOVA ▶ Oneway...**, select RATE in the **Response** text box and BREAST in the **Factor** text box, click on the **Comparisons** button, click on **Tukey's, family error rate** so that an X shows in its check-box, type 1 in its text box, and click **OK**. The result is [The first part of the output will be identical to the output given in Exercise 16.57.]

```
Tukey 99% Simultaneous Confidence Intervals
All Pairwise Comparisons among Levels of BREAST

Individual confidence level = 99.63%
```

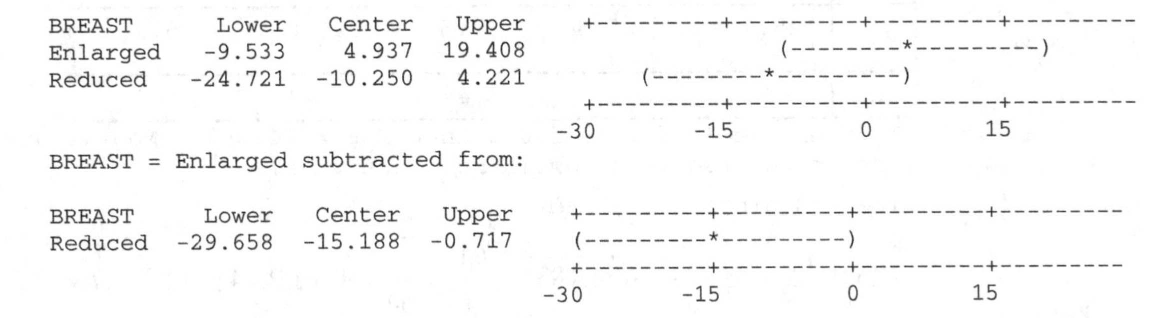

```
BREAST = Control subtracted from:

BREAST      Lower   Center   Upper  +---------+---------+---------+---------
Enlarged   -9.533    4.937  19.408                (--------*---------)
Reduced   -24.721  -10.250   4.221       (--------*---------)
                                    +---------+---------+---------+---------
                                  -30       -15         0        15

BREAST = Enlarged subtracted from:

BREAST      Lower   Center   Upper  +---------+---------+---------+---------
Reduced   -29.658  -15.188  -0.717  (---------*---------)
                                    +---------+---------+---------+---------
                                  -30       -15         0        15
```

Looking at the intervals that do not include zero, we conclude that the mean singing time in the enlarged patch-size group is different from that in the reduced patch-size group. The mean singing time in the control group cannot be declared different from that of either the reduced or enlarged patch-size groups.

16.105 Using Minitab, with all of the data in a column named HARDNESS and the type of artificial tooth material in a column named MATERIAL, we choose **Stat ▶ ANOVA ▶ Oneway...**, select HARDNESS in the **Response** text box and MATERIAL in the **Factor** text box, click on the **Comparisons** button, click on **Tukey's, family error rate** so that an X shows in its check-box, type 5 in its text box, and click **OK**. The result is [The first part of the output will be identical to the output given in Exercise 16.59.]

```
Tukey 99% Simultaneous Confidence Intervals
All Pairwise Comparisons among Levels of MATERIAL

Individual confidence level = 99.62%


MATERIAL = Duracross subtracted from:

MATERIAL    Lower   Center    Upper  --------+---------+---------+---------+-
Duradent  -19.319  -15.617  -11.915  (-----*----)
Endura    -15.802  -12.100   -8.398       (-----*----)
                                     --------+---------+---------+---------+-
                                          -14.0      -7.0        0.0       7.0

MATERIAL = Duradent subtracted from:

MATERIAL   Lower  Center  Upper  --------+---------+---------+---------+-
Endura    -0.185   3.517  7.219                               (----*----)
                                 --------+---------+---------+---------+-
                                      -14.0      -7.0        0.0       7.0
```

Looking at the intervals that do not include zero, we conclude that the mean hardness of Duracross is different from that of Duradent and of Endura, but we cannot declare that the hardness of Duradent is different from that of Endura.

16.107 Step 1: Family confidence level = 0.90.

Step 2: κ = 5 and ν = n - k = 236 - 5 = 231. Consulting the problem, we are given that $q_{0.10}$ = 3.50.

Step 3: Before obtaining all possible confidence intervals for $\mu_i - \mu_j$, we construct a table giving the sample means and sample sizes.

j	1	2	3	4	5
$\bar{x}_j$	25.9	25.8	27.5	26.0	25.9
n_j	46	54	47	42	47

In Exercise 16.61, we found that MSE = 24.183. Now we are ready to obtain the required confidence intervals.

The endpoints for $\mu_1 - \mu_2$ are

$$(25.9-25.8)\pm\frac{3.50}{\sqrt{2}}\cdot\sqrt{24.1834}\cdot\sqrt{\frac{1}{46}+\frac{1}{54}}=0.1\pm2.44=-2.34 \quad to \quad 2.54$$

The endpoints for $\mu_1-\mu_3$ are:

$$(25.9-27.5)\pm\frac{3.50}{\sqrt{2}}\cdot\sqrt{24.1834}\cdot\sqrt{\frac{1}{46}+\frac{1}{47}}=-1.6\pm2.52=-4.12 \quad to \quad 0.92$$

The endpoints for $\mu_1-\mu_4$ are:

$$(25.9-26.0)\pm\frac{3.50}{\sqrt{2}}\cdot\sqrt{24.1834}\cdot\sqrt{\frac{1}{46}+\frac{1}{42}}=-0.1\pm2.60=-2.70 \quad to \quad 2.50$$

The endpoints for $\mu_1-\mu_5$ are:

$$(25.9-25.9)\pm\frac{3.50}{\sqrt{2}}\cdot\sqrt{24.1834}\cdot\sqrt{\frac{1}{46}+\frac{1}{47}}=0\pm2.52=-2.52 \quad to \quad 2.52$$

The endpoints for $\mu_2-\mu_3$ are:

$$(25.8-27.5)\pm\frac{3.50}{\sqrt{2}}\cdot\sqrt{24.1834}\cdot\sqrt{\frac{1}{54}+\frac{1}{47}}=-1.7\pm2.43=-4.13 \quad to \quad 0.73$$

The endpoints for $\mu_2-\mu_4$ are:

$$(25.8-26.0)\pm\frac{3.50}{\sqrt{2}}\cdot\sqrt{24.1834}\cdot\sqrt{\frac{1}{54}+\frac{1}{42}}=-0.2\pm2.50=-2.70 \quad to \quad 2.30$$

The endpoints for $\mu_2-\mu_5$ are:

$$(25.8-25.9)\pm\frac{3.50}{\sqrt{2}}\cdot\sqrt{24.1834}\cdot\sqrt{\frac{1}{54}+\frac{1}{47}}=-0.1\pm2.43=-2.53 \quad to \quad 2.33$$

The endpoints for $\mu_3-\mu_4$ are:

$$(27.5-26.0)\pm\frac{3.50}{\sqrt{2}}\cdot\sqrt{24.1834}\cdot\sqrt{\frac{1}{47}+\frac{1}{42}}=1.5\pm2.58=-1.08 \quad to \quad 4.08$$

The endpoints for $\mu_3-\mu_5$ are:

$$(27.5-25.9)\pm\frac{3.50}{\sqrt{2}}\cdot\sqrt{24.1834}\cdot\sqrt{\frac{1}{47}+\frac{1}{47}}=1.60\pm2.51=-0.91 \quad to \quad 4.11$$

The endpoints for $\mu_4-\mu_5$ are:

$$(26.0-25.9)\pm\frac{3.50}{\sqrt{2}}\cdot\sqrt{24.1834}\cdot\sqrt{\frac{1}{42}+\frac{1}{47}}=0.1\pm2.58=-2.48 \quad to \quad 2.68$$

Step 4: Based on the confidence intervals in Step 3, no two pairs of means can be declared different.

Step 5: We summarize the results with the following diagram.

14mg	210mg	Placebo	100mg	60mg
(2)	(5)	(1)	(4)	(3)
25.8	25.9	25.9	26.0	27.5

Interpreting this diagram, we conclude with 90% confidence, there is not a difference between any two mean body mass indexes in the five denosumab treatment groups.

16.109 Step 1: Family confidence level = 0.99.

Step 2: κ = 6 and ν = n - k = 160 - 6 = 154. Consulting the problem, we are given that $q_{0.01}$ = 4.85.

Step 3: Refer to the problem for the sample means and sample sizes. In Exercise 16.63, we found that MSE = 26.850. Now we are ready to obtain the required confidence intervals.

The endpoints for $\mu_1 - \mu_2$ are

$$(55.1 - 44.6) \pm \frac{4.85}{\sqrt{2}} \cdot \sqrt{26.850} \cdot \sqrt{\frac{1}{46} + \frac{1}{11}} = 10.5 \pm 5.96 = 4.54 \text{ to } 16.46$$

The endpoints for $\mu_1 - \mu_3$ are:

$$(55.1 - 59.1) \pm \frac{4.85}{\sqrt{2}} \cdot \sqrt{26.850} \cdot \sqrt{\frac{1}{46} + \frac{1}{30}} = -4.0 \pm 4.17 = -8.17 \text{ to } 0.17$$

The endpoints for $\mu_1 - \mu_4$ are:

$$(55.1 - 40.6) \pm \frac{4.85}{\sqrt{2}} \cdot \sqrt{26.850} \cdot \sqrt{\frac{1}{46} + \frac{1}{11}} = 14.5 \pm 5.96 = 8.54 \text{ to } 20.46$$

The endpoints for $\mu_1 - \mu_5$ are:

$$(55.1 - 62.6) \pm \frac{4.85}{\sqrt{2}} \cdot \sqrt{26.850} \cdot \sqrt{\frac{1}{46} + \frac{1}{44}} = -7.5 \pm 3.75 = -11.25 \text{ to } -3.75$$

The endpoints for $\mu_1 - \mu_6$ are:

$$(55.1 - 43.0) \pm \frac{4.85}{\sqrt{2}} \cdot \sqrt{26.850} \cdot \sqrt{\frac{1}{46} + \frac{1}{18}} = 12.1 \pm 4.94 = 7.16 \text{ to } 17.04$$

The endpoints for $\mu_2 - \mu_3$ are:

$$(44.6 - 59.1) \pm \frac{4.85}{\sqrt{2}} \cdot \sqrt{26.850} \cdot \sqrt{\frac{1}{11} + \frac{1}{30}} = -14.5 \pm 6.26 = -20.76 \text{ to } -8.24$$

The endpoints for $\mu_2 - \mu_4$ are:

$$(44.6 - 40.6) \pm \frac{4.85}{\sqrt{2}} \cdot \sqrt{26.850} \cdot \sqrt{\frac{1}{11} + \frac{1}{11}} = 4.0 \pm 7.58 = -3.58 \text{ to } 11.58$$

The endpoints for $\mu_2 - \mu_5$ are:

$$(44.6 - 62.6) \pm \frac{4.85}{\sqrt{2}} \cdot \sqrt{26.850} \cdot \sqrt{\frac{1}{11} + \frac{1}{44}} = -18.0 \pm 5.99 = -23.99 \text{ to } -12.01$$

The endpoints for $\mu_2 - \mu_6$ are:

$$(44.6 - 43.0) \pm \frac{4.85}{\sqrt{2}} \cdot \sqrt{26.850} \cdot \sqrt{\frac{1}{11} + \frac{1}{18}} = 1.6 \pm 6.80 = -5.20 \text{ to } 8.40$$

The endpoints for $\mu_3 - \mu_4$ are:

$$(59.1 - 40.6) \pm \frac{4.85}{\sqrt{2}} \cdot \sqrt{26.850} \cdot \sqrt{\frac{1}{30} + \frac{1}{11}} = 18.5 \pm 6.26 = 12.24 \text{ to } 24.76$$

The endpoints for $\mu_3 - \mu_5$ are:

$$(59.1 - 62.6) \pm \frac{4.85}{\sqrt{2}} \cdot \sqrt{26.850} \cdot \sqrt{\frac{1}{30} + \frac{1}{44}} = -3.5 \pm 4.21 = -7.71 \text{ to } 0.71$$

The endpoints for $\mu_3 - \mu_6$ are:

$$(59.1-43.0)\pm\frac{4.85}{\sqrt{2}}\cdot\sqrt{26.850}\cdot\sqrt{\frac{1}{30}+\frac{1}{18}}=16.0\pm5.20=10.80 \textit{ to } 21.20$$

The endpoints for $\mu_4 - \mu_5$ are:

$$(40.6-62.6)\pm\frac{4.85}{\sqrt{2}}\cdot\sqrt{26.850}\cdot\sqrt{\frac{1}{11}+\frac{1}{44}}=-22.0\pm5.99=-27.99 \textit{ to } -16.01$$

The endpoints for $\mu_4 - \mu_6$ are:

$$(40.6-43.0)\pm\frac{4.85}{\sqrt{2}}\cdot\sqrt{26.850}\cdot\sqrt{\frac{1}{11}+\frac{1}{18}}=-2.4\pm6.80=-9.20 \textit{ to } 4.40$$

The endpoints for $\mu_5 - \mu_6$ are:

$$(62.6-43.0)\pm\frac{4.85}{\sqrt{2}}\cdot\sqrt{26.850}\cdot\sqrt{\frac{1}{44}+\frac{1}{18}}=19.6\pm4.97=14.63 \textit{ to } 24.57$$

Step 4: Based on the confidence intervals in Step 3, the pairs of means that can be declared different are: μ_1 is different from μ_2, μ_4, μ_5, and μ_6, μ_2 is different from μ_3 and μ_5, μ_3 is different from μ_4 and μ_6, μ_4 is different from μ_5, and μ_5 is different from μ_6.

Step 5: We summarize the results with the following diagram.

Education	Math & Sciences	Communications	Business	Computer Science	Engineering
(4)	(6)	(2)	(1)	(3)	(5)
40.6	43.0	44.6	55.1	59.1	62.6

(Underlines: Education–Communications; Business–Computer Science; Computer Science–Engineering.)

Interpreting this diagram, we conclude with 99% confidence that the mean starting salaries of Education, Math Sciences, and Communication majors is different than that of Business, Computer Science, and Engineering Majors. Also, the mean starting salary of Business majors is different than that of Engineering majors.

16.111 In Exercise 16.65, we determined that conducting a one-way ANOVA with the weight data was not reasonable. Tukey comparisons will not be carried out for the weight data. However, in Exercise 16.65, we determined that conducting a one-way ANOVA with the length data was reasonable. Using Minitab, with all of the data in a column named LOSS and the group in a column named LENGTH, we choose **Stat ▶ ANOVA ▶ Oneway...**, select LENGTH in the **Response** text box and SPECIES in the **Factor** text box, click on the **Comparisons** button, click on **Tukey's, family error rate** so that an X shows in its check-box, type 5 in its text box, and click **OK**. The Tukey comparison part of the output is

```
        Tukey 95% Simultaneous Confidence Intervals
        All Pairwise Comparisons among Levels of SPECIES

        Individual confidence level = 98.95%

        SPECIES = Lahna subtracted from:

SPECIES    Lower   Center   Upper  ---------+---------+---------+---------+
Siika     -5.804   -1.506   2.793                (----*----)
Saerki   -12.388   -9.661  -6.934     (--*--)
Parkki   -14.941  -11.578  -8.216  (----*---)
                                   ---------+---------+---------+---------+
                                         -8.0       0.0       8.0      16.0

SPECIES = Siika subtracted from:

SPECIES    Lower   Center   Upper  ---------+---------+---------+---------+
Saerki   -12.683   -8.155  -3.627    (-----*----)
Parkki   -15.010  -10.073  -5.135  (-----*------)
                                   ---------+---------+---------+---------+
                                         -8.0       0.0       8.0      16.0
SPECIES = Saerki subtracted from:

SPECIES   Lower  Center  Upper  ---------+---------+---------+---------+
Parkki   -5.570  -1.918  1.734            (----*---)
                                ---------+---------+---------+---------+
                                      -8.0       0.0       8.0      16.0
```

Looking the intervals that do not contain zero, the data provide sufficient evidence to conclude that the mean length of Lahna fish is different from those of Saerki and Parkki. Also, the mean length of Siika fish is different from those of Saerki and Parkii. No other mean lengths can be declared different.

16.113 In Exercise 16.67, we determined that conducting a one-way ANOVA with these data was reasonable, although normality was in question for one sample. Using Minitab, with all of the data in a column named LENGTH and the group in a column named SPECIES, we choose **Stat ▶ ANOVA ▶ Oneway...**, select LENGTH in the **Response** text box and SPECIES in the **Factor** text box, click on the **Comparisons** button, click on **Tukey's, family error rate** so that an X shows in its check-box, type 5 in its text box, and click **OK**. The Tukey comparison part of the output is

```
Tukey 95% Simultaneous Confidence Intervals
All Pairwise Comparisons among Levels of SPECIES

Individual confidence level = 99.55%

SPECIES = Hedge Sparrow subtracted from:

SPECIES          Lower   Center    Upper
Meadow Pipit  -1.6292  -0.8225  -0.0158
Pied Wagtail  -1.1977  -0.2181   0.7615
Robin         -1.5111  -0.5464   0.4183
Tree Pipit    -1.0110  -0.0314   0.9482
Wren          -2.9710  -1.9914  -1.0118
```

```
SPECIES          +---------+---------+---------+---------
Meadow Pipit               (-----*----)
Pied Wagtail                  (------*-----)
Robin                        (-----*------)
Tree Pipit                       (------*-----)
Wren             (------*-----)
                 +---------+---------+---------+---------
              -3.0      -1.5       0.0       1.5

SPECIES = Meadow Pipit subtracted from:

SPECIES          Lower    Center    Upper
Pied Wagtail   -0.1815    0.6044   1.3904
Robin          -0.4912    0.2761   1.0434
Tree Pipit      0.0052    0.7911   1.5770
Wren           -1.9548   -1.1689  -0.3830
SPECIES          +---------+---------+---------+---------
Pied Wagtail                         (----*----)
Robin                              (----*----)
Tree Pipit                            (----*-----)
Wren                     (----*----)
                 +---------+---------+---------+---------
              -3.0      -1.5       0.0       1.5
SPECIES = Pied Wagtail subtracted from:

SPECIES       Lower    Center    Upper
Robin       -1.2757   -0.3283   0.6191
Tree Pipit  -0.7759    0.1867   1.1492
Wren        -2.7359   -1.7733  -0.8108

SPECIES        +---------+---------+---------+---------
Robin                     (------*-----)
Tree Pipit                    (-----*------)
Wren             (-----*------)
               +---------+---------+---------+---------
            -3.0      -1.5       0.0       1.5
SPECIES = Robin subtracted from:

SPECIES       Lower    Center    Upper
Tree Pipit  -0.4324    0.5150   1.4624
Wren        -2.3924   -1.4450  -0.4976


SPECIES        +---------+---------+---------+---------
Tree Pipit                      (-----*------)
Wren                 (-----*------)
               +---------+---------+---------+---------
            -3.0      -1.5       0.0       1.5
SPECIES = Tree Pipit subtracted from:


SPECIES    Lower    Center    Upper     +---------+---------+---------+--------
Wren     -2.9225  -1.9600  -0.9975       (-----*-----)
                                        +---------+---------+---------+--------
  -3.0      -1.5       0.0       1.5
```

Looking at the intervals that do not contain zero, we conclude that the mean lengths of cuckoo eggs laid in the nests of hedge sparrows were different from those laid in the nests of meadow pipits and wrens; those in nests of meadow pipits differed from those laid in the nests of tree pipits and wrens; and those laid in the nests of pied wagtails, robins, and tree pipits differed from those laid in the nests of wrens.

16.115 In Exercise 16.69, we determined that conducting a one-way ANOVA with these data was reasonable. Using Minitab, with all of the data in a column named PRICE and the type of book in a column named SUBJECT, we choose **Stat ▶ ANOVA ▶ Oneway...**, select PRICE in the **Response** text box and SUBJECT in the **Factor** text box, click on the **Comparisons** button, click on **Tukey's, family error rate** so that an X shows in its check-box, type 5 in its text box, and click **OK**. The Tukey comparison part of the output is

```
Grouping Information Using the Tukey Method and 95% Confidence

SUBJECT      N    Mean  Grouping
Science     37  184.67  A
Law         45  168.71    B
Medicine    28  161.67    B
Technology  42  139.07      C

Means that do not share a letter are significantly different.
Tukey Simultaneous Tests for Differences of Means
```

Difference of Levels	Difference of Means	SE of Difference	95% CI	T-Value	Adjusted P-Value
Medicine - Law	-7.04	3.28	(-15.55, 1.47)	-2.15	0.143
Science - Law	15.96	3.02	(8.12, 23.81)	5.28	0.000
Technology - Law	-29.63	2.92	(-37.22, -22.05)	-10.14	0.000
Science - Medicine	23.00	3.41	(14.15, 31.86)	6.74	0.000
Technology - Medicine	-22.59	3.32	(-31.22, -13.97)	-6.80	0.000
Technology - Science	-45.60	3.07	(-53.56, -37.63)	-14.85	0.000

Individual confidence level = 98.96%

Looking at the intervals that do not contain zero, we conclude that the mean price of law books differs from the mean price of science books and technology books. The mean price of medicine books differs from the mean price of science books and technology books. Also, the mean price of science books differs from the mean price of technology books. The mean price of law books cannot be determined to be different than the mean price of medicine books.

16.117 In Exercise 16.71, we determined that conducting a one-way ANOVA with these data was reasonable, although the rule of 2 was violated. Using Minitab, with all of the data in a column named HEMOLEVEL and the type of sickle cell disease in a column named TYPE, we choose **Stat ▶ ANOVA ▶ Oneway...**, select HEMOLEVEL in the **Response** text box and TYPE in the **Factor** text box, click on the **Comparisons** button, click on **Tukey's, family error rate** so that an X shows in its check-box, type 5 in its text box, and click **OK**. The Tukey comparison part of the output is

```
Tukey 95% Simultaneous Confidence Intervals
All Pairwise Comparisons among Levels of TYPE

Individual confidence level = 98.05%

TYPE = HB SC subtracted from:

TYPE     Lower  Center   Upper  -------+---------+---------+---------+--
HB SS  -4.224  -3.048  -1.871  (----*----)
HB ST  -1.405  -0.213   0.978            (----*----)
                                -------+---------+---------+---------+--
                                    -2.5       0.0       2.5       5.0

TYPE = HB SS subtracted from:


TYPE   Lower  Center  Upper  -------+---------+---------+---------+--
HB ST  1.785   2.834  3.883                          (---*----)
                             -------+---------+---------+---------+--
                                 -2.5       0.0       2.5       5.0
```

Looking at the intervals that do not contain zero, we conclude that the mean hemoglobin level of patients with type HB SS sickle cell disease differs than that of patients with types HB SC and HB ST.

16.119 The family confidence level is the confidence we have that all the confidence intervals contain the differences between the corresponding population means. We make our simultaneous comparison of the population means based on these confidence intervals; so the family confidence level is the appropriate level for comparing all population means simultaneously.

16.121 The confidence intervals are the following:

$\mu_2 - \mu_1$ is from -2.43 to 5.43

$\mu_3 - \mu_1$ is from -7.36 to 1.36

$\mu_4 - \mu_1$ is from -7.91 to 0.31

$\mu_3 - \mu_2$ is from -8.69 to -0.31

$\mu_4 - \mu_2$ is from -9.23 to -1.37

$\mu_4 - \mu_3$ is from -5.16 to 3.56.

Exercises 16.5

16.123 The conditions required for using the Kruskal-Wallis test to compare k population means are:

(1) Simple random samples
(2) Independent samples
(3) Same-shape populations
(4) All sample sizes are 5 or greater.

16.125 equal

16.127 The measure of total variation of all the ranks is variance of all the ranks, which can be expressed as SST/(n-1), where SST is the total sum of squares computed for the ranks.

16.129 K has approximately a chi-square distribution with k - 1 degrees of freedom, so for five populations, we use critical values from the right hand side of the chi-square distribution with 4 degrees of freedom.

16.131 One-way ANOVA test

16.133 Neither test

16.135 (a) The defining formula, given in Equation 16.5 is $K = \frac{SSTR}{SST/(n-1)}$, where we will apply the defining formula for SST and SSTR to the ranks of the sample data, not to the sample data themselves. Those defining formulas are $SST = \sum (r_i - \bar{r})^2$ and $SSTR = \sum n_j (\bar{r}_j - \bar{r})^2$.

Sample 1	Rank	Sample 2	Rank	Sample 3	Rank
31	4	49	5	97	7
7	2	17	3	71	6
		1	1		

From this table, we calculate that $\bar{r}_1 = 3$, $\bar{r}_2 = 3$, $\bar{r}_3 = 6.5$, and $\bar{r} = 4$.

$$SST = (4-4)^2 + (2-4)^2 + (5-4)^2 + (3-4)^2 + (1-4)^2 + (6-4)^2 + (7-4)^2 = 28$$

$$SSTR = 2(3-4)^2 + 3(3-4)^2 + 2(6.5-4)^2 = 17.5$$

$$K = \frac{17.5}{28/(7-1)} = 3.75$$

(b) The computing formula, given in equation 16.6 is

$K = \frac{12}{n(n+1)} \sum \frac{R_j^2}{n_j} - 3(n+1)$. We have $R_1 = 6$, $R_2 = 9$, and $R_3 = 13$

$$K = \frac{12}{7(7+1)}(6^2/2 + 9^2/3 + 13^2/2) - 3(7+1) = 3.75$$

(c) The two are the same.

(d) The defining formula and the computing formula for K are equivalent as long as there are no ties in the ranks.

16.137 (a) The defining formula, given in Equation 16.5 is $K = \frac{SSTR}{SST/(n-1)}$, where we will apply the defining formula for SST and SSTR to the ranks of the sample data, not to the sample data themselves. Those defining formulas are $SST = \sum (r_i - \bar{r})^2$ and $SSTR = \sum n_j (\bar{r}_j - \bar{r})^2$.

Sample 1	Rank	Sample 2	Rank	Sample 3	Rank	Sample 4	Rank
14	10	2	1	11	7	9	5
4	2	16	11	10	6	12	8
13	9					8	4
						7	3

From this table, we calculate that $\bar{r}_1 = 7$, $\bar{r}_2 = 6$, $\bar{r}_3 = 6.5$, $\bar{r}_4 = 5$ and $\bar{r} = 6$.

$$SST = (10-6)^2 + (2-6)^2 + (9-6)^2 + (1-6)^2 + (11-6)^2 + (7-6)^2 + (6-6)^2$$
$$+ (5-6)^2 + (8-6)^2 + (4-6)^2 + (3-6)^2 = 110$$

$$SSTR = 3(7-6)^2 + 2(6-6)^2 + 2(6.5-6)^2 + 4(5-6)^2 = 7.5$$

$$K = \frac{7.5}{110/(11-1)} = 0.682$$

(b) The computing formula, given in equation 16.6 is

$$K=\frac{12}{n(n+1)}\sum\frac{R_j^2}{n_j}-3(n+1).$$ We have $R_1=21$, $R_2=12$, $R_3=13$, and $R_4=20$

$$K=\frac{12}{11(11+1)}(21^2/3+12^2/2+13^2/2+20^2/4)-3(11+1)=0.682$$

(c) The two are the same.

(d) The defining formula and the computing formula for K are equivalent as long as there are no ties in the ranks.

16.139 (a) $n=3+3+4=10$ (b) $k=3$

(c)

Sample 1	Rank	Sample 2	Rank	Sample 3	Rank
8	10	2	2	4	6.5
4	6.5	1	1	3	4
6	8.5	3	4	6	8.5
				3	4
	25		7		23

$R_1=25$, $R_2=7$, and $R_3=23$

16.141 (a) $n=4+3+5+5+3=20$

(b) $k=5$

(c)

Sample 1	Rank	Sample 2	Rank	Sample 3	Rank	Sample 4	Rank	Sample 5	Rank
7	14	5	9	6	11	3	1	7	14
4	4.5	9	18.5	7	14	7	14	9	18.5
5	9	4	4.5	5	9	7	14	11	20
4	4.5			4	4.5	4	4.5		
				8	17	4	4.5		
	32		32		55.5		38		52.5

$R_1=32$, $R_2=32$, $R_3=55.5$, $R_4=38$ and $R_5=52.5$

16.143 (a) $n=3+3+3+3=12$

(b) $k=4$

(c)

Sample 1	Rank	Sample 2	Rank	Sample 3	Rank	Sample 4	Rank
11	11	9	8	16	12	5	5
6	6	2	2	10	9.5	1	1
7	7	4	4	10	9.5	3	3
	24		14		31		9

$R_1=24$, $R_2=14$, $R_3=31$, and $R_4=9$.

16.145 The total number of populations being sampled is k = 3. Let the subscripts 1, 2, and 3 refer to the populations in order from left to right on the table. The total number of observations is n = 18. Also, $n_1 = n_2 = n_3 = 6$.

Step 1: H_0: $\eta_1 = \eta_2 = \eta_3$ (median number of recalled brands are equal)

H_a: Not all the medians are equal.

Step 2: $\alpha = 0.05$

Step 3:

Violent	Rank	Sexually Explicit	Rank	Neutral	Rank
0	3	0	3	4	17
0	3	2	12.5	1	7.5
2	12.5	2	12.5	2	12.5
1	7.5	2	12.5	3	16
1	7.5	0	3	2	12.5
0	3	1	7.5	7	18
	36.5		51		83.5

$$K = \frac{12}{n(n+1)}\sum\frac{R_j^2}{n_j} - 3(n+1)$$

$$= \frac{12}{18(18+1)}\left(\frac{36.5^2}{6} + \frac{51^2}{6} + \frac{83.5^2}{6}\right) - 3(18+1) = 6.7747$$

Step 4: df = k - 1 = 2; critical value = 5.991

For the p-value approach, $0.025 < P(\chi^2 > 6.7747) < 0.05$.

Step 5: Since 6.7747 > 5.991, reject H_0. Because the p-value is smaller than the significance level, reject H_0.

Step 6: At the 5% significance level, the data provide sufficient evidence to conclude that the median number of advertisements a viewer is able to recall is different when the TV program is violent, sexually explicit, or neutral.

16.147 The total number of populations being sampled is k = 4. Let the subscripts 1, 2, 3, and 4 refer to Northeast, Midwest, South, and West, respectively. The total number of observations is n = 32. Also, $n_1 = n_2 = n_3 = n_4 = 8$.

Step 1: H_0: $\eta_1 = \eta_2 = \eta_3 = \eta_4$ (median asking rents are equal)

H_a: Not all the medians are equal.

Step 2: $\alpha = 0.10$

Step 3:

Sample 1	Rank	Sample 2	Rank	Sample 3	Rank	Sample 4	Rank
1293	4	1605	11	642	1	694	2
1581	9	1639	12.5	722	3	1345	5
1781	18	1655	15.5	1354	6	1565	8
2130	23	1691	17	1513	7	1649	14
2149	25	2058	20	1591	10	1655	15.5
2286	27	2115	22	1639	12.5	2068	21
2989	30	2413	28	1982	19	2203	26
3182	31	3361	32	2135	24	2789	29
	167		158		82.5		120.5

$$K = \frac{12}{n(n+1)}\sum\frac{R_j^2}{n_j} - 3(n+1)$$

$$= \frac{12}{32(32+1)}(\frac{167^2}{8} + \frac{158^2}{8} + \frac{82.5^2}{8} + \frac{120.5^2}{8}) - 3(32+1) = 6.369$$

Step 4: df = k - 1 = 3; critical value = 6.251

For the p-value approach, $0.05 < P(\chi^2 > 6.369) < 0.10$.

Step 5: Since 6.369 > 6.251, reject H_0. Because the p-value is smaller than the significance level, reject H_0.

Step 6: At the 10% significance level, the data provide sufficient evidence to conclude that a difference exists among the median asking rents in the four U.S. regions.

16.149 The total number of populations being sampled is k = 4. Let the subscripts 1, 2, 3, and 4 refer to the populations in order from left to right on the table. The total number of observations is n = 30. Also, $n_1 = n_3 = 7$ and $n_2 = n_4 = 8$.

Step 1: H_0: $\mu_1 = \mu_2 = \mu_3 = \mu_4$ (mean number of eggs are equal)

H_a: Not all the means are equal.

Step 2: $\alpha = 0.10$ Step 3:

IF & IM	Rank	IF & AM	Rank	AF & IM	Rank	AF & AM	Rank
18	11.5	26	30	11	2	19	17
19	17	21	23	18	11.5	22	26
22	26	19	17	22	26	16	5
19	17	19	17	17	7.5	20	21.5
22	26	20	21.5	13	3	18	11.5
18	11.5	17	7.5	19	17	19	17
25	29	17	7.5	5	1	17	7.5
		22	26			14	4
	139		149.5		68		109.5

$$K = \frac{12}{n(n+1)}\sum\frac{R_j^2}{n_j} - 3(n+1)$$

$$= \frac{12}{30(30+1)}(\frac{138^2}{7} + \frac{149.5^2}{8} + \frac{68^2}{7} + \frac{109.5^2}{8}) - 3(30+1) = 6.0156$$

Step 4: df = k - 1 = 3; critical value = 6.251

For the p-value approach, $P(\chi^2 > 6.0156) > 0.10$

Step 5: Since 6.0156 < 6.251, do not reject H_0. Because the p-value is greater than the significance level, do not reject H_0.

Step 6: At the 10% significance level, the data do not provide sufficient evidence to conclude that a difference exists in mean number of eggs among the four cross combinations of infected and antibiotic treated spider mites.

16.151(a) The total number of populations being sampled is k = 4. Let the subscripts 1, 2, 3, 4, and 5 refer to the five strains. The total number of observations is n = 25. Also, $n_1 = 5$, $n_2 = 5$, $n_3 = 5$, $n_4 = 5$ and $n_5 = 5$.

Step 1: H_0: $\mu_1 = \mu_2 = \mu_3 = \mu_4 = \mu_5$ (mean bacteria counts are equal)

H_a: Not all the means are equal.

Step 2: $\alpha = 0.05$

Step 3:

Strain A	Rank	Strain B	Rank	Strain C	Rank	Strain D	Rank	Strain E	Rank
9	2	3	1	10	3	14	5	33	18
27	12	32	17	47	22	18	8	43	20
22	10	37	19	50	23	17	7	28	13
30	15	45	21	52	24	29	14	59	25
16	6	12	4	26	11	20	9	31	16
	45		62		83		43		92

$$K = \frac{12}{n(n+1)}\sum \frac{R_j^2}{n_j} - 3(n+1)$$

$$= \frac{12}{25(25+1)}\left(\frac{45^2}{5} + \frac{62^2}{5} + \frac{83^2}{5} + \frac{43^2}{5} + \frac{92^2}{5}\right) - 3(25+1) = 7.185$$

Step 4: df = k - 1 = 4; critical value = 9.488

For the p-value approach, $P(\chi^2 > 7.185) > 0.10$.

Step 5: Since 7.185 < 9.488, do not reject H_0. because the p-value is larger than the significance level, do not reject H_0.

Step 6: At the 5% significance level, the data do not provide sufficient evidence to conclude that a difference exists among the mean bacteria counts among the five strains of *Staphylococcus aureus*.

(b) The Kruskal-Wallace test may be used whenever the distributions of the variable have the same shape. If the bacteria counts in all five strains have normal distributions with equal standard deviations, then the five distributions will have the same shape, and the Kruskal-Wallace may be used. If the populations are actually normally distributed with equal standard deviations, then the one-way ANOVA is the better choice since it was designed expressly for this situation and is more powerful than the Kruskal-Wallace test under these circumstances.

16.153 (a) Using Minitab, with all of the data in one column named RENT and four regions of the U.S. corresponding to each of the rent values in a column named CLASS, we choose **Stat ▶ Nonparametrics ▶ Kruskal-Wallis...**, select RENT in the **Response** text box, select REGION in the **Factor** text box, and click **OK**. The result is

```
Kruskal-Wallis Test on RENT

REGION      N  Median  Ave Rank      Z
Midwest     6   734.5       4.5  -2.97
Northeast   5  1005.0      14.4   1.70
South       4   876.0       8.5  -0.76
West        5  1025.0      15.4   2.14
Overall    20              10.5

H = 12.23  DF = 3  P = 0.007

* NOTE * One or more small samples
```

(b) Since the P-value is 0.007, which is less than the 0.05 significance level, we reject the null hypothesis of equal means; the data provide sufficient evidence to conclude that there are differences between the mean monthly rents in the four regions of the U.S.

16.155 (a) Using Minitab, with all of the data in one column named RATE and categories of patch-size manipulations corresponding to each of the singing rates in a column named BREAST, we choose **Stat ▶ Nonparametrics ▶ Kruskal-Wallis...**, select RATE in the **Response** text box, select BREAST in the **Factor** text box, and click **OK**. The result is

```
Kruskal-Wallis Test on RATE

BREAST      N  Median  Ave Rank      Z
Control     8  16.900      15.1   1.29
Enlarged    8  14.900      16.3   1.84
Reduced     8   3.350       6.1  -3.12
Overall    24              12.5

H = 9.86  DF = 2  P = 0.007
```

(b) Since the P-value is 0.007, which is less than the 0.01 significance level, we reject the null hypothesis of equal means; the data provide sufficient evidence to conclude that there are differences between the singing rates for the three patch-size manipulations.

16.157 (a) Using Minitab, with all of the data in one column named HARDNESS and categories artificial teeth materials corresponding to each of the volumes worn off in a column named MATERIAL, we choose **Stat ▶ Nonparametrics ▶ Kruskal-Wallis...**, select HARDNESS in the **Response** text box, select MATERIAL in the **Factor** text box, and click **OK**. The result is

```
Kruskal-Wallis Test on HARDNESS
MATERIAL     N  Median  Ave Rank      Z
Duracross    6   39.45      15.5   3.37
Duradent     6   24.15       3.5  -3.37
Endura       6   27.45       9.5   0.00
Overall     18               9.5

H = 15.16  DF = 2  P = 0.001
H = 15.20  DF = 2  P = 0.000  (adjusted for ties)
```

(b) Since the P-value is 0.001, which is less than the 0.05 significance level, we reject the null hypothesis of equal means; the data provide sufficient evidence to conclude that there are differences between the hardnesses for the three artificial teeth materials.

16.159 (a) In Exercise 16.65, normal probability plots of the lengths and weights of the four different species of fish were linear. The length data had similar standard deviations, thus the Kruskal-Wallis test is reasonable on the length data. However, the weight data had different standard deviations. To check whether the distributions of the weight data are similar in shape, we will obtain box plots using Minitab. Choose **Graph ▶ Boxplot,** select the **Multiple Y's Simple** version and click **OK**. Enter WEIGHT_Lahna, WEIGHT_Siika, WEIGHT_Saerki, and WEIGHT_Parkki in the **Graph variables** text box and click **OK**. The result is

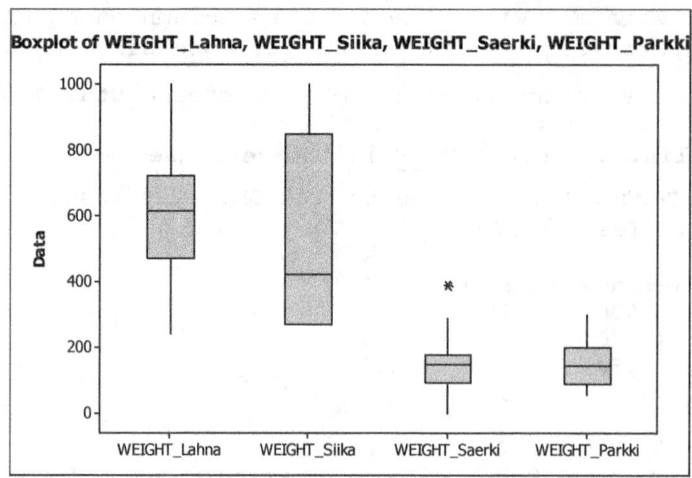

The shapes of the distributions are not close to being the same. The assumptions for Kruskal-Wallis test are not satisfied and the test should not be performed on the weight data.

(b) We will perform the analysis only on the length data. Using Minitab, choose **Stat ▶ Nonparametrics ▶ Kruskal-Wallis...**, select LENGTH in the **Response** text box, select SPECIES in the **Factor** text box, and click **OK**. The result is

```
Kruskal-Wallis Test on LENGTH

SPECIES    N  Median  Ave Rank      Z
Lahna     35   30.40      52.0   6.11
Siika      6   27.05      46.5   1.22
Saerki    20   20.50      19.2  -4.34
Parkki    11   19.00      13.1  -4.02
Overall   72              36.5

H = 47.90  DF = 3  P = 0.000
H = 47.92  DF = 3  P = 0.000  (adjusted for ties)
```

(c) The P-value of the test is 0.00, which is less than the 0.05 significance level. Reject the null hypothesis of equal means. The data provide sufficient evidence to conclude that the mean lengths of the species of fish are different.

(d) The one-way ANOVA in Exercise 16.65 also had a P-value of 0.000 and we reached the same conclusion.

16.161 (a) In Exercise 16.67, normal probability plots of the lengths of cuckoo eggs laid in the nests of six species of birds were linear with approximately the same standard deviations, with the exception of the Meadow Pipit, which had several outliers. To check whether the distributions are similar in shape, we will obtain box plots using Minitab. Choose **Graph ▶ Boxplot,** select the **Multiple Y's Simple** version and click **OK**. Enter Meadow Pipit, Tree Pipit, Hedge Sparrow, Robin, Pied Wagtail, and Wren in the **Graph variables** text box and click **OK**. The result is

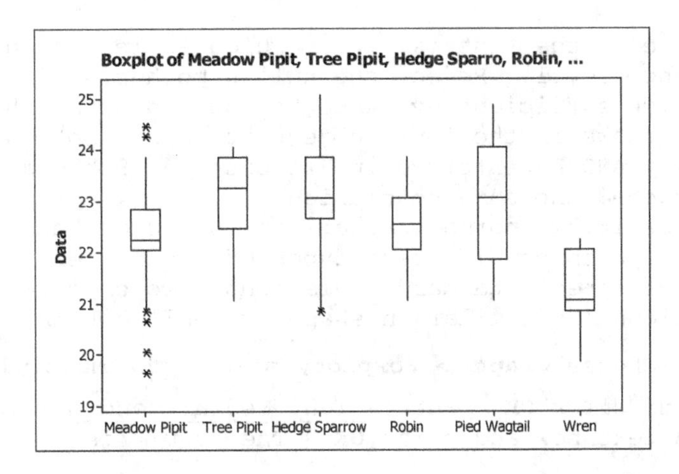

Thus, the six distributions appear to range from symmetric to slightly left skewed. The Kruskal-Wallis test is reasonable.

(b) Using Minitab, choose **Stat ▶ Nonparametrics ▶ Kruskal-Wallis...**, select LENGTH in the **Response** text box, select SPECIES in the **Factor** text box, and click **OK**. The result is

```
Kruskal-Wallis Test on LENGTH

SPECIES           N  Median  Ave Rank      Z
Hedge Sparrow    14   23.05      80.9   2.33
Meadow Pipit     45   22.25      54.8  -1.40
Pied Wagtail     15   23.05      72.6   1.44
Robin            16   22.55      64.7   0.52
Tree Pipit       15   23.25      82.8   2.65
Wren             15   21.05      19.8  -4.85
Overall         120              60.5

H = 34.80  DF = 5  P = 0.000
H = 35.04  DF = 5  P = 0.000  (adjusted for ties)
```

(c) The P-value of the test is 0.000, which is less than the 0.05 significance level. Reject the null hypothesis of equal means. The data provide sufficient evidence to conclude that the mean egg lengths in the six types of nests are not equal.

(d) The one-way ANOVA performed in Exercise 16.67 had a P-value of 0.006, very close to this result, and we reached the same conclusion.

16.163 (a) In Exercise 16.69, normal probability plots of the book prices for the four subject areas were linear with approximately the same standard deviations. Thus, the four distributions appear to be the same shape, and the Kruskal-Wallis test is reasonable.

(b) Using Minitab, choose **Stat ▶ Nonparametrics ▶ Kruskal-Wallis...**, select PRICE in the **Response** text box, select SUBJECT in the **Factor** text box, and click **OK**. The result is

```
Kruskal-Wallis Test on PRICE

SUBJECT       N  Median  Ave Rank      Z
Law          45   167.0      88.1   2.10
Medicine     28   163.1      72.9  -0.48
Science      37   186.5     120.9   7.05
Technology   42   137.7      27.5  -8.49
Overall     152              76.5

H = 93.01  DF = 3  P = 0.000
```

(c) The P-value of the test is 0.000, which is less than the 0.05 significance level. Reject the null hypothesis of equal means. The data provide sufficient evidence to conclude that the mean price for hardcover books in the four subject areas are not equal.

(d) The one-way ANOVA performed in Exercise 16.69 had a P-value of 0.000, and we reached the same conclusion.

16.165 (a) In Exercise 16.71, normal probability plots of the hemoglobin levels of patients with three different types of sickle cell disease were linear, but with different standard deviations. To check whether the distributions are similar in shape, we will obtain box plots using Minitab. Choose **Graph ▶ Boxplot,** select the **Multiple Y's Simple** version and click **OK**. Enter High, Medium, and Low in the **Graph variables** text box and click **OK**. The result is

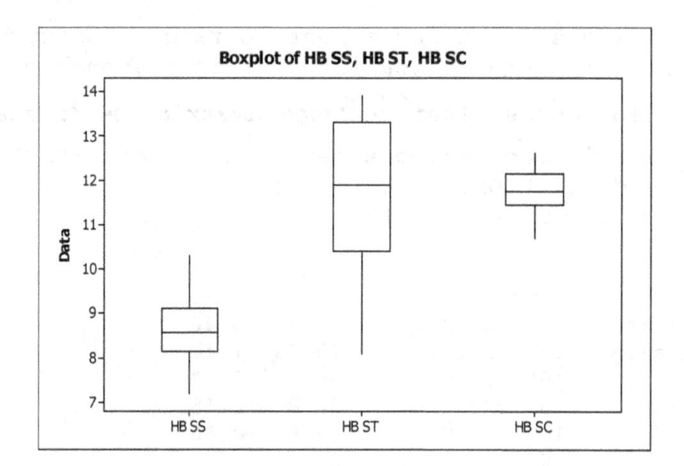

The shapes of the distributions are close to being the same, although with different spreads. The assumptions for Kruskal-Wallis test are satisfied.

(b) Using Minitab, with all of the data in one column named HEMOLEVEL and magazine educational levels in a column named TYPE, we choose **Stat ▶ Nonparametrics ▶ Kruskal-Wallis...**, select HEMOLEVEL in the **Response** text box, select TYPE in the **Factor** text box, and click **OK**. The result is

```
Kruskal-Wallis Test on HEMOLEVEL

TYPE       N   Median   Ave Rank       Z
HB SC     10   11.750       29.4    2.53
HB SS     16    8.550        9.6   -4.88
HB ST     15   11.900       27.6    2.68
Overall   41                21.0

H = 23.92  DF = 2  P = 0.000
H = 23.94  DF = 2  P = 0.000 (adjusted for ties)
```

(c) The P-value is 0.000, which is less than the significance level of 0.05, so we reject the null hypothesis that the mean hemoglobin levels are the same for the three types of sickle cell disease. We conclude that there are differences in the mean hemoglobin levels for the three types of disease. Without a formal analysis, it would appear that the mean hemoglobin level for the HB SS group is lower than either of the other two groups, and the other two groups have approximately equal means.

(d) The one-way ANOVA performed in Exercise 16.71 also had a P-value of 0.000, and we reached the same conclusion.

16.167 (a) The defining formula and the computing formula for K are equivalent as long as there are no ties in the data. The computing formula uses the facts that the sum of the first n integers is n(n+1)/2 and that the sum of the squares of the first n integers is n(n+1)(2n+1)/6. Thus it assumes that all of the ranks are distinct integers. The discrepancy between the formulas results, as in this case, when there are ties in the ranks, invalidating the formula for the sum of squares of the first n integers. The formula for the sum of the first n integers remains valid.

(b) No. In this example, both values of K are greater than the critical value of 5.991.

(c) No. Using Table of chi-square values, either value leads to the P-value being between 0.005 and 0.01, but close to 0.01. While there would be a difference in P-values anytime the K values differ, keep in mind that the distribution of K is only approximately chi-square, so the P-values aren't exact in any case.

Review Problems for Chapter 16

1. One-way ANOVA is used to compare means of a variable for populations that result from a classification by one other variable called the factor.

2. (i) Simple random samples: Check by carefully studying the way that the sampling was done.
(ii) Independent samples: Check by carefully studying the way that the sampling was done.
(iii) Normal populations; Check with normal probability plots, histograms, dotplots.
(iv) All populations have equal standard deviations; this is a reasonable assumption if the ratio of the largest standard deviation to the smallest one is less than 2.

3. F distribution

4. There are n = 17 observations and k = 3 samples. The degrees of freedom are (k - 1, n - k) = (2, 14).

5. (a) Variation among sample means is measured by the mean square for treatments, MSTR = SSTR/(k - 1).
(b) The variation within samples is measured by the error mean square, MSE = SSE/(n - k).

6. (a) SST is the total sum of squares. It measures the total variation among all of the sample data. $SST = \sum (x - \bar{x})^2$

SSTR is the treatment sum of squares. It measures the variation among the sample means. $SSTR = \sum n_i (\bar{x}_i - \bar{x})^2$

SSE is the error sum of squares. It measures the variation within the samples. $SSE = \sum (x - \bar{x}_i)^2 = \sum (n_i - 1) s_i^2$

(b) SST = SSTR + SSE. This means that the total variation in the sample can be broken down into two components, one representing the variation between the sample means and one representing the variation within the samples.

7. (a) One purpose of a one-way ANOVA table is to organize and summarize the quantities required for ANOVA.

(b)

Source	df	SS	MS=SS/df	F-statistic
Treatment	k - 1	SSTR	MSTR=SSTR/(k - 1)	F=MSTR/MSE
Error	n - k	SSE	MSE=SSE/(n - k)	
Total	n - 1	SST		

8. If the null hypothesis is rejected, a multiple comparison is done to determine which means are different.

9. The individual confidence level gives the confidence that we have that any particular confidence interval will contain the population quantity being estimated. The family confidence level gives the confidence that we have that all of the confidence intervals will contain all of the population quantities being estimated. The family confidence level is appropriate for multiple comparisons because we are interested in all of the possible comparisons.

10. Tukey's multiple-comparison procedure is based upon the Studentized Range distribution or q-distribution.

11. Larger. One has to be less confident about the truth of several statements at once than about the truth of a single statement. For example, if one were 99% confident about a single statement, the confidence that two statements were both true must be smaller since there is no way to have 100% confidence in the second statement. Similarly, each time a statement is added to the list, the overall confidence that all of them are true must decrease.

12. The parameters for the q-curve are $\kappa = k = 3$, and $\nu = n - k = 17 - 3 = 14$. [k = number of samples and n = total number of observations.]

13. Kruskal-Wallis test

14. Chi-square distribution with k - 1 degree of freedom [k = number of samples]

15. If the null hypothesis of equal population means is true, then the means of the ranks of the k samples should be about equal. If the variation in the means of the ranks for the k samples is too large, then we have evidence against the null hypothesis.

16. Use the Kruskal-Wallis test. The outliers will have a greater effect on the ANOVA than on the Kruskal-Wallis test since an outlier will be replaced by its rank in the Kruskal-Wallis test and the rank of the most distant outlier is either 1 or n regardless of how large or small the actual data value is. In other words, the Kruskal-Wallis test is more robust to outliers than is the ANOVA.

17. 2

18. 14

19. 3.74

20. 6.51

21. 3.74

22. (a) $q_{0.05} = 3.70$ (b) 4.89

23. (a) The total number of populations being sampled is k = 3. Let the subscripts 1, 2, and 3 refer to A, B, and C, respectively. The total number of pieces of sample data is n = 12. Also, $n_1 = 3$, $n_2 = 5$, $n_3 = 4$. The following statistics for each sample are: $\bar{x}_1 = 3$, $\bar{x}_2 = 3$, and $\bar{x}_3 = 6$. Also, $s_1^2 = 4$, $s_2^2 = 6$, and $s_3^2 = 18$. Note, the mean of all the sample data is $\bar{x} = 4$. We use this information as follows:

(b) SST $= \sum(x - \bar{x})^2 = (1 - 4)^2 + \ldots + (3 - 4)^2 = 110.0$

SSTR $= n_1(\bar{x}_1 - \bar{x})^2 + n_2(\bar{x}_2 - \bar{x})^2 + n_3(\bar{x}_3 - \bar{x})^2$

$= 3(3 - 4)^2 + 5(3 - 4)^2 + 4(6 - 4)^2 = 24.0$

SSE $= (n_1 - 1)s_1^2 + (n_2 - 1)s_2^2 + (n_3 - 1)s_3^2$

$= (3 - 1)4 + (5 - 1)6 + (4 - 1)18 = 86.0$

SST = SSTR + SSE since 110.0 = 24.0 + 86.0

(c) Let T_1, T_2, and T_3 refer to the sum of the data values in each of the three samples, respectively. Thus: $T_1 = 9$ $T_2 = 15$ $T_3 = 24$. Also, the sum of all the data values is $\Sigma x = 48$, and their sum of squares is $\sum x^2 = 302$.

Consequently: $$SST = \sum x^2 - \frac{(\sum x)^2}{n} = 302 - \frac{48^2}{12} = 110$$

$$SSTR = \frac{T_1^2}{n_1} + \frac{T_2^2}{n_2} + \frac{T_3^2}{n_3} - \frac{(\sum x)^2}{n} = \frac{9^2}{3} + \frac{15^2}{5} + \frac{24^2}{4} - \frac{48^2}{12} = 24$$

and SSE = SST − SSTR = 110 − 24 = 86 .

(d)

Source	Df	SS	MS=SS/df	F-statistic
Treatment	2	24	12.000	1.255
Error	9	86	9.556	
Total	11	110		

24. (a) MSTR is a measure of the variation between the sample mean losses for highway robberies, gas station robberies, and convenience store robberies.

(b) MSE is a measure of the variation within the three samples.

(c) The four assumptions for one-way ANOVA, given in Key Fact 16.2: simple random samples, independent samples, normal populations, and equal standard deviations. Assumptions 1 and 2 on simple random independent samples are absolutely essential to the one-way ANOVA procedure. Assumption 3 on normality is not too critical as long as the populations are not too far from being normally distributed. Assumption 4 on equal standard deviations is also not that important provided the sample sizes are roughly equal.

25. (a) The samples and their normal scores are shown in the table below along with the residuals which are $x_i - \bar{x}_j$.

	Hwy	w	Residual	Gas	w	Residual	Conv. St.	w	Residual
	952	-1.18	-27.8	1298	-1.28	129.2	844	-1.28	82.8
	996	-0.50	16.2	1195	-0.64	26.2	921	-0.64	159.8
	839	0.00	-140.8	1174	-0.20	5.2	880	-0.20	118.8
	1088	0.50	108.2	1113	0.20	-55.8	706	0.20	-55.2
	1024	1.18	44.2	953	0.64	-215.8	602	0.64	-159.2
				1280	1.28	111.2	614	1.28	-147.2
Total	4899			7013			4567		
Mean	979.8			1168.8			761.2		
St.Dev.	92.90			126.1			139.0		

We will plot the normal scores w against the data for each type of store. The graphs have been produced using Minitab, but you can easily do them by hand. The standard deviations for each type of robbery are shown in the above table.

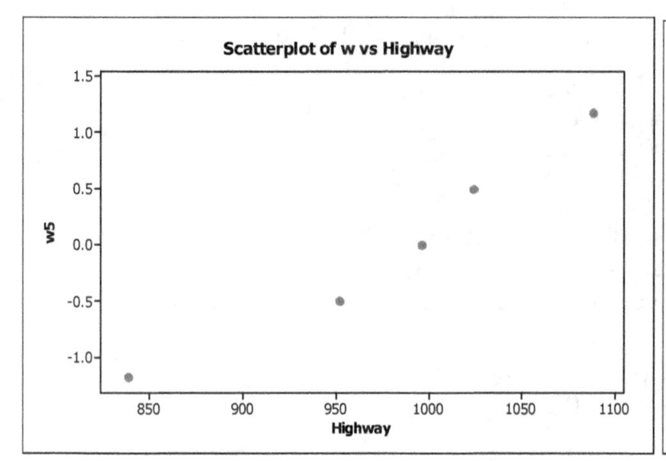

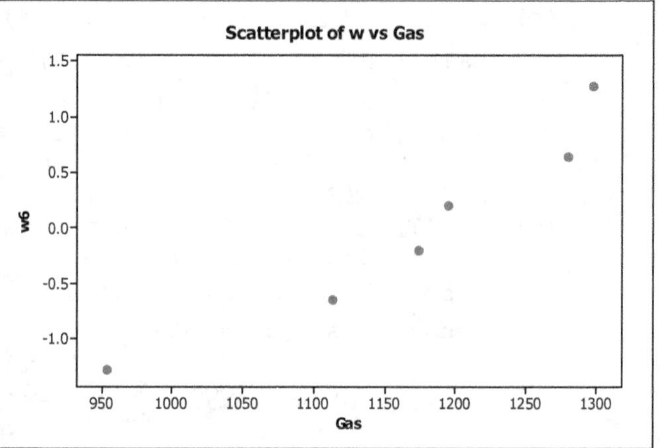

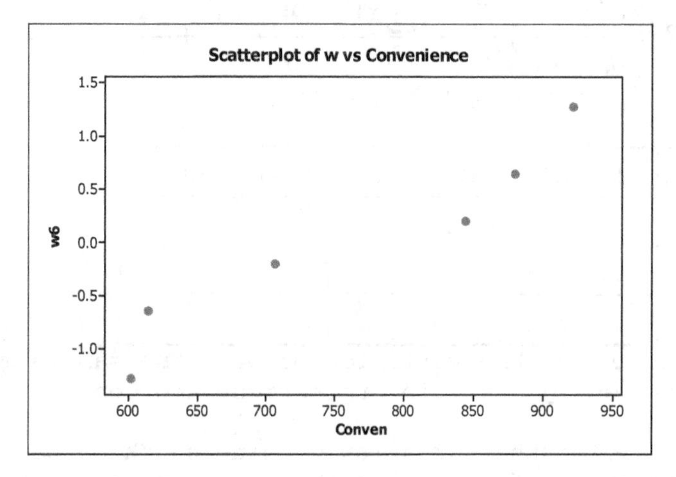

(b) Now order all of the residuals in the previous table from smallest to largest and plot them against the normal scores in Table III for n = 17. Also plot the residuals against the means for each type of robbery. The results are

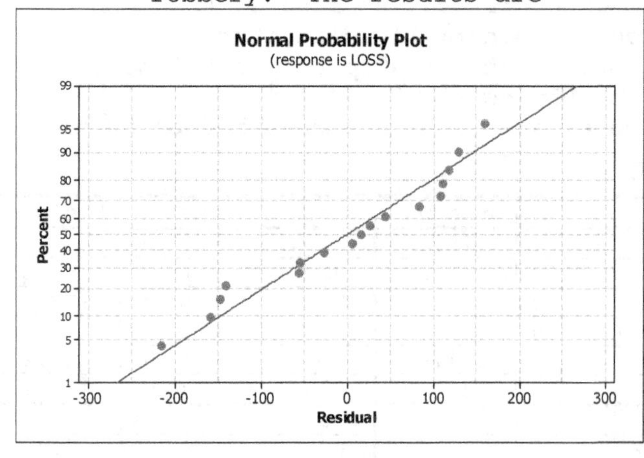

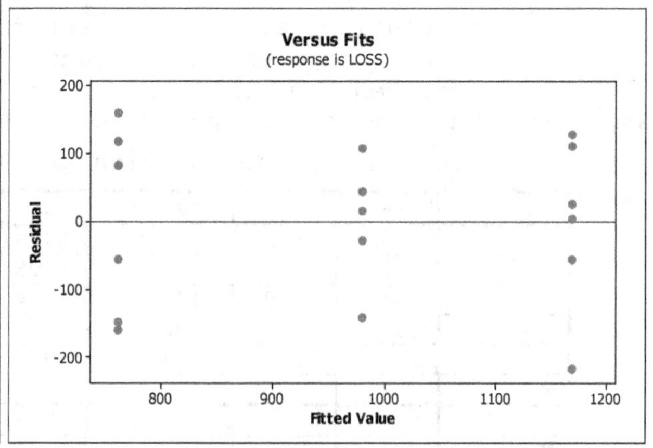

(c) All three of the normal probability plots are close to linear, and the ratio of the largest to smallest standard deviation is 1.50. The residual analysis shows that the normal probability plot of the residuals is close to linear and the last graph shows that the spreads of the residuals are about the same for each type of robbery. Thus the ANOVA assumptions of normality and equal standard deviations appear to be reasonable.

26. The total number of populations being sampled is k = 3. Let the subscripts 1, 2, and 3 refer to Highway, Gas station, and Convenience store, respectively. The total number of observations is n = 17. Also, $n_1 = 5$, $n_2 = 6$, and $n_3 = 6$.

Step 1: $H_0: \mu_1 = \mu_2 = \mu_3$ (population means are equal)

H_a: Not all the means are equal.

Step 2: $\alpha = 0.05$

Step 3: The sums of squares are

$$SST = \sum x^2 - \frac{(\sum x)^2}{n} = 16{,}683{,}857 - \frac{16{,}479^2}{17} = 709{,}889.882$$

$$SSTR = \frac{T_1^2}{n_1} + \frac{T_2^2}{n_2} + \frac{T_3^2}{n_3} - \frac{(\sum x)^2}{n} = \frac{4899^2}{5} + \frac{7013^2}{6} + \frac{4567^2}{6} - \frac{16479^2}{17} = 499{,}349.416$$

and SSE = SST - SSTR =709,889.882- 499,349.416 = 210,540.466

The one-way ANOVA table is:

Source	df	SS	MS = SS/df	F-statistic
Treatment	2	499,349.416	249,674.708	16.602
Error	14	210,540.466	15,038.605	
	16	709,889.882		

The F-statistic is defined and calculated as

$$F = \frac{MSTR}{MSE} = \frac{249{,}674.708}{15{,}038.605} = 16.602$$

Step 4: df = (k - 1, n - k) = (2, 14); critical value = 3.74

For the P-value approach, $P(F > 16.602) < 0.005$.

Step 5: From Step 3, F = 16.602. From Step 4, $F_{0.05} = 3.74$. Since 16.602 > 3.74, reject H_0. Since the P-value is smaller than the significance level, reject H_0.

Step 6: At the 5% significance level, the data provide sufficient evidence to conclude that a difference in mean losses exists among the three types of robberies.

27. (a) Step 1: Family confidence level = 0.95.

Step 2: $\kappa = 3$ and $\nu = n - k = 17 - 3 = 14$. From Problem 22(a), we find that $q_{0.05} = 3.70$.

Step 3: Before obtaining all possible confidence intervals for $\mu_i - \mu_j$, we construct a table giving the sample means and sample sizes.

j	1	2	3
$\bar{x}_j$	979.8	1168.8	761.2
n_j	5	6	6

In Problem 26, we found that MSE = 15,038.605. Now we are ready to obtain the required confidence intervals.

The endpoints for $\mu_1 - \mu_2$ are

$$(979.8 - 1168.8) \pm \frac{3.70}{\sqrt{2}} \cdot \sqrt{15038.605} \cdot \sqrt{\frac{1}{5} + \frac{1}{6}} = -189 \pm 194.279 = -383.279 \text{ to } 5.279$$

The endpoints for $\mu_1 - \mu_3$ are

$$(979.8-761.2)\pm\frac{3.70}{\sqrt{2}}\cdot\sqrt{15038.605}\cdot\sqrt{\frac{1}{5}+\frac{1}{6}}=218.6\pm194.279=24.321 \text{ to } 412.879$$

The endpoints for $\mu_2 - \mu_3$ are

$$(1168.8-761.2)\pm\frac{3.70}{\sqrt{2}}\cdot\sqrt{15038.605}\cdot\sqrt{\frac{1}{6}+\frac{1}{6}}=407.6\pm185.238=222.362 \text{ to } 592.838$$

Step 4: Based on the confidence intervals in Step 3, we declare means μ_1 and μ_3 different and also means μ_2 and μ_3. Means μ_1 and μ_2 are not declared different.

Step 5: We summarize the results with the following diagram.

Convenience store (3)	Highway (1)	Gas Station (2)
761.2	979.8	1168.8

(Highway and Gas Station are joined by a line in the diagram.)

(b) Interpreting this diagram, we conclude with 95% confidence that the mean loss due to convenience-store robberies is different than that due to highway robberies and gas station robberies; the mean loss due to highway robberies and gas-station robberies cannot be declared different.

28. (a) The total number of populations being sampled is k = 3. Let the subscripts 1, 2, and 3 refer to highway, gas station, and convenience store, respectively. The total number of pieces of sample data is n = 17. Also, $n_1 = 5$, $n_2 = n_3 = 6$.

Step 1: H_0: $\mu_1 = \mu_2 = \mu_3$ (mean losses are equal)

H_a: Not all the means are equal.

Step 2: $\alpha = 0.05$

Step 3:

Sample 1	Rank	Sample 2	Rank	Sample 3	Rank
839	4	953	9	602	1
952	8	1113	13	614	2
996	10	1174	14	706	3
1024	11	1195	15	844	5
1088	12	1280	16	880	6
		1298	17	921	7
	45		84		24

$$K=\frac{12}{n(n+1)}\sum\frac{R_j^2}{n_j}-3(n+1)$$

$$=\frac{12}{17(17+1)}(\frac{45^2}{5}+\frac{84^2}{6}+\frac{24^2}{6})-3(17+1)=11.765$$

Step 4: df = k - 1 = 2; critical value = 5.991

For the p-value approach, $P(\chi^2 > 11.765) < 0.005$.

Step 5: Since 11.765 > 5.991, reject H_0. Since the p-value is smaller than the significance level, reject H_0.

Step 6: At the 5% significance level, the data provide sufficient evidence to conclude that a difference exists in mean losses among the three types of robberies.

(b) It is permissible to perform the Kruskal-Wallis test because normal populations having equal standard deviations have the same shape. It is better to use the one-way ANOVA test when the assumptions for that test are met because it is more powerful than the Kruskal-Wallis test.

(c) The P-value for the ANOVA test was less than 0.005. The Kruskal-Wallis P-value was also less than 0.005. Thus, we arrive at the same conclusion when testing at the 0.05 significance level.

29. The total number of populations being sampled is k = 3. The total number of observations is n = 56. Also, $n_1 = 29$, $n_2 = 11$, $n_3 = 16$

Step 1: H_0: $\mu_1 = \mu_2 = \mu_3$ (mean angles of anterior-posterior foot pressure are equal among three groups)
H_a: Not all the means are equal.

Step 2: $\alpha = 0.01$

Step 3: We can determine the mean of all the observations by the following

$$\bar{x} = \frac{n_1\bar{x}_1 + n_2\bar{x}_2 + n_3\bar{x}_3}{n_1 + n_2 + n_3} = \frac{29(54.2) + 11(60.6) + 16(60.7)}{29 + 11 + 16} = 57.3143$$

Then, use the defining formulas to get the SSTR and SSE.

$$SSTR = \sum n_i(\bar{x}_i - \bar{x})^2 = 29(54.2 - 57.3143)^2 + 11(60.6 - 57.3143)^2 + 16(60.7 - 57.3143)^2$$
$$= 583.4286$$
$$SSE = \sum (n_i - 1)s_i^2 = 28(6)^2 + 10(8)^2 + 15(7)^2 = 2383$$

and SST = SSE + SSTR = 2383 + 583.4286 = 2966.4286
MSTR = SSTR/(k - 1) = 583.4286/(3 - 1) = 291.7143
MSE = SSE/(n - k) = 2383/(56 - 3) = 44.9623
F = MSTR/MSE = 291.7143/44.9623 = 6.4880

The one-way ANOVA table is:

Source	df	SS	MS = SS/df	F-statistic
Treatment	2	583.4286	291.7143	6.4880
Error	53	2383.0	44.9623	
Total	55	2966.4286		

Step 4: df = (k - 1, n - k) = (2, 53); From the table in the problem, critical value = 5.03
For the p-value approach, P(F > 6.4880) < 0.005.

Step 5: From Step 3, F = 6.4880. From Step 4, $F_{0.01} = 5.03$. Since 6.4880 > 5.03, reject H_0.
Since the p-value is smaller than the significance level, reject H_0.

Step 6: At the 1% significance level, the data provide sufficient evidence to conclude that there is a difference between the mean angle of anterior-posterior center of foot pressure for patients that have genu valgum, genu varum, or neither condition.

30. Using Minitab, choose **Graph ▶ Probability Plot,** select the **Simple** version and click **OK**. Enter Active, Passive, and Non in the **Graph Variables** text box, click on the **Multiple Graphs** button and select **In separate panels of the same graph**, and click **OK**. The result is

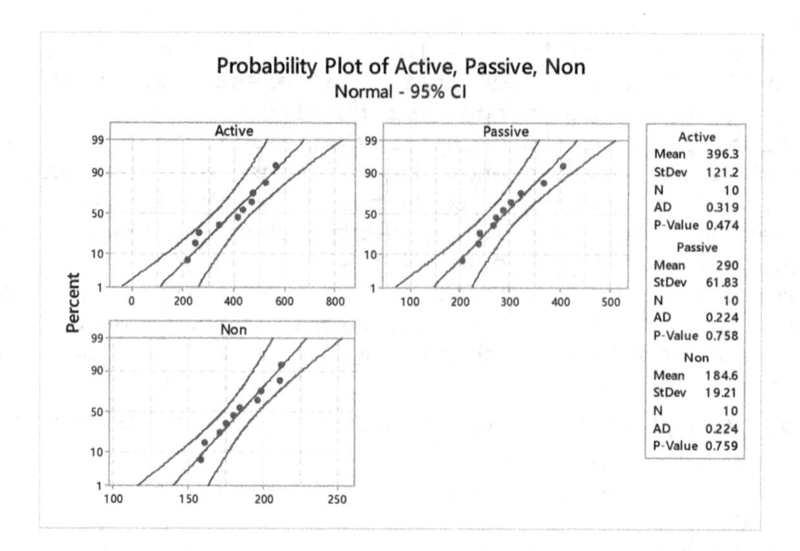

The standard deviations of the three groups are shown in the right hand panel of the graph above as 121.2, 61.83, and 19.21. The variables are approximately normal. However, the rule of 2 is violated since the ratio of the largest to smallest standard deviation is 121.2/19.21 = 6.31. Therefore, the assumption of equal population standard deviations is not met.

31. Using Minitab, choose **Graph ▶ Histogram,** select the **Simple** version and click **OK**. Enter Active, Passive, and Non in the **Graph Variables** text box, click on the **Multiple Graphs** button and select **In separate panels of the same graph**, and click **OK**. The result is

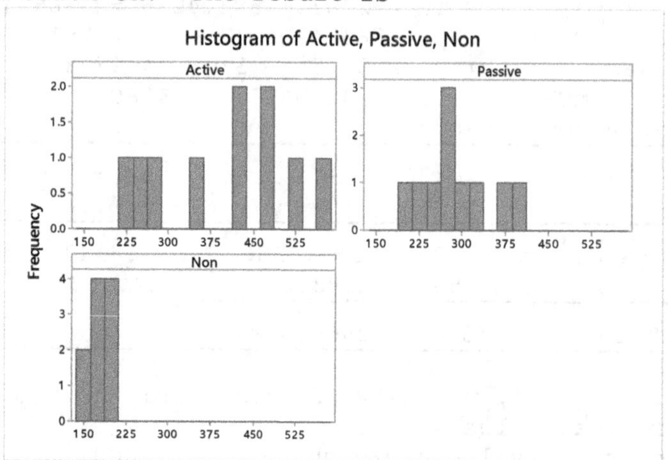

Graphical analysis of the three samples suggest that the distributions are not the same shape. Therefore, performing the Kruskal-Wallis test is not appropriate.

32. (a) Using Minitab, choose **Graph ▶ Probability Plot,** select the **Simple** version and click **OK**. Enter Loss, Stable, and Gain in the **Graph Variables** text box, click on the **Multiple Graphs** button and select **In separate panels of the same graph**, and click **OK**. The result is

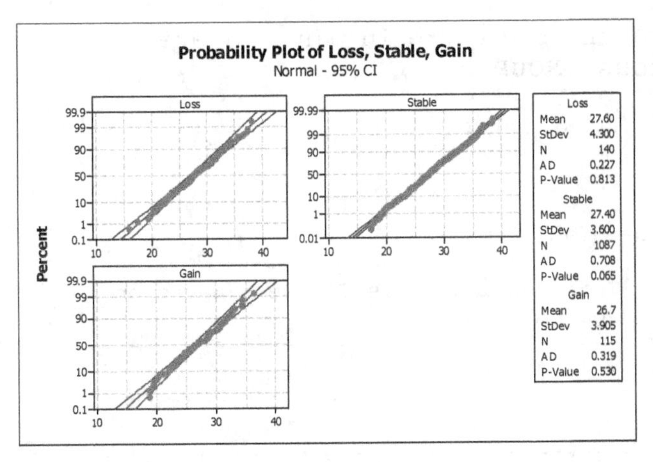

The standard deviations of the three groups are shown in the right hand panel as 4.300, 3.600, and 3.905.

(b) To carry out the Analysis of variance and residual analysis, we will use the data in a single column, which is named BMI. In the next column, which is named GROUP, are the names of each of the regions corresponding to each data value in BMI. The residual analysis consists of plotting the residuals against the means (or fits) and constructing a normal probability plot. This is done at the same time as the analysis of variance in Minitab. To do this, we choose **Stat ▶ Anova ▶One-way...**, select BMI for the **Response:** text box and GROUP in the **Factor:** text box. Enter 95 in the **Confidence level** box. Then click on the **Graphs** button and click to place check marks in the boxes for **Normal plot of residuals** and **Residuals versus fits**, and click **OK**. Click on the **Comparisons** button and select **Tukey's**, and enter 5 in the **family error rate** box, and click **OK**. The graphical results are

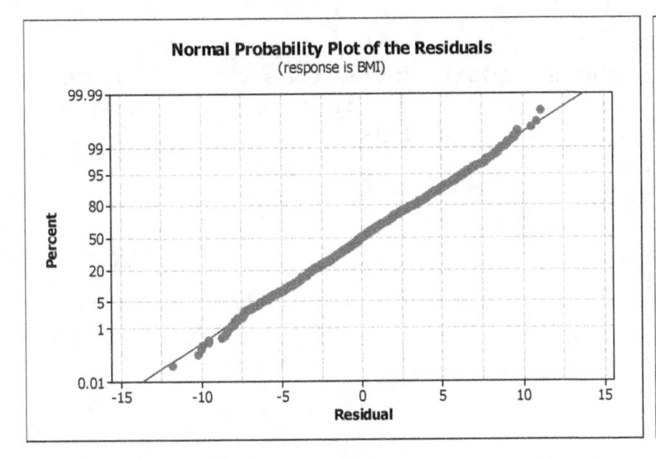

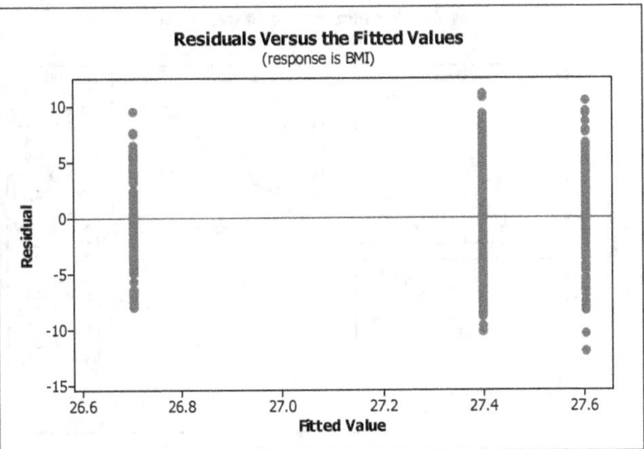

(c) The normal probability plots in part (a) are linear as is the normal probability plot of the residuals in part (b). The plot of the residuals against the fitted values shows approximately equal spreads. The rule of 2 is not violated since the ratio of the largest to smallest standard deviation is 4.3/3.6 = 1.19. A one-way ANOVA is reasonable.

(d) The one-way ANOVA results from the procedure in part (b) are

```
One-way ANOVA: BMI versus GROUP

Source    DF       SS    MS     F      P
GROUP      2     60.0  30.0  2.18  0.113
Error   1339  18379.7  13.7
Total   1341  18439.6

S = 3.705   R-Sq = 0.33%   R-Sq(adj) = 0.18%

                                Individual 95% CIs For Mean Based on
                                Pooled StDev
Level      N    Mean  StDev  ------+---------+---------+---------+---
Gain     115  26.700  3.905  (----------*----------)
Loss     140  27.603  4.300                   (---------*---------)
Stable  1087  27.399  3.600                      (---*--)
                             ------+---------+---------+---------+---
                                 26.40     27.00     27.60     28.20
```

(e) The P-value of the test is 0.113, which is larger than the 0.05 significance level. Do not reject the null hypothesis of equal means. The data do not provide sufficient evidence to conclude that there is a difference in body mass index among the three groups of men with different weight losses.

(f) Since the ANOVA test was not significant, it is not necessary to use Tukey's multiple comparison. None of the means will differ from the others.

33. (a) Using Minitab, choose **Graph ▶ Probability Plot,** select the **Simple** version and click **OK**. Enter Loss, Stable, and Gain in the **Graph Variables** text box, click on the **Multiple Graphs** button and select **In separate panels of the same graph**, and click **OK**. The result is

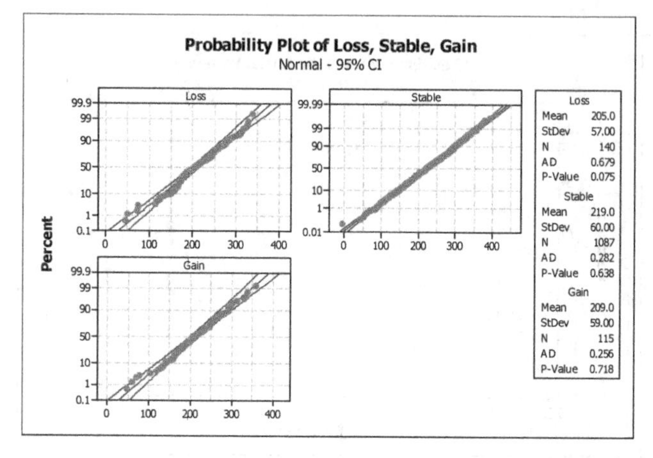

The standard deviations of the three groups are shown in the right hand panel as 57, 60, and 59.

(b) To carry out the Analysis of variance and residual analysis, we will use the data in a single column, which is named POWER. In the next column, which is named GROUP, are the names of each of the regions corresponding to each data value in POWER. The residual analysis consists of plotting the residuals against the means (or fits) and constructing a normal probability plot. This is done at the same time as the analysis of variance in Minitab. To do this, we choose **Stat ▶ Anova ▶One-way...**, select POWER for the **Response:** text box and GROUP in the **Factor:** text box. Enter 95 in the **Confidence level** box. Then click on the **Graphs** button and click to place check marks in the boxes for

Normal plot of residuals and **Residuals versus fits**, and click **OK**. Click on the **Comparisons** button and select **Tukey's**, and enter 5 in the **family error rate** box, and click **OK**. The graphical results are

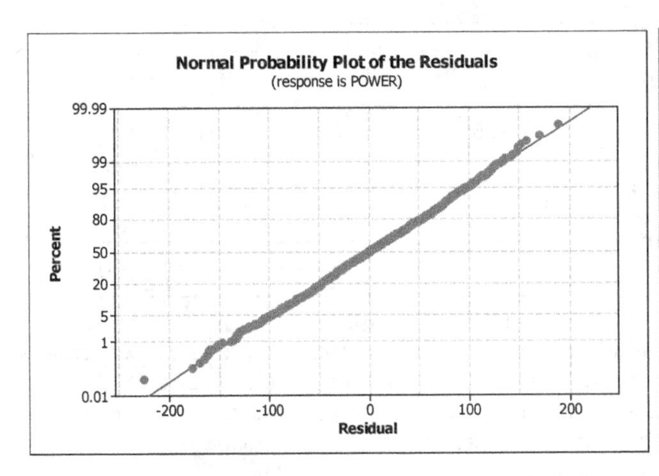

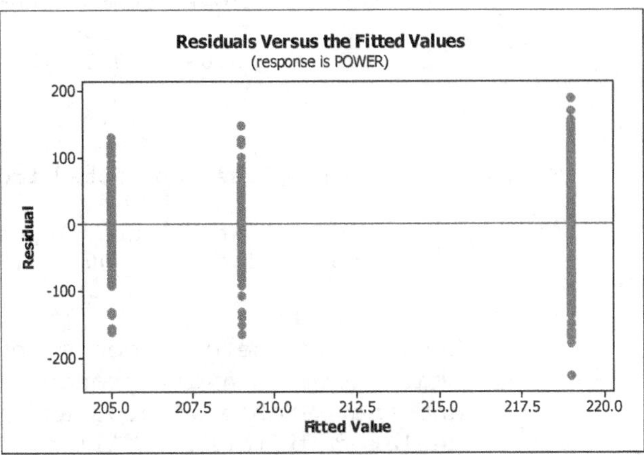

(c) The normal probability plots in part (a) are linear as is the normal probability plot of the residuals in part (b). The plot of the residuals against the fitted values shows approximately equal spreads. The rule of 2 is not violated since the ratio of the largest to smallest standard deviation is 60/57 = 1.05. A one-way ANOVA is reasonable.

(d) The one-way ANOVA results from the procedure in part (b) are

One-way ANOVA: POWER versus GROUP

```
Source    DF       SS     MS     F      P
GROUP      2    31742  15871  4.47  0.012
Error   1339  4758204   3554
Total   1341  4789946

S = 59.61   R-Sq = 0.66%   R-Sq(adj) = 0.51%

                                  Individual 95% CIs For Mean Based on Pooled
                                  StDev
Level      N    Mean  StDev      -+---------+---------+---------+--------
Gain     115  209.00  59.00           (---------------*---------------)
Loss     140  205.00  57.00      (-------------*-------------)
Stable  1087  219.00  60.00                                  (----*----)
                                 -+---------+---------+---------+--------
                                 196.0     203.0     210.0     217.0

Pooled StDev = 59.61
```

(e) The P-value of the test is 0.012, which is smaller than the 0.05 significance level. Reject the null hypothesis of equal means. The data provide sufficient evidence to conclude that there is a difference in leg power among the three groups of men with different weight losses.

(f) The Tukey multiple comparison results produced by the procedure in part (b) are

```
Tukey 95% Simultaneous Confidence Intervals
All Pairwise Comparisons among Levels of GROUP

Individual confidence level = 98.06%

GROUP = Gain subtracted from:

GROUP    Lower   Center   Upper   --------+---------+---------+---------+-
Loss    -21.56    -4.00   13.56       (----------*-----------)
Stable   -3.68    10.00   23.68                  (--------*--------)
                                  --------+---------+---------+---------+-
                                        -15         0        15        30

GROUP = Loss subtracted from:

GROUP    Lower  Center   Upper  --------+---------+---------+---------+-
Stable    1.47   14.00   26.53                    (-------*--------)
                                --------+---------+---------+---------+-
                                      -15         0        15        30
```

Looking at the confidence intervals that do not contain zero, we see that there is a difference in mean leg power between the "Loss" group and the "Stable" group, and that there are no other pairs that can be declared different. All of this can be said with 95% confidence.

34. (a) Using Minitab, choose **Graph ▶ Probability Plot,** select the **Simple** version and click **OK**. Enter 25-34, 35-44, 45-54 and 55-64 in the **Graph Variables** text box, click on the **Multiple Graphs** button and select **In separate panels of the same graph**, and click **OK**. The result is

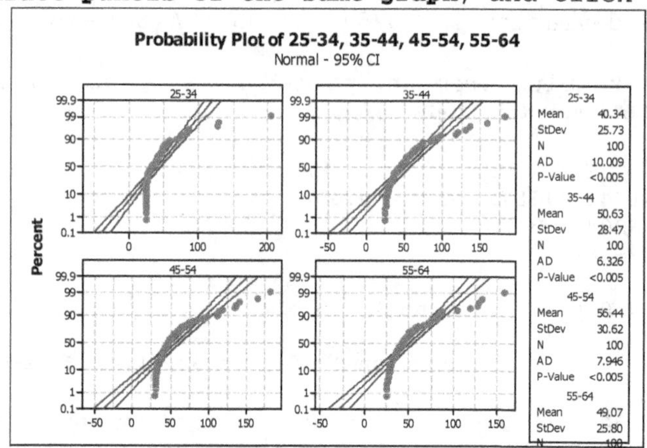

The standard deviations of the incomes in the four age groups are 25.73, 28.47, 30.62, and 25.80.

(b) To carry out the Analysis of variance and residual analysis, we will use the data in a single column, which is named INCOME. In the next column, which is named AGE, are the categories for ages corresponding to each data value in INCOME. The residual analysis consists of plotting the residuals against the means (or fits) and constructing a normal probability plot. This is done at the same time as the analysis of variance in Minitab. To do this, we choose **Stat ▶ Anova ▶One-way...**, select INCOME for the **Response:** text box and AGE in the **Factor:** text box. Enter 95 in the **Confidence level** box. Then click on the **Graphs** button and click to place check marks in the boxes for **Normal plot of residuals** and **Residuals versus fits**, and click **OK**. Click on the **Comparisons** button and select **Tukey's**, and enter 5 in the **family error rate** box, and click **OK**. The graphical results are

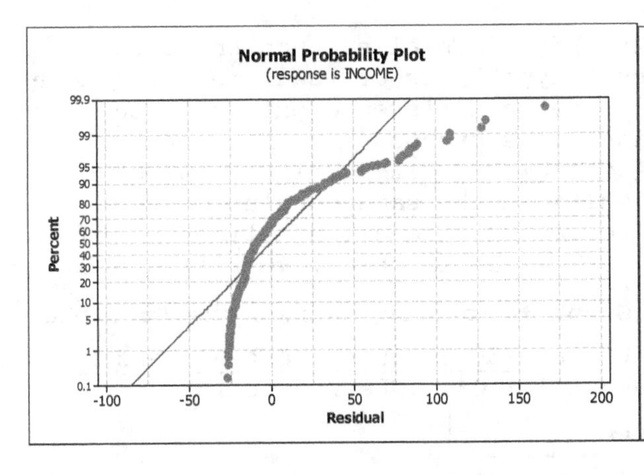

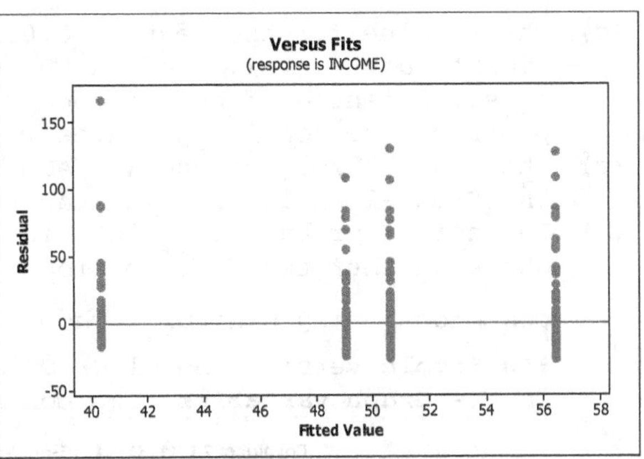

(c) The normal probability plots for each age group are non-linear, as is the residual normal probability plot in part (b). The rule of 2 is not violated as the ratio of the largest to smallest standard deviation is 30.62/25.73 = 1.19. The normality assumption is not reasonable, so parts (d)-(f) are omitted.

35. (a) Since the one-way ANOVA was reasonable, the Kruskal-Wallis test is also reasonable.

(b) Using Minitab, choose **Stat ▶ Nonparametrics ▶ Kruskal-Wallis...**, select BMI in the **Response** text box, select GROUP in the **Factor** text box, and click **OK**. The result is

Kruskal-Wallis Test on BMI

GROUP	N	Median	Ave Rank	Z
Gain	115	26.60	608.3	-1.83
Loss	140	27.55	696.8	0.82
Stable	1087	27.40	674.9	0.67
Overall	1342		671.5	

H = 3.74 DF = 2 P = 0.154
H = 3.74 DF = 2 P = 0.154 (adjusted for ties)

(c) The P-value for the test is 0.154, which is larger than the 0.05 significance level. Do not reject the null hypothesis of equal means. There is not sufficient evidence to conclude that a difference exists in the body mass index of the men in the three weight loss groups.

(d) The P-value for the one-way ANOVA test was 0.113. This is close to the Kruskal-Wallis P-value and results in the same conclusion.

36. (a) Since the one-way ANOVA was reasonable, the Kruskal-Wallis test is also reasonable.

(b) Using Minitab, choose **Stat ▶ Nonparametrics ▶ Kruskal-Wallis...**, select POWER in the **Response** text box, select GROUP in the **Factor** text box, and click **OK**. The result is

Kruskal-Wallis Test on POWER

GROUP	N	Median	Ave Rank	Z
Gain	115	204.4	623.1	-1.40
Loss	140	202.8	594.8	-2.47
Stable	1087	219.6	686.5	2.93
Overall	1342		671.5	

H = 8.90 DF = 2 P = 0.012
H = 8.90 DF = 2 P = 0.012 (adjusted for ties)

(c) The P-value for the test is 0.012, which is smaller than the 0.05 significance level. Reject the null hypothesis of equal means. There is sufficient evidence to conclude that a difference exists in the leg power of the men in the three weight loss groups.

(d) The P-value for the one-way ANOVA test was 0.012. This is the same as the Kruskal-Wallis P-value and results in the same conclusion.

37. (a) The normal probability plots in Problem 34 were all non-linear. To check whether the distributions are similar in shape, we will obtain box plots using Minitab. Choose **Graph ▶ Boxplot,** select the **Multiple Y's Simple** version and click **OK**. Enter 25-34, 35-44, 454-54, and 55-64 in the **Graph variables** text box and click **OK**. The result is

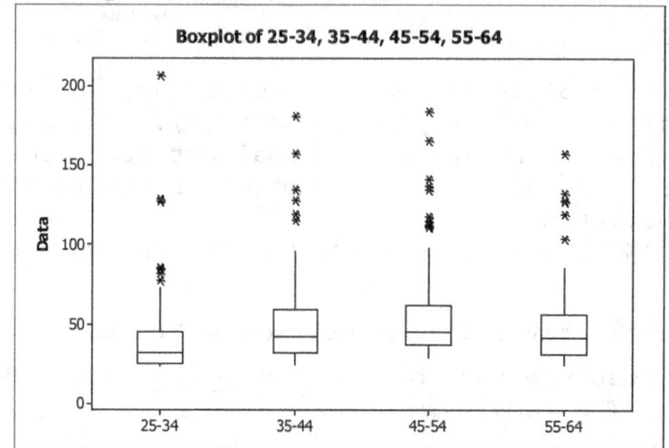

It appears that all four of the distributions have about the same shape, right-skewed, about the same spread. Thus the assumptions for using the Kruskal-Wallis test are reasonable.

(b) Choose **Stat ▶ Nonparametrics ▶ Kruskal-Wallis...**, select INCOME in the **Response** text box, select AGE in the **Factor** text box, and click **OK**. The result is

Kruskal-Wallis Test on INCOME

AGE	N	Median	Ave Rank	Z
25-34	100	32.10	142.0	-5.85
35-44	100	42.60	210.8	1.03
45-54	100	45.70	243.5	4.30
55-64	100	41.50	205.7	0.52
Overall	400		200.5	

H = 40.49 DF = 3 P = 0.000
H = 40.49 DF = 3 P = 0.000 (adjusted for ties)

(c) The P-value of the test is 0.000, which is smaller than the 0.05 significance level. Reject the null hypothesis of equal means. The data provide sufficient evidence to conclude that there is a difference in the mean incomes of the four age groups.

(d) No one-way ANOVA was performed.